T0180686

Lecture Notes in Electrical Engineering

Volume 373

About this Series

"Lecture Notes in Electrical Engineering (LNEE)" is a book series which reports the latest research and developments in Electrical Engineering, namely:

- Communication, Networks, and Information Theory
- Computer Engineering
- Signal, Image, Speech and Information Processing
- Circuits and Systems
- Bioengineering

LNEE publishes authored monographs and contributed volumes which present cutting edge research information as well as new perspectives on classical fields, while maintaining Springer's high standards of academic excellence. Also considered for publication are lecture materials, proceedings, and other related materials of exceptionally high quality and interest. The subject matter should be original and timely, reporting the latest research and developments in all areas of electrical engineering.

The audience for the books in LNEE consists of advanced level students, researchers, and industry professionals working at the forefront of their fields. Much like Springer's other Lecture Notes series, LNEE will be distributed through Springer's print and electronic publishing channels.

More information about this series at http://www.springer.com/series/7818

Doo-Soon Park · Han-Chieh Chao
Young-Sik Jeong · James J. (Jong Hyuk) Park
Editors

Advances in Computer Science and Ubiquitous Computing

CSA & CUTE 2015

 Springer

Editors
Doo-Soon Park
Department of Computer Software
 Engineering
Soonchunhyang University
Chungnam
Korea

Han-Chieh Chao
Institute of Computer Science
 and Information Engineering
National Ilan University
Taiwan
R.O.C

Young-Sik Jeong
Department of Multimedia Engineering
Dongguk University
Seoul
South Korea

James J. (Jong Hyuk) Park
Department of Computer Science
 and Engineering
Seoul University of Science & Technology
Seoul
Korea

ISSN 1876-1100   ISSN 1876-1119   (electronic)
Lecture Notes in Electrical Engineering
ISBN 978-981-10-9121-6   ISBN 978-981-10-0281-6   (eBook)
DOI 10.1007/978-981-10-0281-6

Printed on acid-free paper

This Springer imprint is published by SpringerNature
The registered company is Springer Science+Business Media Singapore Pte Ltd.

Message from the CSA 2015 General Chair

International Conference on Computer Science and its Applications (CSA 2015) is the 7th event of the series of international scientific conference. This conference takes place Cebu, Philippines, Dec. 15–17, 2015. CSA 2015 will be the most comprehensive conference focused on the various aspects of advances in computer science and its applications. CSA 2015 will provide an opportunity for academic and industry professionals to discuss the latest issues and progress in the area of CSA. In addition, the conference will publish high quality papers which are closely related to the various theories and practical applications in CSA. Furthermore, we expect that the conference and its publications will be a trigger for further related research and technology improvements in this important subject. CSA 2015 is the next event in a series of highly successful International Conference on Computer Science and its Applications, previously held as CSA 2014 (6th Edition: Guam, December, 2014), CSA 2013 (5th Edition: Danang, December, 2013), CSA 2012 (4th Edition: Jeju, November, 2012), 1CSA 2011 (3rd Edition: Jeju, December, 2011), CSA 2009 (2nd Edition: Jeju, December, 2009), and CSA 2008 (1st Edition: Australia, October, 2008).

The papers included in the proceedings cover the following topics: Mobile and ubiquitous computing, Dependable, reliable and autonomic computing, Security and trust management, Multimedia systems and services, Networking and communications, Database and data mining, Game and software engineering, Grid and scalable computing, Embedded system and software, Artificial intelligence, Distributed and parallel algorithms, Web and internet computing and IT policy and business management.

Accepted and presented papers highlight new trends and challenges of Computer Science and its Applications. The presenters showed how new research could lead to novel and innovative applications. We hope you will find these results useful and inspiring for your future research. We would like to express our sincere thanks to Steering Chairs: James J. (Jong Hyuk) Park (SeoulTech, Korea), Han-Chieh Chao (National Ilan University, Taiwan) and Mohammad S. Obaidat (Monmouth University, USA). Our special thanks go to the Program Chairs: Aziz Nasridinov

(Chungbuk National University, Korea), Neil Y. Yen (The University of Aizu, Japan), Yu Chen (State University of New York, USA), Hwamin Lee (SoonChunHyang University, Korea), Isaac Woungang (Ryerson University, Canada), all Program Committee members and all the additional reviewers for their valuable efforts in the review process, which helped us to guarantee the highest quality of the selected papers for the conference.

We cordially thank all the authors for their valuable contributions and the other participants of this conference. The conference would not have been possible without their support. Thanks are also due to the many experts who contributed to making the event a success.

CSA 2015 General Chair

Michael Hwa Young Jeong, Kyung Hee University, Korea

Message from the CSA 2015 Program Chairs

Welcome to the 7th International Conference on Computer Science and its Applications (CSA 2015) which will be held in Cebu, Philippines, Dec. 15–17, 2015. CSA 2015 will be the most comprehensive conference focused on the various aspects of advances in computer science and its applications.

CSA 2015 provides an opportunity for academic and industry professionals to discuss the latest issues and progress in the area of Computer Science. In addition, the conference contains high quality papers which are closely related to the various theories and practical applications in Computer Science. Furthermore, we expect that the conference and its publications will be a trigger for further related research and technology improvements in this important subject. CSA 2015 is the next event in a series of highly successful International Conference on Computer Science and its Applications, previously held as CSA 2014 (6th Edition: Guam, December, 2014), CSA 2013 (5th Edition: Danang, December, 2013), CSA 2012 (4th Edition: Jeju, November, 2012), CSA 2011 (3rd Edition: Jeju, December, 2011), CSA 2009 (2nd Edition: Jeju, December, 2009), and CSA 2008 (1st Edition: Australia, October, 2008).

CSA 2015 contains high quality research papers submitted by researchers from all over the world. Each submitted paper was peer-reviewed by reviewers who are experts in the subject area of the paper. Based on the review results, the Program Committee accepted papers.

For organizing an International Conference, the support and help of many people is needed. First, we would like to thank all authors for submitting their papers. We also appreciate the support from program committee members and reviewers who carried out the most difficult work of carefully evaluating the submitted papers.

We would like to give my special thanks to Prof. James J. (Jong Hyuk) Park, Prof. Han-Chieh Chao, and Prof. Mohammad S. Obaidat, the Steering Committee Chairs of CSA for their strong encouragement and guidance to organize the symposium. We would like to thank CSA 2015 General Chair, Prof. Michael Hwa Young Jeong. We would like to express special thanks to committee members for their timely unlimited support.

CSA 2015 Program Chairs

Aziz Nasridinov, Chungbuk Nationl University, Korea
Neil Y. Yen, The University of Aizu, Japan
Yu Chen, State University of New York, USA
Hwamin Lee, SoonChunHyang University, Korea
Isaac Woungang, Ryerson University, Canada

Organization

Honorary Chair

Doo-soon Park   SoonChunHyang University/KIPS President, Korea

Steering Chairs

James J. Park   SeoulTech, Korea
Han-Chieh Chao   National Ilan University, Taiwan
Mohammad S. Obaidat   Monmouth University, USA

General Chairs

Michael Hwa Young Jeong   Kyung Hee University, Korea

Program Chairs

Aziz Nasridinov   Donggguk University, Korea
Neil Y. Yen   The University of Aizu, Japan
Yu Chen   State University of New York, USA
Hwamin Lee   SoonChunHyang University, Korea
Isaac Woungang   Ryerson University, Canada

International Advisory Board

Hsiao-Hwa Chen	Sun Yat-Sen University, Taiwan
Philip S. Yu	University of Illinois at Chicago, USA
Yi Pan	Georgia State University, USA
Jiankun Hu	RMIT University, Australia
Shu-Ching Chen	Florida International University, USA
Victor Leung	University of British Columbia, Canada
Qun Jin	Waseda University, Japan
Vincenzo Loia	University of Salerno, Italy
Frode Eika Sandnes	Oslo University College, Norway
Kyungeun Cho	Dongguk University, Korea

Publicity Chairs

Kwang-il Hwang	Incheon national University, Korea
Jason C. Hung	Overseas Chinese University, Taiwan
Jungho Kang	Sungsil University, Korea
Min Choi	Chungbuk National University, Korea
Deok-Gyu Lee	Seowon University, Korea
Byung-Gyu Kim	Sunmoon University, Korea
Amiya Nayak	University of Ottawa, Canada
Weiwei Fang	Beijing Jiaotong University, China
Kevin Cheng	Tatung University, Taiwan
Sung-Ki Kim	Sunmoon University, Korea

Program Committee

Ahmed El Oualkadi	Abdelmalek Essaadi University, Morocco
Bartosz Ziolko	AGH University of Science and Technology, Techmo, Poland
Bela Genge	University of Targu Mures, Romania
Chang Jerry Hsi-Ya	NCHC, Taiwan
Chao-Tung Yang	Tunghai University, Taiwan
Chen Yuan-Fang	Dalian University of Technology, China
Chia-Hung Yeh	National Sun Yat-sen University, Taiwan
Chih-Lin Hu	National Central University, Taiwan
Cho-Chin Lin	National Yilan University, Taiwan
Dakshina Ranjan Kisku	National Institute of Technology Durgapur, India
Dalton Lin	National Taipei University, Taiwan

Dongkyun Kim	KISTI, Korea
Don-Lin Yang	Feng Chia University, Taiwan
Eunyoung Lee	Dongduk Women's University, Korea
Guan-Ling Lee	National Dong Hwa University, Taiwan
Guo Bin	Institute Telecom & Management SudParis, France
HaRim Jung	Sungkyunkwan University, Korea
Hongsoo Kim	SK Planet, Korea
Hoon Choi	Chungnam National University, Korea
Hsieh Sun-Yuan	National Cheng Kung University, Taiwan
Ivanova Malinka	Technical University of Sofia, Bulgaria
Jaesuhp Oh	Sookmyung Woman's University, Korea
Jemni Mohamed	ESSTT, Tunisia
Jen-Wei Hsieh	National Taiwan University of Science and Technology, Taiwan
Jerzy Respondek	Silesian University of Technology Poland
Jin Wei	Amazon Web Services Seattle, USA
Jun Yong-Kee	Gyeongsang National University, Korea
Jun-Won Ho	Seoul Women's University, Korea
Kuei-Ping Shih	Tamkang University, Taiwan
Kwang Sik Chung	Korea National Open University, Korea
Li-Jen Kao	Hwa Hsia Institute of Technology, Taiwan
Lukas Ruf	Consecom AG, Switzerland
Marco Listanti	DIET, Roma, Italy
Maurizio Lazzari	National Research Council, Italy
Maytham Safar	Kuwait University, Kuwait
Metin Basarir	Sakarya University, Turkey
Muhammad Javed	Dublin City University, Ireland
Muhammad, Younas	Oxford Brookes University, UK
Nader F. Mir	San Jose State University, USA
Paprzycki Marcin	Polish Academy of Sciences, Poland
Pascal Lorenz	University of Haute Alsace, France
Prabu Dorairaj	NetApp, India/USA
Qingyuan Bai	Fuzhou University, China
Schulz Frank	SAP Research, Germany
Sha Kewei	Oklahoma City University, USA
Shingo Ichii	University of Tokyo, Japan
Thepvilojanapong Niwat	Mie University, Japan
Toshihiro Yamauchi	Okayama University, Japan
Tzung-Pei Hong	National University of Kaohsiung, Taiwan
Ventzeslav Valev	Bulgarian Academy of Sciences, Bulgaria
Wang Chien-Min	Academia Sinica, Taiwan
Xubo Song	Oregon Health and Science University, USA

Young-Gab Kim   Sejong University, Korea
Yu Qian   University of Regina, Canada
Yue-Shan Chang   National Taipei University, Taipei
Yutaka Watanobe   University of Aizu, Japan

Message from the CUTE 2015 General Chairs

On behalf of the organizing committees, it is our pleasure to welcome you to the 10th International Conference on Ubiquitous Information Technologies and Applications (CUTE 2015), will be held in Cebu, Philippines on December 15–17, 2015.

This conference provides an international forum for the presentation and showcase of recent advances on various aspects of ubiquitous computing. It will reflect the state-of-the-art of the computational methods, involving theory, algorithm, numerical simulation, error and uncertainty analysis and/or novel application of new processing techniques in engineering, science, and other disciplines related to ubiquitous computing.

The papers included in the proceedings cover the following topics: Ubiquitous Communication and Networking, Ubiquitous Software Technology, Ubiquitous Systems and Applications, Ubiquitous Security, Privacy and Trust. Accepted papers highlight new trends and challenges in the field of ubiquitous computing technologies. We hope you will find these results useful and inspiring for your future research.

We would like to express our sincere thanks to Steering Committees: Young-Sik Jeong (Dongguk University, Korea), Laurence T. Yang (St.Francis Xavier University, Canada), Hai Jin (Huangzhong University of Science and Technology, China), Chan-Hyun Youn (KAIST, Korea), Jianhua Ma (Hosei University, Japan), Minyi Guo (Shanghai Jiao Tong University, Japan), & Weijia Jia (City University of Hong Kong, Hong Kong). We would also like to express our cordial thanks to the Program Chairs & Program Committee members for their valuable efforts in the review process, which helped us to guarantee the highest quality of the selected papers for the conference.

Finally, we would thank all the authors for their valuable contributions and the other participants of this conference. The conference would not have been possible without their support. Thanks are also due to the many experts who contributed to making the event a success.

CUTE 2015 General Chairs

James J. Park, Seoul National University of Science & Technology, Korea
Victor Leung, University of British Columbia, Canada
Shu-Ching Chen, Florida International University, USA

Message from the CUTE 2015 Program Chairs

Welcome to the 10th International Conference on Ubiquitous Information Technologies and Applications (CUTE 2015), will be held in Cebu, Philippines on December 15–17, 2015.

The purpose of the CUTE 2015 conference is to promote discussion and interaction among academics, researchers and professionals in the field of ubiquitous computing technologies. This year the value, breadth, and depth of the CUTE 2015 conference continues to strengthen and grow in importance for both the academic and industrial communities. This strength is evidenced this year by having the highest number of submissions made to the conference.

For CUTE 2015, we received a lot of paper submissions from various countries. Out of these, after a rigorous peer review process, we accepted only high-quality papers for CUTE 2015 proceeding, published by the Springer. All submitted papers have undergone blind reviews by at least two reviewers from the technical program committee, which consists of leading researchers around the globe. Without their hard work, achieving such a high-quality proceeding would not have been possible. We take this opportunity to thank them for their great support and cooperation.

We would also like to sincerely thank the following invited speakers who kindly accepted our invitations, and, in this way, helped to meet the objectives of the conference:

- Prospero C. Naval, Jr., Ph.D., University of the Philippines, Philippines

Finally, we would like to thank all of you for your participation in our conference, and also thank all the authors, reviewers, and organizing committee members. Thank you and enjoy the conference!

CUTE 2015 Program Chairs

Gangman Yi, Gangneung-Wonju National University, Korea
Robert C. H. Hsu, Chung Hua University, Taiwan
Xin Chen, University of Hawaii, USA

Organization

Honorary Chair

Doo-soon Park   SoonChunHyang University, Korea

Steering Committee

Young-Sik Jeong   Dongguk University, Korea (Leading Chair)
Laurence T. Yang   St. Francis Xavier University, Canada
Hai Jin   Huangzhong University of Science and Technology, China
Chan-Hyun Youn   KAIST, Korea
Jianhua Ma   Hosei University, Japan
Minyi Guo   Shanghai Jiao Tong University, Japan
Weijia Jia   City University of Hong Kong, Hong Kong

General Chairs

James J. Park   SeoulTech, Korea
Victor Leung   University of British Columbia, Canada
Shu-Ching Chen   Florida International University, USA

Program Chairs

Gangman Yi	Gangneung-Wonju National University, Korea
Robert C.H. Hsu	Chung Hua University, Taiwan
Xin Chen	University of Hawaii, USA

International Advisory Committee

Jeong-Bae Lee	Sunmmon University, Korea
Yang Xiao	University of Alabama, USA
Bin Xiao	The Hong Kong Polytechnic University, Hong Kong
Seok Cheon Park	Gachon University, Korea
Han-Chieh Chao	National Ilan University, Taiwan
Sanghoon Kim	Hankyong National University, Korea
HeonChang Yu	Korea University, Korea
Hung-Chang Hsiao	National Cheng Kung University, Taiwan
Javier Lopez	University of Malaga, Spain
Nammee Moon	Hoseo University, Korea
Ved Kafle	NICT, Japan
Kuan-Ching Li	Providence University, Taiwan
Im-Yeong Lee	SoonChunHyang University, Korea
Hongli Luo	Indiana University, USA
Min Hong	SoonChunHyang University, Korea
Shanmugasundaram Hariharan	TRP Engineering College (SRM Group), India
Eunyoung Lee	Dongduk Women's University, Korea
Chao-Tung Yang	Tunghai University, Taiwan
Ming Li	California State University, USA

Publicity Chairs

Deqing Zou	HUST, China
Seung-Ho Lim	Hankuk University of Foreigh Studies, Korea
Weili Han	Fudan University, China
Joon-Min Gil	Catholic University of Daegu, Korea
Yunsick Sung	Keimyung University, Korea
Se Dong Min	SoonChunHyang University, Korea

Program Committee

Alfredo Cuzzocrea	University of Calabria, Italy
Antonis Gasteratos	Democritus University of Thrace, Greece
Chen Uei-Ren	Hsiuping University of Science and Technology, Taiwan
Cheonshik Kim	AnYang University, Korea
Chi-Fu Huang	National Chung Cheng University, Taiwan
Hariharan Shanmugasundaram	Pavendar Bharathidasan College of Engineering and Technology, India
Hua Edward	QED Systems, USA
Jiqiang Lu	Institute for Infocomm Research, Singapore
Jong-Myon Kim	University of Ulsan, Korea
KeunHo Ryu	Chungbuk National University, Korea
Klyuev Vitaly	University of Aizu, Japan
Kwangman Ko	Sangji University, Korea
Lai Kuan-Chu	National Taichung University, Taiwan
Lu Leng	Southwest Jiaotong University Emei Campus, China
Pai-Ling Chang	Shih-Hsin University, Taiwan
Pinaki Ghosh	Atmiya Institute of Technology & Science, India
Pit Pichappan	AISB, Hungary
Pyung-Soo Kim	Korea Polytechnic University, Korea
Raylin So	National Chengchi University, China
Ren-Song Ko	National Chung Cheng University, Taiwan
Ruben Rios	Universidad de Malaga, Spain
Ryszard Tadeusiewicz	AGH University of Science and Technology, Poland
Toshiyuki Amagasa	University of Tsukuba, Japan
Xiao Liu	East China Normal University, China
Yang-Sae Moon	Kangwon National University, Korea
Young Ik Eom	Sungkyunkwan University, Korea
Yu-Chen Hu	Providence University, Taiwan
Yunquan Zhang	Institute of Computing Technology, CAS, China

Contents

Contents

Contents

Cryptanalysis of Enhanced Biometric-Based Authentication Scheme for Telecare Medicine Information Systems Using Elliptic Curve Cryptosystem

Jongho Mun, Jiseon Yu, Jiye Kim, Hyungkyu Yang and Dongho Won

Abstract To achieve the security and privacy, many remote user authentication schemes based on cryptography for telecare medicine information systems have been proposed. Recently, Lu et al. proposed an enhanced biometric-based authentication scheme for telecare medicine information system using elliptic curve cryptosystem. They found that Arshad et al.'s scheme was vulnerable to off-line password guessing attack and user impersonation attack, and claimed that their scheme was more secure against various attacks. However, we demonstrate that their scheme is still insecure and vulnerable to outsider attack, user impersonation attack, server impersonation attack and smart card stolen attack in this paper.

Keywords Elliptic Curve Cryptosystem · Remote user authentication · Telecare Medicine Information Systems · Biometric

J. Mun · J. Kim · D. Won(✉)
Colloge of Information and Communication Engineering, Sungkyunkwan University,
Seoul, South Korea
e-mail: {jhmoon,jykim,dhwon}@security.re.kr

J. Yu
Center for Information Security Technologies (CIST), Korea University,
Seoul, South Korea
e-mail: gsun2@korea.ac.kr

H. Yang
Department of Computer and Media Information, Kangnam University,
Yongin, South Korea
e-mail: hkyang@kangnam.ac.kr

© Springer Science+Business Media Singapore 2015
D.-S. Park et al. (eds.), *Advances in Computer Science and Ubiquitous Computing*,
Lecture Notes in Electrical Engineering 373,
DOI: 10.1007/978-981-10-0281-6_1

1

1   Introduction

The Telecare Medical Information Systems (TMISs) provides an efficient method to enhance the medical process between doctors and nurses at a clinical center or home healthcare (HHC) agency, and patients at home. In the TMISs, the patients only need to stay at home, they can still access a convenient and prompt treatment from the medical center over Internet or mobile networks [1]. For enhancing the security and privacy of TMISs, the mutual authentication and session key establishment between a patient and doctor is very important.

Recently, many researchers proposed a remote user authentication schemes [2][3] [4][5]. In 2013, Awasthi et al. [6] proposed a biometric authentication scheme for TMISs using nonce. However, Mishra et al. [7] found that Awasthi et al.'s scheme was vulnerable to off-line password guessing attack and it cannot provide efficient password change method. After that, Tan et al. [8] also found that Awasthi et al.'s scheme cannot resist reflection attack and did not provide three factor security and user anonymity. To overcome the weaknesses of Awasthi et al.'s scheme, Tan et al. proposed a three factor authentication scheme and claimed that their scheme was secure against various attacks. In 2014, Arshad et al. [9] demonstrated that Tan et al.'s scheme cannot resist denial-of-service and replay attack, and proposed an improved elliptic curve cryptosystem (ECC) [10]-based scheme to prevent the vulnerabilities. Recently Lu et al. [11] found that Arshad et al.'s scheme fails to protect against off-line password guessing attack and user impersonation attack and claimed that their scheme was more secure and satisfied many security features. However, we find that Lu et al.'s scheme is still insecure and vulnerable to outsider attack, user imperso-nation attack, server impersonation attack and smart card stolen attack.

The rest of the paper is organized as follows. We begin by reviewing Lu et al.'s biometric-based authentication scheme using elliptic curve cryptosystem in section 2. Then in section 3, we describe security weaknesses in Lu et al.'s scheme. Finally, we conclude this paper in Section 4.

2   Review in Lu et al.'s Scheme

In this section, we review the biometric-based authentication scheme for telecare medicine information systems using elliptic curve cryptosystem by Lu et al. in 2015. As previous researches, Lu et al.'s scheme consists of three phases: registration, login and authentication, and password updating phase which as follows. The notations used in Lu et al.'s scheme are summarized as Table 1.

Table 1 Notations used in Lu et al.'s scheme

Term	Description
$\mathcal{U}, \mathcal{S}$	The patient and the telecare server
$\mathcal{ID}_i, \mathcal{PW}_i, \mathcal{B}_i$	Identity, password, biometric of the patient $\mathcal{U}$
$\mathcal{H}(\cdot)$	Biohash function
$h_1(\cdot), h_2(\cdot)$	Hash function $h_1: \{0, 1\}^* \rightarrow \{0, 1\}^l$ Hash function $h_2: \{0, 1\}^* \rightarrow \mathcal{Z}_{\mathcal{P}}^*$
$\mathcal{P}$	The base point of the elliptic curve $\mathcal{E}$ with order n
$\mathcal{X}$	Private key selected by $\mathcal{S}$
$\oplus, \|$	Exclusive-or operation and concatenation operation

2.1 Registration Phase

In this phase, the patient $\mathcal{U}$ initially registers with the remote server $\mathcal{S}$ as follows:

The patient $\mathcal{U}$ enters his identity $\mathcal{ID}_i$, password $\mathcal{PW}_i$, and biometric $\mathcal{B}_i$. Then, $\mathcal{U}$ computes $\mathcal{MP}_i = \mathcal{PW}_i \oplus \mathcal{H}(\mathcal{B}_i)$ and sends $\{\mathcal{ID}_i, \mathcal{MP}_i\}$ to the server $\mathcal{S}$ over a secure channel.

1. When receiving the registration request message from $\mathcal{U}$, the server $\mathcal{S}$ computes $\mathcal{AID}_i = \mathcal{ID}_i \oplus h_2(\mathcal{X})$, $\mathcal{V}_i = h_1(\mathcal{ID}_i \| \mathcal{MP}_i)$ and issues a smart card $\mathcal{SC}_i$ which contains the information $\{\mathcal{AID}_i, \mathcal{V}_i, h_1(\cdot), h_2(\cdot), \mathcal{H}(\cdot)\}$ to $\mathcal{U}$.

2.2 Login and Authentication

When a patient $\mathcal{U}$ wants to login, $\mathcal{U}$ inserts his/her smart card $\mathcal{SC}_i$ into a card reader and an identity $\mathcal{ID}_i$, password $\mathcal{PW}_i$, and biometric $\mathcal{B}_i$.

$\mathcal{SC}_i$ computes $h_1(\mathcal{ID}_i \| \mathcal{PW}_i \oplus \mathcal{H}(\mathcal{B}_i))$ and verifies whether it is equal to the stored value $\mathcal{V}_i$. If it holds, $\mathcal{U}$ passes through the verification. Hence, $\mathcal{SC}_i$ selects a random number d_u and computes $\mathcal{K} = h_1(\mathcal{ID}_i \| \mathcal{ID}_i \oplus \mathcal{AID}_i)$, $\mathcal{M}_1 = \mathcal{K} \oplus d_u \mathcal{P}$, $\mathcal{M}_2 = h_1(\mathcal{ID}_i \| d_u \mathcal{P} \| \mathcal{T}_1)$, and transmits $\{\mathcal{M}_1, \mathcal{M}_2, \mathcal{AID}_i, \mathcal{T}_1\}$ to $\mathcal{S}$.

1. When receiving the login request message from $\mathcal{U}$, $\mathcal{S}$ first checks whether $|\mathcal{T}_c - \mathcal{T}_1| \leq \Delta\mathcal{T}$, where $\mathcal{T}_c$ is current timestamp of the server $\mathcal{S}$. If it holds, $\mathcal{S}$ uses his private key $\mathcal{X}$ to derive $\mathcal{ID}_i'$ by computing $\mathcal{AID}_i \oplus h_2(\mathcal{X})$, he/she then computes $d_u \mathcal{P}' = \mathcal{K} \oplus \mathcal{M}_1$ and checks $h_1(\mathcal{ID}_i' \| d_u \mathcal{P}' \| \mathcal{T}_1) \stackrel{?}{=} \mathcal{M}_2$. If it is true, $\mathcal{S}$ generates a random number d_s and computes $\mathcal{M}_3 = \mathcal{K} \oplus d_s \mathcal{P}$, $\mathcal{SK} = d_s d_u \mathcal{P}$, $\mathcal{M}_4 = h_1(\mathcal{K} \| d_u \mathcal{P} \| \mathcal{SK} \| \mathcal{T}_2)$, where $\mathcal{T}_2$ is the current timestamp. At last, $\mathcal{S}$ sends the login response message $\{\mathcal{M}_3, \mathcal{M}_4, \mathcal{T}_2\}$ to $\mathcal{U}$.
2. After receiving the login response message from $\mathcal{S}$, $\mathcal{U}$ first checks the freshness of $\mathcal{T}_2$. Then, $\mathcal{U}$ retrieves $d_s \mathcal{P}$ by computing $\mathcal{M}_3 \oplus \mathcal{K}$ and computes $\mathcal{SK}' = d_u d_s \mathcal{P}$, $\mathcal{M}_4' = h_1(\mathcal{K} \| d_u \mathcal{P} \| \mathcal{SK}' \| \mathcal{T}_2)$ to verify whether

$\mathcal{M}_4'$ is the equal to the received $\mathcal{M}_4$. If it holds, $\mathcal{U}$ computes $\mathcal{M}_5 = h_1(\mathcal{K}\|d_s\mathcal{P}\|\mathcal{SK}\|\mathcal{T}_3)$ and then sends the message $\{\mathcal{M}_5, \mathcal{T}_3\}$ to $\mathcal{S}$, where $\mathcal{T}_3$ is the current timestamp.

3. When receiving the authentication response message $\{\mathcal{M}_5, \mathcal{T}_3\}$ from $\mathcal{U}$, $\mathcal{S}$ checks whether $|\mathcal{T}_c - \mathcal{T}_3| \leq \Delta\mathcal{T}$ and computes $\mathcal{M}_5' = h_1(\mathcal{K}\|d_s\mathcal{P}\|\mathcal{SK}\|\mathcal{T}_3)$ to verify whether $\mathcal{M}_5'$ is the equal to the received $\mathcal{M}_5$. If both conditions holds, $\mathcal{S}$ authenticates $\mathcal{U}$ and accepts $\mathcal{SK}$ as the session key for further operations.

2.3 Password Change

When a patient $\mathcal{U}$ wants to change his/her password from $\mathcal{PW}_i$ to $\mathcal{PW}_i'$, $\mathcal{U}$ implores this phase. The password change phase needs to pass the following steps:

$\mathcal{U}$ inserts his/her smart card $\mathcal{SC}_i$ into a card reader and keys an identity $\mathcal{ID}_i$, password $\mathcal{PW}_i$, and biometric $\mathcal{B}_i$. Then, $\mathcal{SC}_i$ computes $h_1(\mathcal{ID}_i\|\mathcal{PW}_i \oplus \mathcal{H}(\mathcal{B}_i))$ and verifies whether it is equal to the stored value $\mathcal{V}_i$. If it is not valid, $\mathcal{SC}_i$ rejects password changing request.

1. $\mathcal{U}$ inputs a new password $\mathcal{PW}_i'$, and hence $\mathcal{SC}_i$ computes $\mathcal{V}_i' = h_1(\mathcal{ID}_i \|\mathcal{PW}_i' \oplus \mathcal{H}(\mathcal{B}_i))$. Finally, $\mathcal{SC}_i$ replaces $\mathcal{V}_i$ with $\mathcal{V}_i'$.

3   Security Flaws in Lu et al.'s Scheme

In this section, we present that Lu et al.'s scheme is still insecure and vulnerable to outsider attack, user impersonation attack, server impersonation attack and smart card stolen attack.

3.1 Outsider Attack

Any adversary $\mathcal{A}$ who is the legal patient and owns a smart card can obtain stored information $\{\mathcal{AID}_\mathcal{A}, \mathcal{V}_\mathcal{A}, h_1(\cdot), h_2(\cdot), \mathcal{H}(\cdot)\}$ from his/her smart card. Then, he/she can easily compute $h_2(\mathcal{X}) = \mathcal{ID}_\mathcal{A} \oplus \mathcal{AID}_\mathcal{A}$. The value $h_2(\mathcal{X})$ is same for each legal patient and the hash result of the server $\mathcal{S}$'s secret key $\mathcal{X}$.

3.2 User Impersonation Attack

Assume that an outsider adversary $\mathcal{A}$ intercepts the login request message $\{\mathcal{M}_1, \mathcal{M}_2, \mathcal{AID}_i, \mathcal{T}_1\}$ between $\mathcal{U}$ with $\mathcal{S}$. Then, $\mathcal{A}$ can easily impersonate a legal patient $\mathcal{U}$ to server $\mathcal{S}$.

$\mathcal{A}$ first computes $\mathcal{U}$'s identity $\mathcal{ID}_i$ by computing $\mathcal{AID}_i \oplus h_2(\mathcal{X})$. Hence, $\mathcal{A}$ generates a random number d_a and computes $\mathcal{K} = h_1(\mathcal{ID}_i\|h_2(\mathcal{X}))$, $\mathcal{M}_1' =$

$\mathcal{K}\oplus d_a\mathcal{P}$, $\mathcal{M}_2' = h_1(\mathcal{ID}_i||d_a\mathcal{P}|\mathcal{T}_1')$, where $\mathcal{T}_1'$ is current timestamp. Finally, $\mathcal{A}$ sends $\{\mathcal{M}_1', \mathcal{M}_2', \mathcal{AID}_i, \mathcal{T}_1'\}$ to $\mathcal{S}$.

1. Upon receiving the login request message from $\mathcal{A}$ who pretends to be $\mathcal{U}$, the login request messages can successfully pass $\mathcal{S}$'s verification. Hence, $\mathcal{S}$ generates a random number d_s and computes $\mathcal{M}_3 = \mathcal{K}\oplus d_s\mathcal{P}$, $\mathcal{SK} = d_s d_a\mathcal{P}$, $\mathcal{M}_4 = h_1(\mathcal{K}||d_a\mathcal{P}||\mathcal{SK}||\mathcal{T}_2)$, where $\mathcal{T}_2$ is the current timestamp. At last, $\mathcal{S}$ sends the response message $\{\mathcal{M}_3, \mathcal{M}_4, \mathcal{T}_2\}$ to $\mathcal{A}$.

2. After receiving the login response message, $\mathcal{A}$ first checks the freshness of $\mathcal{T}_2$. Then, $\mathcal{A}$ retrieves $d_s\mathcal{P}$ by computing $\mathcal{M}_3\oplus\mathcal{K}$ and computes $\mathcal{SK}' = d_a d_s\mathcal{P}$, $\mathcal{M}_4' = h_1(\mathcal{K}||d_a\mathcal{P}||\mathcal{SK}'||\mathcal{T}_2)$ to verify whether $\mathcal{M}_4'$ is the equal to the received $\mathcal{M}_4$. If it holds, $\mathcal{A}$ computes $\mathcal{M}_5 = h_1(\mathcal{K}||d_s\mathcal{P}||\mathcal{SK}||\mathcal{T}_3)$ and then sends the message $\{\mathcal{M}_5, \mathcal{T}_3\}$ to $\mathcal{S}$, where $\mathcal{T}_3$ is the current timestamp.

3. When receiving the message $\{\mathcal{M}_5, \mathcal{T}_3\}$ from $\mathcal{A}$, $\mathcal{S}$ continues to proceed the scheme without detected. Finally, $\mathcal{A}$ and $\mathcal{S}$ "successfully" agree on a session key $\mathcal{SK}$. Unfortunately, the server $\mathcal{S}$ mistakenly believes that he/she is communicating with the legitimate patient $\mathcal{U}$.

3.3 Server Impersonation Attack

Assume that an outsider adversary $\mathcal{A}$ intercepts the login request message $\{\mathcal{M}_1, \mathcal{M}_2, \mathcal{AID}_i, \mathcal{T}_1\}$ between $\mathcal{U}$ with $\mathcal{S}$. Hence, $\mathcal{A}$ can easily impersonate a server $\mathcal{S}$ to the patient $\mathcal{U}$.

$\mathcal{A}$ generates a random number d_a and computes $\mathcal{ID}_i = \mathcal{AID}_i\oplus h_2(\mathcal{X})$, $\mathcal{K} = h_1(\mathcal{ID}_i||h_2(\mathcal{X}))$, $\mathcal{M}_3 = \mathcal{K}\oplus d_a\mathcal{P}$, $\mathcal{SK} = d_a d_u\mathcal{P}$, and $M_4 = h_1(\mathcal{K}||d_u\mathcal{P}||\mathcal{SK}||\mathcal{T}_2)$ where $\mathcal{T}_2$ is the current timestamp. At last, $\mathcal{A}$ sends the response message $\{\mathcal{M}_3, \mathcal{M}_4, \mathcal{T}_2\}$ to $\mathcal{U}$.

1. After receiving the login response message from $\mathcal{A}$, $\mathcal{U}$ first checks the freshness of $\mathcal{T}_2$. Then, $\mathcal{U}$ retrieves $d_a\mathcal{P}$ by computing $\mathcal{M}_3\oplus\mathcal{K}$ and computes $\mathcal{SK}' = d_u d_a\mathcal{P}$, $\mathcal{M}_4' = h_1(\mathcal{K}||d_u\mathcal{P}||\mathcal{SK}'||\mathcal{T}_2)$ to verify whether $\mathcal{M}_4'$ is the equal to the received $\mathcal{M}_4$. If it holds, $\mathcal{U}$ computes $\mathcal{M}_5 = h_1(K||d_u\mathcal{P}||\mathcal{SK}||\mathcal{T}_3)$ and then sends the message $\{\mathcal{M}_5, \mathcal{T}_3\}$ to $\mathcal{A}$, where $\mathcal{T}_3$ is the current timestamp.

2. When receiving the message $\{\mathcal{M}_5, \mathcal{T}_3\}$ from $\mathcal{U}$, $\mathcal{A}$ and $\mathcal{U}$ "successfully" agree on a session key $\mathcal{SK}$. Unfortunately, the patient $\mathcal{U}$ mistakenly believes that he/she is communicating with the legitimate server $\mathcal{S}$.

3.4 Smart Card Stolen Attack

Assume that an outsider adversary $\mathcal{A}$ steals the smart card of any legitimate patient $\mathcal{U}$. Then, an adversary $\mathcal{A}$ can easily obtain the $\mathcal{U}$'s identity $\mathcal{ID}_i$ by computing $\mathcal{AID}_i\oplus h_2(\mathcal{X})$ and impersonate a legal patient $\mathcal{U}$ to server $\mathcal{S}$ using the method as same as user impersonation attack.

4   Conclusion

In this paper, we demonstrated that the security analysis of Lu et al.'s scheme and indicated that their scheme was vulnerable to outsider attack, user impersonation attack, server impersonation attack and smart card stolen attack. Finally, our further research direction ought to propose a secure use authentication scheme which can solve these problems.

Acknowledgments This work was supported by Institute for Information & communications Technology Promotion (IITP) grant funded by the Korea govern-ment (MSIP) (No.R0126-15-1111, The Development of Risk-based Authentication·Access Control Platform and Compliance Technique for Cloud Security).

References

1. Lambrinoudakis, C., Gritzalis, S.: Managing Medical and Insurance Information Through a Smart-Card-Based Information System. Journal of Medical Systems **24**(4), 213–234 (2000)
2. Lee, Y., Kim, J., Won, D.: Countermeasure on Password-Based Authentication Scheme for Multi-server Environments. Multimedia and Ubiquitous Engineering **308**, 459–466 (2013)
3. Choi, Y., Nam, J., Lee, D., Kim, J., Jung, J., Won, D.: Security Enhanced Anonymous Multi-Server Authenticated Key Agreement Scheme Using Smart Cards and Biometrics. The Scientific World Journal **2014**, 1–15 (2014)
4. Jung, J., Jeon, W., Won, D.: An enhanced remote user authentication scheme using smart card. In: Proceedings of the 8th International Conference on Ubiquitous Information Management and Communication (ICUIMC), pp. 62–66 (2014)
5. Mun, J., Kim, J., Lee, D., Jung, J., Won, D.: An improvement of efficient dynamic ID-based user authentication scheme using smart cards without verifier tables. In: Proceedings of the International Conference on Security and Management (SAM), pp. 152–156 (2015)
6. Awasthi, A.K., Srivastava, K.: A Biometric Authentication Scheme for Telecare Medicine Information Systems with Nonce. Journal of Medical Systems **37**(9964), 1–4 (2013)
7. Mishra, D., Mukhopadhyay, S., Kumari, S., Khan, M.K., Chaturvedi, A.: Security Enhancement of a Biometric based Authentication Scheme for Telecare Medicine Information Systems with Nonce. Journal of Medical Systems **35**(41), 1–11 (2014)
8. Tan, Z.: A User Anonymity Preserving Three-Factor Authentication Scheme for Telecare Medicine Information Systems. Journal of Medical Systems **38**(16), 1–9 (2014)
9. Arshad, H., Nikooghadam, M.: Three-Factor Anonymous Authentication and Key Agreement Scheme for Telecare Medicine Information Systems. Journal of Medical Systems **38**(136), 1–12 (2014)
10. Koblitz, N.: Elliptic curve cryptosystems. Mathematics of Computation **48**(177), 203–209 (1987)
11. Lu, Y.R., Li, L.X., Peng, H.P., Yang, Y.X.: An Enhanced Biometric-Based Authentication Scheme for Telecare Medicine Information Systems using Elliptic Curve Cryptosystem. Journal of Medical Systems **39**(32), 1–8 (2015)

Cryptanalysis on Symmetric Key Techniques Based Authentication Scheme for Wireless Sensor Networks

Younsung Choi, Youngsook Lee and Dongho Won

Abstract In wireless sensor networks, user authentication scheme is a critical and important security issue to prevent adversary's illegal approach to wireless sensors. After Das introduce a user authentication scheme for wireless sensor networks, various studies had proceeded to proposed more secure and efficient authentication scheme but many schemes had security problem on smart card attack. So Chem et al. suggested a secure user authentication scheme against smart card loss attack using symmetric key techniques but this scheme does not still resolve some security vulnerability. So by the cryptanalysis, this paper shows that Chem et al's scheme has problems on perfect forward secrecy, session key exposure by gateway node, anonymity, and the password check.

Keywords User authentication scheme · Cryptanalysis · Wireless Sensor Networks

1 Introduction

Recently, wireless sensor networks (WSNs) have been substantially investigated by researches. It can observe various hazardous conditions, such as volcanic

Y. Choi · D. Won(✉)
Department of Computer Engineering, Sungkyunkwan University, 2066 Seoburo, Suwon, Gyeonggido 440-746, Korea
e-mail: {yschoi,dhwon}@security.re.kr

Y. Lee
Department of Cyber Investigation Police, Howon University,
64 Howon University 3 Gil, Impi-Myeon, Gunsan-Si, Jeonrabuk-Do 573-718, Korea
e-mail: ysooklee@howon.ac.kr

This research was supported by Basic Science Research Program through the National Research Foundation of Korea funded by the Ministry of Science, ICT & Future Planning (NRF-2014R1A1A2002775)

© Springer Science+Business Media Singapore 2015

D.-S. Park et al. (eds.), *Advances in Computer Science and Ubiquitous Computing*,
Lecture Notes in Electrical Engineering 373,
DOI: 10.1007/978-981-10-0281-6_2

temperature, battlefield surveillance[1]. To use more secure and efficient communication on WSNs, various studies on user authentication had progressed from previous times. Das introduced a two-factor user authentication scheme firstly using password and smart card[2]. Since then, various researchers analyzed Das's authentication scheme and showed the problems of the scheme, and proposes more secure and efficient user authentication scheme for WSNs. Nyang and Lee[3] pointed out that Das's scheme has vulnerability on an offline password guessing attack and node compromise attack. Khan-Alghathbar showed Das's scheme is insecure to gateway nodes bypass attack[4]. After analyzing Das's scheme and Khan-Alghathbar's scheme, Vaidya et al. point out that they have security problem attacks such as smart card loss attacks[5].

To resolve the smart card attack such as offline password guessing attack, Yuan proposed biometric-based authentication scheme[6], and Yeh et al. improve Yuan's scheme[7] and Choi et al. proposed security enhanced scheme using elliptic curve cryptography for WSNs to solve Yuan and Yeh et al.'s security problem[8].

Chen et al. proposed a secure user authentication scheme against smart card loss attack for WSNs using symmetric key techniques. This scheme improves the security using only password and symmetric key algorithm without biometrics and public key crypto-system[9]. But we review and analyze Chen et al.'s scheme and found out various security problems on this scheme.

The remainder of this paper is organized as follows. Section 2 briefly reviews Chen et al.'s authentication scheme while Section 3 provides a detailed security analysis on perfect forward secrecy, session key exposure by gateway node, anonymity, and password check problem of Chen et al.'s scheme. Section 4 concludes this paper.

2   Review of Chen et al.'s Authentication Scheme

In this section, we describe the phase of Chen et al.'s scheme. This scheme is divided into three phase: the registration phase, password updating phase, and authentication phase. Figure 1 shows the authentication phase of Chen et al.'s scheme. Gateway node GW has secret values x_a and x_s, and share $h(x_s||SID_n)$ with sensors node S_n[9].

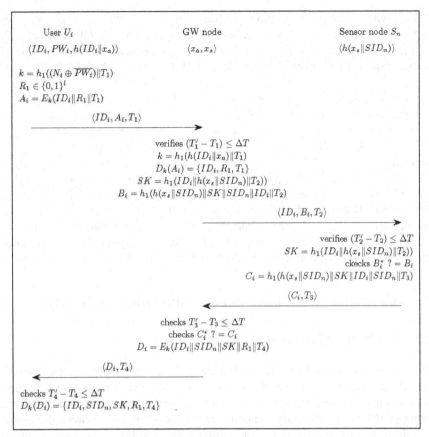

Fig. 1 Authentication phase of Chen et al.'s scheme

[Registration phase] (1) User U_i selects identifier ID_i and password PW_i, generates a random number b, and computes $\overline{PW_i} = h(PW_i \| b)$. Then U_i sends ID_i and $\overline{PW_i}$ to the gateway node GW over a secure channel. (2) GW computes $N_i = h(ID_i \| x_a) \oplus \overline{PW_i}$. GW stores $\{ID_i, N_i, h(\cdot)\}$ into a smart card and sends it to U_i. (3) After U_i receives the smart card, U_i inputs b into smart card and finishes the registration phase.

[Password updating phase] (1) U_i inserts smart card into the terminal and enters ID_i, old password PW_i, new password PW_i^*. (2) U_i's smart card verifies the entered ID_i using stored value ID_i. If ID_i is accurate, then U_i is approved to change the password. (3) U_i's smart-card calculates $\overline{PW_i} = h(PW_i \| b)$, $\overline{PW^*}_i = h(PW_i^* \| b)$, and $N_i^* = N_i \oplus \overline{PW_i} \oplus \overline{PW^*}_i$ and then replaces old value N_i with new value N_i^*, and finishes this phase.

[Authentication phase] (1) U_i inserts smart card to terminal and ID_i and PW_i. Smart card computes $\overline{PW_i} = h(PW_i \| b)$, $k = h_1((N_i \oplus PW_i) \| T_1)$, T_1 is current timestamp of U_i. The smart card generates random number $R_1 \in \{0, 1\}^l$, computes $A_i = E_k(ID_i \| R_1 \| T_1)$,

$E(\cdot)$ is asymmetric key cryptography function; $E_k(*)$ means that "$*$" is encrypted by $E(\cdot)$ with symmetric key k. Then it sends $\{ID_i, A_i, T_1\}$ to GW. **(2)** GW receives $\{ID_i, A_i, T_1\}$. GW verifies T_1. If $(T'_1 - T_1) \leq \Delta T$, then GW continues to the next step, ΔT denotes expected time interval for the delay. GW computes $k=h_1(h(ID_i\|x_a)\|T_1)$ and $D_k(A_i) = \{ID_i, R_1, T_1\}$, where $D_k(*)$ means that "$*$" is decrypted by the key k. Then GW checks whether decrypted messages ID_i and T_1 are equal to received ones. GW computes $SK=h_1(ID_i\| h(x_s\|SID_n)\|T_2)$, $B_i = h_1(h(x_s\| SID_n)\|SK\|SID_n\|ID_i\|T_2)$, SK is the session key between U_i and S_n. Then the GW sends the message $\{ID_i, B_i, T_2\}$ to the sensor node S_n. **(3)** S_n validates T_2, computes the $SK = h_1(ID_i\|h(x_s\|SID_n)\|T_2)$. S_n computes $B_i^* = h_1(h(x_s\| SID_n)\|SK\|SID_n\|ID_i\|T_2)$ and then checks whether $B_i^*= B_i$. S_n computes $C_i= h_1(h(x_s\| SID_n) \|SK \| ID_i\|SID_n\| T_3)$. S_n sends the message $\{C_i, T_3\}$ to GW. **(4)** GW first verifies T_3, computes $C_i^*=h_1(h(x_s\| SID_n)\|SK\|ID_i\|SID_n\|T_3)$ and then checks whether $C_i^* = C_i$. GW computes $D_i= (ID_i\| SID_n\| SK \|R_1\| T_4)$, then sends the message $\{D_i, T_4\}$ to U_i. **(5)** U_i computes $(D_i) = \{ID_i, SID_n, SK, R_1, T_4\}$; then U_i checks whether the decrypted messages ID_i, R_1, and T_4 are equal to the previous ones. If it is same, user U_i establishes trust on GW and establishes SK with sensor node S_n.

3   Cryptanalysis of Chen et al.'s Authentication Scheme

3.1   No Perfect Forward Secrecy

Chen et al.'s authentication scheme does not provide the perfect forward secrecy. Perfect forward secrecy means that a session key derived from a set of long-term keys will not be compromised if one of the long-term keys is compromised in the future [8,10]. So an adversary can compute the session key sk between the U_i and S_j if the adversary knows the one of long-term key $h(x_s\|SID_n)$ in future. Figure 2 describes the lack of perfect forward secrecy on Chen's authentication scheme. Adversary can get ID_i and T_2 in public communication. If adversary know one of user's long-term secret $h(x_s\|SID_n)$, the adversary can compute the session key $SK = h_1(ID_i\| h(x_s\|SID_n)\|T_2)$. Moreover, the adversary can get the previous information ID_{pi} and T_{p2}. Then, the adversary can compute all of previous session key SK_p between user and sensor node.

> - Adversary get ID_i and T_2 in public channel.
> - Adversary know one of long-term secret : $h(x_s \| SID_n)$.
>   - Adversary has ID_i and T_2 , computes SK as follows,
>   - $SK = h_1(ID_i \| h(x_s \| SID_n) \| T_2)$.
> - Adversary can get previous ID_{pi} and T_{p2}, then computes SK_p.
>   - $SK_p = h_1(ID_{pi} \| h(x_s \| SID_n) \| T_{p2})$.
> - Adversary can compute all of previous session key SK_p.

Fig. 2 Lack of perfect forward secrecy on Chen et al.'s scheme

3.2   Session Key Exposure by GW node

In Chen et al.'s scheme, session key is used for secure communication between U_i and S_n after authentication phase. It is important matter that anyone cannot compute SK with the exception of U_i and S_n even if GW node is trust node. However, Chen et al.'s scheme is vulnerable to this problem. Figure 3 describes the session key exposure by GW on Chen et al.'s scheme. GW can have ID_i, x_s, SID_n and T_2, so GW can compute session key $SK=h_1(ID_i\|\ h(x_s\|SID_n)\|T_2)$. Therefore, GW can compute all of session key between U_i and S_n. It means that GW decrypts the secret message between U_i and S_n so GW can gain all of the important information between every U_i and S_n.

- Session key SK is used to communicate securely between U_i and S_n.

- It is important that anyone cannot compute the SK except for U_i and S_n.

- But GW can compute the SK between all of registrated U_i and S_n.

    • GW has ID_i, x_s, SID_n, T_2.

    • $SK = h_1(\ ID_i\ \|\ h(\ x_s\ \|\ SID_n)\ \|\ T_2\ )$.

$\Rightarrow$ GW can compute all session key sk between U_i and S_n.

$\Rightarrow$ GW can decrypt the secret messages between U_i and S_n.

$\Rightarrow$ GW can acquire the important information between U_i and S_n.

Fig. 3 Session key exposure by GW node of Chen et al.'s scheme

3.3   Lack of Anonymity

Figure 4 describes the problem due to the lack of anonymity on Chen et al.'s scheme. User U_i sends own ID_i to GW using public communication, and GW sends ID_i to the sensor S_n without any protection. So an adversary can easily acquire ID_i from public communications. Therefore the adversary can get various information using user's ID_i. For observing GW's receiving communication, the adversary can obtain how many users are registered to GW. Moreover, the

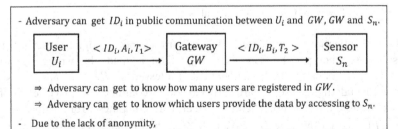

Fig. 4 Lack of Anonymity on Chen et al.'s scheme

adversary can get to know which users communicate with S_n, in other words, which users obtain the data by accessing to S_n. the lack of anonymity in Chen et al.'s scheme cause various problem so it need to be addressed by providing user anonymity through various *ID* protection technique.

3.4   Lack of Password Check

Chen et al.'s scheme cannot provide the password check in user's smart card therefore various problems occur in authentication phase and password updating phase. Figure 5 describes the problems due to the lack of password check on Chen et al.'s scheme.

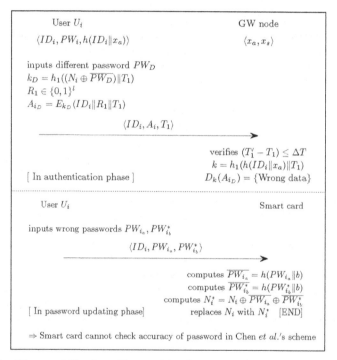

Fig. 5 Lack of Password Check on Chen et al.'s scheme

In authentication phase, user's smart card cannot recognize user's wrong password so encrypts the data using k_D, which is made by wrong password. Moreover, U_i sends the faulty authentication messages $<ID_i, A_{iD}, T_1>$ to GW. In password updating phase, smart card cannot discontinue the phase even if U_i inputs wrong old password. Therefore, smart card computes wrong values such as $\overline{PW_{ia}}$, $\overline{PW^*_{ib}}$, and $N_i^* = N_i \oplus \overline{PW_{ia}} \oplus PW^*_{ib}$, then replace N_i with wrong N_i^*. This problem causes that user cannot execute authentication phase due to wrong N_i^* even if U_i inputs normal new password.

4   Conclusion

To solve various problems such as smart card loss attack, Chen et al. proposed a secure user authentication scheme using symmetric key techniques for WSNs but it still has some problems. In this paper, we first review Chen et al.'s scheme and analyze this scheme using the cryptanalysis. So we have identified that Chen et al.'s scheme designed for WSNs is vulnerable to: lack of perfect forward secrecy, session key exposure by GW node, lack of anonymity, lack of password check. This study on cryptanalysis analysis can be used to make more secure authentication scheme.

References

1. Nam, J., Choo, K.K.R., Han, S., Kim, M., Paik, J., Won, D.: Efficient and Anonymous Two-Factor User Authentication in Wireless Sensor Networks: Achieving User Anonymity with Lightweight Sensor Computation. PLOS ONE **10**(4) (2015)
2. Das, M.L.: Two-factor user authentication in wireless sensor networks. IEEE Transactions on Wireless Communications **8**(3), 1086–1090 (2009)
3. Nyang, D., Lee, M.K.: Improvement of Das's two-factor authentication protocol in wireless sensor networks. IACR Cryptology ePrint Archive (2009)
4. Khan, M.K., Alghathbar, K.: Security analysis of 'two-factor user authentication in wireless sensor networks'. In: Advances in Computer Science and Information Technology. Lecture Notes in Computer Science, vol. 6059, pp. 55–60. Springer, Germany (2010)
5. Vaidya, B., Makrakis, D., Mouftah, H.T.: Improved two factor user authentication in wireless sensor networks. In: IEEE International Conference on Wireless and Mobile Computing, Networking and Communications (WiMob 2010), pp. 600–606, October 2010
6. Yuan, J.J.: An enhanced two-factor user authentication in wireless sensor networks. Telecommunication Systems **55**(1), 105–113 (2014)
7. Yeh, H.L., et al.: A secured authentication protocol for wireless sensor networks using Elliptic Curves Cryptography. Sensors **11**(5), 4767–4779 (2011)
8. Choi, Y., Lee, D., Kim, J., Jung, J., Nam, J., Won, D.: Security enhanced user authentication protocol for wireless sensor networks using elliptic curves cryptography. Sensors **14**(6), 10081–10106 (2014)
9. Chen, L., Fushan, W., Chuangui, M.: A Secure User Authentication Scheme against Smart-Card Loss Attack for Wireless Sensor Networks Using Symmetric Key Techniques. International Journal of Distributed Sensor Networks (2015)
10. Kim, J., Lee, D., Jeon, W., Lee, Y., Won, D.: Security analysis and improvements of two-factor mutual authentication with key agreement in wireless sensor networks. Sensors **14**(4), 6443–6462 (2014)

Proving Sufficient Completeness of Constructor-Based Algebraic Specifications

Masaki Nakamura, Daniel Gaina, Kazuhiro Ogata and Kokichi Futatsugi

Abstract OBJ algebraic specification languages, for example, OBJ3, CafeOBJ and Maude, are formal specification languages which support several sophisticated functions to describe and verify large and complex specifications. Recently, the proof score method, which is an interactive formal verification method for OBJ languages, based on constructor-based algebras has been developed and several practical case studies have been reported. Sufficient completeness is one of the most important properties of constructor-based specifications, which guarantees the existence of initial models. In this study, we give a sufficient condition for sufficient completeness of constructor-based specifications based on the theory of term rewriting.

Keywords Algebraic specification · CafeOBJ · Constructor-based specification · Sufficient completeness · Parameterized specification · Term rewriting

1 Introduction

In algebraic specifications and term rewriting systems, operators may be divided into constructors and defined operators. Defined operators are expected to be defined over terms which are constructed from only constructors. Such a property is called sufficient completeness [4, 5]. Recently, CafeOBJ algebraic specification language [1] supports description of specifications whose models are constructor-based algebras [2]. In constructor-based specifications, sufficient completeness is

M. Nakamura(✉)
Toyama Prefectural University, 5180 Kurokawa, Imizu, Toyama 939-0398, Japan
e-mail: ikasama1974+csa2015@gmail.com

D. Gaina · K. Ogata · K. Futatsugi
Japan Advanced Institute of Science and Technology, 1-1 Asahidai, Nomi, Ishikawa 923-1292, Japan

© Springer Science+Business Media Singapore 2015
D.-S. Park et al. (eds.), *Advances in Computer Science and Ubiquitous Computing*,
Lecture Notes in Electrical Engineering 373,
DOI: 10.1007/978-981-10-0281-6_3

defined over terms constructed from not only constructors but also variables whose sorts are not sorts of constructors. It allows us to describe specifications of a high abstraction level, like generic lists. In this study, we give a sufficient condition for sufficient completeness of constructor-based specifications based on the theory of term rewriting.

2   Preliminaries

We introduce notions and notations related to constructor-based algebraic specifications in the literatures [2, 3]. A constructor-based signature $(S, \leq, \Sigma, \Sigma^C)$ (abbr. a signature Σ) consists of a set S of sorts, a partial order $\leq$ on S, an S-sorted set Σ of operators, and a set $\Sigma^C \subseteq \Sigma$ of constructors. An operator $f \in \Sigma_{ws}$ is indexed with the string $ws \in S^+$ of sorts, called its rank (its arity $w \in S^*$ and sort $s \in S$). For example, for a sort Nat, an operator $+ \in \Sigma_{\text{Nat Nat Nat}}$ is a binary operator whose arity is Nat Nat and sort is Nat. We may use Σ and Σ^C as the unions of all Σ_{ws} and Σ_{ws}^C respectively if no confusion arises. A sort of a constructor is called a constrained sort. The set of constrained sorts is defined as $S^{cs} = \{s \in S \mid f \in \Sigma_{ws}^C\}$. Non-constrained sorts are called loose and the set of loose sorts are denoted by $S^{ls} = S \setminus S^{cs}$. An operator is called constrained if its sort is constrained. The set of constrained operators is denoted by $\Sigma^{S^{cs}} = \{f \in \Sigma \mid f \in \Sigma_{ws}, s \in S^{cs}\}$. The following is an example of signatures written in CafeOBJ specification language, which denotes generic lists:

```
mod* LIST{
  [Elt, List]
  op nil : -> List  {constr}
  op (_;_) : Elt List -> List  {constr}
  op rev : List List -> List  ...
}
```

where LIST is a name of the specification (the module), Elt and List are sorts, operators are declared after the keyword op with the rank, like op f : w -> s for $f \in \Sigma_{ws}$. Constructors are declared with the operator attribute constr. Thus, $S = \{\text{Elt}, \text{List}\}$, $\Sigma_{\text{List}} = \{\text{nil}\}$, $\Sigma_{\text{Elt List List}} = \{\_;\_\}$, $\Sigma_{\text{List List List}} = \{\text{rev}\}$, $\Sigma^C = \{\text{nil}, \_;\_\}$, $S^{cs} = \{\text{List}\}$ and $\Sigma^{S^{cs}} = \{\text{nil}, \_;\_, \text{rev}\}$.

A term is a well-sorted tree structure whose nodes are operators and leaves are variables. In CafeOBJ, variables are declared after var or vars.

```
vars E E' : Elt
var L : List
```

For example, a constant nil is a term of List, and E ; L and rev(E ; E' ; L) are terms of List. Note that underlines in the operator declaration stand for the positions of arguments in CafeOBJ. Let X be an S-sorted set of variables. The set of terms over Σ and X is denoted by $T_\Sigma(X)$. A constructor term is a term constructed from constructors and variables of loose sorts, called loose variables. The set of constructor terms is denoted by $T_{\Sigma c}(Y)$ where Y be a set of loose variables. A constrained term $t \in T_{\Sigma^{cs}}(Y)$ is a term constructed from constrained operators and loose variables.

A (constructor-based) specification consists of a (constructor-based) signature and axioms. We deal with only equations as axioms in this study. An equation, denoted by $l = r$, is a pair of terms l and r of a same sort. In CafeOBJ, an equation is declared after eq. The following is the last half of the specification LIST:

```
mod* LIST{   ...
    var E : Elt   vars L   L' : List
    eq rev(nil, L) = L .
    eq rev(E ; L, L') = rev(L, E ; L') .
}
```

An ordinary specification $((S, \leq, \Sigma), E)$ denotes a set of $(S, \leq, \Sigma)$-algebras satisfying its axioms E, and a constructor-based specification $((S, \leq, \Sigma, \Sigma^C), E)$ denotes its subset where a carrier set of constrained sorts consists of interpretations of constructor terms, which are formally defined as follows: A Σ-algebra M interprets a sort s as a carrier set M_s and an operator $f \in \Sigma_{s_1 \cdots s_n s}$ as a function $M_f : M_{s_1} \times \cdots \times M_{s_n} \to M_s$. For an S-sorted assignment $a : X \to M$, an S-sorted map $a^\# : T_\Sigma(X) \to M$ is defined as the unique extension to a Σ-homomorphism, that is, $a^\#(f(t_1, \ldots, t_n)) = M_f(a^\#(t_1), \ldots, a^\#(t_n))$. A Σ-algebra M satisfies an equation $l = r$ when $a^\#(l) = a^\#(r)$ for any assignment $a : X \to M$. A constructor-based Σ-algebra M is a Σ-algebra which satisfying that for each constrained sort $s \in S^{cs}$ and each element $m \in M_s$, there exists a constructor term $t \in T_{\Sigma c}(Y)$ and an assignment $a : Y \to M$ such that $a^\#(t) = m$. We denote the set of all denotations of a specification SP by $|Mod(SP)|$. For example, let $L \in |Mod(LIST)|$. For any element $l \in L_{List}$, there exists a constructor term $t \in T_{\Sigma c}(Y)_{List}$ and assignment $a : Y \to L$ such that $l = a(t)$. Roughly speaking, each element of L_{List} has a corresponding constructor term $y_1; \cdots; y_n;$ nil.

Sufficient completeness is one of the most important properties of constructor-based specifications since it guarantees the existence of initial models. Roughly speaking, it guarantees all functions are well-defined on constructors. For example, an operator rev should returns a constructor term for any pair of input constructor terms when it is sufficiently complete. Sufficient completeness is defined as follows:

Definition 1. A constructor-based specification $((S, \leq, \Sigma, \Sigma^C), E)$ is sufficiently complete if for all $M \in |Mod((S, \leq, \Sigma, \Sigma^{S^{cs}}), E)|$ we have $M \in |Mod(S, \leq, \Sigma, \Sigma^C)|$.

It is known that sufficient completeness is undecidable. A main purpose of this study is to give a sufficient condition of sufficient completeness which can be proved automatically.

3   A Sufficient Condition of Sufficient Completeness

In [2], a sufficient condition of sufficient completeness is given as follows:

Proposition 1. A constructor-based specification $((S, \leq, \Sigma, \Sigma^C), E)$ is sufficiently complete if for all constrained term $t \in T_{\Sigma^{scs}}(Y)$ there exists a constructor term $t^c \in T_{\Sigma^c}(Y)$ such that $E \vDash t = t^c$.

LIST satisfies the condition of Proposition 1. For example, a constrained term $rev(y; z; nil, x; nil)$ is equivalent to a constructor term: $rev(y; z; nil, x; nil) = rev(z; nil, y; x; nil) = rev(nil, z; y; x; nil) = z; y; x; nil$. However, the following specification of a map function on lists does not satisfy it:

```
mod! MAP{
  [ Elt, List ]
  op nil : -> List {constr}
  op _ ; _ : Elt List -> List {constr}
  op map : List -> List
  op f : Elt -> Elt
  var E : Elt var L : List
  eq map(nil) = nil .
  eq map(E ; L) = f(E) ; map(L) .
}
```

where the operator map takes a list and returns a list where a function f is applied to all elements of the list, e.g. $map(x; y; z; nil) = f(x); f(y); f(z); nil$. The left-hand side of this equation is a constrained term, however it does not have an equivalent constructor term. Note that the right-hand side is not a constructor term since f is not a constructor. We call such an operator whose sort is loose a loose operator, defined as $\Sigma^{S^{ls}} = \{f \in \Sigma \mid f \in \Sigma_{ws}, s \in S^{ls}\}$. From the definition, we have $\Sigma = \Sigma^{scs} \uplus \Sigma^{S^{ls}}$. In this example, the operator map looks like a well-defined on lists. Indeed, while it does not satisfy the condition of Proposition 1, it is sufficiently complete (proved later). To handle the above example, we weaken the condition of Proposition 1 as follows:

Theorem 1. A constructor-based specification $((S, \leq, \Sigma, \Sigma^C), E)$ is sufficiently complete if for all constrained term $t \in T_{\Sigma^{scs}}(Y)$ there exists $u \in T_{\Sigma^C \cup \Sigma^{S^{ls}}}(Y)$ such that $E \vDash t = u$.

Proof.   Let $m \in M_s$ for $M \in \left| Mod\left((S, \leq, \Sigma, \Sigma^{scs}), E\right)\right|$ and $s \in S^{cs}$.   From Definition 1, we need to show $M \in \left| Mod\left((S, \leq, \Sigma, \Sigma^C), E\right)\right|$, that is, there exists a term $t^c \in T_{\Sigma^c}(Y)$ and an assignment $f : Y \to M$ where Y is a finite set of loose variables, such that $f^{\#}(t^c) = m$, where $f^{\#} : T_{\Sigma^c}(Y) \to M$ is the unique extension of f to a Σ^C-homomorphism. From the definition of denotation of constructor-based specifications and Definition 1, for $M \in \left| \text{Mod}\left((S, \leq, \Sigma, \Sigma^{s^{cs}}), E \right) \right|$, we have $m = f^{\#}(t)$ for some $t \in T_{\Sigma^{s^{cs}}}(Y)$ and $f : Y \to M$. By the hypothesis, $M \vDash t = u$ for a term $u \in T_{\Sigma^C \cup \Sigma^{sls}}(Y)$. It follows that $m = f^{\#}(t) = f^{\#}(u)$. Next, from u and f, we make a constructor term $t^c \in T_{\Sigma^C}(Y')$ and $f' : Y' \to M$ such that $f'^{\#}(t^c) = m$. A constructor term t^c is defined by replacing all maximal subterms $u_i = g(\cdots)$ whose root is a loose operator $g \in \Sigma^{sls}$ in t^c with fresh variables y_i. The maximal subterm means that g is the first non-constructor operator from the root position of u to the subterm $u_i = g(\cdots)$. For example, when $u = f(A); f(f(B)); f(f(f(C)); \text{nil}$, we have $t^c = y_0; y_1; y_2; \text{nil}$. Let Y' be $Y \cup \{y_0, y_1, \dots\}$ and define $f' : Y' \to M$ as $f'(y) = y$ for $y \in Y$ and $f'(y_i) = f^{\#}(u_i)$ for $y_i \in Y' \setminus Y$. Then, we have $f'^{\#}(t^c) = f^{\#}(u) = m$.   □

The term $map(x; y; z; \text{nil})$ has no equivalent constructor term as shown above but has a term constructed from constructors, loose variables and loose operators: $f(x); f(y); f(z); \text{nil}$.

4   Proving Sufficient Completeness by Term Rewriting

In this section, we give a way to prove sufficient completeness based on Theorem 1 with term rewriting techniques. Term rewriting is useful for equational reasoning, where bidirectional equations are regarded as directional rewrite relations. A rewrite rule is a pair (l, r) of terms of a same sort, denoted by $l \to r$. A set of rewrite rules is called a term rewriting system (TRS). A term t is written by $l \to r \in R$ into t' when t has an instance of the left-hand side l as a subterm and t' is the result of replacing the subterm with the corresponding instance of the right hand side, denoted by $t \to_R t'$. The reflexive and transitive closure of $\to_R$, i.e. zero, one or more than one rewrite steps, is denoted by $t \to_R^* t'$. An equational specification (or a set of equations) can be regarded as a TRS where all equations are regarded as left-to-right rewrite rules. For example, for the specification MAP, we have $map(A; B; \text{nil}) \to_{MAP} f(A); map(B; \text{nil}) \to_{MAP} f(A); f(B); map(\text{nil}) \to_{MAP} f(A); f(B); \text{nil}$.

Hereafter we may deal with a TRS as a specification (or its axioms) and vice versa if no confusion occurs. We give a definition of sufficient completeness w.r.t. rewriting, written by SCR, as follows:

Definition 2. A constructor-based specification $((S, \leq, \Sigma, \Sigma^C), E)$ is sufficiently complete w.r.t. rewriting (SCR) if for all constrained term $t \in T_{\Sigma^{s^{cs}}}(Y)$ there exists $u \in T_{\Sigma^C \cup \Sigma^{sls}}(Y)$ such that $t \to_E^* u$.

We call the condition of Theorem 1 SCE (sufficient completeness w.r.t. equations). It is trivial that SCR implies SCE since the rewrite relation $\to_E^*$ is included in the equivalence relation $=_E$. Thus, SCR implies sufficient completeness. A TRS R is called terminating if there is no infinite rewrite sequence $t_0 \to_R t_1 \to_R \cdots$. If a specification is sufficiently complete and terminating, for each constrained term $t \in T_{\Sigma^{Scs}}(Y)$ we can compute $u \in T_{\Sigma^C \cup \Sigma^{Sls}}(Y)$ such that $t \to_E^* u$. A constrained and non-constructor operator $g \in \Sigma^{Scs} \setminus \Sigma^C$ is interpreted into a function defined on constrained sorts. A term without such operators $\Sigma^{Scs} \setminus \Sigma^C$ can be regarded as an answer. In this sense, sufficient completeness and termination guarantees that an input term has an answer and it can be computed in finite time. To apply SCR to prove sufficient completeness, we need to find u such that $t \to_E^* u$, however it is a kind of a reachability problem and known as another undecidable property. It is known that under the assumption of termination, to find such a term u is decidable [4, 5] in ordinary specifications. We apply the technique to constructor-based specifications.

Definition 3. For $\Sigma' \subseteq \overline{\Sigma^C}$, a term $f(\bar{t})$ is Σ'-basic iff $f \in \overline{\Sigma^C \cup \Sigma'}$ and $\bar{t}$ are terms constructed from $\Sigma^C \cup \Sigma'$ and loose variables Y, that is, $\bar{t} \in T_{\Sigma^C \cup \Sigma'}(Y)$.

Corollary 1. If a constructor-based specification $((S, \leq, \Sigma, \Sigma^C), E)$ is terminating and all Σ^{Sls}-basic terms are E-reducible, i.e., it can be rewritten by E, then it is SCR.

Consider MAP. Although termination is undecidable in general, there are lots of decidable termination proving methods and tools. MAP is terminating and can be proved by a classical termination method, RPO (recursive path ordering) for example. Let $\Sigma' = \Sigma^{Sls} = \{ f \}$. Then, $\overline{\Sigma^C \cup \Sigma^{Sls}} = \{ \text{map} \}$ and the terms map(nil) and $\text{map}(f(X) ; \text{nil})$ are Σ^{Sls}-basic. Note that $f(X)$ is not. To prove sufficient completeness by Corollary 1, we need to prove E-reducibility of basic terms. Although basic terms are infinite in general, by using techniques of cover sets in [4], E-reducibility of all basic terms can be decidable. For example, in MAP, E-reducibility of the cover set $\{\text{map(nil)}, \text{map}(E ; L) \}$ implies E-reducibility of all basic terms. Therefore, MAP is sufficiently complete.

5   Conclusion

We gave an example of constructor-based specifications whose sufficient completeness cannot be proved by the existing sufficient condition. Through analysing the problem, we proposed a new sufficient condition applicable such examples and a proof method based on term rewriting. The result is expected to be implemented into CafeOBJ system as a pre-processor to check if an input specification is well-defined before doing specification simulations and interactive verifications with proof scores.

Acknowledgments This work was supported by MEXT KAKENHI Grant Number 23220002.

References

1. CafeOBJ. http://cafeobj.org/
2. Futatsugi, Kokichi, Gaina, Daniel, Ogata, Kazuhiro: Principles of proof scores in CafeOBJ. Theorem Computer Science **464**, 90–112 (2012)
3. Gaina, D., Futatsugi, K.: Initial semantics in logics with constructors. Journal of Logic and Computation **25**(1), 95–116 (2015)
4. Kapur, D., et al.: Sufficient-completeness, ground-reducibility and their complexity. Acta Informatica **28**, 311–350 (1991)
5. Schernhammer, F., Meseguer, J.: Incremental checking of well-founded recursive specifications modulo axioms. In: Proceedings of the 13th International ACM SIGPLAN Conference on Principles and Practice of Declarative Programming, pp. 5–16 (2011)

Effect of Zooming Speed and Pattern on Using IPTV by Zoomable User Interface

Samuel Sangkon Lee and Haeng-Suk Chae

Abstract The users preferred faster movement for the zooming speed. The execution performance of zooming interface affects the actual running time. This study was designed for the movement of robot arm so users expected the movement similar to the actual movement of human. It is considered to be natural that, like human's movement, the robot arm slowly adjusts its position toward a target after the rapid burst movement in early stage. It is assumed that users expect the movement different from human's movement for the zooming. Users expect the geometric extension of the object detected through visual organs as the movement by three dimensional approach of the visual object. According to our experiment result, it is shown that users have relatively not bad feeling from Speed 6 to Speed 12 excluding Speed 4 condition which is the lowest speed.

Keywords Zooming speed · Zooming patter · Zoomable UI · User preference · Naturalness · IPTV

1 Introduction

As the IPTV technology allowing bi-directional interaction has emerged recently, the web browsing by a TV is being issued. Since users use remote controls grasping with their hands in comfortable positions for TV contents, they may have

S.S. Lee(✉)
Department of Computer Science and Engineering,
Jeonju University, 303 Chonjam-Ro, Wansan-Gu, Jeonju, South Korea
e-mail: samuel@jj.ac.kr

H.-S. Chae
KT Corporation, 90 Bulljeong-Ro, Bundang-Gu, Seongnam City,
Gyeonggi-Do, South Korea
e-mail: hs.chae@kt.com

© Springer Science+Business Media Singapore 2015
D.-S. Park et al. (eds.), *Advances in Computer Science and Ubiquitous Computing*,
Lecture Notes in Electrical Engineering 373,
DOI: 10.1007/978-981-10-0281-6_4

23

difficulties for information search if hundreds of words such as conventional desktop GUIs and tens of links are displayed on a screen.

Zoomable UI (hereinafter ZUI) is a kind of GUI to help information search in the limited screen area and a good way to maintain the availability when browsing webs in an IPTV. Since, in particular, the web browsing must search meaningful information in comparison with picture browsing or map navigation zooming spatial information, the semantic zooming method is more effective to differentiate the information amount in each extension stage [Cockburn, Savage, 2003]. However, the semantic zooming method must be stepwise, and not continuous, to display different information amount in a constant stage, unlike the physical zooming method. Therefore, a movement occurs when extending each stage, and its speed and pattern make users feel the naturalness of the ZUI.

2   Background

2.1   Zooming Speed

The zooming speed can be determined as size change ratio per second. For example, if the size increases from 2 to 4 for a second, the zooming speed will be $4/2 = 2$. The current zooming UI, Jazz, is using smooth zooming. The speed is, relatively, very slow but is allowed because the contents provide by the current zooming UI are small. If the zooming is applied at this speed in a large screen, the speed will make users feel very slow and cause frustration and errors. Hence, faster speed is required, but too fast speed can cause anxiety and another error [Shneiderman, 1997]. As a result of the study for the speed of the existing zooming UI, the appropriate zooming speed can be calculated with the equation below [Guo et al, 2000]:

$$\text{Speed} = \text{Zoom Rate of 8 x per/sec.} \qquad (1)$$

2.2   Zooming Pattern

When designing the speed of the zooming UI, the zooming progress pattern must be, also, considered. The pattern is called zooming pattern and can be divided as whether the zooming is progressed at a constant speed or accelerated. In addition, the position of the velocity peak can be changed depending on the point the speed is fastest in the whole moving time. The example of the pattern is shown in [Figure 1] below [Shibata & Inooka, 1998].

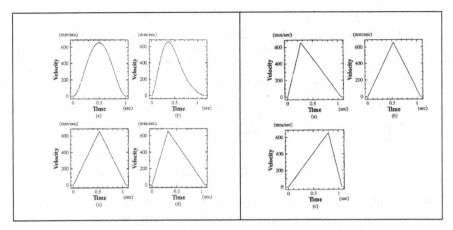

Fig. 1 Velocity Pattern: (Left) Bell-shape 50%, 30%, Triangular 50%, 30%, and (Right) Triangular 25%, 50%, 70%

2.3   Speed and Patterns

In order to feel zooming interface naturally, the zooming speed and the zooming pattern must be harmonized properly. The study of Shibata and Inooka (1998) showed that the change pattern of the zooming speed could change the users' preference. Therefore, this paper will find the speed and the pattern to cause proper sensitivities for users in the zooming user interface (ZUI) of IPTV.

■   **Operational Definition about the Zooming Speed and Pattern**
The ZUI was enabled up to three stages by considering that TV users preferred simple UI and the zooming speed and pattern were defined according to the extension ratio for each stage. The definition of the zooming speed for the experimentation to deduce the optimized zooming speed and pattern of the zooming speed is as follows:

- Zooming   Speed $= \dfrac{\mathrm{Extension\ Ratio}}{\mathrm{Time}}$ (Extension   ratio   basis:
  Longitudinal size of component)
- Extension Ratio $= \dfrac{\mathrm{Target\ Size - Original\ Size}}{\mathrm{Original\ Size}}$

For example, if the target size is increased by 100 % for one second, 100% / 1 sec. is defined as speed 1. Therefore, 200% / 1 sec = speed 2, and 100% / 0.5 sec = speed 2. In this experiment, 3 zooming levels were used as a prototype. Five conditions of the zooming speed used in the experiment can be specified as the following table.

Table 1 Speed and Time according to Level

Speed	Time	
	Level 1 → Level 2(100%)	Level 2 → Level 3(70%)
4	1 / 4 sec.	0.7 / 4 sec.
6	1 / 6 sec.	0.7 / 6 sec.
8	1 / 8 sec.	0.7 / 8 sec.
10	1 / 10 sec.	0.7 / 10 sec.
12	1 / 12 sec.	0.7 / 12 sec.

The zooming pattern was defined before the experiment to deduce the optimized zooming speed and pattern as follows: First, the above five speed conditions were premised as a constant speed condition as shown in the above table. At this time, distance (s) = time (t) × velocity (v). So on the basis of this condition, the zooming speed was defined under the acceleration condition. The acceleration condition was designed as three levels of 0.3t (early-fast), 0.5t (middle-fast), 0.7t (last-fast) depending on the position of the peak speed.

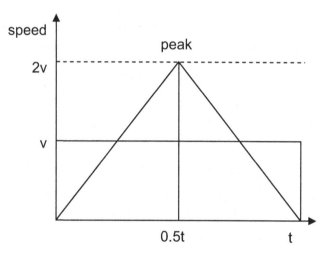

Fig. 2 Constant Speed and Acceleration Graph

3   Method

- Participant: 25 undergraduates of Yonsei University at South Korea (First: 17 students, and Second: 18 students) took part in the experiment. All participants were proficient in Korean and had normal vision.
- Experiment Design: The experiment was performed twice. The first experiment included five conditions to precisely assess the preference of the

zooming speed and was designed as 5 × 3 within-subject design including three pattern conditions. The second experiment was designed as 2 × 3 within-subject design integrating the speed conditions as two conditions to minimize the rating inaccuracy due to the inclusion of the zooming speed conditions and the zooming patterns.

Table 2 Phased Illustrative Screen of the ZUI Prototype used as Zooming Speed & Pattern Stimulus

Zoom Level	Illustrative Picture	Contents	Description
Level 1 overview		Overview of all contents	Both of focused and unfocused areas are same and small.
Level 2 semi view		Intermediate enlargement and context of the focused contents	Enlarging pointed focus block and detailing contents
Level 3 full view		Display of the entire contents like general web pages	Screen view focused on focus block

Table 3 Summary of the Conditions for the First Experiment

Independent variables	Zooming speed (4, 6, 8, 10, 12) Zooming pattern (Constant speed and three acceleration conditions: Early-fast, Middle-fast and Last-fast)
Dependent variables	Subjective measures (Naturalness, Preference)
Experimental tasks	Assess as 7-point scale for 2 questions for each zooming style (Definitely No = 1, Normal = 4, Definitely Yes = 7) 1. Naturalness: How much is it natural? 2. Preference: How much do you prefer?
Experimental design	Within-subject design Manipulating prototype in each condition, Zooming in/out targets within page

4   Results

4.1   Results of the 1st Experiment

- Naturalness: There was no meaningful difference for the naturalness depending on the zooming speed. As shown in the graph below, the participants had more natural feeling as the zooming speed was faster up to 10. However, they showed a tendency that the natural feeling was reduced when the speed was 10 or higher.
- Preference: The preference by the zooming speed showed a meaning difference. In general, the subjects preferred faster speed [$F(4, 75) = 8.053$. P $< .001$], and they showed the highest preference when the zooming speed was 10 and 12 (Speed 4 : 3.250, Speed 6 : 3.688, Speed 8 : 4.563, Speed 10 : 5.188, Speed 12 : 5.188). As with naturalness, there was a tendency that the subjects preferred faster speed until the zooming speed reached 10. However, there was no meaningful difference in terms of preference when the zooming speed was 10 or higher.

4.2   Comprehensive Analysis of the Speed and the Pattern

In order to compare lower speed conditions and the higher speed conditions, Speed 4 and 6 conditions were grouped as lower condition and Speed 8, 10, 12 conditions were grouped as higher speed condition and then naturalness and preference were compared.

Discussion about 1st Experiment

As a result of 1st experiment, the faster speed was given as thumbs-up in terms of naturalness and preference. Among the presented 5 condition, Speed 10 condition was preferred as optimized. On the other hand, the zooming pattern did not show clear tendency. It was shown that users did not clearly assess their preference for the subtle change of the zooming speed but relatively preferred the faster condition.

For the zooming pattern experiment, it seemed to be difficult for the participants to effectively detect the change due to the inclusion of the Speed 5 condition and pattern. Therefore, the simplification of the speed conditions was required for the effective sensitivity assessment about the zooming pattern.

5   Conclusions

In general, the users preferred faster movement for the zooming speed. It is assumed because the execution performance of zooming interface affects the actual running time. According to the results of the 1st experiment, it is shown that users have relatively not bad feeling from Speed 6 to Speed 12 excluding Speed 4 condition which is the lowest speed. ZUI is one of inevitable alternatives.

Therefore, the sensitivity research like this research will be an innovative attempt for the application of ZUI in order that ZUI becomes a UI paradigm to arouse comfort and satisfaction to users and lead in the optimized user experience.

References

1. Baddeley, A.D., Hitch, G.: Working Memory, The Psychology of Learning and Motivation. Advances in Research and Theory, vol. 8, pp. 47–89. Academic Press, New York (1974)
2. Bederson, B., Hollan, J.: Pad++: a zooming graphical interface for exploring alternative interface physics. In: Symposium on User Interface Software and Technology Archive. Proceedings of the 7th Annual ACM Symposium on User Interface Software and Technology, Marina del Rey, California, United States, pp. 17–26 (1993)
3. Benjamin, B.B., James, D.H., Stewart, J., Rogers, D., Druin, A., Vick, D.: A Zooming Web Browser. Human Factors in Web Development (1997)
4. Good, L., Stefik, M., Baudisch, P., Bederson, B.B.: Automatic Text Reduction for Changing Size Constraints. In: Extended Abstracts on Human Factors in Computing Systems (CHI 2002), pp. 798–799 (2002)
5. Hornbæk, K., Bederson, B.B., Plaisant, C.: Navigation Patterns and Usability of Overview+ Detail and Zoomable User Interfaces for Maps. ACM Transactions on Computer-Human Interaction (TOCHI) 9(4) (2001)
6. Punchihewa, A., Malsha De Silva, A, Diao, Y.: Internet Protocol Television (IPTV), Multi-media Research group, School of Engineering and advanced Technology, Massey University (2010)

Malware Similarity Analysis Based on Graph Similarity Flooding Algorithm

Jing Liu, Yongjun Wang, Peidai Xie, Yuan Wang and Zhijian Huang

Abstract Malware is a pervasive problem in computer security. The traditional signature-based detecting method is ineffective to recognize the dramatically increased malware. Researches show that many of the malicious samples are just variations of previously encountered malware. Therefore, it would be preferable to analysis the similarity of malware to determine whether submitted samples are merely variations of existing ones. Static analysis of polymorphic malware variants plays an important role. Function call graph has shown to be an effective feature that represents functionality of malware semantically. In this paper we propose a novel algorithm by comparing the function call graph based on similarity flooding algorithm to analyze the similarity of malware. Similarity between malware can be determined by graph matching method. The evaluation shows that our algorithm is highly effective in terms of accuracy and computational complexity.

Keywords Malware variants · Similarity analysis · Function call graph · Similarity flooding algorithm

1 Introduction

Malware, short for malicious software, is defined as software that fulfills the deliberately harmful intent of an attacker (e.g., viruses, worms, or Trojan horses).The volume of malware is growing at an exponential pace. According to the report of Symantec, 317,000,000 new samples were received in 2014, close to 1,000,000 new pieces of malware each day. Malware poses a major challenge to the security of computer system.

J. Liu(✉) · Y. Wang · P. Xie · Y. Wang · Z. Huang
College of Computer, National University of Defense Technology,
Changsha, Hunan, China
e-mail: {liujing_nudt,wangyuan,zjhuang}@nudt.edu.cn,
        {wwyyjj1971,xpd2002}@126.com

© Springer Science+Business Media Singapore 2015   31
D.-S. Park et al. (eds.), *Advances in Computer Science and Ubiquitous Computing*,
Lecture Notes in Electrical Engineering 373,
DOI: 10.1007/978-981-10-0281-6_5

Researches have shown that the majority of new malware are variations of already well-known malware. They are created through polymorphic engines and code obfuscation techniques. Though they are almost the same functionality but have different syntactic representation. However, many of the current anti-virus programs available rely on a syntactic signature-based approach. As a consequence, malware variants can easily evade the signature-based detection for they possess different signatures. Besides, creating signature for each variant in a timely fashion is time consuming and labor consuming.

To efficiently and effectively detect the enormous amount of malicious variant, it has been proposed that by comparing the similarity of malware instance to detect polymorphic malware variants. It assumes that similar programs have to share similar features. By determining the similarity based on specific quantitative metric to identify variant of a known malware. Similarity analysis is also benefit for malware triage and classification, grouping malware instances into families and generating family signatures to detect variants.

In this paper we propose a malware similarity analysis method based on function-call graph through similarity flooding algorithm. Function-call graph is a high-level structure feature. It is a directed graph created from disassembled code of program, in which each vertex represents a function and edges represent calls relation among functions. Function call graph as a static feature has the advantage of static analysis, such as scalable, covering all possible path of a malicious program. In addition, function-call graph represents the functionality of a program semantically. A program's functionality is mostly determined by the functions it invokes, which make function call graph more resilient to code obfuscation.

Similarity flooding algorithm is a graph matching algorithm based on fix-point computing. We choose it for several reasons. First, it is an inexact graph matching method and has an approximate polynomial time solution. Second, similarity flooding algorithm takes directed graph as input and based on fix-point, which is natural suit for function call graph. The external system function nodes in function call graph can be viewed just as the fixed points in the method. Moreover, similarity flooding algorithm produces a mapping between corresponding nodes in two graphs as output. We can not only determine the degree of similarity based on the result, but also gain the corresponding matching parts between malicious samples.

The main contributions of this paper are as follows;

- We present a method to compare the similarity of malware instance on basis of function-call graph;
- We compute the level of similarity between function-call graph based on the graph similarity flooding algorithm;
- We propose a prototype to verify our method.

The rest of this paper is organized as follows. Section 2 describes the related work in malware similarity analysis. Section 3 describes the design and implementation of our system. Section 4 evaluates our prototype. Finally, Section 5 summarizes and concludes the paper.

2   Related Work

In the past few years, a great number of methods have been proposed in malware similarity analysis field. Kolter et al. [1] proposed using the n-grams opcodes of instruction sequences present in the malware binary to measure the distance of samples. High-level static structural features have also been extracted to assess the similarity between malware. Halvar Flake [2] put up with using the control flow graph (cfg) of a program as feature. Following the idea of matching graph-feature, some improved methods have been presented [3, 4]. The API sequences extracted by static analysis were taken by Shankarapani MK et al. [5]. He proposed to use the normalized API sequences the program has invoked to assess similarity. As pointed by Moser et al. [6], there are some shortcomings of static analysis methods, obfuscation or self-mutating may affect the accuracy of analysis result.

Due to the limitations of static analysis techniques, dynamic analysis was brought into focus. System call sequences, Windows native APIs have been used as a set feature to compare the similarity of malware samples [3]. However, that information can vary significantly, even between programs that exhibit the same behavior. Therefore, a variety of researches pay attention to high-level feature extract from the basic runtime trace. Ulrich Bayer et al. [7] generalized the execution traces into behavioral profiles. The profiles express malware behavior in terms of operating system (OS) objects and OS operations. Bailey et al. [8] used the reduced collection of those user-visible system state changes (e.g., file written, process created) to create a fingerprint of the malware behavior, then chosen the NCD as distance metric.

3   The Overall of Framework

Malware similarity analysis can be divided into two steps: feature extraction and feature-similarity measure. In this work, we choose the static function-call graph as feature of malware samples. It has the advantages of static feature and is resilient to some obfuscation techniques. For two function-call graphs, they may have different number of vertices. Thus it is not possible to find an exact matching between the two graphs. In addition, the key factor of malware similarity analysis lies in finding the best matching between graphs, rather than searching for the exact way of matching vertices. Therefore, we choose the inexact matching method similarity flooding algorithm to measure the similarity of function-call graphs.

3.1   Function-Call Graph

A function-call graph $G = (V, E)$ is composed of a vertex set V and an edge set E, where each vertex in the graph is corresponding to the function included in the program and the edges mean the call relationship between functions. The vertex function can be classified into three categories: local functions, statically-linked library functions, dynamically-imported functions. Local functions are functions written by malware writers. Statically-linked library functions are library functions that are statically linked into the final distributed binary. Dynamically-imported functions are DLL functions.

Reliably identifying function boundaries in binary is a research challenge. Disassembly tool IDA Pro has achieved reasonable accuracy, so in this paper we use IDA to generate the function-call graph of each malware instance. In IDA representation of function-call graph, vertices only have two kinds. The local functions, naming with sub_xxxxxx, "xxxxxx" stands for the local function name in IDA disassemble. The statically and dynamic library functions are merged into one kind external function, which does not distinguish them separately. The vertex name is just the function name in the library, like "WriteFile", "GetParent" and so on.

Figure 1 shows a part of function call graph from IDA Pro.

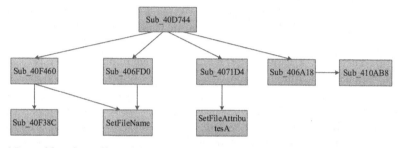

Fig. 1 Part of function call graph

3.2 Similarity Flooding Algorithm

The basic idea of similarity flooding algorithm is that whenever any two nodes in graph G_1 and G_2 are found to be similar, the similarity of their adjacent nodes increases. Thus, over a number of iterations, the initial similarity of any two nodes propagates through the graphs [9]. The algorithm terminates after a fix-point has been reached, i.e. the similarities of all nodes stabilize.

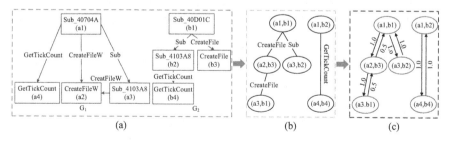

Fig. 2 the pipeline of similarity flooding algorithm

The input of similarity flooding algorithm is two labeled directed graphs. However, the function call graph which we acquire from IDA does not have labels on the arcs, as shown in figure 1. Thus, we define that the arcs from local function to external function are typed the function name of the external function, the edges which connect two local functions are typed with "sub". As shown in figure 2 (a).

The similarity flooding algorithm has four main steps as follows:

a) Initial mapping. As a first step, we need to find an initial mapping of G_1 and G_2. The initial mapping of function-call graph is obtained using a simple string matching method Levenshtein distance. We use this distance to compute the mnemonic instruction sequences difference of local functions and the function name similarity of external functions, then assign a similarity values range from 0 to 1 to indicate how well the corresponding function in G_1 match their counterparts in G_2.

b) Create the connectivity graph. The connectivity graph between two graphs G_1 and G_2 is built this way: $((x_1, x_2), r, (y_1, y_2)) \in CG$ iff $(x_1, r, y_1) \in G_1$ and $(x_2, r, y_2) \in G_2$ [9]. x_1, y_1, x_2, y_2 are the nodes in graph G_1, G_2. r is the label on edges. Figure 2 (b) shows a connectivity graph of G_1 and G_2.

c) Create the propagation graph. For every arc in the connectivity graph, the propagation graph contains an arc going in the opposite direction. We choose the way described in [9] computing the propagation coefficients placed on the arcs of the propagation graph indicate how similar the given map pair propagates into neighbors and back. The propagation coefficients range from 0 to 1. Figure 2 (c) indicates the propagation graph of G_1 and G_2.

d) Similarity propagation. Let σ^i denote the mapping between G_1 and G_2, w denote the weight on each edges. In general, mapping σ^{i+1} is computed from mapping σ^i as follows:

$$\sigma^{i+1}(x+y) = \sigma^i(x, y) + \sum\nolimits_{(v,p,x)\in G_1,(u,p,y)\in G_2} \sigma^i(v,u) \times w((v,u),(x,y)) \qquad (1)$$

The computation process ends whenever the Euclidean distance of (σ^{i+1}, σ^i) is less than the threshold.

We gain a multi-mapping of two graphs, a node of G_1 can be associated to several nodes of G_2. We need to find the best match of each node in the multi-mapping. The Hungarian algorithm [10] was design to solve the assignment problem. The mapping $\sigma_{i,j}$ can be viewed as the cost defined in the algorithm. Assume G_1 has m nodes and G_2 has n nodes. So the cost matrix in bipartite graph matching is:

$$C = \begin{bmatrix} \sigma_{1,1} & \sigma_{1,2} & \cdots & \sigma_{1,n} \\ \sigma_{2,1} & \sigma_{2,2} & \cdots & \sigma_{2,n} \\ \vdots & \vdots & \ddots & \vdots \\ \sigma_{m,1} & \sigma_{m,2} & \cdots & \sigma_{m,n} \end{bmatrix}$$

4   Performance Analysis

To verify the effectiveness of our method, we used the data set from VX Heavens, which contains numbers of malware samples classified by family. Intuitively, the same malware family samples should have higher degree of similarity, different family resemblance lower.

Figure 3 demonstrates the result of our experiment. The highest value 1.0 is similarity score that variants compared with themselves. The families contain Virus.Win32.Adson, Virus.Win32.Afgan. From the figure we can see that the average similarity scores in same family are greater than 0.6, the highest can reach 0.93. The initial mapping and function boundary identification impact the accuracy of the algorithm essentially.

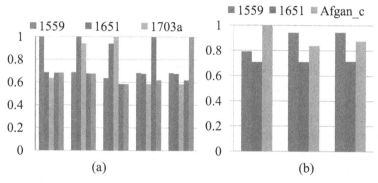

Fig. 3 Similarity scores of each pair of malicious samples

5   Summary and Future Work

In this paper, we propose an efficient malware similarity analysis method using function call graph. Graph matching is based on a well-known schema matching algorithm Similarity Flooding. We choose the function call graph as features of malicious code. Then we apply the schema matching algorithm into function call graph and define the directed labeled function call graph. We construct the connectivity graph and propagation graph with the information and characteristics. Finally we propose using bipartite matching algorithm to filter the mapping result instead of the method used in [9].

In the future we will work on the way assigning the initial mapping value of nodes in connectivity graph and computing the weights of the arcs in the propagation graph. We will also apply other graph matching methods to function call graph for the sake of more accuracy of malware variants detection.

Acknowledgement This work is supported by the National Science Foundation of China (No.61472439, No.61271252).

References

1. Kolter, J.Z., Maloof, M.A.: Learning to Detect and Classify Malicious Executables in the Wild. J. Mach. Learn. Res. **7**, 2721–2744 (2006)
2. Flake, H.: Structural comparison of executable objects. In: Flegel, U., Meier, M. (eds.) DIMVA, vol. 46, pp. 161–173. GI (2004)
3. Gao, D., Reiter, M., Song, D.: Behavioral distance for intrusion detection. In: Valdes, A., Zamboni, D. (eds.) Recent Advances in Intrusion Detection, vol. 3858, pp. 63–81. Springer, Heidelberg (2006)
4. Hu, X., Chiueh, T.-C., Shin, K.G.: Large-scale malware indexing using function-call graphs. In: Proceedings of the 16th ACM Conference on Computer and Communications Security, pp. 611–620. ACM, Chicago (2009)
5. Shankarapani, M., Ramamoorthy, S., Movva, R., Mukkamala, S.: Malware detection using assembly and API call sequences. J. Comput. Virol. **7**, 107–119 (2011)
6. Moser, A., Kruegel, C., Kirda, E.: Limits of static analysis for malware detection. In: 12th Asia-Pacific Computer Systems Architecture Conference, pp. 421–430. IEEE Press, Miami Beach (2007)
7. Bayer, U., Comparetti, P.M., Hlauschek, C., Krügel, C., Kirda, E.: Scalable, Behavior-Based Malware Clustering. NDSS. The Internet Society (2009)
8. Bailey, M., Oberheide, J., Andersen, J., Mao, Z.M., Jahanian, F., Nazario, J.: Automated classification and analysis of internet malware. In: Kruegel, C., Lippmann, R., Clark, A. (eds.) Recent Advances in Intrusion Detection, vol. 4637, pp. 178–197. Springer, Heidelberg (2007)
9. Melnik, S., Garcia-Molina, H., Rahm, E.: Similarity flooding: a versatile graph matching algorithm and its application to schema matching. In: 18th International Conference on Data Engineering, pp. 117–128 (2002)
10. Kuhn, H.W.: The Hungarian method for the assignment problem. Naval Research Logistics Quarterly (1955)

The Effect of Information Quality on User Loyalty Towards Smartphone Applications

Wonjin Jung

Abstract Today's smartphone applications are well equipped with a number of technological innovations and advanced functionalities. At the same time, they have become more complicated in general due to a wide variety of functions. Some are neither easy to use, nor to understand. Even worse, most smartphones have a small display screen that may have a negative impact not only on the quality of information that smartphone applications offer, but also on the perceived usability of the applications. A review of the relevant literature found that past studies have not shown much interest in the impact of the quality of information that smartphone applications provide on the usability of the applications and user loyalty towards the applications. In addition, little research of the relationship between the usability of smartphone applications and user loyalty towards the applications was conducted. Therefore, this study aims to examine: 1) the effects of the quality of information that smartphone applications provide on both the perceived usability of the applications as well as user loyalty towards the applications, 2) the effect of the perceived usability of smartphone applications on user loyalty towards the applications. Structural Equation Modeling (SEM) was employed to analyze data collected through a survey. The results found that information quality significantly influenced the perceived usability of applications as well as user loyalty towards the applications. In addition, the perceived usability of applications also had a positive impact on user loyalty towards the applications.

Keywords Smartphone · Application · Information · Usability · Loyalty

W. Jung(✉)
The School of Business and Economics, Dankook University,
152, Jook-Jun-Ro, Soo-Ji-Goo, Young-In, Kyung-Ki-Do 448-701, Korea
e-mail: jungw@dankook.ac.kr

© Springer Science+Business Media Singapore 2015   39
D.-S. Park et al. (eds.), *Advances in Computer Science and Ubiquitous Computing*,
Lecture Notes in Electrical Engineering 373,
DOI: 10.1007/978-981-10-0281-6_6

1   Introduction

Today's smartphone applications are well equipped with a number of technological innovations and advanced functionalities. These applications provide users with greater functionality and better access to information. At the same time, they have become more complicated in general due to a wide variety of functions. Some are neither easy to use, nor to understand. Even worse, most smartphones have a small display screen. The small display screen of smartphones doesn't provide sufficient space for applications to show various features and information including texts and graphics. If too much features and information are deployed on the same small display screen, their readability and discernability can be reduced. Therefore, the small display screen of smartphones seems to negatively affect not only the quality of information that applications offer, but also the usability of the applications as perceived by smartphone users. In general, complicated applications are considered to be less usable.

A literature review found that information quality is strongly related with issues of usability. Some researchers found that most Web sites that have various usability problems do not provide adequate information [2, 6, 7, 8]. This can be also applicable to smartphone applications. If smartphone applications do not provide adequate quality of information, then they would not be considered to be usable. If smartphone applications are not considered to be usable, it may be difficult for them to secure a loyal user base. In fact, today's popular smartphone applications have secured many loyal users and are maintaining a higher level of usability and information quality. Therefore, it is not too much to say that providing smartphone application users with high-quality information is more imperative than ever.

In sum, the discussion above indicates that the quality of information that smartphone applications provide seems to be critical determinant of not only the usability of the applications, but also user loyalty towards the applications. In addition, the usability of applications may influence user loyalty towards the applications. However, a review of the relevant literature found that past studies have not shown much interest in the impact of the quality of information that smartphone applications provide on the usability of the applications and user loyalty towards the applications. In addition, little research of the relationship between the usability of smartphone applications and user loyalty towards the applications was conducted. Therefore, the following hypotheses are proposed to test those relationships.

H1: The quality of information that smartphone applications offer positively affects the usability of the applications as perceived by smartphone users.

H2: The quality of information that smartphone applications offer positively affects the loyalty of smartphone users towards the applications.

H3: The usability of smartphone applications as perceived by smartphone users positively affects the loyalty of smartphone users towards the applications.

2   Research Methodology, Data Analysis, and Results

The goal of this study is to examine the effects of information quality not only on the perceived usability of applications, but also on user loyalty towards the applications in a smartphone context. In addition, the effect of the perceived usability of smartphone applications on user loyalty towards the applications was also examined. This study collected data by a survey. A total of 236 practitioners and college students participated in the survey. Of the participants, 50.4% were male, and 49.6% were female. 80.1% of the participants were in their twenties. College students majoring in a variety of academic programs including economics, business, computer science at three universities in Korea, made up 66.9% of the participants. On the other hand, practitioners made up 33.1% of the participants. Social networking- and communications-related applications made up 61.9% of the applications that participants had used just before answering the survey questions.

Structural Equation Modeling (SEM) was employed to analyze the proposed theoretical research model and SPSS Statistics and AMOS ver. 18 were the statistical package software used for the analysis. First of all, the reliability of all observable variables was examined to test the measurement model. In order to meet the reliability requirements, the loadings of the observable variables on their respective constructs had to be above 0.6 [1, 3]. The results of the analysis showed that all of the loadings were 0.7 or higher, suggesting that the reliability is adequate (see Table 1).

Table 1 Standardized Regression Weights of Observable Variables, Composite Reliability (CR), and Average Variance Extracted (AVE)

Latent Variables	Estimates	Variance C.R.	Composite Reliability	AVE
Information Quality	.787	7.314	.899	.641
	.799	7.005		
	.816	6.512		
Perceived Usability	.845	7.018	.924	.700
	.854	6.699		
	.810	7.957		
User Loyalty	.900	5.726	.922	.737
	.899	5.767		
	.770	9.266		

Then, the convergent validity of the proposed measurement model was tested by examining the composite reliability (CR) and the average variance extracted (AVE) of the latent variables in the model. AMOS does not have the functions to calculate those values of CR and AVE, so they were calculated manually with the formulas below, as suggested by Fornall and Larcker [4] and Hair et al. [5]. The results of the analyses showed that the values of CR for all latent variables were of 0.8 or higher, well above the recommended cutoff of 0.7 (see Table 1). In addition, all latent variables also had values for AVE greater than the recommended tolerance of 0.5 (see Table 1). Therefore, the proposed measurement model demonstrated a satisfactory convergent validity.

$$\text{CR} = (\sum \text{Standardized Regression Weights})^2 / ((\sum \text{Standardized Regression} \quad [4]$$
$$\text{Weights})^2 + (\sum \text{Variance}))$$

$$\text{AVE} = (\sum \text{Standardized Regression Weights}^2) / N \qquad\qquad\qquad [5]$$

The disrcriminant validity of the proposed measurement model was also examined by comparing the square root of the AVE with the correlations among the constructs. To meet the requirements for the discriminant validity, all variables in the model should have the square root of their VAEs be greater than the correlations with other variables in the model [3]. The results showed that this is the case (see Table 2). Thus, the discriminant validity was also confirmed.

Table 2 Correlation Coefficient Value between Constructs and AVE

Constructs	AVE	$\varnothing^2$	$\varnothing^2$	$\varnothing^2$
Information Quality	.641	0.09	0.061	1.000
Perceived Usability	.700	0.395	1.000	
User Loyalty	.737	1.000		

Next, the goodness of fit was examined to test the structural model. To do so, the indices, such as x^2/df, GFI, AGFI, NFI, TLI, CFI, and RMSEA were examined and the results were as follows: $x^2/\text{df} = 2.028$, GFI = .957, AGFI = .919, NFI = .960, TLI = .969, CFI = .979, and RMSEA = .066. The results of overall fit statistics suggest that the proposed structural model has a fairly good fit.

Finally, the significance and the strength of the relationships between the variables were examined in terms of path coefficients to test the structural model. As expected, information quality had a significant influence on the perceived usability of applications ($\beta = .255$, p = .001) as well as user loyalty ($\beta = .152$, p = .000). In addition, the perceived usability of applications also had a positive impact on user loyalty ($\beta = .565$, p = .017). Therefore, Hypotheses 1, 2, and 3 were all supported. Table 3 below shows the results of the test in detail and Figure 1 presents the results with R^2 values representing the amount of variance.

Table 3 Hypothesis Test

	Paths	Coeff.	Stand. Coeff.	P	Results
H1	Information Quality -> Perceived Usability	.255	.247	.001	Accept
H2	Information Quality -> User Loyalty	.152	.154	***	Accept
H3	Perceived Usability -> User Loyalty	.565	.591	.017	Accept

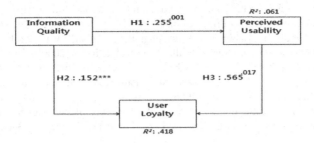

Fig. 1 Structural Model Results

3   Discussion and Conclusion

The results of analyses revealed that information quality significantly influenced the perceived usability of applications as well as user loyalty towards the applications. In addition, the usability of applications as perceived by smartphone users had a positive impact on their loyalty towards the applications.

In practice, application developers may use these findings to improve their applications in terms of information quality and usability. Specifically, the three information quality attributes and the three usability attributes can be used to check whether applications meet smartphone users' information quality expectations and usability expectations. Based on the assessment, application developers may develop information quality and usability management programs to enhance the information quality and the usability of applications and to make their applications more efficiently and effectively. Smartphone users would then not only receive certain types of benefits from high-quality information and usability while using applications, but also show their loyalty towards applications.

In sum, since information quality in smartphone applications is strongly related to the usability of the applications and the loyalty of users towards the applications, effort should be made to improve information quality as an effective strategy for application developers not only to enhance usability for their applications, but also to secure a loyal user base.

References

1. Barclay, D., Higgins, C., Thompson, R.: The Partial Least Squares (PLS) Approach to Causal Modeling: Personal Computer Adoption and Use as an Illustration. Technology Studies **2**, 285–324 (1995)
2. Chang, J.F., Cheung, W., Lai, V.S.: Literature Derived Reference Models for the Adoption of Online Shopping. Information and Management **42**(4), 543–559 (2005)
3. Chin, W.W.: The partial least squares approach for structural equation modeling. In: Marcoulides, G.A. (ed.) Modern Methods for Business Research, pp. 295–336. Lawrence Erlbaum, Mahwah (1998)

4. Fornell, C., Larcker, D.F.: Evaluating Structural Equation Models with Unobservable Variables and Measurement Error. Journal of Marketing Research **18**, 39–50 (1981)
5. Hair, J.F., Black, B., Babin, B., Andersong, R.E., Tatham, R.L.: Multivariate Data Analysis, 6th edn. Pearson Prentice Hall, Upper Saddle River (2006)
6. Tractinsky, N., Cokhavi, A., Kirschenbaum, M., Sharfi, T.: Evaluating the Consistency of Immediate Aesthetic Perceptions of Web Pages. International Journal of Human-Computer Studies **64**, 1071–1083 (2006)
7. Zou, Y., Zhang, Q., Zhao, X.: Improving the Usability of e-Commerce Applications Using Business Processes. IEEE Transactions of Software Engineering **33**(12), 837–855 (2007)
8. Zviran, M., Glezer, C., Avni, I.: User Satisfaction from Commercial Web Sites: The effect of Design and Use. Information and Management **43**, 157–178 (2006)

Appendix

Variables	Questions
Info. Quality	The samrtphone application that I used most recently provided information that is easy enough to understand.
	The samrtphone application that I used most recently provided information that contains adequate detail to achieve my personal goals.
	The samrtphone application that I used most recently provided a sufficient amount of information to achieve my personal goals.
Perceived Usability	The functions that I tried to find in the application were quickly found.
	The information that I tried to find in the application was easily found.
	The application that I recently used provided information quickly.
Loyalty	I intend to use the smartphone application again in the near future.
	I intend to use the smartphone application again at every opportunity.
	I intend to use the smartphone application again to do my personal tasks.

Cryptanalysis of Dynamic ID-Based User Authentication Scheme Using Smartcards Without Verifier Tables

Jaewook Jung, Younsung Choi, Donghoon Lee, Jiye Kim,
Jongho Mun and Dongho Won

Abstract Password-based remote user authentication technique is the most commonly used for secure communication over insecure network environments. Due to its simplicity and efficiency, it is widely used in many fields such as e-commerce, distributed system, remote host login system, etc. In recent years, several dynamic ID-based user authentication schemes using password and smart card have been proposed to provide mutual authentication between the user and server. Recently, Lee proposed an efficient dynamic ID-based user authentication scheme without verifier tables. Lee claimed that his scheme can resist off-line password guessing attack, user impersonation attack and provide user anonymity. In this paper, we demonstrate that Lee's enhanced scheme is not secure against off-line password guessing attack and user impersonation attack in violation of its security claim as well as it fails to preserve user anonymity.

Keywords Authentication · Smart-card · Cryptanalysis · Off-line password guessing attack

1 Introduction

With the rapid increasing need of the internet service and electronic commerce technology, password-based user authentication techniques are an essential security requirement for protecting systems and networks. Due to its simplicity and efficiency, it is used many areas such as E-commerce environment, distributed system, remote host login system, etc.

J. Jung · Y. Choi · D. Lee · J. Kim · J. Mun · D. Won(✉)
Department of Computer Engineering, Sungkyunkwan University,
2066 Seoburo, Suwon, Gyeonggido 440-746, Korea
e-mail: {jwjung,yschoi,dhlee,jykim,jhmoon,dhwon}@security.re.kr

© Springer Science+Business Media Singapore 2015
D.-S. Park et al. (eds.), *Advances in Computer Science and Ubiquitous Computing*,
Lecture Notes in Electrical Engineering 373,
DOI: 10.1007/978-981-10-0281-6_7

45

Since Lamport first proposed a remote user authentication protocol for the insecure channel in 1981 [1], numerous remote user authentication schemes have been proposed [2, 3, 4, 5, 6, 7, 8, 9, 11, 12].

In 2004, Das et al. [2] proposed a dynamic ID-based remote user authentication scheme using smart card. The dynamic ID technique means that prevent the leakage of identity information during login phase, eliminating the risk of ID-theft attack. In their scheme, users are free to choose and change their password. They claimed that their scheme is secure against stolen verifier attack because of there is no verifier table. Besides, Das et al. [2] demonstrated that their scheme can resist replay attack, forgery attack, off-line password guessing attack and privileged insider attack. However, in 2005, Liao et al. [3] showed that Das et al.'s scheme is vulnerable to off-line password guessing attack and proposed an enhanced authentication scheme. Unfortunately, in 2008, Misbahuddin et al. [4] demonstrated that the Liao et al.'s scheme [3] is vulnerable to impersonation attack and reflection attack.

Recently, Lee [9] demonstrated that Lee's scheme [8] cannot withstand off-line password guessing attack, impersonation attack and smart-card theft attack. To surmount this shortcoming, Lee [9] proposed an efficient dynamic ID-based user authentication scheme using smart cards without verifier tables. However, after careful analysis, we find Lee's authentication scheme [9] cannot resist the off-line password guessing attack and user impersonation attack, and it cannot preserve user anonymity. In this paper, we demonstrate this security problem with Lee's authentication scheme [9].

The remainder of the paper is organized as follows: In section 2, we present a review of the Lee's authentication scheme [9]. In section 3, we show the various vulnerabilities of the Lee's authentication scheme. At the end, we draw our conclusion in section 4.

2   Review of Lee's Scheme

In this section, we examine the Lee's authentication scheme [9]. The notations used throughout this paper are summarized in Table 1.

Table 1 Notations used in the scheme of Lee [9]

Notations	Descriptions
U_i	The remote user
ID_i	The user's identity
DID_i	The user's dynamic identity
PW_i	The user's password
T	A time stamp
$h(\cdot)$	A one-way hash function
$\oplus$	The bitwise *Xor* operator

Table 1 (*Continued*)

p, q	The large prime numbers
x	The secret key of remote server S
S	The remote server
R	The random number

2.1   Registration Phase

A new user U_i register with the server S by performing the following steps:

- **Step1.** The user U_i sends his/her PW_i to server S through a secure channel.
- **Step2.** Upon receiving the PW_i, the server S computes $M_i = h(ID_i \oplus x)$ and $N_i = ID_i \oplus h(PW_i)$, where $h(\cdot)$ is a one-way function.
- **Step3.** The server S stores $\{h(\cdot), M_i, N_i, n\}$ into a smart card, where p and q are large prime numbers and $n = p \times q$, and then issues this smartcard to user U_i through a secure channel.

2.2   Authentication Phase

The authentication phase also comprises login phase and verification phase. When user U_i wants to login and authenticate to server S, the following operation will perform:

2.2.1   Login Phase
- **Step1.** The user U_i inserts his/her smartcard into a card reader and inputs ID_i and PW_i.
- **Step2.** The smartcard computes $ID_i = N_i \oplus h(PW_i)$, $b = h(M_i \oplus T)$ and $DID_i = ID_i \oplus b$, where T is the current date/timestamp in the user U_i.
- **Step3.** The smartcard generates random number R and computes $B_i = h(N_i \oplus h(N_i \oplus h(x)) \oplus R)$ and $C_i = b^2 \bmod n$.
- **Step4.** The user U_i sends login request message $M = \{DID_i, C_i, T\}$ to server S.

2.2.2   Verification Phase
- **Step1.** Upon receiving the login request message M from the user U_i, the server S checks the validity of timestamp T.
- **Step2.** The server S utilizes the Chinese Remainder Theorem to solve $C_i = b^2 \bmod n$, since S can derive four roots (b_1, b_2, b_3, b_4) with two large primes p and q.
- **Step3.** The server S checks $h(h(b_i \oplus DID_i \oplus x) \oplus T) =? b$, for $i = 1$ to 4, so that it can obtain the correct value of b. If this is satisfied, the server S accepts the login request; otherwise, the login request is rejected and this phase is terminated.

2.3 Password Change Phase

This phase is invoked whenever U_i wants to change his/her password PW_i to a new password PW_i^{new}. In the password change phase, the user U_i does not communicate with server S.

- **Step1.** The user U_i inserts his/her smartcard into a card reader and inputs PW_i.
- **Step2.** The user U_i chooses a new password PW_i^{new}.
- **Step3.** The smartcard computes $N_i^{new} = N_i \oplus h(PW_i) \oplus h(PW_i^{new})$, where $N_i = ID_i \oplus h(PW_i)$.
- **Step4.** The smartcard replaces the existing value N_i with the new value N_i^{new}.

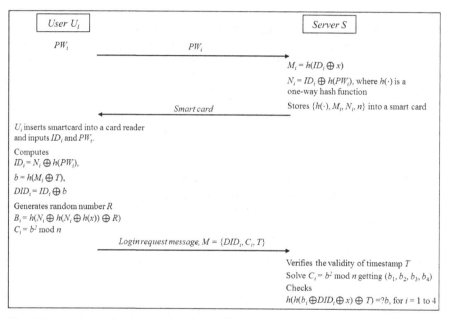

Fig. 1 Registration and authentication phase of Lee's scheme.

3   Cryptanalysis of Lee's Scheme

In this section, we will discuss the flaws of Lee's authentication scheme [9]. We found that Lee's scheme cannot resist the off-line password guessing attack and user impersonation attack in violation of its security claim as well as it fails to preserve user anonymity. The details of these flaws are described follows.

3.1   Failure to Preserve User Anonymity

User anonymity means that the user identity is preserved in secret, and any third party apart from the communicating agents cannot recognize the identity of the user. Now, we show that Lee's scheme fails to preserve the same as follows:

- **Step1.** Attacker extracts the stored parameters $\{h(\cdot), M_i, N_i, n\}$ from the stolen smartcard using the power analysis attack [10].
- **Step2.** Attacker monitors the communication channel and intercepts a login message $M = \{DID_i, C_i, T\}$ in a login session.
- **Step3.** In login phase, using $b = h(M_i \oplus T)$ and $DID_i = ID_i \oplus b$, attacker can easily compute the user's identity, where $ID_i = DID_i \oplus h(M_i \oplus T)$.

Hence, the attacker could easily trace down the relation between the user U_i and the server S by comparing with all of the eavesdropped messages. Consequently, user anonymity is not preserve in Lee's scheme.

3.2   Off-line Password Guessing Attack

In Lee's scheme [9], attacker can obtain the secrets $\{h(\cdot), M_i, N_i, n\}$ in smart card after the attacker has stolen the smart card, and intercept Login request message $\{DID_i, C_i, T\}$ between a user and the server. And then, the attacker tries to off-line password guessing attack by performing the following steps:

- **Step1.** Attacker extracts the stored parameters $\{h(\cdot), M_i, N_i, n\}$ from the stolen smartcard using the power analysis attack [10].
- **Step2.** In login phase, Attacker finds the exact user's identity of the user U_i by executing steps discussed in section 3.1.
- **Step3.** Attacker selects a password candidate PW^*.
- **Step4.** Using obtained user's identity, attacker computes $ID_i^* = N_i \oplus h(PW^*)$
- **Step5.** Attacker repeats above steps from 3 to 4 until the computed result ID_i^* equals the breached information ID_i.

If they are equal, $PW^* = PW_i$, this means that attacker successfully obtain user's password by off-line password guessing attack.

3.3   User Impersonation Attack

In this section, we show that an attacker can successfully login to the server S using the stolen smartcard of a user U_i as follows:

- **Step1.** Attacker monitors the communication channel and intercepts a login message $M = \{DID_i, C_i, T\}$ in a login session.
- **Step2.** Attacker extracts the stored parameters $\{h(\cdot), M_i, N_i, n\}$ from the stolen smartcard using the power analysis attack [10].

- **Step3.** Attacker finds the exact user's identity of the user U_i by executing steps discussed in section 3.1.
- **Step4.** Attacker computes $DID_i = ID_i \oplus h(M_i \oplus T)$, $C_i = b^2$ mod $n = \{h(M_i \oplus T)\}^2$ mod n and selects valid timestamp T^*. Attacker then sends the forged message $\{DID_i, C_i, T^*\}$ to server S for login operation.
- **Step5.** Upon receiving the login request message $M^* = \{DID_i, C_i, T^*\}$, the server S executes the verification procedure using the Chinese Remainder Theorem.
- **Step6.** The server S successfully verifies the login request message because the login request message M^* correctly equals a legitimate user's login request message. With these successful verifications, the server S accepts the forged login request and allows attacker to login to server S.

This attack clearly shows that the attacker can successfully masquerade as a legitimate user using stolen smartcard.

4   Conclusions

In 2014, Lee's proposed an efficient dynamic ID-based user authentication scheme using smartcard and demonstrated its resistance to off-line password guessing attack and impersonation attack. However, in this paper, we point out Lee's scheme is vulnerable to off-line password guessing attack and user impersonation attack in login phase, and it cannot preserve user anonymity.

Acknowledgments This research was supported by the Basic Science Research Program through the National Research Foundation of Korea (NRF) funded by the Ministry of Science, ICT, and Future Planning (2014R1A1A2002775).

References

1. Lamport, L.: Password authentication with insecure communication. Communications of ACM **24**, 770–772 (1981)
2. Das, M.L., Saxena, A., Gulati, V.P.: A Dynamic ID-based Remote User Authentication Scheme. IEEE Transactions on Consumer Electronics **50**, 629–631 (2004)
3. Liao, I.E., Lee, C.C., Hwang, M.S.: Security enhancement for a dynamic id-based remote user authentication scheme. In: Proceedings of the International Conference on the Next Generation Web Services Practices, pp. 22–26 (2005)
4. Misbahuddin, M., Bindu, C.S.: Cryptanalysis of Liao- Lee-Hwang's Dynamic ID Scheme. International Journal of Network Security **6**, 211–213 (2008)
5. Wang, Y.Y., Liu, J.Y., Xia, F.X., Dan, J.: A More Efficient and Secure Dynamic ID-Based Remote User Authentication Scheme. Computer Communications **32**, 583–585 (2009)

6. Ahmed, M.A., Lakshmi, D.R., Sattar, S.A.: Cryptanalysis of A More Efficient and Secure Dynamic ID-Based Remote User Authentication Scheme. International Journal of Network Security & Its Applications 1, 32–37 (2009)
7. Lee, H., Choi, D., Lee, Y., Won, D., Kim, S.: Security Weaknesses of Dynamic ID-based Remote User Authentication Protocol. Proceedings of the World Academy of Science Engineering and Technology 59, 190–193 (2009)
8. Lee, Y.C.: A New Dynamic ID-based User Authentication Scheme to Resist Smart-Card-Theft Attack. Applied Mathematics and Information Sciences 6, 355–361 (2012)
9. Lee, T.F.: An Efficient Dynamic ID-based User Authentication Scheme using Smartcard without Verifier Tables. Applied Mathematics and Information Sciences 9, 485–490 (2014)
10. Kocher, P., Jaffe, J., Jun, B.: Proceedings of Advances in Cryptology (CRYPTO 1999), vol. 1666, pp. 388–397 (1999)
11. Nam, J., Choo, K.K.R., Kim, J., Kang, H.K., Kim, J., Paik, J., Won, D.: Password-Only Authenticated Three-Party Key Exchange with Provable Security in the Standard Model. Sensors 2014 14(4), 6443–6462 (2014)
12. Jung, J., Jeon, W., Won, D.: An enhanced remote user authentication scheme using smart card. In: ICUIMC 2014 (2014)

The Effects of Information Quality on Mental Model and Interactivity in a Smartphone Context

Wonjin Jung

Abstract Based upon the IS literature review, the information quality of applications seems to be related to the mental model that the application users hold as well as the interactivity with the applications. Furthermore, the mental model seems to influence the interactivity with the applications. However, a comprehensive IS literature review found that previous IS research have not shown much interest in those relationships especially in a smartphone context. Therefore, the goal of this study is to suggest a research model that can examine: 1) the effects that the quality of information has not only on the mental model of applications that users form and hold, but also on the interactivity with the given applications, and 2) the effect that the mental model of applications on the interactivity with the applications in a smartphone context. A survey was conducted, and Structural Equation Modeling (SEM) was then used to analyze the data. The results found that the information quality of smartphone applications has not only direct effect on the mental model of the applications that smartphone users have, but also indirect effect on the interactivity with the smartphone applications.

Keywords Smartphone · Application · Information · Mental · Model · Interactivity

1 Introduction

Thanks to a variety of applications, today's smartphones are regarded as versatile for various purposes. Nonetheless, it is still difficult to say that they fully ensure high interactivity due to the restrictions that depend on the hardware features. Most smartphones have a small display screen and this affects the use of

W. Jung(✉)
The School of Business and Economics, Dankook University,
152, Jook-Jun-Ro, Soo-Ji-Goo, Young-In, Kyung-Ki-Do 448-701, Korea
e-mail: jungw@dankook.ac.kr

© Springer Science+Business Media Singapore 2015
D.-S. Park et al. (eds.), *Advances in Computer Science and Ubiquitous Computing*,
Lecture Notes in Electrical Engineering 373,
DOI: 10.1007/978-981-10-0281-6_8

smartphones. Smartphone users usually use their fingers to interact with the smartphones and it is not easy especially for the users with thicker fingers to precisely touch a specific location within the small screen especially.

In addition, the small display screen doesn't offer sufficient space to show a lot of diverse information including text and graphics. If too much information and features are placed on the same small screen, users may have difficulty with interactivity. In order to resolve this problem, smartphone applications usually segment information on several screens or they reduce the size of text and graphics, but the problem still persists.

The small display screen reduces the readability and discernibility, which may have a negative impact on the use of smartphone applications as well as the information quality of the applications. Furthermore, users also have difficulty in generating and developing the mental model of applications. The mental model, known as a cognitive structure that is comprised of specific knowledge and experience, enables smartphone users to interact with applications [5, 6]. Consequently, within the current smartphone environment, the information quality and the interactivity of applications can be affected in a negative way.

In sum, based upon the IS literature review and the discussion above, the information quality of smartphone applications seems to be related to the mental model of the applications that smartphone users form and hold as well as the interactivity with the applications. Furthermore, the mental model of applications seems to influence the interactivity with the applications. However, a comprehensive IS literature review found that previous IS research have not shown much interest in these relationships especially in a smartphone context. Therefore, the goal of this study is to suggest a research model that can examine: 1) the effects that the information quality of applications has not only on the mental model of the applications that users have, but also on the interactivity with the applications, and 2) the effect that the mental model of applications has on the interactivity with the applications in a smartphone context. This study is expected to provide the theoretical and practical recommendations for researchers and application developers through the findings of this research. These recommendations may help to improve the information quality and the interactivity of smartphone applications. The following hypotheses are proposed to test the relationships empirically.

H1: The information quality of smartphone applications positively affects the mental model of the applications that smartphone users form.

H2: The mental model of smartphone applications that smartphone users form positively affects the interactivity with the applications.

H3: The information quality of smartphone applications positively affects the interactivity with the applications.

2 Research Methodology, Data Analysis, and Results

This study aimed to examine the effects that the information quality of smartphone applications has not only on the mental model of the applications that smartphone users form and hold, but also on the interactivity with the applications. In addition, this study also examined the relationship between the mental model of smartphone applications that users have and the interactivity with the applications.

This study conducted a survey to collect data and a total of 236 participants including university students and practitioners volunteered for the survey. The ratio of men to women in the participants was 50.4 to 49.6. Of the participants, 80.1% were in their twenties. 14.8% and 5.1% were in their thirties and forties respectively. University students majoring in various academic programs at three universities in Korea made up 66.9% of the participants. Meanwhile, practitioners made up 33.1% of the participants. Of the smartphone applications that participants had used just before answering the survey questions, 61.9% were social networking- and communications-related applications.

This study used Structural Equation Modeling (SEM) to analyze the proposed research model and SPSS and AMOS ver. 18 were used as the statistical software. First of all, the goodness of fit was checked. To do so, the indices, such as x^2/df, GFI, AGFI, NFI, TLI, CFI, and RMSEA were examined. The results of statistical analyses were as follows: $x^2/df = 1.999$, GFI = .957, AGFI = .919, NFI = .964, TLI = .972, CFI = .981, and RMSEA = .065. The results of overall fit statistics suggest that the proposed model has a fairly good fit.

Next, this study tested the measurement model by examining the reliability of all observable variables. In order to meet the reliability requirements, all observable variables have to have the loadings on their respective constructs higher than 0.6 or ideally 0.7 [1, 2]. The results showed that all of the loadings were 0.7 or higher, which is well above the recommended cutoff level of 0.6 (see Table 1). Therefore, the results suggest that the reliability of all observable variables is adequate.

Table 1 Standardized Regression Weights of Observable Variables, Composite Reliability (CR), and Average Variance Extracted (AVE)

Latent Variables	Estimates	Variance C.R.	Composite Reliability	AVE
Information Quality	.789	7.061	.901	.631
	.796	6.889		
	.798	6.821		
Mental Model	.859	8.049	.948	.780
	.940	4.207		
	.848	8.337		
Interactivity	.841	7.550	.908	.729
	.862	6.885		
	.859	6.979		

Next, this study tested the convergent validity of the measurement model by examining the composite reliability (CR) and the average variance extracted (AVE) of the latent variables. Unluckily, AMOS ver. 18 does not have the functions to calculate the values of CR and AVE, so they were calculated manually. The formulas suggested by Fornall and Larcker [3] and Hair et al. [4] were used to get the values of CR and AVE. The results of the calculations showed that the values of CR for all latent variables were of 0.9 or higher, well above the recommended tolerance of 0.7 (see Table 1). In addition, the values of AVE for all latent variables also had the values of AVE higher than the recommended level of 0.5 (see Table 1). Thus, based upon the analyses, it can be said that the measurement model of this study demonstrated a satisfactory convergent validity.

$$CR = (\sum \text{Standardized Regression Weights})^2 / ((\sum \text{Standardized Regression} \quad [3]$$
$$\text{Weights})^2 + (\sum \text{Variance}))$$

$$AVE = (\sum \text{Standardized Regression Weights}^2) / N \qquad\qquad\qquad [4]$$

Then, discriminant validity of the measurement model was also checked. This analysis was done by comparing the square root of the AVE with the correlations among the latent variables. In order to meet the discriminant validity requirements, all of the variables in the proposed model should have the square root of their AVEs be higher than the correlations with other variables [2]. The results said that this study is the case (see Table 2). Therefore, the discriminant validity was also confirmed.

Table 2 Correlation Coefficient Value between Constructs and AVE

Constructs	AVE	$\varnothing^2$	$\varnothing^2$	$\varnothing^2$
Information Quality	.631	.040	.060	1.000
Mental Model	.780	.340	1.000	
Interactivity	.729	1.000		

Finally, this study tested the structural model in terms of the significance and the strength of the relationships between the variables. As expected, information quality had a significant influence on mental model ($\beta = .252$, $p = .0001$). In addition, mental model also had a positive impact on interactivity ($\beta = .703$, $p = .0001$). However, this study showed that information quality does not have a significant impact on interactivity ($\beta = .079$, $p = .353$). Therefore, only Hypotheses 1 and 2 were supported. Table 3 and Figure 1 below present the results of the test in detail.

Table 3 Hypothesis Test

	Paths	Coeff.	Stand. Coeff.	P	Results
H1	Information Quality -> Mental Model	.252	.076	***	Accept
H2	Mental Model -> Interactivity	.703	.086	***	Accept
H3	Information Quality -> Interactivity	.079	.085	.353	Reject

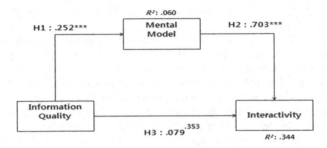

Chi-square=47.988 (df=24) p=.003

Fig. 1 Structural Model Results

3   Discussion and Conclusion

The results of analyses found that the quality of information significantly influenced the mental model of the applications that smartphone users formed. In addition, the mental model of the applications also had a positive impact on the interactivity with the applications. One of interesting findings in this study, however, is that there was no direct relationship between the quality of information and the interactivity with the smartphone applications. This means that the interactivity with smartphone applications may not be enhanced directly by the quality of information that the applications provide. In fact, today's most smartphone applications are easy and simple enough to use. Under these circumstances, high-quality information wouldn't have been necessary for users to interact with the applications.

In practice, mobile application developers can use the findings of this study as a way to improve their applications. More specifically, since this study found that the mental model of smartphone applications and the interactivity with the applications can be secured directly and indirectly by the quality of information, the attributes of information quality that are used in the survey questions including understandability, completeness, and appropriate amount of information can be used to check the quality of information that the applications provide. The three interactivity attributes also can be used to check whether applications meet smartphone users' interaction expectations. Since the improvement of information quality and interactivity can be a powerful strategic advantage for smartphone applications, developers should make an effort to improve the information quality and the interactivity of their applications.

References

1. Barclay, D., Higgins, C., Thompson, R.: The Partial Least Squares (PLS) Approach to Causal Modeling: Personal Computer Adoption and Use as an Illustration. Technology Studies **2**, 285–324 (1995)
2. Chin, W.W.: The partial least squares approach for structural equation modeling. In: Marcoulides, G.A. (ed.) Modern Methods for Business Research, pp. 295–336. Lawrence Erlbaum, Mahwah (1998)
3. Fornell, C., Larcker, D.F.: Evaluating Structural Equation Models with Unobservable Variables and Measurement Error. Journal of Marketing Research **18**, 39–50 (1981)
4. Hair, J.F., Black, B., Babin, B., Andersong, R.E., Tatham, R.L.: Multivariate Data Analysis, 6th edn. Pearson Prentice Hall, Upper Saddle River (2006)
5. Johnson-Laird, P.N.: Mental models. In: Posner, M.I. (ed.) Foundations of Cognitive Science. MIT Press, Cambridge (1989)
6. Rouse, W.B., Morris, N.M.: On Looking into the Black Box: Prospects and Limits in the Search for Mental Models. Psychological Bulletin **100**, 349–363 (1986)

Appendix

Latent Variables	Questions
Information Quality	The smartphone application that I used most recently provided information that is easy enough to understand.
	The smartphone application that I used most recently provided information that contains adequate detail to achieve my personal goals.
	The smartphone application that I used most recently provided a sufficient amount of information to achieve my personal goals.
Mental Model	The way to use the smartphone application that I used most recently was predictable.
	The smartphone application that I used most recently worked as predicted.
	The smartphone application that I used most recently worked as I know.
Interactivity	The smartphone application that I used most recently responded quickly to my commands.
	The smartphone application that I used most recently performed its tasks that I ordered without malfunctioning.
	The smartphone application that I used most recently performed its tasks that I ordered in the way that I expected.

Immersive Dissection Simulator Using Multiple Volume Rendering

Koojoo Kwon, Dong-Su Kang and Byeong-Seok Shin

Abstract Several collections of content representing the human anatomical structure have been introduced. However, most of these contain analog content, such as illustrations, which have difficulty in realistically displaying human organs because they use artificial color and simplified drawings. Digital forms of a human anatomy atlas can provide realistic images from arbitrary viewpoints with real colors. We propose a multiple-volume rendering method for a stereoscopic immersive dissection simulator. It provides a stereoscopic display incorporating a motion recognition sensor, such as the Microsoft Kinect. Users can operate this system without touching or manipulating conventional input devices. To verify the effectiveness of our system, we developed a medical application called virtual dissection. Experimental results show that our system efficiently provides high-quality immersive medical data visualization.

Keywords Virtual dissection · Gesture recognition · Volume rendering · Visible korean

1 Introduction

Visualization methods for human anatomy are widely used in medical imaging [1, 2]. Accurate diagnosis is possible when we use a reconstructed 3D image rather than a 2D image acquired from computed tomography (CT) or magnetic resonance imaging (MRI), because a 3D image provides more accurate spatial information.

Recently, several immersive systems have been proposed to help a user experience simulation of the real world. Some of them have sensors for measuring the distance between the user and the camera to recognize user operation. The Wii

K. Kwon · D.-S. Kang · B.-S. Shin(✉)
Department of Computer Science and Information Engineering,
Inha University, Incheon, Korea
e-mail: mysofs@naver.com, gagalchi@msn.com, bsshin@inha.ac.kr

© Springer Science+Business Media Singapore 2015
D.-S. Park et al. (eds.), *Advances in Computer Science and Ubiquitous Computing*,
Lecture Notes in Electrical Engineering 373,
DOI: 10.1007/978-981-10-0281-6_9

(Nintendo), Playstation Move (SONY), Kinect (Microsoft) and Leap Motion (Leap Motion) are typical examples. These motion recognition devices can increase immersion of user interaction, and they have become compact in size. However, they are dependent on a predefined operating environment, because these sensors cannot accurately identify and interpret all user gestures.

As computer technology advances, a variety of medical devices are being invented and huge volumes of medical data are produced from those devices. The Visible Human, the Chinese Visible Human and the Visible Korean are well-known examples, which enable us to observe the internal structure of the human body with non-invasive methods [3-7]. Since a series of these images is regarded as 3D texture sampled at very high resolution, the size is in the hundreds of gigabytes, even when we scan only a very small region of the human body. These images help to improve the accuracy of a virtual diagnosis. In this paper, we provide an immersive dissection experience system using multiple-volume visualization and gesture recognition techniques.

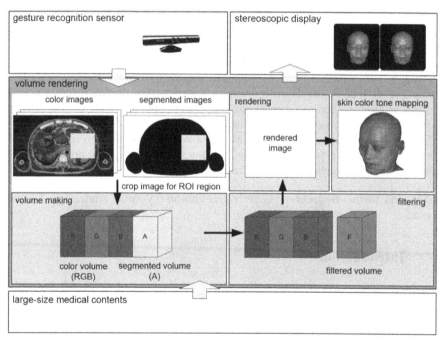

Fig. 1 Framework of our virtual dissection system. It is composed of a gesture recognition sensor, a stereoscopic display and a volume rendering interface.

2   Immersive Virtual Dissection System

Our system is applied to a virtual dissection application that is operated by the user's hand gestures. When performing incision or excision, the volume data should be deformed and moved. That requires some kind internal representation to show the cross-sections made by anatomical operations. For efficient description of the relationship between the part to be incised or excised and the remaining parts, we need a method to produce an exploded view, since both sides of the incised or excised part have important meaning.

We use the framework for recognizing user gestures shown in Fig. 1. Our framework includes a gesture recognition sensor and a volume rendering module. It also includes a large-sized stereoscopic display for providing an immersive environment. The user is positioned on a hot spot in front of the system and experiences the virtual dissection using pre-defined gestures. It provides immersive realism, because the operation is based on a non-contact interface.

In the volume rendering interface, we can use large-size medical content, such as the Visible Male [5], the Visible Female [6] and the Visible Head [7]. We collected volume data using these image archives and cropped the 3D area against the region-of-interest (ROI) in our virtual dissection system. The volume data, which has only color values, additionally needs segmented information, because we cannot recognize shapes of objects in the color values [10]. This segmented data is made manually or semi-automatically. In the filtering step, we can smooth the segmented data using a Gaussian filter. After volume rendering, we adjust the skin color of the result to visualize a realistic human face.

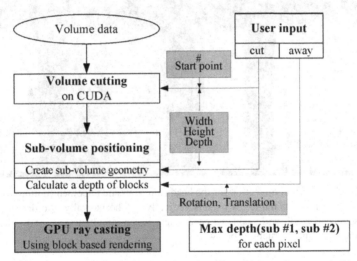

Fig. 2 Overall procedure of the proposed block-based multiple-volume rendering method. The volume is split into several blocks. Each block is rendered by comparing its depth value with the remaining blocks.

Basically, the proposed rendering method uses a graphics processing unit (GPU)-based ray-casting algorithm, and its overall process is depicted in Fig. 2. First, we split the volume data into several sub-volumes with the user-specific parameters under the Compute Unified Device Architecture (CUDA) from NVIDIA. Each sub-volume has its own dimensions (width, height and length). These data are used as scale factors of the corresponding proxy geometry in the rendering phase.

In the vertex shader, the information is used as a parameter to scale a unit-block, the size of which is 1.0×1.0×1.0, into preferred size blocks. The scaled blocks are used as proxy geometry for volume with the scaling operation. Since the proxy geometry can be applied to empty space skipping and early ray termination in GPU ray-casting on the fragment shader, we can accelerate the rendering speed. In the ray sampling step, we render an image in a frame buffer that refers to the multiple 3D textures of sub-volumes separated on CUDA and a pre-calculated opacity transfer function (OTF).

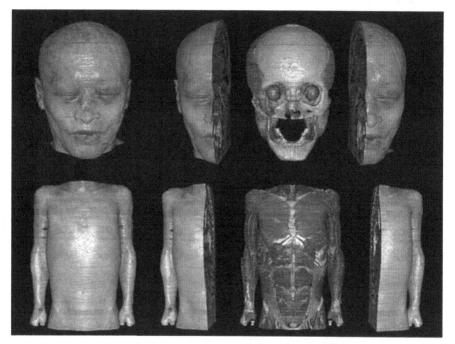

Fig. 3 Example of virtual dissection for the head (top row) and abdomen (bottom row). The method recursively splits the target objects vertically and horizontally in order as follows: skin, muscles and skull, and brain. At each stage, it can simultaneously show the cross-sections and anatomical structures below.

3   Experimental Result

We conducted experiments on a consumer PC equipped with an Intel Core i7-4790 3.6GHz CPU and 16 GB of main memory. The GPU is an NVIDIA GTX 780 with 6GB of video memory. We used the 64-bit Windows 8.1 operating system. Our method is implemented in Direct3D 11 with High Level Shader Language (HLSL).

Figure 3 shows an example of virtual dissection for the head and abdomen. It recursively splits the target objects horizontally and vertically. At each stage, it can simultaneously show the cross-sections and anatomical structure underneath.

It helps us to understand the internal structure of a specific organ and its interrelationships with neighboring organs.

The rendering speed of each screen resolutions is shown in the table 1. Our system includes a high-definition (HD) stereoscopic display device. Our method can provide rendering speed more than 20 *fps* at the HD stereoscopic view. There is no difference between screen resolution 1280×1024 and 1920×1080 since the screen resolution 1920×1080 is includes large empty space.

Table 1 The rendering speed of each screen resolutions (*fps*)

Resolutions	800×600	1280×1024	1920×1080
Rendering speed	60	24.6	24.5

Figure 4 shows our system, which is permanently exhibited at the National Science Museum.

Fig. 4 Left, immersive virtual dissection system exhibited at the National Science Museum, Korea. Right, the system in operation.

4   Conclusion

We developed a stereoscopic immersive experience system for virtual dissection that provides a realistic operation environment using a motion recognition sensor and a stereoscopic display. We also propose a sub-volume based rendering method that minimizes usage of GPU memory and improves rendering speed with parallel processing. Experimental results show that our system efficiently provides high-quality immersive medical visualization.

Acknowledgment This work was supported by the National Research Foundation of Korea (NRF) grant funded by the Korea government (MSIP) (No. 2015R1A2A2A01008248).

References

1. Spitzer, V.M., Scherzinger, A.L.: Virtual anatomy: An anatomist's playground. Clinical Anatomy **19**(3), 192–203 (2006)
2. James, A.P., Dasarathy, B.V.: Medical image fusion: A survey of the state of the art. Information Fusion **19**, 4–19 (2014)
3. Spitzer, V., Ackerman, M.J., Scherzinger, A.L., Whitlock, D.: The visible human male: a technical report. Journal of the American Medical Informatics Association **3**(2), 118–130 (1996)
4. Shin, D.S., Chung, M.S., Shin, B.S., Kwon, K.: Laparoscopic and endoscopic exploration of the ascending colon wall based on a cadaver sectioned images. Anatomical Science International **89**(1), 21–27 (2014)
5. Huang, Y.X., Jin, L.Z., Lowe, J.A., Wang, X.Y., Xu, H.Z., Teng, Y.J., Zhang, H.Z., Chi, Y.L.: Three-dimensional reconstruction of the superior mediastinum from chinese visible human female. Surgical and Radiologic Anatomy **32**(7), 693–698 (2010)
6. Park, J.S., Chung, M.S., Hwang, S.B., Lee, Y.S., Har, D.H., Park, H.S.: Visible korean human: improved serially sectioned images of the entire body. IEEE Transactions on Medical Imaging **24**(3), 352–360 (2005)
7. Schiemann, T., Freudenberg, J., Pflesser, B., Pommert, A., Priesmeyer, K., Riemer, M., Schubert, R., Tiede, U., Höhne, K.H.: Exploring the visible human using the voxel-man framework. Computerized Medical Imaging and Graphics **24**(3), 127–132 (2000)
8. Hwang, S.B., Chung, M.S., Hwang, Y.I., Park, H.S., Har, D.-H., Shin, D.S., Shin, B.-S., Park, J.S.: Improved Sectioned Images of the Female Pelvis Showing Detailed Urogenital and Neighboring Structures. Korean Journal of Physical Anthropologists **23**(4), 187–198 (2010)
9. Park, J.S., Chung, M.S., Shin, D.S., Har, D.H., Cho, Z.H., Kim, Y.B., Han, J.Y., Chi, J.G.: Sectioned images of the cadaver head including the brain and correspondences with ultrahigh field 7.0T MRIs. Proceeding of IEEE **97**, 1988–1996 (2009)
10. Park, J.S., Chung, M.S., Hwang, S.B., Lee, Y.S., Har, D.H., Park, H.S.: Technical report on semiautomatic segmentation by using the Adobe Photoshop. Journal of Digital Imaging **18**, 333–343 (2005)

Business Informatics Management Model

Alena Buchalcevova and Jan Pour

Abstract The aim of this paper is to present a development of the Management of Business Informatics (MBI) model that aims to assist enterprises in managing their business informatics. First, current issues in business informatics management are outlined as well as the results of several surveys conducted worldwide and in the Czech Republic. The MBI model development is described based on design science research methodology.

Keywords Business informatics · Management · Governance · Design science research

1 Introduction

Recent changes in the environment, economy and technology substantially drive a usage of ICT services within companies and the whole society. Businesses are realizing that performance and availability of their technologies are critical to their growth and competitive advantage. These issues are driving business informatics management initiatives. In recent years, standards, frameworks, and best practices addressing different aspects of business informatics management have emerged and matured. Among these, the most quoted are: ITIL [1] and ISO/IEC 20000 [2], ISO/IEC 38500 [3], ISO 27000 [4], the Control Objectives for Information and Related Technology (COBIT) [6] and the Open Group Architecture Framework (TOGAF) [7]. The usage of these standards and frameworks has increased in order to improve IT internal processes, quality, productivity, efficiency and communication with business areas, as well as explore possibilities for innovation [8]. However, several surveys indicate particular challenges linked to their usage. Studies conducted by the IT Governance Institute in 2008 [10] and 2010 [11] show that the vast majority (92%) of respondents are aware of the issues resulting

A. Buchalcevova(✉) · J. Pour
Department of IT, University of Economics,
Prague, W. Churchill Sq. 4, 130 67 Prague 3, Czech Republic
e-mail: {alena.buchalcevova,jan.pour}@vse.cz

© Springer Science+Business Media Singapore 2015
D.-S. Park et al. (eds.), *Advances in Computer Science and Ubiquitous Computing*,
Lecture Notes in Electrical Engineering 373,
DOI: 10.1007/978-981-10-0281-6_10

from the application of these standards and frameworks. While security and compliance are mentioned as important elements, it is people who represent the most critical issue. Results of a survey conducted in 160 SMEs in six Central European countries [11] show an extensive difference between knowledge of ITSM frameworks and their application within companies. Likewise, results of surveys conducted in the Czech Republic [12],[13] demonstrate a low level of usage of these frameworks for management of business informatics. This fact together with limited customization possibilities of these frameworks has led our team to develop a Management of Business Informatics (MBI) model that aims to assist enterprises (including SMEs) in managing their business informatics.

The aim of this paper is to present a development of the MBI model. This paper is organized based on the structure recommended by Gregor and Hevner [14] for presenting design science research. Following introduction, section 2 defines our approach to the research. Section 3 then describes the development of the MBI model and its evaluation. Lastly, the conclusion is presented.

2   Research Method

The MBI model is a methodology for business informatics management and as such it represents an IS artifact according to [15]. Thus, design science research (DSR) can be applied to develop such an artifact. Based on a definition of research contribution maturity levels introduced in [14], the MBI model represents an instantiated artifact, i.e. Level 1 artifact, and within their DSR Knowledge Contribution Framework the MBI model can be classified as an Improvement. The development of the MBI model was based on design science research methodology [16].

3   Business Informatics Management Model

The structure of this section follows individual steps of the design science research methodology [16].

3.1   Problem Identification and Motivation

With the aim to identify the status of business informatics management in the Czech Republic, our team at the Department of Information Technologies at the Prague University of Economics conducted a nationwide survey during 2010 [13] and a subsequent survey in February 2012 [12]. According to these surveys to the most important reasons causing the low utilization of existing business informatics management methodologies and standards belong their complexity, time consuming implementation and high costs. Moreover, existing frameworks do not sufficiently take into account various factors that influence management of business informatics, e.g. sector of the economy, company size, importance of IT for strategic goals, etc. Furthermore, the implementation of such methodologies requires an extensive documentation and high knowledge and skills even in the case of a small enterprise with a simple information system. Consequently, such methodologies are used almost exclusively by larger companies with a significant IT budget [13]. The results of these surveys have led our team to a development of our own tool for business informatics management.

3.2   Solution Objectives

The objective of the Management of Business Informatics (MBI) model is to provide a support for business informatics management activities in companies that figure as users of ICT services. The MBI model aims to provide a solution that suits to specific characteristics of a company which determines the effectiveness of IT governance that cannot be generalized to all types of firms or industries as e.g. Ali and Green point out [17]. The MBI model strives to help organizations to improve the performance of enterprise IT systems, more specifically the quality, availability, security and effectiveness of IT services, and indirectly the overall business performance [13].

3.3   MBI Model Design and Development

Design and development activity is crucial in the whole design science research process. In compliance to Hevner et al. [15], this activity was performed as an iterative search process. The MBI model was defined based on an extensive literature review, analysis of existing standards, methods and frameworks as well as generalized knowledge gained from numerous consulting projects across a wide spectrum of organizations.

The architecture of the MBI model is defined in the UML 2.0 class diagram notation (without methods as on the conceptual level) in Figure 1. A key MBI component represents a Task which describes how to proceed in solving a particular IT management issue. To the examples of Tasks belong: Proposal for Enterprise IT System Sourcing, IT Service Implementation, Service Activation, Security Audit Implementation etc. The MBI model defines a large number of Tasks that are organized in a three-level hierarchy: Management Domain, Task Group and Task level.

Each Task has several attributes. Besides the identification attributes, other attributes represent a specific content of the Task, i.e. Goal, Purpose, Content and Scheme of Activities. An additional content of the Task is represented by relationships to other classes e.g.:

- Document – this class represents a printed or electronic document that is used as a Task input or output.
- Scenario – this class represents a typical issue that needs to be addressed in a business life.

- Application – this class comprises application software that can be utilized for a given Task.
- Metrics – this class constitutes of metrics that are expressed in the context of dimensional modelling as indicators and their analytical dimensions.
- Method – this class describes formalized processes and guidelines to fulfill the goal of a Task.
- Role – this class expresses specific responsibilities of a role holder.
- Factor – this class has a significant impact on the way a particular Task is performed. To the most important Factors belong: Organization Size, Industry Sector (in which an organization operates) and Organization Type (i.e. private company or public institution).

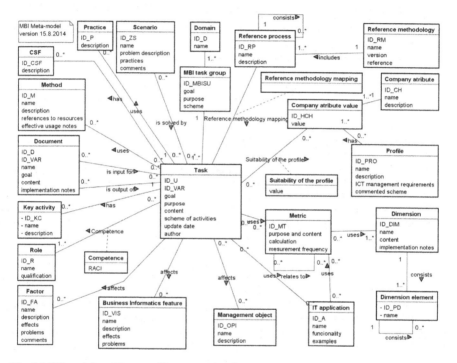

Fig. 1 MBI model architecture (Source: authors)

3.4 Demonstration, Evaluation and Communication

The MBI model was described in detail in [18]. For the purpose of an effective MBI model demonstration, evaluation and usage, the MBI model was implemented as a web application (at the URL mbi.vse.cz). A screenshot of the MBI application shown in Figure 2 demonstrates all Tasks – Roles relationships including the detailed specification based on RACI matrix.

Both the MBI model and the MBI application were presented at the business informatics management conference arranged by the Czech Society for Systems Integration in January 2014 where they received a positive acceptance and ITSMF association expressed its interest in cooperation on the MBI model further development.

The MBI application was in a pilot usage from January 2014 to August 2014. During this time, the MBI application was tested by the MBI team and first MBI users. After this half a year pilot operation, the MBI application was refactored and improvement and enhancement were performed. The MBI 2.0 application came into existence at the end of August 2014. Among the main enhancements are to mention user profiles and application login, full-text search, summary slide with key information related to each object and documents for download. With the aim to enable tracking of the MBI application usage and obtaining feedback from its users, user registration was supplemented. Number of registered users is increasing and has run to 350 currently.

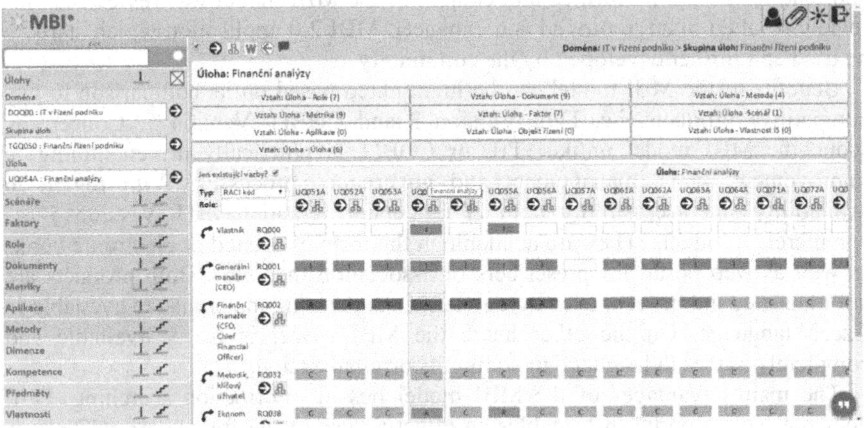

Fig. 2 Screenshot of the MBI application – Task and Roles relationships

The MBI application serves as an information base for companies and supports solution of their IT management problems. Following are examples of companies that use the MBI application:

- iPodnik – the Czech company delivering cloud services for more than 200 clients,
- ITG - the Czech consulting company concentrated on analysis and design of IT systems operating on the Czech and Slovak market,
- Aquasoft – the Czech software company focused on public administration.

Besides use in business the MBI model and application are used in university courses at Prague University of Economics and Czech Technical University in Prague. A MBI community was established which unites MBI content authors and the most active users, organizes meetings, presentations and training and enables to exchange experience. Currently, the MBI application is in Czech, but English version is preparing.

4   Conclusion

In this paper, the Management of Business Informatics (MBI) model was presented focusing on the MBI model development according to design science research methodology. First, current issues in business informatics management were outlined as well as the results of several surveys conducted worldwide and in the Czech Republic. To overcome reported issues, the MBI model was decided to be developed. The MBI model was defined based on an extensive literature review, analysis of existing standards, methods and frameworks as well as generalized knowledge gained from numerous consulting projects across a wide spectrum of organizations. The MBI model was described as to its concepts, architecture, accessibility of information and processes for model manipulation. For the purpose of an effective MBI model demonstration, evaluation and usage, the MBI model was implemented as a web application. The MBI application underwent a pilot usage for half a year when functionality, usability, performance and load tests were performed and content of the MBI model was reviewed. This review resulted in an improved and enhanced MBI 2.0 application which starts to be used and further developed by the community.

Benefits of the MBI model can be highlighted based on its comparison to key competitive products, i.e. ITIL Version 3 and COBIT Version 5. Concerning scope the MBI model unlike ITIL or COBIT is practically driven aiming in improvement of IT management and governance processes. With regard to availability and accessibility both ITIL Version 3 and COBIT Version 5 are commercial products. They are available in the form of printed or electronic books as well as web portal for prescribers or association members. Both products are not translated into Czech, even though certain supporting materials are available in Czech language. On the other hand, the MBI model is freely available and completely in Czech language including its original design.

The main advantages of the MBI model lies in information searching. ITIL does not have a portal that enables an effective use of the framework. COBIT is accessible through a set of electronic or printed guides or COBIT 5 Online but effective information search is not supported. The MBI application primarily focuses on effective information search. Using the MBI application through Scenarios enables solving a concrete management situation very effectively. When finding an appropriate Scenario and opening the page with a Scenario description, it is possible to read through key issues and questions that are connected with the problem to be solved, recommended practices that can be used to solve the issue and a list of all interrelated Tasks. Another way how to direct access to information is to select a particular Role, Task, Document or some other path relevant to the issue you are working on. In case of search for specific term, a full-text searching option is available since the MBI 2.0 version. This way all occurrences of the term within all MBI objects will be obtained.

Lastly, the MBI model is compared in the area of customization. Both ITIL and COBIT have to be adapted to individual needs either in-house or by help of consultants. The MBI model covers the customization requirement within its concept of various MBI model types. The MBI Generic model is intended for all types of organizations and includes generalized best practice guidelines for business informatics management. The MBI Specific model is aimed at organizations that belong to a particular sector of the economy (automotive, banking, public administration, etc.). The content of the MBI Specific model is adapted to particular industry, business, legislative and other conditions that apply to a given sector of the economy. The third type of the MBI model is the MBI model for a concrete organization taking into account particular aspects of an organization.

References

1. ITIL: Introduction to the ITIL Service Lifecycle. TSO, London (2007)
2. ISO/IEC 20000-1 Information technology - Service management - Part 1: Service management system requirements (2011)
3. ISO/IEC 38500 Corporate governance of information technology (2015)
4. ISO/IEC 27001 Information technology – Security techniques – Information security management systems – Requirements (2013)
5. Van Grembergen, W., De Haes, S.: Enterprise Governance of Information Technology: Achieving Strategic Alignment and Value (2009)
6. COBIT 5 - A Business Framework for the Governance and Management of Enterprise IT. Information Systems Audit and Control Association (2015)
7. TOGAF Version 9. The Open Group Architecture Framework. The Open Group (2009)
8. Scheeren, A.W., Fontes-Filho, J.R., Tavares, E.: Impacts of a Relationship Model on Informational Technology Governance: An Analysis of Managerial Perceptions in Brazil. JISTEM - Journal of Information Systems and Technology Management 10(3), 621–642 (2013). doi:10.4301/S1807-17752013000300009. ISSN:1807-1775
9. IT Governance Global Status Report 2008: IT Governance Institute (2009)
10. Global Status Report on the Governance of EnterpriseIT: IT Governance Institute (2011)
11. Küller, P., Vogt, M., Hertweck, D., Grabowski, M.: IT Service Management for Small and Medium-Sized Enterprises: A Domain Specific Approach. Journal of Innovation Management in Small & Medium Enterprises (2012). doi:10.5171/2012.475633
12. Pour, J.: Results of the survey of management of enterprise ICT. Systemova Integrace (1), 49–57 (2012). (in Czech)
13. Pour, J., Vorisek, J., Feuerlicht, G.: Model for management of enterprise IT: considerations of the impact of cloud computing. In: Confenis, pp. 157–168. Trauner Verlag, Linz (2013)
14. Gregor, S., Hevner, A.: Positioning and Presenting Design Science Research for Maximum Impact. MIS Quarterly 37(2), 337–355 (2013)
15. Hevner, A., March, S.T., Park, J., Ram, S.: Design Science in Information Systems Research. MIS Quarterly (2004)
16. Peffers, K., Tuunanen, T., Rothenberger, M.A., Chatterjee, S.: A design science research methodology for information systems research. Journal of Management Information Systems 24(3), 45–77 (2007)
17. Ali, S., Green, P.: Effective information technology (IT) governance mechanisms: An IT outsourcing perspective. Inf. Syst. Front. 14, 179–193 (2012). doi:10.1007/s10796-009-9183-y
18. Vorisek, J., Pour, J., et al.: Management of Business Informatics. Professional Publishing (2012). (in Czech)

Group Awareness in Task Context-Aware E-mail Platform

Masashi Katsumata

Abstract E-mail management applications are among the most used tools for collaborative work in enterprises. However, traditional e-mail management applications lack adequate support to deal with reusing e-mail and e-mail associated resources (attached files, schedule, etc.) management for collaborative work. In our previous study, we developed a task context-aware e-mail platform that helps users send e-mails quickly and efficiently. This platform is based on an ontology-based task context model, that represents the conceptual associations between a task and the task related e-mail process. This study proposes enabling group awareness to facilitate effective collaboration using our system and validates its usefulness.

Keywords Collaborative task · Awareness · Ontology · E-mail management

1 Introduction

E-mail management applications are among the most common tools for collaboration in enterprise environments. However, traditional e-mail management applications lack adequate support to deal with reusing e-mail and e-mail associated resources (attached files, schedule, etc.) management for collaborative work in an enterprise. Knowledge workers can efficiently search and use e-mail messages and corresponding resources by organizing these messages according to individual tasks [1]. Thus, multi-tasking knowledge workers often set up automatic filtering or manually move e-mail messages into folders. In recent times, enhanced existing task management systems and task-centric mail clients have been used for this purpose. Furthermore, several studies that support the

M. Katsumata(✉)
Department of Computer and Information Engineering, Nippon Institute of Technology,
4-1 Gakuendai, Miyashiro-machi, Minamisaitama-gun, Saitama, Japan
e-mail: katumata@nit.ac.jp

© Springer Science+Business Media Singapore 2015
D.-S. Park et al. (eds.), *Advances in Computer Science and Ubiquitous Computing*,
Lecture Notes in Electrical Engineering 373,
DOI: 10.1007/978-981-10-0281-6_11

73

discovery of e-mail messages and related resources by adding meta data to e-mails and related resources have been conducted [1],[2],[3],[4].

In our previous study, we developed a task context-aware e-mail platform that helps users send e-mails quickly and efficiently [5]. Furthermore, this platform extracts data from reply e-mail messages automatically. This platform allows users to automate task-related information classification and user support services using a task context model. We have built the task context model as a semantic representation model of the conceptual associations between a task and the task-related e-mail process. Using the task context model, this platform can provide a context-aware service for mail form composition and automatic mail data extraction. To collaborate effectively, group members must keep all members up-to-date on task progress, status, issues, and date changes. In this study, we introduce a new medium to support group awareness in our task context-aware e-mail platform. This study validates the enabling of group awareness for collaborative work in an enterprise. The remainder of this paper is organized as follows. In Section 2, we review related work. Section 3 describes the task context-aware e-mail platform and group awareness. In Section 4, we validate the enabling of group awareness. In Section 5, we present conclusions and suggestions for future research.

2   Related Work

Much task-centered management tool research has examined adding task management features to e-mail client applications. TaskMaster [2] enhances an e-mail client to function as a task management system and manages resources as e-mail messages and file attachments for each task. Furthermore, a useful user interface with both browsing and operating resources is provided. KASIMIR [3] and OntoPIM [4] are ontology-based personal task-management systems. These systems provide semi-automated functions for retrieving and registering task-related information in e-mail messages according to an ontology-based model. These applications make it easy to move tasks from e-mail clients to dedicated task managers. These studies were primarily concerned with improving the management of and search for task related information. Our research contribution is to provide a support function for reusing managed data to accomplish a collaborative task.

Other studies have explored how computer-mediated tools can be used to share task status information within project teams. In [6], daily project status e-mail messages provide more frequent updates on team members' progress. This approach showed that such updates helped keep team members in the loop regarding the progress of others in the team. In this study, we introduce a new medium to support group awareness in e-mail-based collaborative work environments.

3   Task Context-Aware E-mail Platform

3.1   System Overview

We have implemented a prototype system that executes a service for users in the task context-aware e-mail platform. This prototype includes the following three systems: a task context server, a mail server, and a mail client (Fig. 1). The task context server manages the task context (file path, schedule, contact information, etc.) and its value. The task context server can accept a request command (create, refer, update, and delete) from the mail server and client via TCP/IP. By accepting a request command, the task context server can update the task context model using the Jena API. The mail server is built on Apache James, which is a mail application platform that allows users to code custom applications for e-mail processing. Apache James provides e-mail filtering by way of the Matcher function and provides e-mail processing through the Mailet function. We have introduced extended e-mail headers to realize the concept of our proposed system [5]. After Matcher refers to the extended e-mail headers, Mailet can be executed on basis of the information in the extended e-mail headers. Furthermore, the client can connect to both the mail server and the task context server. In addition to general e-mail operations, the client provides a user interface that manages the task context data. The e-mail message submitted by the client is added to the extended e-mail header automatically. In our prototype, the client displays the structured mail form by referring to the extended e-mail header.

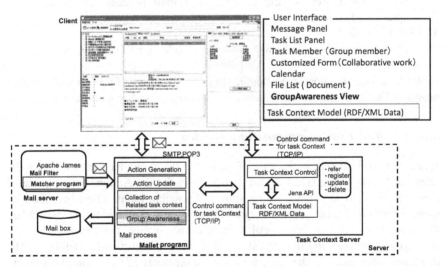

Fig. 1 Task-Context Aware E-mail Platform

3.2   Task Context Model

We have created an ontology-based semantic representation model that represents the conceptual associations between a task and e-mail processes. We call this the task context model. The task context-aware e-mail platform performs the services required by a task user based on a task context model. This model relates the conceptual associations between a task and an e-mail process to physical context entities (e-mail messages, attached files, group members, mail form items, etc.).

The semantic representation of the task context model is based on the Resource Description Framework (RDF). The RDF is a collection of triples, each of which comprises a resource, a property, and a literal. A set of such triples is called an RDF graph.

The task context is represented as a property of a "Task" resource. In the task context model, task contexts are classified as files, schedules, participants, memos, mail form items, etc. Furthermore, we define the concept of Action. The Action concept refers to the type of collaborative task. Such collaborative tasks involve the characteristic used to retrieve the contents from a reply mail for many users within an organization. We have considered the following three Actions for a collaboration task.

1. Event Notification
    This Action represents the concept of a mail process for attendance confirmation for an event being held in an organization.
2. Questionnaire Request
    This Action represents the concept of a mail process for questionnaire requests and collection.
3. File Collection
    This Action represents the mail process for collection of attached files.

We introduce the process portion of the group awareness function in the Mailet program. Furthermore, we implement a GroupAwareness View on the client.

3.3   E-mail Form Composition Service

When a task owner selects the type of Action on the client, an e-mail form for the selected Action is displayed. In the e-mail form for the Action, the task-related data are provided as a list of suggestions of possible input. Therefore, the task owner spends less time typing and querying information related to the task. When the task client receives an e-mail from the task owner, the replying form is displayed by referencing the extended e-mail header X-Action-Model-Type. The displayed input fields in the reply form are the elements of the Action type (event notification, questionnaire request and file collection). A task member can type the value according to the displayed input field on the reply form.

3.4 Data Extraction Service

When the mail server receives an e-mail according to the type of Action, the contents of the reply e-mail are retrieved automatically as the task context. In the task context server, the retrieved task context data are managed as RDF/XML format data based on the conceptual Action model. When the task context server receives a reply e-mail, the value of the task context in the Action is updated, and the state of the Action is displayed on the client's State panel. Thus, the task owner can confirm the state of the task intuitively without checking each reply e-mail. Automated processes such as the generation or update of the Action are performed via Mailet according to the value of the extended mail header. The attendance and the questionnaire data can be written to a comma separated value file on the assumption that the task context data might be used by a spreadsheet application (e.g., Microsoft Excel). For file collections, files are renamed automatically according to a predefined file name from the attached file in the task member's reply e-mail.

3.5 Group Awareness

It is important to support awareness of work in progress to facilitate effective work in an organization. In general, office workers organize and prioritize their own time and manage multiple tasks and schedules (calendar). However, they must adjust task priority and schedules according to the dynamic changes of task related context. In this study, the group awareness function provides task related awareness information to group members in an organization. In general, a knowledge worker has multiple tasks during their daily work. Therefore, they are unlikely to overlook a task status change by focusing on high priority work. In this study, we built an alert function to attract attention by adding tags to e-mail messages. Then, we conducted an experiment to extract user reaction information relative to e-mail messages with these tags.

4 Discussion

To validate the enabling of group awareness with the proposed system, we implemented the logging function to measure awareness in our client and server system. In the logging function, when the tagged mail message is clicked in the message list on the client, the client automatically sends the click information to the context server. We conducted an experiment to verify awareness of tagged mail messages and obtained data from 10 university students (male: 19-22 years) who used the prototype system. The prototype system was installed in our laboratory. We performed an experiment to send e-mail message with tag (changed, attention, etc.) three times. The results indicate that the logging function can identify user interest. We found that task users sense e-mail messages with the tags in the GroupAwareness View on the client.

5   Conclusions and Future Works

This study has proposed a group awareness function that has been added to a context-aware e-mail platform. First, to validate the effect of the group awareness functionality, we built a logging function in our system's client and server. This logging function collects information about when a user clicks an incoming message in the message list on the client. By introducing a Group-Awareness View to the client, we found that users sensed e-mail message with tags and that the logging function could indicate user interest. In future work, as the next step to validate group awareness, we plan to continue investigating the effect of group awareness in practical collaborative work in an organization.

References

1. Krämer, J.-P.: PIM-mail: consolidating task and email management. In: Proceedings of the 28th of the International Conference Extended Abstracts on Human Factors in Computing Systems, pp. 4411–4416 (2010)
2. Bellotti, V., Ducheneaut, N., Howard, M., Smith, I.: Taking email to task: the design and evaluation of a task management centered email tool. In: Proceedings of the SIGCHI Conference on Human Factors in Computing Systems, pp. 345–352. ACM, New York (2003)
3. Grebner, O., Ong, E., Riss, U.V.: KASIMIR—work process embedded task management leveraging the semantic desktop. In: Proceedings of Multikonferenz Wirtshaftsinformatik, Workshop Semantic Web Technology in Business Information Systems, pp. 715–726 (2008)
4. Lepouras, G., Dix, A., Katifori, T., Catarci, T., Habegger, B., Poggi, A., Ioannidis, Y.: On-toPIM: from personal information management to task information management. In: Proceedings of SIGIR Workshop on Personal Information Management, pp. 78–81 (2006)
5. Katsumata, M.: Design and implementation of task context-aware e-mail platform for collaborative tasks. In: Proceedings of the 5th FTRA International Conference on Computer Science and Its Applications (CSA 2013), pp. 233–239 (2013)
6. Brush, A.J., Borning, A.: 'Today' message: lightweight group awareness via email. In: Proceedings of the SIGCHI Conference on Human Factors in Computing Systems (CHI 2003), pp. 920–921. ACM, New York (2003)

An Android-Based Feed Behavior Monitoring System for Early Disease Detection in Livestock

Saraswathi Sivamani, Honggeun Kim, Myeongbae Lee, Changsun Shin, Jangwoo Park and Yongyun Cho

Abstract The objective of this study was to develop an android based mobile application to analyze and prevent the possible disease in the early stage by monitoring the feeding behavior of the livestock. The proposed system secures the information such as feed intake, feeding time, rumination and the feeding rate through the Internet of Things equipped in the livestock farm. Every day feed intake and the feed intake rate are compared with the earlier day, a week before and a month before to find the tardiness or other possible disease in the livestock and an alert is sent to the farmers as text and as well as shown in the mobile application for immediate response. The proposed system is expected to contribute the enhancement of the productivity in the livestock farm, by preventing the disease at its early stage.

Keywords Livestock · Wireless sensor network · Internet of things

1 Introduction

The rate of health disorders in the livestock is rapidly increasing with various changes in the global environment and natural disasters [1]. Increasing health threat in the livestock can lead to major production loss for the farmers which includes the reduced milk production [2], treatment cost or even led to death of the livestock [3], when left untreated. Early detection of the sick animals is one of the solution to and treat with care. In case of individual farmers, the detection of disease is easier as there are a number of interactions between the human and animal. So, when it comes to the large number of herds in the farm, the encounter between the human and animal reduces, where the significance in identification of

S. Sivamani · H. Kim · M. Lee · C. Shin · J. Park · Y. Cho(✉)
Information and Communication Engineering, Sunchon National University,
413 Jungangno, Suncheon, Jeonnam 540-742, Korea
e-mail: {saraswathi,khg_david,lmb,csshin,jwpark,yycho}@sunchon.ac.kr

© Springer Science+Business Media Singapore 2015
D.-S. Park et al. (eds.), *Advances in Computer Science and Ubiquitous Computing*,
Lecture Notes in Electrical Engineering 373,
DOI: 10.1007/978-981-10-0281-6_12

the sickness also minimizes with the advancement in the livestock farm. Technology evolution has given a big hand in surpassing the traditional methods [4], where most of the farm has adopted the concept of sensor networks, RFID technology that helps with the automation of the many services such herd management, health management and environmental control. With this advanced technology, the information of the livestock are obtained precisely and can be used timely, to detect the disease. For this purpose, the tool to monitor the health status to assist the farmer to identify the disease.

As the use of the mobile phones are wide enough, it is more appropriate and perfect for the development of the mobile app to monitor and notify the detection of disease among the livestock farm. In this paper, we design and develop a mobile app for the early detection of the disease with the feed monitoring system. The most common types of disease are easily identified with the temperature of livestock, but some of the disease is rare to identify early and does not show any temperature change. Many researches prove that the change in the pattern of the feed can be a good indicator in the identification of health status [5]. Thus, the proposed system, while maintaining and monitoring the feeding behavior of the livestock, also identifies the possible disease and send an alert message to the farmer, to early health checkup.

2   Related Works

Due to the increased rate of livestock disease, the health factor is more important for the farmers because they result in production losses which can lead to medical expenses [6]. Economic losses are due to premature culling, reduced milk yield and quality, veterinary treatment, and increased labor. Some studies reveal that the clinical mastitis reduces milk yield and cause lameness in the livestock, which is a serious factor in the aspect of livestock's health [7]. Due to increase milk production and stress, there is more possibility of health disorders in the animals, both mentally and physically. If we closely monitor the feeding behavior of the livestock, by noticing the abnormalities, the early detection of disease is highly possible. Likewise, the production cost is also increasing which includes food stock, equipment and energy for the livestock farm.

Automation in livestock farm is developed for various monitoring and control services such as herd management, milk production, environmental control and behavior monitoring [8 - 10]. As we known, taking care of the individual feed status in large farm is not feasible. However, usage of sensor networks, electronic tags, electronic floor scale and RFID reader, can help the feed status to be automatically transferred to the server. A tool to monitor the health status of livestock remotely can assist the farmer to identify disease earlier. In this paper, an android based mobile application is developed to monitor and the status of the feed and also helps in the early detection of the disease by monitoring and comparing the feed rate of livestock with a record of the week and month feed intake rate. Usage of smart phone and its applications within the farm staffs can reduce the labor's burden and also increase the productivity in various ways.

3   Livestock Feed Monitoring System

With the growth of technology, all kinds of data are obtained and transmitted precisely. In livestock farms, most of the services such as environment, weighing, feed intake, milk yielding are converted to automation with sensors, RFID and actuators connected through the wireless sensor network and readers, as shown in the figure 1. In this paper, we plan to focus on developing android based mobile application for the feed monitoring of the livestock, to monitor and detect the abnormalities in cattle. For this, there are three important steps to be noted. Firstly, the identification of the livestock, which is achieved by the RFID tag. Next, the food intake amount, is collected from the electronic scale under the feed bunk which is then transmitted to the server through the wireless network. Lastly, coordinating the data for observation and comparison which can help to early detect the disease.

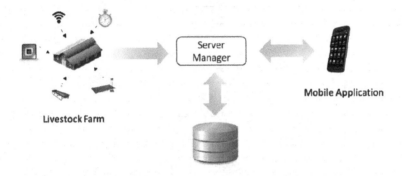

Fig. 1 Overview of Livestock feed monitoring system

Each livestock has RFID tag with unique id that is pierced in the ear, in which their information are stored. The feed bunk is placed over the electronic scale that are controlled by the processor. The intake record starts as soon as the livestock enter the RFID field. Once it leaves the field, the difference in the bunk weight is recorded and sent to the server. With this, the feeding time, feeding rate, amount of intake and the number of times are secured and analyzed. The everyday feed rate is compared with the previous day, a week and a month duration. When the feeding rate is decreased or increased, the possibility of the disease is analyzed and indicated to the farmers. Many studies has undergone for the early detection of the disease using feeding behavior. Gonzalez et al. [5] proved the early detection of the few disease at least one day prior to the normal method. Although the disease prediction are not 100% false proof, it helps to avoid a major crisis on the other hand, by reducing the production loss. Apart from the early detection of the disease, it also helps in the early calculation of the food supply.

3.1 Workflow of the Feed Monitoring System

In the livestock feed monitoring system, the details and the statistics of the feed for each livestock can be viewed through the mobile application. Apart from this, whenever a particular livestock feed rate decreases or increases with the difference of 2.5 SD (Standard Deviation), an alert message containing the details of the livestock and possible disease information are displayed through the mobile application.

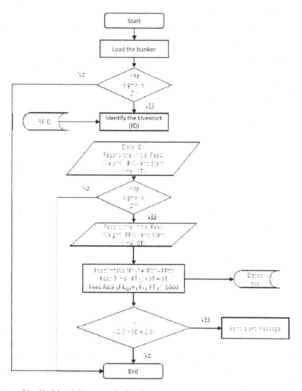

Fig. 2 Workflow of individual livestock feed status

As shown in the figure 2, after each feed of the particular livestock, the feed time, feed rate and the feed intake are logged and compared with the earlier day's log, a week log and a month log. So when the difference appears, an automated message is sent to the farmer, even if doesn't inquire about the livestock.

4   Mobile Application

The application development environment for developing Android OS-based smart application, it operates in the JDK 1.6 version, and Android project studio is used as the basic tool for Android development. The smart application development, as described above, was applied to an actual livestock farm, and it was able to collect the livestock feeding time, feed intake and number of visits through the sensor and RFID readers. The mobile application helps us to provide the details of the livestock where the report is specified. The application gives a wide knowledge of the feeding behavior through a graph view which is compared with the past one week and the past one month system where the status and profile of livestock can be accessed, it also helps to access the alert page directly and the food stock services. The figure 3(b) shows the feeding rate week status of the livestock. Whenever the standard deviation exceeds or decreases than 2.5, then an alert message is sent to the farmer reporting the status and behavior of the livestock, which can also be viewed through the application. As the amount of the feed is calculated precisely, it is easy to find the food stock capacity, which also helps the farmer to make an order on the food stock on right time.

Fig. 3 (a). Home screen of the Livestock Feeding System (b). Weekly status of the livestock feeding rate

When there is a vast number of herds in the barn, identifying the disease on particular cow without any symptom is rather a challenge. Some disease shows symptoms only after a few days of infection. In such rare cases, calculating the feed rate with the record of feed intake by the cow is helpful. When the feed rate of the cow decreases or increases than the normal scope, an alarm message is posted directly to the farmer, after an immediate check-up, the former will be able to distinguish the disease in the cow, at least a day prior than the normal method. With an early detection through the alarm message in the mobile device, proper measure is considered to control or cure the disease precisely.

5   Conclusion

In this paper, we proposed a livestock feed monitoring system based on an Android OS-using mobile application that is useful to check the update on the feed and also can early diagnose the disease with the feeding behavior of the livestock. The proposed system is structured in a mobile application generating abnormal livestock feed behavior by comparing the single day feeding rate with the week and the month of the feeding rate. The application helps to send out an alert to the farmer when the SD of the feeding rate is more or less than 2.5. To test the proposed system, we, in this paper, implemented a server application and android OS-based mobile application. With the proposed livestock disease counseling system in place, we believe livestock farming households can counsel, experts on a livestock abnormal symptom to avoid mishaps and respond to livestock diseases swiftly to minimize its caused damages while safeguarding livestock products.

Acknowledgments This work was supported by the National Research Foundation of Korea (NRF) grant funded by the Korea government. (MEST) (No. 2014R1A1A2059853). This research was supported by the MSIP (Ministry of Science, ICT and Future Planning), Korea, under the C-ITRC(Convergence Information Technology Research Center) (IITP-2015-H8601-15-1007) supervised by the IITP(Institute for Information & communications Technology Promotion).

References

1. Williams, P.E.V.: Animal production and European pollution problems. Animal Feed Science and Technology **53**(2), 135–144 (1995)
2. Ferrari, G., Tasciotti, L., Khan, E., Kiani, A.: Foot-and-Mouth Disease and Its Effect on Milk Yield: An Economic Analysis on Livestock Holders in Pakistan. Transboundary and Eemerging Diseases **61**(6), e52–e59 (2014)
3. Ingvartsen, K.L., Moyes, K.: Nutrition, immune function and health of dairy cattle. Animal **7**(s1), 112–122 (2013)
4. Atzori, L., Iera, A., Morabito, G.: From "smart objects" to "social objects": The next evolutionary step of the internet of things. IEEE Communications Magazine **52**(1), 97–105 (2014)
5. González, L.A., Tolkamp, B.J., Coffey, M.P., Ferret, A., Kyriazakis, I.: Changes in feeding behavior as possible indicators for the automatic monitoring of health disorders in dairy cows. Journal of Dairy Science **91**(3), 1017–1028 (2008)
6. Kaasschieter, G.A., De Jong, R., Schiere, J.B., Zwart, D.: Towards a sustainable livestock production in developing countries and the importance of animal health strategy therein. Veterinary Quarterly **14**(2), 66–75 (1992)
7. Bareille, N., Beaudeau, F., Billon, S., Robert, A., Faverdin, P.: Effects of health disorders on feed intake and milk production in dairy cows. Livestock Production Science **83**(1), 53–62 (2003)
8. Pastell, M., Hautala, M., Poikalainen, V., Praks, J., Veermäe, I., Kujala, M., Ahokas, J.: Automatic observation of cow leg health using load sensors. Computers and Electronics in Agriculture **62**(1), 48–53 (2008)
9. Frost, A.R., Schofield, C.P., Beaulah, S.A., Mottram, T.T., Lines, J.A., Wathes, C.M.: A review of livestock monitoring and the need for integrated systems. Computers and Electronics in Agriculture **17**(2), 139–159 (1997)
10. Firk, R., Stamer, E., Junge, W., Krieter, J.: Automation of oestrus detection in dairy cows: a review. Livestock Production Science **75**(3), 219–232 (2002)

Location-Aware WBAN Data Monitoring System Based on NoSQL

Yan Li and Byeong-Seok Shin

Abstract In order to gather location-aware WBAN data and support efficient and analysis services, we propose a location-aware WBAN data monitoring system based on the NoSQL database system with a spatial index. It can manage JavaScript Object Notation (JSON) and Binary JSON (BSON) document data from mobile gateway devices and, by using the proposed spatial index method, the proposed system can efficiently process location-based requests for medical signal monitoring. We developed our system on MongoDB, which is a document-based NoSQL database system, and evaluated its performance.

Keywords WBAN monitoring system · Location-aware · NoSQL database system

1 Introduction

With the remarkable development and miniaturization of electronic devices and the development of telecommunications technology, human biometric data can be collected anywhere, anytime, by using several kinds of wearable sensors or embedded sensors [1]. Personal vital signs are sent to the monitoring center to check the personal condition. The wireless body area network (WBAN) is constructed with these kinds of devices and signal receivers [2,3]. In a WBAN system, the sensors send the collected data to the gateway node or coordinator node, which filters, samples and aggregates the data to resize it [4]. After that, the cleaned data is sent to monitoring centers and medical experts. In this paper, we call the device that receives all of the patient's vital signs, sending the data to the medical center, a gateway device.

When WBAN sensors collect vital sign data, the location data is also collected, although some sensors do not do so [3]. Because body status will change in

Y. Li · B.-S. Shin(✉)

Department of Computer Information and Engineering, Inha University, Incheon, Korea

e-mail: leeyeon622@gmail.com, bsshin@inha.ac.kr

© Springer Science+Business Media Singapore 2015

D.-S. Park et al. (eds.), *Advances in Computer Science and Ubiquitous Computing*,
Lecture Notes in Electrical Engineering 373,
DOI: 10.1007/978-981-10-0281-6_13

85

different environments, the location factor is very important for health monitoring, and it is used in location-based analyses [5], such as disease statistics based on location, or health status based on location [6]. Health care data and medical data from a WBAN is very large when the medical center covers a specific area with a dense population [7]. And the volume of WBAN data will differ from time to time and day to day. In order to construct this type of large-scale system in a scalable architecture, we need to distribute computing resources across a network [8,9].

In recent years, many researchers have focused on Not Only Structured Query Language (NoSQL) database systems, because NoSQL is a non-relational, schema-free, distributed, easily replicated and horizontally scalable database system that can adaptively support large-scale systems [10,11]. Most of the new NoSQL databases have emerged out of several independent efforts to provide a scalable, flexible database alternative that can effectively address the needs of a high-volume internet application. And many of them are built on extremely solid networking and distribution technologies, although they have diverged significantly from traditional database techniques [12]. Some popular open-source NoSQL databases, such as Cassandra, Redis, MongoDB, VlotDB, Cloudata, Hadoop/Hbase, SimpleDB, CounchDB, and OrientDB, have been widely used in web applications. Although these kinds of database are designed to provide good horizontal scalability for simple read/write database operations distributed over many servers, they entail key-value or document-based storage that does not have spatial support, including neither spatial data type nor spatial queries that most location-based service applications need.

We propose a location-aware WBAN data monitoring system based on a document-based NoSQL database system using MongoDB. First, we analyzed the integrated data model from a WBAN and designed the data model for our system. And in order to support location-aware WBAN data analysis, we propose a spatial data index for NoSQL systems.

The rest of the paper is organized as follows. In Section 2, we present the proposed geospatial WBAN data server on top of MongoDB with a spatial index. In Section 3, we compare and contrast the system with the MySQL for response time and system resource usage. We present the conclusions and future works in Section 4.

2   Location-Aware WBAN Monitoring System

According to the research, location awareness refers to users being offered services based on location information provided by suitable devices or software [2]. So, we think a location-aware WBAN monitoring system should not only monitor and analyze the location based health care data or vital signs, but should also support location-dependant medical services, such as providing local disease rates, the addresses of local healthcare facilities, the successful medical treatment of cases and local health risks, and hazards based on local WBAN data statistics and analysis. Another example is providing users with a personal disease prediction for, say, a specific mountain area because of a specific plant that might be dangerous. In order to provide this kind of service, a location-aware WBAN monitoring system has to quickly store WBAN data, including spatial information, and detect personally dangerous situations in time. In order to provide these kinds of services, we designed a location-aware WBAN monitoring system architecture as follows. Figure 1 shows our proposed architecture.

In Figure 1, the location-aware medical signal monitoring system receives WBAN data from numerous users with the measured location data. The data will be inserted based on MongoDB distribution mechanisms. Before the data are inserted, a location-aware patient manager will determine which MongoDB node could store the data. A spatial query manager finds one inserted query then will collect the person's average body signals and store it with the matched node, and at the same time, the spatial index will be constructed. A location-aware filter is connected with an expert system, gathering real-time local medical and disease information, checking if the person is in trouble or not, and will then send an alarm to that person if the system discovers a possible problem. The spatial index manager is the most important contribution in this paper because, from version 2.0, MongoDB supports a simple spatial index, but it is only limited to point queries and simple nearby queries. So, in this paper, we propose a new spatial index for a NoSQL system. The medical experts system could be connected with other systems to gather real-time medical news to provide predictive information. A map database has to store and update the monitored map information.

Every person has his or her own living area, and every day will live in a specific area only. In a spatial index for a location-aware WBAN, we first focus on how to quickly insert a large volume of data from a WBAN network; second, we determine how to efficiently support personal location-aware services that include aggregation queries or other kinds of query that have to use historical data. So, we propose a new spatial index for location-aware WBAN systems that uses R-tree, which allows nested minimum boundary rectangles (MBRs) and geohash-based B-tree. Scalable

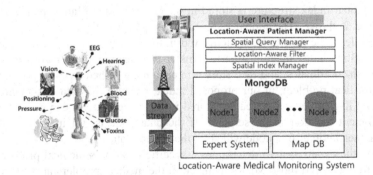

Fig. 1 Location-aware medical monitoring signal system architecture

The NoSQL database data model is a key-value model and cannot support multi-key searches. For example, when we find one location, then the location coordination has to be converted to one-dimensional data first. So, MongoDB uses a geo-hash method to transfer the two-dimensional location data to one-dimensional data, then insert it by using a B-tree index. Also, non-spatial data are also stored based on the B-Tree index.

In our proposed indexing method, in order to support patient location-aware services, we have to separate each patient's living area and store it in different nodes. Figure 2 shows the structure of our spatial index for location-aware WBAN. For tracking one day or one week of a patient's location information, it is easy to detect estimated living area for patients. So every patient's location information for time duration could construct one MBR, which can display the patient's living area. And the patients' data are stored in the same MongoDB node, so that it can support efficient location-based analysis.

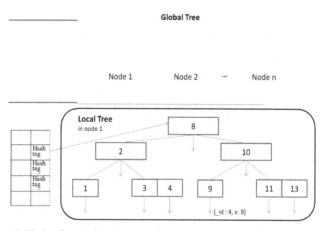

Fig. 2 The spatial index for the location-aware WBAN system.

The global index structure is stored in the Spatial Index Management Module, as Figure 2 shows. The global index decides which patient data are stored in which node in the distributed system. And in the distributed system, it just uses a geohash-Btree index method, first to transfer the location coordinator transferred to geohash coordinates, and then store it in each node. When analyzing a location-based medical problem, in our proposed index, because the data of patients who work and live in the same area are stored in the same node, it can provide quick and efficient location-aware services. If the data node is full in the distributed system, then another free node will store the patient data.

When the expert system determines a specific location's medical problem, the global tree will find those MBRs included in the medical problem area, then send an alert directly to the patients affected. So there is no need to scan distributed nodes, and the system can support quick responses. It fits the needs of emergency medical systems very well.

3   Experimental Results

The proposed system is constructed on MongoDB version 2.2, and the spatial manager is implemented using a multithreaded programming method in Java, combined with MongoDB for the spatial index data. To make the plug-in highly concurrent, this requires a way of avoiding lock contentions when multiple threads or users access the internal index tree at the same time. Thus, we deploy an individual location-aware WBAN spatial index for each insertion thread in order to remove the hotspot resulting from competing locks for one index tree. And each thread shares one process in a modern operating system that is mapped to use one CPU core on the system. Thus, we can exploit the system to be fully loaded by increasing the number of threads. Our geospatial index is designed as a plug-in for MongoDB, so we can use MongoDB's connection model and return a search result format in JSON.

In this evaluation, we sample the medical WBAN data and evaluate our location-aware WBAN service based on the MongoDB NoSQL database using global positioning system (GPS) traces (130 million location data in total) crawled from OpenStreetMap, an open GPS trace project. To evaluate the scalability of our design, we compared the response time with the most popular database (MySQL) only for two basic operations: a check-in and a nearby search. MySQL implements spatial extensions following the specifications from the Open Geospatial Consortium (OGC). This evaluation compares the performance of check-in/nearby, but not a functional comparison.

We measured the performance of the aggregation search of each system with different numberof clients to see the scalability. Figure 3 shows the performance of agreegation query, the query is an SQL expression similar to SELECT AVG(Blood) FROM checkins WHERE ST_DWithin (the_geom, GeometryFromText('POINT (−128.000 45.000)', 4326), 0.01). This means 'get the average blood data within 1 km of alocation coordinate', in this case (−128.000, 45.000). This point location is randomlyselected from the check-in poi data.

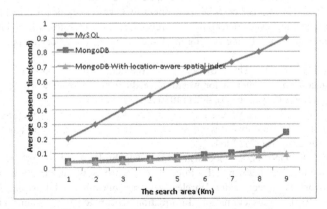

Fig. 3 Average query evaluation performance comparation.

Figure 3 shows that, when the average query evaluated and the search area is larger from 1km area to 9km area, because of the spatial index entented for WBAN data mornitoring, the proposed method is little ourperformed than original MongoDB. And the MySQL is much slower than MongoDB.

4   Conclusions

We propose design and construction method of a NoSQL–based location-aware WBAN data monitoring system. In order to support the location-aware services we propose a new spatial index algorithm for a document-based NoSQL database. We use the algorithm to focus on highly insertion and location-based average searches, which are a common characteristic of mobile applications. We propose and implement a novel location data service based on the MongoDB database using our indexing algorithm. In the future, we will add much more spatial data types, like polygon and multi-polygon, so that our method can propose more functions. Also, we will test more location-aware services for WBAN healthcare monitoring systems to improve our methods.

Acknowledgments This research was supported by a grant of the Korea Health Technology R&D Project through the Korea Health Industry Development Institute (KHIDI), funded by the Ministry of Health & Welfare, Republic of Korea (grant number HI14C0765).

References

1. Pantelopoulos, A.: A Survey on Wearable Sensor-Based Systems for Health Monitoring and Prognosis. IEEE Transactions on Systems, Man, and Cybernetics, Part C: Applications and Reviews **40**(1), 1–12 (2010)
2. Chon, Y., Cha, H.: Lifemap: A smartphone-based context provider for location-based services. IEEE Pervasive Computing **10**(2), 58–67 (2011)
3. Welch, J., Guilak, F., Baker, S.D.: A wireless ECG smart sensor for broad application in life threatening event detection. In: Proceedings of the 26th Annual International Conference of the IEEE Engineering in Medicine and Biology Society, pp. 3447–3449 (2004)
4. Jiang, Z.H., Brocker, D.E., Sieber, P.E., Werner, D.H.: A compact, low-profile metasurface-enabled antenna for wearable medical bodyarea network devices. IEEE Trans. Antennas Propag. **62**(8), 4021–4030 (2014)
5. Xu, J., Zheng, B., Lee, W.-C., Lee, D.L.: The d-tree: An index structure for planar point queries in locationbased wireless services. IEEE Transactions on Knowledge and Data Engineering **16**(12) (2004)
6. Fletcher, R.R., Dobson, K., Goodwin, M.S., Eydgahi, H., Wilder-Smith, O., Fernholz, D., Kuboyama, Y., Hedman, E.B., Poh, M.-Z., Picard, R.W.: iCalm: Wearable Sensor and Network Architecture for Wirelessly Communicating and Logging Autonomic Activity. IEEE Transactions on Information Technology in Biomedicine **14**(2), 215–223 (2010)

7. Ullah, S., Vasilakos, A., Chao, H.-C., Suzuki, J.: Cloud-assisted Wireless Body Area Networks. Information Sciences **284** (2014)
8. Koo, H.-S.: Cloud-assisted Wireless Body Area Networks. Journal of International Journal of Fuzzy Logic and Intelligent Systems **16**(6), 704–709 (2006)
9. Mzila, P., Adigun, M.O.: A Service Supplier Database for Location-Based Mobile Commerce. Distributed Computing Systems Workshops (2007)
10. Hoang, D.B., Chen, L.: Mobile Cloud for Assistive Healthcare (MoCAsH). IEEE Network **27**(5), 56–61 (2013)
11. Wan, J., Zou, C., Ullah, S., Lai, C.-F., Zhou, M., Wang, X.: Cloud-enabled wireless body area networks for pervasive healthcare. In: 2010 IEEE International Conference on Proceedings of Software Engineering and Service Sciences (ICSESS), pp. 431–434 (2010)
12. Somasundaram, M., Gitanjali, S., Govardhani, T.C., Lakshmi Priya, G., Sivakumar, R.: Medical image data management system in mobile cloud computing environment. In: Proceeding of 2011 International Conference on Signal, Image Processing and Applications, IPCSIT, vol. 21, pp. 11–15 (2011)

Optimization of a Hybrid Renewable Energy System with HOMER

Olly Roy Chowdhury, Hong-geun Kim, Yongyun Cho, Changsun Shin and Jangwoo Park

Abstract Renewable energy harvesting is the best option as future energy source and there is no doubt about that. Greenhouse those are far from locality and offshore aqua farms, can make most efficient use of green energy harvesting. One negative approach about green energy that, it is unpredictable. So optimum planning of the resources to be included in a hybrid system and a determination of a perfect working strategy is very important. HOMER is one of the best used software for this optimum planning purpose due to its wide range of analyzing capacity and advanced prediction capability based on the sensitivity of the variables (sensitivity analysis). Moreover HOMER provides options to choose a perfect hybrid system from a large set of results arranged in the order of minimum to maximum cost of energy. This paper represents HOMER based optimum designing of a hybrid system including solar and wind resource for offshore area near about Suncheon, a city of South Korea.

Keywords HOMER · Optimization · System design · Renewable system modeling

1 Introduction

Coastal areas and offshore aqua farms can make the best use of renewable natural resource to meet their power requirements by building hybrid renewable energy systems. HOMER is a computer based system optimization software that simplifies the task of designing and optimization of a hybrid renewable energy system and includes natural resources like solar, wind, hydro as input as well as

O.R. Chowdhury · H.-g. Kim · Y. Cho · C. Shin · J. Park(✉)
Sunchon National University, 413, Suncheon-Si,
Jungangno Suncheon 540-950, Republic of Korea
e-mail: {ollyroy,khg_david,yycho,csshin,jwpark}@sunchon.ac.kr

© Springer Science+Business Media Singapore 2015   93
D.-S. Park et al. (eds.), *Advances in Computer Science and Ubiquitous Computing*,
Lecture Notes in Electrical Engineering 373,
DOI: 10.1007/978-981-10-0281-6_14

fuel based power production option. Homer allows sensitivity analysis based on the availability of resources, quantity of input and also on the cost options. Sensitivity analysis helps assuming the uncertain changes in parameters that cannot be controlled by user. This includes some advanced options like making cost assumptions in future depending on predicted price of some resources such as diesel that varies with time.

In HOMER, the feasibility of different resource combinations can be checked easily based on the Net Present Cost (NPC) and Cost of Energy (CoE) analysis. Homer can model micro power system in both grid connected and off grid condition and can serve electric as well as thermal load factors.

2 Literature Review

Power crisis is the most talked about topics in the world at present. Many developed countries are securing their energy source in future by taking initiatives to create hybrid renewable energy field. On the other hand, developing countries like Bangladesh are also very concerned about this energy issues and are taking steps for harvesting green energy from natural resources. In [1] authors have shown clear comparison between green energy and using IPS. They have mentioned the economic and environmental side effects of using IPS. They have analyzed the impact of using long environmentally friendly and long lasting PV panel systems instead of IPS showed that PV or green energy can be a better replacement of instant power supply. Hybrid wind diesel is a better solution than any one of these to be used alone comparing from economic view [2]. Renewable energy systems are costly for the first time installation but it is long lasting and maintenance is very cheap. Hybrid energy system together with renewable energy can reduce load factor and replacement costs [3]. On the other hand studies of [4, 5, 6] show that hybrid system say wind diesel for example reduce fuel consumption, fuel storage and greenhouse gas emission. For hybrid system design optimization is essential. Optimization of hybrid systems has been studied by [7, 8, 9]. In [10] authors have used HOMER for modeling a hybrid solar-wind-diesel system which can be a great example for modeling standalone power system for coastal areas or offshore industries.

3 Three Principal Tasks of HOMER

3.1 Simulation

The simulation process serves two purposes. First, it checks the system's feasibility. A system is considered to be feasible if it can adequately serve the electric and thermal loads and satisfy any other constraints imposed by the user. Second, it estimates the life-cycle cost of the system, which is the total cost of installing and operating the system over its lifetime and presents as total net present cost (NPC).

3.2 *Optimization*

Optimization determines the optimal value of the variables such as the combination of components that will make up the system and determines the size or quantity of each. In HOMER, the best possible, or optimal, system configuration is the one that satisfies the user-specified constraints at the lowest total net present cost [11].

3.3 *Sensitivity*

Sensitivity analysis helps assess the effects of uncertainty or changes in the variables over which the designer has no control, such as the average wind speed or the future fuel price [11].

4   Modeling System Components

4.1 *Electrical Load Profile*

The load profile is assumed considering the power consumption profile of a fish farm situated in coastal area of Suncheon. HOMER will model the hybrid grid system suitable to meet the power demand of this offshore fish farm. From the load profile shown in fig. (1), it is clear that on two specific time of each day the demand reaches its peak. These two peak hours used to be the feeding time of the fishes.

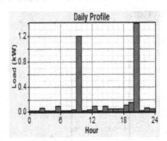

Fig. 1 Daily load profile of the fish farm

4.2 *PV Panel*

PV cells are used to convert solar radiation to electricity. The price of the PV cell is taken keeping in mind the market value and it is fixed as 0.8$ per W power.

Solar panels are placed with derating factor 80%, ground reflectance 20% and without any tracking system. The panels are placed with a slope angle of 40 degree for a lifetime of 20 years.

The annual scaled average for solar irradiation was taken as 4.12 kwh/m^2/d for Suncheon, South Korea with calculated clearness index of 0.489. To do sensitivity analysis, annual scaled average was taken from 3.3 (minimum) to 5.2 (maximum) kWh/m^2/d.

4.3 Turbine

Through a turbine, kinetic energy of wind is converted to electricity. Power conversion depends on wind speed, angle of attack and size of wind turbine i.e., the radius of the rotor, pitch, hub height etc. The wind data for Suncheon was taken from the wind speed record of 2010. Average wind speed was 1.223 m/s. The data was entered as monthly average. The Weibull K factor is 1.06 and the 13th hour of a day is the hour of peak wind speed. Two sensitivity values for wind speed due to sudden change in weather, considered as 1.5 and 5.6 m/s^2. Generic 1kW was chosen from HOMER library for modelling and optimization. The no. of wind turbine is a matter of consideration for best grid configuration with per unit cost of 500 $. The hub height was 25 m and the life time of each turbine is 15 years.

5   Hybrid System Development

For completion of system, battery Surrette was chosen from HOMER library at per unit cost of 700$. And also converter considered for this case is of capacity of 1 kW. The fig. (2) shows HOMER developed hybrid off grid micro renewable power production model consists of solar, wind as renewable source and battery as extra power saver with converter.

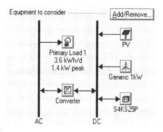

Fig. 2 Hybrid system configuration

	PV (kW)	G1	S4KS25P	Conv. (kW)	Initial Capital	Operating Cost ($/yr)	Total NPC	COE ($/kWh)	Ren. Frac.	Batt. Lf (yr)
	1	1	12	5	$ 12,700	437	$ 18,284	1.089	1.00	12.0
	2		12	5	$ 13,000	432	$ 18,523	1.103	1.00	12.0
	1	2	12	5	$ 13,200	450	$ 18,953	1.128	1.00	12.0
	2	1	12	5	$ 13,500	445	$ 19,193	1.143	1.00	12.0
	3		12	5	$ 13,800	441	$ 19,433	1.157	1.00	12.0
	2	2	12	5	$ 14,000	459	$ 19,863	1.183	1.00	12.0
	3	1	12	5	$ 14,300	454	$ 20,103	1.197	1.00	12.0

Fig. 3 Optimization result

Optimization result in fig. (3) shows that the first combination takes into account both solar and wind but the wind speed responsible here is max among the values considered. It approaches the minimum NPC and lowest COE of all the system. So it is acceptable. But what if the wind velocity is average not maximum.

Sensitivity Results | Optimization Results |

Double click on a system below for optimization results

Solar (kWh/m²/d)	Wind (m/s)		PV (kW)	G1	S4KS25P	Conv. (kW)	Initial Capital	Operating Cost ($/yr)	Total NPC	COE ($/kWh)	Ren. Frac.	Batt. Lf (yr)
4.120	1.220		2			12	5	$ 13,000	432	$ 18,523	1.103	1.00	12.0
4.120	1.500		2			12	5	$ 13,000	432	$ 18,523	1.103	1.00	12.0
4.120	5.600		1	1		12	5	$ 12,700	437	$ 18,284	1.089	1.00	12.0
4.500	1.220		2			12	5	$ 13,000	432	$ 18,523	1.103	1.00	12.0
4.500	1.500		2			12	5	$ 13,000	432	$ 18,523	1.103	1.00	12.0
4.500	5.600		1	1		12	5	$ 12,700	437	$ 18,284	1.089	1.00	12.0
5.200	1.220		1	1		12	5	$ 12,700	437	$ 18,284	1.089	1.00	12.0
5.200	1.500		1	1		12	5	$ 12,700	437	$ 18,284	1.089	1.00	12.0
5.200	5.600		1	1		12	5	$ 12,700	437	$ 18,284	1.089	1.00	12.0
3.300	1.220		2			12	5	$ 13,000	432	$ 18,523	1.103	1.00	12.0
3.300	1.500		2			12	5	$ 13,000	432	$ 18,523	1.103	1.00	12.0
3.300	5.600		1	1		12	5	$ 12,700	437	$ 18,284	1.089	1.00	12.0

Fig. 4 Sensitivity analysis

The fig. (4) shows the optimized configuration together with sensitivity analysis taking solar illumination and wind speed as variables. The first two cases here doesn't involve wind turbine and with two solar panels they approach a system with the minimum cost.

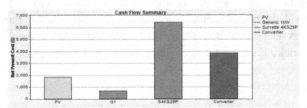

Fig. 5 Cash flow summary

Moreover, from the cash flow summary in fig. (5) it can be seen that the estimated no. of battery is making maximum contribution in huge net present cost and so on cost of energy. So a sensitivity analysis was done considering lesser no. of batteries and to get a better combination of PV panel and wind turbine. The result is shown in fig. (6).

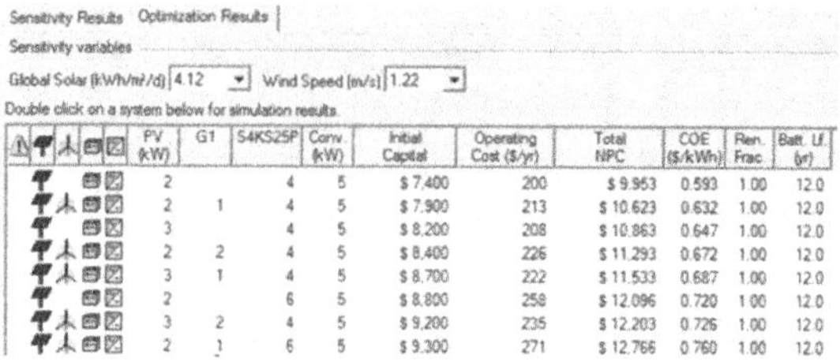

Fig. 6 Sensitivity results with new no. of batteries

Considering less no. of batteries fig. (6) shows that there is a significant change in NPC than before. And also there is a change in system configuration suggestion. The first case here doesn't involve wind turbine and with two solar panels it approach to a system with the minimum cost. But the second result includes wind turbine and in this case both the costs i.e., the NPC and COE has a very little difference with the first case and the difference is negligible in the sense that it is a complete hybrid system with all the renewable sources considered that can show optimum result in both regular average wind speed and solar condition as well as in highly windy and less sunny condition.

6 Conclusion

Renewable energy is the only hope for us as future energy resource. Nature is rich with variety of sources and they are abandoned. But the resource capability varies with time along a day, and with season along a year. The shortcomings of the sources can be minimized and resolved by combining two or more sources together that is by building hybrid energy system. HOMER can do better optimization serving this purpose. This paper represents an idea of building solar and wind hybrid energy harvesting system considering weather particularly for Suncheon. This can be a better idea for near Suncheon coastal or offshore industries for making off grid power source instead of using expensive and short lived batteries.

Acknowledgments This research was supported by IPET (Korea Institute of Planning and Evaluation for Technology in Food, Agriculture, Forestry and Fisheries) through Agri-Bio-industry Technology Development Program, funded by MAFRA (Ministry of Agriculture, Food and Rural Affairs) (No. 315001-5). And This paper was supported by (in part) Sunchon National University Research Fund in 2015.

References

1. Alam, S.M.S.U., Rahman, M.H.: Use of Green Energy Instead of IPS to Lessen Energy Crisis in Bangladesh. In: 2nd International Conference on Green Energy and Technology, pp. 83–86, September 2014
2. Kaldellis, J.K., Kondili, E., Filios, A.: Sizing a Hybrid Wind-Diesel Stand-Alone System on the Basis of Minimum Long-Term Electricity Production Cost. Applied Energy **83**(12), 1384–1403 (2006)
3. Paska, J., Biczel, P., Kłos, M.: Hybrid power systems– An effective way of utilizing primary energy sources. Renewable Energy **34**(11), 2414–2421 (2009)
4. Karki, R., Billinton, R.: Cost-Effective Wind Energy Utilization for Reliable Power Supply. IEEE Transactions on Energy Conversion **19**(2), 435–440 (2004)
5. Liu, X., Islam, S.: Reliability Evaluation of a Wind-Diesel Hybrid Power System with Battery Bank Using Discrete Wind Speed Frame, Analysis. In: Proc. of Int. Conf. on Probabilistic Methods Applied to Power Systems, pp. 1–7. Institute of Electrical and Electronics Engineers, Stockholm (2006)
6. Billinton, R., Bai, G.: Generation Capacity Adequacy Associated with Wind Energy. IEEE Transactions on Energy Conversion **19**(3), 641–646 (2004)
7. Yang, H., Lu, L., Zhou, W.: A novel optimization sizing model for hybrid solar-wind power generation system. Solar Energy **81**(1), 76–84 (2007)
8. Boroy, B.S., Salameh, Z.M.: Optimum photovoltaic array size for a hybrid wind/PV system. IEEE Transaction on Energy Convertion **9**(3), 482–488 (1994)
9. Lal, D.K., Dash, B.B., Akella, A.K.: Optimization of PV/Wind/Micro-Hydro/Diesel Hybrid Power System in HOMER for the Study Area. International Journal on Electrical Engineering and Informatics **3**(3), 307–325 (2011)
10. Zubair, A., Tanvir, A.A., Hasan, M.M.: Off-grid hybrid energy system incorporating renewable energy for a remote coastal area of bangladesh. In: Proceedings of the Global Engineering, Science and Technology Conference, Dhaka, Bangladesh, December 28-29, 2012 (2012)
11. Farret, F.A., Simões, M.G.,: Micropower system modeling with homer. In: Lambert, T., Gilman, P., Lilienthal, P. (eds.) Integration of Alternative Sources of Energy. Wiley-IEEE Press, December 2005. DOI:10.1002/0471755621

Efficient Character Input Scheme Based on Gyro-Accelerometer Sensor for NUI

Hyun-Woo Kim, Boo-Kwang Park, HwiRim Byun, Yoon-A. Heo
and Young-Sik Jeong

Abstract In recent years, as information technology (IT) has advanced rapidly, multimedia smart devices have been designed to provide a variety of services with which users can interact using touch screens. A smart phone is among the typical smart devices and it has provided simple mobility and a convenient interface via touch screens, as well as enabling desktop personal computer (PC) operations thanks to their high performance and miniaturization. In pace with this rapid advancement, a variety of input schemes has also been developed to allow users to enter text conveniently and rapidly. However, despite the development of such various input schemes, learning delay time and typos occur frequently. Furthermore, as touch screen-based multimedia smart devices employ finger-based text inputs, disabled persons or individuals who cannot use their fingers freely may have difficulties. In this paper, a virtual keyboard using a Gyro-Accelerometer sensor (VGA), which is an efficient text input keypad using an accelerometer sensor and gyro sensor, is proposed. The VGA can input text using a gyro sensor and accelerometer sensor embedded in multimedia smart devices. Through the VGA, users whose fingers are not freely available can enter texts easily and comfortably.

Keywords Natural user interface · Multimedia smart device · Gyro sensor · Accelerometer sensor · Character input method · Touch screen

1 Introduction

Thanks to the recent IT technology advancement, multimedia smart devices have been developed. One of the typical multimedia smart devices is a multimedia

H.-W. Kim · B.-K. Park · H. Byun · Y.-A. Heo · Y.-S. Jeong(✉)
Department of Multimedia Engineering, Dongguk University, Seoul, Korea
e-mail: {hwkim,pbg0517,hazzzly,hyagood,ysjeong}@dongguk.edu

© Springer Science+Business Media Singapore 2015
D.-S. Park et al. (eds.), *Advances in Computer Science and Ubiquitous Computing*,
Lecture Notes in Electrical Engineering 373,
DOI: 10.1007/978-981-10-0281-6_15

smart phone, which can provide not only basic functions, such as an alarm, memo pad, contacts list, and camera, but it can also provide other various functions and services with which users can interact. Due to such usefulness, a user can enjoy benefits in terms of improvements in time and space usability and convenience. A multimedia smart phone has provided simple mobility and a convenient interface via touch screens, as well as enabled desktop personal computer (PC) operations thanks to their high performance and miniaturization. To enter texts easily and conveniently by multimedia smart phone users, a variety of virtual keyboards has been developed (e.g., Chun-Ji-In [2, 6], Ez Hangul [2, 6], Neostyle [2, 6], Milgeegle [2, 6], MoAKey [2], and Dingul [2, 6]). However, their input methods are complicated and require a long learning time by users despite fast input. In addition, frequent automatic inputs occur due to incorrect inputs [1, 2, 3, 4, 5, 6, 7, 8, 9, and 10]. As their text inputs are based on finger movements, individuals with disabilities or those whose fingers are not freely available may have trouble entering text using touch and finger movements. In this paper, a text input keypad with a virtual keyboard using Gyro-Accelerometer sensor (VGA), through which a user can enter text easily, is proposed. The VGA is an efficient text input virtual keyboard using an accelerometer sensor and a gyro sensor, through which users can input texts without pushing a virtual keyboard in a smart phone. Through the VGA, users whose fingers are not freely available, including exceptional circumstances can enter texts easily and comfortably.

2   Related Works

In this section, previously developed text input keyboards, such as Chun-Ji-In [2, 6], Ez Hangul [2, 6], Neostyle [2, 6], Milgeegle [2, 6], MoAKey [2], and Dingul [2, 6], are discussed. Chun-Ji-In is one of the generally used text input methods, which can add characters in a similar manner as real writing. However, it requires a long input time to complete a sentence. Ez Hangul is also one of the text input methods generally used in feature phones, and it is similar to Chun-Ji-In. Ez Hangul takes a long time to learn how to input due to difficulties in the consonant and vowel placement. In addition, Neostyle consists of short consonant and short vowel formats in the keyboard placement. It can support text inputs through drag recognition, but it has a high typo rate due to incorrect drags. Meanwhile, Milgeegle can combine vowels by dragging a touch pad in the upper, lower, right, and left directions while touching the pad. MoAKey consists of the keyboard placement of consonants only. When pushing one of the consonants, possible vowels are activated around the touch pad. Finally, Dingul can complete a single text by pushing one button while touching a consonant and dragging it in the upper, lower, right, and left directions. However, it has a high typo rate due to incorrect drags, as well as a long learning time in early use. Chun-Ji-In, Ez Hangul, Neostyle, Milgeegle, MoAKey, and Dingul are all touch-based input schemes and some of them require drags according to their features. Such an input scheme is prone to inconvenience for those who cannot use touch or buttons due to finger disabilities.

3   VGA Input Scheme

The VGA is a consonant input method using a gyro sensor and an accelerometer embedded in a multimedia smart device, which is shown in Fig. 1. Figure 1 also shows an input of "ㄱ" in Korean while facing a smart phone screen upward to change to the consonant input screen and moving it to the right and lower directions.

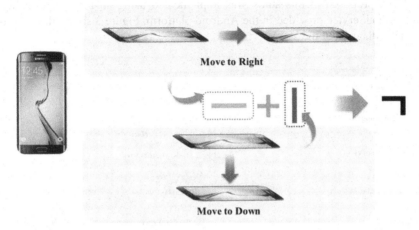

Fig. 1 How to input a consonant in the VGA

Figure 2 shows the vowel input method in the VGA. It shows an input of "ㅓ" in Korean while facing the smart phone screen downward to change to the vowel input screen and moving it to the right and lower directions.

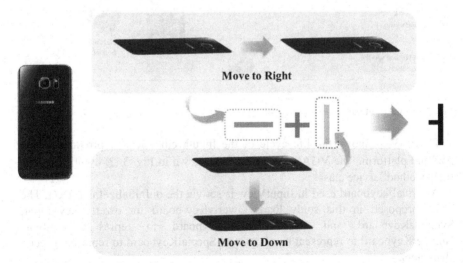

Fig. 2 How to input a vowel in the VGA

As shown in the above figures, the Korean character "거" can be entered. Using the above methods, users can input texts without touching or dragging their smart phones.

4   Design of VGA

The VGA proposed in this study was implemented using an input class called InputMethodService provided in the Android platform. Figure 3 shows the overall schematic diagram of the VGA.

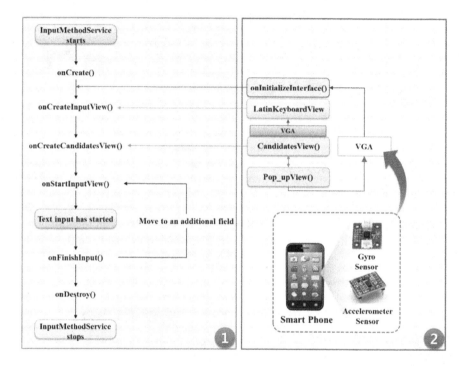

Fig. 3 VGA Architecture

Figure 3 ① shows the life cycle of the InputMethodService provided in the Android platform. The VGA was designed as shown in Fig. 3 ② by utilizing the InputMethodService class.

A virtual keyboard used in InputView is set via the onInitializeInterface(). The VGA proposed in this study loads QwertyKey-board for qwerty key input, SymbolsKeyboard and SymbolsShiftedKeyboard to represent symbols, NumberKeyboard to represent numbers, and SpecialKeyboard to represent special characters.

Once the keyboard is activated for input by onCreateInputView(), a basic keyboard is set to the VGA. Here, a gyro sensor embedded in the smart phone is activated.

Furthermore, onStartInputView() is called from EditBox or inputtable boxes in Android, through which characters are entered. Then, the gyro sensor, activated for text input, senses a smart phone's tilting direction, thereby displaying characters through analysis against the input table. Until the user completes the text inputs, this process is iterated from FinishInput() to StartInputView(). Once the input is entered, a point in the gyro sensor is initialized to move to the center.

Finally, onDestroy() is automatically called once the user finishes the input so that it controls the gyro sensor to be deactivated.

5   Implementation of VGA

The VGA proposed in this paper can change consonant or vowel input modes via the smart phone display's facing direction or shaking according to the user's circumstances, as shown in Fig. 4.

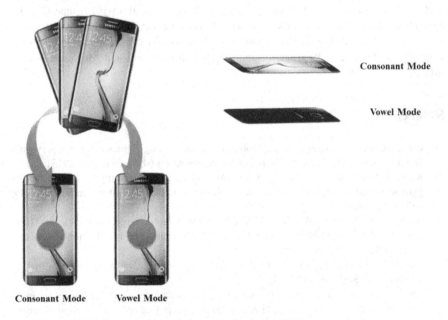

Fig. 4 Consonant and vowel character input modes in the VGA

6   Conclusions

This paper proposed a VGA through which all users can input text, regardless of their finger conditions, into multimedia smart devices. The VGA can support a user's motion-based input mode via the movements of the multimedia smart devices rather than finger-based touch and drag modes, as used in conventional multimedia smart devices. Through the mode proposed in this study, those who wear gloves, who have disabilities, and whose fingers are not available can enter text using this convenient text input method. The input speed via the VGA may be slower than that of Chunjiin, Ez hangeul, neoSTYLE, milgeegle, MoAKey, and dingul due to its user motion-based text input. However, it can enter text efficiently because it does not require frequent pushes or touches by the fingers. Furthermore, it can enter text efficiently when users may have unexpected accidents that make their hands unavailable for text input freely.

In the future, a study on text input methods for various languages other than Korean characters will be conducted by utilizing various sensors embedded in multimedia smart devices.

Acknowledgments This research was supported by Basic Science Research Program through the National Research Foundation of Korea (NRF) funded by the Ministry of Education (NRF-2014R1A1A2053564). And also this research was supported by the MSIP (Ministry of Science, ICT and Future Planning), Korea, under the ITRC(Information Technology Research Center) support program (IITP-2015-H8501-15-1014) supervised by the IITP(Institute for Information & communications Technology Promotion). And also this work (Grants No. C0249205) was supported by Business for Cooperative R&D between Industry, Academy, and Research Institute funded Korea Small and Medium Business Administration in 2014.

References

1. Choi, M., Park, J., Jeong, Y.-S.: Mobile cloud computing framework for a pervasive and ubiquitous environment. Journal of Supercomputing **64**(2), 331–356 (2013)
2. Kim, H.-W., Park, J.H., Jeong, Y.-S.: An efficient character input scheme with a gyro sensor of smartphone on ubiquitous cluster computing. Cluster Computing **18**(1), 147–156 (2015)
3. Song, E.-H., Kim, H.-W., Jeoing, Y.-S.: Visual Monitoring System of Multi-Hosts Behavior for Trustworthiness with Mobile Cloud. Journal of Information Processing Systems **8**(2), 347–358 (2012)
4. Moein, S., Gebali, F., Traore, I.: Analysis of Covert Hardware Attacks. Journal of Convergence **5**(3), 26–30 (2014)
5. Ahn, J., Han, R.: An indoor augmented-reality evacuation system for the Smartphone using personalized Pedometry. Human-centric Computing and Information Sciences **2**(18), 1–23 (2012)
6. Jeong, Y.-S., Yeom, T.-K., Park, J.S., Park, J.H.: Efficient model of Korean graphemes based on a smartphone keyboard. Electronic Commerce Research **13**(3), 357–377 (2013)
7. Lim, Y.-W., Lim, H.: A Design of Korean Input Method using Direction of Vowel on the Touch Screen. Journal of Korea Multimedia Society **14**(7), 924–932 (2011)

8. Sarkar, K.: Automatic Single Document Text Summarization Using Key Concepts in Documents. Journal of Information Processing Systems **9**(4), 602–620 (2013)
9. Yeom, T.-K., Park, J.S., Park, I.-H., Park, J.H.: Toggle Keyboard: Design and Implementation of a New Keyboard Application Based on Android. Ubiquitous Information Technologies and Applications **214**, 807–813 (2013)
10. Kim, H.-W., Song, E.-H., Jeong, Y.-S.: Message input scheme based on gyro sensor in smart embedded devices for NUI. Future Information Technology: Future Tech 2014, vol. 309, pp. 529–534. Springer, July 2014

Ubiquitous Bluetooth Mobile Based Remote Controller for Home Entertainment Centre

Andy Shui-Yu Lai and Ching-Laam Leung

Abstract This paper describes the use of a mobile Android phone in an experimental way to develop a remote controller for multi-player games which are installed and resided in a Home Entertainment Centre in a ubiquitous computing environment. The Home Entertainment Center could be a Smart TV and the Android mobile is programmed to be a remote control device which is used to send sockets via Bluetooth network to Java based Bluetooth server embedded in a Smart TV to control the movements of the game characters playing in the Smart TV with multimedia effects in all directions and actions, such as shootings, in games along its game play. Our investigation focuses on an extended form of ubiquitous computing which game software developers utilize to develop a game remote controller on mobile phone for multi-players. We call this study an experimental ubiquitous computing application in which the Bluetooth embedded in the Home Entertainment Centre which can be discoverable and paired with, the clients, remote mobile devices and instantaneously send the game data via Bluetooth piconnet to the paired remote mobile devices. Currently, mobile computing feeds data information into the game server. However, designing real-time ubiquitous mobile remote controller with Home Entertainment Centre is still a daunting task and much theoretical and practical research remains to be done to reach the ubiquitous computing era. In this paper, we applied the open-source Android Bluetooth and the Java-Based technology in developing a Bluetooth server multi-player mobile game in distributed ubiquitous computing, which strongly focuses on the emergence of technologies that embrace android mobile and Bluetooth latest technology.

Keywords Home entertainment centre · Ubiquitous computing · Mobile programming · Bluetooth technology

A.S.-Y. Lai(✉)
Faculty of Science and Technology, Technological and Higher Education Institute,
Tsing Yi, Hong Kong
e-mail: andylai@vtc.edu.hk

C.-L. Leung
Department of Computer Science and Creative Technologies,
University of the West of England, Bristol, UK
e-mail: jt8486@yahoo.com.hk

© Springer Science+Business Media Singapore 2015
D.-S. Park et al. (eds.), *Advances in Computer Science and Ubiquitous Computing*,
Lecture Notes in Electrical Engineering 373,
DOI: 10.1007/978-981-10-0281-6_16

1   Introduction

The Home Entertainment Centre concept has existed for years. The term Smart Centre or Intelligent Entertainment Centre followed and has been used to introduce the concept of combining the advanced technologies with computer intelligence and game products to form a complete entertainment package at home[1]. Our work is to apply the open-source Bluetooth technology and the Java-Based technology Android in developing a remote controller on mobile phone to control the multi-player PC games resided in Home Entertainment Centre or Smart TV in distributed ubiquitous computing. It strongly focuses on the emergence of technologies that embrace android mobile and Bluetooth open-sources. Our investigation focuses on an extended form of ubiquitous computing which game software developers utilize to develop a game remote controller with mobile phone which is for player of multi-player PC games. We call this study an experimental ubiquitous computing application in which the Bluetooth embedded in the Home Entertainment Centre and can be discoverable and paired with remote device and instantaneously send the data via Bluetooth Piconet to the connected Android mobile device. Currently, mobile computing feeds data information into the game server. However, designing real-time ubiquitous mobile controller with PC games in Smart TV as Home Entertainment Centre is still a daunting task and much theoretical and practical research remains to be done to reach the ubiquitous computing era [2][3]. In this paper, we present the overall architecture and discuss, in detail, the implementation steps taken to create the Bluetooth and Android based remote controller for context-aware multi-player games in PC workstation.

This experimental context-aware application can be viewed as a concrete example of automatic contextual reconfiguration and context-triggered actions in ubiquitous computing. Likewise, the implementation steps in this application can be viewed as micro-architectural elements of a ubiquitous computing framework that documents and motivates the semantics in ubiquitous computing in an effective way [4].

The rest of the paper is organized as follows: Section 2 presents the system architecture of Bluetooth mobile based remote controller for home entertainment centre. Section 3 describes the design and implementation of mobile remote controller application and explains its connection and interaction between home entertainment centre and mobile device. Section 4 we conclude with a note on the current status of the project.

2   System Architecture

One significant aspect of the merging mode of ubiquitous computing is the constantly changing execution environment. This work presents an experimental context-aware application that provides context-aware information to game server and game players in a mobile distributed computing environment. Android is the software activity designed to work with smartphones based on the Android open source operating system and Bluetooth is an open-source wireless technology and prototyping platform based on flexible, easy-to-use hardware and software [5]. Due to the advancement of wireless technology, there are several different connections are introduced such GSM, WIFI, ZIGBEE, and Bluetooth. Among the four popular wireless connection, Bluetooth has its unique specifications and employs it suitable capability for Home Entertainment Centre. Bluetooth with globally available frequencies of 2400Hz is able to provide connectivity up to 100 meters at sped of up to 3 Mbps. In addition, a Bluetooth master device is able to connect up to 7 devices in a Piconet [6]. The capabilities of Bluetooth are sufficient to be implemented the design. It has also been found that most of the current smartphones are come with the built-in Bluetooth adapters that will directly reduce the cost of implementation of the system.

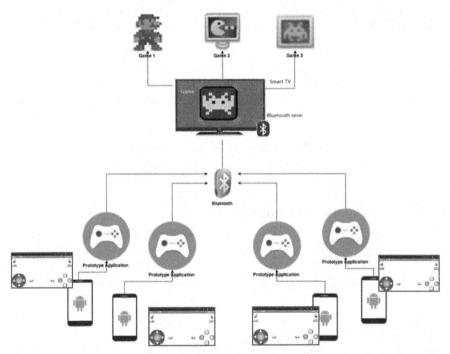

Fig. 1 Network Architecture Bluetooth Based Remote Controllers with a Home Entertainment Centre

Fig. 1 shows a pictorial representation of the network architecture of Bluetooth based remote controllers for a Home Entertainment Centre. It consists of the distributed network components: Android Mobile Remote Controllers, Bluetooth prototype embedded application, Bluetooth devices (Bluetooth slaves), Home Entertainment Centre (Smart TV), Bluetooth device and multi-player games embedded in Smart TV (Bluetooth master).

In Bluetooth, a Piconet is a collection of up to 8 devices that frequency hops together. Each Piconet has one master usually a device that initiated establishment of the Piconet, and up to 7 slave devices. Master's Blue tooth address is used for definition of the frequency hopping sequence. Slave devices use the master's clock to synchronize their clocks to be able to hop simultaneously. When a device wants to establish a Piconet it has to perform inquiry to discover other Bluetooth devices in the range. Inquiry procedure is defined in such a way to ensure that two devices will after some time, visit the same frequency same time when that happens, required information is exchanged and devices can use paging procedure to establish connection [6].

When more than 7 devices need to communicate, there are two options. The first one is to put one or more devices into the park state. Bluetooth defines three low power modes sniff, hold and park. When a device is in the park mode then it disassociates from and Piconet, but still maintains timing synchronization with it. The master of the Piconet periodically broadcasts beacons (warning) to invite the slave to rejoin the Piconet or to allow the slave to request to rejoin. The slave can rejoin the Piconet only if there is less than seven slaves already in the Piconet.

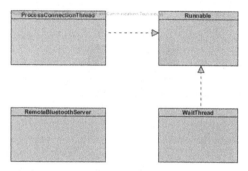

Fig. 2 Class Diagram of Bluetooth Server (Master)

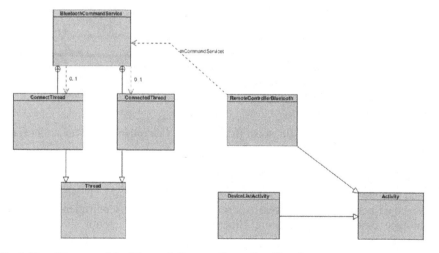

Fig. 3 Class Diagram of the Bluetooth Remote Controller (Slave)

Fig. 2 presents the software architecture with class diagram for the Bluetooth game server (master) installed in a Smart TV and Fig. 3 shows the class diagram for the mobile remote controller (slave) on Android platform, where the master RemoteBluetoothServer class in Bluetooth Server (Home Entertainment Centre - master) manages the messages from RemoteControllerBluetooth class in Bluetooth Remote Controllers (Android mobile devices – slaves) using the Bluetooth applications embedded in both the Smart TV and the Android mobile phones.

3   Android Motion Control Design and Implementation

The Android user interface design is for game players to remotely control the game characters playing in Home Entertainment Centre to move forward and backward at different speeds, turn right and left for different angles with multimedia effects.

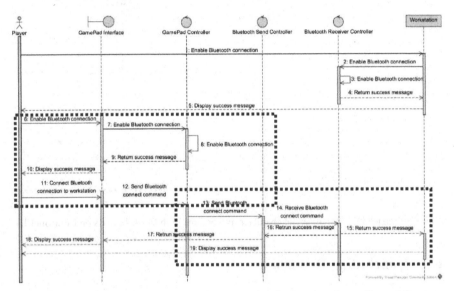

Fig. 4 Sequence Diagram of establishing the connection for Remote Controller and Centre

The sequence diagram in Fig. 4 presents how the BlueCove library is applied in application to establish the connection between the player's mobile phone and the Home Entertainment Centre. In the diagram, on the left hand side, the GamePad provides the Interface and Controller to the Player and, on the right hand side, the Bluetooth establishes the link between the Send Controller and Receiver Controller in system.

The sequence diagram in Fig. 5 shows the implementation of a main function to control the game character movements in game. At times, the Player presses the DPAD buttons on GamePad to send the DPAD data via BluetoothSendController to

Home Entertainment Centre Server which, in turn, returns the data back to GameController to decode and generate the character movement actions for controlling the game characters. Likewise, when Player releases the DPAD buttons on GamePad, the GamePad will send the release DPAD button data signals via BluetoothSendController to Home Entertainment Centre Server which, in turns, will return DPAD data accordingly to the commands of GameController to stop the game character movement in game.

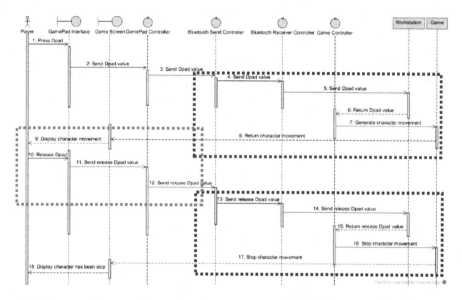

Fig. 5 Sequence Diagram of game character movements with DPAD buttons on Controller UI

4   Concluding Remark

We have presented and implemented an experimental Bluetooth mobile-based remote controller for Home Entertainment Centre in the ubiquitous and mobile computing development environment. The project has successfully applied the Bluetooth and Java Technology to achieve the idea of accessing Home Entertainment Centre by using mobile phone devices as game remote controllers. We have described the overall architecture and discusses, in detail, the implementation steps taken to implement the Bluetooth and Android based remote controller for context-aware multi-player games. We believe that Bluetooth is a good candidate in wireless networking technology and can make a significant contribution towards distributed applications in a mobile remote control game in a ubiquitous system environment.

References

1. Hamed, B.: Design & Implementation of Smart House Control Using LabVIEW. International Journal of Soft Computing and Engineering (IJSCE) **1**(6), January 2012. ISSN: 2231-2307
2. Lai, A.S.Y.: Mobile Bluetooth-Based Multi-Player Game Development in Ubiquitous Computing. Journal of Computational Information Systems (JCIS) **6**(14), December 2010
3. Lai, A.S.Y.: Ubiquitous Mobile Game Development using Arduino on Android Platform, Lecture Notes of Electrical Engineering (LNEE 280). In: 8th International Conference on Ubiquitous Information Technologies and Applications (CUTE2013). Springer-Verlag, December 2013
4. Gu, T., Pung, H.K., Shang, D.Q.: A service-oriented middleware for building context-aware services. Journal of Network and Computer Applications **20**, 1–18 (2005). Elsevier
5. Roh, J.-k., Jin, S.: Device Control Protocol using Mobile Phone. In: 16th International Conference on Advanced Communication Technology (ICACT) (2014)
6. Official Bluetooth website (2015). http://www.bluetooth.com/Pages/Bluetooth-Home.aspx

Robust Feature Design for Object Detection

Woong Hu, Min-su Koo, Jae-Hyun Nam,
Byung-Gyu Kim and Sung-Ki Kim

Abstract Feature matching is a basic process of many computer vision applications such as object detection or image stitching. A lot of state of the art technologies of feature extraction are focusing on the internal features of object. They perform great matching result but some of them spend extensive computational time to extract and match the features. In this paper, we propose a robust method for keypoint detection, description, and matching. The proposed algorithm is based on the point that people can recognize the object through the silhouette, firstly. We use object's contour to extract their features. We can see that the proposed scheme can reduce much time to extract features through experiments.

Keywords Keypoint matching · Feature descriptor · Feature extraction · Object detection

1 Introduction

Most of feature extraction algorithms have considered object's internal keypoints like histogram of gradient (HOG) [1], 3D histogram of gradient location and orientation [2], a lot of corners [3], the maxima of the determinant of the Hessian matrix [4], oriented Features from Accelerated Segment Test (FAST) and Rotated Binary Robust Independent Elementary Features (BRIEF) [5] and binary string by results of simple brightness comparison tests [6].

W. Hu · M.-s. Koo · J.-H. Nam · B.-G. Kim(✉)
Department of Computer Engineering, SunMoon University, Asan-Si, Republic of Korea
e-mail: {huwoong,ms.koo,jh.Nam,bg.kim}@mpcl.sunmoon.ac.kr

S.-K. Kim
Department of Infomation and Communication Engineering,
SunMoon University, Asan-Si, Republic of Korea
e-mail: skkim@sunmoon.ac.kr

© Springer Science+Business Media Singapore 2015   117
D.-S. Park et al. (eds.), *Advances in Computer Science and Ubiquitous Computing*,
Lecture Notes in Electrical Engineering 373,
DOI: 10.1007/978-981-10-0281-6_17

For over a decade, the Scale Invariant Feature Transform (SIFT) keypoint detector and descriptor maintain best of feature matching algorithms in a number of applications using visual features, including object detection, image stitching, etc. However, it spends a large computational time. Likewise other methods also consume not a few time to extract features.

When human recognizes object, firstly human visual system extracts the object's outline. Then, human looks at the object's texture or color. That's why we try to recognize the object through the silhouette as shown in Fig. 1. In plain words, other feature matching algorithms focused on very detail parts of objects instead of looking at the whole of them. So they generally have consumed extensive computational time to extract and match the features.

In this paper, we propose an efficient method for object detection. We extract angles and directions from the divided contour blocks. Then we constitute local descriptor for matching features. The proposed algorithm is attractive because of its good performance and low computational cost.

The organization of paper is following: Section 2 presents the overall object detection system, and the proposed feature extraction method is also defined. Section3 discusses simulation results, and concluding remarks are contained in Section 4.

Fig. 1 Examples of silhouette (cat, car, dog, human).

2 Proposed Algorithm: Oriented Angular Keypoints

We describe the procedure of algorithm and the detailed mechanism. We have five-steps for object matching. First, we find the contour from input image. The result of this stage is very important for overall performance, because this algorithm is basis on the object's outline. Second, we extract a principal outline from the result of the first step. When we extract a contour from an image, there are many noises or holes in an image. So we need to eliminate noise and holes.

Third, we constitute the block region. We divide image into blocks through the contour. This stage is to compute angle and direction of each block. Fourth, we compute angle and direction from divided blocks. We used the "law of cosines" to compute angle and direction. So, set of those will be a descriptor of object. Then, we match the query descriptor with the reference descriptor. Finally, we score matching similarity between query and reference.

Fig. 2 Input images (top) and extracted features from input image (bottom).

2.1 Preprocessing (Outline Detection)

Figure 3 shows the process how to detect outline. We used Gabor filter for smoothing input image and binarize from the smoothed image. Lastly, the contour is extracted by the Canny edge detector.

Fig. 3 Process of outline detection. Original input image (left top), Gabor filtered image (right top), binarized image (left bottom), detected outline image (right bottom).

2.2 Extract Principal Outline

When we try to divide the outline into blocks, some of holes or noise components disturb this process. So we eliminated the holes and noise in advance.

In this stage, we label each group of neighboring pixels. Then, we compare total pixel count of each group. The greatest number of pixel group as is only remained shown in Fig. 4. In labeling process, we explore 8 neighbors of the center point until meet the end of pixels.

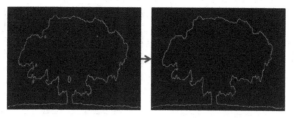

Fig. 4 Detected outline image (left), extracted principal outline (right)

2.3   Constitute Block Region

We constitute block region to compute angle and direction. Each part of the outline will be one feature includes angle, direction and location. We make all blocks that include pixel in their center point. We draw a line from the center point to passing through block outline point as shown in Fig. 5. It makes that we can preserve principal characteristics from the original image and can extract features easily at the same time. The size of block affects both accuracy and speed of the proposed algorithm. In our case, the size of block was selected as width/10 (pixels).

Fig. 5 A block that has pixel in the center point (left), a block that drawn two lines to boundary of block (right)

Fig. 6 Outline image (left), found blocks (center), transformed outline from the given block (right)

2.4   Computation of Angle and Direction

From the divided blocks, we can compute angle and direction using "law of cosines" as Eqs. (1) and (2). Originally, the "law of cosines" generalizes the Pythagorean Theorem, which holds only for right triangles. But its expansion resolves the limitation. If we know length of all three sides, we can compute each angle. So we computed Euclidian distance as Eq. (3) to get angles.

$$
\begin{aligned}
a &= b\cos C + c\cos B, \\
b &= c\cos A + a\cos C, \\
c &= a\cos B + b\cos A.
\end{aligned}
\tag{1}
$$

$$a^2 = b^2 + c^2 - 2bc \cos A. \tag{2}$$

$$\text{distance } d_{p-q} = \sqrt{(p-q)^2}. \tag{3}$$

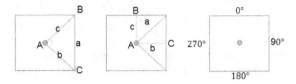

Fig. 7 Compute angle (left), Compute direction (center, right)

Figure 7 shows how to get the angle and direction in the given block. We can calculate angle from points B, A, C in left image and direction from points B, A, C in the center image. Point B of the center image is an anchor point. Point C in center image is in middle of angle. Right image shows the reason why we set point B to anchor point.

2.5 Object Matching

In the matching strategy, we found nearest neighbor between a query image and the reference image. The nearest neighbor is defined as the keypoint with minimum Euclidean distance d as shown in Eq. (4). Minimum Euclidean distance is the sum of difference between query point and its corresponding point. And then we calculated similarity s using Euclidean distance as illustrated in Eq. (5). Based on the measured the similarity s, we verify whether input is the object which we look for or not as Eq. (6). We set a t = 0.25 through experiments.

$$d_{(distance)} = \sum_{i=1}^{n} \sqrt{((p_i - q_i)^2)} \tag{4}$$

$$s_{(similarity)} = 1/(1+d) \tag{5}$$

$$O_{(detected\ object)} = \begin{cases} 1 & s > t \\ 0 & otherwise \end{cases} \tag{6}$$

3 Experimental Results

We have used Caltech101 [8] to evaluate the performance with SIFT [2], SURF [4] that are most popular algorithms. We picked scissors and stapler for positive samples, inline skate for negative samples.

The average of positive dataset's similarity s becomes 0.57 and negative dataset's similarity s becomes 0.08. Therefore, we chose 0.2~0.3 as threshold t in our experiments as shown in Table 1. The both of SIFT [2] and SURF [4] are for matching exactly same object. By the way, the proposed algorithm OAK is focused on generally similar object. That is why the result shows that the performance of SIFT [2] and SURF [4] is not good in these sequences as shown in Table 2.

When we tried to match object with SIFT [2], and SURF [4], they found just trained image. However, the proposed algorithm improved performance and reduced the computational time, because we find object having similar shape not exactly same.

But, if the proposed algorithm could not get enough outlines, then of course it could not find object. If we can find all of object's outlines in the input image, the proposed algorithm may find all similar shape objects.

Table 1 Performance comparison of the proposed algorithm with different threshold.

Sequence	Algorithms (t)	TPR (%)	FPR (%)	Time (ms)
scissors	OAK(0.2)	94.87	6.45	9.27
	OAK(0.25)	82.05	3.22	9.27
	OAK(0.3)	71.79	3.22	9.27
stapler	OAK(0.2)	93.33	9.67	10.27
	OAK(0.25)	88.89	9.67	10.27
	OAK(0.3)	73.33	9.67	10.27

Table 2 Performance comparison of the proposed algorithm with SIFT [2] and SURF [4].

Sequence	Algorithms	TPR (%)	FPR (%)	Time (ms)
scissors	SIFT	51.28	48.38	482.21
	SURF	61.53	45.16	140.01
	OAK(t=0.25)	*82.05*	*3.22*	*9.27*
stapler	SIFT	55.56	38.7	562.27
	SURF	53.33	35.48	177.21
	OAK(t=0.25)	*88.89*	*9.67*	*10.27*

4   Conclusion

In this paper, we have used angles and directions of object as an object detector. The proposed new descriptor is created based on contour of object. The experimental results show that object detection can be accelerated by our approach in fifty times than SIFT matching, even achieves ten times or more than SURF matching while retaining credible accuracy.

References

1. Dalal, N., Triggs, B.: Histogram of Oriented Gradients for Human Detecting. Computer Vision and Pattern Recognition, 886–893 (2005)
2. Lowe, G.G.: Distinctive Image Feature from Scale-Invariant Keypoints. International Journal of Computer Vision **60**, 91–110 (2004)
3. Rosten, E., Drummond, T.: Machine learning for high-speed corner detection In: European Conference on Computer Vision, vol. 3951, pp. 430–443 (2006)
4. Bay, H., Tuytelaars, T., Gool, L.V.: SURF: speeded up robust features. In: European Conference on Computer Vision, vol. 3951, pp. 404–417 (2006)
5. Rublee, E., Rabaud, V., Konolige, K., Bradski, G.: ORB: an efficient alternative to SIFT or SURF. In: International Conference on Computer Vision, pp. 2264–2571 (2011)
6. Leutenegger, S., Chil, M., Siegwart, R.Y.: BRISK: binary robust invariant scalable keypoints. In: International Conference on Computer Vision, pp. 2548–2555 (2011)
7. Mikolajczyk, K., Schmid, C.: A Performance Evaluation of Local Descriptors. Pattern Analysis and Machine Intelligence **27**, 1615–1630 (2005)
8. Fei-Fei, L., Fergus, R., Perona, P.: Learning generative visual models from few training examples: an incremental Bayesian approach tested on 101 object categories. Computer Vision and Image Understanding **106**, 59–70 (2007)

An Efficient Key Management Scheme for Advanced Metering Infrastructure

Min-Sik Kim, Seung-Kyu Park, Hyo-Seong Kim and Kwang-il Hwang

Abstract In this paper we focus on key management scheme for a specific WSN application such as Advanced Metering Infrastructure (AMI), which is based on hierarchical tree network and performs metering data aggregation along the tree. We utilize a novel identity based key generation and propose a hierarchical key management algorithm based on encapsulated encryption. Through experiments, we prove that the proposed key management scheme is suitable for small embedded networked systems which have energy, memory, and processing constraints.

Keywords Key management · Hierarchical network · Security · WSN

1 Introduction

Advances in WSN (wireless sensor networks) have lead to numerable deployment of various sensor nodes over the world. In addition to this, the importance of WSN security is arising as one of the most significant design considerations, because sensor nodes are often deployed in remote, hostile, and unattended environments, and thus they are exposed to various security attacks. So far a number of research relating to WSN security were contributed to protecting WSN sites from a variety

M.-S. Kim · S.-K. Park · K.-i. Hwang(✉)
Incheon National University, Incheon, Korea
e-mail: hkwangil@inu.ac.kr

H.-S. Kim
Leotek Co. Ltd, Incheon, Korea

K.-i. Hwang—This work was supported by the Industrial Core Technology Development Program (10049009, Development of Main IPs for IoT and Image-Based Security Low-Power SoC) funded by the Ministry of Trade, Industry & Energy, and also supported by a grant (12-TI-C01) from Advanced Water Management Research Program funded by Ministry of Land, Infrastructure and Transport of Korean government.

© Springer Science+Business Media Singapore 2015
D.-S. Park et al. (eds.), *Advances in Computer Science and Ubiquitous Computing*,
Lecture Notes in Electrical Engineering 373,
DOI: 10.1007/978-981-10-0281-6_18

of security attacks. The research relating to WSN security are covering from cryptography to key management, and some literatures [1 and 2] provided summary and taxonomy of WSN security. However, WSN is recently being specific to applications or services which require more concrete network and data aggregation model. Therefore, in this paper we focus on some WSN application model which is based on hierarchical and tree-based network. In particular, out target application is focused on AMI (Advanced Metering Infrastructure). To develop a suitable security framework for such WSN model, we utilize a novel identity based key generation and propose a hierarchical key management algorithm based on encapsulated encryption. The proposed key management scheme is basically designed to meet the following requirements: small overhead in enc/decryption time and memory. It is suitable for small embedded networked systems which have energy, memory, and processing constraints.

2   Related Work

For the last few decades, a number of security algorithms, more specifically key management algorithms in WSN, have been introduced. Some literatures [1 and 2] also provided a summary and taxonomy of WSN key management schemes. Therefore, in this paper we intensively investigate some research on hierarchical, tree based key management [3 - 6], identity-based key generation [7 - 9], and polynomial based key management [10 - 15], which are closely related to our work.

First, in tree-based key distribution schemes [3 and 4], sensor nodes are arranged in a tree in which each sensor node communicates with its parent node. So the key establishment has done between neighboring nodes along the tree. In addition, a tree of keys is built for the hierarchical network [5 and 6], where the keys at a certain level are distributed to the corresponding class of nodes. The keys at higher levels can be used to derive the keys at lower levels, but not vice versa. Identity-based key establishment schemes utilizes multivariate key establishment [8] based on unique ID to establish a key. Some literature [7] uses hamming distance greater than one. In addition, an identity based random key predistirubution scheme [9] is proposed. Polynomial-based key establishment scheme is based on pair-wise keys to overcome some of probabilistic predistribution scheme's disadvantages and to reduce communication overhead [10 - 15].

Even though the existing key establishment scheme shows good performance in some environment and network model that they assumed, they do not provide concrete design and implementation details to construct specific WSN based on hierarchical-tree based network and data aggregation model. Therefore, it is necessary to develop a new key establishment scheme to fit for the specific application.

3   Identity-Based Hierarchical Key Management

In this Section, we present our proposed key management scheme. The proposed scheme is based on a cluster tree network model cooperated with hierarchical ID systems, and provides a powerful enc/decryption functionality using a secure key generation method based on unique node ID and key seed generated by server. In order to reduce security overhead, encapsulation encryption is performed at each level. It also results in increasing security strength.

3.1   Network Model

A tree consists of n-level hierarchy, and ID of a node is determined according to the underlying level. The ID is used for pair-wise authentication between a parent and child node, and ID structure is as follows: $ID(i)$ = {Level 1 ID, LEVEL 2 ID, ..., LEVEL n ID}, n = depth of the hierarchy. For example, as shown in Fig. 1, full ID of node 1 at level 4 is {1,1,1,1}. In addition, for intermediate nodes, which are not leaf nodes, lower level ID is all '0'. e.g., ID of node 2 at level 2 is {2,2,0,0}.

3.2   Key Establishment Overview

The procedure of the proposed scheme is largely divided into install phase and runtime phase, and during runtime key refresh (periodic key establishment or by request) is performed. In install phase, a tree can be constructed based on authentication using pre-distributed keys. After completion of tree construction, server propagates key seed to be used for generating pair-wise distributed keys cooperated with unique IDs of nodes.

3.3   Key Generation

One of the critical factors in key management is to use a powerful key generation. In this paper we propose an identity-based key generation using a polynomial based on hierarchical ID and key seed from server. The key generation formula is as follows:

$$f(x) = \sum_{i=1}^{n} Id_i x^{a_i} \tag{1}$$

Where ID_i represents an ID at level i, a is a seed value from server and has different value every key refresh phase. n represents the maximum depth of level.

3.4   Key Management

Initial network construction process is achieved by ascendant nodes in a distributed manner. A parent sends a group key using its own ID, and pair-wise authentication is performed between a parent and children. After completion of a tree construction, BS requests server network key establishment based on the constructed network information, server reply with ACK containing a new key seed. The ACK message is propagated to the whole network along the tree, and thus each node establishes keys for its parent and children, if any, obtaining keys from (1) based on its unique ID and key seed.

4   Experimental Results

In this Section we first introduce our test-bed and then present experimental results based on our Test-bed. Sensor nodes used for Test-bed consist of TI MSP430 16bit microcontroller and CC1120 sub 1-GHz RF modem. NIST AES128 is used for encryption/decryption functions with ECB mode. Test-bed is composed of 1 Base Station (BS), 2 Level-2 nodes, 4 Level-3 nodes, and 3 level hierarchy WSN is tested. To evaluate the performance of the proposed scheme, we compared our scheme with non-encryption and full message encryption method in terms of memory utilization and security overhead.

Fig. 1 shows observation result of security overhead with respect to our scheme (encryption encapsulation) and full message encryption, respectively. We calculated the total overhead as the number of nodes increases from 10 to 100. The total overhead is obtained by sum of Encryption time (Te), propagation time (Tp), and Decryption time(Td). As shown in that figure, while the overhead of full message encryption shows rapid increase as the number of nodes increases, our scheme shows lower rate of increase. That is because our scheme encrypts not all the message but only data area, and the messages are encapsulated at every level.

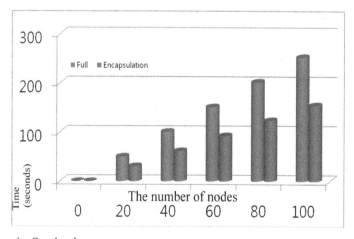

Fig. 1 Security Overhead

5   Conclusion

In order to develop a suitable security framework for hierarchical tree-based WSN model, we proposed a novel identity based key generation and a hierarchical key management scheme based on encapsulated encryption. The proposed key management scheme is basically designed to meet the following requirements: small overhead in enc/decryption time and memory and through experiments based on our Test-bed, we proved that the proposed scheme is suitable for small embedded networked systems which have energy, memory, and processing constraints. We expect that our security framework will contribute to a design guideline of security framework for hierarchical WSN applications.

References

1. Vamsi, P.R., Kant, K.: A taxonomy of key management schemes of wireless sensor networks. In: 2015 Fifth International Conference on Advanced Computing & Communication Technologies (ACCT), vol. 690(696), pp. 21–22 (2015)
2. Bala, S., Sharma, G., Verma, A.K.: Classification of Symmetric Key Management Schemes for Wireless Sensor Network. International Journal of Security and Its Applications 7(2), March 2013
3. Lee, J., Stinson, D.R.: Deterministic key pre-distribution schemes for distributed sensor networks. In: ACM Symposium on Applied Computing 2004, Waterloo, Canada. Lecture Notes in Computer Science, vol. 3357, pp. 294–307 (2004)
4. Blom, R.: Theory and application of cryptographic techniques. In: Proceedings of the Eurocrypt 84 Workshop on Advances in Cryptology, pp. 335–338. Springer, Berlin (1984)
5. Zhu, S., Setia, S., Jajodia, S.: LEAP: efficient security mechanisms for large-scale distributed sensor networks. In: Proceedings of the 10th ACM Conference on Computer and Communications Security (2003)
6. Jang, J., Kwon, T., Song, J.: A time-based key management protocol for wireless sensor networks. In: Proceedings of ISPEC. LNCS, vol. 4464, pp. 314–328 (2007)
7. Liu, D.G., Ning, P., Li, R.F.: Establishing pairwise keys in distributed sensor networks. ACM Trans. Inf. Syst. Secur. 8(1), 41–77 (2005)
8. Delgosha, F., Fekri, F.: A Multivariate Key-Establishment Scheme for Wireless Sensor Networks. IEEE Transactions on wireless communications 8(4) (2009)
9. Das, A.: Improving Identity-based Random Key Establishment Scheme for Large-scale Hierarchical Wireless Sensor Networks. International Journal of Network Security 14(1), 1–21 (2012)
10. Liu, D., Ning, P.: Improving Key Pre-Distribution with Deployment Knowledge In Static Sensor Networks. ACM Transactions on Sensor Networks 1(2), 204–239 (2005)
11. Liu, D., Ning, P., Li, R.: Establishing Pairwise Keys in Distributed Sensor Networks. ACM Transactions on Information and System Security 1(8), 41–77 (2005)
12. Liu, D., Ning, P.: Location-based pair-wise key establishment for static sensor networks. In: Proceedings of 1st ACM Workshop on Security of Ad Hoc and Sensor Networks (2003)

13. Liu, D., Ning, P.: Establishing pair-wise keys in distributed sensor networks. In: Proceedings of 10th ACM Conference on Computer and Communications Security CCS 2003 (2003)
14. Zhang, W., Tran, M., Zhu, S., Cao, G.: A random perturbation-based scheme for pairwise key establishment in sensor networks. In: Proceedings of MOBIHOC 2007, Montréal, Québec, Canada (2007)
15. Eschenauer, L., Gligor, V.: A key management scheme for distributed sensor networks. In: Proceedings of the 9th ACM conference on Computer and Communications Security, CCS 2002, pp 41–47 (2002)
16. Bloom, R.: An optimal class of symmetric key generation systems. In: Proceedings of EUROCRYPT 84, pp 335–338 (1985)

Levenshtein Distance-Based Posture Comparison Method for Cardiopulmonary Resuscitation Education

Yunsick Sung, Ji Won Kim, Hyung Jin Park and Kyung Min Park

Abstract Practice-based educations helps trainees to understand the theorical knowledge by lectures easily. However, given that the limited number of trainers, the effect of practice can be reduced. Therefore, the automatic evaluation approaches of practices is demanded. This paper proposes a Levenshtein distance-based posture comparison method for evaluating the postures during cardiopulmonary resuscitation (CPR) practice. The proposed posture comparison method can be extended to the evaluation of consecutive gestures that are consisted of postures.

Keywords Levenshtein distance · Posture · Gesture · Cardiopulmonary resuscitation · CPR

1 Introduction

As recently cardiopulmonary resuscitation (CPR) educations are one of the essential educations and have conducted for students and the public widely, the research related to CPR is also conducted actively. For an example, the effect of CPR-based knowledge and education attitudes of university students is analyzed [1, 2].

Y. Sung · J.W. Kim
Faculty of Computer Engineering, Keimyung University, Daegu 42601, South Korea
e-mail: {yunsick,jiwon1219}@kmu.ac.kr

H.J. Park
College of Fine Art, Keimyung University, Daegu 42601, South Korea
e-mail: phj@kmu.ac.kr

K.M. Park(✉)
College of Nursing, Keimyung University, Daegu 42601, South Korea
e-mail: kmp@kmu.ac.kr

© Springer Science+Business Media Singapore 2015
D.-S. Park et al. (eds.), *Advances in Computer Science and Ubiquitous Computing*,
Lecture Notes in Electrical Engineering 373,
DOI: 10.1007/978-981-10-0281-6_19

131

The continuous education combining CPR theory with CPR practices is required to increase CPR-related knowledge. In addition, the development of the systematic and professional education programs is also demanded.

As above, the importance of practices is emphasized. However, there is difficulty to evaluate how much the performed gestures/postures during practices are accurate. For an example, when chest compressions are performed on a CPR Annie, it should be checked whether the finger shape is correct or not and whether the speed of the compression is conducted continuously or not. Given that the number of trainers is limited, it is required to evaluate the performed gestures/postures automatically.

This paper proposes a Levenshtein distance-based posture comparison method to evaluate CPR postures accurately without trainers. The difference of two postures is calculated by comparing the properties of the two CPR postures. Users can recognize the amount how much own postures are inaccurate based on the difference of the CPR postures.

The rest of this paper is organized as follows: Section 2 introduces a Levenshtein distance-based posture comparison method. Section 3 explains the experiment by applying the proposed method. Finally, Section 4 conclusions of this paper.

2   Posture Comparison Processes

The traditional Levenshtein distance is a string matrix to measure the difference between two strings [3]. Instead of applying to strings, Levenshtein distance-based comparison approach is also applied to the comparison of movements [5-6]. However, because of the weak concept of an initial movement, two movements could not be compared accurately. This paper compares two postures, one of CPR postures and the other of user postures revising the previous week concept

In this paper, each posture is expressed by multiple properties that can be measured by sensors or can be defined manually in advance. For an example, the properties of a posture is the measured values of magnetometers, gyroscopes, accelerations and electromyograms (EMG). EMG is an electro diagnostic medicine device for evaluating activity produced by muscles.

Two postures are compared as follows. First, three postures are prepared to be compared: a CPR posture (c), a user postures (u) and an initial posture (p). This paper requires one additional posture as an initial posture that is the first posture for CPR to compare a user posture with a CPR posture. The approach how to choose one of CPR postures and one of user postures is not in the ranges in this paper.

Secondly the properties of the three postures are obtained. The properties of the initial posture and the CPR posture are defined before the comparison of the three postures and then utilized. The properties of the user posture are measured by attached sensors and obtained in real-time. Given that multiple sensors can be utilized to measure the properties of a posture, the related properties in a sensor are grouped as a vector. Therefore, the properties of postures are denoted as numeric values, vectors or mixed values.

Finally, Levenshtein distance is calculated based on Levenshtein distance matrix. If a user posture contains m properties and a CPR posture contains n properties, a $m \times n$ matrix will be made. i and j are the indice of properties of a user posture and a CPR posture. Therefore, u_i and c_j are the ith and jth properties of postures, u and c. When i is 0 or j is 0, the element $e_{i,j}$ in Levenshtein distance matrix is assigned by $|u_i - p_i|$ or $|c_j - p_j|$ as Equation (1)

$$f(i,j) = \begin{cases} if\ i = 0\ then\ |c_j - p_j| \\ else\ if\ j = 0\ then\ |u_i - p_i| \\ else \\ \quad min \begin{cases} f(i-1,j) + w_{u',i,j} \\ f(i,j-1) + w_{c',i,j} \\ f(i-1,j-1) + w_{p',i,j} \end{cases} \end{cases} \tag{1}$$

When i or j is not 0, the element $e_{i,j}$ is assigned by the sum of the previous element and the cost $w_{d,i,j}$. $w_{d,i,j}$ is calculated as shown in Equation (2).

$$w_{d,i,j} = \begin{cases} if\ d =\ 'u'\ then \\ \quad if\ i \geq j\ then\ \frac{|c_i-u_i|-min(c_i)}{max(c_i)-min(c_i)} \\ \quad else\ if\ i < j\ then\ \frac{|u_i-p_i|-min(u_i)}{max(u_i)-min(u_i)} \\ else\ if\ d =\ 'c'\ then \\ \quad if\ i > j\ then\ \frac{|c_j-p_j|-min(c_i)}{max(c_i)-min(c_i)} \\ \quad else\ if\ i \leq j\ then\ \frac{|c_j-u_j|-min(c_i)}{max(c_i)-min(c_i)} \\ else \\ \quad if\ i \geq j\ then\ \frac{|c_j-p_j|-min(c_i)}{max(c_i)-min(c_i)} \\ \quad else\ if\ i < j\ then\ \frac{|c_j-u_j|-min(c_i)}{max(c_i)-min(c_i)} \end{cases} \tag{2}$$

where d is defind by the type of the posture corresponding to an added property. For an example, a property of a user posture is added, d is 'u'.

3   Experiments

In the experiments, postures were measured by utilizing two Myo [4]. Myo is the arm band shape-based wearable gesture control device that is put to a wrist. The Myo measures four types of properties: Gyroscope, accelerator, orientation, and muscle data (EMG). When Myo is put to wrist and the arm attached Myo is moved, the movements of muscles in the arm are measured and then the data of the measured movements is transferred through Bluetooth.

In the experiments, all properties are classified into two groups: one for the properties of a left hand u^l and the other for the properties of a right hand u^r. Table 1 shows the properties of the left hand u^l.

Given that all elements of a gyroscope have the most values in the ranges from -500 to 500, all elements in the value of the gyroscope are revised. For an example, $\vec{x}$ is replaced by $\vec{x}'$ that is calculated by Equation (3) and is utilized instead of $\vec{x}$.

$$p^l_{g.\vec{x}'} = \begin{cases} if\left(p^l_{g.\vec{x}} > 500\right) then\ 500 \\ else\ if\ (p^l_{g.\vec{x}} < -500)\ then - 500 \\ else\ p^l_{g.\vec{x}} \end{cases} \tag{3}$$

Table 1 The properties of a user posture are described by the values of sensors such as gyroscope, acceleration, orientation, and EMG data. Each sensor value is expressed by a vector.

			Property	Unit Weight	Element Range
Posture	u^l	Gyroscope	$p^l_g = <\vec{x}', \vec{y}', \vec{z}'>$	$\dfrac{1}{1000\ rad/s}$	-500~+500
		Acceleration	$p^l_a = <\vec{x}, \vec{y}, \vec{z}>$	$\dfrac{1}{10\ g/s^2}$	-5~+5
		Orientation	$p^l_o = <\vec{r}, \vec{p}, \vec{y}>$	$\dfrac{1}{2\ degree}$	$\vec{r}$: -90~90 $\vec{p}$: -180~180 $\vec{y}$: -180~180
		EMG data	$p^l_e = <e_1, e_2, ..., e_7>$	$\dfrac{1}{254}$	-127~+127

In the experiments, CPR practices are described by four postures. Especially the CPR posture and the user posture of the 2nd step were compared with the initial posture of the 2nd step to validate the proposed method. Table 2 shows the measured properties of postures, p, u and c.

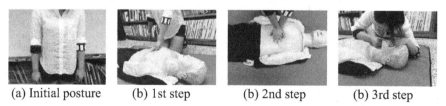

(a) Initial posture   (b) 1st step   (b) 2nd step   (b) 3rd step

Fig. 1 CPR steps are defined by four steps in the experiments. The 2nd step was utilized to validate the propose method.

Table 2 All properties of the posture c, the posture u, and the posture i are measured by utilizing Myo.

	p^l_g p^l_o	p^r_g p^r_o	p^l_a p^l_e	p^r_a p^r_e
c	<0.06,-1.13,-5.88> <-0.83,1.23,-0.74>	<-0.63,0.69,3.38> <-0.77,1.23,-0.82>	<0.30,-0.02,-094> <6,-13,-12,-3,-2,-49,-6,10>	<-0.27,-0.20,-0.91> <48,-7,3,-1,-2,0,13,20>
i	<-0.94,0.44,1.19> <-0.75,1.27,-0.77>	<-0.13,-1.25,8.56> <-0.76,1.30,-0.78>	<0.13,0.28,-0.97> <2,-5,4,-1,1,-2,4,-5>	<0.02,0.26,-0.94> <-2,-2,-4,-3,-5,-6,-3,-3>
u	<-0.50,-0.50,1.94> <-0.82,1.22,-0.75>	<0.81,-0.81,-2.69> <-0.83,1.15,-0.84>	<0.20,-0.29,-0.92> <-5,-5,-5,-10,-11,-2,-14,-5>	<-0.36,-0.12,-0.92> <8,-20,-32,-14,-7,-7,-19,-2>

Table 3 shows the Levenshtein distance of the comparison between the posture u and the posture c. As a result, two postures have the difference about 0.9970.

Table 3 Levenshtein distance matrix is calculated based on the proposed method.

					c				
		p^l_g	p^r_g	p^l_a	p^r_a	p^l_o	p^r_o	p^l_e	p^r_e
		0.0000	0.0096	0.0076	0.0509	0.0354	0.0736	0.0593	0.4134	0.4409
u	p^l_g	**0.0045**	**0.0096**	0.0127	0.0585	0.0405	0.109	0.0644	0.4185	0.8543
	p^r_g	0.0126	**0.0096**	0.0856	0.0675	0.0495	0.1141	0.0734	0.4778	0.8594
	p^l_a	0.0695	0.0791	**0.0186**	**0.0576**	0.0885	0.1231	0.1124	0.4868	0.9187
	p^r_a	0.0417	0.0468	0.0603	**0.0576**	**0.0677**	0.1413	0.1225	0.5258	0.9277
	p^l_o	0.0658	0.0468	0.0558	0.0948	**0.0677**	**0.0834**	0.1427	0.5359	0.9667
	p^r_o	0.1344	0.1812	0.0558	0.0948	0.2021	**0.0834**	**0.1427**	0.5561	0.9768
	p^l_e	0.2165	0.1395	0.2723	0.0948	0.4186	0.2999	**0.1584**	0.5561	0.9970
	p^r_e	0.3425	0.4820	0.1485	0.4373	0.7611	0.6424	**0.5009**	**0.5993**	**0.9970**

4   Conclusion

This paper proposed a CPR posture comparison method based on Levenshtein distance algorithm. Given that the previous Levenshtein distance-based algorithm did not defines an initial posture clearly, this paper defined the initial posture by the first posture of CPR and then utilized when a CPR posture and a user posture were compared. By adding the initial posture, the comparison of the two postures becomes accurate more. In the experiment, Myos were utilized to measure the properties of CPR postures. Based on the measured properties and the proposed method, two CPR postures were compared. Users could confirm the amount how much own postures are different from CPR postures.

Acknowledgement This research was supported by the Keimyung University Research Grant of 2015.

References

1. Kim, H.M.: Effects of Knowledge and Education Experience to Attitude on Cardiopulmonary Resuscitation among University Students. Graduate School of Public Health, Inje University (2014)
2. Lee, M.H., Choi, S.H., Park, M.J.: Effect of CPR Training for Lay Trainees on Their Knowledge and Attitudes. Journal of Korean Academy of Fundamentals of Nursing **14**(2), 1225–9012 (2007)
3. Navarro, G.: A Guided Tour to Approximate String Matching. ACM Computing Surveys **33**(1), 31–88 (2001)
4. Myo. https://www.thalmic.com/myo/
5. Sung, Y., Cho, K.: An expanded levenshtein distance algorithm for action similarity measurement of virtual characters. In: Proceedings of International Conference on Computer and Applications, p. 149 (2012)
6. Sung, Y.: Macro-action generation method using demonstration-based learning. In: Virtual Environments, Doctor Theory, Dongguk University (2012)

HMM Based Duration Control for Singing TTS

Najeeb Ullah Khan and Jung Chul Lee

Abstract In order to develop a HMM based singing TTS system, we need a huge singing voice database to train HMM model parameters. However there is no singing voice database publically available and the construction of it is much more difficult than that of speech database. In this paper we propose a new method to improve the naturalness of singing TTS system using HMM models from speech database. Duration control model based on the syllabic analysis is applied to adapt speech duration model to singing duration model. The proposed method results in better singing voice quality compared to the maximum likelihood generation of durations using the speech database.

Keywords Duration control · HMM · Singing TTS

1 Introduction

Singing voice synthesis is the generation of a song using a computer given its lyrics and musical notes. In recent years singing voice synthesizers have gained a lot of popularity. The Vocaloid singing synthesizer [1] developed by Yamaha has been used for commercial purposes by professional musicians [2]. To get a new voice in singing synthesizers, a large singing voice database is required. Hence there is a need for more flexible approaches to singing voice synthesis.

Recently hidden Markov model based approaches have been applied to the singing voice synthesis. In [3-5], the authors have used a singing voice database for training context dependent models. In [6], pitch and speaker adaptation techniques have been used to cope with the limited size of singing voice database. However these systems utilizes singing voice database. Though singing voice can be generated using HMM models from speech database, synthetic voice is unnatural due to the characteristic differences between speech and singing voice.

In this paper we present a duration control method for singing TTS given context dependent HMMs trained on speech database to improve the naturalness.

N.U. Khan · J.C. Lee(✉)
School of Electrical Engineering, University of Ulsan, Ulsan, South Korea
e-mail: electronicengr@gmail.com, jungclee@ulsan.ac.kr

© Springer Science+Business Media Singapore 2015   137
D.-S. Park et al. (eds.), *Advances in Computer Science and Ubiquitous Computing*,
Lecture Notes in Electrical Engineering 373,
DOI: 10.1007/978-981-10-0281-6_20

We used a speech database for training context dependent HMMs, and combined these context dependent HMMs with a control model devised through the analysis of a small natural singing voice set to solve the limited training data problem.

The rest of the paper is organized as follows. Section 2 discusses the baseline singing voice synthesis system based on HMM. Duration control is discussed in section 3. Section 4 and 5 describes experimental results and conclusions respectively.

2   HMM Based Singing TTS

Our singing TTS system uses open source HTS synthesis system [7]. In this work, the same setup is used for the training part as that used for the text to speech synthesis. In the training phase a labeled speech database is used to train context dependent hidden Markov models, modeling the spectrum, F0 and duration of each context dependent unit. The output of the score editor for each note is a set of phoneme labels, duration and the target F0 value.

In the synthesis part a text string in lyric is processed by the language processing module to output the sequence of context dependent models. The spectral and excitation parameters are generated from the context dependent HMM parameters using a parameter generation algorithm with duration and the target F0 value information from our score editor. Finally the singing voice is synthesized using a MLSA synthesis filter [8]. The speech database consists of labeled speech data. F0 extraction module extracts the log F0 for each voiced frame while the spectral parameter extraction module determines the Mel frequency cepstral coefficients for each frame.

For each phoneme, the spectrum is modeled by a five state left to right HMM, and pitch pattern by Multi-Space probability distribution HMM [8].

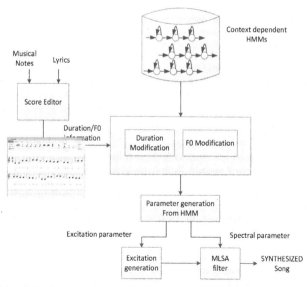

Fig. 1 HMM based Singing TTS

Duration for each state was modeled using single Gaussian distribution. State tying and decision tree clustering was performed using questions about left right contexts of each phoneme for spectrum, excitation and duration models independently.

3 Duration Control

In speech the duration and pitch are determined by the speaker; however, in singing voice the duration, pitch of each phoneme are prescribed in a musical score. This paper is concerned about adaptation rules for modifying duration models trained on speaking voice to generate adequate durations for the singing voice. In subsection 3.1 we formulate the problem of state duration assignment given the HMM models and the note duration using maximum likelihood principle. In subsection 3.2 we describe the modified method for state duration assignment based on the state level analysis of singing voice data while subsection 3.3 describes the duration assignment based on the syllable and phoneme level analysis of the singing voice.

3.1 Maximum Likelihood Based Duration Control Method

This method is a modification of the speech duration control given in [8]. Given a note with duration D and phoneme HMMs corresponding to the note $\lambda_1, \lambda_2, ..., \lambda_N$, the number of frames for each state are determined so as to maximize the log likelihood

$$\log(q_1, q_2, ..., q_D \mid \lambda_1, \lambda_2, ..., \lambda_N, D) = \sum_{i=1}^{N} \sum_{s=1}^{S} \log(p_{is}(d_{is})), \tag{1}$$

with the constraint that the durations should sum to the duration of the musical note

$$D = \sum_{i=1}^{N} \sum_{s=1}^{S} d_{is} \tag{2}$$

where p_{is} is the duration probability distribution for state s of model i and d_{is} is the duration of state s of model i.

For the case of Gaussian duration models,

$$p_{is}(d_{is}) = \frac{1}{\sqrt{2\pi\sigma_{is}^2}} e^{-\frac{(d_{is}-\mu_{is})^2}{2\sigma_{is}^2}} \tag{3}$$

and the above constrained optimization problem can be solved using the Lagrange multiplier as follows

$$L(d_{1.1},...,d_{N.S},\lambda) = \sum_{i=1}^{N}\sum_{s=1}^{S}\log p_{is}(d_{is}) + \lambda\left(\sum_{i=1}^{N}\sum_{s=1}^{S}d_{is} - D\right) \tag{4}$$

where

$$\log p_{is}(d_{is}) = -\frac{1}{2}\log(2\pi) - \frac{1}{2}\log(\sigma_{is}^2) - \frac{1}{2\sigma_{is}^2}(d_{is} - \mu_{is})^2 \tag{5}$$

Setting derivatives of the Lagrange L to zero with respect to d_{is} and λ we get the following set of equations;

$$L_{d_{is}} = -\frac{1}{\sigma_{is}^2}(d_{is} - \mu_{is}) + \lambda = 0 \tag{6}$$

$$L_{\lambda} = \sum_{i=1}^{N}\sum_{s=1}^{S}d_{is} - D = 0. \tag{7}$$

Solving (19) and (20) the duration for each state is given by

$$d_{is} = u_{is} + \rho\sigma_{is}^2 \tag{8}$$

$$\rho = \frac{\left(D - \sum_{i=1}^{N}\sum_{s=1}^{S}\mu_{is}\right)}{\sum_{i=1}^{N}\sum_{s=1}^{S}\sigma_{is}^2}. \tag{9}$$

3.2   State Level Analysis Based Duration Control

To analyze the difference between the durations of segments in speaking and singing voice, two different sets of monophone HMM duration models were trained on a speech database and a small singing voice database. Based on the statistics of HMM parameters, the monophones were categorized into six categories; vowels, semivowels, voiced fricatives, unvoiced fricatives, stops and nasals. The affricate DH was categorized as voiced fricative while CH was categorized as unvoiced fricative. The aspiration HH, silence SP and vowels were grouped into one category. Each state's mean for the trained models was normalized with the total duration of the model. An average value of the duration for each category was calculated using the normalized means. The normalized mean for state j of model p_i is given by $M_{speech}(p_i, j)$ and $M_{singing}(p_i, j)$ for speech and singing voice respectively. The scaling coefficients were computed as follows

$$Coeff(c, j) = \frac{\sum\limits_{p_i \in c} M_{\text{singing}}(p_i, j)}{\sum\limits_{p_i \in c} M_{\text{speech}}(p_i, j)} \tag{10}$$

where c is the phoneme category and j is the state number.

3.3   Syllable Analysis Based Duration Control Method

The basic idea is that each syllable type has different duration distributions in speech and singing voice. In the analysis phase, singing voice was aligned to phonetic transcription and syllable information was added to the label files. Initially syllable types such as CV, VC, CVC, CCVC were considered in the analysis. Further sub-groups within each syllable type were defined with manner of articulation. All the instances belonging to the same syllable type or a subgroup within a syllable type were pooled together and ratio of each phoneme in a syllable was calculated.

During synthesis for each musical note, the duration of each phoneme in a syllable was determined by the corresponding ratio. Given the phoneme durations, maximum likelihood approach was used to get the state durations for each phoneme.

4   Experimental Results and Conclusion

To test the proposed duration control methods described in the previous section a singing database was constructed. Lyrics for a collection of English children songs were collected to be used as reference for the singer and as word level transcription for automatic phonetic alignment. A native amateur male singer sung the songs which were recorded at a sampling rate of 48 KHz. The recorded database of songs comprised of 14 minutes of duration. For the speech database, a male speaker database rms in the CMU arctic speech database was used [9].

A test song, "twinkle twinkle little star" was used for the evaluation of the duration assignment algorithms described in section 3. Since we had only phoneme level alignment available for the natural singing voice, the state durations generated for each phoneme instance were summed up to get the phoneme durations for comparison with the natural singing voice.

Figure 2 shows the histograms of the difference between the natural and assigned phoneme durations using different algorithms in number of frames. As can be seen in figure 2(a) and figure 2 (b), the maximum likelihood and state level analysis methods have many phonemes with duration difference greater than 30 frames. This could mainly be due to the fact that these methods do not use any broad contextual information. In contrast, the syllabic analysis based duration assignment shown in figure 2(c) shows much improved performance with only a few phonemes with error greater than 30 frames. Further classification of the syllabic types into subgroups results in even better results with no phoneme duration difference of more than 20 frames as shown in figure 2(d).

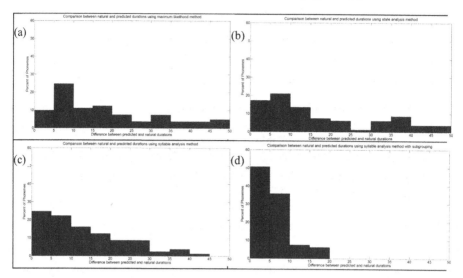

Fig. 2 Histograms of phoneme duration difference between natural singing and proposed methods (a) maximum likelihood method, (b) state level analysis method, (c) syllable analysis, (d) syllable analysis with subgrouping.

5   Conclusion

In this work different strategies for duration control for singing TTS system based on a speech database have been investigated. Among the methods described, the syllable analysis for natural singing voice has improved the assignment of durations to each phoneme. Further classifications of the syllable types have reduced the assignment error of less than 20 frames for most of the phonemes. The singing voice generated using the syllable analysis has relatively better perceptual performance with good intelligibility and naturalness.

Acknowledgments This work was supported by the research fund of the University of Ulsan, Ulsan, South Korea.

References

1. Kenmochi, H., Ohshita, H.: VOCALOID-commercial singing synthesizer based on sample concatenation. In: INTERSPEECH, Antwerp, Belgium, pp. 4009–4010 (2007)
2. Kenmochi, H.: Singing synthesis as a new musical instrument. In: ICASSP, Kyoto, Japan, pp. 5385–5388 (2012)

3. Saino, K., Zen, H., Nankaku, Y., Lee, A., Tokuda, K.: An HMM-based singing voice synthesis system. In: INTERSPEECH, Pittsburgh, Pennsylvania, pp. 2274–2277 (2006)
4. Oura, K., Mase, A., Yamada, T., Muto, S., Nankaku, Y., Tokuda, K.: Recent development of the HMM-based singing voice synthesis system-Sinsy. In: SSW, pp. 211–216 (2010)
5. Nakamura, K., Oura, K., Nankaku, Y., Tokuda, K.: HMM-Based singing voice synthesis and its application to Japanese and English. In: ICASSP, Florence, Italy, pp. 265-269 (2014)
6. Shirota, K., Nakamura, K., Hashimoto, K., Oura, K., Nankaku, Y., Tokuda, K.: Integration of speaker and pitch adaptive training for HMM-based singing voice synthesis. In: ICASSP, Florence, Italy, pp. 2559–2563 (2014)
7. Zen, H., Nose, T., Yamagishi, J., Sako, S., Masuko, T., Black, A.W., et al.: The HMM-based speech synthesis system (HTS) version 2.0. In: SSW, pp. 294–299 (2007)
8. Tokuda, K., Nankaku, Y., Toda, T., Zen, H., Yamagishi, J., Oura, K.: Speech synthesis based on hidden markov models. Proceedings of the IEEE **101**, 1234–1252 (2013)
9. CMU_ARCTIC speech synthesis databases. http://festvox.org/cmu_arctic/

Data Analysis of Automated Monitoring System Based on Target Features

M.S. Islam, J.C. Lee, J.P. Shin and U.P. Chong

Abstract Automated monitoring system provides a potential technique for monitoring the indoor environment. Traditional automated monitoring systems are very expensive. This paper presents a cheap automated monitoring system which is based on the motion sensors and ultrasonic sensor. This automated monitoring system allows the users to access information directly on the web site through browser or Wireless Mobile Terminals at any time. Cross correlation is used to calculate the distance of targets using ultrasonic signal. We demonstrate the effectiveness of our system by analyzing the echo signals. We conducted a series of real-world experiments with different targets and target positions. Additionally, we did experiment and analyzed the ultrasonic signal between transmitter and receiver directly without echo signals. The experimental result shows the reliability of the system.

Keywords Cross correlation · Automated monitoring · Ultrasonic sensor · Mobile monitoring · Targets

1 Introduction

The rapid development of communication and network technology has speeded up the digital control system to replace the pace of analog control system. The increasing requirements of the markets, which include banks, companies and other important institutes [1] are established on three main concepts; low power, less cost and high efficiency. An increasing amount of attention has been paid to

M.S. Islam · J.C. Lee · U.P. Chong(✉)
School of Electrical Engineering, University of Ulsan, Ulsan, South Korea
e-mail: Saiful05eee@yahoo.com, {jungclee,upchong}@ulsan.ac

J.P. Shin
School of Computer Science and Engineering, University of AIZU, Aizuwakamatsu, Japan
e-mail: jpshin@u-aizu.ac

© Springer Science+Business Media Singapore 2015
D.-S. Park et al. (eds.), *Advances in Computer Science and Ubiquitous Computing*,
Lecture Notes in Electrical Engineering 373,
DOI: 10.1007/978-981-10-0281-6_21

improve reliability, availability, and safety in most autonomous control systems such as remote control station.

Since last few years, automated security systems are getting more and more awareness and importance. On the market, currently it already has Intelligent Monitoring System, Video Monitoring System and Sensor Monitoring System, etc. The present security systems are suffering with various issues like security, data analysis.

Researchers, investigators, or maintenance personnel use a variety of methods for monitoring. The authors in [2] introduce a potential technique for monitoring the indoor environment. A web-based building environmental monitoring system is proposed in [3], based on wireless communication scheme. In terms of performance metrics in [4], some applications have been discussed in the building monitoring system. Moreover, a wireless network based monitoring system was deployed in a number of residential and commercial buildings in [5, 6]. For these approaches a human operator is required to monitor the camera images at all times when security is to be maintained, visually scanning the images to determine whether or not suspicious activity is taking place. This is a monotonous task which may lead to operator fatigue and inevitably results in errors [7]. Currently, the data link between cameras onboard an unmanned aerial vehicle and the monitoring terminal [8] and new technological solution for the ocean observation using an unmanned surface vehicle [9] have been researched. Kinematic and dynamic analyses have been done and the mathematical model is developed for the marine platform of the surface vehicle [9].

The objective of this paper is to analyze the echo signals of automated monitoring system achieving the requirement of a real time security system, which is expected to calculate the distance of object at the same time using ultrasonic sensor. We explained the details of this system in our previous paper [11]. The rest of the paper is organized as follows: Section 2 and 3 describes the design of automated monitoring system including motion detection, and ultrasonic signal processing. In Section 4, we describe the cross correlation technique related to target detection. Experiments and performance assessment are given in Section 5. Finally, the conclusions are presented in Section 6.

2   System Scheme Design

Flow chart of overall automated monitoring system is presented in figure 1. This automated monitoring system which can detect the targets in the concerning region using the motion detector and then send ultrasonic signal to the targets. There are four detection regions (I, II, III, IV) which detect the objects from two motion sensors in the figure 1. When the control unit receives detection signal from motion detectors it sends signal to servo motor. Then server motor moves toward the detected region and ultrasonic sensor sends/receives signals from/to target at same time. Then control unit sends the informations to the server computer through wireless communication. The server computer calculates the distance between sensors and target using the ultrasonic signals, and sends information to mobile devices.

Flow chart of overall unmanned monitoring system has been presented in figure 2. This unmanned monitoring system which can detect the targets in the region using the motion detector and then send ultrasonic signal to the targets. Moreover, this proposed approach calculates the distance of targets using ultrasonic sensor with cross correlation. It will be discussed later.

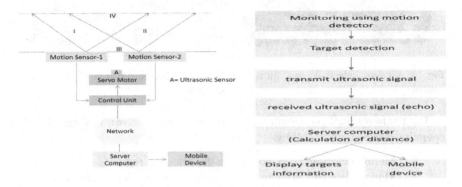

Fig. 1 Overall block diagram for automated monitoring system

Fig. 2 Flowchart of automated monitoring system

3   Motion Detection System

The main disadvantages of existing monitoring systems are to continuously capture or record the video or data. So the system consumes huge power and need large memory to store the database of the captured image or video or data. To overcome these disadvantages we proposed the new system working only motion detection occurred.

Motion detection system consists of motion detector, control board, servo motor, and wireless communication system as shown figure 3. There are two motion detector in our system. The motion sensor-1 keeps monitoring left detection region(region I) and motion sensor- 2 keeps monitoring right detection region(region II) as shown figure 1. The mid region(region IV) is detected   when both motion detectors   are activated. Detection region III is not necessarily for detection because region IV already covers it. When the motion detectors detect the target in any of four region, then control unit sends signal to servo motor to move toward that detected region. Wireless mobile system transfers it to server computer system.

Fig. 3 Motion detection and ultrasonic   **Fig. 4** Experimental setup of ultrasonic
system   sensors

The hardware implementation for motion detection is shown in figure 4. The operating procedure is as follows.

1. Motion detectors keep monitoring the detection region.
2. Motion detectors recognize the region when motion occurred and send detection signal to control board.
3. Control board activates the servo motor.
4. The servo motor with ultrasonic transceiver turn towards the target.
5. The transceiver receive the echo signal from target and sends information to sever computer through wireless communication.

4 Distance Calculation Using Ultrasonic Signal

Ultrasonic distance detection is based on the time of flight of an ultrasonic pulse. An ultrasonic pulse will be partially reflected from the object. The time difference between transmitting a pulse and receiving the reflected signal is a measure for the distance between the transmitter and the object. We use the cross correlation to find the time difference. Correlation is the process to determine degree of 'fit' between two waveforms and to determine the time at which the maximum correlation coefficient or "best fit" occurs. For this system, we correlate between the transmitted signal and the received signal, then we get the time difference between transmitted and received signal. The figure 4 shows experimental setup of ultrasonic sensors.

In this section we will drive the equation of the cross-correlation for distance calculation. If we consider x(n) as the transmitted signal from the ultrasonic sensor, then the returned signal, r(n) may be modeled as:

$$r(n) = \alpha x(n - D) + w(n) \qquad (1)$$

Where $w(n)$ is assumed to be the additive noise during the transmission, α is the attenuation factor, D is the delay which is the time taken for the signal to travel from the transmitter to the target and back to the receiver.

Now the auto-correlation of the transmitted signal $x(n)$ with itself (constant shift l) be [10]

$$C_{xx}(l) = \sum_{n=-\infty}^{n=+\infty} x(n)x(n-l)$$ (2)

The maximum value of the auto-correlation will be at the delay time l. The cross-correlation between the transmitted signal, x (n) and the received signal, r (n) be [10]

$$C_{xr}(l) = \sum_{n=-\infty}^{n=+\infty} r(n)x(n+l)$$ (3)

Now using Equation (2) and (3)

$$C_{xr}(l) = \alpha \ C_{xx}(l\text{-}D) + C_{wx}(l)$$ (4)

Since the noise signal w (n) and the transmitted signal, x(n) are uncorrelated then, $C_{wx}(l)$=0. Therefore equation (4) will be

$$C_{xr}(l) = \alpha C_{xx}(l\text{-}D)$$ (5)

Comparing the equation (5) with equation (2), the maximum value of the cross-correlation will occur at l=D, which is our interest in cross-correlation from which we can detect our target.

5   Experiments and Performance Assessment

In this real system experiment, we use two different type of targets: human target and human target with metal reflector shown in figure 5 and figure 6 respectively. There are 5 target positions are used for each type of reflector from 1 meter to 5 meter shown in figure 11. The center frequency of the ultrasonic sensors is 50 kHz, then ultrasonic detection signal acquisition uses 250 kHz sampling frequency. At the first in our experimental results, we present performances of our automated monitoring system with respect to human target and human target with metal reflector. Then, we will compare received signal strength between these two reflectors. To aim this, we use transceiver (one device used as transmitter and receiver). Average value is computed over a short period (180 ms) of time which is used in deciding echo signal strength. Finally, transmitter and receiver are used to show the characteristics of receiver without targets with 5 different positions.

Figure 9 shows the comparison results of received signal strength for human target and human target with metal reflector. It can be noticed that the signal strength of human target with metal reflector is higher than human target. This implies that our developed system is higher sensitive to metal reflector.

Figure 7 shows strength of received signal with human reflector. Here, received signal strength is decreasing according to distances. As can be seen, maximum signal strength is at target position 1meter and minimum at target position 5 meter. Strength of received signal for human target with metal reflector is shown in figure 8. Signal strength decaying rate is almost linear with respect to target positions except from 1 meter to 2 meter.

Fig. 5 Indoor experiment of human target   **Fig. 6** Human target with metal reflector

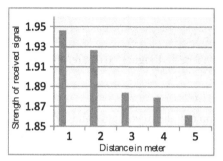

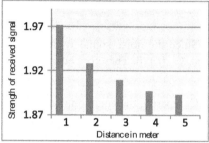

Fig. 7 Ultrasonic received (echo) signal **Fig. 8** Ultrasonic received (echo) signal
with human target   with human target and metal reflector

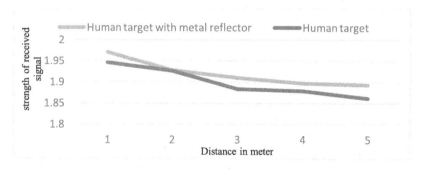

Fig. 9 Comparison between human target and human target with metal reflector

To evaluate our system effectively, we analyze received signal without using targets. In this case transmitter sends the ultrasonic signals and receiver takes the transmitted signal from transmitter directly. The receiver was placed in front of the transmitter shown in figure 11. This signals strength are shown in figure 10. From the compare between figure 9 and figure 10, both cases signal strength is decreased as the target position is increasing except figure 10 where signal strength at 1 meter and 2 meter are same, it is due to the saturation level.

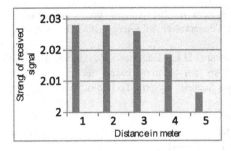

Fig. 10 Ultrasonic received signal without target

Fig. 11 Transmitter and receiver

6   Conclusions

Traditional automated monitoring systems are very expensive. So, in this paper we presents a cheap automated monitoring system that can detect the targets in the concerning region using the motion sensors, that is considerably inexpensive compared with the conventional 3D motion analysis system based on high-speed camera. This automated monitoring system allows the users to access information directly on the web site through browser or Wireless Mobile Terminals at any time. A theoretical analysis is presented on cross correlation related to target distance calculations. An important contribution of this paper is to analyze the echo signals of this automated monitoring system. A series of real-world experiments are conducted with different targets and target positions. It can be noticed from the experimental results that the signal strength of human target with metal reflector is higher than human target. We also analyzed received signal without using targets. The accuracy of this method is demonstrated by comparing target echoes with the received signal (without target). The experimental result shows the reliability of the system.

References

1. Boonsawat, V., Ekchamanonta, J., Bumrungkhet, K., Kittipiyakul, S.: Xbee wireless sensor networks for temperature monitoring. In: ECTI-CARD 2010, pp. 10–14 (2010)
2. Dong, Q.: Design of Building Monitoring Systems Based on Wireless Sensor Networks. Journal of Wireless Sensor Network **2**(9), 703–709 (2010)
3. Jang, W.-S., Healy, W.M., Skibniewski, M.J.: Wireless sensor networks as a part of a web-based building environmental monitoring system. Automation in Construction **17**(6), 729–736 (2008)
4. Jang, W.S., Healy, W.M.: Wireless Sensor Network Performance Metrics for Building Applications. Energy and Buildings **42**(6), 862–868 (2010)
5. Tessa, D., Elena, G., James, B.: Wireless sensor networks to enable the passive house-deployment experiences. In: Smart Sensing and Context, pp. 177–192. Springer, Berlin Heidelberg (2009)

6. Verma, A.: A multi layer bank security system. In: 2013 International Conference on Green Computing, Communication and Conservation of Energy (ICGCE), Chennai, pp. 914–917 (2013)
7. Freer, J.A., Beggs, B.J., Femandez-Canque, H.L., Chevriert, F., Goryashkot, A.: Moving object surveillance and analysis for camera based security systems. In: International Carnahan Conference on Security System Technology, Sanderstead, pp. 67–71, October 18–20, 1995
8. Priyanga, M., Rajaramanan, V.: Unmanned Aerial Vehicle for Video Surveillance Using Raspberry Pi. International Journal of Innovative Research in Science, Engineering and Technology 3(3), 1715–1720 (2014)
9. Gowthaman, D.R., Bhalamurugan, R., Balaji, T., Manoj kumar, V.: Design and modeling of unmanned surface vehicle. In: The Proceeding of 2014 International Conference on Computation of Power, Energy, Information and Communication (ICCPEIC), pp. 283–289 (2104)
10. Islam, M.S., Chong, U.: Detection of Uncooperative Targets Using Cross-Correlation in Oceanic Environment. International Journal of Digital Content Technology and its Applications (JDCTA) 7(12), August 2013
11. Islam, M.S., Lee, J.C., Chong, U.P.: Unmanned monitoring system using motion detection and ultrasonic signals. In: Proc. of the Second Intl. Conf. on Advances in Information Processing and Communication Technology - IPCT 2015, pp. 84–88 (2015). doi:10.15224/978-1-63248-044-6-71

A Persistent Web Data Architecture with Named Data Networking

Euihyun Jung

Abstract Since the Web decay has been considered as an evil to degrade the quality and the persistency of Web contents, a lot of studies have tried to resolve it. However, these studies have not given satisfactory solutions yet due to the inherent feature of HTML. To resolve the Web decay issue, a persistent Web data architecture based on Named Data Networking (NDN) is suggested in this paper. The architecture enables users to manage their data on NDN and achieve the persistency of the data with simple HTML tags. With the architecture, even when Web contents are copied to multiple sites, the persistency of the contents is guaranteed without any human intervention.

Keywords Named Data Networking (NDN) · Web decay · Persistency · Link rot

1 Introduction

Although the Web is a primary information source of human society, the decay of Web contents has been an unresolved headache since it was invented [1][2]. Literally, Web contents start to decay just after they are created, because information units comprising Web contents become easily outdated and inconsistent as time goes by.

From this kind of the Web decay, link rot is a common and well-known phenomenon [3][4]. Since people are used to meet tons of irritating 404 pages during Web surfing, a lot of studies have pointed out this phenomenon and suggested various solutions [4][5]. Needless to say, the link rot promptly affects user experiences and it is a typical proof of the Web decay, but other information units of Web contents such as images, or texts are also decayed. Even worse, this kind of the decay is invisible to users and is hard to be detected. For this reason,

E. Jung(✉)
Department of Computer Science, Anyang University, Jungang-ro Buleun-myeon, Ganghwa-gun, Incheon 602-14, Korea
e-mail: jung@anyang.ac.kr

© Springer Science+Business Media Singapore 2015
D.-S. Park et al. (eds.), *Advances in Computer Science and Ubiquitous Computing*,
Lecture Notes in Electrical Engineering 373,
DOI: 10.1007/978-981-10-0281-6_22

153

people have suffered from the wrong addresses, missing phone numbers, or other outdated Web contents without noticing them.

The main reason of the Web decay is caused from that HTML has been designed to render visual components. Once after a user wrote his/her data into a printed HTML page, the data has been converted to mere text in the page. Therefore, if some changes occur in the data, the changes will not be reflected promptly and the user has to manually find and modify them in the page. Since this process is cumbersome and irritating, most Web contents will go to decay eventually.

To remedy this, the persistency-needed data should be separately managed and be dynamically rendered into a Web page. This idea has been already used in the server-side approach where data are stored in a database and are rendered into HTML on demand. In the approach, if a change is needed, a modification of the data in the database is enough to keep away the Web decay. However, the server-side approach has also some weaknesses, though it looks like a promising solution for the Web decay. First, the approach has poor usability in maintaining Web contents. Whenever people want to add or modify their contents, they have to ask everything to server-side programmers or their administrators. It is neither practical nor sustainable. Second, even if Web contents are maintained well in an original site, the copied contents on the other sites cannot avoid the decay because they are mere text and they will not be reflected from any changes of the original one. Eventually, the copied contents ruin the persistency of the original one around the global Web.

From the case of the server-side approach, two technical issues have been drawn to efficiently resolve the Web decay. First, the persistency-needed data should be maintained on a dedicated data space and the data ought to be easily accessed and indicated over a boundary of the data space. Second, some user-friendly mechanisms are needed for people to easily link the data on the data space to HTML without the help from server-side staffs.

To satisfy these issues, a persistent Web data architecture with Named Data Networking (NDN)[6] is suggested in this paper. The proposed architecture enables users to easily manage their persistency-needed data with NDN and to simply link the data to their Web contents. It can resolve various Web decay including the link rot and it can also support the persistency of the dispersed copies of Web contents.

The rest of this paper is organized as follow. Section 2 describes design considerations, the proposed architecture and the comparison with other existing methods. Section 3 concludes the paper.

2   System Design

2.1   Design Considerations

A Dedicated Data Space. The architecture needs a separated and dedicated data space that keeps the persistency-needed data from mingling with plain Web visual contents. However at the same time, it should support the data to be easily accessed from the outside of the data space. For this purpose, Named Data Networking (NDN) is adopted in the research. NDN is one of network structures based on Information-Centric Networking (ICN) [7] and it is a totally different networking technology from the conventional ones. To transfer data, the conventional packet networking including IP delivers a packet according to communication endpoints contained in the packet. Comparing to this, in NDN, participants fetch data only by a given name without considering a location of the data. For example, if people want to listen to the "sugar.mp3" of Maroon 5, they have to know the location of the song on IP network before accessing the song. Instead, in NDN, they just give a name of the song such as "/music/maroon_5/sugar.mp3" to NDN and then NDN fetches the song from somewhere on NDN without bothering users. Due to its independent communication and distributed manner of data management, NDN can be a solid base for a dedicated data space holding persistency-needed data. Also, the names play an important role to enable users to indicate the data in any Web pages without considering originated servers or data's locations.

Link Between NDN and the Web. From the user's perspective, easy indication of the data on NDN in the HTML is essential to achieve great usability. Moreover, the indication should allow users not to worry about the underlying resolution process for the data. For this purpose, a simple HTML tag should be designed to enables users to easily indicate their persistency-needed data on NDN. Since NDN does not have any mechanism to link data on NDN to HTML pages, a resolver JavaScript is made to render the persistency-needed data into Web pages that contain the indication. Integration of NDN and HTML enables users to easily maintain the persistency of their data with simple HTML.

2.2   Architecture

Conceptual Structure. As shown in Fig. 1, the proposed architecture has three functional parts; a dedicated data space on NDN, NDN-aware HTML tags, and resolver JavaScripts. With the architecture, users can easily manage their persistency-needed data and get persistent HTML pages only by writing simple tags for the data. When an HTML page containing the tags for the data is viewed on somebody's browser, the tagged data are dynamically rendered via the resolver JavaScripts. If some data are changed, the changes are promptly reflected to the corresponding HTML pages when the HMTL pages are loaded. The users can even allow other users to copy their pages without worrying about ruining the persistency of their data.

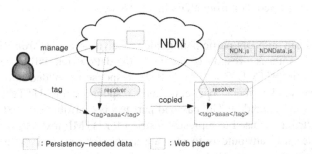

Fig. 1 The proposed architecture consists of three parts; a dedicated data space on NDN, resolver JavaScripts and NDN-aware HTML tags.

The Dedicated Data Space. In the dedicated data space, each data needs to have its own unique name. A hierarchical name space is designed starting from "/organization/ndndata/". Using this namespace, users can allocate a unique name below each organization for their data. For example, a user named "Jeff" of UCLA can allocate his name and email such like "/ucla/ndndata/jeff/name" and "/ucla/ndndata/jeff/email" respectively. Although the allocation is performed locally and independently, the resulted namespace will become a global unique name space.

The NDN-Aware HTML Tags and Resolvers. In the architecture, the most important design consideration in NDN-aware HTML tags is how to minimize introducing new tags to achieve maximum interoperability with current HTML. However, it is difficult to indicate various data such as images, links, and texts consistently without new tags. To resolve the contradiction, HTML attributes are used instead of introducing new tags because HTML tags are allowed to extend their functionality with user-defined custom attributes. By using this approach, the architecture is able to achieve the interoperability.

In the architecture, two attributes are used. First is the existing HTML "class" attribute. When the attribute has "NDNDATA" value, the tag containing the attribute is considered as an NDN-aware tag. The other one is a new attribute "ndn_id" which describes an NDN name. Below sample HTML shows how these attributes are used.

```
<h2 class="NDNDATA" ndn_id="/ucla/ndndata/jeff/name"/>
<img src="http://www.example.org/sample.jpg" class="NDNDATA"
  ndn_id="/ucla/ndndata/jeff/profile_picture"></img>
<a href="http://www.example.org/paper" class="NDNDATA"
  ndn_id="/ucla/ndndata/paper/ndn_overview">NDN Overview</a>
```

In order to resolve the indicated tags into real NDN data, two resolver JavaScripts are used in this architecture. First is NDN.js [8] which is a JavaScript implementation of a client library for NDN and takes charge of resolving an NDN name into the corresponding persistent data. Second is NDNData.js which is newly made in this research. It finds the indicated tags in an HTML page and renders the transferred data from NDN into HTML.

Response. When an NDN participating node receives a request from NDN.js with a data name, it returns a response containing the data which will be rendered into the corresponding HTML. The format of the response is divided into "meta" and "payload" parts. The "meta" part contains a response status. HTTP status codes and messages are extended. The payload part has two attributes; "text" and "url". The "text" attribute is used to represent a string for HTML text tags such as <H1> or <LI>. The "url" attribute is to indicate a link for <a> and <img> tags. The below response shows a sample response.

```
{
  meta: {
    status: 200,
    msg: "SERVER_OK",
    version: 1.0,
  }, payload {
        text : "a persistent string that will be rendered into a Web page",
        url : "a persistent link for an image or a link",
}
```

Operation. Once after a user indicates an NDN-aware tag in a Web page with the proposed attributes, the tag will be fetched from the data space on NDN and then be rendered into the corresponding HTML. When the Web page containing the tags is browsed, the rendering process is started. First, NDNData.js finds an NDN-aware tag in the page and requests NDN.js to resolve an "ndn_id" in the found tag into the persistency data. The resolved data from NDN is returned to NDNData.js as a JSON formatted response and then NDNData.js renders the response into the corresponding HTML. In Fig. 2, "/anyang/ndndata/jung/hero" is rendered into a "lincoln.jpg" initially. However, if a data owner changes the data from "lincoln.jpg" to "kennedy.jpg" on NDN later, the corresponding image on the Web page will be changed instead of breaking the image of the HTML page.

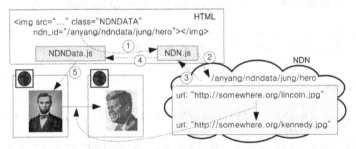

Fig. 2 The operation shows how the changed image on NDN is reflected on the corresponding Web page.

Table 1 The comparison of the proposed architecture with other methods.

Method	Defense against the Web decay	Usability	Multiple sites	Additional components
Static HTML	No	Best	No	No
Server-side approach	Yes	Bad	No	Needed
The proposed one	Yes	Good	Yes	Needed

Comparison. To verify an effectiveness of the proposed structure, the proposed architecture was compared with current Web methods as shown in Table 1. Static HTML is a basic way to make a Web page. It is the best at usability and does not need any scheme except HTML, but it reveals a severe weakness of the Web decay. Although the server-side approach can be a partial solution to the rot of a local site, it shows poor usability because it forces users to get help from server-side staffs to maintain the persistency. Also, it cannot handle the decay of the same Web contents dispersed on multiple sites. The proposed architecture is pretty good at usability and solves the various kinds of the Web decay not only on a local site but also on multiple sites. Although it needs additional scheme such as NDN and the resolver JavaScripts, it is worth when a strong level of persistency is required.

3   Conclusion

The decay of Web contents is a historical headache and has not been resolved yet. The main reason of the decay is caused from that most Web contents mingle their data with visual component in printed HTML. Therefore, when Web contents need to be modified, users will suffer to find the location of the target Web contents and have to correct the changes manually. Even if the contents are copied to other sites, there is hardly to handle it. The server-side approach can partly resolve this problem, but it has poor usability and it also cannot handle the case of the copied contents.

In order to resolve the issues, the NDN-based architecture is suggested. The architecture consists of a dedicated data space on NDN and NDN-aware HTML tags. With the architecture, users can indicate their data with only NDN names in HTML without any complex procedures. The architecture resolves the various kinds of the Web decay that the existing studies have not resolved yet. It also supports the persistency of the Web contents dispersed on multiple sites.

Acknowledgments I would like to show my gratitude to Jeff Burke at UCLA REMAP for his professional comments about NDN that greatly improved the manuscript.

References

1. Spinellis, D.: The Decay and Failures of Web References. Communications of the ACM **46**(1), 71–77 (2003)
2. Nelson, M.L., Allen, B.D.: Object Persistence and Availability in Digital Libraries. D-Lib Magazine **8**(1) (2002)
3. Hughes, B.: Link? rot: URI citation durability in 10 years of ausweb proceedings. In: Proceedings of the 12th Australasian Web Conference (AusWeb 2006), Noosa, Australia (2006)
4. Zittrain, J., Albert, K., Lessig, L.: Perma: Scoping and Addressing the Problem of Link and Reference Rot in Legal Citations. Legal Information Management **14**(2), 88–99 (2014)
5. Goh, D.H., Ng, P.K.: Link Decay in Leading Information Science Journals. Journal of the American Society for Information Science and Technology **58**(1), 15–24 (2006). Wiley
6. Zhang, L., Afanasyev, A., Burke, J., Jacobson, V., Claffy, K., Crowley, P., Papadopoulos, C., Wang, L., Zhang, B.: Named Data Networking. ACM SIGCOMM Computer Communication Review **44**(3), 66–73 (2014). Wiley
7. Bari, M.F., Chowdhury, S.R., Ahmed, R.: A Survey of Naming and Routing in Information-Centric Networks. IEEE Communication Magazine **50**(12), 44–53 (2012)
8. Shang, W., Thomson, J., Cherkaoui, M., Burke, J., Zhang, L.: NDN.JS: a JavaScript client library for named data networking. In: IEEE Conference on Computer Communications Workshops, pp. 309–40 (2013)

Load Shedding for Window Queries Over Continuous Data Streams

Kwang Rak Kim and Hyeon Gyu Kim

Abstract To cope with bursty data arrivals, a stream query processor may perform load shedding to cut the system load by discarding some portion of tuples kept in memory. Load shedding can be conducted in a stateful operator such as join or aggregation in a query plan tree or in a dedicated operator, called *load shedder*, which is typically placed at the entry of the query processor. In this paper, we show that the load shedding can also be performed in window operators. In general, window operators are placed in the initial phase of a query plan and designed to ignore tuples whose arrivals are not in a predefined order. With the functionality to control the number of tuple discards, they can play the role of load shedding. In the proposed method, the conventional load shedder is not necessary and the number of query processing steps can be reduced, which leads to performance improvement of the continuous query processing.

Keywords Data streams · Load shedding · Sliding windows · Disorder control

1   Introduction

There has been substantial interest in applications that monitor the continuous streams of data items such as stock exchanges, network measurements, web page visits, sensor readings and so on [1]. In general, these applications are required to produce their results in a timely manner over rapidly incoming tuples (i.e. records) whose arrival rate may exceed the output rate of queries in a system. To meet the time constraint, many stream query processors such as Aurora [2] and STREAM [3] employ a *load shedder* which discards some portion of input tuples when a system cannot cope with the arrival rate of input tuples.

K.R. Kim · H.G. Kim(✉)
Department of Computer Engineering, Sahmyook University,
Seoul 139-742, Republic of Korea
e-mail: hgkim@syu.ac.kr

© Springer Science+Business Media Singapore 2015   159
D.-S. Park et al. (eds.), *Advances in Computer Science and Ubiquitous Computing*,
Lecture Notes in Electrical Engineering 373,
DOI: 10.1007/978-981-10-0281-6_23

So far, many load shedding techniques have been proposed, which can be classified into the following two groups.

- Operator-level load shedding

  In this scheme, load shedding is performed in each stateful operator such as join or aggregation. Regarding this, Viglas et al. [4] suggested a method to maximize the number of output tuples in a sliding window join. Babcock et al. [5] proposed an algorithm to minimize the degree of inaccuracy in window aggregate queries.

- System-level load shedding

  In this scheme, load shedding is performed in a dedicated operator placed at the entry of a stream query processor. Tatbul et al. [6] presented *semantic load shedding* where tuples are discarded based on values designated by a user-defined QoS (Quality of Service) specification.

In this paper, we show that the load shedding can also be performed in window operators. Our motivation is that input tuples can be discarded from the window operators. More specifically, window operators are designed not to accept tuples whose arrivals are not in a predefined (generally, timestamp) order. Such tuple drops are necessary to identify window boundaries clearly from the out-of-ordered tuple arrivals [7]. From this, by adding the functionality to control the number of tuple discards, window operators can play the role of load shedding.

To support load shedding in window operators, their structure needs to be modified. The operators should be organized to receive various QoS parameters. For example, a stream query processor may request a window operator to drop n tuples due to the memory overflow. The system may also request it to accept only m percent of input tuples for a predefined period of time. To satisfy a given QoS parameter, an elaborate disorder control mechanism must be provided during the window processing. Regarding this, we will discuss how to control the number of tuple drops from out-of-ordered tuple arrivals.

The proposed method can be viewed as an operator-level load shedding mechanism because load shedding is performed in each window operator. On the other hand, it can also be viewed as a system-level load shedding. In general, window operators are placed in the initial phase of a query plan tree. This is similar to the load shedder which is located at the entry of a query processor. Thus, if window operators perform the role of load shedding together, the dedicated load shedder is not necessary and the number of query processing steps is reduced. From this, the proposed method can improve performance of continuous query processing.

2 Preliminaries

2.1 Time-Based Windows

Data streams are inherently infinite and unbounded. If queries on the streams involve joins or aggregates, they cannot be easily answered because the operators cannot start processing until entire input data is ready. Such operators are called *blocking operators* in literature. A common solution to this issue is to restrict the range of stream queries. Sliding windows are used for this purpose, which contain the most recent data of the stream.

There are various types of sliding windows, including *time-based, tuple-based*, and *value-based* windows. Among them, time-based windows are most commonly used in stream applications [8], where the window boundaries are determined by timestamps of input tuples. For example, consider a query to ask for an average value of sensor readings over the latest 30 seconds. This query can be specified as the following SQL-like query, where the parameter RANGE denotes the window size.

Q1. SELECT AVG(value)
     FROM Sensors [RANGE 30 secs]

The above query is manipulated whenever a new tuple arrives from the input stream *Sensors*. When a new tuple with timestamp t arrives, a window interval is determined by $(t - 30, t]$. A window operator organizes a collection of tuples belonging to the window. Then, an average value of the tuples in the window will be returned after computation of the aggregation AVG. To simplify discussion, we assume in this paper that the time-based windows are used for stream query processing.

2.2   Disorder Control

One of the well-known issues when using the time-based windows is that input tuples may not arrive in their (generation) timestamp order. There are various causes of disorder such as network transmission delay, merging unsynchronized streams, data prioritization and so on. The out-of-ordered input tuples are typically discarded by a stream query processor in order to facilitate identification of window boundaries and contents. Such tuple drops may lead to inaccurate query results when aggregations are involved in a query.

To save out-of-ordered tuples and improve accuracy, input tuples can be buffered until they can go out without violating the timestamp order. In this case, we need to be able to decide when and which tuples can go out from the buffer. Regarding this, many existing algorithms including [7, 9] utilizes the maximum network delay seen in the streams. For example, given the max delay m at a certain time t, tuples with timestamps smaller than $t - m$ are moved out from the buffer in this scheme. Such a timestamp $t - m$ is called a *heartbeat* [10] or a *punctuation* [11] in the literature. A newly arriving tuple whose timestamp is smaller than the latest heartbeat value is discarded.

In this way, tuple drops can also occur during the processing of sliding windows. This implies that it is possible to discard some portion of input tuples kept in the window buffer for the purpose of load shedding, while windowing tuples together. To enable this, elaborate disorder control is required. Regarding this, our previous work [12] proposed a method to control the number of tuple discards according to a given QoS parameter. Using the method, load shedding can be performed in the window operators, which will be discussed below.

3   Proposed Method

During the continuous query processing, various types of QoS parameters can
be given from a user or a system. To simplify discussion, we only consider the
following two QoS specifications.

- Discard n tuples from memory immediately.

- Accept only m percent of input tuples (for a predefined period of time)

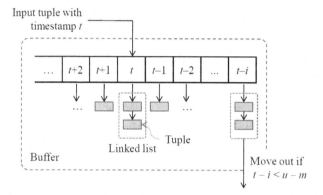

Fig. 1 Window buffer to keep tuples in an increasing order of their timestamps

Given the first specification, we may randomly choose n tuples from the window buffer. But this is not a proper solution because *valid* tuples (i.e., latest tuples
belonging to the current window) can be chosen as victims in this case. In this
case, accuracy of query results can be degraded from the elimination of the valid
tuples (that will be involved in the computation of aggregate functions specified in
a query if not discarded). To minimize its influence on the accuracy, we need to
discard the earliest n tuples (including *expired* tuples that do not belong to the
current window) from the window buffer.

Figure 1 shows the structure of a window buffer to keep tuples in an increasing
order of their timestamps. A circular array is used whose index corresponds to the
tuples' timestamps. Each element of the circular array has a linked list to keep
tuples with the same timestamp. In this structure, by counting the number of tuples
in each array element, the earliest n tuples can easily be identified and discarded
from the buffer.

Satisfying the second specification is not straight-forward. As a naïve approach
to meet the given percentage m, for a predefined period of time, we may accept
the consecutive n input tuples and ignore other tuples arriving after the acceptance. For example, suppose that 100 tuples arrive in a system every second, and
we need to accept 50 percent of input tuples. Then, to meet the constraint, we may
receive the first 50 tuples and ignore the remaining 50 tuple arrivals. But among
the accepted ones, some tuples can be expired (with too early timestamps) that do
not belong to the current window. If the number of expired tuples is large, we

unnecessarily discard input tuples, which lead to inaccurate query results. To fix it, we may accept more α tuples. But in this case, there is no criterion to determine the value of α.

Regarding this, our previous work [12] solved this problem. In the work, the second specification is interpreted as "select and drop $(1 - m)$ percent of tuples from out-of-order tuple arrivals". It then estimates a heartbeat (timestamp) ω which can guarantee that the percent of tuple discards does not exceed $(1 - m)$. To solve the problem, it assumes that tuples are randomly generated from remote sources and their network transmission delays follow a normal distribution. Given the percentage $D = (1 - m)$, the buffer size, (i.e., the number of tuples that can be kept in the window buffer), denoted n, can be estimated as follows.

$$n \approx \sqrt{2} \cdot z(D) \cdot \sigma \lambda$$

Above, $z(D)$ is the z-value of the normal distribution and σ is a standard deviation of the distribution. λ denotes the average arrival rate of input tuples. Note that our previous work assumes some distributions on tuple generation and network transmission delays. When the assumption does not hold, the above estimation measure cannot be used.

On the other hand, in many real-world applications, the distribution of input events may not be known in advance. If the domain of input tuples is discrete in this case, the *empirical distribution function* can be used to estimate the buffer size satisfying the given drop ratio D. For the time-based windows, the function (cumulative distribution function) can be defined as follows.

$$F(\tau) = \frac{Number\ of\ tuples\ whose\ timestamps \le \tau}{Total\ number\ of\ tuples\ in\ memory}$$

Using the function F, the heartbeat satisfying D can be obtained as follows.

$$h = arg\ \max_{\tau}(D \ge F(\tau))$$

The buffer size can also be estimated based on the heartbeat value. Below, $f(t)$ denotes a function that returns the number of tuples with timestamp t from the buffer whose structure is depicted in Figure 1, and t_c is the current wall-clock time (or timestamp of the latest tuple arrival).

$$n = \sum_{h < t \le t_c} f(t)$$

Using the above functions, it is possible to control the number of tuple drops even when the distribution of input tuples is not known. The functions are included in a module called a *heartbeat estimator* in our window operator. The module calculates a heartbeat or a buffer size that satisfies a given drop ratio. Its calculation can be triggered whenever a predefined time interval elapses, or when the system issues a load shedding request.

4   Conclusion and Future Work

In this paper, we discussed how window operators can support load shedding in continuous query processing. To enable this, the functionality to control the number of tuple discards is added to a window operator in the proposed method. In the proposed method, a user or a system can request it to discard n tuples from memory (immediately) or to discard only m percent of input tuples (for a predefined period of time). To support the given QoS parameter, we present the buffer structure of a window operator and discuss estimation functions to calculate the buffer size satisfying the given tuple drop ratio.

The proposed method eliminates the necessity of a dedicated load shedder which is placed at the entry of a stream query processor and discards some portion of input tuples when requested. From this, the number of query processing steps is reduced, which leads to the performance improvement of stream query processing. To verify it, we plan to conduct experiments with various test data sets.

Acknowledgement This paper was supported by the Sahmyook University Research Fund in 2014.

References

1. Babcock, B., et al.: Models and issues in data stream systems. In: Proc. of the ACM PODS, pp. 1–16 (2002)
2. Abadi, D., et al.: Aurora: A new model and architecture for data stream management. VLDB Journal 12(2), 120–139 (2003)
3. Arasu, A., et al.: STREAM: The Stanford stream data manager. IEEE Data Eng. Bull. **26**(1), 19–26 (2003)
4. Viglas, S.D., Naughton, J.F., Burger, J.: Maximizing the output rate of multi-way join queries over streaming information sources. In: Proc. of the VLDB Conference, pp. 285–296 (2003)
5. Babcock, B., Datar, M., Motwani, R.: Load shedding for aggregation queries over data streams. In: Proc. of the ICDE, pp. 350–361 (2004)
6. Tatbul, N., et al.: Load shedding in a data stream manager. In: Proc. of the VLDB Conference, pp. 309–320 (2003)
7. Srivastava, U., Widom, J.: Flexible time management in data stream systems. In: Proc. of the ACM PODS, pp. 263–274 (2004)
8. Kim, H.G., Kim, M.H.: A review of window query processing for data streams. Journal of Computing Science and Engineering 7(4), 220–230 (2013)
9. Maier, D., et al.: Semantics of data streams and operators. In: Proc. of the ICDT, pp. 37–52 (2005)
10. Johnson, T., Muthukrishnan, S., Shkapenyuk, V., Spatscheck, O.: A heartbeat mechanism and its application in Gigascope. In: Proc. of the VLDB Conference, pp. 1079–1088 (2005)
11. Li, J., et al.: Semantics and evaluation techniques for window aggregates in data streams. In: Proc. of the ACM SIGMOD, pp. 311–322 (2005)
12. Kim, H.G., Kim, C., Kim, M.H.: Adaptive disorder control in data stream processing. Computing and Informatics **31**(2), 393–410 (2012)

Improved Location Estimation Method of Trilateration in Ubiquitous Computing Indoor Environment

Jeonghoon Kwak, Hyunseok Jang, Yunsick Sung and Young-Sik Jeong

Abstract Estimating the location of users, Unmanned Aerial Vehicles (UAVs), and devices is the key to provide the diverse kinds of services in ubiquitous computing environments. Given that Global Positioning System (GPS) cannot be utilized in indoor environments, beacon-based and vision-based location estimation approaches are utilized. However, the accuracy of the location does not reaches the demanded amount. This paper propose an indoor location estimation method of beacons based on trilateration. By utilizing three APs and three beacons, the location of one of beacons can be predicted accurately. In the experiment, the proposed method was compared with the traditional trilateration method. Comparing to the result of the traditional trilateration method, the proposed method reduced the distance errors to about 20%.

Keywords Indoor location · UAV · Beacon · Drone · Trilateration

1 Introduction

Global Positioning System (GPS) is the key to estimate the location of users, Unmanned Aerial Vehicles (UAVs), and devices in ubiquitous computing

J. Kwak
Department of Computer Engineering, Graduate School, Keimyung University, Daegu, South Korea
e-mail: jeonghoon@kmu.ac.kr

H. Jang · Y. Sung(✉)
Faculty of Computer Engineering, Keimyung University, Daegu, South Korea
e-mail: {jhs9424,yunsick}@kmu.ac.kr

Y.-S. Jeong
Department of Multimedia Engineering, Dongguk University, Seoul, South Korea
e-mail: ysjeong2k@gmail.com

© Springer Science+Business Media Singapore 2015   165
D.-S. Park et al. (eds.), *Advances in Computer Science and Ubiquitous Computing*,
Lecture Notes in Electrical Engineering 373,
DOI: 10.1007/978-981-10-0281-6_24

environments. Given that GPS cannot be utilized in indoor environments, the diverse kinds of alternative approaches [1-3] have conducted. To estimate the locations of users, UAVs, and devices in indoor location, beacon-based trilateration approaches [4-5] or wireless indoor positioning [6] can be utilized. However, because of indoor environments, the measured distances and locations are usually inaccurate. Therefore, it is required to revise the traditional distance measurement and location estimation approaches.

This paper proposes a location estimation method of beacons. By utilizing Relative Distance Estimation Method (ReDEM) [3] and Trilateration [4], the location of a single beacon can be predicted. In the experiment, the proposed method was compared with the traditional trilateration method. Given that the distance from a beacon to a Access Point (AP) was revised by ReDEM, the accuracy of the location of a becon was improved. As a result, The proposed method reduces the errors of estimating locations to about 20%.

The rest of this paper is organized as follows: Section 2 proposes a beacon location estimation method in indoor environments. Section 3 discusses the validation of the proposed method. Finally, Section 4 concludes of this paper.

2   Trilateration-Based Beacon Location Estimation Method

In this paper, two types of beacons are utilized: one for a target beacon and others for reference beacons. The location of the target beacon is changed and estimated by the proposed method. The reference beacons are utilized to obtain the distances between APs in real-time. There are also two types of APs. One type contains a beacon and the other does not have any beacon. The processes for estimating a target beacon location are divided into nine stages as shown in Figure 1.

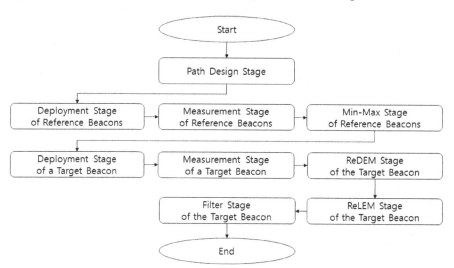

Fig. 1 Beacon location estimation processes are divided into nine stages. Four stages are for reference beacons and five stages are for a target beacon.

Firstly, the moving path of a target beacon is designed. Given that the proposed method is based on trilateration, the moving path is defined by a single edge or multiple edges of a triangle as shown in Figure 2. To simplify the explanation of the proposed method, we limit that the moving path is defined only by a single edge of the triangle.

Second, three APs and two reference beacons are deployed considering the moving path during Deployment Stage of Reference Beacons. Three APs are deployed in advance at the vertices of the triangle. Two reference beacons are located at the two end of the moving path.

Third, the distance from reference beacons to APs are measured and collected n times during Measurement Stage of Reference Beacons. The distance from the ith reference beacon (the ith AP) to the jth AP (the jth reference beacon) at time t is defined as $d_{i,j,t}$ where $i \neq j$. Fourth, the minimum and maximum of the distances from each reference beacon to APs a obtained. The minimum and maximum of $d_{i,j,t}$s are denoted as $\min(d_{i,j,t})$ and $\max(d_{i,j,t})$.

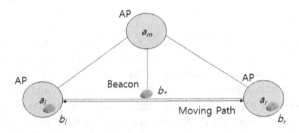

Fig. 2 There are three APs and three beacons. Two APs, the AP a_l and the AP a_r, contains reference beacons, the beacon b_l and the beacon b_r. The beacon b_* moves from the AP a_l to the AP a_r.

Fifth, a target beacon is deployed one of the end of the moving path. Sixth, the current distances between the target beacon and reference beacons are measured that are denoted by $d_{i,*,t}$. Seventh, ReDEM is applied to revise $d_{i,*,t}$ during ReDEM Stage of the Target Beacon. The revised distance of $d_{i,*,t}$ is $d'_{i,*,t}$ as shown in Equation (1).

$$d'_{i,*,t} = (d_{i,*,t} - \varphi_{i,j,r}) \times \frac{\max(d_{i,j,t}) - \min(d_{i,j,t})}{\varphi_{i,j,100-r} - \varphi_{i,j,r}} + \min(d_{i,j,t}) \qquad (1)$$

where $i \neq j$ and b_i and b_j are existed. $\varphi_{i,j,\gamma}$ denotes the top γ% distance among the measured distances from a reference beacon. Eighth, the location l_t is calculated by utilizing trilateration [6] based on the revised distances of the target beacon during ReLEM Stage of the Target Beacon. Finally the locations out of the moving path are adjusted during filter stage. We call the proposed method Relative Location Estimation Method (ReLEM).

3   Experiment

In the experiment, three APs are utilized with the corresponding beacons and a target beacon moved five times as shown in Figure 3. The path was defined as a three meter-line. The distances from each reference beacon to APs were measured about 250 times during 60 seconds. γ was 95%

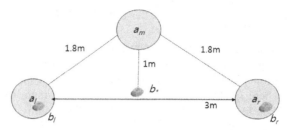

Fig. 3 The distances between a starting point to the location of a target beacon were calculated.

The proposed method-based distances were compared with the result of expected distances and traditional trilateration-based distances as shown in Figure 4. The accumulated difference between the expected distances and the traditional trilateration-based distances were 3916.5 m and the accumulated difference between the expected distances and the proposed method-based distances were 766.74 m. As results, the proposed method reduces the difference to about 20%.

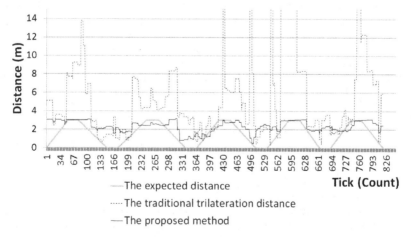

Fig. 4 The distances between a starting point to the location of a target beacon were calculated.

4   Conclusions

The traditional trilateration-based beacon location estimation approaches are based on the distances among beacons. Given that the distances between beacons are inaccurate in indoor environments, the traditional trilateration approaches could not be utilized. This paper proposes an improved trilateration method by applying ReDEM. Given that the accuracy of the distance among beacons is improved by ReDEM, the proposed method becomes accurate.

Acknowledgement This research was supported by Basic Science Research Program through the National Research Foundation of Korea(NRF) funded by the Ministry of Science, ICT & Future Planning (NRF-2014R1A1A1005955).

References

1. Son, B., Kang, S., Lee, H., Lee, D.: A Real Time Quadrotor Autonomous Navigation and Remote Control Method. IEMEK Journal of Embedded Systems and Applications **8**(4), 205–212 (2013)
2. Moo, S., Ha, S., Eom, W., Kim, Y.: 3D Map Generation System for Indoor Autonomous Navigation. The Korean Society for Aeronatical & Space Sciences **13**, 166–173 (2014)
3. Sung, Y., Kwak, J., Jeong, Y., Park, J.H.: Beacon distance measurement method in indoor ubiquitous computing environment. In: The 4th International Conference on Ubiquitous Computing Application and Wireless Sensor Network, Jeju, Korea, July 8–10, 2015
4. Lee, H.C., Lee, D.M.: A Study on Localization System using 3D Triangulation Algorithm based on Dynamic Allocation of Beacon Node. The Journal of Korea Information and Communications Society **36**(4), 378–385 (2011)
5. Lee, H.C., Lee, D.M.: The 3-Dimensional Localization System Based on Beacon Expansion and Coordinate-Space Disassembly. The Journal of Korea Information and Communications Society **38B**(1), 80–86 (2013)
6. Liu, H., Darabi, H., Banerjee, P., Liu, J.: Survey of Wireless Indoor Positioning Techniques and Systems. IEEE Transactions on Systems, Man, and Cybernetics, Part C: Applications and Reviews **37**(6), 1067–1080 (2007)

Performance Comparison of Relational Databases and Columnar Databases Using Bitmap Index for Fast Search of 10Gbps Network Flows

Sunoh Choi, Hyun-Wook Park, Joo-Young Lee,
Jong-Hyun Kim and Ik-Kyun Kim

Abstract Recently a lot of Cyber Attacks are done. In order to prevent these Cyber Attacks, first we should be able to analyze attacks from network traffic like packets and flows. However, in high speed network like 10Gbps, there are millions of packets and tens of thousands of flows per second. So, it is very difficult just to find a flow which a security investigator wants to see within a short time. To make search fast, we have to use indexes used in databases. In this paper, we show loading time and search time in relational database and columnar database using bitmap index.

Keywords Database · Network · Flow · Bitmap index

1 Introduction

Nowadays a lot of Cyber Attacks are done. So, many computers became out of order since they were infected by 320 Cyber Attack in South Korea [9]. On the other hand, critical information about Italian Hacking Team was leaked [10]. To prevent these Cyber Attacks is getting more important. The first step of preventing these Cyber Attacks is to analyze them from network traffics. To analyze Cyber Attacks, we can investigate network packets, network flows, extracted files, and IDS alerts.

However, in high speed networks as 10Gbps, millions of packets and tens of thousands of flows pass every second. So, it is very difficult to find a flow which

S. Choi(✉) · H.-W. Park · J.-Y. Lee · J.-H. Kim · I.-K. Kim
Network Security Research Group,
Electronics and Telecommunications Research Institute,
218 Gajeong-ro, Daejeon, South Korea
e-mail: {suno,j8305gusdnr,joolee,jhk,ikkim21}@etri.re.kr

© Springer Science+Business Media Singapore 2015   171
D.-S. Park et al. (eds.), *Advances in Computer Science and Ubiquitous Computing*,
Lecture Notes in Electrical Engineering 373,
DOI: 10.1007/978-981-10-0281-6_25

an investigator wants to check within a short time. In order to find a flow among a lot of flows, we need indexes used in databases. By using indexes, we can find a flow fast with time complexity $O(log\ n)$ when there are n flows. It is known that traditional RDBMS like MySQL [3] using B+ tree [4] for indexes can't store tens of thousands of flows every second [7]. So, we need other type of databases to store tens of thousands of flows every second and search a flow within a short time. It is InfiniFlux [5] using Bitmap Index. We will show loading time and search time in MySQL and InfiniFlux.

This paper is organized as follows. In section 2 and 3, we will introduce flow and queries used in analyzing Cyber Attacks respectively. In section 4 and 5, we will show loading time and search time of MySQL and InfiniFlux. Finally in section 6, we give conclusion.

2   Network Flows

A network flow [1] is presented as follows.

<SRC_IP, SRC_PORT, DST_IP, DST_PORT, PROTOCOL, START_TIME, END_TIME, PACKETS, BYTES>

It lets us know that how many packets has transmitted from a host to the other host and when the first packet has passed and the last packet has received and what protocol has been used to transmit them.

In high-speed networks as 10Gbps, it has a lot of flows. In [2], an ISP in South Korea has about one billion flows a day. It is very difficult to find a flow which a security investigator wants to see among a lot of flows within a short time. In order to achieve this goal, we have to use *indexes* for flows as it will be discussed in next section.

3   Queries

Queries which are used when a security investigator finds a flow are as follows.

Q1: SELECT * FROM FLOW WHERE SRC_IP='A' and DST_IP='B';

Q2: SELECT PROTOCOL, SUM(BYTES) FROM FLOW GROUP BY PROTOCOL ORDER BY COUNT(*) DESC;

The first query is to find all flows which are transmitted from a host A to a host B. The second query is to see how many bytes have been transmitted using each protocol. Since there are a lot of flows, it is very difficult to get results for these queries within a short time.

This paper's main goal is to find a flow among a lot of flows like one billion flows within a short time. For this goal, we have to use indexes. Indexes are widely used in databases as MySQL [3]. MySQL is a kind of traditional relational database management system (RDBMS). In order to make indexes, RDBMS uses B+ tree [4]. In B+ tree, a node's value is greater than the value of the left child node and less than the value of the right child node and the tree is always balanced by using splitting. So, the search time complexity is $O(log\ n)$. However, the insert (or loading) time complexity is $O(n\ log\ n)$. So, in order to reduce the loading time we have to load data into database tables before making indexes. After that, we have to make indexes. So, it is not appropriate to use relational databases to find a flow in high speed networks as 10Gbps.

In order to meet our requirements, we use a database using another type of indexes. It is InfiniFlux [5] using Bitmap Index [6, 7, 8]. Bitmap index is a structure to make search queries fast. As Figure 1, the data has a range of 0 to 3 and they are mapped to binary arrays respectively. When the value is 1, the second bit of the array is set to 1. When the value is 3, the fourth bit is set to 1. When an investigator finds a flow which has a source A and a destination B, it is done fast by using bitwise operations between two bitmap indexes.

However, a shortcoming of bitmap index is that it requires a lot of storage space. It is mitigated by using run length encoding (RLE) compression as Figure 1. Second shortcoming is that bitmap index can be used for only appending. When data need to be modified or deleted, bitmap index can't be used. However, since Bitmap index has much faster loading time than relational databases, we use InfiniFlux using Bitmap index to find a flow among a lot of flows within a short time. In addition, since network flows do not need to be modified and they are only appended, bitmap index is appropriate for network flows. In next section, we will show loading time and search time in relational database and columnar database using bitmap index.

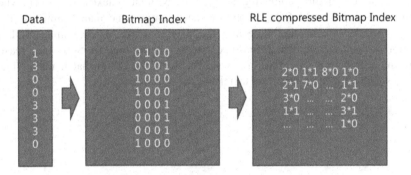

Fig. 1 Data, Bitmap Index and RLE compresses Bitmap Index [8]

4   Loading Time

In order to show loading time and search time, we used a machine having Intel CPU 2.6GHz and 64GB memory and CentOS 7.0. On that system, we installed MySQL and InfiniFlux. In the first experiment, we loaded 100K to 100M flows into the databases. In Figure 2, the loading time of InfiniFlux is about ten times faster than MySQL. When we load flows into MySQL, we didn't make indexes in the table of MySQL since it took too much time when the indexes exist. After we load flows into MySQL, we make indexes. The time to make indexes in MySQL is not shown in Figure 2. If we consider the time for making indexes in MySQL, the loading time of InfiniFlux would be much faster than MySQL. Note that when we load flows into InfiniFlux, indexes are automatically generated.

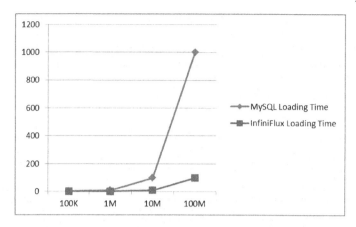

Fig. 2 Loading Time (Seconds) of MySQL and InfiniFlux

5 Search Time

In this section, we show search time of MySQL and InfiniFlux. As shown in Figure 3, when there is no index in MySQL, it takes a lot of time. InfiniFlux shows fast search time compared to MySQL without indexes. However, when there are indexes in MySQL, it shows faster search time than InfiniFlux. But, if we consider the loading time and the time to make indexes in MySQL in Section 4, we conclude that InfiniFlux is totally better than MySQL. One reason why the search time of InfiniFlux is slower than MySQL is that it makes local indexes every 1 million flows for efficiency.

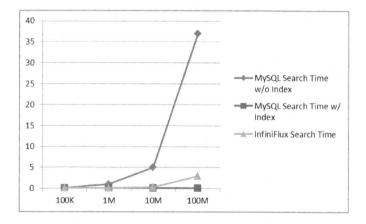

Fig. 3 Search Time (Seconds) of MySQL and InfiniFlux

6 Conclusion

In this paper, we show the loading time and the search time of relational database using B+ tree and columnar database using bitmap index. Columnar database using bitmap index shows much faster loading time than relational database using B+ tree. Even if relational database shows faster search time than columnar database, we conclude that columnar database is much better than relational database in order to find a flow among a lot of flow in high speed networks like 10Gbps since the former has much faster loading time than the latter.

Acknowledgments This work was supported by Institute for Information & communications Technology Promotion (IITP) grant funded by the Korea government (MSIP) (No.B0101-15-0300, The Development of Cyber Blackbox and Integrated Security Analysis Technology for Proactive and Reactive Cyber Incident Response).

References

1. Flow: https://en.wikipedia.org/wiki/Traffic_flow_(computer_networking)
2. Woo, S., Jeong, E., Park, S., Lee, J., Ihm, S., Soo Park, K.: Comparison of Caching Strategies in Modern Cellular Backhaul Networks. ACM MobiSys (2013)
3. MySQL: https://www.mysql.com
4. B+ tree: https://en.wikipedia.org/wiki/B%2B_tree
5. InfiniFlux: http://www.infinibio.com
6. Bitmap Index: https://en.wikipedia.org/wiki/Bitmap_index
7. Deri, L., Lorenzetti, V., Mortimer, S.: Collection and Exploration of Large Data Moniroring Sets Using Bitmap Databases. TMA (2010)
8. Fusco, F., Dimitropoulos, X., Vlachos, M., Deri, L.: pcapIndex: An Index for Network Packet Traces with Legacy Compatibility. ACM SIGCOMM Computer Communication Review (2012)
9. 320 Cyber Attack: https://en.wikipedia.org/wiki/2013_South_Korea_cyberattack
10. Hacking Team: https://en.wikipedia.org/wiki/Hacking_Team

Group ID Issuing Model Using Temporal Explicit Movement in Social Life Logging

Young Ho Jo, Jung Nyun Lee, Hea Jin Kim, Jin Cheol Woo,
Yu Jin Lee and Min Cheol Whang

Abstract We live in an era in which social network services are very active due to the development of technology. People are eager to get connected to other people. The present study is aimed to design a model that a user can determine his connection with others based on the temporal explicit movements among his explict information in daily life and identified the group automatically giving random number according to the determined co-movement. This study filmed and analyzed the explicit movements of participants. The movments was divided them into two modes of behaviors: interaction mode and private mode. The division was necessary because this study was aimed to find out the difference in co-movement and group ID (indentification) by mode of behavior. 17 group IDs were derived from interaction mode and 20 group IDs from private mode. Social connection was observed to be updated and changed more often and changed in private mode than in interaction. This study was successful to determine connected group automatically and issue group IDs.

Keywords Group · Group ID · Temporal explicit movement · Co-movement · Interaction · Life logging

Y.H. Jo · H.J. Kim
Team of Technology Development, Emotion Science Center,
Maebongsan-ro, Mapo-gu, Seoul 03909, South of Korea
e-mail: imzeus05@gmail.com, shaonu@hanmail.net

J.N. Lee · J.C. Woo · Y.J. Lee
Department of Emotion Engineering, Sangmyung University,
Hongji-dong, Jongno-gu, Seoul 03016, South of Korea
e-mail: {blueleen2,mcun}@naver.com, ses2104@nate.com

M.C. Whang(✉)
Department of Digital Media, Sangmyung University,
Hongji-dong, Jongno-gu, Seoul 03016, South of Korea
e-mail: whang@smu.ac.kr

© Springer Science+Business Media Singapore 2015
D.-S. Park et al. (eds.) , *Advances in Computer Science and Ubiquitous Computing*,
Lecture Notes in Electrical Engineering 373,
DOI: 10.1007/978-981-10-0281-6_26

177

1   Introduction

The incredible development of information and communications technology enabled to establish online network infrastructure that connects the world as one. In addition, such mobile devices as smartphone using the online network infrastructure are flooding the market. As a variety of mobile devices such as smartphone, tablet PC and laptop computer are launched in the market, relationship building gets more active online (Wellman & Gulia, 1999). Creating and developing connection is the representative function of social network service as commonly known as 'SNS'. SNS users share their daily lives and give and take feedbacks by uploading comments. SNS is used as a tool to solidify the existing relationship with someone already in it and helps build a relationship with new people with similar interest (Y.S. Cha, et al. 2012).

Developing human relationship is mainly caused from the fact that a human is intrinsic to love building relationship. However, he has to expose himself to others and actively appeal to them who want to make a relationship so they can join relationship building with him together, feeling sympathized with and familiar to him. Ironically, on the other hand, such behavior can be considered as bothering or add pressure on people, too (D. Harrison Mcknight, et al. 1998).

This study is to design a model that can find others who have the same explicit signal[1] pattern in daily life and automatically build a relationship. There are many factors indicating relationship building such as bio signal (e.g. heart beat), consumed contents and behavior pattern, which are often used in life logging (Lijuan Zhou and Cathal Gurrin, 2012). This study chose temporal explicit movement information that can be captured by a camera and determine group. This model analyzed people's temporal explicit movements when they interchange with others and determined how much they were co-movement. Eventually, social groups and their ID was automatically determined. This study is to develop grouping and identifying users in life logging and provide social behavior data to predict social trend and pattern in further study.

2   Method

2.1   *Experimental Method*

This study conducted an experiment on 3 subjects (2 males and 1 female) in the same art gallery by video-recording them in two different modes: when they interacted with each other and when they didn't interact. The mode in which they interacted with each other is named "interaction mode" while the mode in which they didn't interact with each other is named "private mode". Interaction mode means a pattern of behavior that they communicated and interacted with each other and private mode is a pattern of behavior that they used their own

---

[1] Bio temporal explicit movements of a body.

smartphones and surfed on Internet instead of interacting with others. These behaviors were video-taped and their temporal explicit movements were extracted from the recorded video. The camera (GoPro Hero4 Black, USA) was used to capture their temporal explicit movements. The recording device was set to 1920*1080 Full HD and 30 fps mode for capturing images. To facilitate data extraction from the video images, Daum PotEncoder (Daum Kakao, South of Korea) was used to encode the recorded images into one 1280*720 HD, 30 fps. The filming length was 17 minutes and 43 seconds (1003 seconds) for interaction mode and private mode, respectively.

Table 1 Experimental Design

Mode	Task	Description
Interaction	Mutual Interaction	Interact through communication
Private	Internet Surfing	Internet surfing on personal smartphone

Fig. 1 Interaction Mode (left) and Private Mode (right)

2.2 Signal Analysis

Participant's temporal explicit movement data was extracted from the encoded data using LabView System (National Instrument, USA). The standard deviations of the temporal explicit movement were compared to extract determination data of co-movement. Co-movement is determined on time base by comparing the standard deviation of one's temporal explicit movements with that of others (Kelly and Barsade, 2001). For example, supposed that A is the standard deviation of Subject 1, B for Subject 2 and C for the combined standard deviation of Subject A and B. Then when C < A and C < B, they are considered as co-movement. Here, value '1' is assigned to the case of co-movement and '0' to non-co-movement. The determination data for each pair of the three subjects were extracted. Number of cases where each pair is determined as co-movement is 7 as in Figure 2. Then, 3 determination data of co-movement could be extracted: Subject 1 and 2, Subject 2 and 3, and Subject 3 and 1. One determination data can be produced every second. Therefore, a total of 1003 determination data were produced because the total length of the recorded video is 1003 seconds.

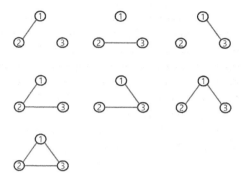

Fig. 2 Three subjects pair's co-movement cases

2.3   Identifying Group

Group is identified on the basis of the determination data of co-movement. Here, it is necessary to set the reference duration of co-movement. It was set to 19.4 seconds, which was obtained from the standard deviation of the combined time of interaction mode and private mode.

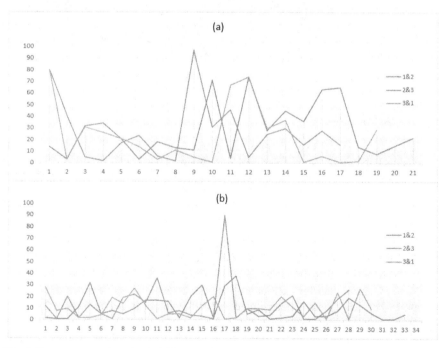

Fig. 3 Co-movement time (sec.) (a) Interaction Mode (b) Private Mode

Table 2 Summary of Co-movement Time

Mode	Interaction Mode	Private Mode	Interaction and Private Mode
Number of Co-movement	57	91	148
Mean	26.4 sec.	11.2 sec.	17.1 sec.
Standard Deviation	24.3 sec.	12.4 sec.	19.4 sec.

The reference duration is necessary to set because co-movement and non-co-movement alternate, so group ID can only be issued automatically after a certain time period from the onset of co-movement. If group ID is allowed to be issued every time co-movement occurs, 57 group IDs and 91 group IDs can be created for interaction mode and private mode, respectively. However, when the reference time is applied, the numbers reduced to 25 group IDs and 34 group IDs will be created for interaction mode and private mode, respectively[2].

3 Result

The reference time for co-movement duration was calculated and co-movement's cases by each pair was categorized. Figure 3 (a) and Figure 4 (a) show the co-movement of each pair over 1003 seconds and the groups made for co-movement. The yellow boxes in Figure 3 (b) and Figure 4 (b) show the groups of each mode.

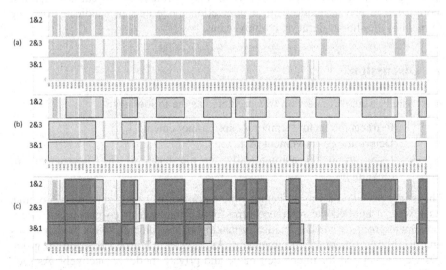

Fig. 4 Grouping of Interaction Mode (a) Co-movement raw data (b) Grouping using co-movement maintenance time (c) Final grouping, Generate 17 group IDs

---

[2] The number simply counted form the beginning of co-movement.

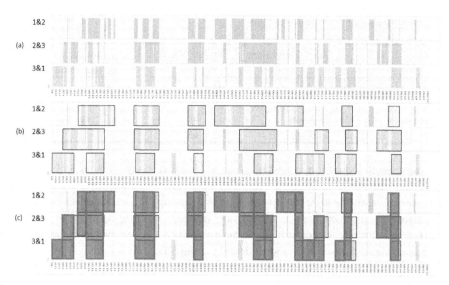

Fig. 5 Grouping of Private Mode (a) Co-movement raw data (b) Grouping using co-movement maintenance time (c) Final grouping, Generate 20 group IDs

And then the co-movement's pairs during the same period were grouped for issuing group ID. The orange boxes in Figure 3 (c) and Figure 4 (c) are the final groups. Eventually, 17 group IDs and 20 group IDs were generated for interaction mode and private mode, respectively. When it was private mode, more group IDs were created than interaction mode, which means connection is cut and resumed more often in private mode than in interaction mode and also indicates that the former has the shorter duration co-movement.

4 Discussion

In this study, social connection was predicted as the followings:

1. Extracts the factor 'temporal explicit movement'.
2. Determine co-movement by it.
3. Issue group ID automatically on the basis of the determined co-movement.

This study provided a way to help people automatically build relationship online. As a way, it applied the reference time for the duration of co-movement to co-movement groups when automatically making co-movement groups, which varies time to time, to maintain co-movement. As a result, 17 group IDs and 20 group IDs were generated for interaction mode and private mode, respectively. As seen in Figure 3 (c) and Figure 4 (c), there are ranges where each of the pairs became co-movement but discarded from the final groups. The groups belonging to these ranges should be determined for inclusion or exclusion based on more precise criteria.

Further study is necessary to define more various factors by which grouping is made and study the weight of each factor. In addition, it is deemed necessary to study on a system that can issue group ID automatically by applying this proposed model.

Acknowledgments This work was partly supported by the ICT R&D program of MSIP/IITP [R0126-15-1045, The development of technology for social life logging based on analyzing social emotion and intelligence of convergence contents] and the National Strategic R&D Program of MOTIE(Ministry of Trade, Industry and Energy) [10044828, Development of augmenting multisensory technology for enhancing significant effect on service industry].

References

1. Cha, Y.S., Kim, J.H., Kim, J.H., Kim, S.Y., Kim, D.K., Whang, M.C.: Validity analysis of the social emotion model based on relation types in SNS. Korean Journal of the Science of Emotion and Sensibility **15**(2), 283–296 (2012)
2. Harrison Mcknight, D., Cummings, L.L., Chervany, N.L.: Initial trust formation in new organizational relationships. Academy of Management Review **23**(3), 473–490 (1998)
3. Kelly, J.R., Barsade, S.G.: Mood and Emotions in Small Groups and Work Teams. Organizational Behavior and Human Decision Processes **86**(1), 99–130 (2001)
4. Zhou, L., Gurrin, C.: A Survey on Life Logging Data Capture. SenseCam Symposium (2012)
5. Wellman, B., Gulia, M.: Net Surfers Don't Ride Alone: Virtual Communities As Communities. Communities in Cyberspace, 167–194 (1999)

An Analysis of Infographic Design for Life-Logging Application

Sojung Kwak, Joeun Lee and Jieun Kwon

Abstract As life-logging technology develops recently, applications which record personal behavior and thought are increasing. This paper intends to analyze infographic design examples related to life-logging applications which are used by people for recording personal information and sharing daily life. About 20 applications are searched with the keyword of life-logging and contents organization centered on human behavior and infographic design such as visual, sound and motion elements are studied. It is judged that this infographic analysis of recent life-logging app design can be used as a guide for designing life-logging applications which are expected to increase in the future.

Keywords Life-logging · Infographic · Application design

1 Introduction

Life-logging means recording and remembering information on personal daily life and sharing or communicating with other people. Recently, life-logging is easily implemented by smart phones and wearable devices. In particular, popularization of smart phones makes it possible to record personal thought and behavior without the limitation of time and space and share with other people by using SNS (Social Networking Service) such as Kakaotalk, Facebook and Tweeter.[1] It is also possible to collect and analyze data which is generated in those services in order to find out the pattern of personal life as well as crowd behavior. This data analysis can be used for trend prediction and marketing purpose. Therefore, infographic design becomes important which helps users recognize meaningful data from vast data and easily and correctly use them. This paper intends to analyze life-logging applications in order to recognize the recent trend of infographic design and discuss about the limitations and possibilities for effective design.

S. Kwak
Department of Game, Graduate School, Sangmyung University, Seoul, Republic of Korea

J. Lee · J. Kwon(✉)
Department of Emotion Engineering, Graduate School, Sangmyung University, Seoul, Republic of Korea
e-mail: jieun@smu.ac.kr

© Springer Science+Business Media Singapore 2015
D.-S. Park et al. (eds.), *Advances in Computer Science and Ubiquitous Computing*,
Lecture Notes in Electrical Engineering 373,
DOI: 10.1007/978-981-10-0281-6_27

For this study, first, the concept and features of life-logging are defined and expression methods of life-logging applications are examined. The general concept of life-logging and theories which have been developed until now are studied by literature search and expression methods of applications actually used on mobile phones are studied. Second, top 10 applications are searched from Android and App Store with the keyword of life-logging and design examples of total 20 (10 for each) life-logging applications are analyzed. Analysis is performed according to the applications type, menu layout and design elements. Third, the infographic design of life-logging applications is discussed based on the detailed analysis.

2 Life-Logging

2.1 Concept and Features of Life-Logging

Life-logging is to store and record personal life and information such as moving path, behavior and quantity of motion in digital space by using smart phones and wearable devices.[3] The aim of life-logging is to record all things which occur in daily life in order to store and extend personal memory and share with other people. Materials for life-logging include pictures, images, texts and voices. Devices for tracking personal life in the digital era include portable cameras, biometric devices, other portable devices and networked systems.[4] Device-created data is collected to analyze personal life and manage self-recorded materials. It can be used for personal development or personalized service. Also, integration of personal data can be developed to a means to understand human and society. This paper will focus on smart phone applications which are popularly used now among life-logging media.

2.2 Life-Logging Information

Life-logging records and stores user behavior based on time and space. Therefore, contents are organized according to human behavior, and time- and space-based data is stored for each content in order to express information. For life-logging information, data is analyzed according to time such as date, month and year or records are integrated according to user behavior or event occurring place. Records are automatically stored by wearable devices or smart phones through the quantification of biometric data or manually stored by the user. For recorded data, infographic through diagram is mainly used for intuitive recognition of statistical figures.

3 Analysis of Life-Logging Applications

3.1 Method and Scope of Life-Logging Application Analysis

This paper analyzes applications which are searched by the keyword of "Life-Logging" in Android Play Store and iPhone App Store. This search selects top 10 applications for each in July 2015 and total 20 applications are selected as shown in Table 1.

For analysis method, each application is directly executed and analyzed focusing on infographic. First, type classification and contents (menu) layout according to the content character of applications are studied. Second, infographic design elements such as Visual elements, Sound elements and Motion elements are analyzed. Visual elements include Color, Tone, Combination concept of color, Contents Style and Icons Shape, Sound elements include Background Music, Sound Effect and Voice, and Motion elements include Basic Motion and Interactivity.

Table 1 Basic information of the Application

no	Android (Play store)		iPhone (App store)	
	Title	Production	Title	Production
1	Lifelog	Sony	iLifeLogger	Natz Soft
2	Universal pedometer	SenseMe	MoneySmart	MoneySmart Inc.
3	Pedometer Walk	ALLTIMESOFT	Optimized	OptimizeMe GmbH
4	Life log Application	Yokotama	Timenote	Katsunobu Ishida
5	ActiFit Pedometer Challenges	MicroMovie Media GmbH	Life Analytics	Media Circuit Co., Ltd.
6	Nike+ Running	Nike+ Running	All of my life.	MW Products
7	Runtastic Pedometer	Runtastic	My Fit Log	PQRS MOBILE (ASIA) LIMITED
8	Saga Automatic Lifelogging	ARO	LifeStats	Placer Labs Inc.
9	Smart e-SMBG Diabetes lifelog	ARKRAY Marketing, Inc.	Time Golden	huang shan
10	Woman Log Calendar	ABISHKKING	Loggo	Dan Haggren

3.2 Results of Life-Logging Infographic Design Analysis

(1) Content Type

Life-logging applications consist of 9 health apps, 9 personal life apps and 2 diaries and most of contents are related health. 1 (minimum) to 24 (maximum) contents (menus) are organized and 7.8 contents are included on average. Among 20 applications for which are confirmed, 15 (6 for iPhone and 9 for Android) applications include a pedometer. The content organizations of each application are shown in Table 2. Step, number of heartbeats, walking and life are mostly confirmed.

As 20 life-logging applications examined in this study focus on health, contents are also limited to measuring bio-signal and behavior and delivering health-related information. But, life-logging targeted for women or applications focusing on certain content such as finance are increasing.

(2) Infographic Elements

Infographic elements are divided into Visual elements, Sound elements and Motion elements for analysis.

Table 2 Application Contents

Android (Play store)

title		Lifelog	Universal Pedometer	Pedometer Walk	Life log Application	ActiFit Pedometer Challenges	nlke+ Running	Runtastic Pedometer	Saga Automatic Lifelogging	Smart e-SMBG Diabetes lifelog	Woman Log Calendar
Global Menu	Calendar	Calendar									Calendar
	Bookmark	Bookmark									
	Setting	Setting	Setting	Setting	Setting	Setting	Setting	Setting	Setting	Setting	Setting
Contents Menu	Calorie	Calorie		Calorie				Calorie			
	Step	Step	Step	Step		Step		Step		Step	
	BPM	BPM	BPM				BPM				BPM
	Walk	Walk	Walk	Walk		Walk		Walk		Walk	
	Run	Run				Run	Run				
	Sleep	Sleep	Sleep		Sleep						
	Bike	Bike				Bike					
	Travel	Travel	Travel			Travel					
	Internet	Internet									
	Camera	Camera									
	Music	Music									
	Movie	Movie									
	Game	Game									
	Book	Book									
	Browsing	Browsing									
	Customize	Customize									
			Place	Mile	Meal	toilet		Speed		Blood	Feeling
			Exercise	Time	Study			Weather		Level	Notice
			Weight		Work					Weight	
										Meal	
Total Number		20	9	6	5	7	3	6	1	7	5

iPhone (App store)

title		iLife Logger	MoneySmart	Optimised	Timenote	Life Analytics	All of my life	My Fit Log	LifeSum	Time Golden	Legge
Global Menu	Setting	Setting	Setting	Setting	Setting	Setting	Setting	Setting	Setting	Setting	Setting
Contents Menu	Calorie			Calorie				Calorie			
	Step	Step		Step				Step			Step
	Walk	Walk	Walk	Walk				Walk			Walk
	Sleep	Sleep			Sleep			Sleep		Sleep	
	Travel		Travel							Travel	
	Internet		Internet								
	Camera					Camera	Camera		Camera		
	Book	Book	Book							Book	
	Browsing	Browsing	Browsing								
	Customize	Customize	Customize	Customize	Customize					Customize	Places
		Work	Beauty	weight	Work		Diary		Diary	Work	Storyline
		Eat	Clothes	Health	Study		Star Score			Study	Journey
		House Work	Hobby	Water	Socialize					Travel	
		Exercise	Meal	Stress	Eat					Eat	
		Shopping	Groceries		Play					Shopping	
		Entertainment	Salary		Exercise					House Work	
		Ect	Bonus							Entertainment	
			Serving							Sport	
			House Work								
			Tour								
			Rent								
			Water								
			Phone								
			Cell Phone								
			Tax								
			Insurance								
			Hospital								
			School								
Total Number		15	24	9	9	1	4	5	3	13	6

First, Visual elements are Color, Tone, Arrangement of Color, Style and Shape for analysis. Color is marked with Munsell color wheel (Red, YellowRed, Yellow, Green Yellow, Green, BlueGreen, Blue, PurpleBlue, Purple, PurpleRed). Tone (Vivid, Strong, Bright, Light, Deep, Pale, Very Pale, Dull, LightGray, Gray, Dark) is marked with overall one. According to color analysis, green and blue are most used based on main color and Vivid, Bright and Deep tones are mainly used. For arrangement of color, light and dynamic concept is expressed. It is judged that it is applied to express contents related to daily life and health. For application color, iPhone and Android show difference where iPhone application colors are dynamic and mellow and Android applications colors are dynamic and solid. They also show difference in the arrangement of color where iPhone apps are designed with light and natural concept and Android apps are designed with natural, dynamic and cute concept.

Second, all designs can be designed into most types of graphic style. Graphic elements are used rather than images to help the visual recognition of information according to time. For icons, most are expressed in Square and Circle and have the simple style of fill & stroke in 2D form. Colors are mainly used to classify contents. As for the form, graphs are used to help easy and correct understanding of accumulated data according to time. Bar graph or broken-line graph is used to show data according to time and accurate amount is expressed in a number.

Third, for sound elements, Background Music is not found, Sound Effect is only found in Time Golden and Voice is only found in 'nike+Running' and 'Universal pedometer'. It seems that sound elements are not essential as applications focus on recording daily life, and are rather not effective for intuitive transfer of much information.

Fourth, Motion elements are classified into Basic Motion and Interactivity and only Tap and Drag are found in all applications. As most applications are to show data created by individuals, they are made to respond to tap interaction with simple and intuitive motion. As life-logging stores and expresses a lot of data in daily life, it does not include many functions and maps one function with one motion, showing the tendency of micro interaction.

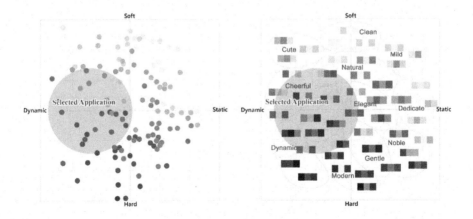

Fig. 1 Color Image Scale & Arrangement of Color Image Scale to Life logging Application

Table 3 Infographic Design for Android Life-Logging Application

no.	App Name	Image	Color	Tone	Combination concept of color	Style	Icon	sort	color	Music	Event	Touch	Voice	Basic Motion	Type of Touch
1	lifelog		all	Bright, light	Cute	Graphic	Square	sans-serif	W	-	-	-	-	Tap, Drag	Tap
2	Universal pedometer		all	bright, ligh	Cute	Graphic	Circle	sans-serif	B, K	-	-	-	0	Tap, Drag	Tap
3	Pedometer Walk		Gray, G, B	Dull	Natural	Graphic	Square	sans-serif	K	-	-	-	-	Tap, Drag	Tap
4	Life log Application		K, B	Vivid	Dynamic	Graphic	Square	sans-serif	W	-	-	-	-	Tap, Drag	Tap
5	ActiFit Pedometer Challenges		G, All	trong, Du	Breezy	Graphic	Circle	sans-serif	B, K	-	-	-	-	Tap, Drag	Tap
6	nike+ Running		R	Strong	Dynamic	Graphic	Square	sans-serif	R, K	-	-	-	-	Tap, Drag	Tap
7	Runtastic Pedometer		P, G	Strong	Elegant	Graphic	Square	sans-serif	K	-	-	-	-	Tap, Drag	Tap
8	Saga Automatic Lifelogging		G	Deep, Dul	Natural	Graphic	Square	sans-serif	K	-	-	-	-	Tap, Drag	Tap
9	Smart e-SMBG Diabetes lifelog		Gray, G	Strong	Natural	Graphic, Image	Square	sans-serif	K	-	-	-	-	Tap, Drag	Tap
10	Woman Log Calendar		R	Light	Breezy	Graphic	Square	sans-serif	K	-	-	-	-	Tap, Drag	Tap

Table 4 Infographic Design for iPhone Life-Logging Application

no.	App Name	Image	Color	Tone	Combination concept of color	Style	Icon	sort	color	Music	Event	Touch	Voice	Basic Motion	Type of Touch
1	iLifeLogger		all	Strong, Dull	Dynamic	Graphic	Square	sans-serif	K	-	-	-	-	Tap, Drag	Tap
2	MoneySmart		B, R	Vivid	Breezy	Graphic	Circle	sans-serif	K, B, R	-	-	-	-	Tap, Drag	Tap
3	Optimized		all	Vivid, Strong	Breezy	Graphic	Circle, Square	sans-serif	K, W	-	-	-	-	Tap, Drag	Tap
4	Timenote		all	Vivid, Strong	Breezy	Graphic	Square	sans-serif	K	-	-	-	-	Tap, Drag	Tap
5	Life Analytics		all	Vivid, Dull	Dynamic	Graphic	Circle	sans-serif	K	-	-	-	-	Tap, Drag	Tap
6	All of my life.		B, G	Dull, Vivid	Natural	Graphic, Image	Square	sans-serif	K	-	-	-	-	Tap, Drag	Tap
7	My Fit Log		B, G	Strong, Bright	Natural	Graphic	Square	sans-serif	K	-	-	-	-	Tap, Drag	Tap
8	LifeStats		B, R, G	Gray, Bright	Natural	Graphic	Square	sans-serif	K	-	-	-	-	Tap, Drag	Tap
9	Time Golden		all	Strong, light	Breezy	Graphic	Circle	sans-serif	K, W	-	O	-	0	Tap, Drag	Tap
10	Loggo		G, B, R	Strong, Bright	Natural	Graphic	Square	sans-serif	K	-	-	-	-	Tap, Drag	Tap

4   Conclusion

This paper aims to analyze how life-logging information is designed in smart phones in order to identify recent trend. Conclusions according to the analysis are as follows.

First, contents of most life-logging applications are related to health. Several applications are extended to SNS, Internet and games but a variety of design styles are not confirmed due to the limited field. Measurement data for primary bio-signals such as health-related motion or heartbeat is statistically shown.

Second, in visual respect, dynamic and mellow colors are most used for design, and as aforementioned, green and blue colors are mainly expressed due to focusing on health. It seems that most use of natural and dynamic arrangement of color is related to the field of health. From the viewpoint of record of daily life, positive and bright tone is used rather than negative one to help the comfortable and stable access of users.

Third, for Sound elements and Motion elements, most of applications did not show special things to be noted. It is judged that this is because sound and motion as an element for correct and intuitive transfer of information are less important than in applications in other fields.

This study has analyzed the infographic design of current life-logging applications from the viewpoint of design. Most applications aimed for information transfer, and as targeted users are extended, a variety of designs has to be developed according to users. It is expected that this analysis can contribute to differentiated design fit for increasingly extended and diversified contents in providing life-logging services.

Acknowledgment This work was supported by the ICT R&D program of MSIP/IITP [R0126-15-1045; the development of technology for social life logging based on analyzing social emotion and intelligence of convergence contents].

References

1. Achilleos, K.: Evolution of Lifelogging. University of Southampton, Hampshire (2010)
2. Jeon, J.H., Yeon, J., Lee, S.-G, Seo, J.: Exploratory Visualization of Smarphone-based Lifelogging Data using Smart Reality Testbed. Seoul National University (2014)
3. Hankyong Economy Glossary. http://s.hankyung.com/dic/
4. Hopfgartner, F., Yang, Y., Zhou, L.M., Gurrin, C.: User interaction templates for the design of lifelogging systems. In: Semantic Models for Adaptive Interactive Systems (2013)
5. Soo Youn, P., Seung, K.: Usability Evaluation of Lifelogging Application on Mobile. Journal of Digitaldesign, 117–127 (2014)

Cardiovascular Synchrony for Determining Significant Group in Social Life Logging

Jincheol Woo, Young-Ju Kim, Seung Seob Shin, Young Ho Jo,
Hojung Choi and Mincheol Whang

Abstract This study is to confirm significant group determination for life logging. Six undergraduate students were participated in experiment. They were divided to two groups consist of three persons. Subjects were presented to sixteen movie clips relating to emotion of four categories. PPIs (Pulse to Pulse Intervals) extracted from PPG were analyzed. Pearson correlation was tested to PPIs between subjects. When the same emotional stimuli were presented, PPIs of twelve pairs were highly correlated significantly. These results showed the possibility to utilize criteria for synchrony for determining significant group.

Keywords Cardiovascular synchrony · PPG · PPI · Emotion

1 Introduction

People desire fundamentally to record and recollect a specific event or an experience of the past. A subjective diary has been used by traditional recording tools. This recorded information was not reliable because it depended on own memory [1]. As sensor technology advanced, a daily life was able to be recorded in real-time not only external data as image, sound, light intensity but also internal data as heart rate, galvanic skin reflex, skin temperature [2]. Most studies in the field of life logging have focused on "How to record the reliable data in various types?" and "How to support reminiscence satisfactorily?". SenseCam is a camera

J. Woo · Y.-J. Kim · S.S. Shin · Y.H. Jo · H. Choi
Department of Emotion Engineering, Sangmyung University,
Hongji-dong, Jongno-gu, Seoul 110-743, South of Korea

M. Whang(✉)
Department of Digital Media, Sangmyung University,
Hongji-dong, Jongno-gu, Seoul 110-743, South of Korea
e-mail: whang@smu.ac.kr

© Springer Science+Business Media Singapore 2015
D.-S. Park et al. (eds.), *Advances in Computer Science and Ubiquitous Computing*,
Lecture Notes in Electrical Engineering 373,
DOI: 10.1007/978-981-10-0281-6_28

193

around the neck that can measure the front image and distance, light intensity. From this equipment, measured data could improve user's reminiscence and recognize human activities [3, 4].

While social network service has emerged as a major issue, in field of life logging, research about sharing method in personal daily life needed with other people. Social concept in life logging defined to detect and record social situations automatically and to recognize social interactions immediately [5, 6]. A seminal study in this area is the work of the collective intelligence formed by social tagging data in web. This study showed that collective intelligence from many and unspecified persons enhanced the reminiscence [7]. Previous researches determined a social connection by sharing time, space or event without consideration about user's feedback [5-8].

Relationship between parents and infant is the first social connection in humans. Mother and infant were linked physiologically during physical and mental interactions. Heart information from ECG (Electrocardiogram) was used to physiological precursor of synchrony [9, 10]. In field of life logging, ECG was not easy to utilize than PPG (Photoplethysmogram). Therefore, the recently developed life logging devices included PPG sensor almost. In many studies pulse information from PPG could be utilized as an alternative measurement of ECG [11-13]. This study is to investigate cardiovascular synchrony between two persons during exposure to emotional stimuli in same category.

2 Materials and Methods

2.1 Participants

Physically and mentally healthy undergraduate students (five males and one female, average age: 24.1 ± 3.3) were participated in this experiment. All participants read and signed an informed consent form. And they were paid nearly 30 \$ as a remuneration.

2.2 Task Procedures and Experimental Task

Three subjects were seated comfortably watching the front at a distance of 290 cm from projection screen of 100 inches shown as Figure 1. The projector (Taiwan, BenQ W1070+) and the speaker were utilized for showing movie clips.

Sixteen movie clips were selected for causing emotion based on two dimensions model [14]. The list of movie clips presented according to number of sequence shown as Table 1. Each category includes four movie clips (Pleasant-Arousal: 1, 3, 7, 13; Pleasant-Relaxation: 4, 12, 15, 16; Unpleasant-Arousal: 5, 6, 10, 14; Unpleasant- Relaxation: 2, 8, 9, 11). Before the start of each movie clips, a grey screen was showed during two minutes thirty seconds in order to stabilize physiological response.

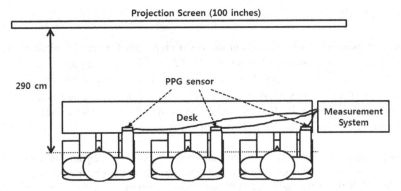

Fig. 1 Experiment Configuration.

Table 1 List of movie clips.

No. of Sequence	Name of movie clips	Categories of emotion	Play times
1	The visitors	Pleasant, Arousal	2m 9s
2	ET	Unpleasant, Relaxation	4m 35s
3	When Harry Met Sally...	Pleasant, Arousal	2m 45s
4	Forrest Gump	Pleasant, Relaxation	2m 1s
5	Scream	Unpleasant, Arousal	6m 33s
6	The Shining	Unpleasant, Arousal	4m 15s
7	There's something about Mary	Pleasant, Arousal	2m 26s
8	City of Angels	Unpleasant, Relaxation	4m 15s
9	A perfect world	Unpleasant, Relaxation	4m 27s
10	Misery	Unpleasant, Arousal	3m 31s
11	Life is beautiful	Unpleasant, Relaxation	2m 7s
12	Dead Poets Society	Pleasant, Relaxation	2m 40s
13	There's something about Mary	Pleasant, Arousal	2m 55s
14	Scream 2	Unpleasant, Arousal	3m 35s
15	Life is beautiful	Pleasant, Relaxation	4m 9s
16	The eighth day	Pleasant, Relaxation	2m 10s

2.3 Analysis Method

Data were measured to 500 Hz sampling rate from PPG sensor and were abstracted for 2 minutes at the beginning of the movie clip. First, each peak of PPG raw data were measured using Labview 2010 (USA, National Instruments). Then PPI (Pulse to Pulse Intervals) was calculated and integrated by four categories of emotion. Pearson correlation were tested to between subjects.

Fig. 2 Analysis Method

3   Result

Table 2 presented values of correlation coefficient and P value between subjects. First rows informed group information. P1, P2, and P3 were group A, and P4, P5, and P6 were group B. Group D means pairs between different groups. First row showed four categories (UA: Unpleasant-Arousal, PA: Pleasant-Arousal, PR: Pleasant-Relaxation, UR: Unpleasant-Relaxation) of two dimensional emotions. And second row indicated the meaning of value.

Table 2 Values of correlation coefficient and P value between participants. Bolded decimal point emphasized to R higher than 0.3, and significant P value. Underlined decimal point indicate to R between 0.1 and 0.3, and P value lower than 0.01.

Group	Pairs	Unpleasant-Arousal (UA)		Pleasant-Arousal (PA)		Pleasant-Relaxation (PR)		Unpleasant-Relaxation (UR)	
		R	P value	R	P value	R	P value	R	P value
A	P1-P2	0.041	0.375	**0.467**	**0.000**	<u>0.283</u>	<u>0.000</u>	0.114	0.013
	P1-P3	<u>0.141</u>	<u>0.002</u>	0.075	0.103	<u>0.222</u>	<u>0.000</u>	0.021	0.651
	P2-P3	**0.314**	**0.000**	<u>0.185</u>	<u>0.000</u>	**0.459**	**0.000**	**0.331**	**0.000**
B	P4-P5	**0.301**	**0.000**	<u>0.169</u>	<u>0.000</u>	**0.331**	**0.000**	<u>0.142</u>	<u>0.002</u>
	P4-P6	0.076	0.097	**0.308**	**0.000**	<u>0.175</u>	<u>0.000</u>	**0.322**	**0.000**
	P5-P6	<u>0.217</u>	<u>0.000</u>	-0.078	0.089	**0.418**	**0.000**	<u>0.185</u>	<u>0.000</u>
D	P1-P4	<u>0.166</u>	<u>0.000</u>	0.065	0.154	<u>0.134</u>	<u>0.003</u>	<u>0.159</u>	<u>0.000</u>
	P1-P5	0.038	0.403	**0.309**	**0.000**	0.023	0.623	0.055	0.227
	P1-P6	-0.048	0.293	-0.347	0.000	-0.005	0.919	0.082	0.072
	P2-P4	<u>0.287</u>	<u>0.000</u>	0.042	0.362	**0.320**	**0.000**	-0.013	0.768
	P2-P5	0.058	0.205	<u>0.151</u>	<u>0.001</u>	-0.010	0.821	-0.127	0.005
	P2-P6	<u>0.226</u>	<u>0.000</u>	-0.144	0.002	-0.110	0.016	0.089	0.050
	P3-P4	<u>0.121</u>	<u>0.008</u>	<u>0.199</u>	<u>0.000</u>	<u>0.108</u>	<u>0.018</u>	<u>0.140</u>	<u>0.002</u>
	P3-P5	<u>0.106</u>	<u>0.020</u>	-0.142	0.002	0.019	0.675	-0.133	0.003
	P3-P6	0.078	0.089	<u>0.134</u>	<u>0.003</u>	0.037	0.418	**0.372**	**0.000**

In case of correlation results in same group, for watching movie clips to UA category, P2-P3 (R: 0.314, P: 0.000) and P4-P5(R: 0.301, P: 0.000) were correlated significantly. In PA category, P1-P2 (R: 0.467, P: 0.000) and P4-P6 (R: 0.308, P: 0.000) pairs were highly correlated. In PR category, P2-P3 (R: 0.459, P: 0.000), P4-P5 (R: 0.331, P: 0.000), and P5-P6 (R: 0.418, P: 0.000) pairs were correlated significantly. In UR category, P2-P3 (R: 0.331, P: 0.000) and P4-P6 (R: 0.322, P: 0.000). Nine pairs were that correlation coefficient were between 0.1 and 0.3, and P value were lower than 0.01.

In case of correlation results between different groups, for watching movie clips to PA category, P1-P5 (R: 0.309, P: 0.000) was correlated. In PR category, P2-P4 (R: 0.320, P: 0.000) was correlated. In UR category, P3-P6 (R: 0.372, P: 0.000) was correlated significantly. And twelve pairs were that correlation coefficient were between 0.1 and 0.3, and P value were lower than 0.01.

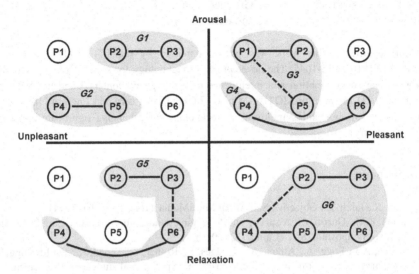

Fig. 3 shown is results of grouping. Criteria for synchronization were correlated that R was higher than 0.3, and P was lower or equal than 0.00. Two dimensional space of emotion were drawn. And six participants were indicated to P1 to P6. G1 to G6 and grey background shape were showed groups according to synchronization. The solid line informed to synchronize between participants in same group. The dot line informed synchronization between different groups.

Grouping results were shown in figure 3. P2-P3 and P4-P5 pairs were grouped to G1 and G2 in UA category. P1, P2, and P5 were grouped to G3 in PA category. P2, P3, P4 and P6 were grouped to G5 in UR category. P2 to P6 were grouped to G6 in PR category.

4   Conclusion and Discussion

Many studies in life logging have been carried out to record and reminisce own daily life [1-6]. Most researchers have not treated criteria of social connection [1-4]. In order to confirm heart rhythm synchronization between two persons showing the same emotional stimuli, PPIs from were extracted from raw PPG and tested to Pearson correlation.

The results from correlation test showed that synchronization take place in the case of pleasant more in the case of unpleasant. The most synchronization emerged in pleasant-relaxation category. Grouping results revealed that it could be grouped in all case of emotion categories. Groups in relaxation case were larger than in arousal case.

These results showed the possibility to utilize criteria for synchrony for determining significant group. Grouped results were expected to be used to estimate the mass of the patterns and trends.

Acknowledgments This work was partly supported by the ICT R&D program of MSIP/IITP [R0126-15-1045, The development of technology for social life logging based on analyzing social emotion and intelligence of convergence contents] and the National Strategic R&D Program of MOTIE (Ministry of Trade, Industry and Energy) [10044828, Development of augmenting multisensory technology for enhancing significant effect on service industry].

References

1. Jain, R., Jalali, L.: Objective Self. IEEE MultiMedia **21**(4), 100–110 (2014)
2. Zhou, L.M., Gurrin, C.: A survey on life logging data capture. In: SenseCam 2012: 3rd Annual Symposium. SenseCam 2012 (2012)
3. Sellen, A.J., Fogg, A., Aitken, M., Hodges, S., Rother, C., Wood, K.: Do life-logging technologies support memory for the past?: an experimental study using sensecam. In: Proceedings of the SIGCHI Conference on Human Factors in Computing Systems, pp. 81–90. ACM (2007)
4. Doherty, A.R., Caprani, N., Conaire, C.Ó., Kalnikaite, V., Gurrin, C., Smeaton, A.F., O'Connor, N.E.: Passively recognising human activities through lifelogging. Computers in Human Behavior **27**(5), 1948–1958 (2011)
5. Groh, G., Lehmann, A., Wang, T., Huber, S., Hammerl, F.: Applications for social situation models. In: Proc. Int'l Conf. Wireless Applications and Computing Conference, Freiburg, Germany (2010)
6. Sueda, K., Duh, H.B.L., Rekimot, J.: Social life logging: can we describe our own personal experience by using collective intelligence?. In: Proceedings of the 10th Asia Pacific Conference on Computer Human Interaction. ACM (2012)
7. Whittaker, S., Kalnikaitė, V., Petrelli, D., Sellen, A., Villar, N., Bergman, O., Clough, P., Brockmeier, J.: Socio-technical lifelogging: deriving design principles for a future proof digital past. Human-Computer Interaction. **27**(1–2), 37–62 (2012)
8. Burns, W., Nugent, C., McCullagh, P., Zheng, H.: Design and evaluation of a smartphone based wearable life-logging and social interaction system. In: 2014 IEEE 27th International Symposium on Computer-Based Medical Systems (CBMS). IEEE (2014)
9. Feldman, R., Magori-Cohen, R., Galili, G., Singer, M., Louzoun, Y.: Mother and infant coordinate heart rhythms through episodes of interaction synchrony. Infant Behavior and Development **34**(4), 569–577 (2011)
10. Feldman, R.: From biological rhythms to social rhythms: Physiological precursors of mother-infant synchrony. Developmental Psychology **42**(1), 175 (2006)

11. Hsu, C.C., Weng, C.S., Liu, T.S., Tsai, Y.S., Chang, Y.H.: Effects of electrical acupuncture on acupoint BL15 evaluated in terms of heart rate variability, pulse rate variability and skin conductance response. The American Journal of Chinese Medicine **34**(01), 23–36 (2006)

12. Posada-Quintero, H.F., Delisle-Rodríguez, D., Cuadra-Sanz, M.B., de la Vara-Prieto, R.F.: Evaluation of pulse rate variability obtained by the pulse onsets of the photoplethysmographic signal. Physiological Measurement **34**(2), 179 (2013)

13. Lin, P.C., Hsu, K.C., Chang, C.C., Hsiao, T.C.: Reliability of instantaneous pulse rate variability by using photoplethysmography. In: Workshop on Biomedical Microelectronic Translational Systems Research (WBMTSR 2014). EPFL (2014)

14. Russell, J.A.: A circumplex model of affect. Journal of Personality and Social Psychology **39**(6), 1161 (1980)

Correlation Between Heart Rate and Image Components

Min Woo Park, Jiwon Im, Jieun Kwon, Mincheol Whang and Eui Chul Lee

Abstract Because the vision-based environmental context awareness technique has become a more important issue for implementing affective computing application, the effects of image components should be investigated in terms of human responses. In this paper, we studied the correlation between heart rate and image components such as hue, saturation, and intensity. We collected heart rate signals from subjects who respectively watched 32 different kinds of images chosen from a public emotion image database. As a result, we confirmed that the heart rate trend decreased in terms of hue and intensity but increased in terms of saturation. In addition, we confirmed that the correlation between heart rate and saturation was greater than that between heart rate and hue and intensity in terms of R^2 value. That is, heart rate trends can change depending on the effect of image components.

Keywords Heart rate · Image components · Hue · Saturation · Intensity

1 Introduction

Recently, the number of research studies performed for measuring heart rate in terms of psychological and clinical services has increased [1][2]. Especially to implement an intelligent life-logging service, heart rate is widely monitored by

M.W. Park · J. Im
Department of Computer Science,
Graduate School, Sangmyung University, Seoul, Republic of Korea

J. Kwon · M. Whang
Department of Emotion Engineering,
Graduate School, Sangmyung University, Seoul, Republic of Korea

E.C. Lee(✉)
Department of Computer Science, Sangmyung University, Seoul, Republic of Korea
e-mail: eclee@smu.ac.kr

© Springer Science+Business Media Singapore 2015
D.-S. Park et al. (eds.), *Advances in Computer Science and Ubiquitous Computing*,
Lecture Notes in Electrical Engineering 373,
DOI: 10.1007/978-981-10-0281-6_29

using wearable devices such as a smart watch and band. Moreover, information about the environmental scene is important for recording human experience and estimating emotion. However, no research has been conducted to validate the correlation between heart rate and quantitative scene information on environmental image.

In fields of marketing and psychology, many studies have been performed for qualitative validation of the psychological effects of color [3]. In previous works, emotion is measured by using physiological signals such as electroencephalography, electrocardiography (ECG), and photoplethysmography (PPG) [4]. In addition, the effect of color light stimulation is analyzed in terms of heart rate variability by using two-color (e.g., red and blue) light panels [5]. Consequently, human emotion can be affected by image components such as color and lightness. Image components can be represented by using the HSI (hue, saturation, and intensity) color model, which is easily extracted from generally used RGB color images.

To validate the correlation between quantitatively represented image components and heart rate, an experiment was performed in this research and the correlation coefficient between heart rate and each component value of HSI was determined. For this purpose, 32 different kinds of images were chosen from the International Affective Picture System (IAPS) [9] emotion image database.

2   Proposed Method

2.1   Heart Rate Measurement

Heart rate is defined as the number of heart beats per minute, or as heart rate per unit time. Generally, heart rate is measured by using sensors such as PPG and ECG. In this study, a PPG sensor of USB (universal serial bus) type was used for acquiring the successive heart rate information, as shown in Fig. 1. Specifically, the R-peak, which is the zero crossing position from the positive to negative gradient, was used for calculating heart rate. Then, the time interval was detected between two consecutive R-peaks [6].

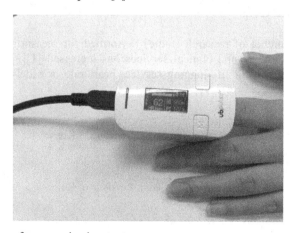

Fig. 1 PPG sensor for measuring heart rate

Thus, heart rate can be calculated by dividing 60 seconds and the RRI, which is the interval between two R-peaks. For example, in case of 1 second of RRI, the heart rate is calculated as 60 times per minute (= 60 seconds/1 second) based on the following equation:

$$\text{Heart Rate} = \frac{60}{t} \qquad (1)$$

2.2 Image Components

The image components contain HSI as shown in Fig. 2.

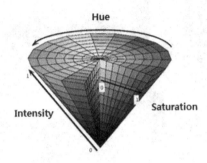

Fig. 2 HSI color model [7]

In our method, the image components are extracted by converting from the RGB to the HSI of the original image by using Eqs. (2)–(4) [8]. As the components of the RGB color model imply both the meaning of brightness and color, using only brightness or colors as a color model for application is difficult. On the other hand, the limitation of the RGB model can be solved by using the HSI color model because it represents separately color, brightness, and saturation. Therefore, in this study, the HSI model is used for analyzing the effects of color.

$$H = \cos^{-1}\left[\frac{0.5*[(R-G)+(R-B)]}{(R-G)^2+(R-B)(G-B)x^{1/2}}\right] \qquad H \in [0, \pi] \text{ for } B \leq G \qquad (2)$$

$$H = 2\pi - \cos^{-1}\left[\frac{0.5*[(R-G)+(R-B)]}{(R-G)^2+(R-B)(G-B)x^{1/2}}\right] \qquad H \in [\pi, 2\pi] \text{ for } B > G$$

$$S = 1 - 3*\min(R, G, B) \qquad S \in [0, 1] \qquad (3)$$

$$I = (R + G + B)(3*255) \qquad I \in [0, 1] \qquad (4)$$

To normalize the generally used scale of values, the H, S, and I values were respectively converted into the ranges of 0–360, 0–255, and 0–255, respectively, according to the following formula: $H = H \times 180°/\pi$; $S = S \times 255$ and $I = I \times 255$.

3   Experiments and Results

3.1   Configuration

The experimental environment consisted of a monitor and a PPG sensor, as shown in Fig. 3. In our experiments, 6 subjects who had no visual impairments participated voluntarily (male = 3, female = 3, mean age = 25.4). The subjects gazed at a monitor plane from a distance of 1.2 m. In addition, the subject wore a PPG sensor on the index finger of the left hand.

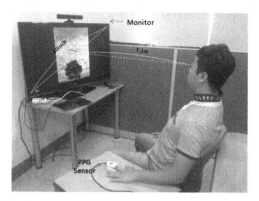

Fig. 3 Experimental Environment

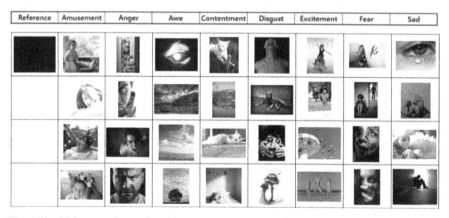

Fig. 4 The 32 images chosen from the IAPS database [9]

Fig. 5 Experimental Procedure

For emotional stimuli, the IAPS database, which contained 32 different kinds of images that represent eight emotions such as amusement, anger, awe, contentment, disgust, excitement, fear, and sad for not deflecting in particular emotion were used [9]. Originally, the IAPS database consisted of 400 images, including 50 images for each of the eight emotions. In this study, 32 images that represented the eight emotions, with 4 images each, were selected randomly among the 50 images, as shown in Fig. 4. Then, the average values of the image components such as hue, saturation, and intensity were extracted from the 32-image data.

Next, the experimental procedure consisted of two stages, as shown in Fig. 5. First, reference data were measured in 20 seconds. After that, the subjects watched the 32 images in 20 seconds per image. All of the 32 images were shown randomly. While the subjects were watching the randomly shown images, their heart rates were simultaneously measured by using a PPG sensor, as mentioned earlier and as shown in Fig. 1.

3.2 Results

For analyzing the correlation between heart rate and HSI, we used the average values of HSI (X-axis) and heart rate (Y-axis) in 6 subjects that were derived from the 32 images. For eliminating individual differences in heart rate, we used normalized heart rate data, which subtracted original experimental data from the reference data. Heart rate responses were extracted from the images by calculating the average of only the latter 10 seconds even though heart rate was recorded in 20 seconds per image. Consequently, the average heart rate measured from watching the same image is determined as the final feature value.

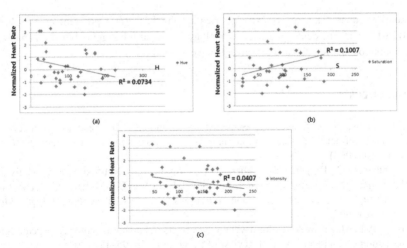

Fig. 6 Results on the correlation between HSI (X-axis) and heart rate (Y-axis): (a) hue, (b) saturation, and (c) intensity

The experimental result can be analyzed in three viewpoints such as hue, saturation, and intensity. First, the heart rate trend decreased with the increment in hue value, as shown in Fig. 6 (a). Second, heart rate trend increased from the scale of 0 to 255 in terms of saturation, as shown in Fig. 6 (b). Last, the heart rate trend decreased from a scale of 0 to 255 in terms of intensity, as shown in Fig. 6 (c). In detail, heart rate gradually decreased from 0° (red) to 240° (blue color) in terms of hue. Moreover, we confirmed that the correlation of heart rate with saturation was greater (0.1007) than its correlation with hue (0.0734) and intensity (0.0407) in terms of R^2 value even though R^2 values are significantly smaller. As a result, we confirmed that the heart rate variability trend can change depending on the effects of image components.

4   Conclusion

In this paper, we studied the correlation between heart rate and image components such as hue, saturation, and intensity. We collected heart rate signals from subjects who respectively watched 32 different kinds of images chosen from a public emotion image database. Based on the results, we confirmed that the heart rate trend decreased in terms of hue and intensity but increased in terms of saturation. However, R^2 values were significantly smaller despite this trend.

To solve the limitations of this study, we applied the weighted average method and focused on gaze point at which the subjects mostly watched the images. In addition, we applied the logistic and polynomial regression methods instead of the linear regression method.

Acknowledgment This work was supported by the ICT R&D program of MSIP/IITP [R0126-15-1045; the development of technology for social life logging based on analyzing social emotion and intelligence of convergence contents].

References

1. Delplanque, S., N'Diaye, K., Scherer, K., Grandjean, D.: Spatial frequencies or emotional effects? discrete wavelet analysis. Journal of Neuroscience Methods **165**(1), 144–150 (2007)
2. Berntson, G.G., Bigger, J.T., Eckberg, D.L., Grossman, P., Kaufmann, P.G., Malik, M., Nagaraja, H.N., Porges, S.W., Saul, J.P., Stone, P.H., Van Der Molen, M.W.: Heart rate variability: origins, methods, and interpretive caveats. Psychophysiology **34**(6), 623–648 (1997)
3. Eliot, A.J., Maier, M.A.: Color psychology: effects of perceiving color on psychological functioning in humans. Annual Review of Psychology **65**, 95–120 (2014)
4. Sohn, J.H.: Measurement of emotion and sensibility using physiological signals. Journal of the Korean Society of Precision Engineering **18**(2), 14–25 (2001)

5. Litscher, D., Wang, L., Gaischek, I., Litcher, G.: The Influence of New Colored Light Stimulation Methods on Heart Rate Variability, Temperature, and Well-Being: Results of a Pilot Study in Humans. Evidence-Based Complementary and Alternative Medicine (2013)
6. Park, M.W., Kim, C.J., Whang, M., Lee, E.C.: individual emotion classification between happiness and sadness by analyzing photoplethysmography and skin temperature. In: Fourth World Congress on Software Engineering (WCSE), pp. 190–194 (2013)
7. Gonzalez, R.C., Woods, R.E., Eddins, S.L.: Digital Image Processing Using MATLAB. Pearson Prentice Hall (2004)
8. RGB to HSI. http://www.cse.usf.edu/~mshreve/rgb-to-hsi
9. Machajdik, J., Hanbury, A.: Affective image classification using features inspired by psychology and art theory. In: Proceedings of the International Conference on Multimedia, pp. 83–92. ACM (2010)

Experimental Verification of Gender Differences in Facial Movement According to Emotion

Yoonkyoung Kim and Eui Chul Lee

Abstract In emotion expression, there are many differences between males and females, such as degree, appearance, and factors arousing emotion. However, the differences are not yet fully verified due to the lack of effective feature extraction methods. Therefore, we developed a facial movement extraction method to perform an experimental verification. To induce four kinds of emotion, both visual and auditory stimuli were simultaneously presented to subjects. Among 121 facial feature points, 31 main points were selected for the analysis of facial movement. The results indicated that males' facial movements were almost identical for all the emotions. On the other hand, females' facial movements showed different tendencies. Moreover, we found that the average movement for males was higher than for females, while discriminative power for different emotions was higher for females than for males.

Keywords Facial expression · Emotion recognition · Micro-movement · Gender difference

1 Introduction

The face not only shows abundant expression, but also indicates immediate responses via various parts, such as the forehead, eyebrows, eyes, mouth, etc. Even though many methods for facial expression recognition have been studied, accuracy and reasonable speed are still weak points.

In the 1960's, facial expressions were identified by Paul Ekman. He defined the movements and relationships of facial muscles [1] and created the Facial Action Coding System (FACS), which was designated as a "first face emotion map."

Y. Kim · E.C. Lee(⌗)
Department of Computer Science, Sangmyung University, Seoul, Republic of Korea
e-mail: eclee@smu.ac.kr

© Springer Science+Business Media Singapore 2015   209
D.-S. Park et al. (eds.), *Advances in Computer Science and Ubiquitous Computing*,
Lecture Notes in Electrical Engineering 373,
DOI: 10.1007/978-981-10-0281-6_30

However, the FACS was only described, but not verified. Therefore, to verify Ekman's definition, it is necessary to actually analyze facial expressions obtained from a camera. Previous research has recognized facial expressions using the Active Appearance Model (AAM) algorithm [2]. The AAM algorithm is a feature extraction method based on Principal Component Analysis (PCA) [3]. Because this method involves a lengthy performance time, it is difficult to implement in real time. Moreover, in cases using an Eigen-point, there are some constraints: face movement should be exaggerated and illumination should be uniformly maintained [4]. Using facial movement features should prove to be a significant method in the aspect of replicating the FACS. Previous research that implemented the FACS used static images, but static images do not sufficiently indicate facial elements and muscle movements [5].

In sensibility ergonomics research, in 1967, Eckard Hess, a psychologist from Chicago University, defined characteristics of pupil accommodation for each gender [6]. He proposed a hypothesis that pupil size could change differently according to gender when looking at an attractive or hostile object under the same lighting intensity. Since this proposal was issued, many researchers have verified gender differences in emotion expression in pupil size, speech signals, and so on [7, 8]. For example, to verify the difference, previous research used three dimensions in its analysis: convert responding, interpersonal expression, and attitude [9]. However, since interpersonal expression was not considered in the study, the results were inaccurate. Another previous research determined that females have a higher rate of distinguishing one emotion from another [10]. This research analyzed two emotions, such as valence and arousal. However, using only two emotions does not provide sufficient data for establishing standards of characteristic facial expression. In addition, to induce emotion, the researchers used an independent visual stimulus. Using a single stimulus is not sufficient for inducing a specific emotion [11]. Other previous research verified characteristics distinguishing between males and females using both facial images and speech signals [12]. They obtained facial feature vectors using Linear Discriminant Analysis (LDA) [13] and the speech signal was obtained by wavelet sub-band. Subsequently, they merged the facial feature vectors and speech signals. The performance of their method was better than using only a single feature. However, LDA, wavelet transform, and merging methods take longer to perform. Thus, these methods are not yet fully efficient.

To overcome these problems, a Kinect camera can be used. In this paper, we chose 31 of the 121 facial feature points obtained by the Kinect Face Tracking SDK that demonstrate significant movement and can be used to analyze facial movement. Moreover, we used a relatively high frequency of 31 Hz. Therefore, we could overcome the limitation of recognizing only extreme expressions. In addition, by using the Kinect camera, it is possible for our program to measure real time performance. We used four emotional states in this study: amused, content, angry, and sad. To induce a specific emotion, both visual and auditory stimuli were presented simultaneously to the subjects.

The next section shows the proposed method we developed and the organization of the experiment. The results of the experiment are presented in section 3.

2   Method of Facial Feature Extraction

In this paper, facial feature points of an active appearance model are obtained via Kinect Face Tracking SDK [14]. The SDK extracts 121 facial feature points. We choose 31 main points that are significant based on Ekman's action units [1]. Fig. 1 shows the numbering of the selected points.

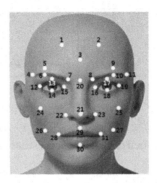

Fig. 1 Numbering of the main points used in the analysis

To analyze pixel changes around the 31 points, we used the pixel distance between two adjacent frames [15]. The method used to calculate the distance includes a shift matching scheme, the local binary pattern, and the Hamming distance [16-18].

To induce emotion, we used two stimuli. In previous research, they estimated an ElectroEncephaloGram (EEG) signal according to a stimulus modality, such as auditory, visual, and combined visual with auditory stimuli [11]. Previous results indicate that using both visual and auditory stimuli is the most efficient method for eliciting emotion. Therefore, we used a combined visual and auditory modality. The visual stimuli we used comprised artistic photography from a photo sharing site [19]. The artistic photographs were verified in terms of emotion classification of low-level features [20]. The auditory stimuli were selected and downloaded from a music sharing site [21], based on Russell's 2D emotion model, with an arousal-relaxation axis and a positive-negative axis. We verified the validity of the auditory stimuli using from four phases: extraction of the candidate sound, Focus Group Discussion (FGD), chi-squared test, and Factor analysis.

Fig. 2 shows Russell's 2D emotion model. The red circles denote the four emotional states that we used.

There are five female and five male subjects. They put on earphones for the auditory stimulus presentation. We presented 50 visual stimuli for each emotion as a slide show. The Kinect camera was located at the center position under the monitor in order to obtain a frontal view of the faces. In addition, to prevent an order effect, the sequence of emotion inducement was randomized [22]. We estimated facial movement during neutral emotion since facial movement differs between individuals. Thus, we could minimize individual variation in the results by removing neutral emotion from the specific emotions induced.

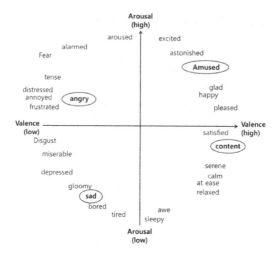

Fig. 2 Russell's emotion model (red circles: emotional states used in this paper)

3   Experimental Results

Results are indicated on a Maya face model. We analyzed the amount of facial movement for each of the four emotions by gender.

Table 1 Results of movement analysis

Emotion	Male	Female
Amused	0.516	0.502
Content	0.513	0.384
Angry	0.442	0.196
Sad	0.598	0.270
Average	**0.517**	**0.338**
Standard deviation	**0.063**	**0.133**

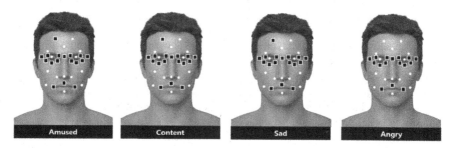

Fig. 3 Results for males' facial movements

The average standard of movement was assessed for each emotional state Table 1. In other words, if movement at some point was higher than average, we considered that point as significant for that emotion. As shown in Table 1, the average movement for males was higher than for females, but the standard deviation was lower. This result is similar to Eckard Hess' hypothesis. These results are displayed in Figs. 3 and 4.

Fig. 3 shows the results of the males' facial movement analysis. The rectangles denote significant points that show a high amount of movement. The results indicate a difference at only one point, that is, the right forehead for the angry emotional state. The movement distributions for the other emotional states were identical.

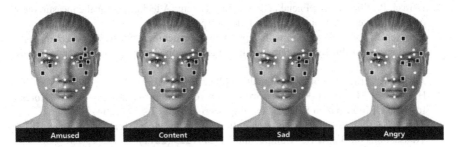

Fig. 4 Results for females' facial movements

Fig. 4 shows the results of the females' facial movement analysis. While the males' facial movement distributions were similar for all emotional states, the females' distributions differed according to each emotion. In particular, for amusement, the movement of the left eye was higher. We can interpret these results two ways. First, among the 31 main points, we found a significant point that can be helpful for recognizing facial emotion. Second, males demonstrate consistent facial expression for changing emotions, and conversely, females demonstrate various facial expressions according to their emotional states.

4 Conclusion

In this paper, we verified differences in the amount of facial movement according to gender. The results indicate that the males' facial movement distribution was similar for all emotional states. On the other hand, the movement distributions for females differed for each emotional state. These findings signify that the ability to distinguish one emotion from another using facial movement is higher for females than for males. Moreover, facial movements were larger when specific emotions were elicited, in comparison to neutral emotion, for both genders.

In future work, we plan to match facial movement with Ekman's action units. Because the FACS has been described, but not verified, we will validate the FACS. Using the algorithm we developed, we could analyze the direction and size of movement in a particular region. In addition, this research would be beneficial for understanding patients with facial nerve palsy.

Acknowledgement This work was the MSIP (Ministry of Science, ICT and Future Planning), Korea, under the ITRC (Information Technology Research Center) support program (IITP-2015-H8501-15-1014) supervised by the IITP (Institute for Information & communications Technology Promotion).

References

1. Ekman, P., Friesen, W.: Facial action coding system: A technique for the measurement of facial movement. Consulting psychologists press (1978)
2. Heisele, B.: Face recognition with support vector machines: global versus component-based approach. In: IEEE International Conference on Computer Vision, pp. 688–694 (2001)
3. Jolliffe, L.: Principal component analysis (2005)
4. Hong, S.-H., Byun, H.-R.: Facial expression recognition using eigen-points. Korean Institute Of Information Science And Engineering **31**, 817–819 (2004)
5. Zhang, L.: Facial expression recognition using facial movement features. IEEE Transactions on Biometrics Compendium **2**, 219–229 (2011)
6. Hess, E.: Attitude and pupil size. Scientific American **212**, 46–54 (1965)
7. Partala, T., Surakka, V.: Pupil size variation as an indication of affective processing. International Journal of Human-Computer Studies **59**, 185–198 (2003)
8. Go, H.-J.: Emotion recognition from the facial image and speech signal. In: SICE 2003 Annual Conference, vol. 3, pp. 2890–2895 (2003)
9. Allen, J.G.: Sex differences in emotionality: A multidimensional approach. SAGE Journals **29**, 711–722 (1976)
10. Montagne, B., Kessels, R.P.C., Frigerio, E., de Haan, E.H.F., Perrett, D.I.: Sex differences in the perception of affective facial expressions: Do men really lack emotional sensitivity? Cognitive Processing **6**, 136–141 (2005)
11. Bos, D O.: EEG-based emotion recognition: The influence of visual and auditory stimuli (2006). http://hmi.ewi.utwente.nl/verslagen/capita-selecta/CS-Oude_Bos-Danny.pdf
12. Thayer, J., Johnsen, B.H.: Sex differences in judgement of facial affect: A multivariate analysis of recognition errors. Scandinavian Journal of Psychology **41**, 243–246 (2000)
13. Mika, S., Ratsch, G., Weston, J., Scholkopf, B., Muller, K.-R.: Fisher discriminant analysis with kernels. In: IEEE Workshop on Neural Networks for signal Processing, pp. 41–48 (1999)
14. Jana, A.: Kinect for windows SDK Programming Guide. Packt Publishing Ltd. (2012)
15. Lee, E.C., Kim, H., Bae, M., Kim, Y.: Analysis of facial movement according to opposite emotions. Journal of Korea Contents Association, Accepted for Publication
16. Kim, Y., Kim, H., Lee, E.C.: Emotion classification using Facial Temporal sparsity. International Journal of Applied Engineering Research **9**, 24793–24801 (2014)

17. Ahonen, T.: Face description with local binary patterns: Application to face recognition. IEEE Transactions on Pattern Analysis and Machine Intelligence **28**, 2037–2041 (2006)
18. Hamming, R.W.: Error detecting and error correcting codes. Bell System Technical Journal **29**, 147–160 (1950)
19. www.deviantart.com (access on 1st August 2015)
20. Machajdik, J., Hanbury, A.: Affective image classification using features inspired by psychology and art theory. In: Proceeding of the International Conference on Multimedia, pp. 83–92 (2010)
21. http://musicovery.com (access on 1st August 2015)
22. Easterbrook, J.A.: The effect of emotion on cue utilization and the organization of behavior. Psychological Review **66**, 183–201 (1959)

Heart Rate Synchronization with Spatial Frequency of Visual Stimuli

Jiwon Im, Min Woo Park and Eui Chul Lee

Abstract Recently, life-logging technology has been adapted to user-supplied data that can be collected from various aspects of daily life. Using this technology, images captured by the camera of a smart device should be continuously processed in order to extract meaningful data applicable to human life and emotion. In this paper, to verify a correlation between emotion synchronization and spatial frequency, images with different spatial frequencies were submitted as visual stimuli to subjects, and heart rate was measured. The results confirmed that heart rate was synchronized with delay in increment and decrement for the spatial frequency of a presented image sequence.

Keywords Heart rate · Spatial frequency · Life-logging · Emotion

1 Introduction

Recently, life-logging technology has been applied to a variety of aspects of daily life through collection and analysis of user-supplied data. The growth of wearable smart devices rapidly pushed life-logging technology into the limelight [1]. Currently, several life-logging technologies are commercially available for healthcare and user experience applications. However, images captured by cameras of smart devices are still not widely used for intelligent life-logging services.

To combine captured images with a user's emotions or experiences, the images should be processed and analyzed to extract meaningful features. Visual information such as color and frequency can affect human emotion [2]. Studies in the fields of psychology and marketing showed that specific color or packaging design patterns can affect a consumer's purchasing needs [3,4]. However, there is no research

J. Im · M.W. Park
Department of Computer Science, Graduate School, Sangmyung University,
Seoul, Republic of Korea

E.C. Lee(✉)
Department of Computer Science, Sangmyung University, Seoul, Republic of Korea
e-mail: eclee@smu.ac.kr

© Springer Science+Business Media Singapore 2015   217
D.-S. Park et al. (eds.), *Advances in Computer Science and Ubiquitous Computing*,
Lecture Notes in Electrical Engineering 373,
DOI: 10.1007/978-981-10-0281-6_31

verifying a correlation between human emotion and the complexity of images of environmental scenes. The complexity of images can be represented as spatial and temporal frequencies.

In this paper, we focus on increment and decrement of spatial frequencies of image sequences. As experimental visual stimuli, 9 checkerboard images of different frequencies were used. To simulate increment and decrement of spatial frequency, 18 checkerboard images were successively presented. Heart rate was also recorded by a photoplethysmography (PPG) sensor on the index finger.

2   Experimental Method

2.1   Visual Stimuli

Nine checkerboard images with different spatial frequencies were used as visual stimuli, as shown in Fig 1. The images were automatically generated as an 8-bit

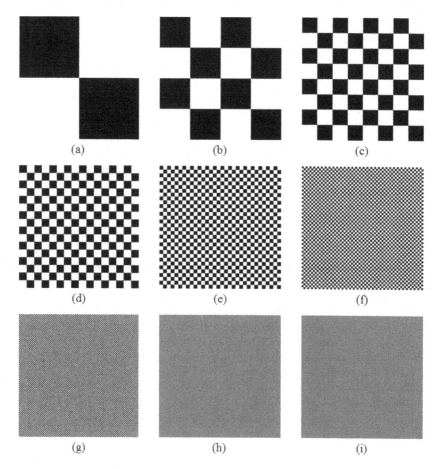

Fig. 1 Sequence of the 9 checkerboard images (a–i) used in experiments. Details of (g), (h), and (i) may be different because of image resolution)

grayscale 1024 × 1024 bitmap by using self-made recursive algorithm-based software. To remove the effect of noise such as semantic content or color on heart rate, no other factors were included in the images. Fig. 1 (a) shows the lowest spatial frequency image used, and is comprised of four squares (each square is 512 × 512). In the sequence progression from Fig. 1 (a) to (i), the spatial frequency is twice that of the previous image horizontally and vertically. Thus, the number of squares in Fig. 1 (i) is 262,144 (= 512 × 512).

2.2 Heart Rate Measurement

Generally, heart rate can be measured by using PPG or electrocardiogram (ECG) sensors. Because ECG sensor attachment is relatively complicated, our experiment used a simple USB-type PPG sensor for estimating successive heart rate information. This device does not need to use a signal amplifier. It acquires and transports 250 blood flow intensity samples per second through the USB interface.

To calculate heart rate, the R-peak (zero crossing position from positive to negative gradient) was detected, followed by the time interval between two successive R-peaks. Consequently, the heart rate can be calculated by dividing 60 seconds by the RRI (Interval between two R-peaks). For example, if RRI is 0.5 seconds, the heart rate is 120 beats per minute (= 60 seconds/0.5 seconds).

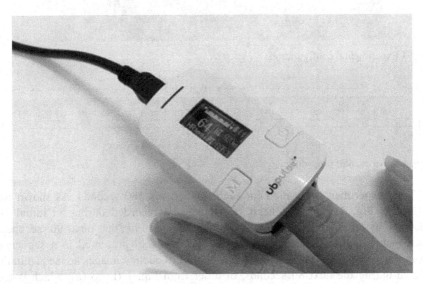

Fig. 2 A commercial PPG sensor used in the experiment [5]

2.3   Experimental Procedure

The experimental configuration including one PPG sensor and a monitor is shown
in Fig. 3. Ten graduate students (male = 6, female = 4, mean age = 25.7)
participated in this experiment. Environmental noise and outdoor sunlight were
blocked.

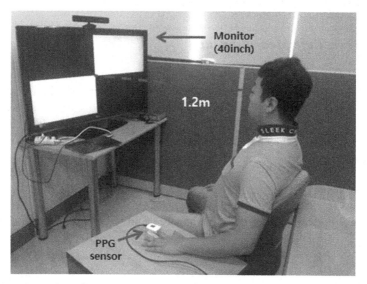

Fig. 3 Experimental configuration

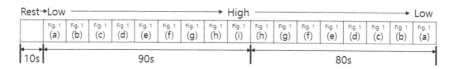

Fig. 4 Experimental procedure

Total experimental duration for each subject was 180 seconds. As shown for
"Rest" in Fig. 4, reference heart rate data were recorded during an initial 10
seconds by showing a completely black image. For the next 90 seconds,
incremental spatial frequency images are successively presented in a sequence
from Fig. 1 (a) to (i). Then, decremental spatial frequency images are sequentially
shown during the next 80 seconds, in order from Fig. 1 (h) to (a), which is the
reverse of the prior 90 seconds. Thus, a total of 170 seconds of heart rate data are
recorded with visual stimuli.

To regularize the size ratio of black and white regions, the visual stimuli (1024
× 1024) are resized to conform to monitor resolution (1920 × 1080), as shown in
Fig. 3.

3 Results

PPG data for one subject were excluded because of motion artifact, but the data for the 9 other subjects were used for heart rate analysis. To correct for individual variation, heart rate values were normalized by subtracting reference data (calculated from the "Rest" stage in Fig. 4) from the values. Then, the normalized data of the 9 subjects were averaged for every stage (10 seconds). The results for the average normalized heart rates are shown in Fig. 5.

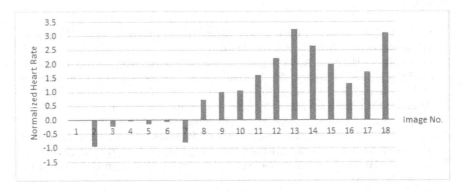

Fig. 5 Experimental results of normalized heart rate tendency according to the visual stimuli presented with different spatial frequencies

Fig. 5 shows an incremental tendency for the heart rate during 110 seconds. After that, the heart rate decreased for 30 seconds, but increased again in the last 20 seconds. Except for the last 20 seconds, the tendency for heart rate change was almost synchronous with the presented image frequencies. The interval between the highest spatial frequency (image No. 10 in Fig. 5) and the highest heart rate (image No. 13 in Fig. 5) may be due to the delay of conversion into heart rate by visual stimuli. Moreover, the heart rate increment in the final 20 seconds may be due to fatigue or stress from prolonged observation of a very high contrast pattern. However, we did confirm delayed synchronization of the heart rate with spatial frequencies of visual stimuli.

4 Conclusion

In this study, images of different spatial frequencies were presented to subjects as visual stimuli, and heart rate was measured. We confirmed that heart rate was synchronized with delay in increment and decrement for the spatial frequency of a presented image sequence. Based on these results, the complexity of edge and motion of environmental images captured by a wearable camera can be used as features to affect and evaluate human emotion and experience.

In future work, we will use scenes of actual environmental images to confirm the correlation with heart rate or other biometric features. In addition, image components such as color and brightness will be considered as affective features.

Acknowledgement This work was supported by the ICT R&D program of MSIP/IITP. [R0126-15-1045, The development of technology for social life logging based on analysis of social emotion and intelligence of convergence content].

References

1. Sellen, A.J., Whittaker, S.: Beyond total capture: a constructive critique of lifelogging. Communications of the ACM **53**, 70–77 (2010)
2. Machajdik, J., Hanbury, A.: Affective image classification using features inspired by psychology and art theory. In: Proceedings of the International Conference on Multimedia, pp. 83–92. ACM (2010)
3. Bradley, M.M., Hamby, S., Low, A., Lang, P.J.: Brain potentials in perception: Picture complexity and emotional arousal. Psychophysiology **44**, 364–373 (2007)
4. Clement, J.: Visual influence on in-store buying decisions: an eye-track experiment on the visual influence of packaging design. Marketing Management, 917–928 (2010)
5. http://www.laxtha.com/ProductView.asp?Model=ubpulse 360&catgrpid=3 (accessed on August 28, 2015)

Effective Similarity Measurement for Key-Point Matching in Images

Sungmin Lee, Seung-Won Jung and Chee Sun Won

Abstract Different similarity measures between the descriptors of the key-points certainly yield different performances in image matching. In this paper we introduce an effective similarity measurement, which considers the distances of each key-point in a query image and its matched key-point with the smallest distance in the test image. Therefore, the distances of all key-points in the query image to the corresponding matched key-points in the test image contribute to the final similarity measurement. On the other hand, the previous method considers only the distances less than a threshold value of all possible key-point pairs, which may ignore a significant part of the key-points in the query image. Our experiments show that the proposed measure yields better performance for image similarity matching and retrieval.

Keywords Image similarity measure · Image retrieval · Key-point detector/descriptor · SIFT

1   Introduction

The similarity between query and test images needs to be measured in image retrieval and recognition problems. Since key-points in images play an important role for the similarity matching, we seek to find the matched key-point pairs between the query and test images. Here, the distance between the descriptors of two key-points is obtained by measuring the differences of the descriptor elements in the feature space [1].

S. Lee · C.S. Won(✉)
Department of Electronics and Electrical Engineering,
Dongguk University-Seoul, 30, Pildong-ro 1gil, Jung-gu, Seoul 100-715, Korea
e-mail: {min4140,cswon}@dongguk.edu

S.-W. Jung
Department of Multimedia Engineering, Dongguk University-Seoul,
30, Pildong-ro 1gil, Jung-gu, Seoul 100-715, Korea
e-mail: swjung83@dongguk.edu

© Springer Science+Business Media Singapore 2015   223
D.-S. Park et al. (eds.), *Advances in Computer Science and Ubiquitous Computing*,
Lecture Notes in Electrical Engineering 373,
DOI: 10.1007/978-981-10-0281-6_32

In [2-3], new key-point descriptors are introduced and evaluated with their similarity measurement. However, the similarity measure used in [2-3] requires a threshold value, which is difficult to fix because of its sensitivity to the performance. Also, not all detected key-points in the query image but only a handful key-point pairs with the distances less than the threshold may contribute to the final (global) similarity, which may distort the global similarity of the two images. Our goal of this paper is to propose a new method to resolve the problems of the previous method.

2   Similarity Measures Between Images with Key-Points

Fig. 1 shows the major steps of the image retrieval by using the image features (descriptors) of the detected key-points. Here, we need to extract the key-points in the query and test images. Then, the feature vector (descriptor) for each key-point is formed. For the key-point detector and descriptor we use the well-known SIFT (Scale-Invariant Feature Vector) method [4]. For the key-point matching all possible combinations of the key-point pairs between the query image and each of the test images are compared for the similarity measurement between the two images. The best matching test image is selected for the retrieval. In this paper we focus on the similarity measure between the two images in terms of the similarity matching of the key-points.

Fig. 1 Procedure of image retrieval with distance metrics

2.1   Related Works

The similarity between the query image I^Q and the test image I^T in [2-3] relies on the selected descriptors with the distances less than a threshold among all possible key-point pairs. Specifically, the similarity $SI(I^Q, I^T)$ is determined as follows:

$$SI(I^Q, I^T) = \frac{\Sigma_{i,j} \Gamma(f_i^Q, f_j^T)}{m \times n} \tag{1}$$

$$\Gamma(f_i^Q, f_j^T) = \begin{cases} 1 & \text{if } dist(f_i^Q, f_j^T) \le th \\ 0 \end{cases} \tag{2}$$

where the feature descriptors $\{f_1^Q, f_2^Q, ..., f_m^Q\}$ and $\{f_1^T, f_2^T, ..., f_n^T\}$ are formed by m detected key-points from the query image I^Q and n from the test image I^T

respectively. $dist(f_i^Q, f_j^T)$ is the Euclidean distance between the two descriptors i and j. The threshold value th in (2) is to be experimentally determined. The performance of this method is sensitive to the selection of the thresholding value (th). So, for a small thresholding value only a handful of key-point pairs may have non-zero values in the numerator of (1), which may distort the global similarity of the images.

2.2 Proposed Similarity Measure

To solve the problems of the previous similarity matching method the distances of each key-point in a query image and its matched key-point with the smallest distance in the test image are used for the global image similarity. That is, the distances of all key-points in the query image to the corresponding matched key-points in the test image contribute to the final similarity measurement between the two images. So, we guarantee a sufficient number of key-points are to be used for the matching. Also, as shown in the following equation, our similarity measure needs no threshold value to be determined. The proposed similarity measure is

$$S2(I^Q, I^T) = \frac{1}{\sum_{i=1}^{m} \min(dist(f_i^Q, f_1^T), dist(f_i^Q, f_2^T), ..., dist(f_i^Q, f_n^T))} \quad (3)$$

where $min()$ represents the minimum value of all elements in the parenthesis. As oppose to (1) our similarity in (3) always finds the shortest distance for each key-point in the query image to those key-points in the test image. All those distances from m key-points are used for the final similarity measurement. Finally, the inverse of the sum of the distances is the score of the similarity between the two images.

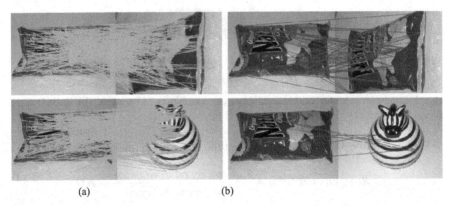

(a)   (b)

Fig. 2 The contributed distances for the similarity calculation: (a) The proposed method: all the detected key-points in the query image contribute to the similarity measure, (b) The existing method: only a part of key-points in the query image are used.

The biggest difference between the existing and our method is the selection of distances to be considered for the global image similarity. Fig. 2 shows the selected key-points used for the image similarity between the proposed method (Fig 2(a)) and the existing method [2-3] (Fig. 2(b)). As shown in Fig. 2(b) with red lines, only partial information (key-points) contributes to the measuring of the image similarity, whereas the distance measures of all the key-points in the query image (the blue lines) are considered in the proposed method.

3   Experiments and Results

The proposed method is tested with 3 databases which are openly available for research purposes. To compare with the result of previous method [2-3], we used *53 objects* [5], *ZuBuD* [5] and *kentucky* [6] in our experiments. Fig. 3 shows representative images in each database, where *53 objects* has images with 53 objects, *ZuBuD* has 265 images of 53 objects and 1005 images of 201 buildings separately. *Kentucky* has 10200 images of 2550 objects, but we used only 1000 objects as the previous method in [2-3].

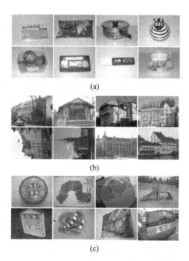

(a)

(b)

(c)

Fig. 3 Database used for our experiments. (a) 53 objects (53 query), (b) ZuBuD (201 query), (c) Kentucky (1000 query)

With the above 3 databases, we can get the recognition rate by the equation (4)

$$Recognition\ rate(\%) = \frac{\#\ of\ correct\ returned\ images}{\#\ of\ returned\ images} \times 100 \qquad (4)$$

The recognition rates with the ground truth in the database are calculated and compared in Table 1. As one can see in the table, by just replacing the previous similarity measure in [2-3] with our method, we obtained significant improvements in the recognition rates. This shows the power of the proposed similarity measure. The results are also shown in Fig. 4. The left-most image surrounded by red lines is the query image and first images are chosen for query images in our experiments.

Table 1 The recognition rate of image retrieval.

Similarity method	DATASET		
	53objects	ZuBuD	Kentucky
Proposed method *S2*	**71.7%**	**81.72%**	**80.03%**
Existing method *S1*	52.45%	75.67%	48.83%

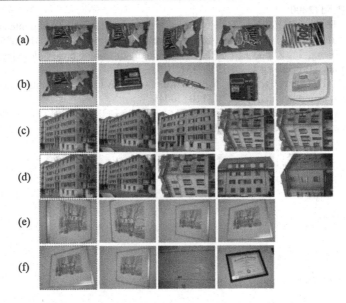

Fig. 4 Result of retrieved images. (a), (c), and (e) are retrieved images with the proposed method. (b), (d), and (f) show the retrieved images with the existing method [2-3].

4   Conclusion

The previous image similarity matching method has a difficulty in choosing the appropriate threshold value. Although the selected threshold finds a sufficient number of key-point correspondences for some images, it may find only a handful key-point pairs for other images, which may distort the global similarity between the two images. To solve this problem we propose a method to consider the best matching key-point in the test image for each key-point in the query image. As a result, the distances of all key-points in the query image to the corresponding matched key-points in the test image contribute to the global similarity measurement between two images, which yields more robust matching performance.

Acknowledgments This work was supported by Basic Science Research Program through the National Research Foundation of Korea (NRF) funded by the Ministry of Education (NRF-2013R1A1A2005024) and by the MSIP(Ministry of Science, ICT and Future Planning), Korea, under the ITRC(Information Technology Research Center) support program (NIPA-2015-H0301-14-4007) supervised by the NIPA(National IT Industry Promotion Agency).

References

1. Aksoy, S., Haralick, R.M.: Probabilistic vs. geometric similarity measures for image retrieval. In: IEEE Conference on Computer Vision and Pattern Recognition, Hilton Head Island (2000)
2. Fan, B., Fuchao, W., Zhanyi, H.: Rotationally invariant descriptors using intensity order pooling. IEEE Trans, Pattern Analysis and Machine Intelligence **34**, 2031–2045 (2012)
3. Xu, X., Feng, L., Zhou, J.: OSRI: A rotationally invariant binary descriptor. IEEE Trans. Image Processing **23**, 2983–2995 (2014)
4. Lowe, D.G.: Distinctive image features from scale-invariant keypoints. International Journal of Computer Vision **60**, 91–110 (2004)
5. http://www.vision.ee.ethz.ch/datasets/
6. http://www.vis.uky.edu/~stewe/ukbench/

Vocabulary Modeling of Social Emotion Based on Social Life Logging

Hea Jin Kim, Young Ho Jo, Young Joo Kim, Hye Sun Kim
and Min Cheol Whang

Abstract Life logging in online space has recently been utilized to share emotion rather than to remember daily life.Therefore, this study is to determine social emotion in life logging. Social emotion was analized and collected from emotion vocabularies of 21 power blogging sites in Korea, such as Naver blog, Kakaotalk, Band, Facebook, Instagram, KaKaostory, and Daum. The collected vocabularies were sorted into morphemes and adjectives were extracted. The frequencies of the adjectives were analyzed, and their similarities was subjectively tested over 99 vocabularies showing high frequency. However, they showed insigniticant. In order to make up for the limition of this frequency analysis, 10 out of total 21 target sites were separately selected and 600 adjectives were extracted again. The extracted 600 vocabularies were sorted into 120 groups by Card sorting, and each group's representative vocabulary was defined as soial emotion demensions.

Keywords Social life logging · Social emotion model · Vocabulary modeling

1 Introduction

Human has been unable to remember valuable experience and daily affairs. Therefore, life logging has been developed to record their history in various forms

H.J. Kim(✉) · Y.H. Jo · Y.J. Kim
Emotion Science Research Center, #405, DMC R&D Center,
37, Maebongsan-ro, Mapo-gu, Seoul 03016, South of Korea
e-mail: shaonu@daum.net

H.S. Kim
Emotion Contents Technology Research Center,
Hongji-dong, Jongno-gu, Seoul 03016, South of Korea

M.C. Whang
Department of Digital Media, Sangmyung University,
Hongji-dong, Jongno-gu, Seoul 03016, South of Korea

© Springer Science+Business Media Singapore 2015
D.-S. Park et al. (eds.), *Advances in Computer Science and Ubiquitous Computing*,
Lecture Notes in Electrical Engineering 373,
DOI: 10.1007/978-981-10-0281-6_33

of text, picture, movie clip, sound, etc. Since human has made an effort to write life record themselves, life logging has been developed to digitizing automatically and to record conveniently.

Online life logging provides to share emotion and interact with others by expressing their own daily lives beyond simple record of daily life. Therefore, life logging expanded its conception into 'Social Life Logging'. Sharing emotion is interactive in life logging and develops social emotion related with others.

Social emotion is formed by social relation, and the reason of social behavior to communicate with others [1][2]. Response to others' emotion can cause embarrassment when someone's excessive care [3]. In many cases, this kind of social emotion is internal emotion condition diversely and delicately expressed by language though it can be shown with direct and instant response like facial expression, movement, voice, etc. [4]. Through the experiments over 2,186 adjectives covering characters and emotions, words chosen by more than 75% of participants were selected as emotion words [5]. Russell arranged 28 emotion vocabularies of pleasant-unpleasant, arousal-relaxation dimension and suggested the correlation between the emotion vocabularies [6]. Korean words to express feelings have been categorized and their frequencies were compared each other according to values of 'prototypicality', 'familiarity', 'pleasant-unpleasant' and 'vitalization' [4]. Vocabulary structure has been established from correlation analysis between categories of social emotion used in SNS(social network service) [7]. However, the emotion words has been restrictedly sampled from dictionaries or reference books [5] and SNS vocabularies [7].

This study was to model Korean social emotion utilized in social life logging, analyzing vocabulary sampled form massive emotion vocabularies in Facebook, Kakaotalk, Blog, etc. Representative vocabularies was determined from extracting valid vocabularies used online, and categorizing them.

2   Method

2.1   Sampling of Vocabularies

The participants was required in this study in logging behavior of continuous update for several years on Naver blog, Kakaotalk, Band, Facebook, Instagram, Kakaostory, and Daum café with high online market share in Korea. The final 21 blogging sites were selected and their vocabularies were collected. In case of chatting on Kakaotalk and Band which reflect users' attitude by keeping records of conversation with others without deletion, they were included as the users of social life logging in terms of record of communication with others. Characteristics of daily life each site are determined in this study as shown in Table 1.

Table 1 Characteristics of Word Collection depending on Target sites

Division			Characteristic
Naver blog	Bulletin board personally operated. Sharing various topics like art, living, know-how, shopping, hobby, leisure, travel, knowledge, trend, etc.	1	Writing daily life and personal thought
		2	Childcare-focused writing
		3	Writing about food and travel
		4	Father's writing daily life of famous restaurant and childcare
		5	famous restaurant-focused contents
		6	Active interaction with blog neighbors about various interests like cooking, girl groups, Starbuck, and games, etc.
Kakaotalk	Representative mobile messenger to support one to one and group chatting	7	Group conversation of friends and coworkers (3 Group)
		8	One to one talk (4 people)
Band	Acquaintance-based exclusive Social Network Service	9	Content of article
		10	Content of comment
Facebook	World' biggest Social Network Service	11	Daily life writing of entertainers(singers) with many followers
		12	Story about family or daily life
		13	Writing about hobby like volunteer work, floral arrangement, and travel
		14	Issues of personal emotion, soccer, social or internet issues, etc.
Instagram	Image-focused Social Network Service	15	Writing daily life with sentimental picture Characteristics of using neologism and hash tag
		16	Writing and sharing daily life with picture Many uses of English, neologism, and hash tag
		17	Sharing one's own design works
		18	Writing date life with girlfriend
		19	Daily life writing in journal style
KaKao story	Korean style Social Network Service to share pictures with acquaintance	20	Writing about current status and hobby of oneself and children
Daum cafe	Membership community site to share interests and useful information	21	Posting café members' one day with pictures and texts

2.2 Analysis of Morpheme

The collected texts from total 21 sites were classified into morphemes, the smallest meaningful unit of a language. If the collected texts were input to morpheme analysis program, the result was suggested as a combination of 'morpheme' and 'part of speech tag'.

열정을 가지고 ⟶ 열정/NNG + 을/JKO + 가지/VV + 고/EC

Morpheme

Parts of speech tag

Morpheme analysis was classified into total 45 parts of speech tag, such as common noun(NNG), verb(VV), adjective(VA), common adverb(MAG). Adjective has been analyzed to express most intense emotion than noun and verb [4]. In this respect, this study was focused to analysis of adjective.

3 Analysis

3.1 Frequency Analysis

Adjectives extracted from each site were integrated and listed. The appearance frequency of relevant vocabularies were analyzed from total 21 integrated text data. 99 vocabularies shown more than 20 times were selected by arranging the order of high frequency.

Table 2 The list of 99 vocabularies extracted by frequency analysis

Vocabulary	Frequency	Vocabulary	Frequency	Vocabulary	Frequency
없다(There is not)	1978	재미있다(fun)	84	얇다(thin)	35
좋다(good)	1805	무섭다(scary)	82	배부르다(full, stuffed)	34
같다(same)	1048	비싸다(expensive)	75	손색없다(Not inferior to)	34
맛있다(delicious, yummy)	889	귀엽다(cute)	72	슬프다(sad)	34
많다(many)	804	수많다(numerous)	71	커다랗다(huge, large)	33
크다(large)	587	가깝다(close)	70	거칠다(rough, coarse)	32
괜찮다(OK, alright)	380	반갑다(glad)	65	고프다(hungry)	32
작다(small)	264	흔하다(common)	65	상관없다(don't care)	32
힘들다(tired, tiring)	232	춥다(cold)	63	시원하다(cool)	32
안되다(impossible)	229	강하다(strong)	59	행복하다(happy, blissful)	32
다르다(different)	219	낮다(low)	57	급하다(urgent, in no time)	31
늦다(late)	184	빠르다(fast)	56	넉넉하다(enough, sufficient)	31
이쁘다(pretty)	182	짧다(short)	56	부끄럽다(shameful, ashamed)	31
고맙다(thankful)	180	낯설다(strange, unfamiliar)	52	곱다(beautiful, fine)	30
멋지다(cool, nice)	155	나쁘다(bad, poor)	50	멋있다(nice)	30
바쁘다(busy, fully occupied)	151	덥다(hot)	50	중요하다(important, significant)	30
즐겁다(pleased, delighted)	144	낫다(better)	49	맛없다(unsavory, unappetizing)	29
쉽다(easy)	130	귀찮다(annoying)	48	자연스럽다(natural)	29
멀다(far)	126	그립다(miss, long for)	48	굵다(thick)	28
새롭다(new, fresh)	125	진하다(thick)	48	과하다(excessive, immoderate)	27
아프다(painful)	125	맑다(clean, clear)	45	순하다(mild)	27
깊다(deep)	123	어리다(young)	45	아깝다(sorry, sad)	27
아깝다(regretful)	123	밝다(bright, brilliant)	44	어둡다(dark, dim)	27
예쁘다(pretty, nice-looking)	111	귀하다(precious, valuable)	43	두껍다(thick, heavy)	26
어렵다(difficult, hard)	100	먹음직스럽다(appetizing, mouth-watering)	42	남다르다(unusual, extraordinary)	25
맵다(hot, spicy)	95	배고프다(hungry)	40	질다(deep, dark)	24
부드럽다(soft, smooth)	95	엄청나다(huge, great)	39	깔끔하다(neat, tidy)	23
넓다(wide)	92	지나치다(excessive, immoderate)	38	깨끗하다(clean, spotless)	22
부럽다(envious)	91	필요하다(necessary)	38	놀랍다(surprising, amazing)	22
높다(high)	88	기쁘다(present)	37	무겁다(heavy, weighty)	22
뜨겁다(hot)	87	드물다(rare, unusual)	37	싱겁다(bland, flat)	21
아름답다(beautiful)	86	똑같다(the same)	37		
편하다(convenience)	86	야무지다(skillful, shrewd)	37		
싫다(hate, dislike)	84	가볍다(light)	36		

3.2 Similarity Evaluation

Similarity was tested on the basis of 99 vocabularies extracted by frequency analysis, and total 5 people site were evaluated including a man and a woman of

20's, a man of 30's, 2 women of 30's. Similarity survey arranged 99 vocabularies into the same lines and columns and made the targeting people evaluate them with scale of 1-7 points. If both word had high similarity from horizontal axis and vertical axis, it was scored close to 7 points. If the similarity was evaluated low, the score was close to 1 point.

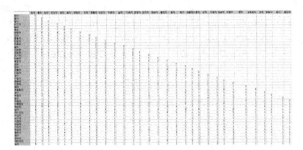

Fig. 1 Similarity Survey Sheet

3.3 Multi-dimensional Scaling

The results of similarity test of 5 participants were applied to multi-dimensional scaling and vocabularies were arranged on dimensional space. Multi-dimensional scaling uses input data to build up multi-dimensional evaluation space by placing relative distances of observing targets. It was aimed to extract social dimension of social life logging and analyze categorized vocabulary groups through producing coordinate of similarity scale between vocabularies input from 1 to 7 points at similarity survey and arranging 99 vocabularies on space.

3.4 Card Sorting

Among 99 vocabularies from the frequency analysis, vocabularies which a specific user repetitively uses in online space can affect the frequency analysis. Therefore, Card sorting was performed, focusing on 600 adjectives shown on 10 out of total 21 sites. Adjectives left out due to morpheme analysis program error were added to 10 selected sites, and so were idioms and compounds with adjective meanings highly used in daily lives.

(a) Arrangement of all adjectives extracted from 10 target sites

(b) Preliminary grouping to define a super ordinate concept vocabularies, like happy and good

(c) Secondary grouping to specify vocabularies at Preliminary grouping

Fig. 2 Card Sorting

4   Results

As the analysis results of multi-dimensional scaling over similarity survey data of 99 vocabularies extracted by frequency analysis, the number of dimension was extracted as 2 dimensions, However it was not secured with adequacy of the number of dimension because stress value was too high with 0.9. Vocabulary dispersion arranged on 2 dimensional space also didn't verify a meaningful result because most of participants' responses over similarity survey were concentrated on 1 point among 1 to 7 points. According to Fig.3. (a) is the result of multi-dimensional

(a) 5 people   (b) 4 people   (c) 1 person

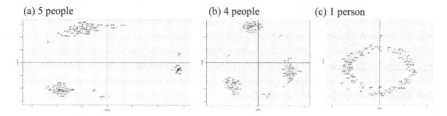

Fig. 3 Multi-dimensional Analysis (The survey result)

Table 3 Representative vocabularies of 120 groups

가깝다(close, nearby)	똑같다(same, identical)	솔직하다(frank)	좋다(good, fine)
가난하다(poor, needy)	똑똑하다(smart, intelligent)	순진하다(naive)	중요하다(important, significant)
가늘다(thin)	많다(many, plenty of)	쉽다(easy)	즐겁다(pleasant, enjoyable)
가능하다(possible)	맑다(clean, pure)	슬프다(sad)	진지하다(serious, earnest)
가지런하다(even, aligned)	맛있다(delicious, yummy)	시다(sour)	짜증나다(irritated, annoyed)
간단하다(simple)	맞다(be correct, be hit)	시시하다(trivial)	차갑다(cold)
강하다(strong, powerful)	매몰차다(cold, harsh)	신성하다(holy)	착하다(good-natured, good-hearted)
건강하다(healthy)	멋있다(nice, wonderful)	싫다(hate, dislike)	철저하다(thorough, exhaustive)
겁없다(unafraid, fearless)	멍청하다(stupid, foolish)	심하다(harsh, severe)	초췌하다(haggard)
경솔하다(rash, hast)	못생기다(ugly, unattractive)	쓸데없다(unnecessary)	촉촉하다(moist)
고맙다(thankful)	무덤덤하다(calm, placid)	아니다(not, hardly)	충분하다(enough, sufficient)
귀찮다(annoying)	무섭다(scary, frightening)	아프다(sick)	치열하다(fierce, intense)
그립다(miss, long for)	무시무시하다(terrible, horrible)	안되다(impossible)	쿨하다(cool, obamaish)
기특하다(admirable, praiseworthy)	미안하다(sorry, feel bad)	안타깝다(regrettable)	크다(large)
나쁘다(bad, poor)	바쁘다(busy)	약하다(weak)	편하다(comfortable, cozy)
낡다(old, worn out)	밝다(bright)	어리다(young)	하고 싶다(would like to, want to)
낮다(low)	배고프다(hungry)	어수선하다(untidy)	하기 싫다(don't want to)
넓다(wide)	별 생각없다(no idea)	없다(no, nothing)	하염없다(blank, dispirited)
느끼다(feel, sense)	부끄럽다(ashamed, shameful)	여전하다(still, no better)	한결같다(unchanging, consistent)
느리다(slow)	부담스럽다(pressured)	예쁘다(pretty)	한심하다(pathetic, pitiful)
늙다(get old)	부지런하다(diligent, hard-working)	외롭다(lonely)	한적하다(secluded)
다르다(be different)	불행하다(unhappy, unfortunate)	우울하다(depressed, gloomy)	행복하다(happy)
다정하다(kind, warmhearted)	비슷하다(similar, alike)	위험하다(dangerous)	허무하다(vain, futile)
달다(sweet)	비싸다(expensive, costly)	이상하다(strange)	화끈하다(bold, hot)
답답하다(stuffy)	비참하다(miserable, pitiable)	작다(small)	확실하다(sure, certain)
당당하다(dignified)	빠르다(fast)	장난 아니다(No kidding, no joke)	황당하다(absurd, ridiculous)
당연하다(natural, no doubt)	빨갛다(red)	재미없다(boring, dull)	흐리다(muddy, cloudy)
대단하다(great, Incredible)	새롭다(new)	재미있다(fun)	흔하다(common, commonplace)
덥다(hot)	생소하다(unfamiliar, strange)	적다(small, little)	흥미롭다(interesting)
따뜻하다(warm, mild)	소중하다(precious, valuable)	적당하다(right, proper)	힘들다(backbreaking, strenuous)

analysis over 5 participants, and (b) is the analysis result of excluding a user to mostly input 1 point on the survey. (c) is the result of 1 person to diversely input 1 to 7 points. Unlike (a) and (b), the case of (c) build up round balance of vocabulary dispersion, but correlations of dispersed vocabularies could be found.

As a result of grouping by card sorting over 600 adjectives extracted from 10 sites, total 120 groups were formed. Representative vocabularies were selected from each of 120 groups. The representative vocabularies were selected among inside of group vocabularies or new vocabularies were defined to represent the relevant group.

Through focus analysis of participatory researchers, the extracted 120 vocabularies were categorized again, and mapped into two-dimensions such as tense and valence dimension. As a result, 61 emotion vocabularies determined 2 dimensional social emotion model according to emotional behavior in social life logging.

The 2 dimensional social emotion model was determined to be composed of tense-calm, positive-negative dimensions in this study. The emotions of 'tense - positive domain' contained 'great/incredible', 'fun', 'would like to/want to', 'nice/wonderful', etc., The emotion of 'calm - positive domain' was characterized by 'good-natured/ good-hearted', 'comfortable/cozy', 'unchanging/consistent' etc., The emotion of 'tense-negative domain' contained 'scary/frightening', 'backbreaking/ strenuous', 'miserable/pitiable' etc., while one of 'calm & negative domain' did 'irritated/ annoyed', 'boring/dull', 'pathetic/pitiful' and others.

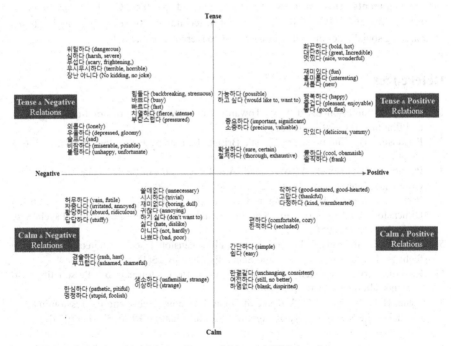

Fig. 4 The 2 dimensional social emotion model in social life logging

5   Discussion

Life logging was in this study considered not only in writing daily life in on-line space into the replay of remembrance, but also in forming social relationship. The proposed social emotion model defined behavior characteristics of social online activity in life logging.

However, since the questionnaire results based on vocabularies extracted from frequency analysis were not generated in a significant level, card sorting was performed for grouping vocabularies and extracting each group's representative vocabularies.

In the further work, the adjectives of 11 sites excluded in this study are also supposed to be included, reconstructed by card sorting, and defined as each new group's representative vocabularies. The extracted representative vocabularies will be also conducted by similarity survey and multi-dimensional analysis to define social emotion dimension based on social life logging. There is the necessity to repetitively conduct vocabulary analysis process so as to obtain the robust adequacy of number of dimensions and extract significant social emotion dimension. In addition, adjective analysis study for the purpose of social emotion modeling will be conducted by expanding it into emotion analysis revealed from verb and noun.

Acknowledgments This work was partly supported by the ICT R&D program of MSIP/IITP [R0126-15-1045, The development of technology for social life logging based on analyzing social emotion and intelligence of convergence contents].

References

1. Oatley, K., Johnson-Laird, P.N.: Towards a cognitive theory of emotions. Cognition and Emotion **1**(1), 29–50 (1987)
2. Parkinson, B.: Emotions are Social. British Journal of Psychology **87**(4), 663–683 (1996)
3. Parkinson, B., Fischer, A., Manstead, A.S.R.: Emotion in Social Relations: Cultural, Group, and Interpersonal Processes, Psychology Pr. (2005)
4. Park, I.J., Min, K.H.: Making a List of Korean Emotion Terms and Exploring Dimensions Underlying Them. Korean Journal of Social and Personality Psychology **19**(1), 109–129 (2005)
5. Bush, L.E.: Individual differences in multidimensional scaling adjectives denoting feelings. Journal of Personality and Social Psychology **25**(1), 50–57 (1973)
6. Russell, J.A.: A Circumplex Model of Affect. Journal of Personality and Socialpsychology **39**(6), 1161 (1980)
7. Hyun, H.J., Whang, M.C.: Valence of Social Emotions' Sense and Expression in SNS. Journal of the Korea Society of Computer and Information **19**(6), 37–48 (2014)

LIDAR Simulation Method for Low-Cost Repetitive Validation

Seongjo Lee, Dahyeon Kang, Seoungjae Cho, Sungdae Sim,
Yong Woon Park, Kyhyun Um and Kyungeun Cho

Abstract Developments in light detection and ranging (LIDAR) have enabled its application in unmanned automotive technology, and various methods using LIDAR are now being proposed. However, it is more difficult to obtain a ground truth dataset to evaluate the performance of algorithms that use a quantity of three-dimensional (3D) points as compared to those that require only 2D images. This paper describes an approach to creating a ground truth dataset for verifying a variety of algorithms by recording the data on detected objects through simulation in virtual space. This approach is able to verify the performance of algorithms in a variety of environments with less cost than the use of actual LIDAR.

Keywords Unmanned vehicle · LIDAR · Multiple sensors · Simulation · Virtual environment

1 Introduction

Research on unmanned driving vehicles has recently accelerated. In particular, research has focused on investigating the technology needed to classify the environment around an automated driving vehicle (in unmanned or safe driving mode) into obstacles or traversable roads. As light detection and ranging (LIDAR) technology has developed, automated driving technology has been studied using three-dimensional (3D) space data in addition to approaches that use 2D images.

S. Lee · D. Kang · S. Cho · K. Um · K. Cho(✉)
Department of Multimedia Engineering, Dongguk University-Seoul,
26 Pildong 3 Ga, Jung-gu, Seoul 100-715, Republic of Korea
e-mail: cke@dongguk.edu

S. Sim · Y.W. Park
Agency for Defense Development, Yuseong P.O.Box 35, Yuseong-gu, Daejeon 34188,
Republic of Korea

© Springer Science+Business Media Singapore 2015
D.-S. Park et al. (eds.), *Advances in Computer Science and Ubiquitous Computing*,
Lecture Notes in Electrical Engineering 373,
DOI: 10.1007/978-981-10-0281-6_34

To completely verify the quality of segmentation or classification results using LIDAR, a ground truth dataset is required to evaluate the results of an algorithm's implementation. However, because the data acquired through LIDAR is 3D data, it is more difficult to generate a ground truth dataset to test these algorithms than it is for those that use 2D images. Furthermore, generating such data incurs significant time and monetary cost because of the expensive LIDAR system that must be purchased and fixed to a vehicle before driving it.

To solve the obstacles mentioned above, this paper proposes a LIDAR simulation approach that reproduces a portable platform and LIDAR in virtual space. This approach is able to obtain a dataset in a variety of environments, within a short period of time, and without the purchase of expensive sensors. The resulting datasets may be used verify the performance of diverse segmentation or classification algorithms.

2   Related Work

Most related research has directly operated LIDAR in an actual environment. Douillard acquired a dataset using various kinds of LIDAR and segmented the 3D point clouds for each object [1]. Behley acquired 3D point clouds from a variety of LIDAR sources and then proposed a classification algorithm that could be generally applied to those point clouds [2].

Some researchers have generated datasets using LIDAR and made the results publicly available. As a result, the other researchers can simply verify their research results and compared the performance against other methods. Geiger disclosed a dataset acquired using a stereo camera, LIDAR, and inertial navigation system [3], and Behley used this dataset to verify the performance an algorithm that classifies the data from LIDAR into pedestrian, car, cyclist, and background classes [4]. While such datasets can be helpful for researchers who require LIDAR data, another process is necessary to acquire further datasets for specific environments not covered by these datasets.

Rohmer developed a robot simulation framework for general purposes [5]. This framework enables users to simulate sensors in a variety of environments. However, because the LIDARs simulated using this framework can have a higher error than actual LIDARs, it is difficult to accurately identify the performance of an algorithm using the relevant LIDAR.

3   LIDAR Simulation System Structure

This paper configured a virtual 3D space for LIDAR sensor simulation and created a mobile platform with sensors mounted in this space. All objects in the 3D space have unique numbers (IDs) that depend on the type of object. A LIDAR sensor exists in the lower hierarchy of the mobile platform and the scope of the sensor changes as the vehicle moves. The sensor mounted on the vehicle acquires the coordinates of collision points on surfaces of objects or terrain by emitting the lasers. As the all objects in the virtual environment are already known, the IDs of collision objects and the flying distances of lasers are acquired based on these coordinates. The position and rotation of the vehicle as well as the distance, ID, and relative coordinates of the object matched to each laser are saved as a binary file. Fig. 1 illustrates the system structure.

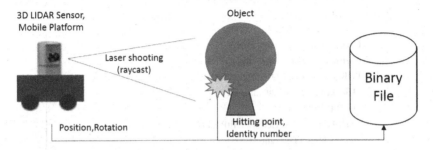

Fig. 1 System Overview.

3.1 LIDAR Sensing Module

The virtual system in this paper simulates the Velodyne-HDL32E sensor. Fig. 2 illustrates the detailed specifications to which the LIDAR sensor scans objects. The module emits the laser and checks for collision objects. The sensor mounted on the vehicle makes 10 360-degree revolutions per s and emits the laser 2,160 times per revolution. Thirty-two beams are aligned vertically and emitted all at once per shot. When the laser is emitted, the angle between the top and bottom beams is 41.34 degrees. The maximum length of each laser is 30 m and the raw data on a terrain or collision object in the scope of beam is extracted. The raw data in this system consists of the coordinates of points where the lasers are reflected by surfaces of objects, relative to the sensor and its distance. If there is nothing that reflects the laser, the x, y, and z values of the coordinates and the distance are set to zero.

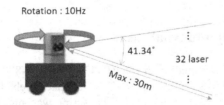

Fig. 2 LIDAR Specifications.

Fig.3 lists the pseudo code of this module. When a laser is emitted in fixed intervals, the number of emissions by the unit can vary, e.g., 2,160 times in intervals of 0.1 s or 216 times in intervals of 0.01 s. The number of emissions is predetermined. When a laser is emitted, the location and rotation values of the vehicle are acquired. Each time the lasers emit, thirty-two beams are shot from the top to the bottom angle. If any terrain or object reflects such beams, the coordinates of that point are calculated. If there is no reflection, the coordinates of the point becomes zero.

```
GetCollisionPoint(Position, Rotation, HitPoint)
    n←GetLaserCount()
    FOR i←0 TO n DO
        CarPosition←GetCarPosition()
        CarRotation←GetCarRotation()
        FOR j←0 TO 32 DO
            HitCheck←LaserHitCollision()
            IF HitCheck = 1
            THEN HitPoint←GetLaserHitPoint(i, j)
            ELSE THEN HitPoint←0
        END FOR
    END FOR
END
```

Fig. 3 Laser simulation pseudo code.

3.2 Binary File Generation Module

After scanning the surrounding terrain, the data are written to a binary file. In the simulation, one data record consists of the coordinates identified in Section 3.1 above, the location and rotation values of the vehicle, and the IDs, distances, and coordinates of the objects colliding with the beams. At this point, the coordinates included in the structure are the local coordinates of the sensor. Because the data record cannot be produced as a file in real time, the records are saved in a linear array and later saved as a binary file. If the time allocated for this is too short, the simulator overwrites the variables too frequently and so it is difficult to efficiently execute the simulation. In contrast, if the allocated time is too long, overflow occurs.

4 Experiment and Analysis

The experiment was executed in an Intel(R) Core(TM) i7-870 CPU, 8 GB RAM, Nvidia Geforce GTX 460 environment. The 3D space, vehicle, and sensor were configured using Unity3D engine [6]. The script was prepared using the C# language.

The virtual vehicle was driven using the keyboard in a 3D space of dimension of 2,000 × 2,000. The sensor mounted on the vehicle emitted a revolving laser (1,080 times per revolution). At this point, the scope of sensor revolution was restricted to 180° in front of the vehicle to improve the scan speed. The sensor revolved 180° per 0.1 s in intervals of 18° per 0.01 s. A total of 108 x 32 laser beams acquired the coordinates of a terrain or collision object per 0.01 s. The coordinates were the local coordinates with respect to the sensor. Fig. 4 presents a visualization of the simulation.

Fig. 4 Visualization of LIDAR simulation.

To test the reliability of a dataset acquired through the implemented system, an application visualizing a binary file was additionally implemented. Using the application, the position and rotation data of the vehicle and the local coordinates of the collision lasers were visualized. The data was verified by drawing 108 x 32 points at intervals of 18° per 0.01 s with the revolution of 180° per 0.1 s. Furthermore, a depth image was generated to verify the depth data. Fig. 5 illustrates the position of the virtual driver and Fig. 6 presents the image after resizing the depth map from 1080 x 32 pixels to 1080 x 160.

Fig. 5 Verification result of simulation.

Fig. 6 Depth-based grayscale image.

To correctly simulate a real sensor, it is necessary to acquire data about potential collisions by emitting the laser every 0.01 s. When the simulation was activated, it took 0.01 to 0.03 s to scan and was slower than the real sensor. However, the vehicle speed was proportionally slowed such that, in accordance with the virtual implementation, data was saved per 0.01 s. This was similar to a true sensor in driving mode.

5   Conclusion

This study executed an experiment to acquire the scan results of a LIDAR sensor in virtual space. The sensing module simulated the Velodyne-HDL32E sensor. The sensing results were saved as binary data for further applications. When the dataset was acquired, the scanning speed was slower than a real sensor. However, the operating speed of the entire program was slowed in proportion to the scanning speed. As a result, the dataset acquired was similar to the data from a true sensor. The results were verified using a player. Further study will implement an environment that is exactly the same as an actual sensing environment using multithreading.

The time, financial, and labor costs of generating a LIDAR ground truth dataset will be significantly reduced because the data are acquired in a virtual space. In particular, datasets in a variety of environments can be easily acquired by implementing diverse kinds of virtual spaces using a computer. Moreover, the performance of the algorithm can be accurately measured because the data on all scanned objects is saved.

Acknowledgements This work was supported by the Agency for Defense Development, Korea.

References

1.   Douillard, B., Underwood, J., Kuntz, N., Vlaskine, V., Quadros, A., Morton, P., Frenkel, A.: On the segmentation of 3D LIDAR point clouds. In: 2011 IEEE International Conference on Robotics and Automation (ICRA), pp. 2798–2805 (2011)
2.   Behley, J., Steinhage, V., Cremers, A.B.: Performance of histogram descriptors for the classification of 3D laser range data in urban environments. In: 2012 IEEE International Conference on Robotics and Automation (ICRA), pp. 4391–4398 (2012)
3.   Geiger, A., Lenz, P., Stiller, C., Urtasun, R.: Vision meets robotics: The KITTI dataset. International Journal of Robotics Research **32**(11), 1231–1237 (2013)
4.   Behley, J., Steinhage, V., Cremers, A.B.: Laser-based segment classification using a mixture of bag-of-words. In: 2013 IEEE/RSJ International Conference on Intelligent Robots and Systems (IROS), pp. 4195–4200 (2013)
5.   Rohmer, E., Singh, S.P.N., Freese, M.: V-REP: a versatile and scalable robot simulation framework. In: 2013 IEEE/RSJ International Conference on Intelligent Robots and Systems (IROS), pp. 1321–1326 (2013)
6.   Unity-Game Engine. http://unity3d.com/

Posture Recognition Using Sensing Blocks

Yulong Xi, Seoungjae Cho, Kyhyun Um and Kyungeun Cho

Abstract Posture recognition has been investigated in a variety of fields including medicine, HCI, and video games. In particular, it is important to improve the generality and recognition speed of posture recognition to improve the user experience in diverse kinds of user environments. This paper proposes a posture recognition algorithm with high generality and recognition speed. The algorithm is able to recognize a variety of postures, regardless of the number of joints recognized in human skeleton data. Furthermore, experimental results show that the method can quickly process a large quantity of data and recognize 22 postures in real time.

Keywords Posture recognition · Human-computer interface · Pattern recognition · Support vector machine

1   Introduction

Posture recognition has been widely used in a variety of fields including medicine, HCI, and artificial intelligence. Serious games based on posture recognition have been applied in diverse fields including rehabilitation practice for patients [1], brain therapy for children [2], psychotherapy for children with autism, and the prevention of dementia. Furthermore, posture recognition enables users to control the surrounding environment, for instance, in smart homes or virtual reality, by recognizing postures. Such applications recognize postures on the basis of skeleton data acquired by an RGB-D camera.

However, when using skeleton data, the coordinates of the body joints are required. Thus, the feature data extracted from these coordinates raise the complexity of the recognition algorithm. Because posture recognition algorithms have been developed for a specific sensor, they cannot be directly applied to environments using other sensors and are hence restricted in usage.

Y. Xi · S. Cho · K. Um · K. Cho(✉)
Department of Multimedia Engineering, Dongguk University-Seoul,
26 Pildong 3 Ga, Jung-Gu, Seoul 100-715, Republic of Korea
e-mail: cke@dongguk.edu

© Springer Science+Business Media Singapore 2015   243
D.-S. Park et al. (eds.), *Advances in Computer Science and Ubiquitous Computing*,
Lecture Notes in Electrical Engineering 373,
DOI: 10.1007/978-981-10-0281-6_35

To solve these problems, we propose a posture recognition algorithm that minimizes the influences from sensors and processes large amounts of data on postures quickly.

2 Related Work

Zhang [3] extracted features from the coordinates of 27 joints using the skeleton data of a human acquired by Kinect and implemented an automatic posture learning function using a support vector machine (SVM) algorithm, using it to recognize 22 postures. Patsadu [4] extracted features from the coordinates of 20 human joints and executed posture learning using four machine learning algorithms: a back propagation neural network (BPNN), SVM, decision tree, and naive Bayes. In experiments recognizing three kinds of pause postures, it was verified that BPNN and SVM were the most effective algorithms.

Le [5] experimentally verified the efficiency of a posture recognition method using various features from human joints. The features included the absolute coordinate values of all joints without scaling, seven joint angles with and without scaling, nine joint angles with and without scaling, 17 joint angles without scaling, and 18 joint angles with scaling. Using these features, the algorithm could recognize four postures.

The main existing approaches acquire human skeleton data and used them to form the features of each joint. For recognizing the postures of the entire human body, a number of joints are used. However, these approaches use a specific number of joints. When the number of joints varies with respect to various postures, the generality of the learned data is reduced.

This paper proposes an approach to express three-dimensional (3D) space, including human beings, as a binary array-type feature without directly applying the human skeleton data. The proposed approach is able to recognize a variety of postures of a human being with fewer data. Furthermore, this approach can recognize postures even if the sensor that recognizes the human skeleton using 3D space data is changed.

3 Posture Recognition

The approach proposed in this paper configures 3D space into $12 \times 12 \times 12$ small sensing blocks to describe the scope of human postures. In this space, the 3D human model expresses the postures of a human being instead of the actual human being. At this point, the number of joints of the skeleton driving the 3D human model can be adjusted. Each sensing block detects an intersection with the 3D human model and expresses this intersection in bit data. The resulting one-dimensional intersection status array is configured by arranging the bit data of all sensing blocks in sequence. When a specific sensing block intersects with the 3D human model, the index value of the relevant sensing block is set to one in the intersection status array. In contrast, when a specific sensing block does not intersect with the 3D human model, the index value of the relevant sensing block is set to zero. Fig. 1 shows the intersection statuses of sensing blocks depending on the position of the 3D human model.

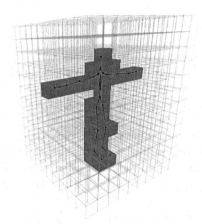

Fig. 1 Human posture encoded in sensing blocks.

When the intersection status of each sensing block is expressed as an element, a single intersection status array can be formed of the 1,728 total elements. When each element value in the intersection status array is expressed as a zero or one, the array has the form of $p = \{0, 0, 0, 1, 0, 1, ..., 0, 0, 0\}$. Using this intersection status array, one pose can be expressed as a single binary number.

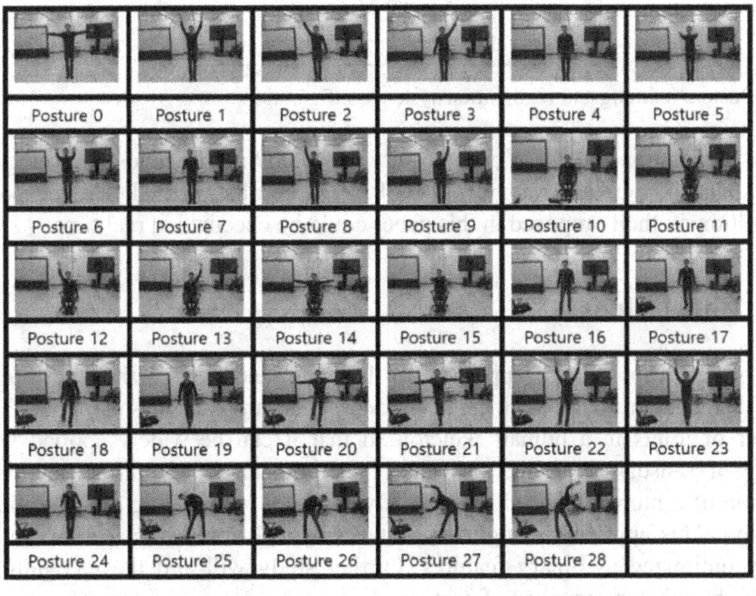

Posture 0	Posture 1	Posture 2	Posture 3	Posture 4	Posture 5
Posture 6	Posture 7	Posture 8	Posture 9	Posture 10	Posture 11
Posture 12	Posture 13	Posture 14	Posture 15	Posture 16	Posture 17
Posture 18	Posture 19	Posture 20	Posture 21	Posture 22	Posture 23
Posture 24	Posture 25	Posture 26	Posture 27	Posture 28	

Fig. 2 Twenty-nine learned postures.

This study learned and recognized 29 postures using the features of the binary numbers generated by the intersection status array using multi-class classification-based SVM. Fig. 2 presents the 29 postures.

4   Recognition Experiment

This study executed the experiments in an Intel (R) Core (TM) i5-4690 CPU, 8 GB RAM, NVidia GeForce GTX 760 environment. Furthermore, the 3D human model and 3D space data collection tool were implemented using the Unity 3D engine [6]. The SVM function were added using C# and the accord.net library [7].

Fig. 3 presents the overall structure of the posture learning and recognition system implemented for this experiment. In this system, the feature extraction module specifies the postures of the 3D human model, and the posture data is then learned in the posture learning module. The posture recognition module can then recognize a posture using the data from the posture database.

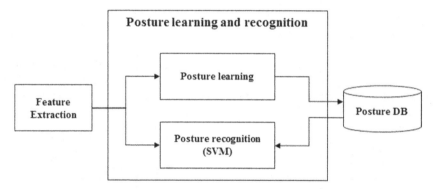

Fig. 3 Posture learning and recognition system configuration.

The recognition speed per each posture was 0.009 milliseconds on average and the average recognition rate was 94.03%. The results verified that the posture recognition method proposed in this paper could be executed in real time.

5   Conclusions

This paper proposed an approach to learning and recognizing postures by collecting human action data in 3D space. This approach has no limit on the number of joints in a human skeleton so that it can be widely applied with a variety of sensors. Moreover, because the approach is not influenced by the direction of a human being, it has the advantage of recognizing a posture in any direction. This approach recognizes postures in real time using features that are simply optimized as binary features. Further study will add the recognition of gestures based on a sequence of postures.

Acknowledgments This research was supported by the MSIP (Ministry of Science, ICT, and Future Planning), Korea, under the ITRC (Information Technology Research Center) support program (IITP-2015-H8501-15-1014) supervised by the IITP(Institute for Information & Communications Technology Promotion).

References

1. Omelina, L., Jansen, B., Bonnechère, B., Jan, S. V.S-, Cornelis, J.: Serious games for physical rehabilitation: designing highly configurable and adaptable games. In: Proceedings of the 9th International Conference on Disability, Virtual Reality and Associated Technologies (ICDVRAT 2012), pp. 195–201 (2012)
2. Pirani, E.Z., Kolte, M.: Gesture Based Educational Software for Children with Acquired Brain Injuries. International Journal of Computer Science and Engineering **2**(3), 790–794 (2010)
3. Zhang, Z., Liu, Y., Li, A., Wang, M.: A novel method for user-defined human posture recognition using kinect. In: The 2014 7th International Congress on Image and Signal Processing, pp. 736–740 (2014)
4. Patsadu, O., Nukoolkit, C., Watanapa, B.: Human gesture recognition using Kinect camera. In: 2012 International Joint Conference on Computer Science and Software Engineering (JCSSE), pp. 28–32 (2012)
5. Le, T.-L., Nguyen, M.-Q., Nguyen, T.-T.-M.: Human posture recognition using human skeleton provided by Kinect. In: 2013 International Conference on Computing, Management and Telecommunications (ComManTel), pp. 340–345 (2013)
6. Unity-Game Engine: http://unity3d.com/
7. Accord.NET-Machine Learning Framework: http://accord-framework.net/

Design of Secure Protocol for Session Key Exchange in Vehicular Cloud Computing

Jungho Kang and Jong Hyuk Park

Abstract Vehicle cloud computing is a technology that converges vehicle networking technology with cloud computing technology. It is expected to contribute to society where vehicles have become mainstream, by combining the vehicle networking technology VANET with existing cloud infrastructure. But on the other hand, issues regarding personal information and privacy have risen. This paper suggests a protocol that ensures the safe transmission of session keys for a safer vehicle cloud environment.

Keywords Cloud computing · Key exchange · Security · VANET · Vehicular

1   Introduction

VANET(Vehicular Ad Hoc Network) showed that by combining vehicle networking technology and cloud computing technology, the following can be a reality [1][2]. The stable internet access in a moving car makes voice recognition possible. This is essential in securing safety in the vehicle. The voice recognition function can be used in real time when one wants to access the cloud services. When the road is frozen, the car's sensor detects this and sends the information to the cloud, making it possible for the cars behind to access the same information. The camera built in the rear view mirror takes photos of the road and sends visual information to the cloud, while vehicles can check in real time such information and know what is several

J. Kang
Department of Computing, Soongsil University,
402 Information Science Building, 369 Sangdo-Ro, Dongjak-Gu, Seoul 156-743, Korea
e-mail: kjh7548@naver.com

J.H. Park(✉)
Department of Computer Science and Engineering, Seoul National University
of Science and Technology, Gongneung 2-dong, Nowon-Gu, Seoul 139-743, Korea
e-mail: jhpark1@seoultech.ac.kr

© Springer Science+Business Media Singapore 2015   249
D.-S. Park et al. (eds.), *Advances in Computer Science and Ubiquitous Computing*,
Lecture Notes in Electrical Engineering 373,
DOI: 10.1007/978-981-10-0281-6_36

hundred meters ahead of them so that they can avoid overly congested areas. Before entering areas that drastically slow down, they can slow down gradually to prevent accidents [3]. Vehicle cloud computing technology that will change our lives uses the internet. But since it uses the internet, issues about personal information leaking have always been raised [4][5]. This paper describes the protocol that allows safe transmission of encrypted keys to protect personal information. Chapter 2 addresses VANET and cloud architecture, while Chapter 3 presents a safe protocol for session key exchange. Chapter 4 evaluates and compares various options and Chapter 5 presents the conclusion.

2   VANET and Vehicular Cloud Architecture

This Chapter describes VANET and the three types of cloud architecture which are vehicle cloud, roadside cloud and central cloud [3].

2.1   VANET

VANET is an evolved form of Mobile Ad-hoc Network(MANET) and is defined as inter-node communication on the move just like MANET. But it is different in that it is limited to nodes in vehicles that are moving at a fast pace in a certain direction. It defines wireless network between vehicles and allows communication without the help of a third part. It makes possible to exchange traffic information or accident information to make driving safer and more convenient. The structure of VANET consists of Vehicle-to-Vehicle (V2V) communication and Vehicle-to-Infrastructure (V2I) communication [6]. In addition, each section forms a cloud to form a collaborative network where services can be exchanged. However, if the communication with the cloud is attacked maliciously, then not only is the service disrupted, but user privacy can be violated. This makes security channels a must [7][8].

2.2   Vehicular Cloud Architecture

Vehicular Cloud Architecture consists of Vehicular Cloud, Roadside Cloud, and Central Cloud.

2.2.1 Vehicular Cloud
Vehicular Cloud is a cloud formed by V2V communication to promote sharing between vehicles. IT supports sharing or computing of information between vehicles in real time so that drivers can be updated consistently. Even if there is a lack of resources, highly sophisticated services can be used. Through the roadside cloud, vehicular cloud can be expanded and the central cloud can be accessed [9].

2.2.2 Roadside Cloud

Roadside Cloud forms a network between Road-Side Units(RSUs). Basically it plays the role of intermediator between the car with its vehicular cloud and the central cloud infrastructure, and ensures inter-vehicle communication broadcasting. Moreover, RSUs share and process information coming from a moving vehicle to provide convenience, and session keys ensure a safe communication [9].

2.2.3 Central Cloud

Central Cloud is a cloud formed by the service provider. It stores the authentication information of the driver and at the driver's request, the roadside cloud provide services to the correct driver. It also plays the role of intermediary so that a trusting relationship can be formed between the vehicle and the RSU [9].

3 Secure Protocol for Session Key Exchange

This chapter suggests a protocol for session key exchange for a safe vehicle cloud environment. Vehicles communicate with the base station through the RSU, and this presumes that not all channels are safe. The terms and symbols used in the proposed security protocol are shown in Table 1.

Table 1 The terms and symbols used in the scheme

CertVehicle	Vehicle's Certificate
N1,N5	Vehicle's Random Number
N2,N3	RSU's Random Number
N4,N6	BS's Random Number
Pri_RSU	RSU's Private Key
Pub_RSU	RSU's Public Key
Req( )	Request
Res( )	Response
Verify( )	Verify
E( )	Encryption
D( )	Decryption
GenSesKey( )	Generate Session Key
B_R_SK	Session Key Between Base station and RSU
R_V_SK	Session Key Between RSU and Vehicle
B_V_SK	Session Key Between Base station and Vehicle

Before exchanging session keys, the Vehicle, RSU and BS exchange vertification and implement verification for mutual authentication. The vehicle requests an authentication certificate from the RSU, and the RSU requests an authentication certificate from BS. RSU transmits the certificate to the vehicle, and the BS to the RSU. The vehicle and the RSU verifies the authentication certificate transmitted through the CA. If the authentication certificate is valid, the vehicle sends its own authentication certificate to the RSU. And then, the vehicle generates a random number N1, encrypts it using RSU's open key and transmits it along with the authentication certificate. The RSU verifies the authentication certificate of the vehicle through CA and decrypts the random number N1 using its own private key. After this, the RSU sends its own authentication certificate to the BS and generates a random number, encryptes it using BS's public key before transmitting. The BS verifies the authentication certificate of RSU through CA and decrypts the random number using its own private key thereafter. RSU then generates a random number N2, encryptes it using an private key of the vehicle and transmits it to the vehicle. BS, too, generates a random number, encryptes it using an public key of RSU and transmits it to RSU. The vehicle and RSU decrypts the random numbers using its own private key and verifies the random number.

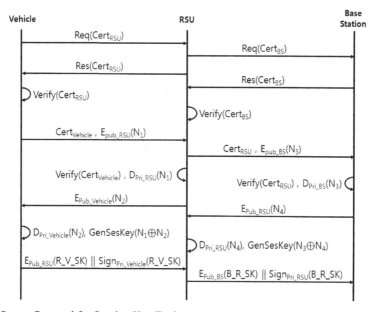

Fig. 1 Secure Protocol for Session Key Exchange

After this process, the vehicle generates a session key using the random number that it generated and the random number it received from RSU. RSU, too, generatees a session key using its own random number and the random number it received from BS. The vehicle encryper the generated session key using the public key of RSU, signs using a private key and transmits it to the RSU. RSU, too, transmits the session key and the signature to the BS through the same calculation as the one done by the vehicle. The session keys generated and trasmitted are used through signature verification. This allows for a safe session key exchange among the vehicle, RSU and the BS.

4   Security Analysis

This Chapter analyzes the security aspects of the suggested protocol.

4.1   Replay Attack

The proposed scheme is safe against replay attacks because it is designed so that the values used for authentication are changed continuously by the random numbers created by the BS, RSU and vehicle. There are the following scenarios in which the attacker makes the replay attack.

1. Session in which the data is collected
   - Collection 1: The attacker collects the authentication certificate and random number that the vehicle transmits to the RSU. Moreover, it collects the encrypted session key and the signature that the vehicle transmits to the RSU.

2. Session where the replay attack is implemented
   - Attack 1: The attacker attempts to receive authentication from the RSU, using the collected authentication certificate and random number value.
   - Verification 1: When the RSU verifies the authentication certificate, the one that the attacker used is passed because it is the authentication certificate from the vehicle. The random number, too, is the value generated by the vehicle and therefore the RSU accepts the random number unconditionally.
   - Attack 2: The attacker transmits the session key it had collected in the previous session as well as the signature to the RSU.
   - Verification 2: RSU first verifies the signature. Since it is the correct value that the vehicle signed, signature verification is a success. However, the attack fails during the authentication of the session key. This is because the RSU's random number which is one of the session key's seed value is different from the random number of the previous session, and thus the session key is different from what the RSU had expected. As a result, the attacker fails in conducting a replay attack.

4.2   Eavesdropping and Brute Force Attack

In the suggested security protocol, all values transmitted through the network are encrypted using the public key. Once the session key is exchanged, the encrypted key is transmitted. Therefore, the suggested security protocol is safe against eavesdropping. If the length of the general open key is assumed to be 2048bits, the likelihood of a brute force attack is about 2^{2048}.

4.3   *Mutual Authentication*

In the suggested protocol, all players conduct mutual authentication using authentication certificates. Through the CA, the protocol is halted if the authentication certificate is invalid. Only when the authentication certificate is valid does the exchange of session keys take place.

5   Conclusion

This paper suggested an authentication and session key protocol in a vehicle cloud environment, to ensure a safe exchange of the driver's personal information and protection of privacy. It suggests the generation of a session key using mutual authentication and random numbers, with a protocol tailored to the VANET environment. The suggested protocol has been confirmed through analysis to be safe against the vulnerabilities in the network environment. The protocol is expected to serve the role as a basic protocol that can be further evolved using security methods such as Zero-Knowledge authentication.

References

1. Heo, M., Rhee, K.-H.: Vehicle Cloud Computing Utilization and Security Requirements. Korean Institute of Information Security 24(2), 42–49 (2014)
2. Cloud Computing Support Center. The Convergence Services Cases on Cloud Computing, August 2014
3. http://ngconnect.org/service-concepts/connected-service-vehicle/
4. Blum, J.J., Eskandarian, A., Hoffman, L.J.: Challenges of intervehicle ad hoc networks. IEEE Transactions on Intelligent Transportation Systems 5(4), 347–351 (2004)
5. Yousefi, S., Mousavi, M.S., Fathy, M.: Vehicular ad hoc networks (VANETs): challenges and perspectives. In: 2006 6th International Conference on ITS Telecommunications Proceedings. IEEE (2006)
6. Al-Sultan, S., et al.: A comprehensive survey on vehicular Ad Hoc network. Journal of Network and Computer Applications 37, 380–392 (2014)
7. Chen, Y.-S., Hsu, C.-S., Cheng, C.-H.: Network mobility protocol for vehicular ad hoc networks. International Journal of Communication Systems 27(11), 3042–3063 (2014)
8. Conti, M., Giordano, S.: Mobile ad hoc networking: milestones, challenges, and new research directions. IEEE Communications Magazine 52(1), 85–96 (2014)
9. Altayeb, M., Mahgoub, I.: A survey of vehicular ad hoc networks routing protocols. International Journal of Innovation and Applied Studies 3(3), 829–846 (2013)

Zero-Knowledge Authentication for Secure Multi-cloud Computing Environments

Hyungjoo Kim, Hyunsoo Chung and Jungho Kang

Abstract A multi-cloud computing refers to the environment where services such as resources and software, etc., can be shared and provided through an agreement between two or more cloud computing service providers. A user requires multiple authentications in order to use the cloud service between the shared service providers. However at such time, users are exposed to vulnerabilities such as their authentication information being exposed to service providers whom they did not sign up with or being exposed during the multiple authentications. Therefore in this paper, the zero-knowledge authentication protocol for ensuring anonymity is proposed. The proposed protocol is safe for all know vulnerabilities and can be used as a safe protocol in multi-cloud environment.

Keywords Anonimity · Authentication · Multi-cloud computing · Security · Zero-knowledge

1 Introduction

A multi-cloud computing refers to the environment where cloud computing service providers with rich resources and cloud computing service providers lacking software can share the advantages and disadvantages to provide the service. In the case of multi-cloud computing based on IaaS(Infrastructure as a Service), since the resources are shared by the service providers, an accurate classification and security on the stored data is required. In the case of PaaS(Platform as a Service) or SaaS(Software as a Service), an accurate measurement such as on usage time is required. In addition, unlike IaaS, since the user must directly access the service providers whom they did not sign up with, association of authentication information and authentications are required [1].

H. Kim · H. Chung · J. Kang(✉)
Soongsil University, Dongjak-Gu, Republic of Korea
e-mail: {hyungjoo.kim,hsj6553}@ssu.ac.kr, kjh7548@naver.com

© Springer Science+Business Media Singapore 2015   255
D.-S. Park et al. (eds.), *Advances in Computer Science and Ubiquitous Computing*,
Lecture Notes in Electrical Engineering 373,
DOI: 10.1007/978-981-10-0281-6_37

To achieve this, SSO(Single Sign On), etc. was proposed but it holds problems for storing excessive authentication information and legal issues. In addition, a problem of authentication information being continuously transmitted to the network and a problem of having one's user history being exposed to the unknown service providers exists [2]. Therefore in this paper, a multi-cloud authentication protocol that supports the anonymity based on zero-knowledge is proposed. In the multi-cloud environment based on brokers, the proposed protocol can verify the suitability of the user based on anonymity without exposing themselves by using a temporary ID and zero-knowledge authentication [3].

2   Multi-cloud Computing

This section will describe the multi-cloud and the public, private and hybrid-cloud which form the multi-cloud [4].

2.1   Multi-cloud

A cloud service as an on-demand type of service providers can be used anytime, anywhere through billing. According to the service types, it is divided into SaaS, PaaS and IaaS, and according to the way the service is provided, it is divided into public, private and hybrid cloud [5]. As such, the cloud service can be provided in a variety of forms and structures depending on the service providers, and already, iCloud, Google Drive, Dropbox and Amazon Workspace are being serviced in several of areas. A multi-cloud has emerged so that depending on the needs of users, more cloud services can be provided than the service being provided by a single cloud. A multi-cloud service is not limited to one service but it is a technology that provides a variety of cloud services utilizing the cloud resources and technologies shared between the service providers. At such time, a third trusted party called the broker is placed to relay the resources and services between the service providers. However, while the services for multi-cloud environment became diverse, the structure became complicated due to the association between clouds, and a security threat such as inheriting various problems of cloud has significantly increased. In addition, when sharing the authentication and its information between each cloud, a threat of exposing personal information exists [6].

2.2   Cloud Type

Cloud can be divided into public cloud, private cloud and hybrid cloud [7].

2.2.1 Public Cloud

Public cloud is the cloud that provides service to various enterprises and individuals through the Internet network according to their demands. Since it is open to public, anyone who can access the Internet can use this service regardless of time and place. However, due to its nature of being public, it has a security problem of being accessed by a malicious attacker [8].

2.2.2 Private Cloud

Private cloud is a closed-type cloud which cannot be accessed from the outside as the services being provided by the service providers are limited to a specific network. With the feature of not being able to access from the outside, it is widely used in the enterprise network but it cannot be used other than the specific place and time. However, unlike the public cloud, it has the advantage of being difficult for malicious attackers to access [9].

2.2.3 Hybrid Cloud

Hybrid cloud is a combination of public cloud and Private Cloud. While providing the service to the users through a public channel, sensitive materials are stored in the private cloud to prevent the leakage and to protect against the malicious attackers. In addition, it is highly extended cloud that shares resources and exchange information between the public cloud and the private cloud [10].

3   Zero-Knowledge Authentication Protocol

In this section, a user anonymity authentication protocol is proposed based on Zero-knowledge authentication for the multi-cloud environment. It is based on SaaS and assumes that cross authentication and session key was distributed between the users; service provider 1(SP1), broker; service provider 2(SP2). In addition, the user is only subscribed to the service provider 1 and assumes the situation of using the software of the service provider 2. The terms and symbols used in the proposed security protocol are shown in Table 1.

Table 1 The terms and symbols used in the scheme

N	Random Number
h()	Hash Function
SK	Session Key
Pub	Public Key
Pri	Private Key
RB	Redundancy Bits from r0 and r1

The user will request the use of service of service provider 2 to the service provider 1, and the corresponding requeset is transmitted to the broker. The broker verifies the contract between the service providers and respond with "continue" to the service provider 1. Using the session key with the broker, the service provider 1 transmits the encrypted public key and a temporary ID of the user. The broker transmits the generated random number, temporary ID and the public key of the user to the service provider 2. In addition, the broker also sends the generated random number to the service provider 1. The service provider 1, hashes the random number and the session key from the service provider 2 and transmits to the user. At this time, the service provider 2 hashes the random number and the session key of the service provider 1 and generates the equal value as the value received by the user.

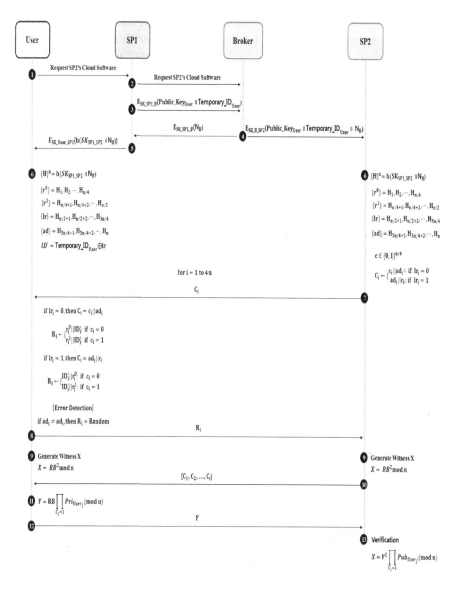

Fig. 1 Zero-Knowledge Authentication Protocol

Hash value of 4n bits is divided into a bit-stream of each n bits. The temporary ID of user is transmitted to service provider 2 through the challenge/response process.

The 4n bit stream is divided into n bits of r0, r1, lr, and ad. Afterwards, user concatenates its own temporary ID, RN and lr creating ID'. Service provider 2 creates bit 'c' to enable the Challenge/Response where it is concatenated with ad to create 'C'. The position of where the ad is concatenated differs depending on the lr value. When the bit value of lr is 0, ad is concatenated behind c, and when the lr bit value is 1, ad is concatenated in front of c. Thus, when lr is expressed in a sequence of n bits, the individual lr value of the lr addresses from first to last determines the ad position. Service provider 2 sends challenge bits 'C' of n bits to the user.

User sends the response value R of service provider 2's challenge bits 'C'. Depending on the lr and 'c' value, a different combination of r0 and r1 value to the ID' value is made. With the challenge/response process ID' is sent to service provider 2. User verifies the ad position and value from the sent 'C' and if the correct 'C' value is not sent a random number is generated and sent. After the ID transmission through the challenge/response process is completed. After a temporary ID is transmitted, the user and the service provider 2 perform an authentication based on Zero-knowledge.

First, in the challenge/response process, the user will generate the value called RB by gathering not used r0 and r1 bit. Afterwards, by multiplying the RB, modular n is computed to generate the witness X. Generated X is transmitted to the service provider 2 and the service provider 2 who knows this RB verifies the X. After verification, the service provider 2 transmits the challenge value configured with 0 and 1 to the user. The user, after the exponentiation of challenge value to his private key, multiplies the RB and generates the modular computed value Y. Generated Y is transmitted to the service provider 2 and the service provider 2 performs exponential operation of the public key and verifies that Y is same as X. When the Y is same as X, the anonymity authentication is successful.

4   Security Analysis

4.1   Replay Attack and Relay Attack

The probability for a successful mafia attack is $2*(1/4)^n$, and the probability for a successful terrorist attack is $2*(1/4)^n$. The proposed scheme is safe against replay attacks because it is designed so that the values used for authentication are changed continuously by the random numbers and challenge bits created by the broker and service provider 2.

4.2   Anonymity and Eavesdropping

In the process for the proposed protocol, all values are encrypted or the location of value changes during the transmission. Therefore, unless the attacker knows the encryption key or the location of all bits, the information cannot be eavesdropped.

In addition, it supports the anonymity since the bit location of the ID is changed during transmission and the authentications are made through Zero-knowledge authentication.

4.3 Forward Security and Error Detection

The user and service provider 2 know the position where the ad value is transmitted according to the lr value in the challenge/response process. That is, the user and service provider 2 are able to detect an error through challenge bits if the service provider 2 and user used the correct hash function. The user and service provider 2 also detect an additional error through response values for challenge bits. The user and service provider 2 detect an additional error through the remainder value and polynomial expression during the second round.

5   Conclusion

In this paper, the anonymous ID transmission method that transmits a hidden temporary ID through a challenge/response process was proposed. In addition, by performing Zero-knowledge authentication using the public key and anonymous ID, a protocol that can process the authentication in anonymity was proposed. The proposed protocol based on anonymity was analyzed to be safe for all known vulnerabilities. It is expected to be used as a safe authentication protocol in the future multi-cloud computing environment.

References

1. AlZain, M.A., Pardede, E., Soh, B., Thom, J.A.: Cloud computing security: from single to multi-clouds. In: 2012 45th Hawaii International Conference on System Sciences (2012)
2. Murukutla, P., Shet, K.C.: Single sign on for cloud. In: 2012 International Conference on Computing Sciences (2012)
3. Maurya, A.K., et al.: Modeling and verification of fiat-shamir zero knowledge authentication protocol. In: Advances in Computer Science and Information Technology. Computer Science and Engineering, pp. 61–70. Springer Berlin, Heidelberg (2012)
4. Lee, J., Son, J., Kim, H., Oh, H.: An Authentication Scheme for Providing to User Service Transparency in Multicloud Environment. Journal of The Korea Institute of Information Security & Cryptology 23(6), December 2013
5. AlZain, M., et al.: Cloud computing security: from single to multi-clouds. In: 2012 45th Hawaii International Conference on System Science (HICSS). IEEE (2012)
6. Ardagna, D.: Cloud and multi-cloud computing: current challenges and future applications. In: Proceedings of the Seventh International Workshop on Principles of Engineering Service-Oriented and Cloud Systems. IEEE Press (2015)

7. Jadeja, Y., Modi, K.: Cloud computing-concepts, architecture and challenges. In: 2012 International Conference on Computing, Electronics and Electrical Technologies (ICCEET). IEEE (2012)
8. Ren, K., Wang, C., Wang, Q.: Security challenges for the public cloud. IEEE Internet Computing 1, 69–73 (2012)
9. Ghanbari, H., et al.: Feedback-based optimization of a private cloud. Future Generation Computer Systems 28(1), 104–111 (2012)
10. Li, J., et al.: A hybrid cloud approach for secure authorized deduplication. IEEE Transactions on Parallel and Distributed Systems 26(5), 1206–1216 (2015)

PUF-Based Privacy Protection Method in VANET Environment

Mansik Kim, Wonkyu Choi, Ayoung Lee and Moon-seog Jun

Abstract The information and communication technology industry has brought about change in everyday lives, including cars, creating a Vehicular Ad-hoc Network (VANET). VANET refers to communication between cars during operation that allows exchanges of messages on traffic information, general safety or responsibility measures to make it safer and more convenient for drivers. Such messages can include personal information and thus, if exposed to malicious attackers can lead to a violation of privacy. Physical Unclonable Functions(PUF) is a challenge-response system where reproduction is impossible due to the inherent physical characteristics gained during the Integrated Circuits (ICs) processing. This paper suggests a protocol that distributes session keys by carrying out the handover and mutual authentication based on PUF without exposing the ID of the car to protect the driver's privacy.

Keywords Vehicular Ad-hoc Network · Physical Unclonable Functions · ID protection · Privacy · Anonymity

1 Introduction

Vehicular Ad-hoc Network (VANET) is a vehicle network that expanded to include even the field of cars with progresses in the ICT industry. It is an evolved form of Mobile Ad-hoc Network(MANET) where communication between vehicles or between the vehicle and the station allows the formation of an autonomous network that enables exchange of information while on the move [1][2]. In particular, traffic information, general safety information and responsibility-related messages can be exchanged over the network to prevent accidents and ease traffic congestion [3]. Such information not only play a critical

M. Kim · W. Choi · A. Lee · M.-s. Jun(✉)
Soongsil University, Dongjak-Gu, Republic of Korea
e-mail: {mansik,gkteehrm,aylee,mjun}@ssu.ac.kr

© Springer Science+Business Media Singapore 2015   263
D.-S. Park et al. (eds.), *Advances in Computer Science and Ubiquitous Computing*,
Lecture Notes in Electrical Engineering 373,
DOI: 10.1007/978-981-10-0281-6_38

role in the vehicle's safety but also include the driver's personal information. Thus, if a malicious attack occurs, this may pose a threat to the life of the driver as well as compose a case of violation of privacy. In order to resolve this issue, Petit et al. suggested a method of generating pseudonyms based on PUF but the issue of needing a fuzzy extractor in addition still remains [4].

Therefore, in order to address these issues, this paper suggests an authentication method that ensures anonymity based on Physical Unclonable Functions (PUF) while still distributing the session keys safely.

2   Vehicular Ad-hoc Network (VANET)

VANET is a wireless ad hoc network that provides communication between On-Board Units (OBUs) and mutual communication between the OBU and the nearby roadside equipment to provide ubiquitous access to drivers on the road and enable the Intelligent Transportation System(ITS) [5]. Moreover, while it offers communication between moving nodes like MANET, unlike MANET the mobility of the driver who is the OBU is limited and has a high speed on the road [2]. AS seen in Fig. 1, VANET consists of Vehicle-to-Vehicle (V2V), road-side infrastructure units (RSUs) and vehicle-to-infrastructure (V2I) [6]. As such, through the information that is exchanged between the channels of VANET ensure a safe and convenient driving experience for the driver, if there is a malicious attack then the privacy of the driver may be violated. In particular, when there is an attack on RSU, a large amount of user data can be taken. Therefore, it is important to establish safe communication channels while ensuring anonymity.

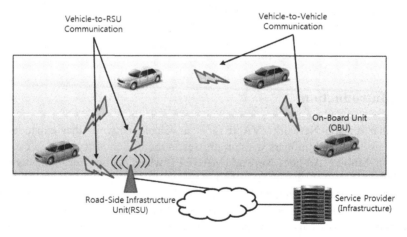

Fig. 1 Vehicular Ad-hoc Network Structure

3 Physical Unclonable Functions(PUF)

Physical Unclonable Functions(PUF) is an innovative physical primitives that generates a unique value that is physically unclonable. There is also no need for secure EEPROMs or any other expensive hardware to safely store the secret value as the PUF can store the secret key or the black-box challenge-response system can be used for mutual authentication [7]. Since PUF has the characteristics of a unique hardware where reproduction cannot be done based on environmental differences or arbitrarily applied physical changes, it offers powerful physical security [8]. The unique challenge-response value generated from the PUF is stored in the DB and later it is compared with the challenge-response value of the device that has installed PUF. Since each challenge-response value is disposed of after being used, it cannot be re-used by a malicious attacker and is thus safer.

4 Proposed Protocol

The suggested protocol consists of the mutual authentication process and the handover process. The notation used throughout the paper is shown in Table 1.

Table 1 Scheme Notation

Notation	Meaning
ID_{OBU}	OBU ID
ID_{RSU}	RSU ID
SK	Session key
C	Challenge value
R	Response value
K	Shared secret key between OBU and SP

4.1 OBU Mutual Authentication Phase

OBU at first registers its own ID_{OBU} and the PUF's Challenge-Response onto the SP and share the secret key K. When there is first contact with RSU1 during operation on the road, mutual authentication is carried out through the protocol as seen in Fig. 2. OBU, once it receives the request, C_1, ID_{RSU1} values that are broadcast from the RSU1, then stores the ID_{RSU1} and calculates R_1 through $PUF(C_1)$. Then a new challenge value C_2 is selected to calculate $E_K(ID_{OBU} \oplus C_2)$ using the secret key K shared with SP. Along with R_1, this is delivered as a mutual authentication response message to RSU1. The RSU1 that receives the value generates the session key SK_{new} and relays as a response message $E_K(ID_{OBU} \oplus C_2)$, as well as R_1, C_1 and SK_{new} to SP. SP searches the ID_{OBU} that matches the R_1 and C_1 to find the secret key K. $E_K(ID_{OBU} \oplus C_2)$ is decrypted to calculate R_2 through $PUF(C_2)$. Once through the secret key K, the $E_K(SK_{new} \oplus R_2)$ is relayed to the

RSU1, RSU1 again relays this value to OBU. OBU decrypts $E_K(SK_{new} \oplus R_2)$ to calculate $PUF(C_2)=R'_2$. If $R_2==R'_2$ is true, the session key SK_{new} is extracted.

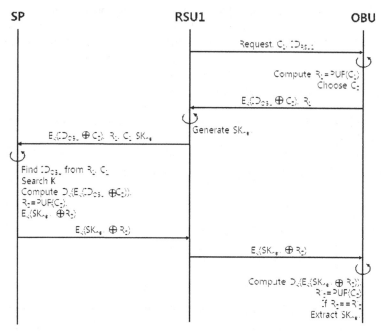

Fig. 2 OBU Mutual Authentication Protocol

4.2 OBU Handover Phase

OBU after carrying out the first mutual authentication with RSU1 then moves into the frequency range of RSU2 to carry out the handover. The handover process is as seen in Fig. 3. OBU carries out the same process as the mutual authentication protocol shown in Fig. 2 except that instead of a mutual authentication response message, it sends a handover request message until SP calculates the $E_K(SK_{new} \oplus R_2)$. Once SP relays $E_K(SK_{new} \oplus R_2)$ to RSU1, then RSU1 encrypts the session key it had shared with OB or SK_{old} to relay it to OBU along with the ID_{RSU2} value. OBU decrypts $E_K(E_{SKold}(SK_{new} \oplus R_2))))$ to calculate $PUF(C_2)=R'_2$. If $R_2==R'_2$ is true, then session key SK_{new} is extracted.

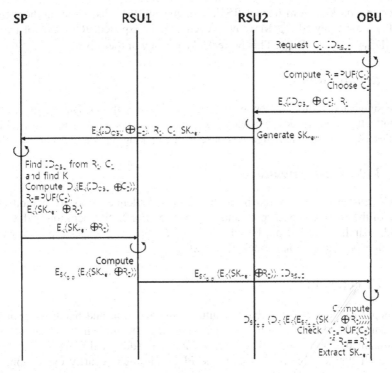

Fig. 3 OBU Handover Protocol

5 Security Evaluation

In this section, we present a security analysis of proposed protocol against various attacks.

5.1 Eavesdropping Attacks

An attacker may eavesdrop on the vehicle's mutual authentication or handover communication. However, the parameters value that are exchanged by the OBU during the RSU communication process has only one response value from the PUF which is not re-used. Moreover, ID_{OBU} is relayed as an encrypted by secret key K after XOR operation with C2. The session keys are not exposed, which makes it safe from attacks or the other RSUs.

5.2 Masquerading Attacks

In a VANET environment, attackers masquerade as RSU during the mutual authentication or handover process to attempt communication with the OBU. However, SK_{new} generated by the RSU is encrypted by secret K which is shared by the SP with OBU, and during handover it is additionally encrypted into the former session key from former RSU. This means RSU that is not authenticated by SP cannot relay SK_{new}. Moreover, even if K is exposed, RSU cannot extract from $ID_{OBU} \oplus C_2$ the ID_{OBU}. This protects the privacy of the driver.

5.3 Replay Attacks

In the suggested protocol, Replay Attacks cannot be a threat. During the mutual authentication and handover process, the message exchanged is not re-used and includes a PUF-based challenge-response value, making reply attacks unfeasible.

5.4 Vehicle Anonymity

Vehicle anonymity is ensured by a PUF-based challenge-response method. The ID_{OBU} of the vehicle is encrypted and transmitted after XOR operation with C_2. At the SP, search is carried out based on the PUF-based challenge-response, which does not expose to external attackers or RSU.

6 Conclusion

This paper suggested a PUF-based mutual authentication and handover protocol that ensures the vehicle's ID is not exposed, so that the anonymity of the user can be maintained. The protocol is realized without any additional functions but using just the challenge-response value of the PUF. Through security evaluation, the safety against various attacks and the validity of privacy protection were verified. This protocol can be applied to various privacy services in the VANET environment where the driver's ID needs to be kept anonymous.

References

1. Conti, M., Giordano, S.: Mobile ad hoc networking: milestones, challenges, and new research directions. IEEE Communications Magazine 52(1), 85–96 (2014)
2. Al-Sultan, S., et al.: A comprehensive survey on vehicular Ad Hoc network. Journal of Network and Computer Applications 37, 380–392 (2014)
3. Raya, M., Hubaux, J.-P.: The security of vehicular ad hoc networks. In: Proceedings of the 3rd ACM Workshop on Security of Ad hoc and Sensor Networks. ACM (2005)
4. Petit, J., et al.: On the potential of PUF for pseudonym generation in vehicular networks. In: 2012 IEEE Vehicular Networking Conference (VNC). IEEE (2012)
5. Wang, Y., Li, F.: Vehicular ad hoc networks. In: Guide to Wireless Ad hoc Networks, pp. 503–525. Springer, London (2009)
6. Calandriello, G., et al.: Efficient and robust pseudonymous authentication in VANET. In: Proceedings of the Fourth ACM International Workshop on Vehicular Ad hoc Networks. ACM (2007)
7. Herder, C., et al.: Physical unclonable functions and applications: A tutorial. Proceedings of the IEEE 102(8), 1126–1141 (2014)
8. Suh, G.E., Devadas, S.: Physical unclonable functions for device authentication and secret key generation. In: Proceedings of the 44th Annual Design Automation Conference. ACM (2007)

Design of Authentication Protocol Based on Distance-Bounding and Zero-Knowledge for Anonymity in VANET

Minjin Kim, Keun-Chang Choi, Hanna You and Moon-seog Jun

Abstract VANET is a communication network for exchanging information between vehicles where accidents and traffic information can be delivered between vehicles or receive services through an infrastructure. Although efficiency of traffic system and convenience of drivers are provided through VAN communication, malicious information can lead to traffic disturbance or risk to drivers. Therefore, a way to prevent an intermediate attack should be considered. For considerations there are mobility of vehicles and the fact that it is multiple objects and the existence of RSU. In this paper, it ensures the integrity and a cross-certification is conducted through Distance-bounding and Zero-Knowledge. The corresponding protocol has verified the safety through a security analysis on the security vulnerabilities in VANET environment. As a result of this study, integrity in VANET environment can be secured with reliability through illegal RSU detection.

Keywords Vehicular networks · Zero-knowledge authentication · Distance-bounding · Diffie-Hellman key exchange · Vehicular security

1 Introduction

The VANET(Vehicular ad hoc networks) is expected to spread as it can efficiently control and provide information such as road conditions and traffic information, etc.

M. Kim · K.-C. Choi · M.-s. Jun(✉)
Soongsil University, Dongjak-Gu, Republic of Korea
e-mail: {minjini57,muziag,mjun}@ssu.ac.kr

H. You
Convergence Laboratory, KT R&D Center, Seoul, Republic of Korea
e-mail: hanna.you@kt.com

© Springer Science+Business Media Singapore 2015
D.-S. Park et al. (eds.), *Advances in Computer Science and Ubiquitous Computing*,
Lecture Notes in Electrical Engineering 373,
DOI: 10.1007/978-981-10-0281-6_39

However, there could be a malicious RSU(Road-Side Units) or eavesdropping attackers. Malicious information can cause confusion to traffic system or becomes a serious threat against the safety of drivers. Therefore, in VANET, a method for sustainable reliability and receiving certification each time the vehicles meet with each RSU is important. A method for authenticating with SP(Service Provider) is required while preventing the exposure of identification of vehicles to the RSU.

Therefore in this study, a method for safely passing the RSU while receiving authentication of SP and receiving certification from each RSU over the road is proposed. Based on Zero-knowledge authentication, a protocol that ensures the anonymity without the exposure of identification is designed. It will be safe from the security threats of corresponding protocol tracking [1].

2 VANET (Vehicular ad hoc networks)

2.1 VANET Overview

With the development of communications technology, a communication between all objects called IoT(Internet of Things) system has been distributed and accordingly, communications between vehicles were made possible. For vehicle network, there is Vehicular ad hoc network [2].

VANET provides the safety and convenience of drivers and the utilization of traffic system. It consists of the vehicle equipped with OBU(On-Board Unit) which allows wireless communication, RSU that provide wire and wireless link with the infra-network installed on the road and the SP who provides the service. VANET supports the communication with ITS(Intelligent Transportation System) using the wireless system of the vehicle or communications between vehicles. In other words, there are V2I(Vehicle-to-Infrastructure) and V2V(Vehicle-to-Vehicle).

2.2 VANET Authentication Security Considerations

VANET which was derived from the mobile ad-hoc network communicates between vehicles through the OBU in each vehicle [3]. A key is allocated from RSU for communication between vehicles and also communicates with other vehicles or infrastructure using the corresponding key. Prior to communication, an authentication that can determine whether the corresponding user is a true user or RSU is required. For authentication, the backbone network called the SP and the vehicle are the key followed by the RSU which connects such backbone with the vehicle network. RSUs are placed on the road at regular intervals. As there are a large number of RSUs, when a malicious RSU is installed which communicates with vehicles without proper authentication, it becomes a fatal security threat. In addition, since one vehicle or one RSU requires a communication for frequent new authentication, if some portion of identification is exposed each time, it becomes a target for hackers. Therefore if possible, identification should not be exposed [4].

3   Zero-Knowledge Authentication

3.1   Zero-Knowledge Authentication Properties

Zero-knowledge authentication satisfies the following 3 properties using the method of proving to the verifier without directly exposing the contents known by the certifier [5].

• Completeness: If the content of authentication is true, an honest certifier should be able to convince the honest verifier.
• Soundness: If the content of authentication is false, no matter whom the dishonest certifier may be, he should not be able to prove true to the honest verifier.
• Zero-Knowledgeness: If the content of authentication is true, the verifier cannot know anything except the true false statement.

Lu et al. [6] in the corresponding paper, Zero-knowledge authentication method in anonymous P2Ps(peer-to-peer networks) was proposed. 3 properties of Zero-knowledge were satisfied and a small traffic amount and overhead was verified. When designing the reliable management system based on anonymity, one-way hash function was used to generate the anonymity rather than using a real identity.

4   Distance-Bounding Protocol

4.1   Distance-Bounding Overview

Distance-Bounding Protocol(DBP) was proposed by Brand and Chaum in 1993 in order to resist the relay attack in wireless environment [7]. Although various DBP variant protocols were proposed, the basic model shares the pre-random values in the order of low speed→high speed→low speed and transmits 0 and 1 bit in a high speed. Through this bit stream, mutual authentication and time is measured.

Response according to the attempted value is delivered in high speed one bit at a time and this response must be made within the upper-bound. The verifier must determine whether the corresponding process was made within a reasonable time, and through the time difference check, the close distance can be roughly verified [8].

5   Proposal

The basic assumptions of protocol proposed in this paper are as follows. First, the vehicle and SP have a unique Seed value. Second, the public key of the SP, RSU and vehicle is opened and known.

The protocol is proposed as the Figure 1 and the parameters of Table 1 are used.

Table 1 Parameters

Parameter	Description
ID_R, ID_C	The identifier of RSU, Car.
N_{RSU}, N_{Car}	The random number generated by RSU, Car.
Pub_SP, Pub_RSU, Pub_Car	The Public key by SP, RSU, Car.
Pri_SP, Pri_RSU, Pri_Car	The Private key by SP, RSU, Car.
r_R, r_C	The random number for distance bounding generated by RSU, Car.
n	The Blum integer n = pq, in which p, q are large primes.
C_i	The challenge number for distance bounding generated by Car.
a, b	The random integer generated by Car, RSU.
S	The Diffie-Hellman Secret Key for RSU, Car.

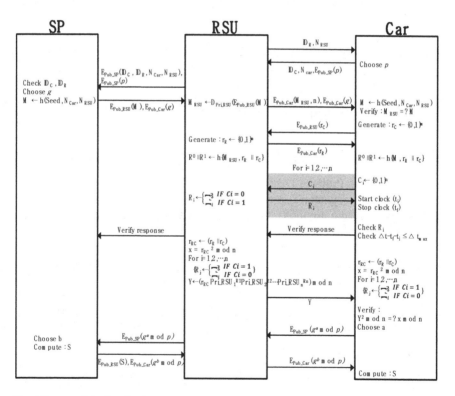

Fig. 1 Proposed Authentication Protocol

5.1   Protocol Process

RSU transmits its ID_R and N_{RSU} to the Car entering his section. At this time, when the Car transmits its ID_C, Ncar and $E_{Pub\_SP}(p)$ value to RSU, the authentication process begins. Here, the P value is chosen for the generation of DH key.

RSU sends the $E_{Pub\_SP}(ID_C, ID_R, N_{Car}, N_{RSU})$ and $E_{Pub\_SP}(p)$ values to the SP and SP verifies the ID value and selects the g. using the Seed value shared with the car and a random number transmitted by the RSU and car, a hash value M is generated and transmitted.

RSU decrypts M using its private key for inputting to MRSU and transmits to the Car by selecting the modular n. At this time, verification is made on the RSU to determine whether it is a legitimate RSU received from the SP by comparing the M_{RSU} with the M value computed from the Car.

The Car and RSU generates random values for DBP in order to perform mutual exchange. Afterwards, R_i is received from RSU through the C_i, and by measuring the starting time and the termination time of transmission, verification is made on whether the communication was made within a valid time with an accurate response. Through this and through DBP process, a cross authentication between the RSU and Car is made. After verification, the confirmed message is sent to RSU for computing the ZKP and transmitted to SP to inform the cross authentication results.

The RSU and Car uses the R_j value left from DBP. So without any challenge for ZKP, the computed resulting value Y is transmitted to the Car by RSU. If the following Y^2 mod n =? x mod n values are satisfied the a value is selected for the DH, and g^a mod p is sent to SP, and SP transmits the g^b mod p and computed secret value S to the RSU. RSU checks the S value and delivers the g^b mod p to the Car and a mutual communication begins using the DH key.

6   Security Analysis

The protocol proposed in this paper has the following characteristics.

6.1   Resistance to Relay Attack

As the bit is exchanged within a valid time using DBP, two nodes which are physically close can be verified. In addition, with validity verification, it is safe from the relay attack.

6.2   Ensures Anonymity

Cross authentication is performed using ZAP without exposing the identification. In the protocol of this paper, the bit array R^0_i, R^1_i generated in DBP is used as a response on the C_i value received from the Car. Only the n bit is used for each

$R^0_{i\,and}\,R^1_j$. In this protocol, the value left at this point is utilized in ZKA. It has the advantage of omitting the exchange process for initial computed value. Conclusively, anonymity can be secured with a fewer number of exchange.

6.3 Forward Security and Error Detection

In the process of exchanging committed R^0_i, R^1_i as the response value on C which is a random value of bit array R^0_i, R^1_i generated during the DBP process, an error detection is possible if the location of even 1 bit is wrong among the response bits, and through this, a forward security and cross authentication is possible.

7 Conclusion

In VANET, there are security vulnerabilities which can be caused by a malicious RSU. Accordingly, a protocol that provides anonymity and cross authentication was proposed. Proposed protocol has verified the safety on the relay attack, anonymity and forward security & error detection through a security analysis. In the future, the proposed protocol that provides anonymity in VANET is expected to be used.

References

1. Maurya, A.K., et al.: Modeling and verification of fiat-shamir zero knowledge authentication protocol. In: Advances in Computer Science and Information Technology. Computer Science and Engineering, pp. 61–70. Springer Berlin, Heidelberg (2012)
2. Sou, S.-I., Tonguz, O.K.: Enhancing VANET connectivity through roadside units on highways. IEEE Transactions on Vehicular Technology 60(8), 3586–3602 (2011)
3. Paul, B., et al.: Vanet routing protocols: Pros and cons. arXiv preprint (2012) arXiv:1204.1201
4. Feiri, M., et al.: The impact of security on cooperative awareness in VANET. In: 2013 IEEE Vehicular Networking Conference (VNC). IEEE (2013)
5. Goldreich, O., Micali, S., Wigderson, A.: Proofs that yield nothing but their validity or all language in np have zero-knowledge proof systems. J. Assoc. Comput. Mach. 38, 691–729 (1991)
6. Lu, L., Han, J., Liu, Y., Hu, L., Huai, J., Ni, L.M., Ma, J.: Pseudo trust: Zero-knowledge authentication in anonymous P2Ps. IEEE Trans. Parallel Distrib. Syst. 19, 1325–1337 (2008)
7. Cremers, C., et al. Distance hijacking attacks on distance bounding protocols. In: 2012 IEEE Symposium on Security and Privacy (SP). IEEE (2012)
8. Avoine, G., Kim, C.H.: Mutual Distance Bounding Protocols. IEEE Transactions on Mobile Computing 12 (2013)

Design of Exploitable Automatic Verification System for Secure Open Source Software

Bumryong Kim, Jun-ho Song, Jae-Pyo Park and Moon-seog Jun

Abstract As more people use IT products, the application extent of software has increased along with demand for it. In addition to commercialized software, open source software is also seeing its market grow rapidly. But open source software is developed by those without expert knowledge in security. As a result, many security vulnerabilities arise and are taken advantage of for attacks. Therefore, in this paper, we suggested the design of an exploitable automatic verification system for secure open source software to address these issues. It is expected that, through the use of this system, the reliabilities of the open source software, the developers of the open source software, and the corporations using can be improved.

Keywords Exploitable · Exploit · Exploitable verification · Open source · Software vulnerability

1   Introduction

With the increase in the use of IT products, software is increasingly applied to wider areas including smartphones, medical devices and vehicles. In addition to commercial software, open source software has also seen a rapid increase in its use. Open source software by nature is developed by those without expert knowledge on security. This leads to many security vulnerabilities that can be taken advantage of. Actually, the number of the cases of the cyber-attacks that took place in 2011 was 5.5 billion. It had increased by 81% compared to 2010. And 75% of the attacks were the attacks that abused the security vulnerabilities of the software themselves. As an example, due to the fatal, weak points of the 'open SSL', the personal information of over 900 people who were the tax payers to the Canada Revenue Agency was leaked. And, in Great Britain, there was an accident

B. Kim · J.-h. Song · J.-P. Park · M.-s. Jun(✉)
Soongsil University, Dongjak-Gu, Republic of Korea
e-mail: {gflawer,jhsong,pjerry,mjun}@ssu.ac.kr

© Springer Science+Business Media Singapore 2015
D.-S. Park et al. (eds.), *Advances in Computer Science and Ubiquitous Computing*,
Lecture Notes in Electrical Engineering 373,
DOI: 10.1007/978-981-10-0281-6_40

275

in which the computer network of the Mumsnet, which possesses about 1.5 million members [1].

Thus, this paper suggests the design of an exploitable automatic verification system for secure open source software to address these issues.

2   Preliminaries

2.1   Open Source Software(OSS)

Open source software refers to software whose source code is open, allowing anyone to use, edit, distribute and re-distribute it. The global OSS market has seen an annual growth rate of 18.8% [2].

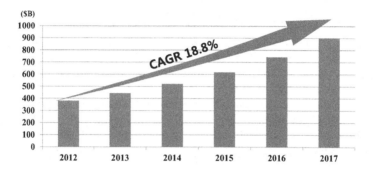

Fig. 1 Growth rate of the Software market

As more people use open source software such as Linux, many victims have arisen, too. For example, there is the vulnerability of GNU Bash which was discovered in September, 2014. This vulnerability is called "ShellShock". It allows malicious activity such as system control by interesting a specific code using code injection method to an arbitrary environmental variable in GNU Bash. This led to many secondary damages such as the cost to maintain patches or conduct urgent updates, or the violation of private information.

2.2   Web Crawling

Web Crawling is also called Crawling or Spidering. It refers to the method of constantly searching and copying data in the web environment through automatic methods. The software called Web Crawler visits a website, searches all or selective data and generates a copy. The data can be processed according to the preferred format of the user and saved as a data base or file [3].

2.3 Software Testing Methodologies

A White Box Testing method is a way to test the internal logical structure and behavior based on source code. It allows one to judge the validity of a code error or activity and can view the process of an activity. This method has the advantage of clarifying all source codes and enabling test scenario drafting, but it is impossible to implement without a source code. There is also the danger of undermining the objectivity of the test as a result of conducting it according to the activity of the code.

A Black Box Testing method is one of the SW test methods where the internal structure and operation principle don't have to be known to test the software operation. Usually the function is tested from the outside by looking at what kind of outputs a certain input generates. It does not require a source code and has many advantages for large scale software but it cannot detect logical error and cannot make various test cases [4]

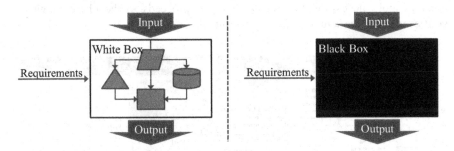

Fig. 2 Black box testing and White box testing

2.4 Information Related to Global Vulnerabilities

CWE(Common Weakness Enumeration) is a compilation of software weaknesses gathered by MITRE with the support of the National Cyber Security Division within the U.S. Department of Homeland Security. It is provided for free to public sectors around the world. CWE mainly defines the weakness and explains the platform for weakness and example code and describes the way to lessen the weaknesses [5][6].

CVE(Common Vulnerabilities and Exposures) is a list of security vulnerabilities detected over time. If CWE can be called a categorization system of general vulnerabilities, then CVE can be seen as the chronological history of discovered security vulnerabilities. CVE can be sorted into two states: entry and candidate. Entry is a state which permits the inclusion of a CVE identification number and candidate is a state in which it is being reviewed in order to be included in the list. The form of CVE is shown in the form of CVE-year shown-identification number [7][8].

CAPEC(Common Attack Pattern Enumeration and Classification) does not categorize vulnerabilities but attack patterns. It was sponsored by the U.S. Department of Homeland Security and developed by the company Cigital to manage the list that compiles various attack patterns, actual cases, vulnerability test and detection methods for each attack pattern, response methods and level of importance. Such information can be helpful in reducing security vulnerability [9][10].

3   Design of Exploitable Automatic Verification System

In this chapter, the organization of the system that automatically verifies whether the weak points are exploitable and the detailed functions are explained. The system proposed is organized into the pre-execution function tool and the exploitable verification part. Based on the results and others of the like of analyzing the information and the source codes collected with the pre-execution function tool, the exploitable verification part proceeds to the phase in which verifies whether the source codes are exploitable or not.

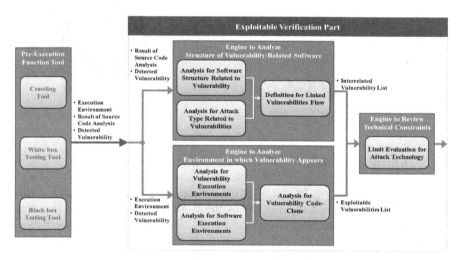

Fig. 3 Exploitable Automatic Verification System

3.1   Pre-Execution Procedure

The pre-execution function tool is organized into the crawling tools, the black box testing tools, and the white box testing tools. And the information and the database to be used for the exploitable verification part are provided. Crawling Tool collects information on the execution environment of open source software, its source code and the international vulnerability categorization system, then provides such information to other tools or engines. The Black Box Testing Tool and White Box Testing Tool conduct test on open source software and conclude the suspected vulnerable code-clone. The detected vulnerability (code suspected of vulnerability) is then provided to the Exploitable Verification Part.

3.2 Exploitable Verification Part

Exploitable Verification Part consists of the analysis engine for the environment in which vulnerability appeared, an engine to analyze the software structure related and the engine to establish a violation scenario. Using these, it verifies whether the vulnerability is exploitable.

3.2.1 Engine to Analyze Structure of Vulnerability-Related Software

This engine is an engine for judging what kind of an interrelationship that the offensive actions that can appear according to the weak points have. This engine is organized into the analysis module for the software structure related to the vulnerability, the analysis module for the attack type related to the vulnerabilities, and the definition module for the linked vulnerabilities flows.

The analysis module for the software structure related to the vulnerability analyzes the interrelationship between the structure of the software and the vulnerabilities. By extracting the information that traces the flow of source code, it traces back to the section where vulnerability exists or identifies the degree of reliance by functions that were discovered to have vulnerabilities. This is done by analyzing the calling relationship between functions. This helps identify the likelihood of a vulnerability spreading. The structural analysis of software defines and lists a series of characteristics that can be seen as vulnerabilities.

The Analysis module for attack type related to vulnerabilities is based on source code vulnerability analysis such as CWE. It reviews whether follow-up attacks have been detected and whether risk factors exist for future attacks.

Based on the standards of the resultant lists of the precedent, two modules, the definition module for the linked vulnerabilities flows examines the crossing items and reconstitutes by merging the flows of the offensive actions possible.

3.2.2 Engine to Analyze Environment in Which Vulnerability Appears

This engine is an engine that plays the role of judging how the sources and patterns known to be weak points will move in the actual operation environment. This engine is organized into the analysis module for the vulnerability execution environments, the analysis module for the software execution environments, and the analysis module for the Vulnerability Code-Clones.

The module to analyze the environment in which vulnerability appears analyzes the execution environment where vulnerability might appear and maps it by considering the vulnerability environment through white box testing or black box testing.

The software execution environment analysis module generates a while list of execution environments based on the recommended environment for open source software that were searched through web crawling.

The module to analyze the vulnerability code-clone generates a list of exploitable vulnerabilities and assesses the code-clone of source code as well as the source code of the open source software based on the execution environment white list and execution environment analysis.

3.2.3   Engine to Review Technical Constraints

This engine, through the module that reviews technical constraints of attacks, tests whether any constraint conditions exist in the environment for the attack to detect an error in detection due to such constraints.

4   Conclusion

In addition to the increase of the utilizations of the IT products recently, the application scope of the software has been widened to become a diverse environment. Accordingly, the usage scope and the usage frequency, too, of the open source software has been increased. But in many cases such software have security vulnerabilities that are exploited.

In this paper, by automatically verifying whether the security vulnerabilities of the open source software are exploitable, a system for securing the stability was proposed to the developers and the security experts. By distinguishing the indispensable functions and the roles needed for the system, each organizational element was defined. And the flowchart of the organic system between the organizational elements that have been defined was designed. It is highly expected that, through this system that is proposed, the secondary damages due to the future development costs, the maintenance and repair costs, and the attacks on the security vulnerabilities will be prevented.

Acknowledgments This work was supported by the ICT R&D program of MSIP/IITP. [R0112-14-1061, The analysis technology of a vulnerability on an open-source software, and the development of Platform]

References

1. Bodhani, A.: Feeling lucky [cybersecurity]. Engineering & Technology **10**(1), 44–47 (2015)
2. Korea OSS Promotion Forum. Business Guide of Open Software (2014)
3. Cothey, V.: Web-crawling reliability. Journal of the American Society for Information Science and Technology **55**(14), 1228–1238 (2004)
4. Patton, R.: Software testing. Sams Pub. (2006)
5. CWE - Common Weakness Enumeration. http://cwe.mitre.org

6. CWE/SANS Top 25 Most Dangerous Programming Errors (2011). http://cwe. mitre.org/top25/
7. CVE - Common Vulnerabilities and Exposures. http://cve.mitre.org
8. Mell, P., Scarfone, K., Romanosky, S.: Common vulnerability scoring system. IEEE Security & Privacy **4**(6), 85–89 (2006)
9. CAPEC - Common Attack Pattern Enumeration and Classification. http://capec. mitre.org
10. Barnum, S.: Common attack pattern enumeration and classification (capec) schema description. Cigital Inc., http://capec.mitre.org/documents/documentation/CAPEC_ Schema_Description_v1 **3** (2008)

User Authentication Method Design Based on Biometrics in a Multi-cloud Environment

Seokhwa Song, JaeSeung Lee, Jaesik Lee and Moon-seog Jun

Abstract Users in a multi-cloud environment use two or more cloud services and each cloud has an authentication module and user management module. However, in order to use a variety of cloud, users must perform a user authentication for each cloud. At such time, as the user authentication information is stored in various clouds, there are problems of being captured or stolen. In this paper, a user authentication method based on biometrics using biometric information is proposed in order to provide a safe user authentication in a multi-cloud environment.

Keywords Multi cloud · Biometric authentication · Security · Mobile cloud · Privacy

1   Introduction

Recently with a rapid spread of cloud environment, a variety of cloud services are being provided. At the same time, a multi-cloud service where two or more different cloud interact with each other based on a specific contract is catching attention. Through such, users use only one cloud service but services from various clouds can be used. In addition, to the service providers lacking resources such as computing ability and storage, etc. an opportunity of being able to borrow and use virtual resources from other cloud services is provided [1][2]. But in multi-cloud, there are problems of having to perform cross authentication with

S. Song · J. Lee · M.-S. Jun(✉)
Soongsil University, Dongjak-Gu, Republic of Korea
e-mail: {shsong,Ijs0322,mjun}@ssu.ac.kr

J. Lee
Korea Internet & Security Agency, Seoul, Korea
e-mail: j30231@kisa.or.kr

© Springer Science+Business Media Singapore 2015   283
D.-S. Park et al. (eds.), *Advances in Computer Science and Ubiquitous Computing*,
Lecture Notes in Electrical Engineering 373,
DOI: 10.1007/978-981-10-0281-6_41

each cloud in order for the users to use a variety of cloud services. Not only that, since user authentication information is stored in each cloud, there is a threat of being captured or stolen by malicious attackers. In order to solve these problems, an integrated authentication method SSO(Single Sign-On) which allows the use of various cloud services using single authentication information was proposed. However in SSO, there is a problem of its privacy being violated or the entire system being halted by overloading when the SSO server is attacked [3]. User authentication method proposed by Lee et al is a 2-Factor authentication method combined with PKI(Public Key Infrastructure) authentication and OOB(Out-of-Band) authentication technique in the cloud computing environment. However, the authentication method proposed in that paper has the inconvenience of going through two authentication processes [4].

Therefore, this paper has proposed a user authentication method based on biometrics using a variety of biometric information in a multi-cloud environment. In Chapter 2, it deals with multi-cloud and biometric authentication, and in Chapter 3, a detailed explanation of authentication protocol proposed in this paper will be given, and in Chapter 4, an assessment will be made, and in Chapter 5, presents the conclusion

2   Learning System

2.1   Multi Cloud

Although users are only using one service, multi-cloud can allow the use of multiple cloud services and provide a wider and a variety of cloud services. This can solve the problem of user authentication information stored in the cloud from being leaked in the case of security accidents due to a system error in a specific cloud, natural disaster or hacking [5][6].

2.2   Biometric Authentication

Biometric recognition is a technique which verifies a person using their unique biometric characteristics such as fingerprints, iris, retina, vein and faces, etc. Biometric information which is unique to each person can be easily distinguished without any threat of being lost or stolen as long as the body is not compromised. After finding a lifelong individual characteristics and registering this using an automated means, it is compared and determined with the proposed information. Currently, FIDO Alliance has proposed a UAF(Universal Authentication Factor) security protocol based on biometric authentication technique [7].

FIDO Alliance is a global consortium which was established back in 2012 in order to define the standard method of FIDO technology. Internet enterprises such as Google, PayPal, Alibaba, Lenovo, Microsoft, Discover Financial Service, Nok Nok Labs, Qualcomm and Samsung, etc. are its participating members. In addition,

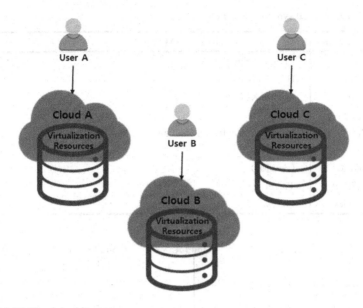

Fig. 1 Multi Cloud Architecture

FIDO Alliance has separated the authentication protocol and authentication method and developed the new FIDO authentication system based on various biometrics such as fingerprints and iris recognition, etc. instead of the existing ID and password authentication method. This is a new authentication technique which can be applied in the smart mobile environment while increasing the authentication strength and the convenience of users. FIDO standard consists of UAF(Universal Authentication Factor) and U2F(Universal 2nd Factor). UAF is a technique that authenticates the user in conjunction with the online service where biometric information such as fingerprints and iris, etc. is used in the authentication process instead of ID and password method. U2F is a secondary authentication factor along with the existing password using online services where a separate double authentication devices are used such as USB dongle or smart cards which stores information [8].

3   Proposed Protocol

In this section, a user authentication method based on biometrics is proposed using biometrics information in a biometric multi-cloud environment. Cloud Broker having fiduciary relationship with Cloud A and Cloud B, assumes that secret values SB1 and SB2 has been exchanged with Cloud A and Cloud B. The terms and symbols used in the proposed security protocol are shown in Table 1.

Table 1 The terms and symbols used in the scheme

Notation	Meaning
R	Random Number
H()	Hash Function
Pri	Private Key
Pub	Public Key
K	Shared Key between User and Cloud A
S	Server's Key
CID	Cloud ID
Ticket	Ticket
Policy	Policy

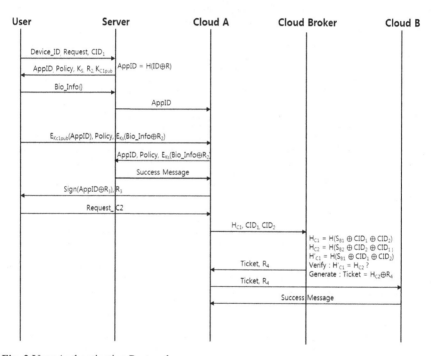

Fig. 2 User Authentication Protocol

Users send Device_ID and CID_1 along with a requesting message to the server in order to undergo an authentication process. Server generates the Device_ID and a random number and computes against XOR for hashing and generates the AppID, and the key value of the server K_S and a random number R_2 which will be used during the user authentication and policy is generated. The user finally sends the biometic information Bio_Info{} to the server and the server after confirming the Bio_Info{}, sends the AppID to the Cloud A.

The user, in order to approach the Cloud A, computes the random number R_2 and XOR with the $E_{ks}(AppID)$, Policy and Bio_Info{} which was encrypted with K_S a key value of the server and transmits to Cloud A. Cloud A, after computing the random value R_2 and XOR with the AppID, Policy and biometric information Bio_Info{} which was obtained by decrypting the AppID with its private key, it is transmitted to the server after encrypting it with the K_S a key value of the server. The server then verifies the biometric information of the user which came from Cloud A. When the correct information is verified, the server sends the success message to the Cloud A. Cloud A, after computing the AppID, random number R3 and XOR, signs with its own private key and send to the user. The user, in order to approach the Cloud B, sends the Request_C2 message to Cloud A.

Cloud A in order to access the Cloud B, hashes and computes the secret value SB1 exchanged with Cloud A with the CID_1, CID_2 and XOR and sends to Cloud Broker along with CID_1 and CID_2. Cloud broker, after hashing and computing the secret value S_{B2} exchanged with Cloud B with the CID_1, CID_2 and XOR, generates H_{C2}, and when the H_{C1} received from Cloud A having the same value as H_{C2} is verified, computes the H_{C2}, random number R_4 with XOR and generates the Ticket. Generated Ticket is sent to Cloud A along with the random number R_4. Cloud A sends the Ticket and the random number R_4 received from the Cloud Broker to the Cloud B. Cloud B, after verifying the Ticket sent by Cloud A, sends the success message to Cloud A.

4   Security Analysis

4.1   Replay Attack

Attackers can eavesdrop on the communication message between the user and Cloud A with an attempt for reuse. However, all communications between the user and Cloud A includes a new random number each time, it is safe from the reuse by the attackers.

4.2   Brute Force Attack

Attackers can attempt indiscriminate attacks on the key in order to find the AppID or Device_ID of the user which is encrypted with the public key of the Cloud A. However, assuming that the length of the public key is 2048bit, the chance of the public key being captured is very slim and even if captured, Device_ID cannot be found from AppID.

4.3   Eavesdropping Attacks

Attackers can overhear the communication messages between the server and Cloud A and the messages between the user and Cloud A. However, the value which is disclosed among the communication messages between the user and Cloud A is a random number value R which changes each time and the disclosed variable policy. And since the AppID is encrypted with the public key of Cloud A, and the Bio_Info $\oplus$R2 is encrypted with a secret key between the server and user, the attacker cannot find the AppID of the user or the biometric information Bio_Info.

5   Conclusion

In this paper, a user authentication method using a variety of biometric information in a multi-cloud environment was proposed. Proposed environment was found to be safe through a security analysis against Replay Attack, Brute Force Attack and Eavesdropping Attacks. In the future, it is expected to be used as a safe user authentication protocol in a multi-cloud environment.

References

1. Chow, R., et al.: Authentication in the clouds: a framework and its application to mobile users. In: Proceedings of the 2010 ACM Workshop on Cloud Computing Security Workshop. ACM (2010)
2. Lee, S., Ong, I., Lim, H., Lee, H.: Two Factor Authentication for Cloud Computing. International Journal of KIMICS **8**(4), August 2010
3. Graser, T., et al.: Single Sign On. U.S. Patent Application 12/094,858
4. Chang, H., Choi, E.: User authentication in cloud computing. In: Ubiquitous Computing and Multimedia Applications, pp. 338–342. Springer Berlin, Heidelberg (2011)
5. Ahn, H., et al.: User authentication platform using provisioning in cloud computing environment. In: Advanced Communication and Networking, pp. 132–138. Springer Berlin, Heidelberg (2011)
6. Choudhury, A.J., et al.: A strong user authentication framework for cloud computing. In: 2011 IEEE Asia-Pacific Services Computing Conference (APSCC). IEEE (2011)
7. Chen, N., Rui, J.: Analysis and improvement of user authentication framework for cloud computing. In: Advanced Materials Research, vol. 756 (2013)
8. FIDO: Fast Identity Online Alliance Privacy Principles. The FIDO Alliance : Privacy Principles Whitepaper FIDO Alliance, February 2014

A Study on Framework for Developing Secure IoT Service

Miyeon Yoon and Jonghyun Baek

Abstract This paper proposes IoT security framework including protection of key value in a device. Also, we suggest seamless key protection in spite of heterogeneous IoT framework by making 'initial secret value' on IoT device information and secret value made by result value through authentication mechanism through FIDO mechanisms.

Keywords IoT security · Architecture · Key management · Authentication

1 Introduction

This paper proposes a key management and authentication for IoT security. We suggest protection method of key materials in spite of heterogeneous IoT security methods of IoT framework[2,3,4] by making 'initial secret value' on IoT device information and secret value made by result value through authentication mechanism on FIDO mechanisms[1]. In chapter 2, we introduce related works. Chapter 3 proposes IOT security principles, key value protection method in an unveiled device. In conclusion, we suggest further works in near future.

2 Related Works

The security for IoT has been researched for a long time. Initially, RFID/USN was emerged and discussed. For larger understandings, we has been studying IoT and CPS to apply to our real life and traditional industries. OneM2M[4] is a major group to discuss this issue. They published IoT security architectures and its functionalities. However, they only suggest security mechanisms based on

M. Yoon(✉) · J. Baek
KISA, Seoul, Republic of Korea
e-mail: {myyoon,jhbaek}@kisa.or.kr

© Springer Science+Business Media Singapore 2015
D.-S. Park et al. (eds.), *Advances in Computer Science and Ubiquitous Computing*,
Lecture Notes in Electrical Engineering 373,
DOI: 10.1007/978-981-10-0281-6_42

TLS/DTLS[5,6] and cryptography such as ECC[7] and standard symmetric algorithm. And Allseen[2] and OIC[3] also published IoT framework and its securities. We proposes IoT security principles for all connected IoT parties on any IoT platforms. And So we need methods for a secret value, only known as a device itself on each connected IoT service, to protect authentication and encryption key or intial secret value owned by a device over any IoT framework.

3 Proposal for IoT Security Framework

We assume there is an IoT Platform performs management of IoT Gateways, Devices and service platforms. And a security controller provides all procedures for key management and authentication. Here, IoT device(Dp) is a very lightweight to provide sensing and forwarding, and another device(GDk) is enough to perform security facilities. The GDk is not only an IoT device, but also IoT gateway for sensing and collecting and routing.

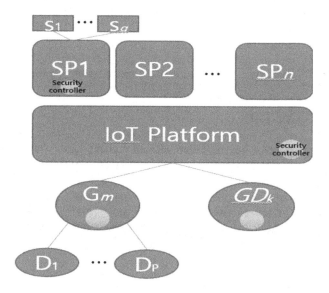

Fig. 1 IoT Security controllers over IoT systems

Since there exists so many IoT platforms, we define a common IoT platform to manage a device, gateway and service platform that is an IoT platform provided by major vendors. And the common IoT platform is deployed on global standard such as oneM2M.

We propose IoT security principles from security design and deploying IoT, and especially, suggest a protection of secret value of key management and authentication for IoT security setup.

3.1   IoT Security Principles

For security by design for IoT, we suggest seven principles. Each principles is for designed for three phases-design, development/implementation and service management.

(1) Security and Privacy by design
- All entities of Fig1 should consider security and privacy by design.
- Providing security methods for confidentiality, integrity/authentication and availability with preventing misuse of IoT device on lightweight characteristics.
- Proper access control, end-to-end communication security and integrity/authentication on service environment.
- Consideration for S/W or H/W security mechanisms, and applying security standards.
- Privacy considerations for IoT device and service policy.
- Establishment of secure channels, anonymous savings, access control and authentication, confidentiality, secure storage.
- Establishment for privacy policy

(2) Apply and Verification for secure S/W and H/W
- Especially IoT device(D, GD) should implement these factors.
- Applying Secure coding mechanisms
- Verifying S/W security
- Access control and authentication between security S/W and H/W

(3) Provision of initial security by default
- Especially IoT device(D, GD) should implement these factors.
- Setting up of Security modules and parameters
- Strong password
- Secure management of encryption and authentication key
- Establishment of secure channels between devices

(4) Configuration of secure parameters and correct implementation of security protocols
- All entities on IoT security framework should be reflected.
- Appling security protocols
- Securities for heterogeneous protocols and networks
- Developing IoT Security platform
- Consideration of securities for applications

(5) Continuous update and patch of vulnerabilities for S/W and Firmware
- IoT device(D, GD) should implements these factors in lightweight ways

(6) Security and Privacy management system for secure deploying service
   - All entities is related to secure management in service aspects
(7) Corresponding Cyber-incident and confirming responsibility for cyber attacks
   - Especially IoT device should develop some methods for cyber security
   - Sharing for zero-day or security vulnerabilities among IoT parties.

3.2 Protection for Key Value of Key Management and Authentication

Initially, the Dp or GDk authenticate by itself using FIDO[1] mechanism. While authenticating, it has ID value and its biometric or a secret value from the dongle device. And then it stores its hashed value as below:

$D_{p\_initial}$:= $\mathbf{H}$ ($auth\_value, K_{\text{ID} \oplus \text{H(SerialNumber)}}$), where $\mathbf{H}$ denotes a hash function and the hash key $K_{\text{ID} \oplus \text{H(SerialNumber)}}$ is created by **XOR** calculating ID, which is an identification value as a result of FIDO authentication, to a serial number of IoT device. The $K_{\text{ID} \oplus \text{H(SerialNumber)}}$ is made by arbitrary value as the key material for $\mathbf{H}(SerialNumber)$. The value $D_{p\_initial}$ will be periodically updated by event signal from Gm by the hash function, $\mathbf{H}(D_{p\_i})$. Then the device makes SID(Service ID) to receive IoT service after authenticating with the user information and the value $\mathbf{H}(D_{p\_i})$ between the device and the service platform.

The IoT platform and service platform only knows $\mathbf{H}(D_{p\_i})$. So they cannot get any original $auth\_value$, by one-way characteristics by a hash function. The key information $K_{\text{ID} \oplus \text{H(SerialNumber)}}$ is calculated by ID and the hashed serial number through a registration phase.

The Gm performs the same procedure with Dp. So Gm can use the value as an identifiable value with IoT platform and Service platform.

Gm and GDk is capable of encrypting and routing sensed and collected messages. So based on the identifiable value, they initially calculate the key value $K_{D_{p\_i}}$:= $\mathbf{H}$ (D_{p_i}, $SecretValue$) where $SecretValue$ is initially shared with the Service platform encrypted with $\mathbf{H}(D_{p\_i})$. Between Gm and GDk to the SPj, $K_{D_{p\_i}}$ is used as an initial secret value to make an encryption key for secure communication on the defined network protocol. However, any strong encryption key can be used from Gm and GDk to Sq through IoT platform and the service platform.

Since we create a secret value with unique information of IoT device. This value satisfies uniqueness. Because we apply a hash function to the unique value, IoT platform to its service cannot derive the serial number from the key information. The initial key material of Gm and GDk encrypts key materials defined by each IoT framework. Also, the key should be updated in enough short term. And previous key material should be deleted. So the all key material should be protected by periodically continuous changed key values on each service ID and serial number of its device.

4 Conclusions

We have a plan to develop for verification test elements for IoT security as a continuous works. So the proposed IoT security principals, protection method of key materials defined on each IoT framework will be base methods to perform (2), (3) and (4).

References

1. FIDO Authentication (2015). https://fidoalliance.org/specifications/download/
2. Security 2.0. https://allseenalliance.org/developers/learn/core/security2_0/hld. Allseen alliance, January 2015
3. OIC, OIC_Security_Specification_Project_B_v0.9.9. OIC alliance (2015)
4. oneM2M, TS-0003-Security_Solutions-V1_0_1. oneM2M (2015)
5. DTLS, Datagram Transport Layer Security Version 1.2. RFC 6347 (2012)
6. TLS, The Transport Layer Security (TLS) Protocol Version 1.2. RFC 5246 (2008)
7. ISO/IEC 15946, Information technology – Security techniques – Cryptographic techniques based on elliptic curves, November 2011
8. Internet of Things: Privacy & Security in a Connected World, FTC, January 2015
9. Zhang, W., Qu, B.: Security Architecture of the IoT Oriented to Perceptual Layer. IJ3C 2(2) (2013)
10. Opinion 8/2014 on the on Recent Developments on the IoT. EU Data Protection Working Party, September 2014
11. Jha, A., Sunil, M.C.: Whitepaper: Security Considerations for Internet of Things. L&T Technology Services (2015)
12. Zhao, K., Ge, L.: A survey on the internet of things security, In: 2013 Ninth International Conference on Computational Intelligence and Security. IEEE (2013)
13. Zhang, B., Zou, Z., Liu, M.: Evaluation on security system of internet of things based on fuzzy-AHP method. In: ICEE, May 2011
14. Babar, S., Stango, A., Prasad, N., Sen, J., Prasad, R.: Proposed embedded security framework for internet of things (IoT). In: 2011 2nd International Conference on Wireless VITAE (2011)
15. Roman, R.: Securing the Internet of Things. IEEE Computer 44(9), 51–58 (2011)
16. Heer, T.: Security Challenges in the IP-based Internet of Things. Journal on Wireless Personal Communications 61(3), 527–542 (2011)
17. Whitehouse, O.: Security of Things: An Implementers' Guide to Cyber-security for Internet of Things Devices and Beyond, NCC Group (2014)
18. Babar, S., et al.: Proposed security model and threat taxonomy for the IoT. In: CNSA 2010, pp. 420–429 (2010)
19. Ukil, A.: Embedded security for internet of things. In: NCETACS 2011, pp. 1–6 (2011)
20. Ora, L.: Safety standards in the ARM ecosystem. ARM Whitepaper, January 2015
21. Kempf, J., et al.: Thoughts on reliability in the internet of things. In: Interconnecting smart objects with the Internet workshop, vol. 1 (2011)
22. Linpner, S.: The trustworthy computing security development lifecycle. In: ACSAC 2004, pp. 2–13 (2004)

23. Security Considerations in the System Development Life Cycle (SP800-64 Revision2). NIST (2008)
24. Bormann, C., et al.: Terminology for Constrained-Node Networks (RFC7228). IETF, May 2014
25. Internet of Things Research Study 2014 Report, Hewlett Packard (2014)
26. Pescatore, J.: Securing the Internet of Things Survey. SANS Institute, January 2014
27. Ministry of Science, ICT and Future Planning, South Korea. http://www.msip.go.kr
28. 10 Specialized Action Plans for IoT Development (2013-2015), Ministry of Industry and Information Technology, China

Device Dedication in Virtualized Embedded Systems

SooYoung Kim, Hyunwoo Joe, Dongwook Kang, Jin-Ah Shin, Vincent Dupre, Taeho Kim and Chaedeok Lim

Abstract In embedded systems, hypervisors allow a virtual machine to access directly native hardware devices without sharing with other virtual machines because of conflictions of sharing devices between the virtual machines. In this paper, we present the issue being occurred to dedicate peripheral devices and how to solve it. With the device dedication technique on an embedded hypervisor, we identify the development considerations based on our observation. For proof-of-concept of our solutions, the partial virtualization technique about the clock controller is implemented to a real embedded hypervisor for safety-critical systems.

Keywords Device dedication · Direct access · Embedded hypervisor · Partial virtualization

1 Introduction

For industry area recently, the industrial systems demand to perform not only a simple control of the device as the role of conventional embedded systems, but also a complex and smart functions at the same time. They are evolving to add a new improved functionality and performance while maintaining legacy systems as much as possible rather than to change the systems. However, we must consider the collision problem between existing functions and a newly added one, because in many cases the legacy system was designed to be able to operate only for a particular target. A conflict may cause a big problem in the safety of the entire system.

S.-Y. Kim(✉) · H. Joe · D. Kang · J.-A. Shin ·V. Dupre · T. Kim · C. Lim
Embedded SW Research Department, Electronics and Telecommunications
Research Institute (ETRI), Daejeon, Republic of Korea
e-mail: {sykim,hwjoe,dkang,jashin,vdupre,taehokim,cdlim}@etri.re.kr

D.-S. Park et al. (eds.), *Advances in Computer Science and Ubiquitous Computing*,
Lecture Notes in Electrical Engineering 373,
DOI: 10.1007/978-981-10-0281-6_43

In order to satisfy these trends, high-performance hardware-assisted virtualization technology may be a considerable option to improve safety and add new functionalities. Virtualization technology can perform a variety of missions with just existing software. In addition, it can also support to isolate between them physically and logically, hence it is possible to correspond independently to the failure status of each other from an entire system safety point of view.

Many studies for improving software dependability have been conducted to isolate between the systems by virtualization technology [1][2][3]. The loosely-synchronized, redundant virtual machines architecture is a typical approach. It proposed the architecture using logically isolated redundant copies of the operating system and the application with the virtualization based on XEN [4]. It is able to build the high levels of reliability at extremely low cost systems.

In embedded systems however, consolidation of multiple systems into single hardware is more difficult than server area because constraints for embedded system such as deterministic execution behavior should be accepted on a hypervisor. Especially, I/O virtualization is the most important to maintain deterministic executions for safety-critical systems. Therefore, device dedication technique among various I/O virtualization is the most suitable solution. However, it is not simple to dedicate some peripheral devices to a specific virtual machine. We observe a situation that affected the operation of the device in the guest OS when mapping the register's address with not-aligned page to another guest OS. It is because the mapping makes to approve the control of the device to another guest OS. For example, the device has also a clock which is controlled by clock controller with the different segmented address.

In this paper, with our observation, we present the issue during dedicated peripheral device. In best our knowledge, embedded hypervisors do not consider this issue although they support device dedication features [5]. We focus on some peripheral devices' clock in clock management unit with not-aligned page. For the practical implementation, we solve this issue by applying trap-and-emulate technique for those clocks on a real embedded hypervisor.

The contributions of this paper are as follows:

- We clarify why we need to choose a device dedication technique in safety critical systems.
- We identify the considerations of embedded hypervisor when device dedication technique is applied.
- We observe the possible problems in an embedded hypervisor and suggested the solution.

2   IO Virtualization in Embedded Systems

Typically, the device virtualization techniques are classified into three techniques. They are trap-and-emulate techniques, dedication (direct access), and para-virtualization. The dedication technique among the above ones is the most reasonable way for ensuring the real-time property in safety-critical system.

The hypervisor needs an additional processing layer and the non-constant response time of IO devices in when applying the trap-and-emulate or para-virtualization technique. Most of SoC chips usually offers several same devices, and they can be dedicated to each guest OS. It is possible to take advantage of the performance to the maximum while minimizing the cost. Then when applying the dedication technique, as compared to applying the trap-and-emulate and para-virtualization technique, each guest OS can directly use the existing device driver without the development of device drivers for the hypervisor.

We can easily distinguish the device's control register from others in the memory-mapped I/O of ARM processors [6]. However, the registers for the operating clock of the devices are managed in the different section. Thus, those registers should be virtualized by the trap-and-emulate technique for ensuring the isolation of the device from the other guest OSes. Therefore, the hypervisor must consider how to manage the registers with not-aligned page for applying dedication technique when designing the hypervisor.

For instance, the registers related on controlling device power or clock source in clock management unit are candidate. As shown in Figure 1, The UART1 and UART2 each by mapping to Guest 0 and Guest 1, the control of each device itself has each guest OS, In addition to allow the clock source control and power control of each UART for two guest OSes in clock management unit also describes the situation that has been mapped, respectively.

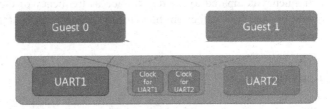

Fig. 1 UART1/UART2 Dedication

As shown in Figure 1, each guest OS can unavoidably control the clock of UART assigned to other guest OS in case of the clock management unit that is not divided into page unit. We resolve that other guest OS do not control the clock of UART assigned to other guest OS by using the trap-and-emulate technique.

3   Implementation

For practical experiments, we try to implement the dedication on Qplus-HYPER[7] which is an embedded hypervisor for running a safety-critical RTOS and a GPOS with ARMv7 virtualization extensions feature. We run a Linux as a GPOS and Qplus-AIR[8] as a RTOS on Qplus-HYPER. Our simple target of I/O devices for proof-of-concept is UART. The target board as arndale8, consists of 4 UART devices. We dedicate the UART1 to the Qplus-AIR and the UART2 to the Linux and on the hypervisor. The clock controller is assigned to the Linux.

For the observation, we set the message from the UART1 and the message from the UART2 at the initial time. Then, while preceding the initialization of the Linux, the UART1 of the Qplus-AIR is crashed without applying our solution. We investigate what is going on the scenario. This is related to the power management feature of a Linux. The Linux blocks clocks for unused devices by itself because of saving the system power. Although the configuration of the Linux does not set the UART2, Linux can access the clock controller. In order to solve this problem, we just add the virtualization part for the clock controller Qplus-HYPER. However, this solution leads the bigger codes and poor performance of the hypervisor. Therefore, we propose to partially virtualize the clock controller. In general rich OSes such as Linux, the most of devices on a target board is assigned by kernel. On the other hands, RTOS for safety-critical systems takes only specific devices. We should consider this situation on an embedded hypervisor while running a GPOS and a RTOS side-by-side. In this scenario, therefore, the hypervisor can map the clock controller's address except for the specific region to the Linux directly and not map the specific region to any guest OS. We need to only virtualize the specific region which needs just one entry of page tables.

This approach can affect to real-time property to RTOS because of extra virtualization part even in the dedication technique. Since the access frequency in the case of the clock management unit is very low, it is the level that should be ignored. Also, similar other registers, since it is expected to exhibit a significantly lower access frequency compared to the register access frequency of the device is actually in operation, and little effect on the real-time property and performance it is expected that.

4 Conclusion

We present the issue being occurred when dedicating a device in an embedded hypervisor. We propose the method of not fully virtualizing the clock controller, but partially virtualizing it. Thus, while using the maximum performance of the device assigned to a specific guest OS, it is possible to minimize the decrease in overall system performance due to the virtualization of the rest of the clock controller. However, it is necessary to ensure that there is the effect of this method.

Later in our research, we will analyze and show the effect of the dedication with partial virtualization of the clock, experimentally. And we will try to apply this method to other devices and make the general solution for the embedded hypervisor.

Acknowledgment This work was supported by the ICT R&D program of MSIP/IITP[R0101-15-0081, Research and Development of Dual Operating System Architecture with High-Reliable RTOS and High-Performance OS].

References

1. Goldberg, R.P.: Survey of Virtual Machine Research. IEEE Computer **7**(6) (1974)
2. Barham, P., Dragovic, B., Fraser, K., Hand, S., Harris, T., Ho, A., Neugebauer, R., Pratt, I., Warfield, A.: Xen and the Art of Virtualization. In: SOSP 2003 (2003)
3. Cereia, M., Bertolotti, I.C.: Virtual machines for distributed real-time systems. Computer Standars & Interfaces (2009)
4. Cox, A.L., Mohanram, K., Rixner, S.: Dependable ≠ unaffordable. In: The 1st Workshop on Architectural and System Support for Improving Software Dependability (2006)
5. Masmano, M., Ripoll, I., Crespo, A., Metge, J.: XtratuM: a hypervisor for safety critical embedded systems. In: 11th Real-Time Linux Workshop, Dresden, Germany (2009)
6. ARM: Principles of ARM® Memory Maps White Paper. ARM (2012)
7. Kim, T., Kang, D., Kim, S., Shin, J., Lim, D., Dupre, V.: Qplus-hyper : a hypervisor for safty-criticla systems. In: The 9th International Symposium on Embedded Technology (2014)
8. Kim, T., Son, D., Shin, C., Park, S., Lim, D., Lee, H., Gim, B., Lim, C.: Qplus-AIR: a DO-178B certifiable ARINC 653 RTOS. In: The 8th International Symposium on Embedded Technology (2013)
9. Arndale Board. http://www.arndaleboard.org/ (accessed on September 2015)

An Estimation Filtering for Packet Loss Probability Using Finite Memory Structure Strategy

Pyung Soo Kim, Eung Hyuk Lee and Mun Suck Jang

Abstract A finite memory structure (FMS) strategy is applied for a packet loss probability (PLP) estimation filtering to estimate the packet rate mean and variance of the input traffic process in real-time, which is called the FMS-PLP estimation filter. The proposed FMS-PLP estimation filter uses only the finite traffic measurements on the most recent window and thus has several inherent properties such as unbiasedness and deadbeat. The proposed FMS-PLP estimation filter can be represented in either a matrix form or a simple summation form. Using computer simulations, the proposed FMS-PLP estimation filter is shown to outperform the Kalman filter with the infinite memory structure for the temporarily uncertain system.

Keywords Packet loss probability estimation · Finite memory structure · Estimation filtering · Kalman filter

1 Introduction

Since measuring packet loss ratio at the intermediate nodes in high speed networks does not seem applicable in real time for real-time and throughput-sensitive Internet services such as multimedia streaming, VoIP, IPTV and Internet of things (IoT), several recent researches have been performed for the estimation of packet loss probability (PLP) [1]-[4]. Among them, due to the compact representation and the efficient manner[1][2], the optimal estimation filtering approach using the Kalman filter has been adopted for the PLP estimation. Due to its inherent good properties, the Kalman filter has been used widely for the optimal estimation of a

P.S. Kim(✉) · E.H. Lee · M.S. Jang
System Software Solution Lab, Korea Polytechnic University,
237 Sangidaehak-ro, Siheung-si, Gyeonggi-do 429-793, Korea
e-mail: pskim@kpu.ac.kr

© Springer Science+Business Media Singapore 2015
D.-S. Park et al. (eds.), *Advances in Computer Science and Ubiquitous Computing*,
Lecture Notes in Electrical Engineering 373,
DOI: 10.1007/978-981-10-0281-6_44

301

noisy measurement using the estimation error obtained from all past measurements to fix the one-step prediction and thus has been applied successfully for various areas[5]-[7]. However, the Kalman filter tends to accumulate the filtering error as time goes and can show even divergence phenomenon for temporary modeling uncertainties and round-off errors due to its infinite memory structure that utilizes all past packet measurements accomplished by equaling weighting and has a recursive formulation[8]-[10].

Therefore, this paper proposes an alternative PLP estimation filter with finite memory structure (FMS) strategy to estimate the packet rate mean and variance of the input traffic process in real-time while removing undesired system and measurement noises. The proposed estimation filter is called the FMS-PLP estimation filter. The proposed FMS-PLP estimation filter is developed under a weighted least square criterion using only the finite traffic measurements on the most recent window. On the other hand, the Kalman filter with infinite memory structure in [1][2] uses past all measurements as well as recent ones. The proposed FMS-PLP estimation filter can be represented in either a matrix form or a simple summation form. It is shown that the proposed FMS-PLP estimation filter has inherent properties such as unbiasedness and deadbeat. Computer simulations show that the proposed FMS-PLP estimation filter can outperform the Kalman filter with infinite memory structure for the temporarily uncertain system.

2   PLP Estimation Filter with Finite Memory Structure Strategy

As shown in [1][2], the packet rate mean and variance of the input traffic are required to be estimated in real-time for the estimation of PLP. The mean μ_i and variance σ_i of the instant packet rate of the input traffic process can be represented by the following state-space model:

$$\begin{bmatrix} \mu_{i+1} \\ \sigma_{i+1} \end{bmatrix} = A \begin{bmatrix} \mu_i \\ \sigma_i \end{bmatrix} + Bu_i + w_i, \ z_i = C \begin{bmatrix} \mu_i \\ \sigma_i \end{bmatrix} + v_i, \tag{1}$$

where

$$A = \begin{bmatrix} 1 - \dfrac{1}{N} & 0 \\ 0 & 1 \end{bmatrix}, \ B = \begin{bmatrix} \dfrac{1}{N} & 0 \\ 0 & 1 \end{bmatrix}, \ C = I, \tag{2}$$

N is the sample size and and u_i is the instant packet rate input. The system noise w_i is a white noise source with zero mean and covariance matrix Q and is considered uncorrelated with the input. In the network, the packet rate mean and variance can be measured and expressed as a measurement v_i. The measurement noise v_i is also a white noise source with zero mean and covariance R that is uncorrelated with the input.

In the current paper, an alternative PLP estimation filter is designed with finite memory structure (FMS) to estimate mean μ_i and variance σ_i of the instant packet rate of the input traffic process. The proposed estimation filter is called the FMS-PLP estimation filter and developed under a weighted least square criterion using only the finite traffic measurements and inputs on the most recent window $[i - M, i]$, denoted by Z_i and U_i respectively, as follows:

$$Z_i \equiv \begin{bmatrix} z_{i-M}^T & z_{i-M+1}^T & \cdots & z_{i-1}^T \end{bmatrix}^T, \quad U_i \equiv \begin{bmatrix} u_{i-M}^T & u_{i-M+1}^T & \cdots & u_{i-1}^T \end{bmatrix}^T. \tag{3}$$

These finite traffic measurements and inputs Z_i and U_i can be represented in the following regression form from the discrete-time system (1):

$$Z_i - \Omega U_i = \Gamma \begin{bmatrix} \mu_i \\ \sigma_i \end{bmatrix} + \Lambda W_i + V_i, \tag{4}$$

where W_i and V_i have the same form as (4) for w_i and v_i, respectively, and matrices Ω, Γ, Λ are as follows:

$$\Omega \equiv \begin{bmatrix} CA^{-1}B & CA^{-2}B & \cdots & CA^{-M+1}B & CA^{-M}B \\ 0 & CA^{-1}B & \cdots & CA^{-M+2}B & CA^{-M+1}B \\ \vdots & \vdots & \cdots & \vdots & \vdots \\ 0 & 0 & \cdots & 0 & CA^{-1}B \end{bmatrix}, \quad \Gamma \equiv \begin{bmatrix} CA^{-M} \\ CA^{-M+1} \\ \vdots \\ CA^{-1} \end{bmatrix}, \tag{5}$$

$$\Lambda \equiv \begin{bmatrix} CA^{-1}G & CA^{-2}G & \cdots & CA^{-M+1}G & CA^{-M}G \\ 0 & CA^{-1}G & \cdots & CA^{-M+2}G & CA^{-M+1}G \\ \vdots & \vdots & \cdots & \vdots & \vdots \\ 0 & 0 & \cdots & 0 & CA^{-1}G \end{bmatrix}. \tag{6}$$

The noise term $\Lambda W_i + V_i$ in (4) is zero-mean white Gaussian with covariance Π given by

$$\Pi \equiv \Lambda \begin{bmatrix} diag \overbrace{\begin{pmatrix} Q & Q & \cdots & Q \end{pmatrix}}^{M} \end{bmatrix} \Lambda^T + \begin{bmatrix} diag \overbrace{\begin{pmatrix} R & R & \cdots & R \end{pmatrix}}^{M} \end{bmatrix}. \tag{7}$$

Now, to get the FMS-PLP estimation filter from the regression form (4), the following weighted least square cost function must be minimized:

$$\left\{Z_i - \Omega U_i - \Gamma \begin{bmatrix} \mu_i \\ \sigma_i \end{bmatrix} \right\}^T \Pi^{-1} \left\{Z_i - \Omega U_i - \Gamma \begin{bmatrix} \mu_i \\ \sigma_i \end{bmatrix} \right\}. \tag{8}$$

Taking a derivation of (9) with respect to $\begin{bmatrix} \mu_i & \sigma_i \end{bmatrix}^T$ and setting it to zero, the proposed FMS-PLP estimation filter $\begin{bmatrix} \hat{\mu}_i & \hat{\sigma}_i \end{bmatrix}^T$ can be obtained with the matrix form as follows:

$$\begin{bmatrix} \hat{\mu}_i \\ \hat{\sigma}_i \end{bmatrix} = \left[(\Gamma^T \Pi^{-1} \Gamma)^{-1} \Gamma^T \Pi^{-1} \right] \left[Z_i - \Omega U_i \right]. \tag{9}$$

The proposed FMS-PLP estimation filter $\begin{bmatrix} \hat{\mu}_i & \hat{\sigma}_i \end{bmatrix}^T$ in (10) for the mean μ_i and the variance σ_i is shown to have several inherent properties such as unbiasedness, deadbeat, robustness. In addition, the proposed FMS-PLP estimation filter has an unbiasedness property when there are noises and a deadbeat property when there are no noises. When there are noises on the window $[i - M, i]$,

$$E[Z_i - \Omega U_i] = \Gamma E \begin{bmatrix} \mu_i \\ \sigma_i \end{bmatrix}.$$

since $\Lambda W_i + V_i$ is zero-mean in (5). Therefore, the following is true:

$$E\begin{bmatrix} \hat{\mu}_i \\ \hat{\sigma}_i \end{bmatrix} = \left[(\Gamma^T \Pi^{-1} \Gamma)^{-1} \Gamma^T \Pi^{-1} \right] E[Z_i - \Omega U_i] = \left[(\Gamma^T \Pi^{-1} \Gamma)^{-1} \Gamma^T \Pi^{-1} \Gamma \right] E \begin{bmatrix} \mu_i \\ \sigma_i \end{bmatrix} = E \begin{bmatrix} \mu_i \\ \sigma_i \end{bmatrix},$$

which shows the unbiasedness property. When there are no noises on the window $[i - M, i]$ for the discrete-time state space model (1) as follows:

$$\begin{bmatrix} \mu_{i+1} \\ \sigma_{i+1} \end{bmatrix} = A \begin{bmatrix} \mu_i \\ \sigma_i \end{bmatrix} + B u_i, \quad z_i = C \begin{bmatrix} \mu_i \\ \sigma_i \end{bmatrix}, \tag{10}$$

measurements Z_i is determined from (5) as follows:

$$Z_i - \Omega U_i = \Gamma \begin{bmatrix} \mu_i \\ \sigma_i \end{bmatrix}.$$

Therefore, the following is true:

$$\begin{bmatrix} \hat{\mu}_i \\ \hat{\sigma}_i \end{bmatrix} = \left[(\Gamma^T \Pi^{-1} \Gamma)^{-1} \Gamma^T \Pi^{-1}\right]\left[Z_i - \Omega U_i\right] = \left[(\Gamma^T \Pi^{-1} \Gamma)^{-1} \Gamma^T \Pi^{-1} \Gamma\right] \begin{bmatrix} \mu_i \\ \sigma_i \end{bmatrix} = \begin{bmatrix} \mu_i \\ \sigma_i \end{bmatrix},$$

which shows the deadbeat property.

The deadbeat property means that the proposed FMS-PLP estimation filter $[\hat{\mu}_i \ \ \hat{\sigma}_i]^T$ tracks exactly the mean μ_i and the variance σ_i at every time for noise-free systems. This property indicates finite convergence time and fast tracking ability of the proposed FMS-PLP estimation filter. Thus, it can be expected that the proposed FMS-PLP estimation filter might be appropriate for fast detection. The window length M can be a useful design parameter for the proposed FMS-PLP estimation filter. Thus, the important issue here is how to choose an appropriate window length M to make the estimation performance as good as possible. Since M is an integer, fine adjustment of the properties with M is difficult. In applications, one way to determine the window length is to take the appropriate value that can provide enough noise suppression. Therefore, it can be stated from above discussions that the window length M can be considered as a useful parameters to make the estimation performance of the proposed FMS-PLP estimation filter as good as possible.

The matrix form (10) of the proposed FMS-PLP estimation filter can be represented by the simple summation form, which might be familiar to digital signal processing (DSP) engineer. The matrices in the matrix form (10) can be represented by

$$(\Gamma^T \Pi^{-1} \Gamma)^{-1} \Gamma^T \Pi^{-1} \equiv [\phi_{M-1} \quad \phi_{M-2} \quad \cdots \quad \phi_1 \quad \phi_0],$$
$$(\Gamma^T \Pi^{-1} \Gamma)^{-1} \Gamma^T \Pi^{-1} \Omega \equiv [\varphi_{M-1} \quad \varphi_{M-2} \quad \cdots \quad \varphi_1 \quad \varphi_0].$$

Then, the proposed FMS-PLP estimation filter $[\hat{\mu}_i \ \ \hat{\sigma}_i]^T$ is represented by the following summation form:

$$\begin{bmatrix} \hat{\mu}_i \\ \hat{\sigma}_i \end{bmatrix} = \sum_{j=0}^{M-1} \phi_j z_{i-j} - \sum_{j=0}^{M-1} \varphi_j u_{i-j}. \tag{11}$$

3 Computer Simulations

The proposed FMS-PLP estimation filter with finite memory structure is compared with the Kalman filter with infinite memory structure in [1][2] via computer simulations. The window length is taken by three cases, $M=5$, $M=10$, $M=15$. The sampling time is set by $N=5$ and noise covariances are taken by $Q=0.05^2$ and $R=\text{diag}(0.02^2 \ 0.02^2)I$, respectively. To make a clearer comparison of estimation performances, simulations of 40 runs are performed using different system and measurement noises, and each single simulation run lasts 500 samples. Two scenarios are considered for simulations. The first scenario considers the nominal system which has no temporarily model uncertainties. In contrast, the second scenario considers the temporarily uncertain system which has temporarily uncertain model parameters such as $\Delta A = \text{diag}(0.01\ 0.01)I$ for the system matrix A and $\Delta C = \text{diag}(0.01\ 0.01)I$ for the measurement matrix C on the interval $100 \leq i \leq 150$. For both nominal and temporarily uncertain systems, the estimation performance is compared by the mean of root-squared estimation error for simulations of 40 runs. Table 1 shows the total average for simulations of 40 runs according to diverse window lengths. For the nominal system, the Kalman filter is shown to outperform the proposed estimation filter for all cases of Ms. In contrast, for the temporarily uncertain system, the proposed estimation filter is shown to outperform the Kalman filter for all cases of Ms. One possible explanation for this is that the proposed FMS-PLP estimation filter can have greater noise suppression and thus the estimation performance can be improved as the window length M increases, while the convergence time of a filtered estimate becomes long. Thus, these simulation results can provide practical guidance on the choice of M.

Table 1 Total average for simulations of 40 runs.

	FMS-PLP estimation filter			Kalman filter
	$M = 5$	$M = 10$	$M = 15$	
Nominal system	0.0783	0.0600	**0.0501**	**0.0337**
Temporarily uncertain system	**0.0946**	0.1120	0.1244	**0.1354**

4   Concluding Remarks

In this paper, to estimate the packet rate mean and variance of the input traffic process in real-time while removing undesired system and measurement noises, the FMS-PLP estimation filter has been proposed. The proposed FMS-PLP estimation filter is developed under a weighted least square criterion using only the finite traffic measurements on the most recent window. The proposed FMS-PLP estimation filter has shown to have several inherent properties such as unbiasedness, deadbeat, robustness. A guideline for choosing appropriate window length is described since it can affect significantly the estimation performance. Computer simulations show that the proposed FMS-PLP estimation filter with finite memory structure can outperform the Kalman filter with infinite memory structure for the temporarily uncertain system. One possible explanation for this is that the proposed FMS-PLP estimation filter can have greater convergence time of a filtered estimate as the window length decreases.

Acknowledgments This research was supported by the MSIP(Ministry of Science, ICT and Future Planning), Korea, under the C-ITRC(Convergence Information Technology Research Center) (IITP-2015-H8601-15-1003) supervised by the IITP(Institute for Information & communications Technology Promotion).

References

1. Zhang, D., Ionescu, D.: A new method for measuring packet loss probability using a Kalman filter. IEEE Transaction on Instrumentation and Measurement **58**(2), 488–499 (2009)
2. Zhang, D., Ionescu, D.: Reactive estimation of packet loss probability for IP-based video services. IEEE Transaction on Broadcasting **55**(2), 375–385 (2009)
3. Parker, B.M.: Measurement of packet loss probability by optimal design of packet probing experiments. IET Communications **3**(6), 979–991 (2009)
4. Vakili, A., Gregoire, J.C.: Real-time packet loss probability estimates from IP traffic parameters. International Journal on Advances in Networks and Services **5**(1), 34–42 (2012)
5. Grewal, M.S.: Applications of Kalman filtering in aerospace 1960 to the present. IEEE Control Systems **30**(3), 69–78 (2010)
6. Simon, D.: Kalman filtering with state constraints: a survey of linear and nonlinear algorithms. IET Control Theory & Applications **4**(8), 1303–1318 (2011)
7. Faragher, R.: Understanding the basis of the Kalman filter via a simple and intuitive derivation. IEEE Signal Processing Magazine **29**(5), 128–132 (2012)
8. Bruckstein, A.M., Kailath, T.: Recursive limited memory filtering and scattering theory. IEEE Trans. Inform. Theory **31**(3), 440–443 (1985)
9. Kim, P.S.: A computationally efficient fixed-lag smoother using recent finite measurements. Measurement **46**(1), 846–850 (2013)
10. Zhao, S., Shmaliy, Y.S., Huang, B., Liu, F.: Minimum variance unbiased FIR filter for discrete time-variant systems. Automatica **53**(2), 355–361 (2015)

Cryptanalysis of User Authentication Scheme Preserving Anonymity for Ubiquitous Devices

Dongwoo Kang, Jongho Mun, Donghoon Lee and Dongho Won

Abstract As the mobile network such as using cell phone, tablet PC, notebook services are gradually increased, a smart card comes to one of the useful thing, because of its convenience and portable. Contemporary, smart card-based authentication also can be one of the most generally authentication method. In 2015, Djellali et al. proposed user authentication scheme with preserving user anonymity and mutual authentication. Also, it provides light and profitable mechanism which can be easily applied to limited power or resources. They claimed their scheme is resisted many networks threat. Unfortunately, we discover some vulnerable weakness. In this paper, we demonstrate that their scheme is still unstable to some network threats, such as insider attack, offline-password guessing attack, impersonation attack and replay attack.

Keywords Anonymity user authentication · Smart card · Network security · Transition probability · Stationary distribution

1 Introduction

Since Lamport proposed the first password-based authentication scheme in 1981[1]. Over the past few years, several studies have made on smart card-based user authentication[2][3][4][5][6][7]. Smart card is very secure and efficient for storing secure information. The primary purpose of the authentication scheme using smart card is to verify and diagnose the reasonable user.

Therefore, Smart card-based authentication scheme was proposed and developed continuously. Center of development, the protection of the user's information is also emerged as important problem. At the moment, there are some

D. Kang · J. Mun · D. Lee · D. Won(✉)
College of Information and Communication Enginnering, Sungkyunkwan University,
2066 Seobu-ro, Jangan-gu, Suwon-si, Gyeongki-do 440-746, Korea
e-mail: {dwkang,jhmoon,dhlee,dhwon}@security.re.kr

© Springer Science+Business Media Singapore 2015
D.-S. Park et al. (eds.), *Advances in Computer Science and Ubiquitous Computing*,
Lecture Notes in Electrical Engineering 373,
DOI: 10.1007/978-981-10-0281-6_45

following security requirements when we proposed authentication scheme. Such as, resistant network threat, anonymity[8][9], mutual authentication[10][11], confidentiality, identification also cheap computation cost, communication cost.

Djellali et al[12]. in 2015 proposed that the new type of authentication scheme which use transition probability matrix and stationary distribution and preserve user anonymity. Djellali proved his scheme can resist various network security threats such as insider attack, replay attack, guessing attack, stolen-verifier attack, forgery attack, impersonate attack and so on. However, unfortunately, we found that there scheme still uncomfortable. It cannot resist insider attack, offline-password guessing attacks, impersonation attack and replay attack.

The rest of this paper is organized as follows: review Djellali et al's user authentication with preserving anonymity scheme in Section 2. In Section 3, we describe security threats of Djellali et al.'s scheme. At the conclusion, we conclude this paper in Section 4.

2   Review in Djellali et al.'s Scheme

This section reviews the user authentication scheme for ubiquitous devices proposed by Djellali et al. In 2015, Djellali et al.'s scheme consists of four phases: setup, registration, authentication and password changing phases which as follows. The notations used in this paper are summarized as Table 1.

Table 1 Notation used in this paper

Notations	Description
U_i	A qualified user i
ID_i, PW_i	User i's identity and password
S	A remote server
K_s	Server secret key
TID_i, TPW_i	User i's temporary identity and password
AK_i	The authenticated session key
nounce	The number of users that could be generated randomly
M_{n*n}	A transition probability matrix
π	A stationary limit distribution of M_{n*n}
h(.)	A collision resistant one-way hash function
$\oplus$	The bitwise XOR operation
||	String concatenation
$E_{SKi}\{M\}$	Encrypted message by SK_i

2.1   Setup Phase

The remote server generates a random transition probability matrix $P = M_{n*n}$ which satisfied irreducible and ergodic Markov chain. After that, server computes stationary limit distribution π which satisfied $\pi = \pi P$.

2.2 Registration Phase

When user U_i wants to access server for a service, U_i initially registers to the server S and is described as follows.

Step 1: U_i chooses his/her identity and password $ID_i, h(PW_i)$, and submits them to the server via a secure channel.

Step 2: After server receives $ID_i, h(PW_i)$ from U_i, the server S performs the following

      (1) Generate a random number y_i between 1 and n
      (2) move forward π_{yi} with d digit numbers and truncate the rest places:
        $\pi'_{yi} = Shift_d(\pi_{yi})$.
      (3) Generate temporary identity $TID_i = h(ID_i||K_s)$
      (4) Generate temporary password $TPW_i = h(PW_i) \oplus \pi'_{yi}$
      (5) $C_i = h(k_s) \oplus y_i$

Step 3: Create a smart card containing $<TID_i, TPW_i, C_i, \pi'_{yi}, h(.)>$ to U_i

2.3 Login and Authentication Phase

To start any communication, the user must first login using smart card. The user inserts his/her smartcard into card-reader and input ID_i^*, PW_i^*. Then, the smart card calculates the following sequence of steps.

Step 1: Compute $TPW_i^* = h(PW_i^*) \oplus \pi'_{yi}$ and verify the validity of the input password. If TPW_i^* is not equal to TPW_i, the login request is rejected.

Step 2: Generate a random number x and a and compute following
      $X_i = TPW_i^* \oplus x, A_i = TID_i \oplus a, L_i = A_i \oplus \pi'_{yi}$

Step 3: Encrypt $(TID_i, TPW_i, \pi'_{yi}, X_i)$ with A_i and send the login message
      $M_1 = < L_i, E_{Ai}(TID_i, TPW_i, \pi'_{yi}, X_i), C_i >$ to the server

After receiving the login message from user, the server performs the following sequence of operations to generate same session key.

Step 1: Calculate $y_i = C_i \oplus h(K_S)$, $\pi = \pi P$, $\pi''_{yi} = Shift_d(\pi_{yi})$ and $A_i = L_i \oplus \pi''_{yi}$.

Step 2: Decrypt $E_{Ai}(TID_i, TPW_i, \pi'_{yi}, X_i)$ with A_i and verify π''_{yi} equals to π'_{yi}, if different, rejects the login request.

Step 3: Retrieve $x = TPW_i \oplus X_i$ and generate random number z and compute $Z = z \oplus \pi'_{yi}$.

Step 4: Construct session key $AK_s = x.z$ and encrypt Z with A_i and send the message $M_2 = < E_{Ai}(Z) >$ to U_i .

Step 5: After receiving the message from the server, U_i decrypt Z with A_i and compute $z = Z \oplus \pi'_{yi}$.

Step 6: Generate the session key $AK_i = x.z$.

2.4 Password Change Phase

In the password-change phase, when a user wants to change his/her password, she/he inserts his/her smart card in the reader and inputs a password changing request. He/She inputs his/her old PW_i and new PW_i^* passwords. The smartcard calculates the following steps.

Step 1: Calculate the new temporary password $TPW_i^* = TPW_i \oplus h(PW_i) \oplus h(PW_i^*)$

Step 2: Change TPW_i with TPW_i^* on the smart card.

3 Security Analysis of Djellali's Scheme

In this section, we point out security weakness of Djeallal's scheme. There are some assumptions are made analysis and design of the scheme.

(1) An illegal attacker can intercept all messages communicated among smart card and remote server with common channel.
(2) An illegal attacker can steal smart card and extract the container of smart card by examining the power consumption of the card
(3) An illegal attacker can alter, delete or resend the captured message

3.1 Insider Attack

According the Djellali's scheme, the password of user $h(PW_i)$ will be reveal to server in the registration phase. If the illegal attacker in the server get this value. He/She can directly obtain the password PW_i by password-guessing attack.

3.2 Smart Card Stolen and Offline-Password Guessing Attack

Smart card stolen attack means an illegal attacker who has with smart card, and performs some illegal action with obtained information in smart card. If an outsider attacker U_a steals the smart card and obtains parameters, $<TID_i, TPW_i, C_i, \pi'_{yi}, h(.)>$. Then U_a can easily do offline-password guessing attack by following step.

1) Outsider illegal attacker U_a calculates $h(PW_i) = TPW_i \oplus \pi'_{yi}$
2) Then, the attacker selects random password PW_i^*, calculate $h(PW_i^*)$ using hash function which contained in also smartcard.
3) If 2)'s result is equal to $h(PW_i)$, the attacker infers that PW_i^* is user U_i's password PW_i.
4) Otherwise, attacker selects another password nominee and performs same steps, until he/she find password.

3.3 Impersonation Attack

Suppose there exists a legitimate but malicious attacker U_a with a smart card containing $\{ TID_i, TPW_i, C_i, \pi'_{yi}, h(.) \}$, U_a can easily intend to launch a user impersonation attack for a right user U_i to performs the following steps:

1) Outsider illegal attacker U_a can get PW_i from his own smartcard by using offline-password guessing attack.

2) The adversary randomly generates a nonce x' and a'

3) The adversary computes following:

$$X'_i = TPW_i^* \oplus x'$$
$$A'_i = TID_i \oplus a'$$
$$L'_i = A'_i \oplus \pi'_{yi}$$

4) The adversary encryption $(TID_i, TPW_i, \pi'_{yi}, X_i')$ with A_i' and transmit the login request message $\{L'_i, E_{A'_i}(TID_i, TPW_i, \pi'_{yi}, X_i'), C_i\}$

5) When receiving the message from adversary who pretends to be U_i, the messages can successfully pass server's verification check following:

Compute $\pi''_{yi} = Shift_d(\pi_{yi})$ and $A'_i = L_{i'} \oplus \pi''_{yi}$

Server decrypt $E_{A'_i}(TID_i, TPW_i, \pi'_{yi}, X_i')$ with A_i'

Check π'_{yi} equal to π''_{yi}

6) Adversary and server can successfully agree on a session key $SK = x'.z$

Based on the foregoing, we can deduct that the resistance to user impersonation attack cannot be assured in Djeallal's scheme.

3.4 Replay Attack

Replay attack is a type of network threat in which valid data transmission is maliciously repeated or delayed. In this scheme, an outsider attacker U_a eavesdrop a message between a user and the server which communicate in common channel. U_a can try to use these messages for opening new communication to a server coming to performs the following steps:

1) Outsider illegal attacker U_a eavesdrop a login messages for performing replay attack $\{L_i, E_{Ai}(TID_i, TPW_i \pi'_{yi}, X_i), C_i\}$.

2) In future, illegal attacker U_a resend a login messages to server

3) Resend messages can successfully pass server's verification check following:

Compute $\pi''_{yi} = Shift_d(\pi_{yi})$ and $A_i = L_i \oplus \pi''_{yi}$

Server decrypt $E_{A_i}(TID_i, TPW_i, \pi'_{yi}, X_i)$ with A_i

Check π'_{yi} equal to π''_{yi}

4) Server constructs the authenticated session key $AK_s = x.z$

The adversary cannot achieve the session key only use replay attack. It just disguise the user. But, when the adversary extract some information from smartcard additionally, also adversary can get session key:

1) Outsider illegal attacker U_a extract π'_{yi} from smartcard.
2) U_a also extract L_i from login message which using replay attack.
3) U_a successfully derive encryption key A_i by $L_i \oplus \pi'_{yi}$. Because it is symmetric key, U_a can use A_i also decryption.
4) U_a decryption $M_2 < E_{Ai}(Z) >$ by A_i, and extract TPW_i from smartcard and compute $x = TPW_i \oplus X_i$, $z = Z \oplus \pi'_{yi}$
5) Illegal attacker U_a can derive session key $AK_i = x.z$

4   Conclusion

In 2015, Djeallal et al. proposed a new type of authentication scheme which uses new secret key making method, transition probability and stationary distribution. Djeallal opinioned his scheme is resistance to famous attacks such as insider attack, replay attack, guessing attack, stolen-verifier attack, reflection attack, forgery attack, parallel-session attack and provide user anonymity. But, Djeallal et al.'s scheme is still unsecure. We showed to this paper, Djeallal et al's scheme cannot resist insider attack, offline-password guessing attack, impersonation attack and replay attack. And to conclude, our future research is to proposed more secure user authentication which preserving anonymity scheme, also it will be able to resist these threats.

Acknowledgement This research was supported by the Basic Science Research Program through the National Research Foundation of Korea (NRF) funded by the Ministry of Science, ICT, and Future Planning (2014R1A1A2002775)."

References

1. Lamport, L.: Password authentication with insecure communication. Communications of the ACM **24**(11), 770–772 (1981)
2. Xu, J., Zhu, W.-T., Feng, D.-G.: An improved smart card based password authentication scheme with provable security. Computer Standards & Interfaces **31**(4), 723–728 (2009)
3. Song, R.: Advanced smart card based password authentication protocol. Computer Standards & Interfaces **32**(5), 321–325 (2010)
4. Shiuh-Jeng, W., Jin-Fu, C.: Smart card based secure password authentication scheme. Computers & Security **15**(3), 231–237 (1996)
5. Jeon, W., Lee, Y., Won, D.: An efficient user authentication scheme with smart cards for wireless communications. International Journal of Security & Its Applications **7**(4), 1–5 (2013)
6. Juang, W.-S.: Efficient password authenticated key agreement using smart cards. Computers & Security **23**(2), 167–173 (2004)

7. Choi, Y., et al.: Security Enhanced Anonymous Multiserver Authenticated Key Agreement Scheme Using Smart Cards and Biometrics. The Scientific World Journal 2014 (2014)
8. Chien, H.-Y., Chen, C.-H.: A remote authentication scheme preserving user anonymity. In: 19th International Conference on Advanced Information Networking and Applications, AINA 2005, vol. 2. IEEE (2005)
9. Yang, J.-H., Chang, C.-C.: An ID-based remote mutual authentication with key agreement scheme for mobile devices on elliptic curve cryptosystem. Computers & Security 28(3), 138–143 (2009)
10. Peris-Lopez, P., et al.: EMAP: an efficient mutual-authentication protocol for low-cost RFID tags. In: Otm 2006 Workshops On the Move to Meaningful Internet Systems. Springer, Heidelberg (2006)
11. Djellali, B., et al.: User authentication scheme preserving anonymity for ubiquitous devices. Security and Communication Networks (2015)

Feasibility and Reliability of a Mobile Application for Assessing Unilateral Neglect

Hyun Lee, Hyeon-Gi Lee and Jiheon Hong

Abstract The healthcare focused on fusion of ICT and mobile device. Conventionally, the tests were used to elicit unilateral neglect from the number of distracter by manual control. We attempted to elucidate reliability of mobile application for assessment of unilateral neglect. Four patients and 30 right-handed normal subjects were recruited in this study. Subjects were required to be examined using the analog and developed digital testing method (Albert, line-bisection, and star-cancellation test). All 30 normal subjects and 4 patients showed full markers for the Albert (18 point) and star-cancellation (26 point) test both analog and digital tests. We found that the Albert, line-bisection, and star-cancellation test are suitable for classification of unilateral neglect using mobile application. We believed that the digital tests are able to be used for assessment of neglect with a paperless measurement, reduction of the observation time, and eliminating the error of the calculation process.

Keywords Unilateral neglect · Mobile application · Albert test · Line bisection · Star cancellation tasks

1 Introduction

The development of information and communication technologies has an important role to play in the medical industry. Medicine has embraced mobile healthcare systems that utilize mobile devices with the widespread introduction of handheld wireless devices [1]. The paradigm of healthcare has evolved from the diagnosis and treatment of disease to its prevention. Patrick K et al. (2008)

J. Hong
Department of Physical Therapy, Sun Moon University, A-San, South Korea

H. Lee(✉) · H.-G. Lee
Department of Computer Science and Engineering, Sun Moon University, A-San, South Korea
e-mail: mahyun91@sunmoon.ac.kr

© Springer Science+Business Media Singapore 2015
D.-S. Park et al. (eds.), *Advances in Computer Science and Ubiquitous Computing*,
Lecture Notes in Electrical Engineering 373,
DOI: 10.1007/978-981-10-0281-6_46

reported that technologies for health-related purposes should be useable by all types of individuals, including the elderly, people with low literacy, and those with permanent or temporary disability. Recently, this convergence of technology and medicine has made a particularly relevant contribution to the neuropsychological rehabilitation of patients with stroke [2].

The World Health Organization estimates that every year, 15 million people have a stroke, causing the permanent disability of 5 million people [3]. Unilateral neglect (ULN) is a common behavioral syndrome in patients with stroke [4]. ULN is defined by a person's neuropsychological inability to process and perceive stimuli on one side of the body or environment [4,5]. ULN is divided into two main categories (modality of behavior and distribution of the abnormal behavior) and five subcategories (sensory, motor, representational, personal, and spatial) [4]. Patients show one type of neglect or a combination of neglect behaviors because UNL is a heterogeneous condition, with different combinations of deficits occurring in different patients [6,7]. Therefore, previous studies have recommended that assessment of neglect include a test battery [8,9].

Traditionally, the assessment of ULN used pencil-and-paper tests such as the Albert, line-bisection, and star-cancellation tests [4,6]. In the past, people thought that pencil-and-paper tests were simple and quick to administer. However, modern technology has a number of advantages over analog methods (such as pencil-and-paper tests). Mobile applications have already been developed for assessing neglect [10]. The digital assessment of neglect is paperless, reducing observation time and eliminating errors in the calculation process. Moreover, digital assessment is conveniently performed for data management.

The validity and reliability of traditional neuropsychological assessments, namely, paper-and-pencil tests, are well established. However, digital assessments need reliability for a clinical approach. The purpose of the present study was twofold. First, we attempted to demonstrate the reliability of a mobile application for assessing ULN. The second aim of this study was to identify differences between the digital and analog approaches.

2   Methods

The study was conducted in the order of each client's information-testing-evaluation. The stage of the study and the content and outcome of each stage are shown in Figure 1. Patients created their account at the system. Then, they performed randomly an each test. Our application automatically calculated and diagnosed the ULN on the basis of previous research data. Figure 2 shows three ULN assessment tools. The data from this application at each stage are stored in the database of the mobile device.

Four patients and 30 right-handed normal subjects participated in this study. The four consecutive stroke patients were recruited according to the following criteria: (1) first ever stroke, (2) more than three months after cerebrovascular accident onset, and (3) no evidence of other serious associative problems (e.g., aphasia, attention deficits, Mini-Mental State Examination). The normal subjects were 30 current students at S University who had no earlier history of neurological, psychiatric, or physical illness. All participants understood the purpose of this study, and all provided written informed consent before participation in this experiment.

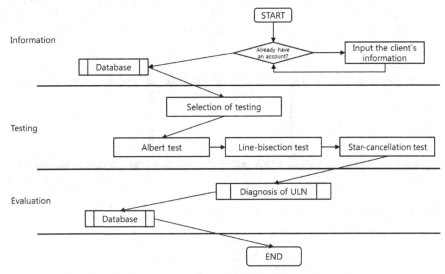

Fig. 1 Main algorithm for the unilateral neglect (ULN) testing application

2.1   The Proposed Digital Testing Methods

2.1.1   Albert Test

For the Albert test, subjects drew a line in the center of the presented lines [11]. A device divided the screen from right to left and left to right, and the device calculated the correct drawing. Finally, unilateral neglect was determined by the number of correct drawings. If the number of correct drawings was not more than 70% of the total presented lines, the patient displayed unilateral neglect. In particular, we utilized a line crossing algorithm in order to distinguish the line that was checked by the patient as shown in Figure 3. This algorithm was utilized in a line-bisection test.

2.1.2   Line-Bisection Test

The line-bisection test was composed of 18 lines (100, 120, 140, 160, 180, and 200mm) [5]. The subject drew a line in the center of the presented lines. A device calculated the deviations between the two lines such as actual center of the line and the line selected by the subject. The value of the line-bisection test score is the average of the 18 lines, and the determined value is presented as %. The deviation of the bisection mark from the true center of the line was measured in mm with left-sided errors scored as negative deviations and right-sided errors scored as positive deviations.

2.1.3 Star-Cancellation Test

On the star-cancellation test, the total number of stars—28 large stars and 28 small ones—are randomly placed on the screen [12]. Subjects are required to select the small stars. When a patient touches a small star, the color of the star is changed by the device, and the device then calculates the number of changed colors by separating the left and right of the screen.

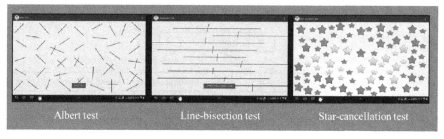

Albert test   Line-bisection test   Star-cancellation test

Fig. 2 Developed digital testing methods for unilateral neglect

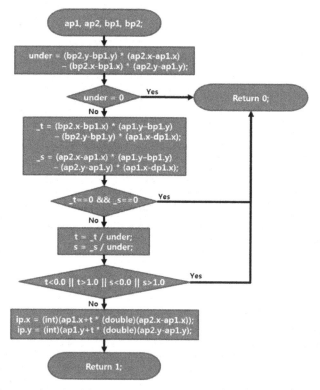

Fig. 3 The line crossing algorithm used in the Albert test and the line-bisection test

3   Results

All subjects were required to be examined using analog and digital testing methods. Each evaluation (the Albert, line-bisection, and star-cancellation tests) was performed two times. For the analog Albert and star-cancellation tests, the number of detected targets was counted manually. Average percent deviation scores on the line-bisection test were manually computed for the sets of lines at each position on the page and for the whole page combined. In contrast, the digital testing methods counted and calculated automatically.

Statistical analysis was performed using SPSS 12.0 for Windows (SPSS Inc, Chicago, IL, USA). To evaluate inter-observer (two times) and intra-observer (analog and digital tests) reliability, we used intra-class correlation coefficients (ICCs). Following Shrout et al. [13], we used the following ICC categories for interpretation: 0.00–0.10=virtually no agreement; 0.11–0.40= slight; 0.41–0.60=fair; 0.61–0.80=moderate; and 0.81–1.00= substantial. We did not perform any statistical analysis of the reliability of the Albert and star-cancellation tests because all subjects obtained a perfect score. The independent t-test was used to determine the differences between the line-bisection test and the analog and digital tests.

Table 1 summarizes the results of the analog and digital testing (Table 1). In terms of the number of correct drawings, all 30 normal subjects and 4 patients showed full markers for the Albert (18 points) and star-cancellation (26 points) tests. The relative average of the line-bisection test was $0.97(\pm2.04)$ for the analog test and $-1.45(\pm1.22)$ for the digital test ($p<0.05$). The results of this study showed moderate inter-observer reliability (analog test; ICC=0.71, digital test; ICC=0.70). In contrast, there was no intra-observer reliability (ICC=0.31).

Table 1 Results of the analog and digital testing

			Albert	Line-bisection	Star-cancellation
Normal Subject	Analog Testing	First	18	0.99($\pm$2.67)	26
		Second	18	0.94($\pm$1.92)	26
	Digital Testing	First	18	-1.42($\pm$1.39)	26
		Second	18	-1.48($\pm$1.40)	26
Patient	Analog Testing	First	18	-2.13($\pm$2.94)	26
		Second	18	0.67($\pm$5.12)	26
	Digital Testing	First	18	-2.81($\pm$1.64)	26
		Second	18	0.19($\pm$3.20)	26

4 Conclusion

In the current study, we investigated the reliability of a mobile application for assessing ULN. We found that conventional tests for ULN, such as the Albert, line-bisection, and star-cancellation tests, are suitable for classifying ULN using a mobile application. The star cancellation and Albert tests have been well established for eliciting ULN [11,14]. In this study, there was no intra-observer reliability with the line-bisection test (analog and digital). However, the relative average of the line-bisection test was in the normal range on both the analog and digital tests [5]. Previous studies reported that cancellation tests have greater test-retest reliability and are often more sensitive for detecting ULN than the line-bisection test [15].

We demonstrated the feasibility and reliability of a mobile application for assessing ULN. We believe that paper-and-pencil testing by hand can be replaced by digital testing methods. This study had limitation that could not demonstrate sensitivity to detection of ULN because of relatively small number of patients. Therefore, further studies including a larger number of patients to elucidate the sensitivity should be encouraged. In the future, we will improve the developed system model for purposes of networked-based rehabilitation.

Acknowledgement This work was supported by the Sun Moon University Research Grant of 2014.

References

1. Patrick, K., Griswold, W.G., Raab, F., Intille, S.S.: Health and the mobile phone. American Journal of Preventive Medicine **35**, 177–181 (2008)
2. Edmans, J., Gladman, J., Hilton, D., Walker, M., Sunderland, A., Cobb, S., Pridmore, T., Thomas, S.: Clinical evaluation of a non-immersive virtual environment in stroke rehabilitation. Clinical Rehabilitation **23**, 106–116 (2009)
3. The world health organization, the world health report 2002 - reducing risks, promoting healthy life, who (2002)
4. Plummer, P., Morris, M.E., Dunai, J.: Assessment of unilateral neglect. Physical Therapy **83**, 732–740 (2003)
5. Schenkenberg, T., Bradford, D.C., Ajax, E.T.: Line bisection and unilateral visual neglect in patients with neurologic impairment. Neurology **30**, 509–517 (1980)
6. Robertson, I.H., Halligan, P.W.: Spatial neglect: A clinical handbook for diagnosis and treatment. Psychology Press, Hove (1990)
7. Parton, A., Malhotra, P., Husain, M.: Hemispatial neglect. Journal of Neurology, Neurosurgery, and Psychiatry **75**, 13–21 (2004)
8. Bailey, M.J., Riddoch, M.J., Crome, P.: Evaluation of a test battery for hemineglect in elderly stroke patients for use by therapists in clinical practice. NeuroRehabilitation **14**, 139–150 (2000)
9. Halligan, P.W., Marshall, J.C., Wade, D.T.: Visuospatial neglect: Underlying factors and test sensitivity. Lancet **2**, 908–911 (1989)

10. Lee, H., Lee, H.G., Yang, S.M., Hong, J.H.: A testing and treatment system model for unilateral neglect patient based on visual feedback training. In: Computer Science and its Applications. LNEE, vol. 330, pp. 683–689 (2015)
11. Albert, M.L.: A simple test of visual neglect. Neurology **23**, 658–664 (1973)
12. Fullerton, K.J., McSherry, D., Stout, R.W.: Albert's test: A neglected test of perceptual neglect. Lancet **1**, 430–432 (1986)
13. Shrout, P.E., Fleiss, J.L.: Intraclass correlations: Uses in assessing rater reliability. Psychological Bulletin **86**, 420–428 (1979)
14. Diller, L., Weinberg, J.: Hemi-inattention in rehabilitation: The evolution of a rational remediation program. Advances in Neurology **18**, 63–82 (1977)
15. Glynda, K., Sue, P., Kim, N.G., John, O.R.: Continuing issues in the assessment of neglect. Neuropsychological Rehabilitation: An International Journal **5**, 239–258 (1995)

A Context-Aware Framework for Mobile Computing Environment

Hoon-Kyu Kim, Choung-Seok Kim and Kyung-Chang Kim

Abstract Context awareness is a computer application that uses context, that is usually used when a person communicates or acts. Context aware computing, a computing paradigm related to context awareness, provides quality service by being aware of users and its neighboring contexts in mobile computing environment. A main research issue in context aware system is to collect and transmit appropriate context data among related nodes in time. In addition, it is very important to design a framework to clearly define functions and roles in the system to provide scalability, reliability, flexibility and reusability. In this paper, we design a logical integrated model for context aware system. We then propose a context aware framework for mobile computing environment based on environment monitoring using a sensor network. Comparison with other context managing framework architectures shows that our framework provides security and privacy lacking in other architectures.

Keywords Context · Context-awareness · Context aware computing · Context aware framework · Mobile computing · Sensor network · Environment monitoring

H.-K. Kim
Agency for Defense Development, Seoul, Korea
e-mail: hunk@add.re.kr

C.-S. Kim
School of Computer Information Engineering, Silla University, Pusan, Korea
e-mail: cskim@silla.ac.kr

K.-C. Kim(✉)
Department of Computer Engineering, Hongik University, Seoul, Korea
e-mail: kckim@hongik.ac.kr

K.-C. Kim—This work was supported by 2015 Hongik University Research Fund.

D.-S. Park et al. (eds.), *Advances in Computer Science and Ubiquitous Computing*,
Lecture Notes in Electrical Engineering 373,
DOI: 10.1007/978-981-10-0281-6_47

1   Introduction

Compared to the well-defined and static traditional computing environment, ubiquitous computing environment has dynamic and open characteristics. Smart space, like intelligent home, smart office and smart car, is an implementation of ubiquitous computing technology and is getting a lot of attention these days. In practice, many of the devices in smart space communicate with each other using wireless sensor network.

In smart space, users must make great efforts to interact with numerous devices and the services provided by these devices that exist in its environment. For the users and devices to interact in the smart space, the devices and the services they provide must be well organized and the efforts to control and manage them must be minimized. Context aware computing is a computing paradigm that recognizes users and surrounding contexts to provide appropriate services in smart space.

In comparison with mobile and ubiquitous computing, that have received a lot of attention with high research interests, research interest in context aware computing is still in the early stages. Even though prototype context aware applications like context aware tourist guide, reminder application and decision tools have emerged [1][2], these applications use restricted context information in a very simple manner. A few applications were acknowledged for having practical usage scenarios, but in reality there are not many success cases for context awareness. To solve this problem, an important research issue that has received a lot of attention recently is collecting needed information from appropriate sources like sensors, and sending and managing right information to application.

In this paper, we design a logical integrated model for context aware system expanding on concepts and requirements for existing context aware application architectures. We then propose a context aware framework for mobile computing environment, more specifically environment monitoring with sensor network, based on the integrated model.

The rest of the paper is organized as follows. We discuss related works on related context aware architectures in Section 2. In Section 3, we present the logical integrated model for context aware system framework. The application of the framework for monitoring system using a sensor network is discussed in Section 4. The conclusion is given in Section 5.

2   Related Works

A variety of architectures for context aware applications exist. These include context toolkit [3], context managing framework architecture [2], service-oriented context aware middleware (SOCAM) [4], context broker architecture (CoBrA) [5] and Gaia [6].

In context toolkit, the division is made between low level sensing and high level application. They proposed a middleware layer that provides function to collect sensor source information, interpret it to a form that is understood by application and then send it to applications that need it. The middleware contains context widget, interpreter and aggregator. The widget sends the collected source context information to either the interpreter or the aggregator for further processing.

The context management framework architecture is layered and consists of four functional elements and two additional functions. The four functional elements are context manager, resource server, context recognition service and application. Two additional functions are change detection service and security. The context manager manages the blackboard and acts as the central server.

The SOCAM is a framework to develop context aware mobile service easily. It consists of context sensing layer, context middleware layer and context application layer. It collects context information from distributed context information providers and interprets it in an appropriate form. The context interpreter is the central server that interprets the context information and sends it to context-aware service. The context-aware service lies in the context application layer and provides appropriate service to users.

CoBrA is a framework that supports context aware computing in smart space based on agents. The context broker is the central component in the framework that helps collect context information from various source, manages the context information model and performs function to share knowledge among agents. The context broker consists of context knowledge base, context reasoning engine, context acquisition module and privacy management module.

Gaia is a middleware developed for user-centric environment and defines user activity area as the active space. It is a framework that is developed for mobile applications that can use diverse devices, uses context information efficiently and enables resource recognition. The architecture consists of Gaia kernel and application framework. The Gaia kernel corresponds to context aware framework and includes management and deployment system for basic services used by all applications and distributed objects.

3 Design of Framework for Context Aware System

In this Section, based on research of existing context aware systems, we analyze basic concepts and requirements for the design of context aware framework, derive design requirements and then design a logical integrated model for context aware system.

The main characteristics of the model must include the following. It should apply in static, dynamic and wireless network environment. It must support context aware middleware solutions, context data collection and transmission between nodes, night situations and uncertainty management. In addition, it must overcome system overhead and hindrance to usability and scalability. The basic design concept include distinction between generating/consuming context data, application to mobile and heterogeneous environments, visible areas for context data, restrictions based on quality of context (QoC) and life cycle of context data.

The design of the integrated architecture model for context aware system is shown in Fig. 1. The proposed integrated model is composed of three sub-systems having a hierarchical structure. The lowest layer is the context data source layer that access context data through physical sensors. The highest layer, which is the context data sink layer signifying the application, is the service level that uses processed context data. The middle layer, which is the middleware of the system, is the context data distribution layer comprising of context data transmission layer and context data management layer. To guarantee system scalability and

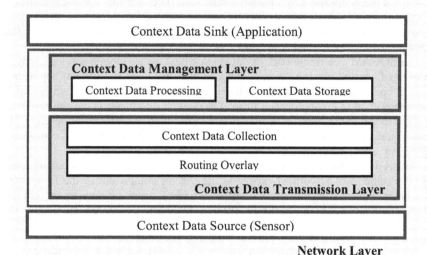

Fig. 1 Logical Integrated Model

reliability, the efficiency of distributing context data is very important and is the key in system design.

The context data transmission layer must present methods (or algorithms) to transmit context data within the system and may be affected by network topology. It consists of context data collection and routing overlay functions. The routing overlay defines the structure of network path. In context data collection, the context data from sensor sources is collected and identified and transmitted to context data management layer. A policy decision must be made to decide which nodes transmits or receives context data.

The context data management layer is the layer that generates high-level abstracted context data through context data processing and storage. It includes context data storage function and provides formal context model to define context aware system based on the model. Various storage methods are possible depending on representation methods, memory cost, overhead and more. In context data processing, it contains operations necessary to formalize context data depending on required system levels.

4   Applying the Framework for Monitoring System

Based on the integrated model, we then design a software architecture framework applicable for environment monitoring with sensor network.

The system has a sensor network structure which is based on wireless network with no human intervention and the sensor nodes are deployed in remote areas not easily accessible by humans. The process in the framework is divided into sensor node tasks and server tasks. In sensor node, the sensor signal is detected, the detected signal is preprocessed, the context information is merged (or fused) and the information is sent to the server. In server, context information received from sensor node is processed and the sensor field information is merged and displayed.

The functional requirements for the sensor node include signal processing for sensed data, preprocess detected signal for each sensor, fuse/inference detected information, establish processing rules and eliminate uncertainty embedded in context information. In server area, merge (fuse) context information collected from each node and standardize context information through alignment, estimation, gating and association. In addition, produce final context based on fusion rules, display final context information and eliminate derived uncertainty.

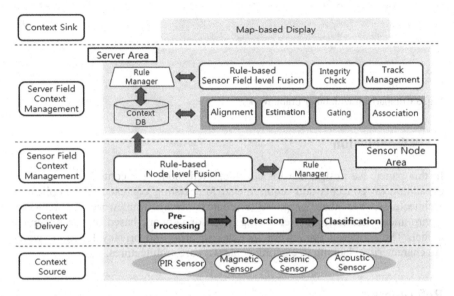

Fig. 2 Framework for Environment Monitoring System

The context aware framework for environment monitoring system using sensor network is shown in Fig. 2. It is based on the integrated model presented in the previous Section. It also has a hierarchical structure composed of context source, context delivery, sensor field/server field context management and context sink layers. The context source and context sink layers corresponds to context data source and context data sink layers respectively in the integrated model. The context delivery layer corresponds to context data transmission layer in the integrated model. The sensor field/server field context management layer corresponds to the context data management layer in the integrated model.

A comparison is made between proposed framework and existing frameworks for context aware systems. As shown in Fig. 3, our proposed framework contains more functions for managing and transmitting context data. In particular, our framework uses history, aggregation and filtering functions in context data processing while other frameworks just uses one or two of the functions. The proposed framework also includes security and privacy functions which are characteristic lacking in existing context aware architectures.

System	Context Data Model	Context Data Management			Context Data Transmission						Network		
		Context Data Processing			Context Data Broadcasting				Routing Overlay				
		History	Aggregation	Filtering	Sensor Access	Flooding-based	Selected-based	Gossip-based	Centralized	De-Centralized	Hybrid	Ad-Hoc	Fixed
Proposed Monitoring System	K-V. P	√	√	√	√			√	√		√		
Context Toolkit	K-V. P		√		√					√	√		
SOCAM	On.B	√	√		√				√				√
CoBrA	On.B		√						√				√
Gaia	F.O.L	√	√					√					√

K-V. P : Key-Value Pairs,
O.B : Object-based, F.O.L : First-Order Logic,
On.B : Ontology-based

Fig. 3 Comparison with other Frameworks

5 Conclusion

In this paper, we designed a logical integrated model for context aware system expanding on concepts and requirements for existing context aware application architectures. We then proposed a context aware framework for real-time environment monitoring system using a sensor network based on the integrated model. The proposed framework includes security and privacy functions which are characteristic lacking in existing context aware architectures.

References

1. Ranganathan, A., Campbell, R., Ravi, A., Mahajan, A.: ConChat: A context-aware chat program. IEEE Pervasive Computing, Special Issue on Context Aware Computing **1**(3), 51–57 (2002)
2. Cheverst, K., Davies, N., Mitchell, K., Friday, A.: Experience of developing and deploying a context-aware tourist guide: the GUIDE project. In: 6th Int'l Conf. on Mobile Computing and Networking, pp. 20–31, August 2000
3. Dey, A.K., Abowd, G.D.: The context toolkit: aiding the development of context-aware applications. In: Proc. of Workshop on Software Engineering for Wearable and Pervasive Computing, pp. 1–3 (2000)
4. Chen, H., Finin, T., Joshi, A.: An Ontology for Context-Aware Pervasive Computing Environments. Knowledge Engineering Review **18**(3) (2003)
5. Ranganathan, A., Campbell, R.H.: A middleware for context-aware agents in ubiquitous computing environments. In: ACM/IFIPUSENIX Int'l Middleware Conference (2003)
6. Park, I., Lee, D., Hyun S.J.: A dynamic context-conflict management scheme for group-aware ubiquitous computing environments. In: 29th Annual Int'l Computer Software and Application Conference (COMPSAC 2005), pp. 359–364 (2005)

The Design of Log Analysis Mechanism in SDN Quarantined Network System

Tae-Young Kim, Nam-Uk Kim and Tai-Myoung Chung

Abstract Due to the development of information and communication technologies, there is a growing dependence on the computer. Based on the accidents occurred in the country, this dependence means that if people have a problem in terms of cyber security, it will cause great confusion nationally. Even though there are a lot of security solutions to prevent it, security incidents have continued to occur because the number of solutions confined to a specific function have duplication and vulnerability. In order to overcome the limitation of the information security environment, it has been proposed an inspection system using the SDN of the new concept. In this paper, we will propose log analysis system to be applied within the inspection system using these SDN.

Keywords Software Defined Networks · Security · Log analysis

1 Introduction

Technology advancement and economic development makes higher reliance on computers. This dependence means that problems occurred in the cyber security can lead to great confusion nationally. This dependence means that problems occurred in the cyber security can lead to great confusion nationally. For example, security threats such as cyber-terrorism events for domestic broadcasters and banks which occurred on March 20, 2013 are being issued to social problems beyond technical problems. So as to prevent these incidents, many companies are introducing a variety of security solutions such as IDS and firewall software. In the same vein, many academic researchers are actively studying the research on intrusion detection and prevention techniques, resulting in a variety of information security solutions.

T.-Y. Kim(✉) · N.-U. Kim · T.-M. Chung
College of Information and Communication Engineering,
SungKyunKwan University, Suwon, Korea
e-mail: {tykim,nukim}@imtl.skku.ac.kr, tmchung@ece.skku.ac.kr

© Springer Science+Business Media Singapore 2015
D.-S. Park et al. (eds.), *Advances in Computer Science and Ubiquitous Computing*,
Lecture Notes in Electrical Engineering 373,
DOI: 10.1007/978-981-10-0281-6_48

However, security incidents are still occurring on an ongoing basis. This is because security techniques used with the complexity of security management have duplication and vulnerability in that these are limited to a specific function. To overcome this limitation of existing information security environment, we propose the systematic and integrated security system of new concept, as shown in Figure 1 [1].

Specifically, the inspection system utilizing SDN starts the detection on the basis of the rules defined in the dictionary when the packet came in on the SDN switch. Then it will be processed by separating them with normal, suspicious and harmful.

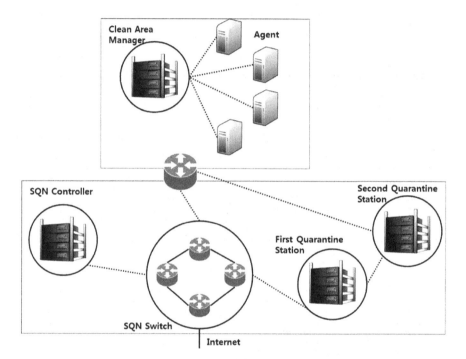

Fig. 1 Composition of inspection system utilizing SDN

Packets pass through the inspection system are stored in a log format. Stored log files are contained a variety of information. Analysis of log files makes it possible to determine the cause of security incidents. This may help to develop a technology that can prevent security threats becoming more complicated and more sophisticated.

But, with the development of technology, improved network transmission speed and suddenly increased amount of packets flowing from the inspection system are made to extend the amount of the log exponentially, which is to store them in log analysis. Existing RDBMS(Relation Data Base Management System) exists problem of performance degradation with regard to process the increased log files efficiently. System of Map-Reduce way based on technology of NoSQL(Not Only

SQL), which is effective to process the data and has recently emerged in connection with the big data, can be a solution.

In this paper, we propose a log analysis system applied Map-Reduce for analyzing log files which has been growing exponentially in the inspection system using the SDN.

2 Security Threats Detectable Through Log Analysis

The inspection system using the SDN is based on the security threats defined by three level and 10 types rooted in Source IP, Destination IP and Destination Port belonging to NASA(Network Attacks Situation Analysis) which is intrusion detection method based on situational awareness developed by Electronics and Telecommunications Research Institute(ETRI) [2]. The inspection system converts the information into the logs and saves it after the system adds information of Source Port, Protocol, and Time that packet is coming in the three factors mentioned above. The stored log data can generate much more subspecialized pattern. Generated pattern can also make a rule to detect attacks. This rule will enable us to execute strong log analysis since it can not only detect attack of Port Scanning, Host Scanning and DoS(Denial of Service) detectable through existing log analysis method, but also discover various network attacks which can be found further through the inspection system.

Table 1 Information of packet detection policy

Order	Input Time	Src IP	Dst IP	Src Port	Dst Port	Protocol	Category	Payload
1	2015.05.24 .17:20:15	54.x.x.x	115.x.x. x	55788	135	TCP	suspicious	..
2	2015.05.24 .17:20:16	54.x.x.x	115.x.x. x	55792	139	TCP	suspicious	..
3	2015.05.25 .16:37:42	112.x.x. x	115.x.x. x	59032	3389	RDP	normal	..
..	..	..	..	..	..	..	..	..
99	2015.05.28 .13:54:22	47.x.x.x	123.200. x.x	59185	3389	TCP	harmful	..
100	2015.05.28 .13:54:24	47.x.x.x	123.200. x.x	4974	3389	TCP	harmful	..

3 Features of NoSQL and Map-Reduce Technology

The existing RDBMS is difficult to extend through Scale-Out method in order to save large amounts of data. Of course, the expensive solution such as Oracle can be extended to more than a petabyte of storage space. However, Internet Services supplied by the service provider such as Google or Facebook tend not to require expensive solutions. Also, these services are not required to join operation in regard to several tens of billion or more data. Rather than strong data integrity, it needed the functions to immediately inquiry data in service failure situations by replicating and keeping data on file safely as needed. Furthermore, in regard to table holding the several tens of billions or more data, the operation such as reorganization of the index or addition of columns is not feasible in situations where the system is being operated. Thus, index and data is being separately operated and requirements that there is no defined schema or free have been added. NoSQL has received attention as solutions tailored to the requirements [3].

NoSQL is possible to access the various types of data by utilizing the key value without existing schema determined by the RDBMS, thereby easily implementing Log Parsing phase in the log analysis mechanism. Also, The Map-Reduce as a framework designed to process large amounts of data in parallel is handled by dividing the data processing across multiple servers. Looking at the model of the Map-Reduce program, when key(k1) and value(v1) is passed to the input of a map function, it processes user's logic by using delivered key-value and then outputs a list of the new key(k2) and value(v2). This function is performed repeatedly resulting in several pieces of output data are generated. There is number of data for each key, causing from sorting the output data as the key. Then, key(k2) and list of value(v2) are entered in the Reduce-function. The reduce-function receives the list of key-value and then outputs the number of values as the result after dealing with the logic defined by users.

map (k1, v1) -> list (k2, v2)

Reduce (k2, list (v2)) -> list (v2)

Fig. 2 Basic model of Map-Reduce

This Map-Reduce model is good for parallel processing because the map-function is capable of parallel processing the input data independently at the same time, meaning that it can be better performance by distributing workload, than when processed by the one server.

4   Log Analysis Mechanism

In the inspection system using the SDN, Log Analysis Mechanism consists of the several phases shown in Figure 2.

First of all, in the Log-Collection phase, it collects log with respect to the packets coming to the quarantine through the SQN switch. Second, it is entered in the Log-Parsing phase in order to facilitate the analysis. Third, Based on the log been parsing, it extracts the pattern and characteristics of the specific security threat in the Feature-Create phase. Fourth, in the Anomaly-Detection phase, it generates a temporary detection policy and determines whether or not properly detected by applying the policy. Finally, if it can correctly detect security threats, it will be added to the inspection system to detect the threats through Policy-Creation phase.

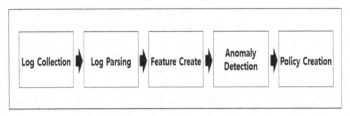

Fig. 3 Process of Log Analysis Mechanism

Port-Scan attack as an example, it can be detected and defined as Port-Scan attack when Dst Port, arbitrarily changed more than a certain number of times at a particular time, is continuously introduced on the same Src IP, Src Port, Dst Ip, Protocol.

Fig. 4 Definition of Port-Scan Detection Policy

5   Conclusion

In this paper, we propose the log analysis process using the Map-Reduce based on NoSQL to apply to the inspection system utilizing the integrated SDN, a new concept. This is expected to be further classified and detected a number of network attacks compared to the existing log analysis system. We will identify the problem not mentioned in the paper by studying the implementation of a log analysis system continuously and proceed with the study through comparison analysis for better performance development.

Acknowledgments This research was funded by the MSIP(Ministry of Science, ICT & Future Planning), Korea in the ICT R&D Program 2014[2014044072003, Development of Cyber Quarantine System using SDN Techniques].

References

1. Nam, K., et al.: The design of SDN quarantined network. In: The 2014 Fall Conference of the KIPS, pp. 559–560 (2014)
2. Jung, Y., et al.: ESM technology trends for the security of the network. Electronics and Telecommunications Trends (2001)
3. Kim, H., et al.: Cloud Computing Implementing Technology. Acornpub (2011)
4. Yoon, S., et al.: A Study on the Intrusion Detection Method using Firewall Log. Journal of Information Technology Applications & Management (2006)
5. McKeown, N.: Software-defined networking. INFOCOM keynote talk (2009)
6. Junho, C., Dong-Gyu, S., Kun-Won, J., Moon-Seog, J.: Analysis of Security Vulnerability on Firewall Logging Mechanism against DDoS Attack. Journal of KIISC **17**(6) (2007)
7. Dae-Soo, C., Gil-Jong, M., Yong-Min, K., Bong-Nam, N.: An Analysis of Large-Scale Security Log using MapReduce. Journal of KIIT **9**(8) (2011)
8. Dean, J., Ghemawat, S.: MapReduce: simplified data processing on large clusters. Communications of the ACM **51**(1) (2008)
9. Chen, W.-Y., Wang, J.: Building a cloud computing analysis system for intrusion detection system. In: CLOUD SLAM (2009)
10. White, T.: Hadoop Definitive Guide. O'Reilly Media (2009)
11. Lee, Y., Kang, W., Son, H.: An internet traffic analysis method with MapReduce. In: The 1st IEEE/IETP Workshop on Cloud Management (2010)

A Study on the Defense MITM with Message Authentication in WLAN Environments

Ji-Hoon Hong, Nam-Uk Kim and Tai-Myoung Chung

Abstract Recently smartphone users is growing and AP are installed in public places while increasing. But AP's have been shared without the most secure in public areas. In addition, a free smartphone application that uses the Internet are being developed to communicate with the server without a page advertising the security process. In this environment, Attacker can obtain the personal information of the number of users of the user and the AP interrupt and a MITM attack, connect to a shared AP. In this paper, we try a MITM attack against the real application report and the result is analyzed. In addition, the application corresponding to the MITM attack in unreliable AP Environment - offers a secure communication between the server and the model evaluation by applying them to the attack scenario.

Keywords MITM attack · ARP spoofing · WLAN message authentication

1 Introduction

The smartphones is very popular and distribution rate increasing rapidly. The wireless Internet service using wireless internet technology and has developed rapidly to. Wireless network map service number of the wireless AP with a unique SSID (Subsystem Identification) according to irradiation of WIGLE to provide from about 140 million worldwide and said to be increased by every 100,000 [1]. The AP installed in public places has become one of the most public for their customers and users. However, while a number of users to access the shared AP may also include an Attacker. Which passes the authentication process for all

J.-H. Hong(✉) · N.-U. Kim · T.-M. Chung
College of Information and Communication Engineering,
SungKyunKwan University, Suwon, Korea
e-mail: {jhhong88,nukim}@imtl.skku.ac.kr, tmchung@ece.skku.ac.kr

© Springer Science+Business Media Singapore 2015　　　　　337
D.-S. Park et al. (eds.), *Advances in Computer Science and Ubiquitous Computing*,
Lecture Notes in Electrical Engineering 373,
DOI: 10.1007/978-981-10-0281-6_49

users connected to the AP is the situation requires the solution because it is impossible in practice. Only public AP is not a problem. Google Play Store applications are being deployed are then registered in the store without a rigorous verification process to applications developed by developers. However, this makes them indiscriminately developed applications or other private information of users several security-related issues. In this paper, when the smart phone users are connected to a shared application AP is exposed to MITM. This paper is organized as follows. The second section reproduces the environment of a building ARP Spoofing attacks based MITM attacks. In the Chapter 3 and 4 offers a secure communication model of the server. And in the Chapter 5, when applying the model proposed in the wireless LAN environment using a proven formula is safe to MITM attacks, and finally Chapter 6 concludes.

2 MITM Attack Scenario

2.1 Environment Construction

To reproduce the MITM attack is described for the deployment. Configure the AP using a LINKSYS-WRT54G router as shown in Figure 1 below and were connected to Target and Attacker. Target is an Android-based smartphone model number is LG-160K is the Attacker is a laptop and installed OS is OS Backtrack5 to the open-source Linux-based hacking tools are installed by default. An important point in the built environment is the same environment that points shared with the AP in use in real life. AP Top, create a duplicate of the RogueAP users is Evil-Twin attacks and other physical wireless LAN environment that taps the send and receive data when connected to RogueAP.

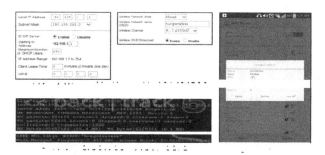

Fig. 1 Environment for MITM attacks

2.2 Attack Process

The MITM attack in progress in the same order as shown in Figure 2 below.

1) Attacker attempts to ARP Spoofing to modulate the ARP table of the Target (Target ARP Attacker's AP MAC address table and the like Jim)

2) Target are to the AP via the Attacker before all the packets that are transmitted and received, this time using the command Attacker fragrouter called for in the normal packet forwarding (packet intercepted in any medium)

3) Attacker attempts to generate random DNS Cache to a local DNS server role, DNS Spoofing attack

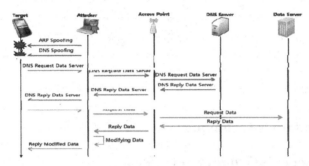

Fig. 2 MITM attack scenarios

Figure 3 shows the page viewed before and after the attack Target applications for city driving. Top left side of the page, right page is a forgery. Attacker attacks that can be modified to have not merely to modify images, such as Application B, C in Figure HTML source such as Application A redirected to Target. User trust is visible in the pages of the application which he is running, I do not know at all that has been modified.

Fig. 3 Damage caused by the attack MITM attack

2.3 Riskiness Analysis of Application

Targeted at applications that are registered in South Korea and saw the Google Play Store to try MITM attack. Specific URL or the HTML page of the application server that communicates with a MITM attack, images, etc. Figure 4 has confirmed that the application 100 applications from the URL, such as the exposure of 53 exposure, application HTML page impressions 32, applications where the image is exposed is 42. According to the details of each of the applications it examined applications are recorded on the number of downloads 50000-10000000. URL, HTML page, the image is exposed to the application are given the number of downloads not only affect the 50%. It can be seen that the number of users is at risk. As a result, most of the application examined is transmitted on a secure communication such as communication using the SSL (Secure Socket Layer) or TLS (Transport Layer Security) for important data.

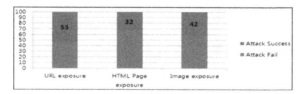

Fig. 4 Identification of the application registered in the Google Play Store

3   Foundation Technique

3.1 Message Authentication Code

HTML page to communicate with the server, Image, URL, etc. are exposed - reas ons discussed earlier in this MITM attack is dangerous WLAN applications. When t he two communication parties A and B may share the secret information K in adva nce, MAC is A message to be sent to B M and the secret information K, it can b e obtained using a function (1), and verify the integrity of the message M encryptio n technique used is [3]. Description of the functions and the terms that are used he rein as follows:

$$MAC = MAC(S, M).\qquad\qquad(1)$$

Where

M = input message
C = MAC function
S = shared secret key
MAC = message authentication code

Figure 5 is generated in the MAC with the MAC source and destination are shown in the process of verifying the integrity. It creates a data block of a fixed size of the original message M using K and function of the pre-shared secret information synapses in the original message.

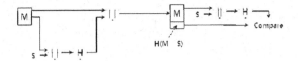

Fig. 5 Generating the MAC and the integrity verification process [3]

MAC is appended after the message was delivered to the recipient, the recipient creates a new MAC for verification using the same K and the sender. If the match is newly generated and received MAC MAC receivers are confident the message has not been changed. Attacker not know if K is-way transmission and reception shared Attacker can not be changed, because the message.

3.2   Security Communication Model Based on MAC

MAC is used for a variety of communication models. Depending on the operating method can be divided into several types. That type is based on the Hash function general, to use the block cipher mode of operation, it is typical to use an encryption algorithm [4]. It is the method of verifying the integrity of the Preneel and Oorschot proposed based MAC Hash function operation method and William sent and received by the Hash value to obtain an IP DHCP server using the suggested data based on the common Hash function [5]. Bellare also has developed the HMAC using the NMAC and Hash Functions Based on this, using a secret key initial value of the Hash Hash function as a black box type HMAC was established in RFC2104 [6][7]. Prasithsangaree offered a way to improve the function of the Radius authentication server, apply the HMAC the AP, Radius server, the communication between the terminal and the authentication server operators in the 3G network environment [8].

4   Proposed Model

In this paper, a MAC-based client for MITM attack defense in the wireless LAN environment - offer server communication model. The proposed model and generates the authentication code for integrity using the shared secret information a URL or HTML, Image exposure of the process of communication, the server and the client. Here, the shared secret information using the unique device identification information of the client smartphone IMEI (International Mobile Equipment Identity). When you enter into the server and the client is connected for the first time to exchange a lot of important information, where the IMEI can be used to exchange key information. Which is a method for verifying the integrity of MITM attack scenario used in the proposed model is applied to the proposed model uses a MAC dealt with in the third section is shown in Figure 6.

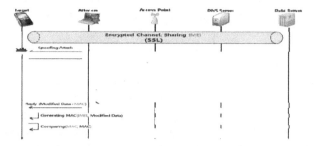

Fig. 6 MITM attack scenario, applying the proposed model

Attacker and forgery while the landscape in the middle of the Data to Target is stuck in the middle of the Target and holding request to the AP Spoofing attacks. But because the MAC is generated using the IMEI and the original data can not modify the MAC Attacker in the state do not know the key. Target generates a new MAC using the data Data and the IMEI received MAC is received, as compared to the received MAC knows that the Data Modified Data.

5   Evaluation on the Proposed Model

Attacker can modify its contents, as shown in Figure below 7 (a) obtain the source code of the HTML page, such as picture 7 (b).

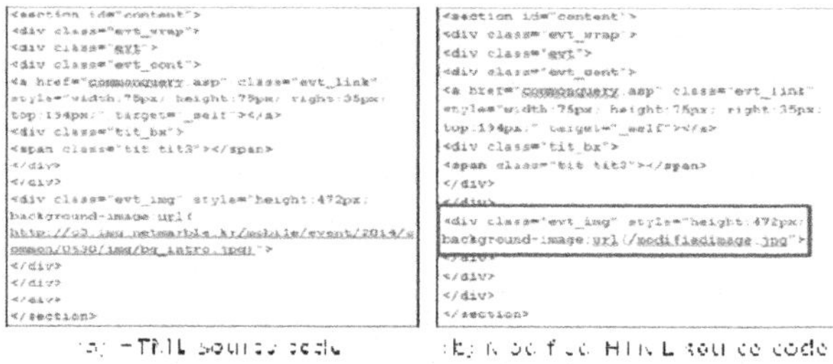

Fig. 7 HTML source code and modify the HTML source code

If the source code is modified without the Attacker it is sent to Target Target proposed model is applied to see the original server is sent Page Image Image Unlike the Attacker is modulated.

If the apply to proposed model to the application and server in such a situation, such as the code in Figure 8 below it is added.

```
public static String calculateRFC2104HMAC(String data, String key)
    throws SignatureException, NoSuchAlgorithmException, InvalidKeyException
{
    SecretKeySpec signingKey = new SecretKeySpec(key.getBytes(), HMAC_SHA1_ALGORITHM);
    Mac mac = Mac.getInstance(HMAC_SHA1_ALGORITHM);
    mac.init(signingKey);
    return toHexString(mac.doFinal(data.getBytes()));
}
```

Fig. 8 HMAC algorithm for the MAC generation source

When the application is to generate a MAC using the source Fig. 7 (a) and Figure 7 (b) is different MAC is generated. This is the same as Figure 9 below.

```
<terminated> MessageAuthentication [Java Application] C:\Program Files\Java\
HTML source code: 0def62838d3e8d020b225497a620262f7a771033
Modified HTML source code: 545975b2c06438797bec0307cd0bc9d439459569
```

Fig. 9 HMAC algorithm generated by different source MAC

The safety of the proposed model, the characteristics associated with the Hash function used to generate the MAC. Attacker can obtain the application data and the server and received by tapping with a MITM attack. In this case, Attacker scenarios, depending on whether any data is divided into two gained.
(1) If the attacker application the Hash function used by the server,
    ① Attacker to find the M corresponding to H (M) having the same value as the Hash value h.
    ② Attacker must attack the enterprise to find a M. (Any M 'is substituted to produce a value to h (M') is repeated until the process for calculating the same as the Hash value h)
    ③ However, it attempts Attacker for m-bit Hash value is proportional to 2^m.
    ④ On average should try to assign a 2^{m-1} of M 'values to find the original message M with a given Hash value h.
    ⑤ Therefore, to find the H (M) = h, M h for any Hash value of M is not known when the computation is not possible.
(2) If the attacker has gained additional M knows the Hash function used by applications and servers,
    ① Attacker is M and the other M 'to produce a new Hash value H (M using the key, and they are randomly generated, and generates a).
    ② Attacker sends the M and H (M ') to the Target. However, the key is used by the Attacker can not produce M 'to H (M') is different to the key of the Target.
    ③ Therefore, even if the Al M to generate a H (M '), such as H (M) when it is not know K can not computationally.

6   Conclusion

A smart phone user and the AP continues to increase, but the threats to the AP that has been disclosed in accordance with the user's request is still present. In this paper, the threats against wireless LAN environment based recreate the ARP Spoofing attack against an MITM attack and try to deploy applications from the

Google Play Store and identified the risks for the top 100 applications. In addition, application-proposes a communication model with a MAC to ensure secure communication, the server and verify the safety from the Attacker. We expect to be able to protect your personal information by connecting to an untrusted shared AP applying the proposed model when using the wireless LAN. However, there is the issue of keys and their length to be used In an application of the real-world environment. Due to a trade-off relationship is established between the security and communication speed to the length of the key. Therefore, future application proposed model for application and study the length of the key and the right key exchange at the time of the initial communication with the server. The communication speed of using SSL with the proposed model and the measure if the SSL also apply to all transmitted and received packets. Considering the trade-off between the measured communication speed and security will be provided a framework for secure communication model in future studies to the wireless LAN MITM attack.

Acknowledgments This research was funded by the MSIP(Ministry of Science, ICT & Future Planning), Korea in the ICT R&D Program 2014[2014044072003, Development of Cyber Quarantine System using SDN Techniques].

References

1. Wireless Geographic Logging Engine. https://wigle.net
2. Android Open Source Project. http://source.android.com
3. Stallings, W.: Cryptography and Network Security: Principles and Practice, 5th edn. PEARSON (2011)
4. Preneel, B., Van Oorschot, P.C.: MDx-MAC and building fast MACs from hash functions. In: Cryptology CRYPT0 1995. Springer, Heidelberg (1995)
5. Bellare, M., Canetti, R., Krawczyk, H.: Keying hash functions for message authentication. In: Cryptology CRYPTO 1996. Springer, Heidelberg (1996)
6. Arbaugh, W.A.: Your 80211 wireless network has no clothes. IEEE Wireless Communications **9**(6) (2002)
7. Bellare, M., Canetti, R., Krawczyk, H.: Message authentication using hash functions: The HMAC construction. RSA Laboratories' CryptoBytes (1996)
8. Prasithsangaree. P., Krishnamurthy, P.: A new authentication mechanism for loosely coupled 3G-WLAN integrated networks. Vehicular Technology Conference (2004)

Interactive Activity Recognition Using Articulated-Pose Features on Spatio-Temporal Relation

Thien Huynh-The, Dinh-Mao Bui, Sungyoung Lee and Yongik Yoon

Abstract A success progress of pose estimation approaches motivates the activity recognition used in CCTV-based surveillance systems. In this paper, a method is proposed for recognizing interactive activities between two human objects. Based on articulated joint coordinates obtained from a pose estimation algorithm, the distance and direction feature are extracted from objects to describe both the spatial and temporal relation. The multiclass Support Vector Machine is finally employed for activity classification task. Compared with existing methods using the public interaction dataset, the proposed method outperforms in overall classification accuracy.

Keywords Interactive activity recognition · Articulated-pose feature · Spatio-temporal relation

1 Introduction

In recent years, human activity recognition has been received more attentions from computer vision and artificial intelligence community due to its important role in applications of video-based surveillance, video annotation, somatic game and human-computer interaction. However, developing an effective system is a challenge due to many issues, such as illumination variation and object occlusion.

T. Huynh-The(✉) · D.-M. Bui · S. Lee
Department of Computer Engineering, Kyung Hee University, Seoul,
Gyeonggi-do 446-701, Korea
e-mail: {thienht,mao,sylee}@oslab.khu.ac.kr

Y. Yoon
Department of Multimedia Science, Sookmyung Women's University,
Seoul 140-172, Korea
e-mail: yiyoon@sookmyung.ac.kr

© Springer Science+Business Media Singapore 2015
D.-S. Park et al. (eds.), *Advances in Computer Science and Ubiquitous Computing*,
Lecture Notes in Electrical Engineering 373,
DOI: 10.1007/978-981-10-0281-6_50

Many researches were proposed for activity recognition with one actor, such as running, walking, waving hands [1]. Some methods considered the interaction activities between the actor and object, i.e., the daily life activities as eating, drinking [2] by using the local features. Few works [3] faced with activities involving two or more human objects in interaction likes as hand shaking, hugging, and punching.

One of the most important issues explored in the activity recognition is the feature extraction. Motivated by existing object recognition approaches using local features such as SIFT descriptor [4], which captured the spatial relations between the points of interest. Recently, authors have been attracted by research direction of pose estimation [5], in which the relative locations of body components are identified. Sparse coding and dictionary learning are two advanced techniques for modeling activities based on extracted features. Unlike the existing dictionary learning approaches utilized the spatio-temporal features, Cai et al. [6] recommended a framework of learning pose dictionary for the human body representation.

In this paper, we focus on the interactive activities between two persons based on the successful outcome of the pose estimation approach [5]. From the articulated-pose coordinates, two kinds of feature, joint distance and joint direction, are extracted to describe the spatial relationship between body components that belong to each object and two objects. Furthermore, the objects are consider not only within a same frame but also in two adjacent frames to enhance the distinctness of benchmarked activities. Finally, a multiclass Support Vector Machine (SVM) classifier is performed for training and classification. In the experiments, we implement the proposed scheme under the different extracted feature categories and compare with the state-of-the-art methods.

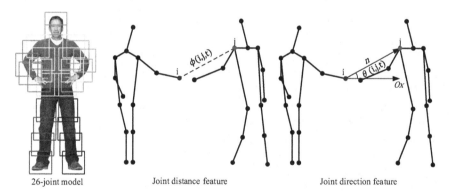

| 26-joint model | Joint distance feature | Joint direction feature |

Fig. 1 An illustration of: (a) 26 articulated joint model and (b) distance and direction feature.

2   Methodology

2.1   Joint Detection

One of the most well-known pose estimation methods was introduced by Yang et al. [5], in which, the body key joint were modeled into a tree structure and a score function was identified to dynamically search human object and effectively detect articulation. Based on capturing the dependence of local appearance on the spatial geometry, Yang's model outperformed classic articulated models in speed and accuracy. The impressive results on several real-life datasets were reported in that research to demonstrate the preeminence of this algorithm when compared with existing algorithms in the task of human pose estimation. In this research, we employed Yang's algorithm for the joint estimation purpose on the interaction dataset [7]. Concretely, the PARSE model with full-body estimation has been used to locate 26 articulated points, shown in Fig. 1, due to the highest accuracy in comparison of 14-point and 51-point model. For enhancing performance of locating articulated points, multiple estimators are trained on the interaction dataset to control huge variance among activities. Each model is tried to detect joints and the best one is selected with maximum score.

2.2   Feature Extraction

The input data in this stage contains the coordinates of body joints estimated from Yang's algorithm. Each joint is denoted as $p_i(x,y)$. Two types of feature calculated from the joint coordinates are the distance and the direction, illustrated in Fig. 1.

Joint Distance: The joint distance feature is defined as the Euclidean distance between all pairs of joints in each person and between two person in a frame and in two adjacent frames. The distance between two joints i of an object X and j of an object Y in the frame t is calculated as follows:

$$\phi^{X,Y}(i,j,t) = \left\| p_{i,t}^X - p_{j,t}^Y \right\| = \sqrt{\left(x_{p_{i,t}^X} - x_{p_{j,t}^Y}\right)^2 + \left(y_{p_{i,t}^X} - y_{p_{j,t}^Y}\right)^2} \tag{1}$$

The above equation is also applied for the same object $(X=Y)$ and developed for two objects in two consecutive frames t and $t'=t-1$:

$$\phi^{X,Y}(i,j,t',t) = \left\| p_{i,t'}^X - p_{j,t}^Y \right\| \tag{2}$$

The joint distance features extracted from two objects are organized into a vector:

$$\Phi = \begin{bmatrix} \Phi^X(t) & \Phi^Y(t) & \Phi^{X,Y}(t) & \Phi^X(t',t) & \Phi^Y(t',t) & \Phi^{X,Y}(t',t) & \Phi^{Y,X}(t',t) \end{bmatrix} \quad (3)$$

where $\Phi^X(t) = \{\phi^{X,X}(i,j,t)\}$, $\Phi^Y(t) = \{\phi^Y(i,j,t)\}$, and $\Phi^{X,Y}(t) = \{\phi^{X,Y}(i,j,t)\}$ for the same frame; and $\Phi^X(t',t) = \{\phi^{X,X}(i,j,t',t)\}$, $\Phi^Y(t',t) = \{\phi^{Y,Y}(i,j,t',t)\}$, $\Phi^{X,Y}(t',t) = \{\phi^{X,Y}(i,j,t',t)\}$, and $\Phi^{Y,X}(t',t) = \{\phi^{Y,X}(i,j,t',t)\}$ for consecutive frames.

Joint Direction: The joint direction feature is identified as the angle between the vector $\vec{n}(i,j)$ and X-axis $\overrightarrow{Ox}$ as an illustration in Fig. 1. Generally, the direction feature between two joints i of an object X and j of an object Y within the frame t is calculated as follows:

$$\theta^{X,Y}(i,j,t) = \angle\left(p_{i,t}^X, p_{j,t}^Y\right) \quad (4)$$

The above equation is also applied for the same object $(X = Y)$ and generally developed for two objects in two frame t and $t' = t-1$:

$$\theta^{X,Y}(i,j,t',t) = \angle\left(p_{i,t'}^X, p_{j,t}^Y\right) \quad (5)$$

The joint distance features extracted from two objects are organized into a vector:

$$\Theta = \begin{bmatrix} \Theta^X(t) & \Theta^Y(t) & \Theta^{X,Y}(t) & \Theta^X(t',t) & \Theta^Y(t',t) & \Theta^{X,Y}(t',t) & \Theta^{Y,X}(t',t) \end{bmatrix} \quad (6)$$

where $\Theta^X(t) = \{\theta^{X,X}(i,j,t)\}$, $\Theta^Y(t) = \{\theta^{Y,Y}(i,j,t)\}$, and $\Theta^{X,Y}(t) = \{\theta^{X,Y}(i,j,t)\}$ for the same frame; while $\Theta^X(t',t) = \{\theta^{X,X}(i,j,t',t)\}$, $\Theta^Y(t',t) = \{\theta^{Y,Y}(i,j,t',t)\}$, $\Theta^{X,Y}(t',t) = \{\theta^{X,Y}(i,j,t',t)\}$, and $\Theta^{Y,X}(t',t) = \{\theta^{Y,X}(i,j,t',t)\}$ for the difference frames. Although, features are formed in distance and direction categories in (3) and (6), they can be structured in spatial and temporal relation as follows:

$$S = \begin{bmatrix} \Phi^X(t) & \Phi^Y(t) & \Phi^{X,Y}(t) & \Theta^X(t) & \Theta^Y(t) & \Theta^{X,Y}(t) \end{bmatrix} \quad (7)$$

$$T = \begin{bmatrix} \Phi^X(t',t) & \Phi^Y(t',t) & \Phi^{X,Y}(t',t) & \Phi^{Y,X}(t',t) & \Theta^X(t',t) & \Theta^Y(t',t) & \Theta^{X,Y}(t',t) & \Theta^{Y,X}(t',t) \end{bmatrix} \quad (8)$$

(a) Hugging   (b) Kicking   (c) Pointing

(d) Punching   (e) Pushing   (f) Hand Shaking

Fig. 2 UT-Interaction Dataset examples after using articulated joint estimation.

3 Experiment

In the experiment stage, we benchmarked the method on the public UT-Interaction Dataset [7] that includes two sets of video data. Six interactive activities are presented in UT-dataset: hugging, kicking, pointing, punching, pushing, and hand shaking. Some snapshots of them with the articulated joint coordinates were represented in Fig. 2. For SVM classifier, the authors used LibSVM [9] with RBF kernel to solve the multiclass classification problem with 10-fold cross validation.

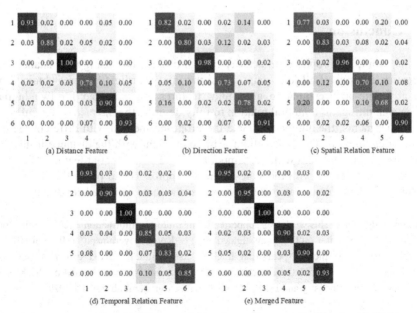

Fig. 3. Confusion matrices of different articulated-pose features. (1: Hugging, 2: Kicking, 3: Pointing, 4: Punching, 5: Pushing, 6: Hand shaking). The average classification rates are 90.56%, 83.89%, 80.83%, 89.44%, and 93.89% corresponding to the case of distance feature, direction feature, spatial relation feature, temporal relation feature and combination.

In the first experiment, we evaluated the proposed scheme on different feature categories. The classification results were represented through the confusion matrices in Fig. 3. Compared with the direction feature, distance feature reported a greater accuracy at most testing activities as in Fig. 3a and 3b. Furthermore, the temporal feature was also represent a more impressive results when compared with the spatial feature, shown in Fig. 3c and 3d., because of more useful interaction information contained in the temporal relation feature category. The highest accuracy was achieved when merging distance and direction feature in the consideration of spatio-temporal relation as Fig. 3e. In the second experiment, we make a comparison of performance in classification accuracy between the proposed method and existing approaches using the mean Accuracy (mAcc) and mean Average Precision (mAP). Based on the results in Table 1, the proposed method outperform others in the same experimental condition with 10-fold validation [3].

Table 1 Method comparison on UT-Interaction Dataset.

Method	mAcc (%)	mAP (%)
PSR [2]	49.09	45.90
BoF [8]	77.12	79.95
SSR [3]	87.42	91.81
Our method	93.89	93.89

4 Conclusion

In this work, we proposed an activity recognition method using the distance and direction features of all joint-pairs. From the joint coordinate dataset obtained by an articulate-pose estimation algorithm, features are extracted for two human objects in the spatio-temporal relation to fully describe the interactive activities. In the experiments, the authors evaluate the proposed method with different feature categories. In particular, our method is more efficient with the distance instead of the direction feature and with the temporal instead of the spatial feature. We extra compare our method with the state-of-the-art approaches to prove the impressive results is task of interaction recognition under the same validated condition.

Acknowledgments This work was supported by Institute for Information & communications Technology Promotion (IITP) grant funded by the Korea government(MSIP) (B0101-15-1282-00010002, Suspicious pedestrian tracking using multiple fixed cameras) and also supported by the Industrial Core Technology Development Program, funded by the Korean Ministry of Trade, Industry and Energy (MOTIE), under grant number #10049079.

References

1. Qiu, Q., Jiang, Z., Chellappa, R.: Sparse dictionary-based representation and recognition of action attributes. In: 2011 IEEE International Conference on Computer Vision (ICCV), pp. 707–714 (2011)
2. Matikainen, P., Hebert, M., Sukthankar, R.: Representing pairwise spatial and temporal relations for action recognition. In: European Conference on Computer Vision: Part I (ECCV 2010), pp. 508–521 (2010)
3. Meng, L., Qing, L., Yang, P., Miao, J., Chen, X., Metaxas, D.N.: Activity recognition based on semantic spatial relation. In: 2012 21st International Conference on Pattern Recognition (ICPR), pp. 609–612 (2012)
4. Moghimi, M., Azagra, P., Montesano, L., Murillo, A.C., Belongie, S.: Experiments on an RGB-D wearable vision system for egocentric activity recognition. In: 2014 IEEE Conference on Computer Vision and Pattern Recognition Workshops (CVPRW), pp. 611–617 (2014)
5. Yang, Y., Ramanan, D.: Articulated Human Detection with Flexible Mixtures of Parts. IEEE Transactions on Pattern Analysis and Machine Intelligence 35(12), 2878–2890 (2013)
6. Cai, J.X., Tang, X., Feng, G.: Learning pose dictionary for human action recognition. In: 2014 22nd International Conference on Pattern Recognition (ICPR), pp. 381–386 (2014)
7. Ryoo, M.S., Aggarwal, J.K.: Spatio-temporal relationship match: video structure comparison for recognition of complex human activities. In: 2009 IEEE 12th International Conference on Computer Vision, pp. 1593–1600 (2009)
8. Delaitre, V., Laptev, I., Sivic, J.: Recognizing human actions in still images: a study of bag-of-features and part-based representations. In: British Machine Vision Conference, pp. 97.1–97.11 (2010)
9. https://www.csie.ntu.edu.tw/~cjlin/libsvm/

Development of Learner-Centric Teaching-Learning Application Model for Ubiquitous Learning

Ji-Hye Bae and Hyun Lee

Abstract In the era of rapidly changing ubiquitous environment, smart devices have developed and penetrated together with numerous educational applications. Ubiquitous learning enables people to study regardless of time and space in a learning environment of efficient communication between instructors and learners. In this study, for more externally expanded learning activities, a learner-centric teaching-learning application was presented, instead of the teaching-learning process in the actual field of education, and the application was applied to the field. A main program used for this research development is App Inventor tool. It was tested in the field to see if it had any effect on learning motivation. The pre/post-experiment program tests of this study found significant differences in all of the four main learning motivation factors – attention, relevance, confidence and satisfaction.

Keywords Ubiquitous learning · Android app · App Inventor · Learner-centric teaching-learning model

1 Introduction

Supported by the rapid information-oriented development, e-Learning has been activated so far by means of desktop computer sets to practice learning in the cyber space. However, given the nature of the physical media of desktop PC hardware, learners can study only in a specific place under the temporal and

J.-H. Bae(✉)
Department of History and Culture Contents, Sun Moon University,
A-san City, South Korea
e-mail: jhbae327@gmail.com

H. Lee
Department of Computer Engineering, Sun Moon University, A-san City, South Korea

© Springer Science+Business Media Singapore 2015
D.-S. Park et al. (eds.), *Advances in Computer Science and Ubiquitous Computing*,
Lecture Notes in Electrical Engineering 373,
DOI: 10.1007/978-981-10-0281-6_51

spatial limitations [1]. But the recent technological trend has made it possible to realize ubiquitous learning where learners can study anytime anywhere through smart devices. In line with the smart device commercialization and penetration such as smart phones or smart pad, users have quickly adapted to such devices. As study on new teaching-learning paradigm in this ubiquitous era has been actively researched, the quick and wide penetration of smart devices accelerates the development of diversified educational applications appropriate for the new educational environment [2]. In the education area, smart devices are highly useful thanks to their accessibility, mobility, convenience, etc. Indeed, educational applications have been actively developed. As educational contents are frequently accessed in this mobile environment based on mobile devices, naturally, the present physically-limited education framework has moved to the ubiquitous environment converging time and space. In additions, compared to the usual educational practice led by instructors for the most part, education in the ubiquitous environment allows learners to lead own study based on own goals and motivation, realizing a learner-oriented teaching-learning practice. In the modern information-based society, it was pointed out that, to change the educational paradigm into a learner-centered constructivism-based one respecting learners' personality, the class instruction system should be structured in terms of team-centered organization, autonomy beyond responsibility, cooperative relationship, shared decision making, initiation, diversity, etc. while introducing more learner-centered instructions for proactive learning process [3]. Accordingly, the education field in such a ubiquitous environment can be also viewed as an appropriate educational environment for the definition of learner-centered education. The purpose of this study is to develop a teaching-learning application expandable to the outside of the actual education field to help instructor-learner and learner-learner interaction and proactive learning practices ultimately for ubiquitous learning process beyond time and space. The main study media of this study is smart devices. The study also seeks that, in the process of using such smart devices, learners acquire updated digital skills and experience learning in the ubiquitous environment. The teaching-learning application development tool herein is the cloud-based Android App Inventor developed by Google and MIT Media Lab. This study is structured as follows; Section 2 examines the features and related studies on ubiquitous learning and App Inventor. Section 3 deals with the development of learner-centric teaching-learning application proposed in this study. Section 4 presents the field application and test results of the proposed teaching-learning application. Section 5 concludes this study.

2   Theoretical Background

2.1   Ubiquitous Learning

Ubiquitous education or u-learning can be defined to realize more creative and learner-centered educational programs by helping education-related organizations and objects in the physical space become smarter and mutually connected to establish a learning environment favorable for learners to study anytime, anywhere, whatever, and through whichever devices. To realize this, the ubiquitous learning environment should be established first so that learners can gain access to any kind of contents, anytime, anywhere, with any device. Second, education service should be provided, which helps learners study according to their own individual needs and promotes more natural learner-teacher interaction. U-learning contents for ubiquitous learning environment building need to be customized to each learner and provided regardless of time, place and device in a creative and natural manner. Study devices in the u-learning environment should be friendlier to people and provide human-centered convenience of use instead of machine-orientation, mobility and easy wearability to realize communication efficiency and mutually interacting environment responsive to user demand anytime anywhere [4]. In this study, by considering such characteristics of the ubiquitous learning, an application for smart devices will be developed, which can support learner-centric customized educational contents and ensure access to education without temporal and spatial restrictions.

2.2   App Inventor

Usually when developing apps, Google's Android SDK is utilized and it supports specialized environment and tool for app development. Since the Java language is utilized for programming therein, anyone other than professional programmer or app developer may have difficulty. App Inventor was developed by Google in 2010 and is currently operated and supported by the MIT Center for Mobile Learning. It is a highly useful cloud service tool for educational purposes as provided free of charge regardless of high-performance hardware or operational systems. It supports block-based programming and component-based object-oriented programming in a similar form to MIT Scratch tool for anyone without app development expertise to easily develop an application. App Inventor provides block programming-based visualized structure to check a learner's process of cognitive thinking in colors and frames from as early as the programming process. In this sense, the programming language is helpful for education on creativity-enhancing way of thinking [5]. App Inventor supports cloud service, enabling online programming and production at its main homepage (http://appinventor.mit.edu). So anyone with a Google account can use it free of charge without going through an additional installation process to own desktop. Thanks to these benefits, App Inventor-based education programs have been more often provided recently along with active study efforts on application development for elevated educational effectiveness. In study [6], university students were instructed to build a utility operable by SMS in real life to observe qualitative changes in their learning process. As a result, the study found the students could establish such a utility usable in real life and explained the program was very useful for that reason. In study [7], elementary school students were investigated regarding App Inventor-based Android app production education. In this study, the participants were found to be motivated for app production. In study [8], App Inventor was employed to develop a real-time/non-real time question-answer application and learners used the app in the real educational field to test its educational effectiveness. The study was limited in developing a simple app centered solely on one instructing person and providing feedback to learners in non-real time. In this present study, App Inventor was used as a main tool to develop the u-learning teaching-learning application. The educational app developed herein is operable in Android OS-based smart devices. This study sought to find out how much learning effect learners could receive in the field of education.

3   Development of Learner-Centric Teaching-Learning Application Based on Ubiquitous Learning

This section presents the design and establishment process of the learner-centric teaching-learning application developed based on App Inventor. Figure 1 shows the main design structure of the learner-centric teaching-learning application proposed herein. The propose application is largely divided into two parts – instructor's activity and learner's activity. The instructor's activity area supports the event activity of registering learning contents and quiz for a specific study theme after log-in with authority. The study materials can be in forms of text, image or web-posted information. Once the learning event is registered, a message is sent to learners and the learners log-in to follow the study program and solve quiz. Then the instructor performs the feedback activity, which is to check the learners' study activities informed back to the instructor as well as their answers to the quiz to make subsequent events such as making comments and scoring. These series of procedures are regular teaching-learning practices between instructors and learners, which can be implemented anytime anywhere without outside the conventional field of teaching. In the learner's activity area, in addition to the function to view the unilaterally-provided study contents from the instructor, there is another function for learners to create a study event for themselves and share study details together for effective learner-to-learner interaction. Such an environment does not rely solely on instructors for study contents but encourages learners to actively participate in contents generation, opening and sharing to show their own initiation in learning process in line with the Web 2.0 paradigm for participation, openness and sharing. Through such learner-centered activities, each individual learner can be a mentor to one another and deepen their learning while playing as a mentor.

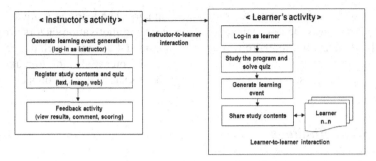

Fig. 1 Main design of learner-centric teaching-learning application

Figure 2 below shows an example of the learner-centric teaching-learning application establishment proposed in this study. The instructor's activity area deals with learning event registration and quiz provision. And accordingly, the learner's activity area shows corresponding pages. The learner's activity page shows a data input box to key in the answers to an instructor's quiz and supports functions such as Send, View Results, Share learning contents and Create learning event.

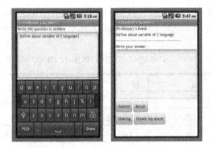

Fig. 2 Example of learner-centric teaching-learning application

4 Application to Education Field

4.1 Study Method and Research Tool

In this study, the above proposed teaching-learning application was practiced in the actual field of education to see if it had any significant effect on learners' learning motivation. This research looked at 30 university students in their early or mid-20s at 4-year Korean universities, who are familiar with smart device manipulation. For this study design, the single group pre/post-test method was employed. In the early education stage, the pre-test was conducted with questionnaire on learning motivation. For the experiment, a 20-hour education program was performed for 4 hours a day and 5 days in total. After the program, post-test was conducted with the same questionnaire. The learning motivation test utilized herein was the re-organization of the learning motivation factors proposed in study [9]. The questionnaire was structured around 4 elements including 5 items on attention; 5 on relevance; 5 on confidence; and 5 on satisfaction to make 20 in total. All of the question items are is 5-point Likert scale and the Cronbach's alpha was .82.

4.2   Study Findings

As a result of analyzing the questionnaire, the average scores of all of the four learning motivation sub-elements increased as in Table 1. It seems because the learners in this research experiment gave more positive answers in generally more items in the post-test than the pre-test. Larger difference was found especially in confidence-related factors. The *t-test* was conducted for attention, relevance, confidence and satisfaction. As a result, significant differences were found in all items as shown in Table 1 ($p < .05$). This finding means that learning by utilizing the learner-centric teaching-learning application proposed herein has a positive effect on learning motivation. The students were found to use the application program more actively outside the class hours mainly to review the class. It is deemed that, amid the high interest in instruction, the smart device-based self-directed learning method helped further enhance their learning motivation.

Table 1 Result of *t-test* on pre/post-experiment learning motivation factors

factor	Group	n	M	SD	t	p
Attention	Pre	30	3.39	.372	-9.402	.000
	Post	30	4.28	.360		
Relevance	Pre	30	3.14	.483	-8.591	.000
	Post	30	4.13	.402		
Confidence	Pre	30	2.69	.455	-15.540	.000
	Post	30	4.52	.453		
Satisfaction	Pre	30	3.80	.495	-3.726	.001
	Post	30	4.25	.430		

5   Conclusion

In this study, a teaching-learning application was presented to help realize ubiquitous learning free of space and time constraints by inspiring instructor-learner and learner-learner interaction and learning motivation through externally expanded learning activities instead of the conventional teaching-learning process in the educational field. Also the application was tested in the actual education process. The main development program used herein was App Inventor which is useful not only for app production but also for app use. The developed app was test-used in actual learning process to see if it had any effect on learning motivation. The pre/post-test herein found significance differences in all of the four main areas of learning motivation – attention, relevance, confidence and satisfaction. In terms of study motivation, the application can be viewed to have an effect on learning motivation regarding learning activities outside the class hours. However, in this study, simply the learning motivation was tested while excluding learning flow or tool usability, etc. From the UI design perspective, due to the limited component support of App Inventor, the page design seems less user-centered while focusing more on functionality as of now. As data capacity and DB design are also limited, building a large-scale DB is difficult. Follow-up study is expected to design a more favorable UI for user environment to help expand user convenience in functionality.

Acknowledgement This work is supported by NRF-2013R1A1A1075980.

References

1. Kim, E.-G., Hyun, D.-L., Kim, J.-H.: Design and Implementation of Learning Content Authoring Framework for Android-based Three-Dimensional Shape. Journal of the Korean Association of Information Education **15**(1), 67–76 (2010)
2. Hye-Ki, M., Hyunji, Y., You-na, S., Se-Eun, L., Young, R.A., Seungho, P.: Designing Smart Pad Application as an Assistant Tool of College Lecture for Cooperative Learning. Design Convergence Study **33.11**(2), 59–71 (2012)
3. Charles, M.R.: A new paradigm of ISD? Educational Technology **36**(3), 13–20 (1996)
4. Nam, H.S.: A Study on the key success factors of teaching and learning model ubiquitous IT for u-learning system. In: The 2010 Fall Conference of Korean Institute of Industrial Engineers, pp. 824–825 (2010)
5. Shin, S., Choi, I., Bae, Y.: Development of STEAM Program using App Inventor: Focusing on the Concept of Speed in Elementary Science Education. The Journal of Korea Contents Association, 530–544 (2015)
6. Wolber, D.: App Inventor and Real-World Motivation. ACM Special Interest Group on Computer Science Education, 601–606 (2011)
7. Rim, H.: Android App Implementation Teaching using App Inventor for Elementary School Students. The Journal of Korea Multimedia Society **16**(12), 1495–1507 (2013)
8. Park, C.-J., Kang, J.-H., Kim, M.-J., Yu, Y.-R., Kim, H.-S., Koh, J.-W.: Design and implementation of real-time/non-real-time question-answer apps by using appinventor. In: The 2012 Summer Conference of Korean Association of Computer Education, pp. 63–66 (2012)
9. Keller, J.M.: Development and use of the ARCS model of instructional design. Journal of Instructional Development **10**(3), 2–10 (1987)

EAP: Energy-Awareness Predictor in Multicore CPU

Dinh-Mao Bui, Thien Huynh-The, YongIk Yoon,
SungIk Jun and Sungyoung Lee

Abstract To deal with inference and reasoning problems, Gaussian process has been considered as a promising tool due to the robustness and flexibility features. Especially, solving the regression and classification, Gaussian process coupling with Bayesian learning is one of the most appropriate supervised learning approaches in terms of accuracy and tractability. Because of these features, it is reasonable to engage Gaussian process for energy saving purpose. In this paper, the research focuses on analyzing the capability of Gaussian process, implementing it to predict CPU utilization, which is used as a factor to predict the status of computing node. Subsequently, a migration mechanism is applied so as to migrate the system-level processes between CPU cores and turn off the idle ones in order to save the energy while still maintaining the performance.

Keywords Proactive prediction · Bayesian learning · Gaussian process · Energy efficiency · CPU utilization

1 Introduction

Research in energy consumption is one of the hot trends in recent years. From time to time, this topic has been the major issue in most of the computing systems.

D.-M. Bui(✉) · T. Huynh-The · S. Lee
Computer Engineering Department, Kyung Hee University, Suwon, Korea
e-mail: {mao,thienht,sylee}@oslab.khu.ac.kr

Y. Yoon
Department of Multimedia Science, SookMyung Women's University,
Seoul, Republic of Korea
e-mail: yiyoon@sm.ac.kr

S. Jun
HPC System Research Section/Cloud Computing Department/ETRI, Daejeon, Korea
e-mail: sijun@etri.re.kr

© Springer Science+Business Media Singapore 2015
D.-S. Park et al. (eds.), *Advances in Computer Science and Ubiquitous Computing*,
Lecture Notes in Electrical Engineering 373,
DOI: 10.1007/978-981-10-0281-6_52

Coming up with the higher performance that the computing node achieves, a large amount of energy is also taken in to account as significant costs. Theoretically, energy is used to conduct the computation, especially in servers. When the servers are up and running, even when they are idle, the energy is still wasted to maintain the power-on status. For that reason, enhancing the energy efficiency in computing system concerns reducing the energy waste on idle computing facilities. However, this effort is challenged by the problem to concurrently maintain the performance.

In this paper, we propose a proactive solution for energy saving in CPU level. By applying the prediction technique which is the Gaussian process regression, the energy application predicts the multicore CPU utilization and activates the process migration between the cores in order to achieve the overall energy efficiency while maintaining a high performance.

This paper is organized as follows. We detail our energy efficiency approach in Section 2. In Section 3, we conduct the performance evaluation of the energy saving application. Our conclusion is summarized in Section 4.

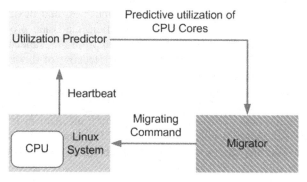

Fig. 1 The architecture of energy saving application.

2   Proposed Method

2.1   Target Process

The energy efficiency architecture for a multi-cores CPU is proposed in this section and described in detail in Fig.1. Basically, the purpose of this architecture is to pro-actively reduce the energy consumption of the CPU cores. Thus, the main functionality of this application is to empty the workload of the CPU cores and then deactivate or stand-by the idle cores to save energy. To do this effectively, the migration procedure on the Migrator component needs the predictive information of the CPU core utilization to determine the source and the destination in order to migrate the target process (from this point, the system process being considered migration will be known as the target process). Primarily based on the current value of the CPU core's utilization, known as the heart-beat, the predictive utilization is calculated in the Utilization Predictor component and plays a critical role in subsequent decision making.

2.2   Prediction Model

In our application, the object of the prediction is to anticipate the utilization of the CPU cores. Bayesian learning and Gaussian [1,2,3,4] process regression are employed as the inference technique and probability framework, respectively. Because the input data for this model is the time series utilization [5], curve-fitting is preferred over function mapping for the mapping approach. It is important to note that the curve-fitting is more flexible with regard to the time series data and non-stationary model.

Assuming that the input data is a limited collection of time location $x=[x_1, x_2, x_3,...x_n]$, a finite set of random variable $Y=[Y_1, Y_2, Y_3,...Y_n]$ represents the corresponding joint Gaussian distribution of incoming processes with regard to the time order. This set over the time constraint actually forms up the Gaussian process:

$$f(y|x) \sim GP(m(x), k(x, x')) \tag{1}$$

with

$$m(x) = E(f(x)) \tag{2}$$

and

$$k(x, x') = E((f(x) - m(x))(f(x') - m(x'))) \tag{3}$$

in which, $m(x)$ is the mean function, evaluated at the location x variable, and $k(x,x')$ is the covariance function, also known as the kernel function[6,7,8]. By definition, the kernel function is a positive-definite function, used to define the prior knowledge of the underlying relationship. Basically, the kernel function is only a mandatory requirement when there is a lack of finite dimensional form of the feature space. Otherwise, it can be dropped by directly calculating the sample. However, this feature space dimension is frequently infinite, which means that the kernel function cannot be directly calculated. For this reason, the kernel function technique is often chosen to tackle the Gaussian process regression. In addition, the kernel function comprises some special parameters that specify its own shape. These parameters are referred to as hyper-parameters.

Generally, the Square-Exponential (SE) kernel, also known as the Radial Basis Function (RBF) kernel [9,10], is chosen as a basic kernel function. In reality, the SE kernel is favored in most Gaussian process applications, because it requires the calculation of only a few parameters. Moreover, there is a theoretical reason to choose this method, as it is a universal kernel appropriate for any continuous function when enough data is given. The formula for SE kernel is described as follows:

$$k_{SE}(x, x') = \sigma_f^2 (-\frac{(x - x')^2}{2l^2}) \tag{4}$$

in which, σ_f is the output-scale amplitude and l is the time-scale of the variable x from one moment to the next. l also stands for the bandwidth of the kernel and, thereby, the smoothness of the function in the model. In addition, l also plays the role of judgement for Automatic Relevance Detection (ARD) [11,12], to discard the irrelevant input dimension. In the next step, we evaluate the posterior distribution of the Gaussian process. Assuming that the incoming value of the input data is (x_*, y_*), the joint distribution of the training output is y, and the test output is y_*, as below.

$$p\binom{y}{y^*} = GP\begin{pmatrix} m(x) & K(x,x')K(x,x_*) \\ m(x_*)' & K(x_*,x)K(x_*,x_*) \end{pmatrix}$$ (5)

here, $K(x_*,x_*)=k(x_*,x_*)$, $K(x,x_*)$ is the column vector made from $k(x_1,x_*)$, $k(x_2,x_*)...,k(x_n,x_*)$. In addition, $K(x_*,x) = K(x,x_*)^T$ is the transposition of $K(x,x_*)$. Subsequently, the posterior distribution over y_* can be evaluated with the below mean m_* and covariance C_*.

$$m_* = m(x_*) + K(x_*,x)K(x,x')^{-1}(y - m(x))$$ (6)

$$C_* = K(x_*,x_*) - K(x_*,x)K(x,x')^{-1}K(x,x_*)$$ (7)

then

$$p(y_*) \sim GP(m_*, C_*)$$ (8)

The best estimation for y_* is the mean of this distribution:

$$\overline{y_*} = K(x_*,x)K(x,x')^{-1}y$$ (9)

In addition, the uncertainty of the estimation is captured in the variance of the distribution as follows:

$$var(y_*) = K(x_*,x_*) - K(x_*,x)K(x,x')^{-1}K(x,x_*)$$ (10)

3   Performance Evaluation

3.1   Experiments

For the performance evaluation, our experiments are aimed at investigating the performance of the proposed application in terms of energy efficiency and execution time. In the initial experiments, the workload is generated *via* the CPU intensive benchmark for one hour to determine the energy savings. In this test, in order to more easily control the number and the intensiveness of the workload, a benchmark software, namely *stress*-1.0.4, is used to simulate the incoming processes. To aggregate the results, the *powerstat* and the *sysstat* software are used to log the power consumption and workload statistics, respectively. All of the information of the benchmarking system is described in Table 1.

Table 1 System configuration.

	Configuration
Platform	64bit
CPU	Intel Core i7-3770, 3.40GHz, Quad core
Storage	800GB
Memory	16GB
OS	CentOS 6.5 kernel: 2.6.32-431.el6.x86_64
Benchmark	stress-1.0.4
Power stat	powerstat-0.01.30-1
System stat	sysstat-9.0.4-27.el6

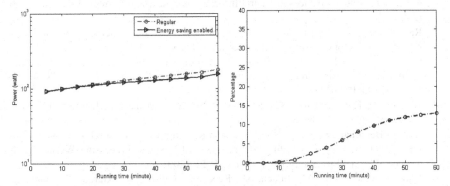

(a) Power consumption over one hour running time (lower is better). (b) Power saving over one hour running time.

Fig. 2 Performance evaluation of proposed method.

3.2 Results

As seen in the Fig.2a, both systems begin with the stand-by mode, which costs 91.49 watts to maintain. Both systems had simultaneous stress tests for duration of 60 minutes. Subsequently, the system with energy saving enabled ends the benchmark test with a consumption of 154.93 watts, in comparison with 177.96 watts for the regular system. Therefore, 23.03 watts are saved (which is equivalent to an energy savings of 12.94%) (Fig.2b). In processor architecture, an energy reduction of 12.94% is significant.

4 Conclusion

The proposed method proves the capability in improving the power consumption of the computing node. To do that, the strategy is to predict the utilization of CPU cores, migrate the target processes and stand-by the idle cores to save the energy.

For further development, as previously mentioned, some parts of the source code that are developed to test this method would be made available under the terms of the GNU general public license (GPL).

Acknowledgements This work was supported by Institute for Information & communications Technology Promotion(IITP) grant funded by the Korea government(MSIP) (No.R0101-15-237, Development of General-Purpose OS and Virtualization Technology to Reduce 30% of Energy for High-density Servers based on Low-power Processors).

References

1. Lawrence, N.D.: Gaussian process latent variable models for visualisation of high dimensional data. Advances in Neural Information Processing Systems **16**(3), 329–336 (2004)
2. Rasmussen, C.E.: Evaluation of Gaussian processes and other methods for non-linear regression (Doctoral dissertation, University of Toronto) (1996)
3. Chalupka, K., Williams, C.K., Murray, I.: A framework for evaluating approximation methods for Gaussian process regression. The Journal of Machine Learning Research **14**(1), 333–350 (2013)
4. Williams, C.K., Rasmussen, C.E.: Gaussian processes for machine learning. The MIT Press, vol. 2(3), p. 4 (2006)
5. Brahim-Belhouari, S., Vesin, J.M.: Bayesian learning using Gaussian process for time series prediction. In: Proceedings of the 11th IEEE Signal Processing Workshop on Statistical Signal Processing, pp. 433–436. IEEE (2001)
6. Roberts, S., Osborne, M., Ebden, M., Reece, S., Gibson, N., Aigrain, S.: Gaussian processes for time-series modelling. Philosophical Transactions of the Royal Society of London A: Mathematical, Physical and Engineering Sciences **371**(1984), 20110550 (2013)
7. Petelin, D., Kocijan, J.: Evolving Gaussian process models for predicting chaotic time-series. In: 2014 IEEE Conference on Evolving and Adaptive Intelligent Systems (EAIS), pp. 1–8. IEEE, June 2014
8. Shewchuk, J.R.: An introduction to the conjugate gradient method without the agonizing pain (1994)
9. Grande, R.C., Chowdhary, G., How, J.P.: Nonparametric adaptive control using Gaussian processes with online hyperparameter estimation. In: 2013 IEEE 52nd Annual Conference on Decision and Control (CDC), pp. 861–867. IEEE, December 2013
10. Banerjee, A., Dunson, D.B., Tokdar, S.T.: Efficient Gaussian process regression for large datasets. Biometrika, ass068 (2012)
11. Hensman, J., Fusi, N., Lawrence, N.D.: Gaussian processes for big data. arXiv preprint arXiv:1309.6835 (2013)

Design and Implementation of LMS for Sharing Learning Resources in e-Learning

Dae Hyun Lee and Yong Kim

Abstract In the information based society, e-learning is being magnified as an education environment where produced knowledge is disclosed and shared and new knowledge is produced through cooperation. For e-learning, interaction tools have been extensively conducted to increse learning effects. However, learning related interaction functions are insufficient in learning windows where actual learning occurs. In the present study, a content learning window (CS, web) where users can effectively share learning resources and opinions was designed and implemented. The presented LMS functions is considered to be capable of increasing learners' interactions in e-learning. The designed LMS will offer better learning environments through reinforcement of users' interactions, increasing target consciousness, and increasing communication.

Keywords e-Learning · LMS · Content · Learning window · Learning resource

1   Introduction

The development of information communication technology has been providing not only teaching-learning methods but also diverse learning opportunities to learners. e-Learning as a learning method is used globally with its advantages such as reduced educational costs and the fact that it enables everybody to receive education dynamically any time anywhere [1]. Learning management systems (hereinafter, LMS) are systems that manage learners, deliver content in Internet based learning, and manage overall learning activities to improve learners' capability. Therefore, since learning is done online through LMS, to satisfaction[2], convenience[3] and enhance learning effects[4], LMS should be sufficiently supported [5][6].

D.H. Lee · Y. Kim(✉)
Department of e-Learning, Graduate School, Korea National Open University,
Seoul, South Korea
e-mail: eogus0709@gmail.com, dragonknou@knou.ac.kr

© Springer Science+Business Media Singapore 2015   367
D.-S. Park et al. (eds.), *Advances in Computer Science and Ubiquitous Computing*,
Lecture Notes in Electrical Engineering 373,
DOI: 10.1007/978-981-10-0281-6_53

To improve learning effects and provide environments similar to off-line learning environments in e-learning, learning tools that will enable interactions in content, LMS, and 3rd Party Solutions have been studied. However, methods of interactions are different by area and tools to support interactions are insufficient in content learning windows where learning occurs. In those LMS learning windows that have been already developed, only those interaction functions provided by the content being operated were generally used and to support interaction functions between teachers and teachers, between teachers and learners, and between learners and learners, an interaction learning tool operated separately from the content learning window should be utilized or the user should move to a 3rd Party Solution to utilize functions. This cause inconvenience to e-learning users because to exchange information with each other, they should move to another LMS during learning.

Therefore, such LMS learning windows have problems such as resultant inconvenience in communication, the lack of concentration on learning, the lack of motivation for learning, learning effects lower than off-line learning, and learner's passive participation [7]. In addition, by enabling learners to participate in diverse kinds of communications, e-learning can increase the degree of learners' participation in education and training processes and improve learners' satisfaction thereby bringing about high learning outcomes [8]. In addition, studies indicating that when interactions are supported more through learning management systems, the effects of learning and satisfaction with learning can be higher have been presented [9][10] [11].

In this study, a content learning window where users can share opinions with learning materials while they undergo content learning was designed and implemented. Through the content learning window, the present study was intended to overcome the 'difficulties in interactions', provide 'motivations for learning', and solve the problem of the 'lack of presence' by providing functions that will enable users to share learning resources, learning data and learning opinions while they undergo learning through e-learning windows instead of the previous function that simply deliver the table of content and the contents.

2   Design and Implementation

Interactions that are implemented on the content learning window were determined as a learning progress sharing function, a file sharing function, and a learning opinion sharing function and these functions were designed as follows.

First, the improve the low learning efficiency due to the impossibility of user's (teacher, learner) interactions during learning, a function was designed that enables users to freely share data such as documents, texts, images, videos and URL.

Second, to support UIs that enable easy sharing of learning resources on the content learning window, learning windows were designed in two types; CS and web. The CS version supports UIs through which learner can easily share learning resources through Drag & Drop and the web version supports functions for sharing learning resources through file searching functions.

Third, among the learning resources shared on the content learning window, texts are provided through UIs in the form of chatting and a function to synchronize text input time and content reproducing time was designed and implemented. Through the foregoing, a function to share learning opinions in non-real time while learning contents was implemented. It was intended to provide interaction based learning environments to users in order to support learning environments with high learning engagement, motivation, and presence. A block diagram of the content learning window service is as shown in Fig.1.

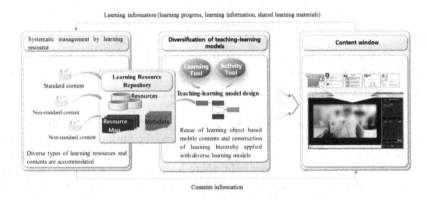

Fig. 1 Block diagram of the content learning window service.

The learning windows provide a CS version and a web version with the same UI (user interface). Although the two versions provide the same UI/UX, they have functional differences. The CS version automatically changes the IE (internet explorer) version of content learning window used on users' PC into the IE version optimized to content development environments. Through this, this version supports users to undergo learning without setting web compatibility review resulting from differences in IE versions. In addition, this version also provides users with convenience such as Drag & Drop. However, this version involves inconvenience because the learning window should be installed before learning through a separate installation file and this causes problems in users' accessibility. Therefore, the web version that does not require separate installation was provided.

As functions for sharing of learning materials on learning windows, an average learning progress checking function, an attached file sharing function, an opinion sharing function, and a content evaluation function were provided and detailed descriptions of the functions are as shown in Table 1 on the content learning window.

Table 1 Learning window function statement

Division	Content
Table of learning contents	Shows the table of learning contents composed on the LMS or LCMS
Attached file	File upload function for teachers or learners to memorize or share learning related data (ex: Text, documents, image, Zip File, URL, etc.)
All opinions	A function to store learners' opinions at certain points of the content(ex: page 3 of the WBT, 2 min. and 30 sec. of image content)
My opinion	A function to inquire into only the opinion prepared by the learner among registered opinions
Good	A function to evaluate functions helpful for learning as "good"
See more comments	A function to show all opinions regardless of time points
X times speed replay	A function to replay video content at x times speeds(0.2x, 0.5x, 1.0x, 1.5x, 2.0x)

To provide services optimized for content development environments, the Microsoft .Net Framework based CS technology and the Spring Framework's web technology were applied to the learning window. The central field of the learning window was designed as a content field where learning is to be done and the right field was constructed so that learning resources can be checked and interactions can be made. The right uppermost field has a function to check my progress rate and the average progress rate and on the right top field, a table of learning contents function and a learning attached file sharing function that can be selected using tab buttons were provided. On the right bottom field, a function to see my learning opinion and a function to see all learning opinions were organized and on the right lowermost field, a function to evaluate content and a function to see at x times of the speed were organized. These functions were provided with switches so that they can be activated or inactivated by users when necessary.

With this configuration, learners can identify their learning information while learning contents without moving to another screen, share their learning resources and opinions with other learners, and see data and opinions shared with other learners. Major screens of the learning window are as shown in Fig. 2.

Fig. 2 A screen where the learning window was implemented

The learning window can be seen on web browsers (IE, Chrome, Safari, Firefox). On the web learning window, files can be shared through file searches and UIs similar to the Windows explorer.

The right top of the learning window was constructed to enable identifying my average progress rates and the mean rates and the right bottom was constructed to have functions to register, check, and download the table of learning contents and attached files. The table of learning contents is a function to inquire into the table of learning contents of the learning contents reproduced in the central field and the attached file function was configured to enable users to upload data, check uploaded files, and download uploaded files.

Files that can be uploaded are documents, images, videos, compressed files, and URLs. The file checking function enables users to check data uploaded by them in real time while they are learning and check files uploaded by other users when restarting content reproduction after terminating learning. The file download function enables users to download files necessary for them among the attached file documents of the learning window and check data through the viewer installed in users' PC.

Users can enter their opinions on the content learning window through comment entries. As major functions, see all opinions, see my opinions, see all comments, and see comments at all times functions were configured. In addition, at the side of the see opinions function, a button to evaluate the content was organized.

3   Suitability Review

To verify the effectiveness of the design and implementation of functions for sharing learning materials on the content learning window, the suitability was reviewed with 15 experts in e-learning. The suitability was reviewed by examining satisfaction with items improved compared to the content learning window currently in use or experienced.

Improved items for which opinions were established were increased concentration on learning, learning efficiency increase, presence increase, interaction increase, UI/UX convenience increase, sense of belonging and bond of sympathy increase, learning achievement and satisfaction increase, and learning environment improvement. According to the results of suitability review conducted with experts, the gross average was 79 points indicating that the developed system is highly suitable. To review by detailed item, the score of "learning environment (UI/UX) improvement (84points)" was shown to be the highest followed by "learner's learning efficiency increase(81 points)", "inter-learner interaction increase (81points)". On the other hand the scores of "increase in interactions between learners and teachers (76 points)" and "increase in learners' sense of belonging /bond of sympathy (76 points)" were shown to be the lowest indicating that these should be considered as matters to be improved for the system.

The e-learning expert group mentioned that in the content learning window, the learning resources sharing function improved users' learning convenience UI/UX, increased interactions and improved the limitations of e-learning. However, the expert group mentioned that when the content learning window is operated in actual e-learning processes, technical stability and the efficiency of operation are important. In addition, the expert group also mentioned that if real time chatting and SNS functions are grafted; interaction functions can be improved a little more.

4   Conclusion and Suggestions

The present study was conducted to define interactions in past e-learning environments and improve 'the difficulties in interactions', 'the lack of learner presence' and 'the lack of learning motivation' in e-learning through utilization of learning resources sharing functions on the content learning window. In addition, this study was also conducted to construct learning environments that can motivate learners for learning and enhance accessibility to interactions through interactions between learners.

Existing learning windows where the largest part of e-learning classes is provided supported only the role of a medium that delivers only the table of contents and the contents. The present study was intended to improve the following through studies of interaction functions on the content learning window.

First, improve difficulties in interactions by designing and implementing UI/UX where learning information, learning resources and learning opinions can be exchanged. Second, provide environments that can motivate learners for learning through functions for sharing learning data, learning resources and learning opinions. Third, provide social presence to learners through functions for sharing learning data, learning resources while learners are undergoing learning and functions for non-real time learning opinion sharing thereby also contributing to relieving the sense of isolation which is a shortcoming of Internet classes.

The results of this study are considered to be utilized as basic data for development of learning management systems for learners' effective learning in e-learning.

References

1. Michaels, J.W., Smillie, D.: Web education some smart investors are betting big bucks that Peter Drucker is right about the brilliant future of online adult education. Forbes MAGAZINE **165**, 92–94 (2000)
2. Kim, J., Park, K., Kim, Y.: Diversity and Satisfaction: Analysis of Learners' Satisfaction According to the Online Learning Interaction. International Journal of Applied Engineering Research **9**(21), 9157–9166 (2014)
3. Oblinger, D., Hawkins, B.L.: The myth about e-Learning: We Don't Need to Worry about e-Learning Anymore. EDUCAUSE Review **40**(4), 14–15 (2005)

4. Rosenberg, M.J.: e-Learning: Strategies for delivering knowledge in the digital age. McGraw-Hill (2000)
5. Becta: Progress and impact of technology in education. Harnessing Technology Review, Becta (2007)
6. Khan, B.H.: Web-based instruction(WBI): What is it and why is it? In: Khan, B.H. (ed.) Web-Based Instruction, pp. 5–18. Educational Technology, Englewood Cliffs (1997)
7. Young, C.Y.: Social Learning Platform Design on Facebook. The Graduate School of Educatio, Ewha Womans University, Major in Computer Science Education (2010)
8. Jin, G.: Development a Monitoring Instrument for Firm's e-Learning, The Graduate School, Ewha Womans University, Dept. of Business Administration (2001)
9. Lee, D.H., Shon, J.G., Kim, Y.: Design and implementation of OSMD based learning management system for mobile learning. In: ICCT 2004 The 4th International Conference on Convergence Technology, vol. 4(1), pp. 154–160 (2014)
10. Eom, W., An, B.-G.: Analysis of fuctions for interactions in learning management system for cyber home learning system. The Journal of Educational Information and Media 15(2), 47–66 (2009)
11. Son, K.A., Woo, Y.: The development strategy of LMS to improve teaching and learning activities in distance education of Korea. Journal of Lifelong Leaning Society 6(2), 127–149 (2010)

A Zero-Watermarking Scheme Based on Spread Spectrum and Holography

De Li, DaYou Jiang and Jong Weon Kim

Abstract This paper proposes a zero-watermarking scheme based on spread spectrum and holography. The method makes no changes to cover images while embedding the owner information of images so as to achieve high transparency. Discrete Wavelet transform, matrix norm and holography are used to generate the feature image in order. While Arnold transformation and spread spectrum technique are employed to disturb and spread the watermark. Finally Visual Cryptography scheme is applied to generate secret image by using a codebook. The experimental results show that the algorithm is effective and robust against most conventional attacks.

Keywords Zero watermark · Spread spectrum · Log-polar map · Matrix norm · Holography · Visual cryptography

1 Introduction

Digital watermarking technique is an effective method to solve the copyright protection problems of digital media. Zero-watermark is a new digital watermarking technology without modifying the data of original cover image. Q. Wen first proposed a zero watermark method by using high-order cumulates [1]. Z. Wang proposed a method based on Zernike moments, the rectangular revolves invariability of Zernike moments was used to withdraw picture characteristic point, and quantified it for the binary sequence as the zero watermarking [2]. D.F. Yu proposed a method (YDF method) based on Fourier-Mellin transformation

D. Li · D. Jiang
Department of Computer Science, Yanbian University, Yanji 133002, China
e-mail: leader1223@ybu.edu.cn, ybdxgxy13529@163.com

J.W. Kim(✉)
Department of Contents and Copyright, Sangmyung University, Seoul 110743, Korea
e-mail: jwkim@smu.ac.kr

© Springer Science+Business Media Singapore 2015   375
D.-S. Park et al. (eds.), *Advances in Computer Science and Ubiquitous Computing*,
Lecture Notes in Electrical Engineering 373,
DOI: 10.1007/978-981-10-0281-6_54

(FMT). In the algorithm, the geometrical-invariant domain is gotten by FMT to the image, and the zero-watermark is constructed by image feature [3]. H. H. Tsai proposed a scheme with geometrical invariants using support vector machine and particle swarm optimization against geometrical attacks. The method is superior to other existing methods under considering the secure channel to deliver the watermark to receivers [4]. P.L. Liu proposed a method (LPL) based on no-subsample contour-let transform (NSCT) and QR-decomposition. The algorithm can effectively resist geometric attacks with high correction accuracy [5]. M. S. Dhoka proposed a method using Hessian Laplace Detector and Logistic map [6]. S. Lin proposed a method based on generated Arnold Transform, spread spectrum and despreading (SSD) techniques [7]. W. J. Zhou proposed a zero-watermarking based on objective reduced-reference stereoscopic image quality assessment (RR-SIQA) method, two kinds of zero-watermarks are constructed according to the characteristics of image structure and stereoscopic perception [8].

In the paper, we adopt DWT and matrix norm to generate the character of the cover image, employ Log-polar transform and Holography to resist rotation and crop attack. To ensure the security, Arnold transform and visual cryptography are also used. The spread spectrum applied to watermark can improve the robustness for adding noise and filtering attacks. Experiments results are shown that our scheme has good robust and high security.

2 Related Work

2.1 Spread Spectrum

Suppose the sequence m of digital binary image of N × M. Then we employ the quadrature amplitude modulation to the sequence to generate a binary polarity sequence of $\{-1, 1\}$.

$$m = \{b_k \mid b_k \in \{-1,1\}\}k = 1, 2, \cdots N \times M \tag{1}$$

Suppose the pseudo-random noise pattern is defined as follows:

$$P = \{P_{i,j}(k) \mid k=1,2,\cdots,N \times M\} \tag{2}$$

Where $P_{i,j}(k)$ is 2-D pseudo-random binary sequence of $\{-1,1\}$ with zero mean generated using the key as the seed. Then the spread spectrum sequence will be

$$w = P \cdot m^{'} = \sum_{k=1}^{N \times M} P_{i,j}(k) b_k^{'} \tag{3}$$

In this paper, we adopted Walsh-Hadamard transformation to generate $P_{i,j}(k)$ for spread spectrum coding.

2.2   Holography

Holography is usually used to generate hologram and embedded the hologram into the cover image as watermark, for improving the performance of a conventional watermarking system against some attacks such as noise, cropping, and so on. The lens-less Fourier transform hologram is employed in this paper. Suppose that the object point source and the reference point source are on the same plane, namely $x_0 y_0$ plane. The CCD plane is named xy plane. The recording distance between two planes is d. We suppose the complex amplitude of an object is $O_0(x_0, y_0)$ and the reference point source is $(-b, 0)$.

According to the Fresnel diffraction theory and ignoring the constant factor, the complex amplitude of the object wave in the CCD plane is defined below in Equation (4).

$$O(x, y) = \iint_{\infty} O_0(x, y) \exp\{\frac{jk}{2d}[(x - x_0)^2 + (y - y_0)^2]\} dx_0 dy_0, \tag{4}$$

The complex amplitude of the reference wave in the CCD plane is defined below in Equation (5).

$$R(x, y) = \exp[\frac{jk}{2d}(x^2 + y^2)] \exp[-\frac{jk}{d}(xx_r + yy_r)], \tag{5}$$

The intensity distribution of the interference pattern in the hologram plane is

$$H(x, y) = |O + R| = |O|^2 + |R|^2 + O^*R + OR^*, \tag{6}$$

Where * denotes the complex conjugate.

2.3   Visual Cryptography

It is a secret sharing scheme that allows a secret to be shared among a set of participants. Considering the convenience and security of the scheme, in this paper, we adopted a (2, 2) visual cryptography method. A secret image is just divided into two shares and each secret pixel of an image is replaced by 1×2 pixels. In addition, the secret image can be recovered by stacking the two sharing images. To generate a secret image, the critical step is to make a codebook of VC technique. The codebook is shown in Figure 1. Combining the feature image and watermark image, the secret image can be generated by them according to the codebook.

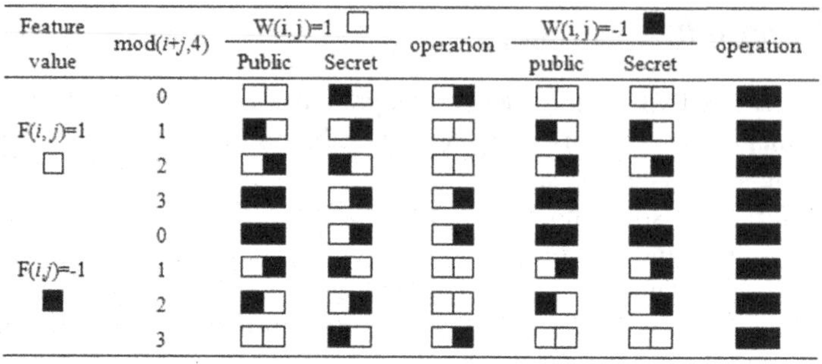

Feature value	mod(i+j,4)	W(i, j)=1 ☐		operation	W(i, j)=-1 ■		operation
		Public	Secret		public	Secret	
F(i, j)=1 ☐	0						
	1						
	2						
	3						
F(i,j)=-1 ■	0						
	1						
	2						
	3						

Fig. 1 Codebook used in the visual cryptography scheme.

3   Zero Watermark Scheme

3.1   Zero Watermark Embedding

The main steps of the embedding procedure are described as follows:

Step1: Watermark image preprocessing

(1) Select a binary image W of 64 × 64 as a watermark,
(2) Apply Arnold transformation with key1 to disturb image W, generate image A_W,
(3) Use Spread spectrum to spread A_W across a wide-band channel, generate image S_W.

Step2: Cover image preprocessing

(1) Select a cover gray image I of 512 × 512,
(2) Use Log-polar map to the cover image, generate image LP_I,
(3) Decompose the image LP_I into seven sub-bands by 2-level DWT and extract the sub-band LL of the image, generate image LL_I,
(4) Divide the image LL_I into 4 × 4 blocks, compute the Frobenius norm of each block matrix and the mean value of them, if it's value is bigger than the mean value, we define the value of that block as 1, otherwise, define that block as 0. So a 32 × 32 binary image MN_I is generated.
(5) Generate a hologram image of 128 × 128 by using hologram to the image MN_I, then convert it into another binary digits (-1 or 1) and take it as the feature image F_I.

Step3: Employ the VC scheme to generate the secret image from image F_I and image S_W according to the codebook, and then register the secrete image S_I to certified authority (CA).

3.2   Zero Watermark Extraction

The watermark extraction process is similar to the embedding process. The main steps of the extraction procedure are described as follows:

Step1: Suspected image processing

(1) Use Log-polar map to the suspected image I', generate image LP_I'.
(2) Decompose the image LP_ I' into seven sub-bands by 2-level DWT and extract the sub-band LL of the image, generate image LL_I',
(3) Divide image LL_I' into 4 × 4 blocks, compute the Frobenius norm of each block matrix and the mean value of them, if it's value is bigger than the mean value, we define the value of that block as 1, otherwise, define the block as 0. So a 32 × 32 binary image MN_I' is generated.
(4) Generate a hologram image of 128 × 128 by using hologram to the image MN_I', then convert it into another binary digits (-1 or 1) and take it as the feature image F_I'.
(5) Generate a public image P_I' by feature image F_I' according to the codebook.

Step2: Obtain the image S_W' from secret image S_I and image P_I'.

Step3: Do the inverse process of spread spectrum and Arnold transformation with key1 in order to get the extracted watermark W'.

4   Experimental Results and Analysis

In experiments, the binary image of 32 × 32 is taken as the owner information. The watermark and cover gray images of 512 × 512 are used as shown in Table 1. The attacks are listed in Table 2.

Table 1 Cover images and watermark image

| Watermark | Lena | Airplane | Peppers | Baboon |

Table 2 A summary for the attacks

Types	Contents
JPEG Compression(JP)	Quality = 80 & 2
Salt & Pepper Noise(SP)	Noise density = 0.01
Multiplicative Noise(MN)	Factor = 0.01
Median Filter (MF)	Size 3 × 3 & 5 × 5
Rotation (RC)	Angle = 30° & 45°
Scaling (S)	Scaling factor = 0.5
Gaussian Filter (GF)	Size 5 × 5 & 8 × 8

380 D. Li et al.

In this paper, the bit correction rate (BCR) measures the similarity between an original watermark W and the extracted watermark W'. The BCR is defined by

$$BCR(W, W') = 1 - \frac{\sum_{k=1}^{m} |W_k - W'_k|}{m} \qquad (7)$$

A series of experiments are conducted, including JPEG compression attack, add Salt & Pepper noise, add multiplicative noise, rotation, image scaling, image filtering. The experimental results after performing multiple attacks on the images of Lena, Airplane, Baboon and Peppers are shown in Figure 2. Figure 3 shows the comparison results for the LPL method [5] and our method.

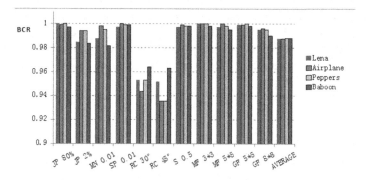

Fig. 2 The BCR values corresponding to various attacks

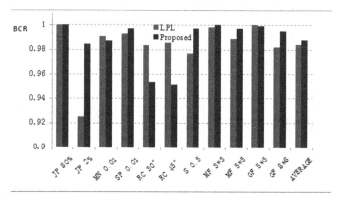

Fig. 3 Comparison of the BCR values for LPL method

From the comparison results in Figure 3, we can see that our proposed algorithm has a better robustness in generally. The BCR values of our method are lower than LPL method under the rotation attacks, because of LPL method used NTSC (non-subsampled contourlet transform) to extract stable SIFT feature points, which is more invariant to rotation attacks than log-polar map transform. The figure shows that our method is robustness for adding noise and filtering attacks due to using the spread spectrum technique.

5   Conclusions

In this paper, a zero-watermarking scheme is proposed which makes good use of the features of discrete wavelet transform, matrix norm, spread spectrum, log-polar and visual cryptography. The robustness of our scheme is validated by common signal processing such as add noise, filtering, JPEG Compression and geometric attacks, including rotation and scaling. Although the proposed scheme still has limitations such as a need for a trusted third party to store the secret image, it is blind and also has high imperceptibility and robustness.

Acknowledgments This research is supported by Ministry of Culture, Sports and Tourism(MCST) and Korea Creative Content Agency(KOCCA) in the Culture Technology(CT) Research & Development Program 2015.

References

1. Wen, Q.: Concept and Application of Zero Watermark. Acta Electronica Snica **31**(2), 214–216 (2003)
2. Wang, Z., Sun, Y.: Zero watermarking algorithm based on Zernike moments. Computer Applications **28**(9), 2233–2235 (2013)
3. Yu, D.F., Li, X.B.: Zero-watermark algorithm based on Fourier-Millin transforming. Computer Engineering and Design **30**(10), 2566–2569 (2009)
4. Tsai, H.H.: A zero-watermark scheme with geometrical invariants using SVM and PSO against geometrical attacks for image protection. Journal of Systems and Software **86**, 335–348 (2013)
5. Liu, P.L.: Zero-watermarking algorithm based on NSTC and QR-decomposition. Electronic Design Engineering **22**(16), 163–170 (2014)
6. Dhoka M.S.: Robust and dynamic image zero watermarking using Hessian Laplace detector and logistic map. In: IEEE International Advance Computing Conference (IACC), Banglore, pp. 930–935 (2015)
7. Sun, L.: A novel Generalized Arnold Transform-based Zero-Watermarking Scheme. Applied Mathematics & Information Sciences **4**, 2023–2035 (2015)
8. Zhou, W.J.: Reduced-reference stereoscopic image quality assessment based on view and disparity zero-watermarks. Signal Processing: Image Communication **29**, 167–176 (2014)

A Novel Seamless Redundancy Protocol for Ethernet

Nguyen Xuan Tien and Jong Myung Rhee

Abstract High availability is crucial for industrial Ethernet networks and Ethernet-based control systems, such as automation networks and substation automation systems. As standard Ethernet does not support fault tolerance capability, the high availability of Ethernet networks can be increased by using redundancy protocols. In this paper, we present a novel seamless redundancy protocol for Ethernet networks called the Redundancy Protocol for Ethernet (RPE). Our proposed RPE not only provides seamless communications with zero switchover time in case of failure, but it also supports any topologies. The RPE is transparent and compatible with standard Ethernet nodes. These features of the RPE make it very useful for time-critical and mission-critical systems, such as substation automation systems, automation networks, and other industrial Ethernet networks.

Keywords Ethernet fault tolerance · Seamless redundancy · Industrial ethernet

1 Introduction

Ethernet has become a de facto standard in today's networks. From private home networks to office networks to automation networks, Ethernet is used on a large-scale basis. Automation networks require a high availability to ensure continuous plant operation. As the failure of the Ethernet automation network, e.g., in a production plant, can result in a halt in production and possibly a significant loss of money; thus, increasing the fault tolerance of Ethernet networks is of high importance.

As Ethernet, standardized by the Institute of Electrical and Electronics Engineers (IEEE) in IEEE 802.3 [1], does not support the fault tolerance capability, various

N.X. Tien · J.M. Rhee(✉)
Information and Communications Engineering Department,
Myongji University, Seoul, Republic of Korea
e-mail: nxtien@gmail.com, jmr77@mju.ac.kr

© Springer Science+Business Media Singapore 2015
D.-S. Park et al. (eds.), *Advances in Computer Science and Ubiquitous Computing*,
Lecture Notes in Electrical Engineering 373,
DOI: 10.1007/978-981-10-0281-6_55

383

redundancy protocols for Ethernet have been developed and standardized, such as the rapid spanning tree protocol (RSTP) [2], media redundancy protocol (MRP) [3], parallel redundancy protocol (PRP) [4], high-availability seamless redundancy (HSR) [4], and others. RSTP and MRP provide redundancy in the network, whereas PRP and HSR provide redundancy in the end nodes. The RSTP and MRP have a drawback on switchover delay. The PRP and HSR provide zero recovery time, but PRP requires a duplicated network infrastructure, whereas the HSR is mainly used in ring topologies.

In this paper, we propose a novel seamless redundancy protocol for Ethernet networks called the Redundancy Protocol for Ethernet (RPE). The RPE provides seamless redundancy for Ethernet networks, is compatible with standard Ethernet nodes, and can support any topology.

2   The Proposed RPE

2.1   RPE Concepts

The RPE is a seamless redundancy protocol for Ethernet networks that provides zero switchover and recovery time in case of network failure. The RPE applies redundancy at layer 2 of the open system interconnection (OSI) reference model. The RPE operates independently of upper layer protocols.

Like the RSTP, the RPE provides redundant links and switches in networks. However, unlike the RSTP, which blocks redundant links between the switches to avoid switching loops in networks, the RPE allows redundant links to be active and used to forward frames. This creates network loops that can cause serious problems, such as a broadcast storm and traffic circulation. To solve problems caused by network loops, RPE uses a duplicate discarding scheme. The RPE switches forward frames that have been received for the first time and discards duplicates of the frames. When a RPE switch receives a frame, it checks whether the frame has previously been received and forwarded. If so, it discards the frame. To identify frame duplicates, the RPE switches insert a RPE tag containing a sequence number into every Ethernet frame received so that a RPE-tagged frame can be uniquely identified by a combination of its source MAC address and the RPE tag's sequence number. By forwarding or flooding a frame for the first time and discarding duplicates of the frame, the RPE provides zero switchover time in case of network failure. In other words, the RPE provides seamless redundancy for Ethernet networks.

The RPE is implemented in RPE switches (SW) to provide seamless redundancy for Ethernet networks. The RPE switches perform the following functions.

- Inserting the RPE tag into Ethernet frames received from Ethernet nodes.
- Forwarding RPE-tagged frames based on the destination MAC address.
- Removing the RPE tag from RPE-tagged frames before sending them to Ethernet nodes.

Fig. 1a shows redundancy in the network under the RSTP and Fig. 1b shows seamless redundancy provided by the RPE. In Fig. 1b, when RPE SW1 receives a frame sent by the source ETH node, it inserts a RPE tag into the frame and forwards the RPE-tagged frame to other RPE switches. RPE SW3 receives some copies of the same RPE frame from links connected to it and it then removes the RPE tag from the first received the RPE frame, sends it to the destination ETH node, and discards duplicates of the RPE frame. If a network failure occurred in the network, such as a link failure between the two switches RPE SW2 and RPE SW3, only one frame is lost and the other frame still reaches RPE SW3. Therefore, there is no communication interruption in the network in case of a network failure.

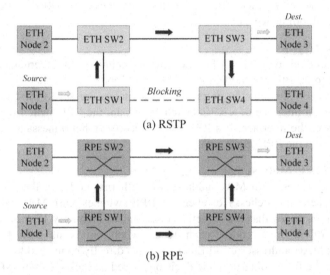

Fig. 1 Redundancy in the network provided by the RSTP and RPE.

2.2 RPE Frame Structure

The RPE operates based on forwarding/flooding RPE-tagged frames received for the first time and discarding duplicates of the frame in RPE networks. The RPE frame is an Ethernet frame that is inserted into a 4-octet RPE tag. The RPE tag consists of a RPE EtherType field and a sequence number field, as shown in Fig. 2.

A RPE frame is uniquely identified by a combination of its source MAC address and the RPE tag's sequence number. Each time an RPE switch inserts a RPE tag into an Ethernet frame sent by an ETH node, it increases the sequence number associated with the source MAC address. The RPE switches can detect duplicates of a RPE frame based on the combination of the source MAC address and sequence number.

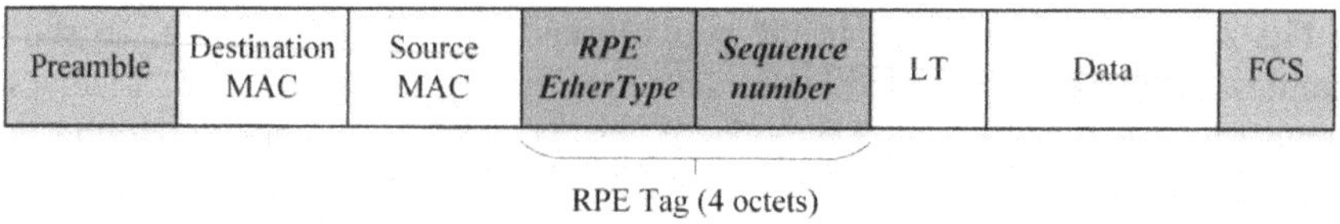

Preamble	Destination MAC	Source MAC	*RPE EtherType*	*Sequence number*	LT	Data	FCS

RPE Tag (4 octets)

Fig. 2 RPE Frame Format.

2.3 RPE Operations

To provide seamless redundancy, the RPE switches first configure their port types and then build a MAC table that contains MAC addresses of Ethernet nodes directly connected to them. Based on the MAC table, the RPE switches make forwarding/flooding decisions.

Setting Port Type

The RPE defines two types of switch ports, including an Ethernet port that is connected to an Ethernet node and a RPE-trunk port that is connected to another RPE switch. By default, the ports of RPE switches are set to the Ethernet port type. When an interface of a RPE switch comes up, the RPE switch sends a hello message over the interface. If a RPE switch receives a hello message on a port, it sets the port to the RPE-trunk type.

Learning MAC Address

The RPE switches learn MAC addresses of Ethernet nodes in the same way as standard Ethernet switches. However, the RPE switches learn MAC addresses of only Ethernet nodes that are directly connected to them. By learning the MAC addresses of Ethernet nodes, each RPE switch builds its own MAC table containing MAC addresses of all Ethernet nodes directly connected to it. The RPE switches make forwarding/flooding decisions based on looking up the MAC table.

Forwarding Frames

Multicast and Broadcast Frames. When a RPE switch receives a multicast or broadcast frame for the first time, it floods the frame to all its ports, except the port on which the frame was received. If the RPE switch receives any duplicates of the frame, it then discards the duplicates.

Unicast Frames. RPE switches make forwarding/flooding decisions based on looking up the destination MAC address in the MAC table. When a RPE switch receives an Ethernet frame, it looks up its MAC table to find an entry that matches the destination MAC address of the frame. If it finds a matched entry, it then forwards the Ethernet frame unchanged to the output port of the matched entry. Otherwise, it inserts a RPE tag to the frame and then floods the RPE-tagged frame over all its active ports, except the port on which the frame was received. When a RPE switch has received a RPE frame for the first time, it looks up its MAC table to find an entry that matches the destination MAC address of the frame. If it finds a matched entry, it removes the RPE tag from the RPE frame and then forwards the corresponding Ethernet frame to the output port of the matched entry. Otherwise, it floods the frame over all its active ports, except the port on which the frame was received. The process of forwarding unicast frames of RPE switches is shown in Fig. 3.

Discarding Duplicates

The RPE provides seamless redundancy in the network for Ethernet networks. Unlike the RSTP, which blocks redundant links to avoid switching loops in networks, redundant links in RPE networks are active and can be used to forward frames. This creates network loops that can cause serious problems, such as a broadcast storm and traffic circulation. To solve problems caused by network loops, the RPE uses a duplicate discarding scheme. The principle of the duplicate discarding rule is that RPE switches discard a frame that they have already received and forwarded. When a RPE switch receives a RPE-tagged frame, it checks whether the frame has previously been received and forwarded. If so, it discards the frame.

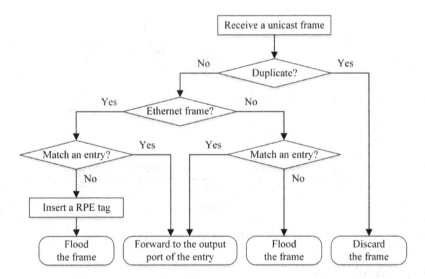

Fig. 3 The process of forwarding unicast frames.

3   Redundancy Performance

In this Section, we analyze the redundancy performance of the RPE compared to that of the RSTP. We consider two similar Ethernet and RPE networks, as shown in Fig. 1a and Fig. 1b, respectively.

In a failure-free case, the RSTP blocks the redundant link between Ethernet SW1 and SW4 to avoid the switching loop. Source ETH node 1 sends unicast frames to destination ETH node 3 through a path of SW1–SW2–SW3. When the link between SW2 and SW3 fails, the RSTP activates corresponding blocked ports in SW1 and SW4 to enable the blocked link between these switches, as shown in Fig. 4a. This process takes from a few hundred milliseconds to a few seconds. In other words, the RSTP suffers a switchover delay in case of network failure.

In the RPE network, as shown in Fig. 1b, when receiving unicast frames sent by source ETH node 1 to destination ETH node 3, RPE SW1 inserts the RPE tag into the frames and floods them into the RPE network. RPE SW2 and RPE SW4 also flood the frames when they receive them for the first time. RPE SW3 receives two copies of each frame from two links connected to it, removes the RPE tag from the first received copy, then forwards the Ethernet frame to the destination ETH node 3, and discards the duplicate. When the link between RPE SW2 and RPE SW3 fails, only one copy of each frame is lost and the other copy still reaches the destination node without switchover delay, as shown in Fig. 4b. Therefore, the RPE provides zero switchover time in case of a network failure.

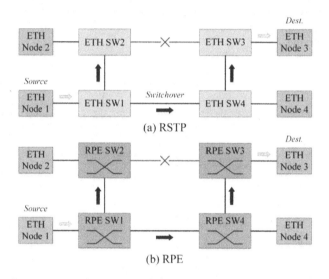

Fig. 4 Redundancy provided by the RSTP and RPE in a failure case.

4 Simulations

To evaluate our proposed RPE protocol, we conducted a simulation using the OMNeT++ simulation tool [5]. We consider the RPE network with four RPE switches (RPE SWs), as shown in Fig. 1b. Source ETH node 1 connecting to RPE SW1 sends unicast frames to destination ETH node 3 connecting to RPE SW3. There is a link failure that occurs during the simulation's data communications, as shown in Fig. 4b. Frame statistics were recorded to evaluate the RPE protocol.

Table 1 shows network frame statistics of the simulation. The simulation results demonstrate that the RPE provides seamless redundancy for standard Ethernet nodes. Unlike the RSTP, which suffers a switchover delay, the RPE provides zero recovery time in the case of a network failure.

Table 1 Network frame statistics of the simulation.

Number of frames sent by source	Number of frames received and processed by destination	Number of frames lost
10	10	0
20	20	0
30	30	0
40	40	0
50	50	0

Table 2 shows a comparison of the RPE and existing redundancy protocols including RSTP, MRP, PRP, and HSR.

Table 2 Comparison of redundancy protocols.

Protocol	Standard	Topology	Seamless Redundancy
RSTP	IEEE 802.1D-2004	Any	No
MPR	IEC 62439-2	Ring	No
PRP	IEC 62439-3	Dual networks	Yes
HSR	IEC 62439-3	Ring	Yes
RPE		Any	Yes

5 Conclusion

In this paper, we proposed a novel seamless redundancy protocol for Ethernet networks called the RPE. Unlike the RSTP, which suffers a switchover delay, the RPE protocol provides seamless redundancy with zero recovery time in case of a network failure. The RPE also supports any network topologies, such as ring and mesh. Additionally, the RPE is transparent and compatible with standard Ethernet nodes. These features make the RPE very suited for Ethernet-based applications that demand high availability and zero recovery time, such as automation networks, substation automation systems, and other industrial Ethernet networks.

Our future work will implement the RPE in hardware devices.

Acknowledgments This work was supported by the 2015 MPEES Advanced Research Center Fund of Myongji University.

References

1. IEEE 802.3 – IEEE Standards for Ethernet. http://www.ieee802.org/3/
2. IEEE 802.1D-2004 – IEEE Standard for Local and Metropolitan Area Networks: Media Access Control (MAC) Bridges. http://www.ieee802.org/1/pages/802.1D-2003.html
3. IEC 62439-2: Industrial Communications Networks: High-Availability Automation Networks, Part 2: Media Redundancy Protocol (MRP), Geneva, Switzerland (2010)
4. IEC 62439-3: Industrial Communications Networks: High-Availability Automation Networks, Part 3: Parallel Redundancy Protocol (PRP) and High-Availability Seamless Redundancy (HSR), Geneva, Switzerland, (2012)
5. OMNeT++ Version 4.6 Simulator. http://www.omnetpp.org/

Mutual Authentication Scheme Based on GSM for NFC Mobile Payment Environments

Sung-Wook Park and Im-Yeong Lee

Abstract Recently, smart devices for various services have been developed using converged telecommunications, and the markets for near field communication (NFC) mobile services is expected to grow rapidly. In particular, the realization of mobile NFC payment services is expected to go commercial, and it is widely attracting attention both on a domestic and global level. However, this realization would increase privacy infringement, as personal information is extensively used in the NFC technology. One example of such privacy infringement would be the case of the Google wallet service. In this paper, we propose mutual authentication scheme based on GSM for protecting user information in NFC mobile payment systems.

Keywords NFC mobile payment · GSM · Authentication · Key agreement

1 Introduction

Near Field Communication (NFC) is a short-range wireless communication standard defined in the ISO/IEC 18092 standard. NFC operates at 13.56 MHz and it can be used for communication between two active devices, or an active device and a passive device. Active devices are powered by a battery, whereas passive devices obtain their energy from the electromagnetic field of the active device.

S.-W. Park(✉) · I.-Y. Lee
Department of Computer Software Engineering,
Soonchunhyang University, Asan-si 336-745, Republic of Korea
e-mail: {swpark,imylee}@sch.ac.kr

This research was supported by the MSIP(Ministry of Science, ICT and Future Planning), Korea, under the ITRC(Information Technology Research Center) support program (IITP-2015-R0992-15-1006) supervised by the IITP(Institute for Information & communications Technology Promotion)

© Springer Science+Business Media Singapore 2015
D.-S. Park et al. (eds.), *Advances in Computer Science and Ubiquitous Computing*,
Lecture Notes in Electrical Engineering 373,
DOI: 10.1007/978-981-10-0281-6_56

Communication always occurs between an NFC initiator and an NFC target. An active device can have both roles, whereas a passive device is always an NFC target. The initiator sends requests to a target and the target answers these requests. NFC operates in a manner that is highly intuitive for the user. Two NFC devices start communicating after bringing them close together (known as "touching" each other). The NFC(Near Field Communication) application service fields that are currently being discussed now can be divided largely into mobile payment service and application services. First, anyone can issue a mobile card and download information onto a card reader that has the required function. Payment is possible just by tagging the card reader during the payment period. According to documents released by KISA, these problems are attributable to the following: partially unsupported privacy encryption, spill concerns affecting corporate internal privacy, excessive privacy requirements, and collection and storage. A privacy disclosure case related to the Google Wallet service can be used as evidence to support these issues. However, there are growing issues with privacy infringement due to increasing personal information exchange via NFC technology. This paper is organized as follows.

2   Related Work

NFC is a short-range wireless communication standard defined in the ISO/IEC 18092 standard. NFC operates at 13.56 MHz and it can be used for communication between two active devices, or between an active device and a passive device. NFC devices facilitate two-way communication between "intelligent devices." They also ensure security by applying NFC-SEC. Communication between devices can reach a speed of up to 424 kbps but the maximum range of NFC communication is 4 cm. NFC services can be used as a payment method, ID card, coupon, and so on. NFC allows the authentication of corporate access rights, while NFC devices can also be used as contactless cards and for reading RF tags. NFC can operate in the following three modes. If the Card Emulation Mode is enabled, a device can be used as a contactless card. When the Reader/Writer Mode is enabled, a device can interact with RF tags. The P2P (Peer-to-Peer) Mode is enabled during bi-directional device communication. NFC Mobile architecture consists of an NFC device, SP, TSM, and a card issuer. Each NFC mobile device contains a MIDlet, Secure Element(SE), and NFC chipset. The NFC chipset supports proximate communications via any of the three NFC modes. The SE is managed by card issuers and used to store payment information.

3   Security Requirements

NFC mobile payment services to the proposed in the present research must meet the following security requirements[1-4].

. Confidentiality: The data used in communication should be shared only with legitimate communication objects and, even if it is exposed during communication, the value of data cannot be inferred.

. Integrity: Data providing on communication are the basis for cash transactions, such as billing. Therefore, The data should not be forged and modulation on communication.

. Mutual Authentication: Each object should provide mutual authentication to verify the legality of objects.

. Replay Protection: An attacker should not use the exposed user payment information multiple times.

. Secrecy: The proposed scheme should maintain a high level of security about payment information.

. Efficiency: The proposed scheme should be computationally efficient in a limited device environment.

4   Proposed Scheme

In this chapter, our proposed NFC payment scheme for payment information detection for security requirements of chapter 3.

4.1   System Parameters

The system parameters in the proposed scheme are as follow:

. $*$: object (A : MNO(TTP) B : Mobile Client(User), C : Merchant(POS))
. TMSI: Temporary Mobile Subscriber Identity
. KC: $E_{Ki}(R)$ using A8 algorithm based on GSM
. Ki: SIM specific key. Stored at a secure location in SIM and at MNO
. KP: Shared key between MNO and shop POS terminal
. LAI: Local Area Identity
. H(): Hash function
. R: Random value(RAND)
. S: generated SRES(Signature Response) using A3 algorithm based on GSM from authentication key Ki and Random Number(RAND)

4.2   Mutual Authentication Phase (Key Agreement)

Step 1: The user attempts to tagging to POS for product payment. And then, POS send to request message to about mobile device ID to the user.

   C→B : Request Mobile ID

Step 2: The user transmits its own information(TMSI, LAI, Mobile ID Infomation) to the shop. POS can be verified that such mobile network of user via LAI code. Because the network code is available in LAI in the form of mobile country code and mobile network code.

Step 3: The POS searches the corresponding MNO using LAI. And then, The POS transmits own information(TMSI, LAI, Shop ID, Mobile ID) to the MNO.

B→A : TMSI, LAI, Shop ID, Mobile ID

Step 4: The MNO identifies the user and look-up table for corresponding triplet of GSM. At this time, In case of incorrect TMSI, a declined message is sent to the user. The MNO re-generate the R and $H(R \oplus K_C)$ using the generated Triplet(R, S, K_C) is transmits to the user.

A→B→C : R, $H(R \oplus Kc)$

Step 5: The user then proceeds to step for verification of R and $H(R \oplus K_C)$. The user can use SIM specific key K_i for generation K_C for using A8 algorithm. Because K_C generated from share key K_i between the MNO and the user.

Step 6: The user generates random value R_S for authentication of MNO. And then, the user transmits key K_C to the MNO. And then, each object performs the following steps:

C→B→A : $E_{Kc}(R\|R_S)$
A→B→C : $E_{Kc}(R_S\|R\|PU_{POSKey})$
C : $K_c = A8(R, K_i)$
C : Generate $K_{c1} = H(K_c)$
C→B→A : Successful Authentication
A→B : $E_{KP}(PU_{ClientKey}\|K_{C1})$

5   Conclusion

The development of IT technology is leading to the development of a wide range of services based on personal information; accordingly, a variety of authentication technologies have emerged for protecting personal information. However, the efficiency and payment information protection of these technologies must be guaranteed if mobile payment services based on NFC are to be widely used. In this paper, we proposed enhanced secure payment authentication scheme for protection of user payment infomation in payment services based on NFC. Our scheme satisfies the necessary requirements. Therefore, our scheme could be effectively applied in an NFC mobile payment environment.

References

1. Chen, W.D., Lien, Y.H., Chiu, J.H.: NFC mobile transactions and authentication based on GSM network, pp. 83–89. IEEE, April 20, 2010
2. Pourghomi, P., Saeed, M.Q., Ghinea, G.: A Proposed NFC Payment Application. IJACSA **4**(8), December 2013
3. Chen, W.D., Lien,Y.H., Chiu, J.H.: NFC Mobile Payment with Citizen Digital Certificate, pp. 120–126. IEEE, June 21–23, 2011
4. Kranz, M., Murmann, L., Michahelles, F.: Research in the Large: Challenges for Large-Scale Mobile Application research. IJMHCI (2013)

Necessity of Incentive System for the First Uploader in Client-Side Deduplication

Taek-Young Youn and Ku-Young Chang

Abstract Client-side deduplication technique is a useful tool for efficient storage management in DB-related services. The technique permits a server to save his storage and a client to store his data without actually uploading it. In this paper, we review client-side deduplication and show that selfish clients can make the technique useless if there is no incentive system by examining some possible scenarios where selfish clients do not follow pre-defined procedure for client-side deduplication. We also give a simple example of incentive system which can be the beginning of the research for the above described issue.

Keywords Storage service · Incentive system · Client-side deduplication

1 Introduction

In these days, various services are supported based on remote storage maintained by a service provider. In the services, the service provider may want to reduce the size of storage since he can increase his gains by reducing the size of storage required for supporting the service. For that reason, deduplication techniques are widely studied.

There are two kinds of deduplication techniques: server-side deduplication and client-side deduplication. In server-side deduplication technique, clients upload their data to storage server's storage and the server receives all data uploaded by clients. Duplicated data will be eliminated periodically for pre-defined time interval. In this case, the server can reduce the cost of storage, but the client cannot reduce any cost. On the other hands, in client-side deduplication technique,

T.-Y. Youn · K.-Y. Chang(✉)
Electronics and Telecommunications Research Institute (ETRI), Daejeon, Korea
e-mail: {taekyoung,jang1090}@etri.re.kr

K.-Y. Chang—This research was supported by ETRI R&D program(15ZS1500).

© Springer Science+Business Media Singapore 2015
D.-S. Park et al. (eds.), *Advances in Computer Science and Ubiquitous Computing*,
Lecture Notes in Electrical Engineering 373,
DOI: 10.1007/978-981-10-0281-6_57

the client does not always store the file. Instead, the client and the server perform a protocol to test if the same file is already stored in the server's storage or not. If the file already exists in the server's database, the client can obtain the ownership of the file without uploading it. To be sure, the client uploads his file when the same file is not stored yet. In short, server-side deduplication improves the cost of storage for the server, but client-side deduplication improves not only the cost of storage for the server but also the cost of communication overhead between the server and the client. Therefore, client-side deduplication is more attractive than the server-side deduplication.

Client-side deduplication should be carefully used since the technique has some vulnerability. When client-side dedpulication technique is used in storage service, the service provider has to verify the ownership of the client since he will obtain the ownership of a file without uploading it. Hence, until now, most of researches in this field focus on the way to proof the ownership of the client regarding the file to be stored.

One of fundamental requirement for client-side deduplication is the existence of a file. In other words, a client can reduce the cost of communication only if the same file is already stored in the server's storage. The file would be firstly stored by a client, and the first client who has tried to store the file to the server's storage would have performed uploading procedure. For a file, only the first uploader actually uploads the file which implies that the first uploader has no gain compared with other clients who will store the file after the first uploader already uploads it. If all clients are not selfish, the above described issue is not a matter to the storage service. However, if all clients are selfish, the merit of client-side deduplication would be diminished. Clients can be selfish since the first uploader is more vulnerable to privacy threat compared with other clients, and thus the above mentioned problem is not a needless anxiety. In this paper, we will discuss the above issue in detail, and give some simple examples of incentive system which can be used to solve the above problem.

2   Brief Review of Client-Side Deduplication

In this section, we briefly review client-side deduplication techniques. To upload a file F with client-side deduplication, the client C and the server S perform a protocol execution as following:

> Step 1. To check the existence of the file F, C generates a tag tag_F for the file. In general, we use a cryptographic hash function $h(\cdot)$ to generate the tag, and the tag is simply computed as $tag_F = h(F)$. Then C sends the tag tag_F to S.

Step 2. S checks the existence of F using given tag. If the same file is not stored in the server's storage, S sends N/A to C. Otherwise, S assigns the ownership of C regarding the file F and informs this fact to C.

Step 3. If C receives N/A from S, he uploads the file F and deletes the file F from his storage. Otherwise, C deletes the file F without uploading it.

Note that, the proof of ownership procedure is also described in the following figure even though it is not a main interest of this paper. (The ownership proof part is bounded by a dotted line.) We did not give detained explanation for it but we include it in the above figure to see that the use of proof of ownership does not influence on our points.

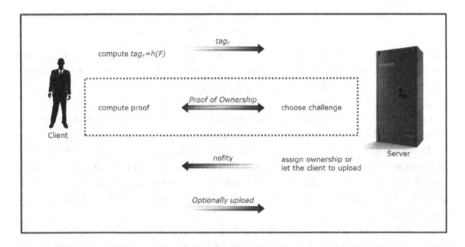

Fig. 1 This figure is the simple client-side deduplication procedure with proof of ownership procedure (which is bounded by a dotted line).

3 Disadvantage for the First Uploader in Client-Side Dedup

In this section, we examine client-side deduplication in terms of the disadvantage to the first uploader in storage service with client-side deduplication. Our main opinion is that a legitimate but selfish user would not follow pre-defined procedure of client-side deduplication to protect his assets. There are two issues regarding this view point. In the following, we discuss each issue in detail.

3.1 Privacy Leakage

One of disadvantages for the first uploader is the privacy leakage. When a client-side deduplication protocol is executed, a client who participates in the protocol execution can figure out the existence of a file. In other words, any client can determine whether a file is exists in the server's storage or not by running the client-side deduplication protocol. This feature is already known in the literature []. Concretely, a client can test the existence of a file F as followings:

> Step 1. The client C generates a tag tag_F for the file F. Then he sends the tag to the server S.

> Step 2. The server will search the file from his storage by using given tag. Then S informs the existence of the file to C.

> Step 3. If the file F is not stored in the server's storage yet, the client stops the protocol without uploading it.

Note that, in the third step, the client does not upload the file when the target file is not stored in server's storage. By performing the above procedure, the client can test the existence of a file without altering the existence of the file. Using this fact, an adversarial client can test if a target user uploads a specific file or not by examining the existence of the file before and after the target user uploading a file.

However, the above described attack scenario works only if the target user is the first uploader. When a client uploads a file to the storage service provider, the existence of the file is changed, and the fact can be used to determine that the user upload the file. Except the first uploader, the above attack scenario cannot be applied, which implies that the first uploader should suffer such privacy vulnerability.

It seems uneasy to mount the above described attack, since an adversary has to choose a target user and a file and correctly guess the upload timing. However, the above attack works correctly, and thus the privacy leakage problem described in this section is still one of significant security holes of client-side deduplication.

3.2 Expensive Service for Devoted Behavior

Recall that the main concern discussed in this paper is disadvantages imposed to the first uploader. In client-side deduplication, there are two burden imposed to the first uploader. The first disadvantage is the privacy leakage described in Section 3.1. The other disadvantage is the cost of service. Note that, in client-side deduplication, a client can store its file without uploading the file only if the same file is already stored in the server's storage. The first copy would be stored by someone, and the client may transmit the file to the server. All clients would pay the same cost for storage service, but the first uploader and other users pay different cost for storing the same file.

The main problem is that the first uploader is more vulnerable to privacy leakage even though he pays much cost than other users. The first uploader actually uploads a file by performing a communication protocol such as TCP, but he can be a victim of an adversarial client's attack. On the other hands, other clients can store their data without actually uploading it, but they are more secure than the first uploader in terms of the privacy.

4   Incentive System in Client-Side Deduplication

In Section 3, we examine two disadvantages for the first uploader. If the first uploader is selfish, he can easily get away from the problem by changing his file. For example, he can attach dummy data to the file, and store the manipulated file instead of the original one. There are various ways to get away from the disadvantages, but they have a feature in common that they will not store the file in the original form. In the client's view point, there is no reason to be the first uploader since there is no merit. Since the service provider can reduce the size of storage for supporting storage service, he should compensate the first uploader's disadvantages. One way of compensation is to make an incentive system for the first uploader.

There are various incentive systems which are already used in online services, and the systems are also suitable to the storage service. For example, the server can discount the cost of service for the first uploader. Various discount policies can be considered but it is not main issue in this paper to give fine policy for storage service. We want to emphasis that to inform the necessity of incentive system in client-side deduplication for the first uploader is the main topic of this paper.

5   Conclusion

In this paper, we reviewed client-side deduplication and show that selfish clients can make the technique useless if there is no incentive system by examining some possible disadvantages imposed to the first uploader. In this paper, we only examine necessity of incentive system in client-side deduplication, but we expect that our research can be a seed for further researches for the design of good incentive system for client-side deduplication.

Acknowledgments This research was supported by Next-Generation Information Computing Development Program through the National Research Foundation of Korea(NRF) funded by the Ministry of Education, Science and Technology (Grant No. 2011-0029925).

References

1. Bellare, M., Keelveedhi, S., Ristenpart, T.: DupLESS: server-aided encryption for deduplicated storage. In: Proc. of the 22nd USENIX Conference on Security, pp. 179–194. USENIX Association (2013)
2. Bellare, M., Keelveedhi, S., Ristenpart, T.: Message-locked encryption and secure deduplication. In: Proc. of EUROCRYPT 2013. LNCS, vol. 7881, pp. 296–312. Springer (2013)
3. Duan, Y.: Distributed key generation for encrypted deduplication: achieving the strongest privacy. In: Proc. of the 6th Edition of the ACM Workshop on Cloud Computing Security, pp. 57–68. ACM press (2014)
4. Harnik, D., Pinkas, B., Shulman-Peleg, A.: Side channels in cloud services: Deduplication in cloud storage. IEEE Security Privacy 8(6), 40–47 (2010)
5. Puzio, P., Molva, R., Onen, M., Loureiro, S.: Cloudedup: secure deduplication with encrypted data for cloud storage. In: Proc. of the 2013 IEEE International Conference on Cloud Computing Technology and Science, pp. 363–370. IEEE Computer Society (2013)
6. Stanek, J., Sorniotti, A., Androulaki, E., Kencl, L.: A secure data deduplication scheme for cloud storage, In: Proc. of Financial Cryptography and Data Security. LNCS, vol. 8437, pp. 99–118. Springer (2014)

Design and Implementation of Disaster Information Alert System Using Python in Ubiquitous Environment

Jae-Pil Lee, Jae-Gwang Lee, Eun-su Mo, Jun-hyeon Lee and Jae-Kwang Lee

Abstract With the rising occurrence of disasters at home and abroad, the disasters threatening the public's life and health, such as natural disasters, have become larger and more complicated. It is very essential to establish the disaster management cooperation system joined by citizens and civic groups. In conventional disaster management system, Disaster Information (DI) is provided by 3G mobile and DMB service. However, it is urgent to build a disaster management system which can respond to disasters effectively depending on their patterns. This study proposed a DI alert system in ubiquitous environments, which provides DI in ubiquitous technology and the network technology supporting Beacon and smart mobile. To do that, this study used the Python-based to design DI web service and implemented the system that collects DI generated in disaster area and Beacon information around the area and provides a DI alert to the smart mobile devices of DI service users.

Keywords Ubiquitous · Disaster · Beacon · Flask · SNS · Crowd sensing

1 Introduction

Globally, natural disasters, such as volcanic eruptions, earthquakes, seismic waves, typhoons, floods, and forest fires, have been on the rise. As a result, a lot of personal injuries and property damage have occurred [1]. An accident occurring naturally has led into safety ignorance which has caused a large scale accident. Therefore, since such a accident was recognized to be a social issue, the interest of the government and the voluntary participation of each civic group have resulted in much active research on disaster prevention [2]. According to Emergency

J.-P. Lee · J.-G. Lee · E.-s. Mo · J.-h. Lee · J.-K. Lee(✉)
Department of Computer Engineering, Hannam University, Daejeon, Korea
e-mail: {jplee,jglee,esmo,jhlee}@netwk.hnu.kr, jklee@hnu.kr

© Springer Science+Business Media Singapore 2015
D.-S. Park et al. (eds.), *Advances in Computer Science and Ubiquitous Computing*,
Lecture Notes in Electrical Engineering 373,
DOI: 10.1007/978-981-10-0281-6_58

Events Database (EM-DAT Database) of the Centre for Research on the Epidemiology of Disasters (CRED) in 2013 at least 330 natural disasters claimed 21,610 lives, made 96.5 million victims, and caused the property losses worth 118.6 billion dollars in 108 countries. The biggest single source disaster that took the most number of lives was Haiyan, the typhoon hitting the Philippines, which killed 7,354 persons. The next biggest one was an Indian flood that had 6,054 persons dead [3]. In the aspect, it is very significant to find the characteristics of temporal and spatial distributions of disasters and analyze their determinants in terms of preparation and strategy of disasters. In the US, big data about weather and climate have been applied more to the public fields, including establishment of climate prediction and disaster response real-time network system using satellite image data [4]. In Korea, since the big accidents including collapses of Seongsu Bridge and Sampung Department Store, disaster and safety management system has been enhanced, and safety culture movement has been expanded [5]. Ubiquitous originated from Lain words mean 'existing everywhere.' The concept of ubiquitous was suggested first by Mark Weiser in 1998. According to his definition, ubiquitous computing was aimed at helping people access information communication network and use diverse information communication services in their everyday life or work environments regardless of time and space [6]. Since the activation of ubiquitous computing technology, the PC based internet access ratio has fallen, but the smart mobile and tablet device based access ratio has increased in the world [7]. In the circumstance, it has been more important to establish a disaster management system that provides disaster information to smart mobile devices in real time through ubiquitous communication and sensor network technology. In previous studies, most safety services focused on big data based crime analysis, and thus there are many problems in applying them to the service of finding disaster conditions accurately and evacuating citizens from relevant locations.

This study proposed a DI alert system that collects and analyzes DI immediately and provides an alert in unpredictable disaster conditions with the use of ubiquitous communication and sensor network technology. To do that, it used the Python based framework 'Flask' to design an alert web service system, and implemented the DI alert system that provides an alert to DI service users in the combined web service of DI in a disaster area and of Beacon information around the area. In this thesis, chapter 2 explains previous studies of disaster management information system; chapter 3 describes the designed and implemented system that collects and analyzes disaster alerts; the last chapter presents conclusions of this study and the future study.

2   Research of Disaster Based Ubiquitous

The research on disaster response system [8] suggested the necessity of Ubiquitous Technology (UT) based disaster response system and the cases of advanced countries in order to cope with national crises of complicated disaster safety. In the disaster response cases of advanced countries, Ubiquitous Government Service Model is suggested as an alternative for UT based service in rejection of human experience. A disaster can be either predictable or unpredictable and thus it is more necessary to provide DI in real time accurately in each region. The overall model of national safety management integrated system to respond to disasters is illustrated in Fig.1.(A).

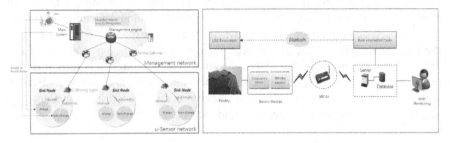

Fig. 1 Technical structure for disaster response(A), u-Disaster Prevention System(B)

The research on Extraction Mashup for Reported Disaster [9] was intended for extracting DI most quickly from numerous media through the use of DI registered in Twitter in real time. In disaster sites, it is very important to report and distribute information quickly. Considering the point that there are not many means to collect information on disaster conditions in real time, the study designed information collection system through social networking service (SNS). In this way, it looked into a method of extracting and using a variety of DI, including disaster occurrence time, locations, and images through crowdsourcing of SNS. However, it failed to provide the service for a communication range area. Therefore, this thesis suggested a study of using the information of both social networking service and Beacon sensor devices with the use of smart mobile devices.

The research on u-Disaster Prevention System [10] suggested u-Disaster Prevention System in ubiquitous environments in order to overcome the problems of conventional fire prevention system. The u-Disaster Prevent System sends and receives sensor data independently in wireless networks, and can provide the optimal evacuation route to reduce human victims even when power is shut off by fire. As shown in Fig.1.(B), depending on the characteristics of facilities, proper evacuation route information is provided in real time with the use of LED electronic displays on walls through Ubiquitous Sensor Network (USN) technology including temperature sensing and smoke sensing. In addition, it established ubiquitous technology based fire evaluation process to suggest the applicability and application plan of the system in real time circumstances. The u-Disaster Prevention System research conducted a field test in a limited environment according to a possible scenario on the assumption that a fire occurred in a particular location. The test revealed that the fire evaluation processes of u-Disaster Prevention System was more effective than general fire evaluation processes in terms of evacuation move distance and evaluation time.

The research on mobile Beacon based oceanographic sensor network [11] proposed a location detection technique using mobile Beacon and geometric constraints method in marine environment. One of the most important oceanographic sensor network technologies is location detection technique. Most location detection techniques are based on the condition that sensor nodes are located on the ground, and thus it is hard for them to be applied to marine environment. To solve the problems of conventional location detection techniques, the proposed one calculated the areas where nodes are expected to exist, detected the locations of the nodes, and thereby it improved accuracy performance from 18% to 82%. Accordingly, to reduce a location detection error, it is recommended to choose the possibly longest points as the minimal distance between Beacon points. However, the technique should receive many Beacon messages and lets sensor nodes consume more energy to find their own location.

3   Disaster Information Alert System

In this thesis, a range of the alert service was limited to service users around the ubiquitous technology based apartment district. The users around a disaster area are able to take disaster detection and judgment service by using their smart mobile devices, neighboring Beacon sensors and DI web service information. In addition, the DI alert system was designed and implemented to provide user participation typed sensing and public monitoring services. This thesis proposed a DI alert system that provides DI service to users in disaster conditions. To prepare for disaster circumstances in a ubiquitous environment, users with Beacon sensor and smart mobile devices subscribe to the DI alert system. Fig.2.(A), illustrates the architecture of the DI alert system based on Beacon sensor information and big data information of SNS. The DI alert system consisted of Smart Mobile Device (SMD), Disaster Web Server (DWS), and Disaster Analysis Center (DAC).

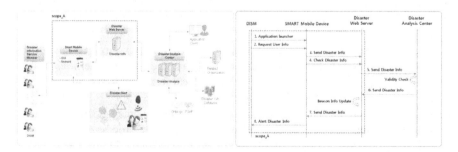

Fig. 2 DI Alert System Model(A), DI Alert System Process(B)

Fig. 2. (B) shows the system process to provide DI alerts efficiently in disaster conditions. The whole system architecture is comprised of Disaster Information Service Member (DISM), Smart Mobile Device (SMD), Disaster Web Server (DWS), and Disaster Analysis Center (DAC). The range suggested in this thesis is scope_A of Fig. 2.(B), and its process is presented as follows: (1) In Application Launcher step, a DISM executes disaster alert APP in SMD. (2) In Request User Info step, the APP accesses DWS to request authentication information of the DISM's ID and PW. (3) In Send Disaster Info step, the DISM sends DI including time, content, and location, and Beacon information to DWS by using Mobile APP in its SWD. (4) In Check Disaster Info step, the DISM monitors DI of DWS on its SMD. (5) In Save Disaster Info step, the DI saved in DWS is sent to DAC. In Validity Check step, the communication state between DWS and DAC is checked regularly. (6) In Send Disaster Info step, DAC receives DI from meteorological center and relevant organizations. After that, DAC sends the information on a disaster area to DWS, and updates the Beacon in the disaster area with the latest data. (7) In Send Disaster Info step, based on the updated Beacon information, the DISM receives the disaster alert information on the disaster area. Lastly, (8) in Alert Disaster Info step, the DI on the disaster area and the Beacon information around the disaster area, received via SMD, is sent to the DISM.

3.1 Implementation of DI Alert System

To make the test environment for the DI alert system, this study implemented the Mobile Application Program (Fig. 3.(A)) , DWS (Fig. 3.(B)) on Windows 7, as shown in Fig. 3., To develop the application for DI alert service, this study employed Python(ver.3.4.3)[12] and Flask(ver.0.10.1). The early Python version was developed for Macintosh, but is now applied to all platforms. Flask, Python based micro framework announced in 2010, is used for web framework based development. It is relatively intuitive and understood easily. As alternatives to conventional web development, Python and Flask provide a simple core engine necessary for web development, and features flexible development environments depending on development domains [13]. To implement the Mobile Application Program, this study used Eclipse (ver.4.5.0) and Android Studio (ver.1.2.2), Integrated Development Environments (IDE) to increase development convenience. And Samsung Galaxy Tab 4, Android SDK(Software Development Kit) Android 5.0.1(API 21) was applied to implement DWS and App.

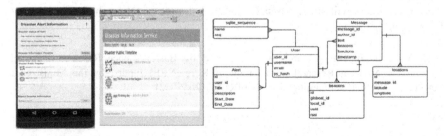

Fig. 3 DI Mobile App(A), Web Service(B), Database Diagram(C)

App function was divided into three types. The first one is to display a DI alert message at the time of accessing a disaster area. The second one is to display the current state of DI web service. The third one is message box and 'send' button to post a message about disaster conditions. In addition, in order for DI service users to post a message about DI, the function of user registration for DI users and the function of posting a message about DI were implemented as web application functions. As such, in the DI alert system, DI service users with smart mobile devices register messages about DI and Beacon information in DWS; DAC verifies the DI and sends the DI to each Beacon device in a disaster area to alert the service users in the disaster area of the disaster. To set up the database of DI web service system, sqlite3 was employed to create DI table for DI alert system table and schema (Fig. 3.(C)) and to generate such fields as message_id, author_id, text, beacons, location, and timestamp, as shown in Fig. 4 in order to save the data generated by web service into the database.

	message_id	author_id	text	beacons	locations	timestamp
	Filter	Filter	Filter	Filter	Filter	Filter
1	13	2	it raining day~	000110001000	36.354483,127.421148	201507150150
2	17	1	The Fire was in the Daejeon	000110001001	36.354483,127.421148	201507150146
3	18	3	It's hot, today	000110001010	36.354483,127.421148	201507130839

Fig. 4 Disaster alert Database diagram and Table queries

4   Conclusions

With the increasing natural disasters in the world, a lot of personal injuries and property damage have occurred. It has been required to establish a system that can immediately and efficiently collect and analyze DI in a disaster area and can respond to the disaster in real time. For disaster management, it is important to have a mutual and efficient communication system in civil society. To overcome the limitations of conventional DI alert service, this study implemented a DI alert system with the use of Beacon devices, Python based DI alert web service, and Android App, and proposed a method of providing DI service users with the information generated by the Beacon device installed in a disaster area. Therefore, it is expected that the system using ubiquitous information technology and smart mobile devices based network technology helps users detect disaster alerts and evacuate from disaster areas in unpredictable disaster circumstances. In the future, these researchers will study classification and storage of DI on the basis of social big data information.

Acknowledgment This research was supported by Basic Science Research Program through the National Research Foundation of Korea(NRF) funded by the Ministry of Science, ICT & Future Planning(NRF-2014R1A1A2055522).

References

1. Centers for Disease Control and Prevention, Emergency Preparedness and Response (2015). http://emergency.cdc.gov/
2. National Emergency Management Agency(South Korea), Disaster Yearbook 2013 (2013)
3. Guha-Sapir, D, Hoyois, P.H., Below, R.: Annual disaster statistical review 2013: The numbers and trends (2014)
4. Korea Meterological Administration, Meteorological Technology and Policy, Meteorological Big Data and the economy (2014)
5. KISTEP: MPSS Disaster Safety R&D direction of the new policy system improvements, ISSUE PAPER 2014-14 (2014)
6. Weiser, M.: The Computer for the 21st Century. Scientific American Magazine, 94–104 (1991)
7. StatCounter (2015). http://gs.statcounter.com/#desktop+mobile-comparison-ww-monthly-201006-201506
8. Young-chul, C., Yong-Guen, B.: JKIICE, Research of UT utilized the services for a disaster response system as role of u-Government (2015)
9. Seo, T.-W., Park, M.-G., Kim, C.-S.: Design and Implementation of the Extraction Mashup for Reported Disaster Information on SNSs (2013) doi:10.9717/kmms.2013.16.11.1297
10. Moon, S.-W., Seong, H.-J.: u-Disaster Prevention System based Real-Time Fire Monitoring in a Building Facility (2011) doi:10.6106/KJCEM.2011.12.1.107
11. Lee, S.-H., Kim, E.-C., Kim, C.-S.: Localization Scheme with Mobile Beacons in Ocean Sensor Networks (2007)
12. Python, Python Software Foundation. https://www.python.org/
13. Grinberg, M.: O'Reilly Media, Flask Web Development: Developing Web Applications with Python (2014)

A Study on the Development of Real-Time Analysis Monitoring System and Its Application of Medical Ins

Mi-Jin Kim, Jong-Wook Jang and Yun-Sik Yu

Abstract By the period of Big Data, the share of open public data and result for bio has actively moved around advanced country, and policy of promote big data development and utilization have been pushed forward for improvement of market competitive power. The most receive attention industry is health and medical treatment.

In this research, to eliminate analysis big data system to analyze information based on patient and medical device data for usage of big data in health and medical industry push forward from policy can activate in small and medium-size medical institution for efficiently business manage to patient management and hospital management.

Keywords Big data · Monitoring system · Real-time analysis

1 Introduction

To increase propagating smart device and usage of social network, user's location and private information has been daily saved in the web[1]. It is well-known as the main cause to create big data. By the period of big data, the share of open public data and result for bio has actively moved around advanced country, and policy of

M.-J. Kim · Y.-S. Yu(✉)
Convergence of IT Devices Institute Busan(CIDI), Dong-Eui University,
176 Eomgwang-ro, Busanjin-gu, Busan, Korea
e-mail: {agicap,ysyu}@deu.ac.kr

J.-W. Jang
Department of Computer Engineering, Dong-Eui University, Busan 614714,
Republic of Korea
e-mail: jwjang@deu.ac.kr

© Springer Science+Business Media Singapore 2015

411

D.-S. Park et al. (eds.), *Advances in Computer Science and Ubiquitous Computing*,
Lecture Notes in Electrical Engineering 373,
DOI: 10.1007/978-981-10-0281-6_59

promote big data development and utilization have been pushed forward for improvement of market competitive power. Big data has created and utilized in various industry especially receive attention in health and medical industry.

From the population aging, increasing of chronic disease and degenerative disease in health and medical industry try various researches on using big data to reduce medical expense, prevent infectious disease, and improve medical service quality and it suggest effective method to efficient diagnosis or searching appropriate treatment, prognosis prediction.

In this research, to eliminate analysis big data system to analyze information based on patient and medical device data for usage of big data in health and medical industry push forward from policy can activate in small and medium-size medical institution for efficiently business manage to patient management and hospital management.

2   Related Research

2.1   Hadoop

Hadoop is a solution with its key focused on distributed processing technology, which is currently most favorable for Big Data processing. It is Java-based framework with Apache open source process to process massive data, using a relatively simple program model[2]. Hadoop is used as a core technology by Yahoo and Facebook, while being applied to many other companies' own solution. Hadoop is composed of the distributed file system called HDFS (Hadoop Distributed File System) and the distributed processing system called MapReduce. The operation method is HDFS, but MapReduce is composed of the masters called Namenode and Datanode as well as multiple slaves in its structure.

2.2   Complex Event Processing

CEP(Complex Event Processing) is complex event processing technology to extract meaningful data in real time basis from events from various event sources, thereby performing the corresponding actions[3]. Event data herein refers to stream data, which are data of continuous massive inputs, with important time sequences and endless data. It is impossible to process and analyze such stream data in real time basis into a conventional Relational Database. CEP is an event data processing solution that can provide real time analysis of such stream data. That is, it is possible to do real time processing of hundreds/millions of various high speed event stream based on In-Memory without saving it to database, file or Hadoop.

2.3   Hadoop and CEP for Big Data Approach

The Big Data-based methodologies have been mainly the Hadoop-oriented approaches from the prospects of saving. More recently, however, the approach to Big Data becomes more important. Indeed, the Hadoop ecological system-based perspectives of batch analysis is important, but the approach from the CEP-based perspectives of real time distributed is growing increasingly more important as in the Gartner report.

With respect to the conventional DB, Hadoop-based batch processing and various high speed event streams, the differences in analytics mechanisms with the In-Memory-based real-time processing are shown in the Fig. 1[4]. In this study, we adopted the CEP technology using the In-Memory-based analytics processing mechanisms where data saved after analysis.

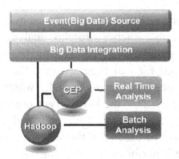

Fig. 1 Big data analysis from the viewpoint of Hadoop and CEP

3   System Compositions and Implementation

CEP technology is currently emerging across business fields, as the simplest and strongest method to implement the real time business intelligence based on timely analysis, providing new values including real time monitoring, with an early alarm and production field management by processing and analyzing various events.

In this study, we aimed at providing a systematic and organized business environment for efficient patient management and administrative management at hospitals, by using CEP-based advantages in consideration of the low cost of Big Data and then establishing the Big Data-applicable real time analytics system in combination with hospital ERP systems with yet a insufficient number of cases. In addition, the main Adaptor and data publisher/customizing functions were implemented, to allow to identify and develop UI screens - depending on the needs of each hospital.

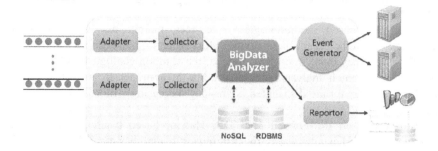

Fig. 2 CEP-based real-time analysis system flow

Real time analytics system flow is composed as shown in Fig. 2. Incoming and outgoing real time data in varieties from/to analytics through Adaptor are converted into internal/external event types such as data protocol and type. While mapping with streams used for real time analysis, the Collector also saves the incoming system from each even onto NoSQL, which allows for time-series analysis as well. For NoSQL DB, the batch event processing was done through Cassandra. In the BigData Analyzer, data collected from the Collector are analyzed using the open source-based real time analysis engine CEP and the Hive-based batch layer, before it performs the mapping function in Reporting-enabled form. The Event Generator performs the functions to convert the real time analyzed results into types, and protocols as user wants, as well as to save the analyzed results to database in real time basis. The Reportor provides the Web-based analysis functions that allows for a user to have visualized approach to the CEP-based real time analytics system, including Dashboard, Alarm, Analysis scheduling for real time analyzed results.

To be brief on the processing of the entire system, Legacy saves log details generated in its own process as in Fig. 3, while performing real time analysis processing as far as major tasks among such processes. Batch layer collects data through data aggregation and extracts data through analysis works. Depending on such extracted results, there may be either storage to middle repository or request for real time analysis. Real time processing layer processes the requests from Legacy and from batch layers, requesting for data as necessary for processing and saving the results to the middle repository. For such processed results, there will be monitoring system API calls. Dashboard visualizes the data depending on the user's request, and then saved the Dashboard-processed data to the middle repository.

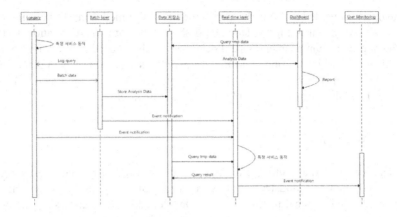

Fig. 3 The entire system process definition

Fig. 4 Result screen of full data   **Fig. 5** Result screen of the selected filter

Fig. 4 is the screenshot of monitoring analyzed of all data on Output Adaptor settings on Dashboard. Through filtering functions, Server/Adaptor/Event/Hour can also be set up in order to analyze necessary parts alone. Fig. 5 is the screenshot of analysis, when filtering Server selected as 210.109.9.177, Output Adaptor selected as Event, Event selected as event_10 and Hour selected as 07:00 ~ 24:00.

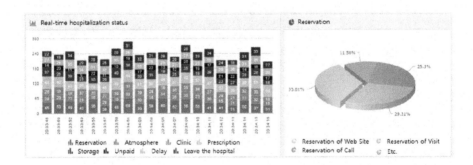

Fig. 6 Real-time hospitalization status screen

Analysis screen for real-time hospitalization status in medical institute using real-time analysis system is Fig. 6. Dashboard shows the status of reservation/waiting/treatment/prescription/payment/unpaid/delay/leave the hospital etc. information looking at once as monitoring real-time. Reservation status represents in graph and it can monitor categorized as website/visiting/phone/others as reservation. It will analyze about disease analysis as collect the data from disease.

4   Conclusion

Big data has created and utilized in various industry especially receive attention in health and medical industry. In this research, big data analysis system has eliminated using medical institution data. Developed big data analysis system can categorize patients by disease case; it can manage by different diseases. In the case of emergency patient, it should be manage patient as reality and efficiently using treat method in real case, not only with diagnosis from developing real-time analysis to reduce time for treat method. Developed big data analysis system should develop as customizing for not only for medical institution, also in ship, distribution, etc. It lasts to keep researching.

References

1. 7 Top Mega Trends in IT Industry for 2013. The Federation of Korean Information Industries, p. 1, 'Big Data Policy' (2013)
2. Park, K.-W., Ban, K.-J., Song, S.-H., Kim, E.-K.: Cloud-based Intelligent Management System for Photovoltaic Power Plants. J. of the Korea Institute of Electronic Communication Sciences 7(3), 591–596 (2012)
3. Luckham, D.C., Frasca, B.: Complex event processing in distributed systems. Computer Systems Laboratory Technical Report CSL-TR-98-754. Stanford University, Stanford, 28, 1998
4. http://www.dbguide.net/knowledge.db?cmd=specialist_view&boardUid=180895&boardConfigUid=108&boardStep=0&categoryUid. Lee Ho-cheol, No. 249, July 29, 2014

Green Treatment Plan Selection Based on Three Dimensional Fuzzy Evaluation Model

Fei Hao, Doo-Soon Park and Sang Yeon Woo

Abstract The advancement of Smart Medical enriches the treatment plans for various diseases. However, how to launch an efficient treatment plan for reducing the pollution, lowering side-effect as well as the treatment cost is becoming a challenge. To cope with this challenge and evaluate the treatment plans during the whole life cycle of it, this paper firstly presents a three dimensional evaluation model and then proposes a green medical treatment plan evaluation and selection strategy based on three dimensional fuzzy evaluation model; Further, an illustrative example is conducted for demonstrating the feasibility of the proposed approach.

Keywords Smart medical · Treatment plan · Degree of membership · Fuzzy evaluation

1 Introduction

The rapid development of modern medical technologies and the deep expansion of the medical knowledge enable the potential treatment plans variety for a certain disease of the patients. For example, for a patient *Joe* who has the "*computer autism*" disease [1], many existing treatment plans might be selected for him. However, some of these treatment plans cannot satisfy the social and economical sustainability, *i.e, higher medicine cost, expensive treatments, more pollution.* Therefore, how to efficiently select the best treatment plan is a quite complicated

F. Hao · D.-S. Park (✉)
Department of Computer Software Engineering,
Soonchunhyang University, Asan, South Korea
e-mail: {fhao,parkds}@sch.ac.kr

S.Y. Woo
Department of Sport Science, Soonchunhyang University, Asan, South Korea

© Springer Science+Business Media Singapore 2015
D.-S. Park et al. (eds.), *Advances in Computer Science and Ubiquitous Computing*,
Lecture Notes in Electrical Engineering 373,
DOI: 10.1007/978-981-10-0281-6_60

problem. The existing literature attempt to evaluate the level of hospital with multi-level fuzzy evaluation approach [2][3][8]. Unfortunately, these researches did not take the life cycle of the treatment at hospital and associated items into account. To tackle the above shortages, this paper presents a green medical treatment plan selection strategy which meets the following requirements: a) low-side effect; b) low-cost; c) low-pollution; d) sustainability. Concretely, the major contributions of this paper are twofold: 1) we devise a three dimensional fuzzy evaluation model for selecting the green medical treatment plan. 2) An illustrative example is used for our model evaluation.

The remainder of this paper is structured as follows: Section 2 presents the three dimensional evaluation model by taking into the relevant evaluation criteria of treatment plan account; The proposed green medical treatment plan selection strategy is provided in Section 3. Section 4 shows an illustrative example and Section 5 concludes this paper.

2   Three Dimensional Evaluation Model

This section is devoted to presenting the conceptual structure of three dimensional evaluation model. As described in the Introduction section, the objectives of the green medical treatment plan is to provide a economical and social sustainable treatment scheme. A good treatment plan should save the design cost, reduce the medical pollution, lower the side-effect for patients during the different life cycle of the selected treatment plan, such as medicines selection, diagnosis, and recovery processing. Based on these motivations, a conceptual structure of three dimensional evaluation model is depicted as follows,

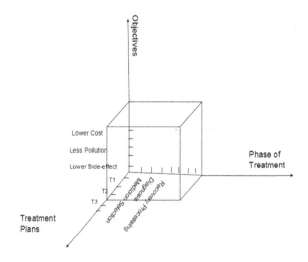

Fig. 1 The Conceptual Structure of Three Dimensional Evaluation Model for Green Medical Treatment Plan Selection

Clearly, Figure 1 shows a three dimensional array including 3 dimensions: objectives, phase of treatment as well as treatment plans. In our paper, we consider 3 attributes, *i.e.*, the lower cost, less pollution and lower side-effect of the potential treatment plans. Regarding to the life cycle of medical treatment plan, it is simple to divided into medicine selection, diagnosis and recovery processing which basically cover the whole life cycle of it; During the real-life treatments, these different life cycle have different weights, denoted as *a,b,c*. For example, the weight for diagnosis phase is obviously greater than others; and a set of treatment plans $T_1, T_2, \ldots T_k \ldots T_n$ are all possible medical treatment schemes that will be evaluated.

3   Green Medical Treatment Plan Selection Strategy Based on Three Dimension Fuzzy Evaluation Model

Smart medical industries emphasize the users' intelligence and experiences, an efficient treatment plan evaluation is facilitating these industries sharply [9][10]. During the evaluation of treatment plan, users usually describe and evaluate them in natural language, such as

> "T_1 generates a little bit pollution, costs much money, rare side-effect during diagnosis";
>
> "T_2 generates a lot of pollution, costs a little money, relative-high side-effect during diagnosis".

Here is a natural question: *which treatment plan is the green one and how to evaluate them?* Aiming to address this question, this section details the proposed green medical treatment plan selection strategy based on three dimensional fuzzy evaluation model. Our strategy is composed of the following steps: 1) establish the membership functions of the fuzzy linguistic terms expressed for objectives evaluation at different phase; 2) calculate the degrees of membership of Step1 and obtain the aggregated degrees of membership for characterize the most green one; 3) rank the treatment plans in terms of maximum aggregated degrees of membership.

3.1   Establishing the Membership Functions

We explore the commonly used fuzzy linguistic terms *"worst"*, *"worse"*, *"average"*, *"better"*, *"best"* which are used to evaluate the treatment plans in terms of different objectives at different phase of it with users' objective opinion.

Let $L=\{l_1, l_2, \ldots l_5\}$ be the set of fuzzy linguistic terms, $P=\{p_1, p_2, p_3\}$ be the different phases of treatment plan, $O=\{o_1, o_2, o_3\}$ be the evaluation objectives; and the corresponding numerical evaluation matrix for a given treatment plan Tk is expressed $E_{ij}^{\ k} = \{e_{ij}\}^k$. Consequently, the membership functions for all possible treatment plans are characterized with triangle membership functions [4][5][11] as shown in Figure 2.

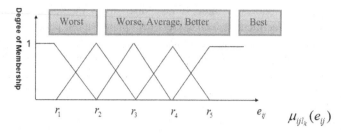

Fig. 2 Construction of Membership Functions for Evaluation of Treatment Plans

Remark: The parameters $r_1, r_2, ... r_5$ are given by users.

3.2 Degree of Membership Aggregation

Up to now, it is not enough to obtain the overall evaluation of the potential treatment plans due to the multiple dimensions criteria. To aggregate these degrees of membership, we firstly aggregate the degrees of membership in terms of treatment phase, that we called **Vertical Aggregation**; then, the overall degrees of membership are aggregated by a linear aggregation formula, termed **Horizontal Aggregation.**

3.2.1 Vertical Aggregation

Vertical aggregation is mainly to aggregate the degrees of membership from the dimension of objectives. Therefore, for a given treatment plan O_i, and a given treatment phase P_i, the vertical aggregation VA_i at P_i, is defined as follows,

$$VA_i = \prod_{j=1}^{3} \max_{l_k \in L} \mu_{ijl_k}(e_{ij}) \tag{1}$$

Eq.(1) indicates that the treatment plans at phase P_i, adopts the approach of maximum degree of membership multiplication to realize the aggregation and obtain VA_i.

3.2.2 Horizontal Aggregation

The objective of horizontal aggregation is to aggregate all the degrees of membership obtained from vertical aggregation. Differs from the vertical aggregation, horizontal aggregation is operated from the dimension of treatment phases for a completed treatment plan about a disease. Since the different phase has different weights for the final evaluation, a linear aggregation formula is constructed like,

$$HA_i = aVA_1 + bVA_2 + cVA_3 \tag{2}$$

now, for any given treatment plan T_i, the HA_i is used to evaluate the advantages/disadvantages of T_i.

3.3   Ranking the Treatment Plans

As a core step of our proposed selection strategy, ranking the treatment plans is in charge of re-ranking of treatment plans according to the HA_i. That is to say, the higher of HA_i is, the most green of the treatment plan is. Formally, the most green medical treatment plan is mathematically described as follows,

$$T_{Green} \leftarrow \max\{HA_i \mid i = 1,2,3\} \qquad (3)$$

obviously, in our analytic case, the most green medical treatment plan T_{Green} has the largest horizontal aggregated degree of membership. It satisfies the economical and social sustainability, therefore it is the most green one among the possible treatment plans.

4   An Illustrative Example

This section presents an illustrative example of green medical treatment plan selection service for diabetes in smart medical which includes three dimensions of treatment plan, treatment phase, and objectives.

Similar with Figure 1, suppose there are 3 possible treatment plans, i.e., T_1, T_2, T_3 and the evaluation matrix for each treatment plan T_i is a 3*3 matrix, these numerical evaluation values can be stored with the following 3-order tensor [6][7] (as shown in Figure 3).

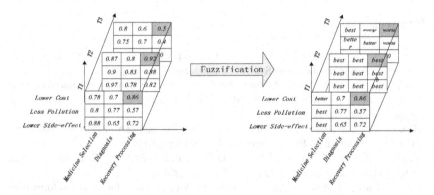

   (a) Numerical-expressed evaluation   (b) Linguistic-expressed evaluation

Fig. 3 Tensor-expressed numerical evaluation and its fuzzification

For better illustration purpose, we set the parameters $r_1, r_2, ... r_5$ be 0.2,0.4, 0.6,0.8, 1 in this example. After fuzzification, we can easily get the reasonable linguistic-expressed evaluation as shown in Fig 3.(b). As a matter of fact, no matter what we set the weights for the treatment phases, the treatment plan T_2 is selected as the most green one for diabetes patients by using the vertical and horizontal aggregation. It is coincided with the user's intuition since the treatment plan T_2 exhibits the best benefits on the aspects of environment, economy and health of patients.

5   Conclusions

This paper aims at evaluating and selecting the most green one from all possible treatment plan for a given disease in order to provide more valued services for patients. Specifically, this paper firstly presents a three dimensional evaluation model and then proposes a green medical treatment plan evaluation and selection strategy based on three dimensional fuzzy evaluation model; Further, an illustrative example is provided for demonstrating the feasibility of the proposed approach. The results analysis reflects a quite good consistence with users' intuition.

Acknowledgments This research was supported by Basic Science Research program through the National Research Foundation of Korea (NRF) funded by the Ministry of Education (No. NRF-2014R1A1A4A01007190) and was partly by supported by the MSIP(Ministry of Science, ICT and Future Planning), Korea, under the C-ITRC(Convergence Information Technology Research Center) (IITP-2015-IITP-2015-H8601-15-1009) supervised by the IITP(Institute for Information & communications Technology Promotion).

References

1. Dickerson, P., Dautenhahn, K.: Interaction between the Autism Children with Computer. Artificial Intelligent Systems and Machine Learning **7**(7), 205–211 (2015)
2. Zhang, Q., Song, Q., Yuan, Z.: Plow Plane Multi-level Fuzzy Evaluation Based on Gray Level Correlation Decision Model and Entropy Value Law. Science & Technology Review **30**(8), 55–60 (2012)
3. Hao, L., Qiu, C.F., Zhao, X.L.: Multi-Level Fuzzy Evaluation Method for Radar Anti-Jamming Effectiveness. Radar Science & Technology (2012)
4. Verma, O.P., Jain, V., Gumber, R.: Simple Fuzzy Rule Based Edge Detection. Journal of Information Processing Systems **9**(4), 575–591 (2013)
5. Pei, Z., Ruan, D., Liu, J., Xu, Y.: Linguistic Values based Intelligent Information Processing: Theory, Methods, and Application, Atlantis press World Scientific (2009)
6. Kolda, T.G., Bader, B.W.: Tensor Decompositions and Applications. SIAM Review **51**(3), 455–500 (2009)

7. Symeonidis, P., Nanopoulos, A., Manolopoulos, Y.: A Unified Framework for Providing Recommendations in Social Tagging Systems Based on Ternary Semantic Analysis. IEEE Transactions on Knowledge and Data Engineering **22**(2), 179–192 (2010)
8. Akdag, H., Kalayci, T., Karagoz, S., Zulfikar, H., Giz, D.: The evaluation of hospital service quality by fuzzy MCDM. Applied Soft Computing **23**, 239–248 (2014)
9. Siddiqui, Z., Abdullah, A.H., Khan, M.K., Alghamdi, A.S.: Smart Environment as a Service: Three Factor Cloud Based User Authentication for Telecare Medical Information System. Journal of Medical Systems **38**(1), 1–14 (2014)
10. Gravenhorst, F., Muaremi, A., Bardram, J., Grunerbl, A., Mayora, O., Wurzer, G., Frost, M., et al.: Mobile Phones as Medical Devices in Mental Disorder Treatment: An Overview. Personal and Ubiquitous Computing **19**(2), 335–353 (2015)
11. Hao, F., Min, G., Lin, M., Luo, C., Yang, L.T.: MobiFuzzyTrust: An Efficient Fuzzy Trust Inference Mechanism in Mobile Social Networks. IEEE Transactions on Parallel and Distributed Systems **25**(11), 2944–2955 (2014)

Performance Analysis of Format-Preserving Encryption Based on Unbalanced-Feistel Structure

Keonwoo Kim and Ku-Young Chang

Abstract Format-Preserving Encryption (FPE) is used to produce a ciphertext with the same length and format as a plaintext. Some approaches to do that are a prefix cipher, a cycle walking, a feistel network and so on. In this paper, we present format-preserving encryption using the unbalanced-feistel structure. We apply our construction to FFX, VAES3 and BPS schemes, which are modes of operation for FPE. We also analyze the performance of three unbalanced-feistel FPE schemes over block ciphers. VAES3 using 128-bit AES encrypts credit card numbers at 5,924,000bps.

Keywords Format-preserving encryption · FPE · Unbalanced-feistel · FFX · VAES3 · BPS

1 Introduction

Block cipher using ECB, CBC, OFB and CFB modes converts an input bit string to an output bit string. These modes are not suitable for encrypting a digit string such as a credit card number and a social security number. Since the data format to be saved after encrypting is not same to the original format before encrypting, the legacy modes require changes to the database schema and associated applications.

In contrast, a ciphertext to be enciphered by format-preserving encryption (FPE) has exactly the same format and length with a plaintext. So, the encrypted data fits into the existing fields unlike the traditional cryptosystem. For example, a FPE-encrypted 16-digit credit card number seems to be another 16-digit credit card number, not a hexadecimal value.

Feistel network is one of well-known and extensively-studied FPE modes. In this paper, we implemented FFX[1,2], VAES3[3] and BPS[4] schemes using the

K. Kim(✉) · K.-Y. Chang
ETRI, 218 Gajeong-ro, Yuseong-gu, Daejeon, Korea
e-mail: {wootopian,jang1090}@etri.re.kr

© Springer Science+Business Media Singapore 2015   425
D.-S. Park et al. (eds.), *Advances in Computer Science and Ubiquitous Computing*,
Lecture Notes in Electrical Engineering 373,
DOI: 10.1007/978-981-10-0281-6_61

unbalanced-feistel construction. Furthermore, we simply compare the performance of three FPE modes over AES, SEED and ARIA.

This work is organized as follows. Section 2 briefly introduces feistel network and format preserving encryption. In Section 3, we present how to efficiently implement FPE modes using the unbalanced-feistel structure. And then, Section 4 analyzes the performance of FFX, VAES3 and BPS over block ciphers. Finally, we conclude in Section 5.

2   Format-Preserving Encryption with Feistel Structure

The feistel-based FPE algorithm takes in a key K, a plaintext P and a tweak T, and produces a ciphertext C. By [1], plaintexts and ciphertexts are regarded as strings over an alphabet Chars = $\{0, 1, ..., radix-1\}$. Members of the alphabet are called *characters*. The number of characters radix in Chars is referred to as the *radix* of the alphabet. Example radix values are 2, 10, and 26, corresponding to bits, digits, and lowercase English letters.

Figure 1 shows the unbalanced- and alternating-feistel structure, respectively. After a character string consisting of plaintext is converted into a numeral string, the encoded plaintext is split into two pieces. On every round, F_K is applied to one part of the data in order to modify the other part of the data. The round function F_K is constructed from a block cipher or a hash function. In the alternating-feistel network FPE, right and left part is alternately used as input value of F_K with a tweak to

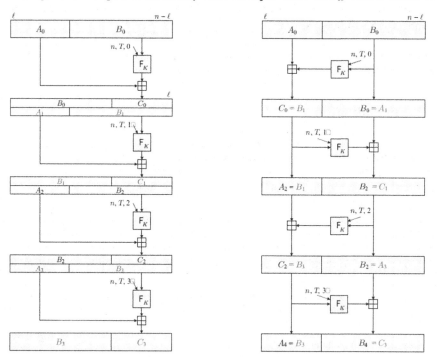

Fig. 1 Unbalanced-Fiestel(Left) and Alternating-Feistel(Right) structure [1]

generate a sub key on each round. In contrast, in the unbalanced-feistel FPE, right input data is always used to F_K on every round. Output of each round is produced by performing block-wise modular addition of the input data under the sub key.

Output of the previous round function becomes the right input of the next round. The left and right output numeral strings of the last round are transformed to the character strings to preserve the original format. The number of rounds depends on the FPE modes that will be applied.

NIST has chosen FFX[radix] [2], VAES3 [3] and BPS-BC [4] as modes of operation for its format-preserving encryption. In addition to the key and designated cipher function, each mode has prerequisites about the choices of the base, radix, and of the range of lengths [5]. Table 1 presents parameters for three FPE modes.

FFX[radix] and VAES3 have a prerequisite for the choice of the maximum tweak length, and VAES3 has an additional parameter, tweakradix, for the choice of the base for tweak strings. BPS-BS permits only 64-bit tweak.

Table 1 Parameters of NIST FPE modes (draft status)

	FFX[radix]	VAES3	BPS-BC
Radix	$2 \sim 2^{16}$	$2 \sim 2^{8}$	$2 \sim 2^{16}$
Input length to be permitted	$2 \sim 2^{32}$	$2 \sim 2[120/\text{LOG}_2(\text{radix})]$ if radix is a power of 2 $2 \sim 2[98/\text{LOG}_2(\text{radix})]$ if radix is not a power of 2	$2 \sim 2[\text{LOG}_{\text{radix}}(2^{96})]$
Tweak length to be permitted	$0 \sim 2^{32}$	$0 \sim [104/\text{LOG}_2(\text{tweakradix})]$	64-bit
Tweak radix	-	$2 \sim 2^{8}$	-
The number of rounds	10	10	8

3   Implementation of Unbalanced-Feistel FPE

3.1   FFX[radix] with Unbalanced-Feistel Structure

Our FFX[radix] uses a maximally-unbalanced Feistel network and a round function based on the CBC-MAC of block cipher. Figure 2 presents an algorithm of unbalanced-feistel FFX[radix]. Following actions need to be taken to enhance the encryption performance.

-   Operation from step 1 to step 4 is performed only once regardless of the rounds.
-   In Step 5-i, the tweak, T, the substring, B, and the round number, i, are encoded as a binary string, Q.
-   In Step 5-i, T and $[0]^{(-t-b-1) \bmod 16}$ consisting of Q can be set prior to every round.

- In Step 5-v, $radix^m$ are able to be precomputed before round computation.
- Size of the value P in Step 4 is always 16-byte. And, the concatenation value of P and Q in Step 5-ii is a multiple of 16-byte all the time. Therefore, in order to calculate CBC-MAC of $P\|Q$ in Step 5-ii, CBC-MAC output of the initial block, P, can be achieved before each round. This job is done only once during 10 rounds of FFX[radix].
- When radix is 10 to encrypt a 16-digit credit card number, values of b and d become 4 and 8 in Step 3, respectively. And then, Step 5-iii is not executed since S is a string for 1 block. That is, first 8-byte of the result of Setp 5-ii is set to S of Step 5-iv. Above description permitted up to 56-digit input. At this time, block cipher is invoked 11 times.

---

1. Let $u = n/2$; $v = n - u$.
2. Let $A = X[1 .. u]$; $B = X[u + 1 .. n]$.
3. Let $b = ceil(ceil(v *LOG_2(radix)) / 8)$; $d = 4*ceil(b/4)+ 4$
4. Let $P = [1]^1 \| [2]^1 \| [1]^1 \| [radix]^3 \| [10]^1 \| [u \bmod 256]^1 \| [n]^4 \| [t]^4$.
5. For i from 0 to 9:
    i. Let $Q = T \| [0]^{(-t-b-1) \bmod 16} \| [i]^1 \| [NUM_{radix}(B)]^b$.
    ii. Let $R = PRF(P \| Q)$.
    iii. Let S be the first d bytes of the following string of $ceil(d/16)$ blocks:
        $R \| CIPH_K (R \ xor \ [1]^{16}) \| CIPH_K (R \ xor \ [2]^{16}) \| ...$
    iv. Let $y = NUM_2(S)$.
    v. Let $c = (NUM_{radix}(A) + y) \bmod radix^m$.
    vi. Let $C = STR^m_{radix} (c)$.
    vii. Let $A = B$.
    viii. Let $B = C$.
6. Return $A \| B$.

Fig. 2 FFX[radix] round function [2,5]

3.2 VAES3 and BPS with Unbalanced-Feistel Structure

VAES3 is treated as a set of parameters of FFX. It features to have a sub-key step that enhances security and lengthens lifetime of the key. Before round computation, the sub-key is derived from the encrypted output of P using the master key. And then, the sub-key is used on every round. Block cipher mode used in VAES3 is ECB not CBC. VAES3 enciphers strings over an alphabet of an arbitrary radix from 2 to 256. In practice, input and tweak length to encrypt a credit card number with a radix of 10 and a tweak radix of 10 is limited to 58 and 31 characters, respectively.

BPS is built upon two components: an internal length-limited block cipher and a mode of operation in order to handle long strings [4]. NIST employs only the first component, names BPS-BC, as its FPE modes. BPS recommends using 8 rounds unlike other two schemes.

4   Performance Analysis

To implement the unbalanced-feistel FFX[radix],VAES3 and BPS-BC, we modified three NIST draft FPE modes to the unbalanced-feistel structure based on AES, SEED and ARIA with 128-bit key. The test was conducted using Microsoft Visual Studio 2010 on Windows PC with Intel i7-3770K CPU@3.40GHz and 4GB RAM.

In order to measure the encryption performance, we have created randomly 1,000,000 sixteen-digit credit card numbers and checked their encryption time. The performance was calculated as the average of the encryption time. Figure 3 presents the performance of three unbalanced-feistel FPE modes over three block ciphers to encrypt credit card numbers while preserving format and length.

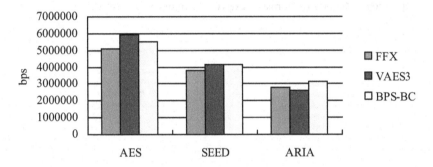

Fig. 3 Performance of unbalanced-feistel FPE over block ciphers

By the result of the experiment, using AES in the round function gives better performance than SEED or ARIA regardless FPE modes. Especially, VAES3 using 128-bit AES has been observed to encrypt credit card numbers at 5,924,000bps. With FPE mode and its structure, the performance also depends on the core block cipher and its optimization.

5   Conclusion

FPE makes it possible to store the encrypted data without changing the existing database schema. And, feistel-based encryption is one of some approaches to achieve FPE. In this paper, we give useful description to implement the unbalanced-feistel FFX[radix], VAES3 and BPS-BC and to enhance encryption/decryption speed. We also compared the performance of three unbalanced-feistel FPE schemes over AES, SEED and ARIA.

Acknowledgement This work was supported by ETRI R&D program. (15ZS1500).

References

1. Bellare, M., Rogaway, P., Spies, T.: The FFX Mode of Operation for Format-Preserving Encryption, Draft 1.1, NIST, February 2010
2. Bellare, M., Rogaway, P., Spies, T.: Addendum to "The FFX Mode of Operation for Format-Preserving Encryption": A parameter collection for enciphering strings of arbitrary radix and length, Draft 1.0, NIST, September 2010
3. Vance, J.: VAES3 scheme for FFX: An addendum to "The FFX Mode of Operation for Format-Preserving Encryption": A parameter collection for encipher strings of arbitrary radix with subkey operation to lengthen life of the enciphering key, Draft 1.0, NIST, May 2011
4. Brier, E., Peyrin, T., Stern, J.: BPS: a Format-Preserving Encryption Proposal
5. NIST Special Publication 800-38G Draft: Recommendation for Block Cipher Modes of Operation—Methods for Format-Preserving Encryption, July 2013

Energy Consumption Reduction Technique on Smart Devices for Communication-Intensive Applications

Jonghee Youn, Doosan Cho, Yunheung Paek and Kwangman Ko

Abstract In these days, mobile devices as smartphone offer users more applications, more communication bandwidth, and more processing, which together put an increasingly heavier burden on its energy usage, which advances in battery capacity do not keep up with the requirement of the modern user. In this paper, we intended to enhance the battery lifetime of mobile devices such as smartphone through communication-intensive offloading technique, we called EnPiler, that can be profile and offload energy intensive method units of communication-intensive applications.

Keywords Mobile computing · Communication offloading · Energy overhead · Static and dynamic code analysis

1 Introduction

Mobile devices are a rapidly growing fragment of the global computing device market, whose processing capabilities are also incredibly increasing for the last few years. In line with continuous increase of use of mobile devices such as

J. Youn
Computer Engineering, Youngnam University, Gyeongsan, South Korea

D. Cho
Electrical and Electronic Engineering, Sunchon National University,
Suncheon-si, South Korea

Y. Paek
Electrical and Computer Engineering, Seoul National University, Seoul, South Korea

K. Ko(✉)
Computer and Information Engineering, SangJi University, Wonju, South Korea
e-mail: kkman@sangji.ac.kr

© Springer Science+Business Media Singapore 2015
D.-S. Park et al. (eds.), *Advances in Computer Science and Ubiquitous Computing*,
Lecture Notes in Electrical Engineering 373,
DOI: 10.1007/978-981-10-0281-6_62

431

smartphone and tablet PC with limited computing resources and capability, research on how to cope with mobile apps which need high performance computing resources for complicated computation has come to the forefront as an important issue. Particularly along with steady efforts for performance improvement of processing speed, storage capacity, is emphasized the importance of technical management in the area of software to prolong the use hours of battery with limited capacity. Under the circumstances, research on how to minimize and manage battery consumption of applications processed in mobile devices must be an important research subject in the related field. There have been previous studies on a diverse range of offloading frameworks with a view to reducing of computing load on mobile system through offloading. That said, most of them focused on how to overcome computing burden by means of computation offloading of units requiring complicated computation or long time of processing by determining in a static or dynamic way, and consequently to reduce burden of battery consumption. However, according to recent study reports, communication-intensive movement becomes the main cause of battery consumption not less than complicated and repeated computation, as one of characteristics of mobile apps[1].

In this paper, we intended to enhance the battery lifetime of mobile devices as smartphone through communication-intensive offloading framework with energy profiling techniques, we called EnPiler, that can be profile and offload energy intensive method units of communication-intensive applications. The rest of this paper is organized as follows. Section 2 describes the previous and related works. Section 3 describes the communication-intensive offloading framework, EnPiler. Finally, Section 4 describes the conclusions and works in progress.

2   Related and Previous Works

In order to make mobile cloud computing more feasible and adaptable for the mobile device environment, recent execution offloading researches dynamically offload their computation to the remote server, based on various runtime performance factors such as execution time, energy consumption, and network latency. As shown in prior works [2][3], which uses the resources of remote server only if it is profitable, is very effective for usual applications in most cases. CloneCloud[2] suggests dynamic execution offloading approach by modifying the mobile execution environment, Dalvik VM, to capture the current application state. Because of their approach, CloneCloud do not need to modify the application code. MAUI[3] is a RPC based offloading architecture which decides at runtime which methods should be offloaded based on the best energy savings possible under the current runtime performance factors. MAUI requests special user annotations on the application code to mark REMs. Although the basic architecture of our offloading framework is inspired by MAUI and ThinkAir[4], these studies target general mobile applications while our work focuses on energy reduction of mobile applications with communication offloading.

More recently, several approaches have been proposed to improve the performance of execution offloading. ThinkAir suggests a dynamic resource allocation scheme, which allocates more than one clone VM for the offloaded application to exploit parallelism and to relieve the lack of memory space. [5] proposes a distributed Energy Efficient Computational Offloading Framework (EECOF) for the processing of intensive mobile applications in MCC. The framework focuses on leveraging application processing services of cloud datacenters with minimal instances of computationally intensive component migration at runtime. As a result, the size of data transmission and energy consumption cost is reduced in computational offloading for MCC. Analysis of the experimental results show that by employing EECOF the size of data transmission over the wireless network medium is reduced by 84 % and energy consumption cost is reduced by 69.9 % in offloading different components of the prototype application.

SorMob[6] is an AOP-based source-level offloading framework, which moves part of computation to the cloud, based on source code profiling. It calls the methods placed at the offloading server through the remote procedure calls so that a programmer does not need to explicitly code the details for this interaction. The use of AOP in SorMob makes the implementation of the offloading framework simpler and more flexible. For the implement offloading for Android applications: either to modify Dalvik Virtual Machine and make the offloading process automated or manually modify each application adding the code for offloading to it. Both of ways are painful compared to AOP, which also provides automated offloading without any need to modify target applications. With AOP adding offloading functionality to an Android application is as easy as adding SorMob library to applications[7].

3   An Energy Consumption Reduction Technique

Figure 1 shows the communication-intensive offloading architecture that consists of EnPiler Local-end and EnPiler Server-end. EnOffload Profiler detects offloading candidate methods with tele-immersive application with the compiler static source analysis technique and generates Energy-aware Profiles & classes appropriate for the model that can measure consumption. To this end, the profiler analyzes the source code by comprising DFG and CFG for the application and collects flow information on calculation, communication and application running. Especially, the key factor to decide offloading is for a local device, e.g. smartphone, to capture all communication IPs and monitor information on the number and volume of communications as well as analyze the communication packet and profiling relating methods.

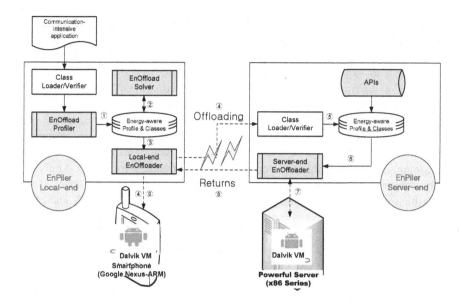

Fig. 1 EnPiler: Communication-intensive Offloading Framework and its workflow.

The EnOffload Solver enables offloading by internally modifying or changing the source code (serialization form) without additional efforts for the offloading candidate methods by the developer. In this process, the study considered communication and server condition to suggest the method to deal with the error that can occur during offloading. It classified the offloading candidate methods into hardware (computing resources) restraint, software restraint, communication and server condition restraint, developed the source code changing method to address each of them and adopted them properly for Java serialization API during offloading.

Local-end EnOffloader (Server-end EnOffloader) performs actual offloading by taking into account external environment (communication and server condition) and data volume needed for offloading based on the offloading candidate methods information. When offloading is determined impossible or unrealistic, the smartphone, local-end mobile device, performs methods. The study suggested the ways to identify and overcome the causes that hamper offloading in the process. The server-end EnOffloader performs tele-immersive operation at the high performance server by receiving optimal offloading information and returns the final running result to the local-end EnOffloader.

4   Conclusions

In this paper, we introduced and designed a conceptual solution, Enpiler, for reducing the energy consumption overhead of local-end mobile devices. For several years many offloading frameworks have been introduced, but there is no a framework that practically used to reduce and minimize of energy consumption since the just focused on computational overheads. In this paper, we describes the architecture and workflow of Enpiler and we anticipate that it may be effective for reducing the overload of local-end smart mobile devices in the situation that needs for video streaming are explosively increasing. Our next step is to realize the proposed key concepts and prototypes in the real environment like a smartphone and present the results of variable experimentals.

Acknowledgments This work was supported by the Engineering Research Center of Excellence Program of Korea Ministry of Science, ICT & Future Planning (MSIP) / National Research Foundation of Korea (NRF) (Grant NRF-2008-0062609), the IT R&D program of MSIP/KEIT. [K10047212, Development of homomorphic encryption supporting arithmetics on ciphertexts of size less than 1kB and its applications], the Brain Korea 21 Plus Project in 2014, Basic Science Research Program through the National Research Foundation of Korea (NRF) funded by the Ministry of Education (NRF-2010-0024529), National Research Foundation of Korea(NRF) grant funded by the Korea government(MSIP) (No. 2014R1A2A1A10051792), Brain Korea 21 Plus Project in 2015 and the MSIP(Ministry of Science, ICT and Future Planning), Korea, under the ITRC(Information Technology Research Center) support program (IITP-2015-R0992-15-1006) supervised by the IITP(Institute for Information & communications Technology Promotion).

References

1. Kwangman, K., Yunheung, P.: Energy-aware profiler: an energy consumption analysis techniques for offloading communication-intensive mobile apps. In: Ubiquitous Information Technologies and Applications(LNEE), vol. 280, pp. 699–703 (2014)
2. Chun, B.-G., Ihm, S., Maniatis, P.: CloneCloud: elastic execution between mobile device and cloud. In: International Conference on Mobile Systems, Applications, and Services (2011)
3. Cuervo, E., Balasubramanian, A., Cho, D.-k., Wolman, A., Saroiu, S., Chandra, R., Bahl, P.: MAUI: making smartphones last longer with code offload. In: International Conference on Mobile Systems, Applications, and Services (2010)
4. Kosta, S., Aucinas, A., Hui, P., Mortier, R., Zhang, X.: ThinkAir: dynamic resource allocation and parallel execution in the cloud for mobile code offloading. In: Proc. of INFOCOM 2012 (2012)
5. Shiraz, M., Gani, A., Shamim, A., Khan, S., Ahmad, R.W.: Energy Efficient Computation Offloading Framework for Mobile Cloud Computing. Journal of Grid Computing **13**(1), 1–18 (2015)
6. Cho, Y., Cho, D., Paek, Y.: Sormob: Computation Offloading Framework based on AOP. KIPS Transactions on Computer and Communication Systems **2**(5), 203–208 (2013)
7. Kwon, D., Yang, S., Paek, Y., Ko, K.: Optimization techniques to enable execution offloading for 3D video games. In: MTAP Special Issues(UCAWSN2015) (2015). (to be published)

Framework of Service Accountability and Policy Representation for Trustworthy Cloud Services

Hyejoo Lee, Hyun-il Kim, Changho Seo and Sang Uk Shin

Abstract In this paper, we propose an architecture for the service accountability in order to provide the trustworthy cloud services for the Cloud End Consumers and also define the structure of document to represent the policy of service accountability. As a result, we show the proposed scheme makes more effective system performance by compared with some exist schemes.

Keywords Cloud computing · Data accountability · Provenance · Cloud service accountability

1 Introduction

The trustworthiness of cloud service as to whether the data of users are safely saved, transferred, and processed becomes more important problem due to the characteristic of the cloud services that the users do not possess the physical computing resources in person. Hence many researches have conducted on the data accountability to provide the trustworthy cloud services [1-9]. It contains

H. Lee · S.U. Shin(✉)
Department of IT Convergence and Application Engineering, Pukyong National University, Busan, Korea
e-mail: hyejoo2010@gmail.com, shinsu@pknu.ac.kr

H.-i. Kim
Department of Convergence Science, Kongju National University, Gongju-si, Korea
e-mail: hyunil89@kongju.ac.kr

C. Seo
Department of Applied Mathematics, Kongju National University, Gongju-si, Korea
e-mail: chseo@kongju.ac.kr

© Springer Science+Business Media Singapore 2015
D.-S. Park et al. (eds.), *Advances in Computer Science and Ubiquitous Computing*,
Lecture Notes in Electrical Engineering 373,
DOI: 10.1007/978-981-10-0281-6_63

various processes such as compliance with service policy establishment, monitoring and management of logging information, auditing, and so on.

In general, some services provided through IaaS (infrastructure as a service) or PaaS (platform as a service) are the service of cloud infrastructure and platform, and it is expected that the cloud services on IaaS or PaaS can be trustworthy if the computer resources or platform is exactly provided in accordance with SLA (service level agreement). However, the type of services which are provided to the end consumers must be SaaS (software as a service), and the trustworthiness of cloud services through SaaS on various business service environment has to consider the service accountability based on some relationship and semantics between the cloud services. In this paper, therefore, we propose an architecture which establishes the service accountability based on the relationship and semantics of processes related to the cloud services. The organization of paper is as follows. In the section 2, we describe some related researches about the data accountability. In section 3, we propose the architecture to fulfill the service accountability on physical and virtual machines and also representational scheme to describe the service accountability policy. In the section 4, we compare the proposed scheme with some exist schemes of data accountability. In the section 5, we describe some future studies for the proposed scheme as a conclusion.

2   Related Studies

Most of studies for the cloud accountability are file-central schemes, which record some log data after monitoring some system calls related to file processing. DataPROVE framework of HP [6] traces some applications, processes, events on VM (virtual machine), but it is basically the file-central logging mechanism using 'Flogger' which monitors file access and transfer on PM (physical machine) and VM, and records a variety of related data such as the accessed filename and path, file access time and date, and so on. Besides this, PASS (provenance-aware storage system) [3] is a prototype for the automatic provenance collection and management, which consists of DAPI (disclosed provenance API) for provenance and seven components. YAO proposed a scheme to provide the accountability as a service domain within one more business service domains in order to provide trust SOA (service oriented service) [13]. Also, the project A4Cloud as EU project for the cloud accountability has been started since 2012 [9-10]. Besides these, there are various studies as to the compliance for service policy on environment of cloud service, secure logging, and monitoring and auditing [11-13]. Most of researches deal with some mechanism to monitor file processing related to data within the cloud. The services on the cloud are made up as the chain of one or more services. If the end consumers want to identify whether the composite services are properly provided, some way to prove it is needed.

3 Framework for Service Accountability Between Cloud Services

3.1 Data Accountability vs. Service Accountability

The independent or combined SaaS, PaaS, and IaaS service models compose the cloud service chain. Within this service chain, the data of users are traced to provide them with the trustworthiness related to the usage, processing and storage of data. In aspect of the data accountability, the data are originated in the end user, and then finally stored to the cloud. Thus, the data accountability has to be applied to some actions related to the files stored to the cloud. Compared with the data, the service, which processes the user's data for specific purpose, starts from the cloud by the processes or by the user's request. The provision of service includes the data as well as the processes which perform some process logic to provide the service. The Cloud End Users expect that the processes are executed according to accurate service flows provided by the service providers. If the users want to confirm their expectation, the service providers must adduce some evidence to satisfy their demand. This is the chief aim of the service accountability. The service encompasses the process execution, data processing and transfer, data storage, and so on. In the end, the service accountability has to be applied to data and processes in close association with each other. In the next section, we explain the service accountability for the cloud to do this.

3.2 Framework of the Service Accountability for the Cloud

Fig. 1 presents the architecture of framework to perform the service accountability. There are two parts as the main modules, one part consists of three tools for the service accountability to monitor some processes which run in user spaces of VMs and the other part within VMM (virtual machine manager) is the controller for service accountability to handle the tools for monitoring processes.

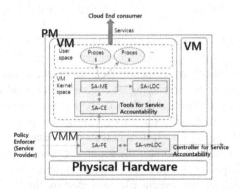

Fig. 1 Cloud service chain which is based on the cloud service model SaaS, PaaS, and IaaS.

The controller for service accountability controls the tools for service accountability according to the policy of service accountability and the tools for service accountability trace some active processes on user's space within VM and collect a variety of data for accountability. The sub modules of the controller for service accountability include SA-PE (service accountability policy engine) and SA-vmLDC (service accountability VM-log data collector). Under this control, the tools for service accountability interpreter the policy and monitor some related processes and data according to this policy. The data collected from VMs are stored as the log data for service accountability. These functions are performed by three sub-modules, SA-CE (service accountability configuration engine), SA-ME (service accountability monitoring engine), and SA-LDC (service accountability log data collector).

3.3 Representational Scheme of Service Accountability Policy

The provision of services consists of the execution of processes, the processing of data by the processes, the offering of services, the transfer of data, and so on. The service accountability is performed by monitoring various actions of processes, data, and network and by recording these log data. An automatic enforcement scheme is required in order to fulfill the service accountability policy. As shown in Fig. 2, we define the structure of XML document to represent the service accountability policy for automatic enforcement.

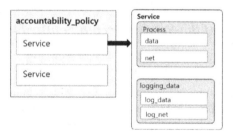

Fig. 2 The structure of XML document for the service accountability policy.

For easy understanding, we give an example that the CSP (content service provider) utilizes the cloud to provide the content services for the end consumers who requests the playback of contents such as video, audio, and so on. We suppose a simple case of the streaming of content which works on the VM. The end users want to keep on managing their content usage information safely. Also they want to confirm whether the service providers have provided the contents as demanded. The CSP has to prove it to gain their confidence. For this, the tools for service accountability have to monitor the actions of process 'streaming' when the end user requests the playback of content. The 'streaming' process consists of two main actions, the reading the content file requested by the end user and network communication to stream it to the user. The various processes must be involved in the service accountability, but we cover only the reading of the content file and network communication for simple example. For this, the service accountability policy is determined as follows:

- The process 'streaming' should monitor the read of content file to prove when the streaming service is provided and what kind of content is provided, recording required log data as follows:
  - Information about the content file: a requested content name, location of it, etc.
  - Information about the process: PID, GID, UID, execution time, date and time when the content file is read, and hash value of log data, etc.
- The process 'streaming' should monitor network communication from the request of the end user to the end of streaming, and recodes the log data as follows:
  - PID, GID, UID of the process, respective date and time when the request of streaming, the start, stop and end of streaming, hash value of packet data, IP Address, MAC Address, and Protocol

This policy is represented by the following XML document.

```xml
<accountability_policy target="streaming-server-01">
  <service name="streaming-content'">
    <process name="streaming" >
      <data type="file" activity="read">
        <name>test.webm</name>
        <logging_id name="data-streaming-test.log">
          12345</logging_id>
      </data>
      <net activity="all">
        <logging_id name="net-streaming.log">
          12345</logging_id>
      </net>
    </process>
    <logging_data>
      <log_data id="12345" target_name="on" uid="on"
        gid="on" pid="on" datahash="on" time="on"/>
      <log_net id="12345" uid="on" gid="on"  pid="on"
        time="on" ip_mac="on" datahash="on"/>
    </logging_data>
  </service>
</accountability_policy>
```

4   Comparison

As mentioned before, DataPROVE and PASS are based on file-centric mechanism among the mechanism to provide the data accountability. The 'Flogger' of DataPROVE is installed on Kernel space of PM and VM, other components are located on Kernel and User space of VM. Each component of PASS is located on Kernel space. The components which locate in Kernel space are influenced by the change of Kernel, they have to be changed if the system's Kernel is changed. In this case where the system's Kernel is frequently changed, the cloud accountability is not cost effective. But the proposed scheme can locates in Kernel or User Space with no restriction. As a result, the components can be located in an of job space according to service environment. Addition to this, in DataPROVE and PASS, the target objects are separately traced and then the log data are recorded without relevance to each other. As a result, it needs post-processing to apprehend the relevance of logged data. However, the proposed method traces data and network communication with relevance to processes and record those log data together, it results in immediately understanding the relevance between the log data. Also, DataPROVE and PASS impose the burden on the performance of system due to monitoring all processing of file. But the proposed scheme can reduce the burden with system performance because it performs the monitoring of specific file, data processing or network processing according to importance of services.

5   Conclusion

The service accountability has to associate the process with data to prove the trustworthiness of services within the cloud, that is, focuses on the process. In this paper, we proposed the architecture and the structure of XML document to represent the service accountability policy. With compared with other schemes, the proposed scheme can locate in any job space of the cloud and has lighter burden than the others. Also the representational scheme allows the automatic enforcement of the service accountability policy.

The future research is to define the architecture for the service accountability more details and the elements of XML for the service accountability policy to be applicable various cases for services, and to implement the prototype to test the performance of the proposed scheme.

Acknowledgments This research was supported by Next-Generation Information Computing Development Program through the National Research Foundation of Korea(NRF) funded by the Ministry of Science, ICT & Future Planning (2011-0029927), Ministry of Education(MOE) and National Research Foundation of Korea(NRF) through the Human Resource Training Project for Regional Innovation (2013H1B8A2032077).

References

1. Macko, P., Chiarini, M., Seltzer, M.: Collecting provenance via the xen hypervisor. In: Proceedings of 3rd USENIX Workshop on the Theory and Practice of Provenance (TaPP 2011) (2011)
2. Muniswamy-Reddy, K., Macko, P., Seltzer, M.: Provenance for the Cloud (2010). https://www.usenix.org/legacy/event/fast10/tech/full_papers/muniswamy-reddy.pdf
3. Muniswamy-Reddy, K., Braun, U., Holland, D.A., Macko, P., Maclean, D., Margo, D., Seltzer, M., Smogor, R.: Layering in provenance systems. In: Proceedings of the 2009 USENIX Annual Technical Conference (USENIX 2009) (2009)
4. Ko, R.K.L.: Data accountability in cloud systems. In: Security. Privacy and Trust in Cloud Systems, pp. 211–238. Springer-Verlag (2014)
5. Ko, R.K.L., et al.: TrustCloud: a framework for accountability and trust in cloud computing. In: 2011 IEEE World Congress on Services, pp. 584–588 (2011)
6. Zhan, O.Q., Krichberg, M., Ko, R.K.L., Lee, B.S.: How to track your data: the case for cloud computing provenance. In: 2011 IEEE International Conference on Cloud Computing Technology and Science, pp. 446–453 (2011)
7. Zhang, O.Q., Ko, R.K.L., Kirchberg, M., Suen, C.H., Jagadpramana, P., Lee, B.S.: How to track your data: rule-based data provenance tracing algorithms. In: 2012 IEEE 11th International Conference on Trust, Security and Privacy in Computing and Communication, pp. 1429–1437 (2012)
8. Ko, R.K.L., Jagadpramana, P., Lee, B.S.: Flogger: a file-centric logger for monitoring file access and transfers within cloud computing environments. In: 2011 IEEE 10th International Conference of on Trust, Security and Privacy in Computing and Communications(TrustCom), pp. 765–771 (2011)
9. Pearson, S., et al.: Accountability for cloud and other future internet services. In: IEEE 5th International Conference on Cloud Computing Technology and Science, pp. 629–632 (2012)
10. A4Cloud Project. http://www.a4cloud.eu/
11. Papanikolaou, N., Pearson, S., Mont, M.C., Ko, R.K.L.: A Toolkit for Automating Compliance in Cloud Computing Services. International Journal of Cloud Computing 3(1), 46–68 (2014)
12. Accorsi, R.: A secure log architecture to support remote auditing. Mathematical and Computer Modelling 57, 1578–1791 (2013)
13. Gehani, A., Ciocarlie, G.F., Shankar, N.: Accountable clouds. In: 2013 IEEE International Conference on Technologies for Homeland Security(HST), pp. 403–407 (2013)

A New Visual Pet Activities Monitoring System Design

Yahya Imad Mohammed and Jong Myung Rhee

Abstract This paper presents a new pet activities monitoring system via video recording in an automated way. It provides a solution for tracking pets and identifying their location through the use of a digital camera without the need for any supplementary hardware equipment. The system uses a modern object recognition algorithm with a motion detection technique. It makes use of a minimum amount of processor time in normal use. Motion detection alongside object recognition techniques increases the system reliability and reduces the response time to track a pet when the pet changes its location. Without using motion detection, the object recognition technique provides a higher error rate and a late response as compared with our utilized technique. This paper makes the most use of the object recognition technique and provides a high frame/second video recording and more accurate pet tracking.

Keywords Motion detection · Object recognition · Haar-like features

1 Introduction

Pets were once considered a leisure interest best kept at home. Nowadays, pets are a serious business for marketing. The demand for pet health care and monitoring systems has increased tremendously, as nine out of ten pet owners consider their pet a member of the family. The desire to provide electronic devices that monitor pets and continues reporting to the owner about their pet's condition has become a field of interest in research.

Having a video recording of a pet can provide information on the pet's condition to the owner. Capturing a long video of a pet with a fixed position camera gives the chance for the pet to move away from the scene and the video, which would not be that beneficial anymore. If the camera can detect the pet's location and rotate its angle to track the pet, it may maintain a reliable video recording.

Y.I. Mohammed · J.M. Rhee(✉)
Myongji University, Seoul, Republic of South Korea
e-mail: yahyamohamd@yahoo.com, jmr77@mju.ac.kr

© Springer Science+Business Media Singapore 2015
D.-S. Park et al. (eds.), *Advances in Computer Science and Ubiquitous Computing*,
Lecture Notes in Electrical Engineering 373,
DOI: 10.1007/978-981-10-0281-6_64

This paper proposes a visual pet activities monitoring system that has the ability to identify a pet and its location through the use of image processing techniques. It suggests an algorithm to recognize a pet's location through real-time video capturing. The system keeps tracking the pet's location and makes sure that the pet is located at the center of the camera lens. The user does not need to control the camera physically, while the system can process the video and recognize a location fast with a low error rate.

Applying image processing techniques to identify a pet's location can be a difficult task for the processor and might take a long time to execute. Human faces approximately share the same features (eyes, mouth, nose, …etc), one classifier might be enough to recognize human faces. On the contrary, pets might have many shapes and colors, therefore it is necessary to build a classifiers for each type of pets. Our proposed algorithm enhances the object recognition using Haar-like features [5] technique that adopted in this paper, and reduces the time required to identify and update the pet's location while it is moving.

2   System Design

Processing a video stream and maintaining a high frame/second ratio are challengeable tasks, as each frame requires a long time to be processed. To reduce the amount of time spent on each frame, the frame size could be reduced, but this method degrades the video quality. To avoid this, an alternate method was suggested, which identifies the location of interest and processes only a certain amount of pixels. The used algorithm has been developed in this system based on motion detection and object recognition techniques. The motion detection pinpoints the location of interest and the object recognition identifies the source of motion. The object recognition does not work unless there is a motion to decrease the processing time needed by the recognition process.

2.1   Motion Detection

Motion detection is a process of detecting changes in the position of an object relative to its surroundings and vice versa. There are many motion detection techniques suggested in image processing, such as background subtraction, also known as foreground detection [1], spatio temporal entropy [2], temporal difference, and optical flow analysis [3]. In this paper, the absolute difference method was used [4]. It has low overhead and does not require any previous learning; on the other hand, learning the background at each position would not be useful, as the background is changed by the camera rotation. The absolute difference method is achieved by comparing two consecutive frames in gray scale mode and determining the motion location, as illustrated in figure 1.

2.2   Object Recognition

Humans have the ability to recognize objects by sight using little effort, despite the fact that the image may have in illumination, rotation, scale, background, etc.

A number of object recognition approaches were proposed to recognize objects within digital images.

In this paper, Haar-like features, suggested by Viola and Jones [5], were selected for object recognition. Haar-like gained its name from the intuitive similarity with the Haar wavelet transform. It is the first successful real-time face detection approach. Haar-like is faster than any previous approach for image recognition and in contrast, it provides accurate results; therefore, we adapted the algorithm in our system.

Haar-like recognizes objects depending on the intensity distribution within the images. An example of Haar-like face detection features is the detection of the eye region. It is obvious that eyes are darker than cheeks, so we can specify the area of the eyes by calculating the darkness difference in the face. On the other hand, Haar-like is robust, because it is not highly effected by light variation.

Haar-like features consider two or three black and white adjacent rectangular regions at a specific location in a detection window (extended Haar-like features, as suggested by Lienhart and Maydt [6], are shown in Fig. 2). By computing the sum of the pixels at each triangular window and then comparing it to the adjacent windows,

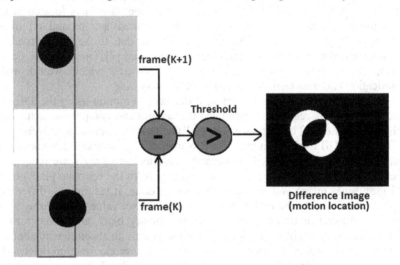

Fig. 1 Absolute difference technique.

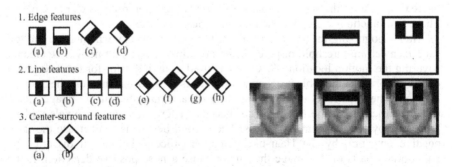

Fig. 2 Haar-like features and a detection sample

subsections of the image can be categorized. These subsections are used to identify the object by comparing it to a learned threshold. In our system, we used a large number of the pets' images for training a Haar-like features classifier to identify pets.

3   System Structure

This project contains three main elements: a camera, a PC, and an Arduino board. The camera is attached to a pan tilt base. The pan tilt has two servo motors to allow the camera to rotate in different directions. The camera is connected to the PC via a USB data bus to transfer the captured frames to the PC. The PC receives the frames and conducts all the processing needed. The obtained results contain the pet location. The PC transfers these results to the Arduino board via the RS232 bus. The Arduino board works as an interface between the PC and the servo motor.

After recognizing the pet's location by the PC, a command is sent to rotate the camera to the proper angle. The system ensures the pet is always located at the center of the screen. Each time the pet moves, the PC sends a command to redirect the camera to the pet's location. The camera sends raw data to the PC. To read the camera data and convert it to a frame, OpenCV was used [7]. OpenCV has a class that supports reading frames from the USB camera, which would make communication and transferring data to the PC an easy task.

The algorithm works as follows. First, it detects the motion using the absolute difference method by reading frames from the camera and comparing each frame with the previous one. The frames are in an RGB color model, so a conversion is made from the RGB format to a gray scale format. After conversion has ended, the absolute difference is calculated between the resulted two gray scale images by subtracting each pixel on the first gray scale image from the opposite pixel on the second gray scale image. When the subtraction is done, it results in a difference frame that contains only the difference between the pixel values of the gray scale frames. The difference frame is converted to a binary black and white format, where zero means an absolute black color and one means an absolute white color.

Now, the erode and dilate processes take place. The erode function deletes the noise that makes the pixel differences in the captured video frames, such as those caused by the camera's small mechanical vibration. The dilate function expands the real motion on the scene so that the object recognition process can take place. These two functions are called the morphological operation.

After recognizing the motion location, the system checks whether the motion exceeds a certain threshold of pixel difference; then, a mask frame will be created covering the motion location. This mask is used as an indicator for the next phase of the algorithm, the object recognition process, to focus only on the motion location. The first benefit of focusing only on a certain part of the image yields that the processing time is decreased, because a fewer amount of pixels will be processed by the recognition process. The second benefit is to decrease the false positive detection by the Haar-like detection process, because it discards the unimportant parts of the image that might cause a false positive detection. If the object recognition process detects a pet on the motion location, the center pixels

(X,Y) of the motion location are saved. Then, this location is calculated to measure the distance between the center of the motion and the center of the camera direction. After that, this distance is converted to the difference angle that the camera needs to be redirected. These results are then sent to rotate the camera to the direction of the pet. The whole system algorithm is shown in Fig. 3.

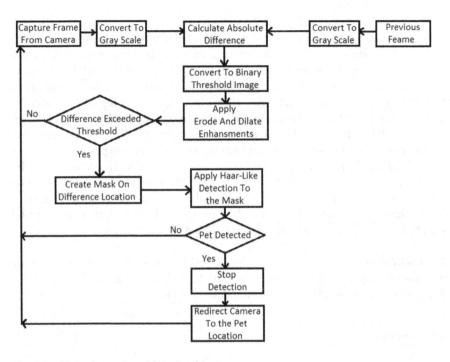

Fig. 3 Pet Detection and tracking algorithm

Table 1 Image recognition and motion detection versus image recognition only.

Image recognition and motion detection (Our proposed approach)	Image recognition only (Haar-like features)
Provides video recording service with range of 20-30 frame/second	Approximately provides 10 frame/second
Applied only to the moving objects therefore, static objects will be ignored to avoid any false positive detection.	Applied to whole image frame, which increases the chances of getting a false positive detection.
Low demand for the processor due to the minimized number of pixels that required to be processed.	Process all the pixels within an image frame, resulting in a high demand for the processor.

However, table 1 shows a comparison between image recognition alongside motion detection (our proposed approach) versus image recognition without motion detection (Haar-like features).

4   Conclusion

Appling motion detection alongside Haar-like features yields a lower error rate and faster detection. A faster detection decreases the time needed from the processor, so that the processor can execute other sets of instructions, such as capturing more frames from the camera to provide a smooth video recording. A low error rate gives the system the reliability to keep tracking objects and the ability to locate the pet among other unwanted objects, which is considered an unwanted false positive detection. On the other hand, the low resource consumption makes it possible to construct the system with cheaper hardware.

However, if there is a large number of moving objects within the sight, the error rate will be increased due to an occurrence of a false positive detection. As a future work, to decrease the error generated by unwanted moving objects, a motion sensor can be installed on the pet to recognize its current motion's state and match it with the sensor's readings that will be detected by the camera.

Acknowledgments This work (Grants No. C02370020100422266) was supported by Business for Cooperative R&D between Industry, Academy, and Research Institute funded Korea Small and Medium Business Administration in 2014.

References

1. Piccardi, M.: Background subtraction techniques: a review. In: IEEE International Conference on Systems, pp. 3099–3104 (2004)
2. Ma, Y.-F., Zhang, H.-J.: Detecting motion object by spatio-temporal entropy. In: IEEE International Conference on Multimedia and Expo ISBN, pp. 265–268 (2001)
3. Shuigen, W., Zhen, C., Hua, D.: Motion detection based on temporal difference method and optical flow field. In: Second International Symposium on Electronic Commerce and Security, pp. 85–88(2009)
4. Shraddha, G., Varma, S., Nikhare, R.: Visual surveillance using absolute difference motion detection. In: Technologies for Sustainable Development, pp. 1–5 (2015)
5. Viola, P., Jones, M.: Robust real-time object detection. Cambridge Research Laboratory Technical Report Series, pp. 1–24 (2001)
6. Lienhart, R., Maydt, J.: An extended set of haar-like features for rapid object detection. In: Proc. IEEE International Conference on Image Processing, pp. 900–903 (2002)
7. Culjak, I., Abram, D., Pribanic, T., Dzapo, H., Cifrek, M.: A brief introduction to OpenCV. In: MIPRO, pp. 1725–1730 (2012)

A Study of Security Level in Cloud Computing

Eric Niyonsaba and Jongwook Jang

Abstract Cloud computing is one of recent technologies paradigm that attracted many computer users. Among these are big, small enterprises and individual users. Cloud computing brought a lot of advantages where users can access computer services through internet. With cloud computing, there is no need of physical components as hardware that will support the company's computer system. Therefore, Cloud computing enables a set of entrepreneurs to start their business with zero investment on IT infrastructure. Data are stored and accessed in a remote server over internet. However, along with the good benefits of cloud computing has to offer, there are some security concerns which make users anxious about the safety, reliability and efficiency of migrating to cloud computing. So, before implementing cloud computing, organizations need to be aware of some security metrics. We will discuss about security concerns and analyze its level based on some of the proposed solutions.

Keywords Cloud computing · Security level · Security metrics · Security issues

1 An Overview on Cloud Computing

Cloud computing is a method of providing a set of shared computing resources that includes applications, computing, storage, networking, development, and deployment platforms as well as business process[1].

Based on cloud computing taxonomy defined by the national Institute of Standards and Technology (NIST), there are three service models [2] in cloud computing which are:

Software as a Service (SaaS): most known and first service, it is a major service given as a part of cloud computing. In this case, the application itself is provided by the service provider, typically via Web browser.

E. Niyonsaba(✉) · J. Jang
Department Computer Engineering, Dong-Eui University,
176 Eomgwangro, Busanjin-Gu, Busan 614-714, Korea
e-mail: niyonsabaeric@gmail.com, jwjang@deu.ac.kr

© Springer Science+Business Media Singapore 2015

D.-S. Park et al. (eds.), *Advances in Computer Science and Ubiquitous Computing*,
Lecture Notes in Electrical Engineering 373,
DOI: 10.1007/978-981-10-0281-6_65

Platform as a Service (PaaS): Providers deliver middleware (databases, messaging engines and so on) and solution stacks for application build, development and deploy.

Infrastructure as a service (IaaS): it is the delivery of computing infrastructure as a service. Provider offers capacity for rent basically hosted data center and servers. In Cloud Computing, the management of the technology and services has moved from user to the service provider's end. The cloud computing provides virtualized resources to the customers using various technologies, for example, Web services, virtualization, and multi-tenancy. The cloud services are delivered to the customer through the Internet. The Web applications are used to access and manage cloud resources that makes Web applications an important component of the cloud computing. The customers' processes are executed in virtualized environment that in turn utilize the physical resources. Multiple virtual processes of various users are allocated to same physical machines that are segregated logically. This gives rise to a multi-tenant environment in the cloud. Despite the provided advantages, the cloud computing is not exclusive of risks with security being the key risk. Security is one of the biggest obstacles that hamper the widespread adoption of cloud computing. Several business and research organization are reluctant in completely trusting the cloud computing to shift digital assets to the third-party service providers. Lack of security is only hurdle in wide adoption of Cloud Computing.

2   Literature Survey

Some of the proposed methods have been discussed in the literature survey for handling security issues in cloud computing. R.Velumadhava Rao and K.Selmani, presented data security challenges and provided some solutions for those challenges to overcome the risk involved in cloud computing [3]. Mazhar Ali, Samee U.khan and Athanasios V.Vasilakos, elaborated security issues from many authors and concluded that the cloud not only retains orthodox security concerns but also entails the novel issues arising due to the use of new technologies and practices. The issues for example of Web services and applications, communication and network, data privacy are the conventional issues that were present in the respective technologies even before the appearance of the cloud computing paradigm [4]. Keiko Hashizume, David G Rosado, Eduardo Femandez-Medina and Eduardo B Fernandez presented security issues for cloud models: IaaS, PaaS, and SaaS which vary depending on the model. They mentioned also a difference between vulnerabilities and threats. Threat is a potential attack that may lead to a misuse of information or resources and vulnerability refers to the flaws in a system that allows an attack to be successful [5].

3   Some of Security Metrics of Cloud Computing

According to the Cloud Security Alliance (CSA), an organization dedicated to ensuring security best practices in the cloud, significant areas of operational risk in the cloud include the following [6]. (1)*Physical security*: Covers security of IT equipment, network assets, and telecommunications infrastructure. (2)*Human resource security*: Appropriate controls need to be in place for the staff working at the facilities of a cloud provider, including any temporary or contract staff. (3)*Business continuity and Disaster recovery*: Cloud providers must have business continuity and data recovery plans in place to ensure that service can be maintained in case of a disaster or an emergency and that any data loss will be recovered. (4)*Incident handling changes in a cloud*: Working with your service provider to control at least part of infrastructure. (5)*Application security changes in the cloud*: covers application security into different areas including securing the software development lifecycle. (6)*Encryption and key management*: Ensures that only intended recipients receive data and can decrypt it. The recipient of an encrypted message uses a key that triggers the algorithm to decrypt the data and provide it in its original state to the authorized user.

4   Security Issues and Solutions

Based on literature review, here are the following security issues and solutions found.

4.1   Security Challenges

Data Security Challenges: Cloud computing presents an added level of risk because essential services are often outsourced to a third party, which makes it harder to maintain data security and privacy, support data and service availability, and demonstrate compliance. In multi-tenancy model, a single instance of application serves all customers. This approach enables more efficient use of resources but scalability is limited. Since data from multiple tenants is likely to be stored in the same database, the risk of data leakage between these tenants is high.

Virtualization Issues: Virtualization allows users to create, copy, share, migrate and roll back virtual machines (VM) which may allow them to run a variety of applications. Virtualized environments are vulnerable of attacks. The virtual Machine Monitor (VMM) or hypervisor is responsible for virtual machines isolation; therefore, if the VMM is compromised, its virtual machines may be compromised. Virtualization introduces the ability to migrate virtual machines between physical servers for fault tolerance, load balancing or maintenance. An attacker can compromise the migration module in the VMM and transfer a victim virtual machine to a malicious server. VM located on the same server can share CPU, memory, I/O, and others. Sharing resources between VMs may decrease the security of each VM.

Web Application and Application Programming Interface (API) Issues: Web applications are typically delivered via the internet through a web browser. Attackers have been using the web to compromise user's computers and perform malicious activities such as steal sensitive data. Cloud providers offer services that can be accessed through APIs (SOAP, REST, or HTTP with XML/JSON).The security of the cloud depends upon the security of interfaces.

Identity Management and Access Control Issues: There are many issues that can arise in cloud due to weak identity management and access control.

Challenges at Contractual and Legal Levels: Adopting the cloud computing, results in moving the organizations data and applications to the administrative control of CSP. This brings many issues to the front, for instance, performance assurance, regulatory laws compliance, geographic jurisdictions, monitoring of contract enforcement.

4.2 Some Solutions on Security Issues

Countermeasures for Data/Storage Issues: Encryption is suggested as a better solution to secure information (e.g. Advanced Encryption Standard).To identify unauthorized users, using of credentials or attributed based policies are better. Digital signatures and homomorphic encryption are also technologies used to secure data. Digital signatures proposes to secure data using digital signature with RSA (Rivest-Shamir- Adleman) algorithm while data is being transferred over the internet. Homomorphic encryption allows performing arbitrary computation on ciphertexts without being decrypted. A technique named Fragmentation-redundancy-scattering (FRS) is also used. This technique aims to provide intrusion tolerance and, in consequence, secure storage. This technique consists in first breaking down sensitive data into insignificant fragments, so any fragment does not have any significant information by itself. Then, fragments are scattered in a redundant fashion across different sites of the distributed system.

Countermeasures of Virtualization Issues: Trusted cloud computing platform (TCCP) which enables providers to offer closed box   execution environments, and allows users to determine if environment is secure before launching their VMs. Trusted virtual datacenter TVDc which insures isolation and integrity in cloud environments. It groups virtual machines that have common objectives into workloads named Trusted Virtual Domains (TVDs). TVDs provides isolation between workloads by enforcing mandatory access control, hypervisor-based isolation, and protected communication channels such as VLANs.

Countermeasures of Web Application and APIs Issues: Web application scanners is a program which scan web applications through the web front-end in order to identify security vulnerabilities. There are also other web applications security tools such as web application firewall. To protect the cloud applications from unauthorized access, the authors proposed the use of Diameter- AAA protocol. The diameter-AAA employs network based access control to filter the illegitimate access request to the cloud applications.

Countermeasures of Identity Management and Access Control Issues: CSA has issued an identity and access Management guidance which provides a list of recommended best practiced to assure identities and secure access management.

This report includes centralized directory, access management, identity management, role-based access control, user access certifications, privileged user and access management, separation of duties, and identity and access reporting.

Countermeasures of Contractual and Legal Levels Issues: The (Web services agreement) ws-agreement defines the syntax and semantics of publicizing the competences of the services providers and to create the template based agreements, and to monitor the agreement acquiescence. The ws-agreement mainly captures the agreement mainly based on quality of service. The SecAgreement extends the template of the ws-agreement to incorporate security constraints and metrics into the terms of SLA. The extended template also integrates the elements that quantify the risks of using specific cloud services.

4.3   Security Level Analysis

Table 1 Security metrics and some techniques used (solutions found in literature review) to improve them.

Security metrics	Techniques used
Incident handling changes in cloud	Virtualizations solutions: Trusted cloud computing platform (TCCP) and Trusted virtual datacenter (TVDc), Mirage,
Application security changes in cloud	Web applications and APIs solutions: Web applications scanners, Diameter-AAA protocol, Open Authorization Platform.
Identity and access management	Role-based access control, user access certifications, privileged user and access management, separation of duties, and identity and access reporting.
Encryption and key management	To identify unauthorized users, using of credentials or attributed based policies are better, use of encryption techniques such as AES (Advanced Encryption Standard) and homomorphic encryption, and Digital signatures method
Business continuity	Contractual and legal levels solutions which comprise Web services agreement and SecAgreement systems.

The purpose of this research is to let customers to understand the level risk they may be assuming when adopting a cloud computing. This level risk can be analyzed relying on solutions listed above to cloud challenges from literature review and cloud metrics. According to largest scope of cloud computing, we have selected relevant metrics and solutions. So far, cloud computing presents a good number of benefits for its users; however, it also raises some security problems which may slow its use. Even if solutions don't fit cloud security on hundred percent, understanding what vulnerabilities exist in cloud computing will help organizations to make the shift towards the cloud. Since cloud computing leverages many technologies, it also inherits their security issues. Solutions are needed in order to mitigate security issues. Security metrics provide visibility into how well the organization is executing on cloud security strategies. Because cloud computing represents a relatively new computing, there is a great deal of uncertainty about how security at all levels (network, host, application and data levels) can be achieved and how application security is moved to cloud computing. Based on security metrics relating to cloud can help an organization to measure, analyze and manage risk. This research differs from others in terms to emphasize on security metrics and solutions before adopting any cloud service.

5   Conclusion and Future Work

The specific solution would be to develop security tools covering all security metrics. But instead of this, users may rely on existing solutions and analyze security level according to security metrics and their expectations. In this paper, we present relevant security requirements and try to assign some existing solutions. It is a kind of a qualitative analysis. In the future a thorough analysis for measuring level security can be developed in order to avoid ambiguity. So far, even though cloud computing remains having security issues, it has more advantages for users to adopt it.

Acknowledgment This work was supported by the Human Resource Training Program for Regional Innovation and Creativity through the Ministry of Education and National Research Foundation of Korea (NRF-2015H1C1A1035898).

References

1. Hurwitz, J., Kaufman, M., Halper, F.: Cloud Services for Dummies, p. 6
2. Smoot, S.R., Tan, N.K.: Private Cloud Computing Consolidation, virtualization, and Service-oriented Infrastructure, pp. 38–41
3. Velumadhava Rao, R., Selvamani, K.: Data security challenges and its solutions in cloud computing. In: International Conference on Intelligent Computing, Communication and Convergence (ICCC-2015), pp. 204–209
4. Ali, M., Khan, S.U., Vasilakos, A.V.: Security in Cloud computing: Opportunities and Challenges. Information Sciences **305**(1), 357–383 (2015)
5. Hashizume, K., Rosado, D.G., Femandez-Medina, E., Fernandez, E.B.: An analysis of Security issues for cloud computing. Journal of Internet Services and Applications (2013). http://www.jisajournal.com/content/4/1/5
6. Hurwitz, J., Kaufman, M., Halper, F.: Cloud Services for Dummies, pp. 46–47

The Dynamic Unequal Clustering Routing Protocol Based on Efficiency Energy in Wireless Sensor Network

Nurhayati

Abstract A problem of unbalanced energy consumption exist and it primarily depend role and on location of a particular node in the network. The in order to make use of their potential, researchers must find solution to some difficulties that are slowing down the wide spread use these networks. The other case the problem faced by wireless sensor networks is the energy efficiency that are used during routing occurs, it affects the lifetime of sensor and residual energy of wireless network. Cluster-based routing protocols are not the same in wireless sensor network (UCR), nodes are grouped into clusters that are not the same size or Unequal Cluster Routing Protocol (UCR) has disadvantages in choosing the head of the cluster if the distance from the node away from the base station and the higher energy from other nodes cause trouble hot spots. In this research will offer a solution efficiency energy and balance energy consumption to prolong lifetime of sensor nodes based on Dynamic Unequal Clustering Routing Protocol namely by focusing the cluster will be done with consideration of dynamic routing on the cluster node are unequal cluster. The next step can do re-clustering node to change between nodes as cluster member or cluster head, so they can save energy on the condition of the hot spots on the node closest to the base station. The routing protocol algorithm called The Dynamic Unequal Cluster Routing Protocol Based on Efficiency Energy in Wireless Sensor Network (DUCRBEE). The research methodology used is primary and secondary data collection and simulation methods. And the results will be shown by the simulation comparing DUCR new algorithm that will extend the lifetime of sensor nodes when compared with BCDCP and UCR.

Keywords Wireless Sensor Network · Clustering · Unequal Clustering Routing (UCR)

Nurhayati(✉)
Department of Informatics Engineering, Science and Technology Faculty,
Syarif Hidayatullah State Islamic University (UIN) Jakarta,
Jl. Ir H Juanda N0. 95, Ciputat, Tang Sel, Banten, Indonesia
e-mail: nurhayati@uinjkt.ac.id

© Springer Science+Business Media Singapore 2015   457
D.-S. Park et al. (eds.), *Advances in Computer Science and Ubiquitous Computing*,
Lecture Notes in Electrical Engineering 373,
DOI: 10.1007/978-981-10-0281-6_66

1   Introduction

Wireless sensor network (WSN) is a collection of sensor nodes in a particular area. Sensor nodes are often used to monitor the environment such as air, lakes, and forests, and others. Each sensor node has the ability to collect data and communicate with other sensor nodes, the data captured by the sensor nodes will be processed and used as a reference for what to do to prevent something or improve something in the monitored area.

This research based the UCR. Consider the hot spots problem when multi-hop forwarding model is adopted while cluster head transmits their data to the base station. The UCR [1, 2] an unequal cluster-based routing protocol in wireless sensor networks, Cluster head closer to base station have smaller cluster size than those farther away from the base station; therefore they can preserve some energy to forward inter cluster data. UCR has weakness in choosing cluster head if the distance of the node is far from base station and the energy is higher than another node. At this condition it is difficult for UCR to choose cluster head and hot spot problem still persists.

2   Related Works

Base Station Control Dynamic Clustering Protocol (BCDCP) [7] every node has similar clustering like LEACH [8] and It have disadvantage when there is a large number of sensor nodes and cluster heads. Due to the large number, sensor nodes would need more energy for intra and inter cluster data transmission. This creates an unbalance in energy consumption and decreases network lifetime.

In [1,2] an unequal cluster-based routing protocol in wireless sensor networks (UCR), it groups node into cluster of unequal size. Cluster head closer to base station have smaller cluster size than those farther away from the base station; therefore they can preserve some energy to forward inter cluster data. UCR has weakness in choosing cluster head if the distance of the node is far from base station and the energy is higher than another node. At this condition it is difficult for UCR to choose cluster head and hot spot problem still persists.

We proposed solution based on related works. The new algorithm Dynamic Unequal Routing Protocol Based on Energy Efficient in Wireless Sensor Networks (DUCRBEE) has the same basic idea of an unequal cluster size (UCR) [1, 2].

3   The System Model and Assumption

3.1   The System Model

The network model of routing protocol system is based on following assumptions:

1. The Base Station located far from the sensing field. Sensors and the Base Station are stationary after deployment.

2. Sensors are homogeneous and have the same capability, and each node is assigned with a unique identifier (ID).
3. Sensors are capable of operating in an active node or low power sleeping mode.
4. Sensors are able to use power control to vary the amount of transmission power, which depends on the distance to the receiver.

3.2   The Energy Model

We also used and adopted a simple model [7 and 8] a known formula that is commonly used in many research [1, 2, 3, 4 , 5 ,6, 7, and 8] to calculate energy on each node. The energy spent for transmission of a 1-bit packet from the transmitter to the receiver at a distance d is defined as follow:

$$E_{Tx}(l,d) = E_{Tx-elec}(l) + E_{Tx-amp}(l,d)$$

$$E_{Tx}(l,d) = E_{elec} + l\varepsilon d^{\alpha} \tag{1}$$

$$E_{Tx}(l,d) = \begin{cases} lE_{elec} + l\varepsilon_{fs} d^2, d < d_0 \\ lE_{elec} + l\varepsilon_{tg} d^4, d \geq d_0 \end{cases}$$

E_{Tx} is Energy Transmission when energy is dissipated in the transmitter of source node. Elec is Energy Electronic where the per bit energy dissipation for running the transceiver circuitry $\varepsilon_{fs}.d^2$ or $\varepsilon_{tg}.d^4$ is equal to the Amplifier Energy depends on transmission distance and acceptable bit-error rate. Where α is the propagation exponent and basically dependent on factor as digital coding and modulation.

The cross over distance d_0 can be obtained from:

$$d_0 = \sqrt{\frac{\varepsilon_{ft}}{\varepsilon_{mp}}} \tag{2}$$

The energy expended to receive message is:

$$E_{Rx}(l) = lR_{elec} \tag{3}$$

E_{Rx} is energy received when energy is needed to received message.

When received data cluster head performs data fusion on received data packets, it assumes that the sent information is highly correlated, thus the Cluster Head can always aggregate the data gathered from members into single length-fixed packet. The energy consumed by cluster Head to receive, E_{Rx} and data fusion or aggregate data (EDA) is derived in equation below. To receive this message the radio expends energy:

$$E_{Rx}(l) = (l)E_{Rx-elec} = l * E_{elec} = l * E_{elec} + E_{DA} \tag{4}$$

Each node consumed the energy for transmission data. The total cluster's energy consumption ($E_{cluster}$) is the sum of the energy consumed by member nodes used to send data to the CH and the energy consumed by CH for receiving an l-length packet data for aggregating and forwarding l-length packet data to the next hop.

4 Dynamic Unequal Clustering Routing Protocol Based on Efficiency Energy in Wireless Sensor Network

The DUCRBEE Algorithm has three phases will be described:

4.1 Set-up Phase

Set-up phase is the process where all algorithms make routing protocol in the sensor area. In this routing, sensor is homogenous and deployed in random area. After the deployment stage, each sensor becomes static and then each sensor node will send a message with all information, which contains of ID Node (ID), Residual Energy (RE) node, and distance (d) from node to base station. That information would be stored in Table-Information-Messages on the Base Station (BS). After all nodes send the information, a few nodes, which have the highest energy, will be selected as the cluster head (CH). The CH can be the number of cluster head depends on the number of cluster to be created and the rest will be a cluster member. Then the base station will process the data in table-information-message to determine the distribution of the cluster. WSN will be divided into several clusters with unequal size and cluster closer to the base station has a smaller size.

4.2 Steady-Set Phase

Steady-Set Phase is a process of data transmission of all nodes member in the clusters that send data until it reaches the base station. The sensor node sends data to the cluster head in each cluster. Then it will be divided in groups with maximum three clusters in each group. Once the group is formed, one of CH in clusters in each group will be selected as cluster head leader (CHL). Then CH will send data to CHL in each group. CHL will send data to the other CHL that closest to the base station. This process of re-clustering will continue to repeat until the entire CH connected, which then be forwarded to the base station.

4.3 Reconstruction Phase

Reconstruction phase is usually used for maintenance to modify routing in accommodations for increased use of energy, which means the use of energy must be spread among the nodes is WSN area and generate the extension of the network. If energy in CHL becomes lower than the energy in neighboring node, it can change and choose another CH as the CHL to send data to base station. The CHL rotated is done to balance energy consumption and prolong the network lifetime.

5   Analysis and Simulation Result

The sensor node set is S ={s1, s2, ..., s_N}, where |S|=N is a random number of sensor nodes. Sensing area within the WSN is divided into several clusters in order to decide whether this node will be a cluster head (CH) or cluster member (CM). After the division of the cluster, will be grouped, and will be decided based on the testing of the entire CH. CH group members will be selected by the energy value of the highest node are in the same group will be CH. Between CH in the same group will be selected CHL which has the highest value as a receiver node CH in sending data from other CH.

Table 1 SIMULATION PARAMETERS

Parameter	Value
Network field	$(0,0) - (100,100)$
Base Station location	$(150,50)$m
N	100
Initial Energy	1 J
E_{elec}	50 nJ/bit
ε_{fs}	10 pJ/bit/m2
d_0	87 m
EDA	5 nJ/bit/signal
Data Packet size	400 bits

The paper focuses on balancing energy consumption and efficiency of a network's lifetime, the performance of DUCRBEE with cluster head characterized by unequal clustering algorithm was evaluated as well as the parameter setting and the energy efficiency of DUCRBEE in term of a network life time. The DUCRBEE protocol was compared to that of the BCDCP and UCR. The simulation model of the network consists of 50 nodes and randomly scattered on the WSN area of 100 mx 100 m with a base station located at (150 m, 50 m). Each node in the beginning of the simulation has energy of 1 Joule. Each node transmits 4000-bit data packets each revolution. Calculation of energy reduction in this simulation is based on the equation (1) to equation (3) on the radio model. The parameters used in this simulation can be seen in table 1.

The following is the result of the comparison algorithm DUCRBEE with UCR and BCDCP in network lifetime (number of sensor nodes alive over time).

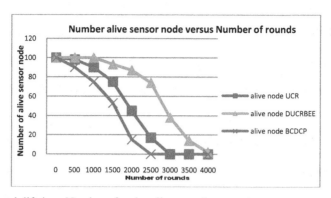

Fig. 1 Network lifetime: Number of nodes alive over time

Figure 1 shows that DUCRBEE clearly improve the lifetime the network compared to the UCR. This can be seen in the number of sensor nodes that are still alive, both of the first die and the sensor node to the last node sensor die. At UCR, CH will be selected randomly based on the amount of residual energy and the distance between the sensor nodes to the base station, but on DUCRBEE besides using the same components as the UCR, the base station cluster will have a cluster size smaller than the distance clusters with base station. DUCRBEE have added value, namely, clusters have been formed are grouped into several groups with maximum three clusters in one group. The UCR died faster 0.35% then DUCR and BCDCP died faster 0.45%. It proves DUCRBEE is more efficient for improving energy efficient than the UCR and BCDCP.

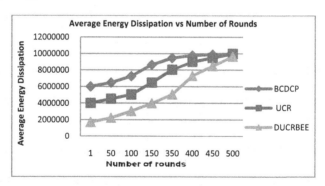

Fig. 2 Average Energy Dissipation vs Number of Rounds

Figure 2 shows that DUCRBEE clearly improve the energy dissipation of sensor network compared to the UCR and BCDCP. This can be seen in the number of sensor nodes that are still having amount of residual energy on the number of sensor node. The DUCREE have residual energy 0.22% then UCR and 0.35% then BCDCP. It proves DUCRBEE is more efficient for improving energy efficient than the UCR and BCDCP.

6   Conclusion

This paper focus on efficiency energy and balance energy consumption to prolong lifetime of sensor nodes based on Dynamic Unequal Clustering Routing Protocol in WSN. Based on the current argument, under normal circumstances Residual Energy and Node Location are not sufficient in maintaining balanced energy consumption within a network. This is due to the fact that when a node is located close to the base station, its' Residual Energy would be low. And normally, a node would have a high value residual energy only when it is placed far away from the Base Station. Hence we have proposed the Dynamic Unequal Clustering Routing Protocol Based on g Energy Efficient Wireless on Sensor Network. Simulation Result demonstrated that DUCRBEE achieve significant efficiency energy and enhance network lifetime compare to BCDCP and UCR. We had shown that DUCRBEE achieved better performance than other clustering based routing protocol.

References

1. Li, C.F., Ye, M., Chen, G., Wu, J.: An energy-efficient unequal clustering mechanism for wireless sensor networks. In: IEEE International Conf. Mobile Adhoc and Sensor Systems, p. 8, November 2005
2. Chen, G., Li, C.F., Ye, M., Wu, J.: An unequal cluster-based routing protocol in wireless sensor networks. Wireless Network. 15, 193–207 (2009)
3. Jazyah, Y., Hope, M.D.: A new routing protocol for UWB MANET. In: Proceeding of the 10th WSEAS International Conference on Eroupean Conference System (ECS 2010), and Signal Processing, Cambridge, UK, pp. 24–28, February 2010
4. Nurhayati: Energy efficient based on mechanism unequal clustering routing protocol wireless sensor networks. In: Interrnational Conference In Recent Researchers in Communications, Signal and Information Technology (SITE 2009), Saint Malo, Italia, pp. 114–119 (2012)
5. Nurhayati, : Re-cluster Node on Unequal Clustering Routing Protocol Wireless Sensor Networks for Improving Energy Efficient. International Journal of Computer and Communication 6(3), 157–166 (2012)
6. Nurhayati, Choi, S.H., Lee, K.O.: A Cluster Based Energy Efficient Location Routing Protocol in Wireless Sensor Networks. International Journal of Computer and Communications 5(2), 67–74 (2011)
7. Muruganathan, S.D., Ma, D.C.F., Bhasin, R.I., Fapojuwo, A.O.: A Centralized Energy-Efficient Routing Protocol for Wireless Sensor Networks. IEEE Radio Communications, March 2005
8. Heinzelman, W., Chandrakasan, A., Balakrishnan, H.: Energy-efficient communication protocol for wireless microsensor networks. In: Proceedings of the 33rd Hawaii International Conference on System Sciences (HICSS 2000), January 2000

A Study on the Localization Algorithm Using RSSI and Directional Antennas Between Sensor Nodes for the DV-HOP Algorithm

MunSuck Jang, ByungChul Kim, PyungSoo Kim and EungHyuk Lee

Abstract This study proposes an algorithm that estimates locations of sensor nodes using directional antennas, which radiate or receive electromagnetic waves effectively in a specific direction for the RDV-HOP algorithm. It shows that the wireless communication range is presented as a maximum value for the directivity value of 6 in 2-way antennas. Alto, it is verified that localization errors are measured as the smallest value.

Keywords Localization algorithm · DV-HOP · RSSI · Directivity · Directional antenna

1 Introduction

Wireless sensor networks become an essential technology for implementing near field communication in different sensor nodes, which represent small, low cost, and multi-function, and for supporting ubiquitous environments. It consists of lots of sensor nodes in order to get physical data. For estimating locations of sensor nodes including estimations of user locations in recent smart environments, several methods such as AOA (Angle Of Arrival)[1], TOA (Time of Arrival)[2], TDOA (Time Difference Of Arrival)[3], RSSI(Receiced Signal Strength Indication), and so on are introduced. Also, studies on the algorithms of APIT[4], APS, and DV-HOP [5,7] have been conducted.

In particular, [7] poposes an RDV-HOP algorithm that estimates locations of sensor nodes by classifying the sections, which allow or disallow to identify distances using RSSI in the communication between sensor nodes for the DV-HOP algorithm that measures the distance between sensor and reference

M. Jang(✉) · B. Kim · P. Kim · E. Lee
Department of Electronic, Korea Polytechnic University, Siheung-si, Gyeonggi-Do, Korea
e-mail: msjang@kpu.ac.kr

© Springer Science+Business Media Singapore 2015
D.-S. Park et al. (eds.), *Advances in Computer Science and Ubiquitous Computing*,
Lecture Notes in Electrical Engineering 373,
DOI: 10.1007/978-981-10-0281-6_67

nodes through calculating the number of HOPs between sensor nodes based on the one-hop distance calculated by the reference node. This proposal has been investigated through isotropic antennas in which antennas in all nodes represent even radiations and no losses in all directions and the isotropic antennas show directivies of practical antennas. However, there exists no such antenna practically and most antennas represent its directivities.

In this study an algorithm that estimates locations of sensor nodes using a directional antenna, which has characteristics of radiating and receiving electromagnetic waves effectively in a specific direction for the RDV-HOP algorithm, is proposed. The simulation implemented in this study was performed using different antennas such as Omnidirectional, which represents a directivity of whole directions, 1-way, 2-way, and 4-way antennas.

2   Localization Algorithm Model

RSSI in wireless sensor networks is an index of representing the reliability of signals using the intensity of received waves in the transmission of wireless data between nodes. Although the measurement of RSSI shows a variation in data due to the radiation pattern and performance in an antenna, the distance between nodes can be measured using linear RSSI data as nodes show small distances between sensor nodes, i.e. nodes are near to each other. Thus, the RDV-HOP algorithm uses a method that recognizes locations of sensor nodes through classifying the distance as estimative and none estimative by measuring the RSSI data.

In this study an algorithm that measures accurate locations of sensor nodes in the reference node using antenna radiation patterns (Omni-, 1-way, 2-way, and 4-way directional radiations) is proposed based on the RDV-HOP algorithm. Fig. 1 shows a network model of the localization algorithm using RSSI information.

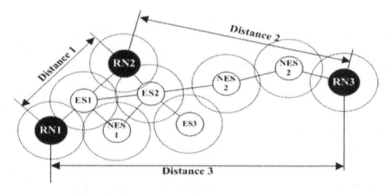

RN : Reference Node(AP)
ES   : Estimative Sensor node(Smart Phone)
NES : None Estimative Sensor node(Smart Phone)

Fig. 1 Network model for the localization algorithm using RSSI

In addition, the antenna has a characteristic that the antenna radiation property is presented by an antenna pattern (radiation pattern) with a mathematical function or graph using a function of space coordinates. The radiation property includes power flux density, radiation intensity, field strength, directivity, phase, and polarization and is identified as isotropic, directional, and omnidirectional.

Antennas represent directivities according to its directional properties as shown in Fig. 2. The directivity is defined as the ratio of the radiation intensity (x_1) for a given direction for the average radiation intensity (x_0) in all directions.

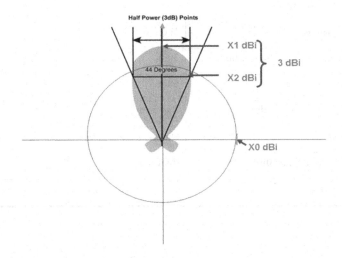

Fig. 2 Directivity and HPBW in an antenna

$$D_0 = 10 \log \frac{x_1}{x_0}$$

Also, the angle between two points (x_2), which becomes a half of the maximum radiation in a plan that includes the maximum direction, is defined as the Half-Power BeamWidth, HPBW. Thus, the directivity is determined according to antenna radiation patterns.

In this study, an approximate equation of Pozar was used. The equation is defined as follows.

$$D_0 \simeq -172.4 + 191 \sqrt{0.818 + \frac{1}{HPBW\,(deg\,ree)}}$$

3   Simulation and Analysis

The configuration of the wireless sensor network simulation implemented in this experiment is presented in Table 1.

M. Jang et al.

Table 1 Configuration of the wireless sensor network simulation

Parameter		Value
Network (Field) scale		500m X 500m
Network model		General, '⌐', '⊏', and '□' shapes
Number of reference node		10
Number of sensor node		500
Data transmission coverage	Ref. node	100m
	Sensor node	20m
RSSI effective range		60m
Distribution of sensor nodes		Random
Antenna type		Omni-, 1-way, 2-way, and 4-way directional
Distribution of antenna direction		Random
Antenna directivity		4,6,8, and 10

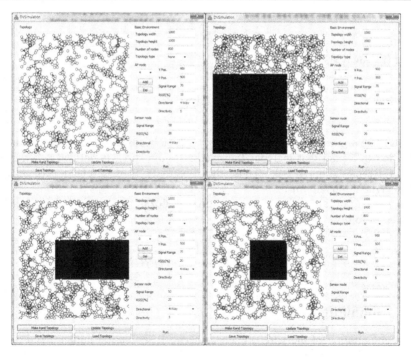

Fig. 3 Different network models

The simulation was performed for obtaining both the average number of HOPs and the location error distance data in the entire sensor nodes according to antenna types, such as Omni-, 1-way, 2-way, and 4-way directional, based on the RSSI effective distance of 60m, which is 60% of the maximum transmission range. In addition, the location error distance data was compared by obtaining it changing the directivity as 4, 6, 8, and 10 in a 2-way directional environment.

The simulator applied in this study used different network shapes (¬, ⊏, and ⊐ shapes) in order to apply it for various environments. Fig. 3 shows different network models.

4 Conclusion

Fig. 4 shows the distance error distribution in using a 2-way directional antenna. It reveals that there is the smallest error as the directivity is determined by 6 and the error is increased according to directivity values of 8. 10, and 4. Although the maximum transmission distance is increased according to increases in the directivity, the wireless communication range is rather decreased as the directivity exceeds a specific value because the HPBW is decreased. Thus, it is verified that the maximum wireless communication range is presented by the directivity of 6 and it also shows small localization errors.

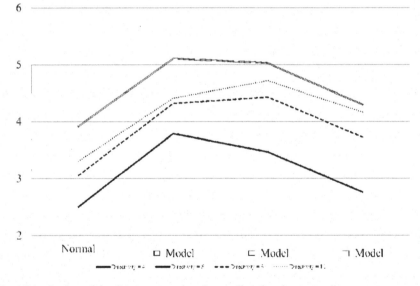

Fig. 4 Distribution of the distance error in a 2-way directional antenna

Acknowledgement This research was supported by the MSIP(Ministry of Science, ICT and Future Planning), Korea, under the C-ITRC(Convergence Information Technology

Research Center) (IITP-2015-H8601-15-1003) supervised by the IITP(Institute for Information & communications Technology Promotion).

This research was supported by the KIAT(Korea Institute for Advancement of Technology), Korea, under Bio-Product GMP Technical Education Program. (N0000961).

References

1. Bahl, P., Padmanabhan, V.N.: RADAR: an in-building RF-based user location and tracking system. In: Proc. of IEEE INFOCOM, vol. 2, pp. 775–784 (2000)
2. Caffery, J.: A new approach to the geometry of TOA location. In: Proc. of IEEE Vehicular Technology Conference (VTC), pp. 1943–1950 (2000)
3. Bahl, P., Padmanabhan, V.N.: An high performance distributed computing. In: Proc. of IEEE INFOCOM, vol. 2, pp. 775–784 (2000). IEEE Press
4. He, T., Huang, C., Blum, B.M., Stankovic, J.A., Abdelzaher, T.: Range-free localization schemes for large scale sensor networks. In: Proc. of the International Conf. on Mobile Computing and Networking, pp. 81–95 (2003)
5. Niculescu, D., Nath, B.: Ad-hoc positioning system. In: Global Telecommunications Conference (GlobeCom), vol. 5, pp. 2926–2931 (2001)
6. Niculescu, D., Nath, B.: DV Based Positioning in Ad hoc Networks. Journal of Telecommunication Systems 22, 267–280 (2003)
7. Jang, M.S., Song, I.S., Kim, P.S.: A Study on Localization Algorithm using Hop Count and RSSI. International Journal of Control and Automation 6(3), 83–94 (2013)

Estimation of Human Social Relation Based on Device Connectivity

Dong-oh Kang and Changseok Bae

Abstract People in these days use personal smart device as a typical device to communicate with others. Talks, texts, SNS, and even e-mails are available over the smartphones. Thus, we can consider the smartphones are able to represent personal closeness. This paper proposes an estimation model of human social relationship based on analysis of device connectivity. We have collected device-level communication data, and analyze them to find out correlation with human social relationships. Experimental results show that the proposed estimation model has linear relationship between device connectivity and human social relations. We can employ these results to construct sociality between devices.

Keywords Human social relation · Device social model · Device connectivity

1 Introduction

Ubiquitous computing environments are advancing into IoT (Internet of Things) worlds. IoT means an infrastructure of interconnected objects, people, systems, and information resources together with intelligent services to allow them to process information of the physical and virtual world and react [1]. Recent smart sensors and smart devices are making it possible to implement the IoT world. IoT worlds usually consider smart sensors as a platform device. Smart devices such as smartphones are much more important in IoT world, because modern smartphones embed a lot of sensors and process collected data to provide intelligent services.

D.-o. Kang
SW Contents Research Lab., Electronics and Telecommunications Research Institute,
Daejeon, Korea
e-mail: dongoh@etri.re.kr

C. Bae(✉)
Department of IT Convergence, Daejeon University, Daejeon, Korea
e-mail: csbae@dju.kr

© Springer Science+Business Media Singapore 2015   471
D.-S. Park et al. (eds.), *Advances in Computer Science and Ubiquitous Computing*,
Lecture Notes in Electrical Engineering 373,
DOI: 10.1007/978-981-10-0281-6_68

Intelligent and autonomous collaboration of smart things is one of the most essential parts in the IoT world. Agent-based collaborations are intuitive approaches for context awareness and intelligent services [2, 3]. Another approach which comes from a relatively recent paradigm is a method to assign social relationship among things. This approach is known as Social Internet of Things (SIoT) or device socialization [4-6].

Modern personal smart devices provide lots of communication methods with other people. We can talk and text with our friends and family members over the smartphones. In addition, we can do SNS and e-mails using them as well. The personal smartphones are a typical device to communicate with others. Thus, we are able to assume communication history on the personal smartphones reflect personal closeness. This means we can estimate human social relationship by analyzing device connectivity. Relationship between device connectivity and human sociality enables to assign sociality to smart devices. This device socialization is the key technology of the intelligent device collaboration.

This paper consists as follows. Section 2 briefly describes related works, and Section 3 explains the proposed scheme. Then, in Section 4, we present experimental results. Finally, we provide conclusions and future works.

2   Related Works

In this section, we discuss a few researches regarding to device social network. According to sociological research result, social closeness between people depends on mutual interaction time and frequency [7]. This result encourages our research for investigating the relationship between human sociality and device connectivity.

Social network of intelligent objects, called Social Internet of Things (SIoT) [4] is the most representative work for device social network services. This work introduces social network concepts among things similar to human SNS (Social Network Services). Device social network based on SIoT framework is employed to implement intelligent service composition. Our work can be a basic part for assigning sociality of smart device based on the relationship between device connectivity and human sociality model. Another attempt is the integration of the existing SNS infrastructure to smart home devices [8]. This work presents a Facebook application for adding and sharing smart home devices with a predefined group by users' intervention. Our work can provide a scheme for autonomous collaboration among smart home devices based on human sociality model.

In addition, we can consider a group clustering method by suggesting friends based on implicit social graph [9]. In this work, email interactions of users are a key component for constructing implicit social graph. This work only considers email interactions which provides less semantic information compared with our scheme. Our work employs phone calls, texts, SNS, and emails as well.

3 Human Sociality Based on Device Connectivity

As described in the previous section, the motivation of our research mainly comes from sociological research result about the relationship between human closeness and level of mutual interactions. This paper considers device connectivity as mutual interactions in sociological research, and investigates social closeness between people from their device connectivity. In this paper, device connectivity consists of emails, texts, phone calls, and intercepts. Similar to the sociological research, we can assume linear combination of these device connectivity parameters depends on social closeness between people. Based on this assumption, we can estimate social closeness of people i and j from degrees of device connectivity by using linear regression model, as shown in Eq. (1).

$$Social\ closeness_{ij} \approx \alpha E_{ij} + \beta M_{ij} + \gamma C_{ij} + \varepsilon,\tag{1}$$

where, 'E' means email interactions, 'M' means instant message interactions, and 'C' means phone call interactions between people i and j. In addition, α, β, and γ are linear regression parameters for each interaction and ε is error term for linear regression model.

We can easily collect times and frequencies for emails, instant messages, and phone calls between other people from a smartphone. Then, we can find the estimation of social closeness between other people by applying linear regression analysis shown in Eq. (1).

As a reference data for social closeness between people, we use direct survey for people by asking to make measure closeness between other people. For example, we can make subjects measure social closeness from score 1 (low) to score 5 (high). This paper considers two kinds of human closeness such as private closeness and business closeness.

We compare this reference and estimated values of social closeness to investigate validity of our assumption. This paper employs Pearson correlation coefficient for the comparison shown in Eq. (2).

$$r = \frac{S_{xy}}{\sqrt{S_{xx}S_{yy}}},\tag{2}$$

$$where\ S_{xx} = \sum_{i=1}^{n} x_i^2 - \frac{\left(\sum_{i=1}^{n} x_i\right)^2}{n},\ S_{yy} = \sum_{i=1}^{n} y_i^2 - \frac{\left(\sum_{i=1}^{n} y_i\right)^2}{n},\ S_{xy} = \sum_{i=1}^{n} x_i y_i - \frac{\left(\sum_{i=1}^{n} x_i\right)\left(\sum_{i=1}^{n} y_i\right)}{n}.$$

If errors or deviations between the reference and estimated values are quite small, we can find social closeness of people from the linear combination of device connectivity parameters such as emails, instant messages, and phone calls times and frequencies.

4 Experimental Results

In order to find device connectivity, this paper collects interactions among 25 devices of 10 subjects for 7 days. Age distribution of 10 subjects is from late twenties to early fifties to make sure diversity of subjects. Data structure of collected device interactions is shown in Table 1.

Table 1 Device connectivity data structure

Data element	Meaning of data element
Interaction ID	Sequential number (integer type)
Send/Receive	Attribute for identifying send interaction or receive interaction
Sender ID	ID of sender
Receiver ID	IDs of receivers
Time stamp	YY-MM-DD HH-MM-SS type time stamp of interaction
Device ID	ID of device used for interaction
Interaction type ID	ID of interaction type (1: email, 2: text, 3: phone call, 4: SNS)

Our experiment consists of 3 steps which include data collection, personal closeness survey, and Pearson correlation analysis.

In the first step, we have collected total 397 interactions such as emails, text messages, phone calls and SNS posts. Figure 1 (a) and (b) illustrate interactions among devices and subjects respectively. The thicker lines are the heavier interactions.

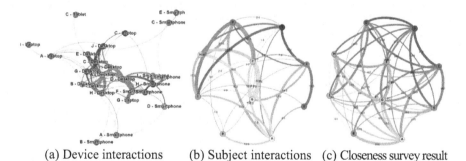

(a) Device interactions   (b) Subject interactions   (c) Closeness survey result

Fig. 1 Interactions among devices and subjects, and closeness survey result

The second step is personal closeness survey. We investigate personal closeness among subjects using direct survey. Figure 1 (c) shows the reply of subjects about the personal closeness among subjects, whereas Figure 1 (b) shows human interactions from device connectivity. We can find intuitively these two pictures are quite similar. In the next step, we can compare the similarity quantatively.

The last step is Pearson correlation analysis. We have compared the relationship between subjects' interactions using their smart devices and personal closeness survey results. Figure 2 shows correlation coefficients and p-values for subjects. We can find correlation coefficient values between device connectivity and personal closeness survey are quite high for most subjects except subject J. On the contrary, p-values are very low for most subjects except J.

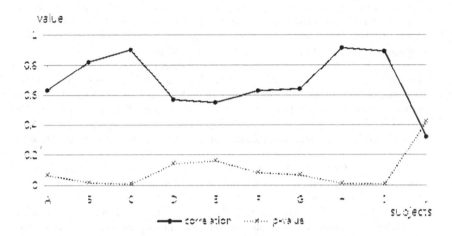

Fig. 2 Relationship between device connectivity and personal closeness

5 Conclusion

In this paper, we have proposed a novel estimation model for human social relationship based on the analysis of device connectivity history. We can consider device-level communication data represent owner's personal relation between others. We have collected talks, texts, SNSs, and e-mails over a personal smartphone, and analyzed them to find out their relationship with human sociality. Based on the experimental results, we have found human social relationship has linear correlation with device connectivity.

These results represent we can construct a socialization model among devices based on human social relationship. Not only sociality among devices but also machine socialization is considered as a key concept for implementing future intelligent IoT (Internet of Things) world such as smart factories. For the future work, we are going to extend our work to implement machine socialization.

Acknowledgement. This research was supported by the ICT R&D program of MSIP/IITP, [B0101-15-0239, Human Friendly Devices (Skin Patch, Multi-modal Surface) and Device Social Framework Technology].

References

1. ISO/IEC JTC 1/SWG 5, Study Report on Internet of Things (IoT). ISO/IEC JTC 1/SWG 5 N00238 (2014)
2. Talia, D.: Clouds Meet Agents: Toward Intelligent Cloud Services. IEEE Internet Computing **16**, 78–81 (2012)
3. Tapia, D., Rodríguez, S., Corchado, J.: A distributed ambient intelligence based multi-agent system for Alzheimer health care. In: Pervasive Computing Computer Communications and Networks, pp. 181–199 (2010)
4. Atzori, L., Iera, A., Morabito, G.: SIoT: Giving a Social Structure to the Internet of Things. IEEE Communications Letters **15**, 1193–1195 (2011)
5. Kosmatos, E., Tselikas, N., Boucouvalas, A.: Integrating RFIDs and smart objects into a unified internet of things architecture. Advances in Internet of Things **1**, 5–12 (2011)
6. Kang, K., Kang, D., Bae, C.: Novel approach of device collaboration based on device social network. In: The 2013 Int. Conference on Consumer Electronics, pp. 248–249 (2013)
7. Granovetter, M.: A Theoretical Agenda for Economic Sociology, CCOP UC Berkeley (2000)
8. Kamilaris, A., Pitsillides, A.: Social networking of the smart home. In: Proc. of International Symposium on Personal Indoor and Mobile Radio Communications, pp. 2632–2637 (2010)
9. Roth, M., Ben-David, A., Deutscher, D., Flysher, G., Horn, I., Leichtberg, A., Leiser, N., Matias, Y., Merom, R.: Suggesting friends using the implicit social graph. In: Proc. of the 16th ACM SIGKDD International Conference on Knowledge Discovery and Data Mining, pp. 233–242 (2010)

Secure Framework for Software Defined Based Internet of Things

SeongHo Choi and Jin Kwak

Abstract Recently, the Internet of Things (IoT) has come into the spotlight. Therefore, these network environments require methods to manage efficiency and security. SDN technology is one of the various realms of study that has been applied to these methods. The present study presents an environment that controls a network using an application. On the basis of this method, software-defined technology can provide efficient network control and security, which is defined using the application. In this paper, we studied the configuration of a software-defined IoT environment and its secure framework.

Keywords SDN · SDP · SDSec · IoT · SDIoT

1 Introduction

Recently, IoT (Internet of Things) technology has been developing, creating many smart devices, combining many network technologies, and providing many services to users. In the process, much attention has been paid to software-defined technology for management such as for network technology, traffic, packets, and so on. Software-defined technology allows for the control of the network through applications, and provides a variety of services. The present method separates network control functions from the network device on which they are implemented into various network functions using applications in the control layer. This environment, in addition to controlling the various network technologies, is able to provide device-specific security details to the software. This property can be flexibly managed by the network as a whole.

S. Choi
ISAA Laboratory, Department of Computer Engineering, Ajou University, Suwon, Korea
e-mail: csha123@ajou.ac.kr

J. Kwak(✉)
Department of Information and Computer Engineering, Ajou University, Suwon, Korea
e-mail: security@ajou.ac.kr

© Springer Science+Business Media Singapore 2015   477
D.-S. Park et al. (eds.), *Advances in Computer Science and Ubiquitous Computing*,
Lecture Notes in Electrical Engineering 373,
DOI: 10.1007/978-981-10-0281-6_69

Y. Jararweh et al., proposed a software-defined IoT framework. This framework was proposed to simplify IoT management processes and provide a software-defined security control model [2]. In this paper, we build upon previous studies in terms of security. The proposed analysis method examines security requirements and secure frameworks for software-defined IoT.

This paper is organized as follows: In Section 2, we analyze software-defined IoT. In Section 3, we analyze security requirements for frameworks within SDIoT (software-defined Internet of Things). In Section 4, we propose secure frameworks for SDIoT. In section 5, we present our conclusions.

2   Software-Defined Internet of Things

The SDIoT environment sends control messages to many network management technology controllers. IoT refers to smart devices in all areas, creating large amounts of data and using various services. In this environment, many researchers have studied technologies for processing such large numbers of smart devices, amounts of traffic, numbers of packets, and so on. Software-defined systems are one applicable technology; they are include such as SDN, SDP, and SDSec. It is possible to control all of the devices connected to networks; this technology can be used in an IoT environment to enable efficient network management. SDIoT is a network structure that can be simplified more easily than a legacy network, because it provides network management and security functions in the application within the controller. The network infrastructures require only OpenFlow-based network devices for networking. The network can receive data to provide secure channels, access control, authentication, integrity, and confidentiality in a software-defined control system.

Fig. 1 shows the SDIoT network layout.

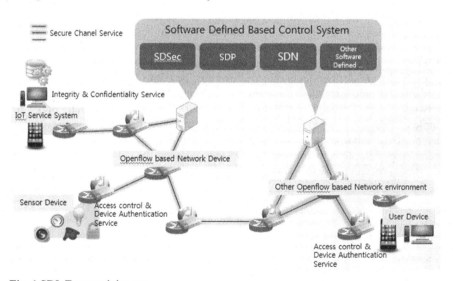

Fig. 1 SDIoT network layout

3   SDIoT Secure Framework Requirements

In the IoT environment, a variety of smart devices are connected using networking technologies; therefore, the network control environment can be complex. However, SDIoT does not use a variety of network control devices; instead, it provides a controller in the control layer; therefore, the network environment is simple. The controller uses a variety of software-defined systems.

The IoT environment have security requirements for Device, network, service and interface. They are included that Access control, Authentication, integrity and confidentiality. Table 1 shows the IoT security requirements.

Table 1 IoT security requirements

Index	Security requirements
Device	Access control, authentication, device integrity, data integrity, and confidentiality
Network	Access control, authentication, data/signal integrity, and confidentiality
Service interface	Access control, authentication, data integrity, confidentiality, privacy, and anti-virus

For IoT security requirements, the controller must meet the following requirements:

❐ Authentication and identification

The IoT environment requires technology for device identification and authentication. However, sensor devices use low-end hardware, unlike typical smart devices. If sensor devices require many calculations, it will result in a large overhead. We required a lightweight technology for the authentication and identification of sensor devices.

❐ Network and system security

SDIoT requires a technique for detecting a threat that might be generated by a variety of devices. IoT will be the occurrence of new types of attacks because of the increase in the number and variety of devices. For example, this environment is expected to see an increased number zero-day attacks. Detecting such an attack requires big-data-based security analysis. This has the advantage of being able to quickly detect the abnormal symptoms of an attack. This method is to be developed as a distributed system for overload decreases.

❐ Integrity, confidentiality, and privacy

The controller must guarantee the confidentiality of collected data and service data. In addition, the integrity of the data should be guaranteed. To achieve this, the controller can be used to encrypt data using integrity verification technology for collected data and service data.

4   SDIoT Secure Frameworks

4.1   Infrastructure

The infrastructure layer configures devices for a variety of networking tasks. The infrastructure layer is controlled by a software-defined controller that includes a variety security technologies and network management methods. Smart devices including sensor devices were connecting by a variety of open-flow-based networks.

4.2   Control Layer

The control layer is an area that provides security services to each device. In this layer, authentication, secure communication, and access control technologies are required. The details are as follows.

❏ Authentication
Device authentication is a feature that provides authentication services for IoT devices. This framework is classified into two techniques for authentication.

- Certificate-based authentication
The smart device can use certificate-based authentication. to achieve this, the controller can request verification of a certificate from a third-party certification authority. However, this technique can difficult to use for sensor devices.

- SPA (single packet authorization)
Sensor devices can use authentication based on SDPs (software-defined perimeters). SDPs use SPA for device authentication. SPA can authenticate the device via an initial seed value.

- Access control
Access control determines access to services and networks for authenticated devices; it allows access to the devices via a variety of policies.

❏ Network security
The purpose of network security is to reduce network threats arising from IoT devices. This framework is classified into two techniques for authentication.

- Secure channel
Secure channels are used to create a secure path for end-to-end communication. For this purpose, we can use technologies such as VPN, TLS, and IPSec. However, the technique used must be lightweight.

- Big-data security analysis

Capabilities for the detection of attacks should be applied to big-data security analytics platforms. Additionally, the controller should implement this function in a distributed system for overhead reduction.

❒ Integrity and confidentiality

Data collected from the IoT should be provided with integrity. Additionally, IoT service data should be received confidentially. For this purpose, the controller should be used to create lightweight cryptography systems and lightweight hash functions.

❒ Privacy policy

The IoT service should be able to ensure the user's privacy. For this purpose, the controller must perform policy support to protect user privacy. And, it can use to encryption function.

4.3   Application Layer

The application layer is the area of the program that uses the services of the controller; it can utilize the various functions of the controller according to the development. This area must be managed using user authentication and access control technology.

Fig. 2 shows the secure framework for SDIoT.

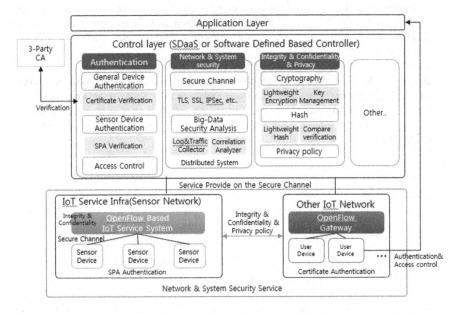

Fig. 2 Secure framework for SDIoT

5   Conclusion

In this paper, we studied a secure SDIoT framework. We configured the system using the network layout and analyzed the SDIoT. In addition, a study of the security requirements for the proposed SDIoT framework was conducted. And, we analyzed the security requirements to develop a security framework for SDIoT. We were proposed a framework is based on the need for authentication, network security, integrity, and confidentiality. Our focus was sensor devices, service systems, and user smart devices. In future studies, we will analyze the detailed operation procedures of this framework and suggest extensions of the security frameworks.

Acknowledgments This research was supported by the MSIP (Ministry of Science, ICT and Future Planning), Korea, under the ITRC (Information Technology Research Center) support program (IITP-2015-R0992-15-1006) supervised by the IITP (Institute for Information & Communications Technology Promotion)

References

1. ITU-T: Overview of the Internet of Things. Recommendation Y.2060 (2012)
2. Jararweh, Y., et al.: SDIoT: A Software Defined based Internet of Things Framework. Journal of Ambient Intelligence and Humanized Computing, 1–9 (2015)
3. McKeown, N.: Software-defined networking. INFOCOM Keynote Talk **17**(2), 30–32 (2009)
4. Huang, H., Zhu, J., Zhang, L.: An SDN-based management framework for IoT devices. In: 25th IET Irish Signals & Systems Conference and China-Ireland International Conference on Information and Communications Technologies (ISSC 2014/CIICT 2014), pp. 175–179 (2014)
5. Cloud Security Alliance: Software defined perimeter. Whitepaper (2013)
6. Benkhelifa, E., et al.: SDSecurity: a software defined security experimental framework. In: IEEE International Conference on Communications (CCSNA 2015), June (2015)
7. Cecchinel, C., Jimenez, M., Mosser, S., et al.: An architecture to support the collection of big data in the internet of things. In: 2014 IEEE World Congress on Services (SERVICES), pp. 442–449. IEEE (2014)

Secure Data Access Control Scheme for Smart Home

Ho-Seok Ryu and Jin Kwak

Abstract Owing to the development of information technology, smart home service providers now provide various services such as remote services and monitoring services through network-based smart home devices. In particular, a variety of services are provided for smart homes via smart devices irrespective of time and place. However, because smart devices are connected to a base network, it is possible for unauthorized devices to access the smart homes. Such indiscriminate data access may result in data leakage, data falsification, and invasion of privacy. Therefore, in this paper, we propose a secure data access control scheme for smart homes.

Keywords Smart home · Data access control · Data management · Internet of things

1 Introduction

Owing to the development of Internet communication technology (ICT), today's society can be considered a ubiquitous society accelerated by the Internet of Things (IoT). Further, the fact that all the smart devices in a smart home are connected to each other has led to the development of various types of smart home technologies and services [1][2].

Smart homes now have access to remote meter reading, remote education, remote control, and many other such services. Further, through green home energy, smart grid and other such technology, a user can manage his/her home

H.-S. Ryu
ISAA Laboratory, Department of Computer Engineering, Ajou University, Suwon, Korea
e-mail: ryuhs@ajou.ac.kr

J. Kwak(✉)
Department of Information and Computer Engineering, Ajou University, Suwon, Korea
e-mail: security@ajou.ac.kr

© Springer Science+Business Media Singapore 2015
D.-S. Park et al. (eds.), *Advances in Computer Science and Ubiquitous Computing*,
Lecture Notes in Electrical Engineering 373,
DOI: 10.1007/978-981-10-0281-6_70

utilities such as gas and electricity. However, unauthorized devices may access smart home data through the network-based smart devices, leading to a variety of security incidents.

Therefore, a smart home server should be managed efficiently to enable the smart home devices to access the smart home internally; therefore, smart homes need an access control technology for preventing unauthorized access by unidentified devices and users [3~6].

The rest of this paper is organized as follows: Section 2 presents an analysis of the security requirements of smart homes, and Section 3 proposes a secure data access control scheme. Section 4 presents a security analysis of the proposed scheme, and Section 5 concludes our findings.

2   Analysis of Security Requirements

Data access by unauthorized devices and users can pose security threats such as data leakage, data falsification, illegal authentication, and invasion of privacy. Therefore, in this section, we analyze the security requirements for secure data access control in smart homes [7].

2.1   Data Confidentiality

In the communication environment of a smart home, personal data, control messages, and sensitive data are transmitted through a network. Therefore, during such communication, an unauthorized third person does not have any idea about these data. Further, in order to prevent an invasion of privacy, data transmission and reception have to be encrypted, and the data have to be stored securely by using certain data encryption methods.

2.2   Device Authentication

Many smart home devices do not have any security functions, and therefore, unauthorized devices/users can access the abovementioned devices through a wireless network. Such unauthorized access may lead to data leakage, injection of malicious code, or malicious contamination of the smart home communication environment. Therefore, certain security certifications are required for smart home devices.

2.3   Access Control

The communication environment of smart homes may allow indiscriminate access by unauthorized devices and/or users, resulting in data leakage, data falsification, invasion of privacy, and many other security threats. Therefore, access of data should be classified into data access control categories such as data download and data upload. Because a smart home environment contains sensitive data, it must be have data access control for regulating data access through appropriate permissions. Therefore, you will be able to access the data only if you use an authorized device.

3   Proposed Scheme

In this section, we propose a secure data access control scheme for smart homes. It authenticates all devices registered to a smart home and provides safe access control of the data.

Smart home devices are divided into public devices and personal devices for effective management. The data acquired or created by smart devices are classified as public data or confidential data by the smart home server in the data upload phase. Public data storage stores temperature data, humidity data, and common device-created data for the smart home. Further, confidential data storage stores sensitive privacy data. The classified data can be accessed through the user information and context-aware data. Through context awareness, the access control module of the smart home server is recognized for preventing unauthorized devices from accessing such user location and context-aware data. These context-aware data include data such as user location, application data, SNS data, and lifestyle data.

3.1   Proposed Scheme Component

• *Use*: When a user wants to use smart home services through external smart devices, the user requests the smart home server to authenticate the user and devices in order to provide access to the smart devices.

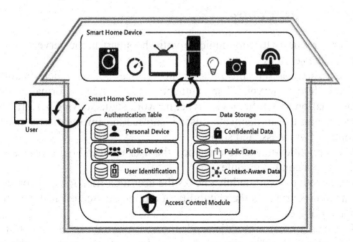

Fig. 1 Access control in the proposed scheme

• *Smart home devices*: Smart home devices generate or collect data in a smart home environment. Smart home devices include televisions, refrigerators, Wi-Fi devices, and smart phones and tablets that can access the smart home remotely. The generated and collected data are transmitted to the smart home server by the smart devices.

• *Smart home server*: The smart home server manages authentication, data management, access control, and the overall system in a smart home environment. The smart home server is composed of an authentication table, database, and access control module; their description is as follows:

- *Authentication table*: The authentication table authenticates the devices and the user's identification to allow access to the smart home network. It is composed of a personal device, the public device, and user identification. After authentication is completed, it transmits the data used for the authentication to access control module.
- *Data storage*: Data storage stores data that are generated and collected in a smart home. It is composed of public data, confidential data, and context-aware data. Further, the data are classified using rules and stored by using an encryption method.
- *Access control module*: The access control module allows access to the database through data transmitted by the authentication table. The access module manages the access control through context-aware data. Through user identification and context-aware data, the access module allows an authorized device/user to access the data storage.

3.2   Data Access Control

1) A user requests to communicate with the smart home server for access to the home services.
2) The authentication table authenticates devices and user identification in the smart home server. The authenticated device information includes information about the device type, network status, device performance, and so on, and the authenticated user identification information consists of id, context awareness, and so on.
3) The authentication table transmits the user data and context-aware data to the access control module.
4) The access control module checks the user identification and context-aware data and determines the suitability of a policy in accordance with the smart home policies.
5) The access control module sends the result of data access suitability to the data storage. Further, the access control module transmits the new context-aware data to the data storage. The data storage then stores the new context-aware data in the context-aware data storage.

6) In accordance with the abovementioned suitability, the data storage transmits the requested data to the user.

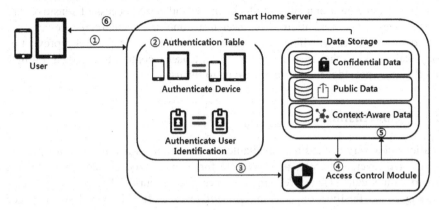

Fig. 2 Data access control in the proposed scheme

4   Security Analysis of the Proposed Scheme

In this section, we analyze the security analysis of our proposed secure data access control scheme in smart home environment.

4.1   Data Confidentiality

The communication environment of smart homes may allow data leaked by an authorized third person. It is able to data leakage such as personal data, control data and sensitive data. In this paper, the smart home server of the proposed scheme stores the encrypted form of the data generated or collected by the smart devices. This suggested scheme can block data leaked through data encrypted.

4.2   Authentication

Authenticated smart devices can insert malignant code. In this situation, the smart device will be malignant smart device. It is able to execute distributed denial of service attack and data falsification. This paper proposed scheme ensures that the stored data are not leaked by providing access to these data only to authenticated devices. Further, Primary authentication is done through device authentication and user authentication.

4.3   Access Control

The smart home server classifies the data into public data and confidential data in the smart home environment and blocks the unauthorized access of sensitive data and privacy data. In addition, if the user loses one of the authorized devices, the smart home server will block the access of the lost device, thereby preventing physical access to the smart home through the user's context-aware data. Further, it efficiently provides the appropriate data to the user.

5   Conclusion

Various new services are now available for smart homes irrespective of time and place. However, currently, it is possible for unauthorized devices to access the smart home data. Therefore, smart homes face security threats such as data leakage, data falsification, and invasion of privacy. In this paper, we analyzed some of these security threats and proposed a secure data access control scheme for smart homes. The proposed scheme can classify securely stored data by using device authentication, user authentication, and context-aware data. Therefore, this paper is expected to contribute to the safe and effective management of smart devices through secure data access control in smart homes.

Acknowledgment This research was supported by the MSIP(Ministry of Science, ICT and Future Planning), Korea, under the ITRC(Information Technology Research Center) support program (IITP-2015-R0992-15-1006) supervised by the IITP(Institute for Information & communications Technology Promotion)

References

1. Li, Y.: Design of a key establishment protocol for smart home energy management system. In: 2013 Fifth International Conference on Computational Intelligence, Communication Systems and Networks, pp. 88–93 (2013)
2. Suh, H.-j., Lee, D.-g., Choe, J.-s., Kim, H.-w.: IoT security technology trends. The Korea Institute of Electromagnetic Engineering and Science **24**(4), 27–35 (2013)
3. Raj, S.V.: Implementation of pervasive computing based high secure smart home system. In: 2012 IEEE International Conference on Computational Intelligence &Computing Research (ICCIC), pp. 18–20 (2012)
4. Han, D.-M., Lim, J.-H.: Smart Home Energy Management System using IEEE 802.15.4 and ZigBee. IEEE Transactions on Consumer Electronics **56**(3), 1403–1410 (2010)
5. Wright, A.: Cyber security for the power grid: cyber security issues & Securing control systems. In: ACMCCS (2009)
6. Son, J.-Y., Lee, J.-H., Kim, J.-Y., Park, J.-H., Lee, Y.-H.: RAFD: Resource-aware fault diagnosis system for home environment with smart devices. IEEE Transactions on Consumer Electronics **58**(4), 1185–1193 (2013)

A Combination of PSO-Based Feature Selection and Tree-Based Classifiers Ensemble for Intrusion Detection Systems

Bayu Adhi Tama and Kyung Hyune Rhee

Abstract Due to the numerous attacks over the Internet, several early detection systems have been developed to prevent the network from huge losses. Data mining, soft computing, and machine learning are employed to classify historical network traffic whether anomaly or normal. This paper presents the experimental result of network anomaly detection using particle swarm optimization (PSO) for attribute selection and the ensemble of tree-based classifiers (C4.5, Random Forest, and CART) for classification task. Proposed detection model shows the promising result with detection accuracy and lower positive rate compared to existing ensemble techniques.

Keywords Particle swarm optimization · Anomaly detection · Ensemble of tree-based classifiers

1 Introduction

With the rapid growth of computer networks, the number of users connected to the Internet has increased year by year. Severe disasters might be risen due to the excessive escalation of malicious intrusion or attack over the Internet. Therefore, the need for providing secure and safe security systems through the use of intrusion detection systems (IDS), encryption, or firewall is required. An IDS plays a vital role to analyze the network events occurring in a computer networks for indication of intrusion presence.

Intrusion aims at attempting to violate computer security policies such as confidentiality, integrity, and availability [1]. To date, significant research concern

B.A. Tama · K.H. Rhee(✉)
Lab. of Information Security and Internet Applications,
IT Convergence and Application Engineering,
Pukyong National University (PKNU), Busan, South Korea
e-mail: {bayuat,khrhee}@pknu.ac.kr

© Springer Science+Business Media Singapore 2015
D.-S. Park et al. (eds.), *Advances in Computer Science and Ubiquitous Computing*,
Lecture Notes in Electrical Engineering 373,
DOI: 10.1007/978-981-10-0281-6_71

in information security is intrusion detection and prevention. Intrusion detection can be considered as a classification analysis which is given to computer network traffic whether as normal or anomaly [2].

A first IDS was proposed by Denning [3], since then, numerous detection techniques including statistical methods, machine learning and data mining have been deployed in order to improve their performance. However, the deployment of efficient model to identify such malicious activity is still challenging task. An IDS must not have high computational burden and can perform intelligently so as to recognize previously unknown attacks. Specifically, an IDS must meet a low false positive rate and high detection rate [4].

In the recent work, a combination of PSO-based feature selection technique and multiple of tree-based classifiers system are proposed. PSO [5] is chosen to reduce computational cost since its capability to automatically search good features. We also consider the fusion of tree-based classifiers such as C4.5 [6], Random Forest [7], and CART [8] for classification analysis in order to increase detection accuracy. The performance result of proposed model is compared with the aforementioned base classifiers as single classifier and the state-of-the art ensemble techniques such as Bagging [9], Real Adaboost [10], MultiBoost [11], and Rotation Forest [12].

The objective of this paper are as follows: firstly, to select the most relevant features for intrusion detection systems using PSO and correlation-based feature selection (PSO-CFS); and secondly, to introduce the fusion of tree-based classifiers to maximizing the classification accuracy.

2   Related Work

Biology-inspired methods have tremendous impact to design computer security systems. They have developed novel and effective protection mechanism. Due to the increased deployment and widespread use of computer systems, traditional approaches often suffer from scalability problems to cope with [13]. Thus, it is important to consider biologically systems as sources of inspiration when designing new approaches [14].

PSO as one of many existing biology-inspired methods have been widely applied in IDS. It has been adopted for core functionality of IDS such as classification task or for secondary functions such as feature selection [4]. For instance, Zainal et al. [15] proposed the integration of rough set theory and particle swarm (Rough-DPSO) for feature selection process in IDS. From the experiment, proposed method offers better representation of data and they are robust. The most recent research regarding the use of PSO for feature selection in intrusion detection is a method called dynamic swarm based rough set (IDS-RS) [16]. IDS-RS is proposed to select the most relevant features that can represent the pattern of the network traffic.

2.1   *PSO and Correlation-Based Feature Selection*

PSO firstly proposed by Kennedy and Eberhart [17], is one of computation technique which is inspired by behavior of flying birds and their means of information exchange to solve the problems. Each particle in the swarm represents possible solution. A number of particle is located in the hyperspace, which has random position φ_i and velocity ϑ_i. The basic update rule for the position and the speed is depicted in Eq. (1) and (2), respectively.

$$\varphi_i(t + 1) = \varphi_i + \vartheta_i(t + 1) \tag{1}$$

$$v_i(t + 1) = \omega\vartheta_i(t) + c_1 r_1 (p_i - x_i) + c_2 r_2 (g - x_i) \tag{2}$$

Where ω denotes inertia weight constant, c_1 and c_2 denotes cognitive and social learning constant, respectively, r_1 and r_2 represent random number, p_i is personal best position of particle i, and finally, g is global best position among all particles in the swarm.

Correlation-based feature selection (CFS) is one of leading subset selection method in machine learning and pattern recognition [18]. CFS uses entropy and information gain theory to measure the uncertainty or unpredictability of a system. The lack of computation using information gain is symmetrical uncertainty and biased of feature with more values. Hence, CFS adopts a coefficient to compensate information gain's bias toward attribute with more values and to normalize its value to the range [0,1].

In this paper, the integration of PSO and CFS is employed. An open source data mining tool, Weka, allows us to combine PSO and CFS as search and evaluation method, respectively. We consider to compare the number of selected features by varying the number of particle and its influence to the performance of classifier.

2.2   Fusion of Tree-Based Classifiers

Nowadays, the fusion of several base classifiers in parallel has been widely applied in many applications. Parallel approach organizes classifiers in parallel. All classifiers are applied for the same input in parallel, and then the result from each classifier is then combined to yield the final output. Moreover, in parallel approach, a combination rule is needed in order to incorporate the output of each classifiers. Once the base classifiers have been trained, a classifier fusion is formed by the rule of voting. In this current work, we consider to compare the accuracy of classifiers fusion using majority voting [19] and average of probabilities rule [20].

Our approach is based on the hypothesis that the use of classifiers fusion, an accurate detection can be obtained. Our base classifiers are C4.5, Random Forest, and CART. The selection of base classifiers, although we could choose other classifiers, is based on the fact that these classifiers are widely applied in many today's applications and show successful performance.

3   Experimental Setup

In this section, the scheme for intrusion detection using PSO-based feature selection and multiple classifier systems is presented. Firstly, we present a framework of intrusion detection using multiple classifiers systems as depicted in Figure 1. Then, each part of the framework such as dataset description, parameter for feature selection process, and the strategy for combining multiple tree-based classifiers will be briefly discussed.

Our experiments were performed using NSL-KDD dataset [21]. It consists of selected records of older and well-known dataset, KDD Cup. NSL-KDD possesses 41 attributes plus one class label. It has 12973 records with 53.3% of normal class and anomaly class (represents 23 attacks) for the rest. All attributes were labeled from A to AO. Dataset is divided into 2 parts. One part which consists of two-third (66%) of dataset will be used for training, and the rest will be used for evaluation.

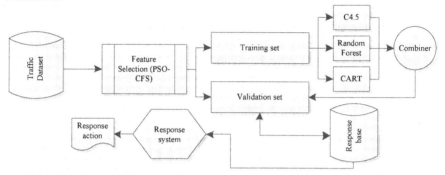

Fig. 1 Framework of intrusion detection systems

The parameter for the feature selection process using PSO-CFS the learning parameter of base classifiers is shown in Table 1. A different number of particle for feature selection are 50, 100, and 200 are denoted by PSO-50, PSO-100, and PSO-200, respectively. We considered the same parameters for base classifiers either as part of MCS or as an individual classifier. We ran the experiments using Weka, running on a system with an Intel Core i5 3.65GHz, 16GB RAM, and Windows 7 Professional.

The performance of classifiers are measured by accuracy and false positive rate (FPR). Accuracy represents the percentage of correctly classified of samples for different number of attributes, whilst FPR denotes the number of incorrectly classified of samples as belonging to positive class.

Table 1 Parameter for feature selection and learning process

PSO-CFS	C4.5	Random Forest	CART
Number of particles: 50, 100, and 200	Confidence factor, $C = 0.25$	Number of trees: 100	Heuristic process
$\omega = 0.33$	Number of folds = 3	Number of slots = 1	Number of pruning: 5
$c_1, c_2 = 0.34$	Min. instance per leaf = 2		Pruning
	Pruning		

4   Result and Discussion

Table 2 summarizes our experimental results. Our proposed scheme with respect to classifiers ensemble using tree-based classifiers performed better than single classifier and existing ensemble classifiers. Based on the feature selection experiment by varying the number of particles, as the number of particle increases, the number of selected features continue to decreases significantly. Nevertheless, the fewer of selected features, the lower performance of classifier has.

Moreover, proposed feature selection scheme PSO-50 with average of probability voting ensemble scheme shows higher accuracy rate than other classifiers. It can be said that in the future, ensemble of tree-based classifiers might become a promising solution to detect anomaly in computer network. With reference to single classifier, RF always performs better than other classifiers with 99.78%, 99.67%, and 99.43% of predictive accuracy for PSO-50, PSO-100, and PSO-200, respectively. Surprisingly, among the ensemble technique, Rotation Forest with C4.5 as base

Table 2 Cross comparison results

Method	Selected Features	Accuracy (%)	FPR
PSO-50			
C4.5		99.71	0.003
RF		99.78	0.002
CART	D, E, F, L, Z, AC, AD, AG, AK, AL, AM	99.72	0.003
Bagging-C4.5		99.76	0.002
Real Adaboost-C4.5		99.77	0.002
Multiboost-C4.5		99.78	0.002
Rotation Forest-C4.5		98.98	0.011
Maj. Voting (C4.5+RF+CART)		99.76	0.002
Average of Prob. (C4.5+RF+CART)		**99.80**	**0.002**
PSO-100			
C4.5		99.62	0.004
RF		99.67	0.004
CART	D, E, F, L, Z, AC, AD, AK, AM	99.60	0.004
Bagging-C4.5		99.62	0.004
Real Adaboost-C4.5		99.64	0.004
Multiboost-C4.5		99.64	0.004
Rotation Forest-C4.5		98.39	0.018
Maj. Voting (C4.5+RF+CART)		99.64	0.004
Average of Prob. (C4.5+RF+CART)		**99.76**	**0.002**
PSO-200			
C4.5		99.39	0.007
RF		99.43	0.006
CART		99.35	0.007
Bagging-C4.5	D, E, F, L, Z, AD	99.38	0.007
Real Adaboost-C4.5		99.40	0.007
Multiboost-C4.5		99.38	0.007
Rotation Forest-C4.5		98.12	0.019
Maj. Voting (C4.5+RF+CART)		**99.44**	**0.006**
Average of Prob. (C4.5+RF+CART)		99.39	0.007

classifier continue to show unsatisfactory performance compared to other ensemble technique. Finally, a good performance could not be obtained when we employed a well-known decision tree method C4.5 as single classifier, yet C4.5 tends to show good performance when we incorporate it in ensemble technique.

5   Conclusion

In this paper, the performance of particle swarm optimization feature selection based and classifiers ensemble are thoroughly studied. The classifiers ensemble was built by tree-based classifiers such as C4.5, Random Forest, and CART, while feature selection was carried out by varying the number of particles. The proposed scheme showed good performance compared to other ensemble techniques. An incorporation of fifty particles of PSO and an average probability voting rule gave us promising performance with 99.8% of accuracy. As future work we will study the performance of classifiers fusion by combining other machine learning technique with different dataset.

Acknowledgement This work was supported by the National Research Foundation of Korea (NRF) grant funded by the Korea government (MSIP) (No. NRF-2014R1A2A1A11052981).

References

1. Liao, H., Lin, C., Lin, Y., Tung, K.: Intrusion detection system: A comprehensive review. Journal of Network and Computer Applications **36**(1), 16–24 (2013)
2. Catania, C., Garino, C.: Automatic network intrusion detection: Current techniques and open issues. Computers & Electrical Engineering **38**, 1062–1072 (2012)
3. Denning, D.: An intrusion-detection model. IEEE Transactions on Software Engineering **2**, 222–232 (1987)
4. Kolias, C., Kambourakis, G., Maragoudakis, M.: Swarm intelligence in intrusion detection: A survey. Computers & Security **30**(8), 625–642 (2011)
5. Moraglio, A., Di Chio, C., Poli, R.: Geometric particle swarm optimisation. In: Genetic Programming. LNCS, vol. 4445. Springer (2007)
6. Quinlan, J.: C4. 5: programs for machine learning. Elsevier (1993)
7. Breiman, L.: Random Forests. Machine Learning **45**(1), 5–32 (2001)
8. Breiman, L., Friedman, J., Stone, C., Olshen, R.: Classification and regression trees. CRC Press (1984)
9. Breiman, L.: Bagging predictors. Machine Learning **24**(2), 123–140 (1996)
10. Friedman, J., Hastie, T., Tibshirani, R.: Additive logistic regression: a statistical view of boosting. Annals of Statistics **95**(2), 337–407 (2000)
11. Webb, G.: MultiBoosting: A Technique for Combining Boosting and Wagging. Machine Learning **40**(2), 159–196 (2000)
12. Rodriguez, J., Kuncheva, L., Alonso, C.: Rotation forest: A new classifier ensemble method. IEEE Transactions on Pattern Analysis and Machine Intelligence **28**(10), 1619–1630 (2006)

13. Williamson, M: Biologically Inspired Approaches to Computer Security. Technical Report, HP Laboratories, Bristol (2002)
14. Twycross, J., Aickelin, U.: An immune-inspired approach to anomaly detection. In: Handbook of Research on Information Security and Assurance. IGI Global (2008)
15. Zainal, A., Maarof, M., Shamsuddin, S.: Feature selection using rough-DPSO in anomaly intrusion detection. In: Computational Science and Its Applications. LNCS, vol. 4705, pp. 512–524 (2007)
16. Chung, Y., Wahid, N.: A hybrid network intrusion detection system using simplified swarm optimization (SSO). Applied Soft Computing 12(9), 3014–3022 (2012)
17. Kennedy, J., Eberhart, R.: A discrete binary version of the particle swarm algorithm. In: IEEE International Conference on Systems, Man, and Cybernetics, pp. 4104–4108 (1997)
18. Hall, M.: Correlation-based feature selection for machine learning. The University of Waikato, Hamilton (1999)
19. Kuncheva, L.: Combining pattern classifiers: methods and algorithms. John Wiley and Sons (2004)
20. Kittler, J., Hatef, M., Duin, R., Matas, J.: On combining classifiers. IEEE Transactions on Pattern Analysis and Machine Intelligence 20(3), 226–239 (1998)
21. Tavallaee, M., Bagheri, E., Lu, W., Ghorbani, A.: A detailed analysis of the KDD CUP 99 data set. In: Second IEEE Symposium on Computational Intelligence for Security and Defence Applications (2009)

Secure Traffic Data Transmission Protocol for Vehicular Cloud

Lewis Nkenyereye and Kyung Hyune Rhee

Abstract One of the main goals of Intelligent Transportation System (ITS) is to provide safety for both the drivers and passengers of vehicles. Collecting and analyzing road traffic data seem to be the first step towards safety related application in ITS. However security issues while collecting and transmitting traffic data need to be properly handled before the full adoption of those systems. Vehicular Cloud (VC) which combines vehicle ad hoc networks (VANETs) and cloud computing (CC) presents a promising paradigm for ITS. In this paper, we present a secure traffic data transmission protocol for vehicular cloud. We make use of ID-based signature and anonymous credential techniques to allow registered vehicles to anonymously transmit the traffic data within the vehicle's sensors.

Keywords Vehicular cloud · Traffic data · Identity based cryptosystems · Digital signature

1 Introduction

Intelligent Transportation system (ITS) seeks to provide safer transportation environment through a variety of applications from safety to infotainment services. ITS has attracted all developed countries because of its direct effect to economic growth. For instance, the Center for Disease Control and Prevention (CDC), affiliated to the of U.S. Department of Health and Human Services indicated that more than 2.3 million adult drivers and passengers were treated in emergency departments as a result of being injured in road crashes in 2009 [2] due to lack of proper maintenance, monitoring of vehicle condition and negligence of

L. Nkenyereye · K.H. Rhee(✉)
Department of IT Covergence and Application Engineering,
Pukyong Nation University, Busan, Republic of Korea
e-mail: {nkenyele,khrhee}@pknu.ac.kr

© Springer Science+Business Media Singapore 2015
D.-S. Park et al. (eds.), *Advances in Computer Science and Ubiquitous Computing*,
Lecture Notes in Electrical Engineering 373,
DOI: 10.1007/978-981-10-0281-6_72

drivers or passengers. VANETs was introduced as a self-organized platform where the vehicles are equipped with wireless communication devices called on board unit (OBU) and road side units (RSUs) fixed along the roads would enable both vehicle to vehicle communication (V2V) and vehicle to infrastructure (V2I) communications. However, in the near future vehicles are predicted to be equipped with a range of sensors such as Vehicle Speed Sensor (VSS), Carbon Monoxide Sensor (CMS), Sulphur Dioxide Sensor (SDS), Nitrogen Oxide Sensor (NOS), Alcohol Sensor (ALS), Sound Sensor (SOS), Vehicle Noise Sensor (VNS), Occupancy Sensor (OCS), Fuel System Leak Sensor (FLS) and Gas Leakage Sensor (GLS) [2]. Those sensors would generate a lot of data which could be of great importance to regional transportation centers. The limited computational capabilities of VANETs is relieved by cloud computing to form vehicular cloud in which vehicles devices are connected to cloud servers to perform complicated computational operations.

Olariu et al. [4] presented a classification for vehicular cloud-based services and their respective potential security threats. The authors pointed out the feasibility and practicability of VC compared to VANETs. VANETs has received much attention on secure navigation related protocols [1,7,10]. The majority of the proposed protocols suggest security mechanisms which could be adopted in order for the vehicles to securely receive road services such as navigation service. In [8], the authors presented a system which captures the GPS recommended minimum specific (RMC) traffic data and transmits it securely to probe land center by means of a communication service provider, however the protocol is not built on vehicular cloud-based environment. In [5], the authors proposed a secure and privacy-aware traffic information protocol in which each vehicle is authenticated to get a symmetric key zone K_Z whenever it enters a new zone. Note that a zone is assumed to cover around 3 RSUs. The vehicle also get a specific RSU key K_{RSU} as it enters RSU coverage zone. Those keys are subject to change for security reasons, which require a heavy key management protocol. In this paper, we propose a secure traffic data transmission protocol for vehicular cloud which insures the secure transmission of traffic data within the vehicle's sensors. We make the following contributions: we present an application model for secure traffic data transmission protocol which would allow a transportation center sever located in the cloud to securely collect traffic information from vehicle's sensors for data mining purposes; we construct our protocol based on ID-based signature scheme [9] and anonymous credential [6] and we provide the security and performance analysis of the proposed protocol.

2   System Architecture

In this section we describe the communication entities within our protocol which are made of TA, TC, RSU and vehicles which communicate through the on board unit (OBU) as shown in fig. 1:

- Trusted Authority (TA): It is in charge of the registration of all entities (TC, RSU and vehicle) inside our system and issues cryptographic materials during the system initialization.
- Transportation Center (TC): It is a server located in the cloud belonging to the regional transportation authority. TC receives traffic data from the vehicles on the road through RSU and broadcast it back to the vehicles after data mining operations.
- Road Side Units (RSU): RSU are databases located along the roads and accessible by the vehicles. They play a middle field role between the vehicles and the TC. Traffic information from vehicle's sensors are transmitted to TC through RSUs and analyzed traffic information from TC are broadcasted to vehicles through RSUs.
- Vehicles: They move within the city and collects information through the sensors installed within the vehicles such as GPS sensor. For simplicity in this work, we will only consider the GPS/RMC sensor data and neglect other embedded sensors such as Vehicle Noise Sensor (VNS), Occupancy Sensor (OCS) [2].

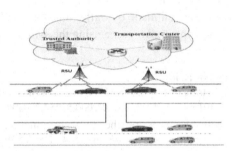

Fig. 1 System Architecture

Our protocol should satisfy the following security requirements:

- Authentication and Authorization: Each vehicle should be authenticated before it transmits traffic data to TC from its sensors.
- Identity privacy preservation: The real identity of a vehicle should be kept secret from other vehicles and TC.
- Unlinkability: An attacker should not be able to link two different traffic data transmission messages sent by the same vehicle.
- Traceability: TA should be able to reveal the real identity of each vehicle which is necessary in case of disputes or suspicious behavior.

3 Protocol Description

In this section, we design a secure traffic data transmission protocol for vehicle cloud which is made of three main sub protocols as follows:

a. System Setup

TA generates two groups G and G_r of order q. Let $e : G \times G \to G_r$ be a bilinear map. Let P denotes a generator in G. TA selects a random $x \in \mathbb{Z}_q^*$ and calculates $P_{TA} = xP$ where x is a private key and P_{TA} is a public key.

For each v_i real identity VID_i, TA generates its private key as $SK_i = xH(PID_i)$ where $PID_i = Enc_{P_{TA}}(VID_i)$ is its pseudo identity and VID_i its real identity [9]. TA publishes the public parameter $\{G, G_r, q, e, P, P_{TA}, H\}$ where $H : \{0,1\}^* \to G$ is a cryptographic hash function. TC chooses a random $t \in \mathbb{Z}_q^*$ as its secret key and publish its public key as $P_{TC} = tP$. TA also regularly sends a revocation list to TC made by v_i pseudo-identities $RCList = \{PID_i, ..., PID_n\}$ [7].

b. Credential Generation

Upon meeting the RSU_j for the first time, v_i sends a request for sensor traffic data transmission credential $tr\_req = Enc_{P_{TC}}(PID_i, mre, k)$ where mre is a request message keyword and k a symmetric key to be used at later stage.

TC decrypts the message with its secret key and check if PID_i is not in the recovation list $RCList$. TC sends $tr\_res = Enc_k(cred_i)$ to v_i through $RSUj$ where $cred_i = tH_1(TR1 \mid T)$ is a credential made by traffic transmission keyword $TR1$ and T the duration period. v_i decrypts the message using the symmetric key k and recover the credential $cred_i$

c. Traffic Data Transmission

When v_i wants to transmit the RMC/GPS sensor information, it sets $M = \{RMC, 225446, A, 4916.45, N, 12311.12, W, 000.5, 054.7, 050915, 020.3, E * 68\}$ representing respectively the type of the GPS information (RMC), the UTC time(22:54:46), the navigation reciever warning (A=OK,V=Warning), the latitude (49 degree,15.45 min North), the longitute (123 degree, 11.12 min West), the speed over ground (000.5 knots), the course stastus (057.7 true), the fix date (05 auguts 2015), the magnetic variation (020.3 degree East) and the mandatory chescksum (*68) as shown in Table 1 [8]. v_i sends a traffic message from its GPS sensor as $\{C, \sigma\}$ where $C = Enc_{P_{TC}}(M)$ and $\sigma = (U_1, U_2)$ is an signature generated on as follows:

$U_1 = cP$ for $c \in \mathbb{Z}_q^*$ $U_2 = SK_i + cred_i + cH(M)$. Upon receiving the data traffic message, TC decrypts C using its private key and verifies if $e(P, U_2) = e(P_{TA}, H(PID_i))(P_{TC}, H(TR1 \mid T))e(U_1, H(M))$. The correctness of the verification can be done as follows:

$e(P, U_2) = e(P, SK_i + cred_i + cH(M))$

$e(P, U_2) = e(P, SK_i)e(P, cred_i)e(P, cH(M))$

$e(P, U_2) = e(P, xH(PID_i))e(P, tH(TR1 \mid T))e(cP, H(M))$

$e(P, U_2) = e(xP, H(PID_i))e(tP, H(TR1 \mid T))e(cP, H(M))$

$e(P, U_2) = e(P_{TA}, H(PID_i))(P_{TC}, H(TR1 \mid T))e(U_1, H(M))$

It the equations holds, TC can be sure of the validity of the requesting v_i. Then TC checks again if the the pseudonym corresponfing to $cred_i$ is not in the recovation list $RCList$. TC can then trust the transmitted traffic data.

Table 1 GPS recommended minimum specific (RMC) data

Time(UTC)	Status	Latitude	Longitude	Speed/ground	Course	Date	MagneticV	Checksum
225446	A	4916.45,N	12311.12,W	0.005	054.7	050815	020.3,E	*68

4   Security Analysis and Performance

According to the aforementioned security objectives, we analyze and discuss the security of the proposed protocol:

Authorization: Before v_i can upload traffic information, it has to get a daily credential generated by TC $cred_i = tH_1(TR1 \mid T)$. Only a vehicle with a valid credential can upload the traffic information from its sensors.

Authentication: During the credential issuance, v_i sends $tr\_req = Enc_{P_{Tc}}(PID_i, mre, k)$ which contains its pseudo-identity. TC checks whether the PID_i is still a valid peudonym from the revocation list $RCList = \{PID_i, ..., PID_a\}$ before generating a credential. Only a vehicle with a valid credential and not within the revocation list can upload traffic data.

Identity Privacy Preservation: During the registration of vehicles by the TA, each vehicle is given a pseudo-identity $PID_i = Enc_{P_{TA}}(VID_i)$ from its real identity. All the operations done by the vehicle's OBU can not reveal the real identity of the vehicle.

Unlinkability: Before v_i can transmitt its traffic information, it generate a signature on the message with a fresh a random value as $\sigma = (cP, SK_i + cred_i + cH_1(M))$ which does not even reveal the pseudo identity of the vehicle. Thus, an attacker can not link the traffic messages sent by a same vehicle.

Traceability: Even though it is hard for an attacker and TC to know the real identity of a vehicle, TA has the capability of revealing vehicles real identity in case of disputes. The pseudo identity of v_i is based on the vehicle real identity.

We further evaluated the performance of the proposed protocol in accordance with M/D/1 queuing model in order to evaluate the efficiency in terms of processing delay of an RSU for authenticating vehicles connecting to RSU on the road as shown in figure 2. We took into consideration the authentication processing time T_{Auth} for a vehicle which is determined by the computational cost of the proposed protocol versus Hussain et a.l [5] in Table 2. We assume that the arrival of vehicles connecting to the RSU occurs at rate β of exponential distribution according to a Poisson process [7] and RSU processes one vehicle at a time with first-come-first-serve principal. The density of vehicles D in RSC's range, while a vehicle gets out of RSC's area, can be estimated by $\beta \times R/v$ where $v=40m/s$ is the vehicle speed and $R=1200m$ is the RSU coverage and RSU's processing time can be calculated as $\alpha=\beta x T_{Auth}$.

Fig. 2 Authentication process delay of [5] and proposed protocol

Table 2 Computation Cost of [5] and proposed protocol

Phase	Hussain et al, [5]	Proposed
OBU	$5T_{mul}+2\,T_{as-enc}$	$T_{mul}+T_{as-dec}$
RSU	$T_{mul}+T_{as-enc}+T_{as-dec}+4T_{pair}$	$T_{mul}+2T_{pair}+T_{as-dec}$

5   Conclusion

In this paper, we proposed a secure traffic data transmission protocol for vehicular cloud. The protocol ensures that the data from the vehicle sensor is anonymously transmitted to the transportation center. The proposed protocol use ID-based signature scheme and anonymous credential to achieve the security objectives. The security and performance analysis of the proposed protocol confirms its efficiency.

Acknowledgments This work was supported by the National Research Foundation of Korea (NRF) grant funded by the Korea government (MSIP) (No. NRF-2014R1A2A1A1 1052981).

References

1. Nkenyereye, L., Tama, B.A., Park, Y., Rhee, K.H.: A Fine-Grained Privacy Preserving Protocol over Attribute Based Access Control for VANETs. J. Wireless Mobile Netw. Ubiquitous Comput. Dependable Appl. **6**(2), 98–112 (2015)
2. Mallissery, S., Pai, M.M., Ajam, N., Mouzna, J., Pai, R.M.: Transport and traffic rule violation monitoring service in ITS: a secured VANET cloud application. In: 12th IEEE (CCNC), pp. 213–218 (2015)

3. Choon, J.C., Cheon, J.H.: An identity-based signature from gap Diffie-Hellman groups. In: Public Key Cryptography-PKC, pp. 18–30. Springer, Heidelberg (2003)
4. Olariu, S., Hristov, T., Yan, G.: The next paradigm shift: from vehicular networks to vehicular clouds. In: Mobile Ad hoc Networking: Cutting Edge Directions, 2nd edn. John Wiley & Sons, Inc., Hoboken (2013)
5. Hussain, R., Rezaeifar, Z., Lee, Y.H., Oh, H.: Secure and privacy-aware traffic information as a service in VANET-based clouds. Pervasive and Mobile Computing (2015). http://dx.doi.org/10.10.16/j.pmcj.2015.07.007
6. Verheul, E.R.: Self-blindable credential certificates from the weil pairing. In: Advances in Cryptology-Asiacrypt. LNCS, vol. 2248, pp. 533–551 (2001)
7. Chim, T.W., Yiu, S.M., Hui, L.C., Li, V.O.: VSPN: VANET-based secure and privacy-preserving navigation. IEEE Transactions on Computers $63(2)$, 510–524 (2014)
8. Babu, P., Rekha, N.R.: Secured GPS based traffic monitoring system in pervasive environment. In: 5th IEEE International Conference on In Software Engineering and Service Science (ICSESS), pp. 775–779 (2014)
9. Gentry, C., Silverberg, A.: Hierarchical ID-based cryptography. In: Advances in Cryptology-ASIACRYPT, pp. 548–566. Springer, Heidelberg (2002)
10. Cho, W., Park, Y., Sur, C., Rhee, K.H.: An improved privacy-preserving navigation protocol in VANETS. J. Wireless Mobile Netw. Ubiquitous Comput. Dependable Appl. $4(4)$, 80–92 (2013)

A Short-Range Tracking System Based on Bluetooth Radio Identification: An Algorithm and Its Implementation

Ibraheem Raed Altaha and Jong Myong Rhee

Abstract The paper describes the application of radio signal in the development of a short-ranged camera-tracking system. Object tracking can enable the static or mobile objects owners to observe their objects in a local position or remotely via the Internet. The tracking system uses Bluetooth (BT) radio frequency to track an object fitted with a radio transmitter. The use of BT receiver and motor-tracking system makes it possible to monitor many BT transmitters at once. Thus, the movements of a number of pets that carries a BT transmitter can be tracked. In this paper, we also present a location-tracking algorithm, which utilizes the tracking process.

Keywords Pet-tracking system · Tracking-system algorithm · Tracking-system implementation

1 Introduction

The use of radio frequency identification (RFID) technology has become widespread in many areas of industry. In RFID, a simple low-cost antenna is attached to an item, which can then be tracked [1]. RFID technology is based on the exchange of data between two entitles, namely a reader/writer and a tag, as shown in Fig. 1. Previous studies have focused on optimizing a variety of tracking systems. Related research has been conducted on developing a smart integrated tracking system based on the internet of things, real time location systems (RTLSs), and RFID [2]. The Korean Institute of Maritime Information and Communication Sciences investigated the use of tracking systems in monitoring the delivery and distribution of goods (e.g., the type of load, product information,

I.R. Altaha · J.M. Rhee(✉)
Myongji University, Seoul, Republic of Korea
e-mail: Ibrahimgate@gmail.com, jmr77@mju.ac.kr

© Springer Science+Business Media Singapore 2015
D.-S. Park et al. (eds.), *Advances in Computer Science and Ubiquitous Computing*,
Lecture Notes in Electrical Engineering 373,
DOI: 10.1007/978-981-10-0281-6_73

delivery status, and missing item status). The goods were tracked with Global Broadcast System (GPS), and the information was then linked with Google maps to determine the location of the goods and distribution paths [3].

Amon and Jamsid et al. studied the application of a tracking system in container storage yards in harbors [4]. They utilized an interface module as a location-tracking technique and GPS and wireless networks to enable efficient container management.

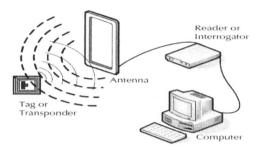

Fig. 1 An RFID tracking system based on a cloud computing system

The rest of this paper is organized as follows. In Section 2, we briefly describe our proposed approach. In Section 3, we illustrate the implementation of our tracking system.

2   The Proposed Approach

In this section, we present our approach for a short-range camera-tracking system based on Bluetooth (BT) transmission. The tracking system consists of a BT transmitter and a stationary BT receiver. As shown in Fig. 2, the transmitter works as a tag, and it has a unique ID. The stationary receiver consists of three parts: the BT receiver, a microcontroller, and a camera set above a motor. The tracking system detects the BT signal of the transmitter (the tagged object) and sends a signal to the microcontroller. The last rotates the motor in the direction of the location of the transmitter. The microcontroller utilizes an embedded algorithm, which filters the tag ID and detects the location of the transmitter.

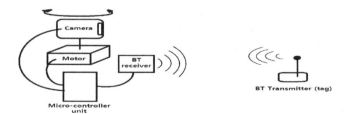

Fig. 2 Set up of the short-range BT tracking system

Our design is based on multiple BT receivers, as shown in Fig. 3. The multiple BT receiver design consists of three receivers: A, B, and C. Receivers are placed in the same plane with an angle separation between them, and each is set to receive a signal within a narrow area of detection. They read and measure the signal ID and strength of the transmitter signal respectively. Based on the microcontroller algorithm, the motor rotates the camera to face the object. Receiver C corrects the direction of the camera in cases where the object is located outside of the field of view. Fig. 3 shows three examples of the application of the proposed short-range BT tracking system. The first case shows the BT transmitter in front of the tracking system. As the transmitter is closer to receiver A than it is to B and C, the microcontroller algorithm rotates the motor to move the camera in the direction of the BT receiver A, so it is facing the BT transmitter. The second case shows what happens when the BT transmitter is moved closer to BT receiver B. In this case, the microcontroller rotates the motor in the direction of BT receiver B, so the camera is facing the BT transmitter. In the final case, if the BT transmitter is moved to a point behind the setup (i.e., closer to BT receiver C), the strength of receiver C will be higher than that of receivers A and B. Thus, the microcontroller will rotate the setup 180° in any direction and apply the same routine as in case 1 and case 2.

3   Implementation

In our implementation, we used an (Arduino UNO Atmega328p) microcontroller [5]. There are two microcontrollers' used in our design, the first one is used to control the BT transmitter and the second one is used to control the BT receivers A, B and C. The microcontroller in general works on making decisions based on the detection algorithm to facilitate the tracking process.

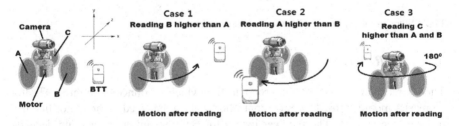

Fig. 3 Setup of the proposed short-range BT tracking system

3.1   BT Transmitter

The BT transmitter consists of an (Aruino Mini) microcontroller, which is connected to a BT HC-06 slave module [6]. Both the microcontroller and the BT module are operated by battery power. To optimize the power saving, two sensors are attached to the transmitter complex: a motion sensor and a where are you (WAY) signal. These two sensors are incorporated in the microcontroller algorithm. Fig. 4 shows the steps involved in checking and decision making.

To save power, the BT transmitter module works in the listening mode without transmitting. A signal is transmitted after one of two conditions:

1. The BT transmitter receives a WAY signal from the sender, indicating that the tracking system has been activated manually.

2. The BT transmitter begins moving. If the object (e.g., a vehicle, an animal, or a person) that is supposed to be tracked is mobile, the motion sensor will indicate that the object is moving.

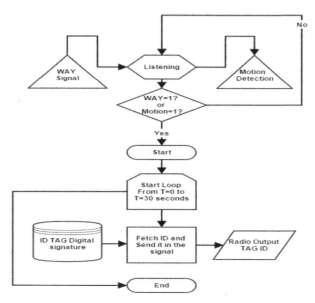

Fig. 4 Steps involved in the BT transmitter algorithm

3.2 BT Receiver

The BT receiver consists of a microcontroller, which is connected to three BT HC-05 master modules [6]. The microcontroller is also connected to the motor that use to rotate the camera. The receiver listens for BT transmitter signals and decodes the BT ID. In addition, the receiver module works as a sender to send a WAY signal to the BT transmitter. The WAY signal is activated either automatically at the tracking system startup or manually. Fig. 5 shows the steps involved in the BT receiver algorithm. After sending a WAY signal, the system receives a signal from sensors A, B, and C. After comparing these readings, the microcontroller applies the cases shown in Fig. 3.

4   Detection Analyses

In this paper, we utilized optimized signal power calculations for the sake of simplicity. As radio frequency waves are propagated through air, and distance will act as a passive transmission medium. As in common with any other type of radiation, the intensity at any point in the medium is inversely proportional to the square of its distance from the sender, where I is the intensity of the signal and d is the distance between the sender and the receiver [7-9].

If we assume an ideal scenario for linear transmission, then the transmission intensity in our proposed will be calculated using the distance between the transmitter and the receiver, multiplied by the radio power coefficient T_C, as shown below:

$$I\alpha \frac{1}{d^2} \qquad (1) \qquad\qquad I = P \times T_C \times \frac{1}{d^2}, \qquad (2)$$

where P is the transmitted power, T_C is the transmission power coefficient and d is the distance between the transmitter and the receiver. The distance plays a major role in the tracking process. When d is large, then the received radio intensity will be small.

Under the assumed ideal environment, the transmitting signal power is estimated based on finding the distance between the BT transmitter and receiver. Since, the BT sender is a mobile object, then it doesn't have a fixed location and it can be expressed by the probability of the existence function $p(x,y)$. Fig. 6 presents a top view of the implementation of the tracking system, facilitating the estimation of the distance between the BT receiver and $P(x,y)$.

In Fig. 6, BT A, B, and C are BT receivers that work together in the tracking process. If we assume that the tracked object has a circular pattern of movement, three major cases can cover the mathematical derivation equation of the tracking process. In the figure, BT Sender 1, 2, and 3 show the probable location of the object in the tracking system. The movement pattern can be summarized as a circular function:

$$r = \sqrt{x^2 + y^2}, \qquad (3)$$

where r is the radius of the circular pattern from the camera's location to tracked object, as shown in Fig. 6, and x and y are the coordinates of the probable location of the tracked object $P(x,y)$. If we considered that the camera's location as a frame of reference that lies at (0, 0), and that any of the BT receivers on (Xn,Yn), then the distance d between the BT receiver and transmitter can be found by measuring the direct short line between them, as below:
111111

$$\overline{N} = \sqrt{(Xn-x)^2 + (Yn-y)^2}, \qquad (4)$$

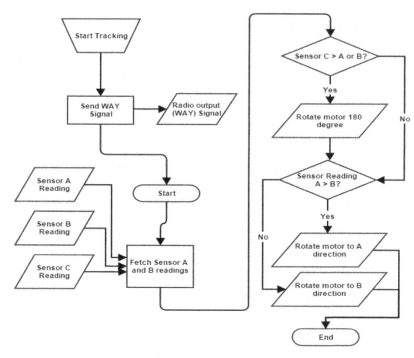

Fig. 5 Steps involved in the BT receiver algorithm

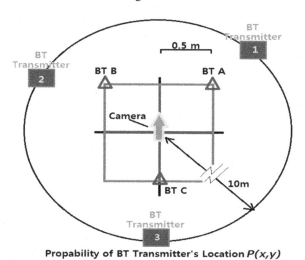

Fig. 6 Overview of the calculation of the distance between the BT receiver and transmitter with the proposed tracking system

where $\overline{N}$ is the distance between the N BT receiver to the BT transmitter from any point in $P(x,y)$. Applying Eq. 4 to the cases shown in Fig. 7 will yield the following:

$$\overline{A} = \sqrt{(Xa-x)^2 + (Ya-y)^2} \text{ ,(5a)} \qquad \overline{B} = \sqrt{(Xb-x)^2 + (Yb-y)^2} \text{ , (5b)} \qquad \overline{C} = \sqrt{(Xc-x)^2 + (Yc-y)^2} \text{ . (5c)}$$

If we assume that r is fixed to 10 m and that the distance between the camera and each sensor is 0.5 m and apply Eqs. 5a, 5b, and 5c, the distance between the BT transmitter and receiver can be calculated, as shown in Fig. 7.

Fig. 7 shows an estimation of the distance between the BT senders and the receiver for a 360° circular pattern and the response of the microcontroller based on the distance. The x-axis represents the angle between the camera and the location of the BT transmitter φp(x,y)=tan-1(y/x). The y-axis represents the distance between the BT transmitter and the BT receiver. The microcontroller selects the lowest value at a specific angle and tilts the camera in the direction of the specific BT receiver.

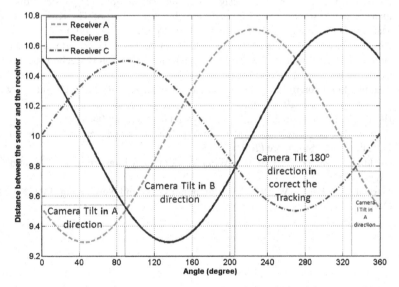

Fig. 7 Distance between the BT transmitter and receiver as function of the slope of $p(x,y)$.

5   Conclusions

In this paper, we presented a tracking algorithm and described its implementation. The proposed tracking system uses BT transmission to locate an object equipped with a tag, which transmits a specific BT signal. Fig. 7 shows the expected results of our approach based on three BT receivers with any tracked object or BT transmitter. The wide viewing angle of the camera (i.e., as wide as 60°) allows for low resolution tracking method to be applied using less BT receivers. Based on our estimations, by increasing the number of the detection points or the number of BT receivers, we can improve the resolution and accuracy of the detection. The latter would make it possible to apply the proposed tracking system in a number of different applications.

Acknowledgments This work was supported by the 2015 MPEES Advanced Research Center Fund of Myongji University.

References

1. Resources and analysis for electronics engineers. http://www.radio-electronics.com/info/wireless/radio-frequency-identification-rfid/technology-tutorial-basics.php
2. Kim, J.-S., Lee, H.-J., Oh, R.-D.: Smart Integrated Multiple Tracking System Development for IOT based Target-oriented Logistics Location and Resources Services. International Journal of Smart Home, 195–204 (2015)
3. Hurm, D.-C., Lee, K.-Y.: Design and implementation of physical distribution management system using RFID and GPS. In: Proceeding of The Korean Institute of Maritime Information Communication Sciences, pp. 441–444 (2007)
4. Amon, O., Jamsid, U., Yusupov, S., Lee, C.-U., Oh, R.-D.: The smart utility monitoring system for industrial safety and enhancement on the IOT technology. In: Int'l Conference on Smart Media Applications (2014)
5. Arduino Microcontroller Guide, Minnesota University (2011). URL: http://www.me.umn.edu/courses/me2011/arduino/
6. HC-06 bluetooth module. http://www.Lanwind.com/files/hc-06en,pdf
7. Rappaport, T.S.: Wireless Communications: Principles and Practice, 2nd edn. Pearson Education, Inc., Singapore (2002)
8. Haykin, S., Moher, M.: Modern Wireless Communications. Pearson Education, Inc., Singapore (2002)
9. Mark, J.W., Zhuang, W.: Wireless Communications and Networking. PHI, New Delhi (2005)

Design and Implementation of a Collaboration Messenger System Based on MQTT Protocol

Hyun Cheon Hwang, JiSu Park and Jin Gon Shon

Abstract Many office working systems have been changed from desktop based systems to mobile device based systems. A communication system which is one of important components in the office working system widely that has been widely used is the push message service via SMS and polling service. However, push message has vulnerability to maintain ordering between messages. In this paper, we design and implement a collaboration messenger system using MQTT protocol to maintain ordering between messages for a reliable office working system.

Keywords Push service · MQTT protocol · Collaboration messenger system · Messages ordering

1   Introduction

Most of data information sharing has developed from a wired communication into a wireless communication with the development of wireless technology. Also, many data communications are changing from the desktop devices to the mobile devices. The mobile data traffic is constantly increasing and the data expected to increase to about 11 times and the expected data capacity is about 15.9EB worldwide in 2018[1] compared to 2013. Many IT services such as groupware services and business services are being provided by the mobile devices. The services for office communication among mobile office systems are using the

H.C. Hwang · J.G. Shon(✉)
Department of Computer Science, Graduate School, Korea National Open University, Seoul, South Korea
e-mail: andy74.hwang@gmail.com, jgshon@knou.ac.kr

J. Park
Department of Computer Science and Engineering, Korea University, Seoul, South Korea
e-mail: bluejisu@korea.ac.kr

© Springer Science+Business Media Singapore 2015   513
D.-S. Park et al. (eds.), *Advances in Computer Science and Ubiquitous Computing*,
Lecture Notes in Electrical Engineering 373,
DOI: 10.1007/978-981-10-0281-6_74

instant messaging system to send and receive messages. And the messaging system, via the SMS and polling system is currently being used a lot of push message transmission.

The push message services were implemented various protocol forms such as XMPP, MQTT, HTTP and CoAP. These various protocols are used according to each situation. In particular MQTT protocol is designed to work on low-power devices nicely as a lightweight protocol and been used in many instant messaging systems. The MQTT protocol is also designed to have a mechanism to guarantee the QoS of each short message, and one of QoS level supports to ensure that only once transmitted through them. However, it can be hard to guarantee the exact order of these messages when a number of them are transmitted from the messenger. In this paper, we discuss how we can guarantee the order of messages from the system which ensures reliable message ordering. Lastly, we create an experiment for reliable message delivery system and reviewed the result.

2 Related Research

2.1 Message Transmission Protocol

A messaging communication system can be distinguished by SMS, polling, push messaging protocol. SMS system's advantage can be used in most mobile devices without separate special mobile application and most mobile communication networks. However, the SMS is short message service and every large message is sent to the client after being divided into short messages. Also, one of SMS's disadvantages is that it can't send multimedia content and that it costs per message as a billing system. In general, the cost of the American area is about 0.07 USD per message [2]. On the other hand, the polling message system's advantage is its cost-efficient system and sends multimedia messages but the polling system still has a disadvantage as it connects to the message server to check if new message occurred periodically thus causing data traffic waste. Such as the inconvenience of both the SMS and polling system, the messaging communication system is changed to gradually transfer a protocol push protocol scheme.

2.2 MQTT Protocol

MQTT protocol is the message push protocol released by IBM[4]. MQTT protocol was designed to transfer the messages reliably under the low-bandwidth network condition and long network delay. Currently Facebook instant messenger application for mobile device also uses the MQTT protocol.

The MQTT push protocol defines the transmission QoS (Quality of Service) level which is an agreement between sender and receiver of a single message regarding the guarantees of delivery of message. There are 3 levels of QoS in MQTT [3][4].

QoS 0: At most once. Send a message at most once and do not guarantee of deliver a message.

QoS 1: At least once. Send a message at least once and it is possible to deliver a message more than once.

QoS 2: Exactly once Send a message exactly once with 4 level handshaking.

A system which needs reliable message delivery uses QoS 2 for delivery a message exact once even if there is time delay. However, the QoS of MQTT push protocol does not guarantee messages ordering when it sends various messages but only guarantee single message delivery. Each message can be guaranteed to deliver to a receiver but message ordering cannot be guaranteed when time delay has occurred because of network condition as shown Figure 1.

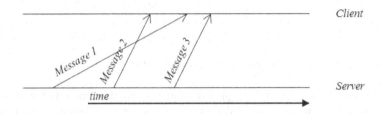

Fig. 1 Reversed messages ordering in message sending flow

3   Design of MQTT Based Collaboration Messenger

3.1   Reliable Messages Delivery Technique

It is very important that keeping message transaction guarantee in collaboration messenger for such things as the financial transaction system. In order to maintain the transaction, messages are required to have atomicity, consistency, and performance [5]. In addition, the collaboration messenger system must be able to view the previous messages for collaboration or the communication history. However, MQTT protocol based message communication system does not guarantee of message ordering and store the message; it uses only the transmission purposes. Therefore, the collaboration messenger system has to keep the messages for reliable collaboration, and has to retransmit when a message is missing or requesting again from the client.

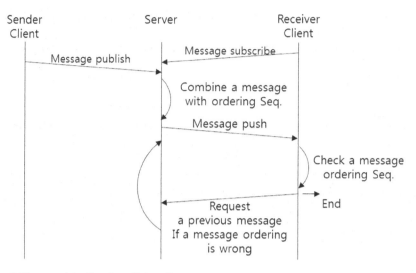

Fig. 2 Message data flow in collaboration messenger system

All messages have to be assigned a sequence number for reliable message delivery and each message are assigned a sequence number based on the time of arrival to the collaboration server, the messages go through the process of being transmitted to each collaboration messenger client that want to subscribe to the topic which is related to the messages. Each collaboration messenger client confirms whether or not the message is missing by checking the sequence number of the received message. Figure 2 shows the message data flow in collaboration message system.

3.2 Design of MQTT Based Collaboration Messenger Protocol

MQTT protocol is composed of MQTT fixed header, variable header and payload parts. The header part is composed of 2bytes fixed size and an additional byte for subscription and connection [4]. The payload area is an area that will be sent to the message. In this paper, we extend the payload area in order to verify the data in a collaborative messenger system. It is designed to expand the payload, as shown in table 1.

Table 1 Payload configuration for collaboration messenger system

Item	Order flag	SEQ	Original message
bit	1	15	n

In table 1, the order flag determines the transmission mode of a message. This order flag is the flag which determines whether a client received missed messages before receiving a current message or a client ignores missed messages while receiving an arriving message. SEQ is the sequence number given by the time arrives to a topic server represents a sequence of messages that occur, in the subject. SEQ consists of 15 bits, which can represent 2^{15} sequence numbers. The original message is sent by the client. Figure 3 shows how the order flag and SEQ is added to original message.

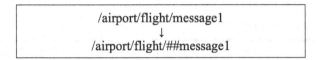

/airport/flight/message1
↓
/airport/flight/##message1

Fig. 3 A message with 2byte collaboration payload

3.3   Collaboration Messenger System Configuration

The client who receives messages subscribes to the topic, and the other client which issues messages publishes the topic to the server. The server reconstructs a message format to verify in the client by using the order flag set in the server and the SEQ of the received message. The client checks SEQ and order flag of the arrival message, and compare with a previous message SEQ. client request a missed message if there is a missed message after comparing SEQs and server resend the missed message to the client. Figure 4 shows the data flow in collaboration messenger system based on MQTT protocol.

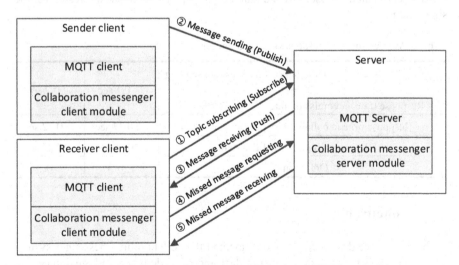

Fig. 4 Data flow in collaboration messenger system based on MQTT protocol

4   Experiment

4.1   Experiment Environment

We verify a collaboration messenger described in this paper through this experiment. The experiment environment for this is in below table 2. And we experiment messages from the client send without missing.

Table 2 Experiment environment

Item	Server	Client
OS	CentOS v5.1	Microsoft Windows 7
RDBMS	MySQL v5.0.95	-
message broker	mosquitto v1.2.3	mosquitto v1.2.3

4.2   Result of Experiment

We measure the result after sending and receiving 1000 messages which is composed with 20byte from client to client at time intervals of 0.1 and 0.2 seconds. Consequently, the delivery time is delayed as compared to the system which does not check the message sequence. However, this delay time is not enough people notice the difference so, it is not the problem between the messengers which is used by human being [6]. Table 3 shows result of the simulations.

Table 3 Average message response time and omission rate

Time interval for message sending / Response time and omission rate	0.1 sec	0.2 sec
Response time in collaboration messenger system	0.37 sec	0.32 sec
Response time in MQTT system	0.20 sec	0.23 sec
Omission rate	0 %	0 %

5   Conclusion

In this paper, we described and experimented the collaboration messenger system based on MQTT protocol to ensure fast message delivery and guarantee the ordering of messages for the real working environment. We confirmed the collaboration messenger system can deliver the messages sequentially without a long time delay.

In future research, we will design and implement the collaboration messenger system to be operational on a mobile device.

Acknowledgement This research was supported by Basic Science Research Program through the National Research Foundation of Korea (NRF) funded by the Ministry of Education (NRF-2014R1A1A2058888).

References

1. Cisco Visual Networking Index: Global Mobile Data Traffic Forecast Update, 2013–2018.   http://www.cisco.com/c/en/us/solutions/collateral/serviceprovider/visual-networking-index-vni/white_paper_c11-520862.html
2. Union, I.T.: Facts and Figures: The World in 2010. ed: International Telecommunication Union, p. 3 (2010)
3. Lee, S., Jung, I., Kim, H., Ju, H.: The design of integrated mobile SNS gateway structure. In: 2013 15th Asia-Pacific Network Operations and Management Symposium (APNOMS), pp. 1–3 (2013)
4. IBM MQTT Protocol Specification.   http://public.dhe.ibm.com/software/dw/webservices/ws-mqtt/mqtt-v3r1.html
5. Schiper, A., Raynal, M.: From group communication to transactions in distributed systems. Communications of the ACM **39**, 84–87 (1996)
6. Nah, F.F.-H.: A study on tolerable waiting time: how long are web users willing to wait? Behaviour & Information Technology **23**, 153–163 (2004)

Financial Security Protocol and Service Model for Joint Account Banking Transaction

JungJin Ahn and Eunmi Choi

Abstract Under todays' popular financial services via Internet banking process, this paper studies on the banking procedures and security protocols that aims to allow banking transactions of joint accounts to proceed within cyberspace. We propose the security protocol design of joint account banking transaction based on Diffie–Hellman protocol and station-to-station protocol. Regardless of whether members of the joint account are in the same bank or other banks, our proposed protocol is verified to be processed according to the transactional operation. This paper shows safety verification against possible threats. Finally, we suggest two types of financial service model for joint accounts in the both cases of a single-bank and multiple-banks joint account.

Keywords Internet banking protocol · Joint account · Diffie–Hellman protocol · Station-to-station protocol

1   Introduction

Financial services mainly use a non-face-to-face service channel through Internet activation, and the existing range of financial services has become available through such non-face-to-face service channels. However, joint account transactions are not yet available for financial service through the non-face-to-face service channels. A joint account is a contract whereby the account holder is two

J. Ahn and E. Choi(✉)
Department of Financial Information Security/Graduate School of Business IT,
Kookmin University, Seoul, Republic of Korea
e-mail: {ajj0602,emchoi}@kookmin.ac.kr

This research was supported by Basic Science Research Program through the National Research Foundation of Korea (NRF) funded by the Ministry of Education, Science and Technology. (Grant Number: 2011-0011507)

© Springer Science+Business Media Singapore 2015   521
D.-S. Park et al. (eds.), *Advances in Computer Science and Ubiquitous Computing*,
Lecture Notes in Electrical Engineering 373,
DOI: 10.1007/978-981-10-0281-6_75

or more people. [4] In the case of deposit transactions, any single account holder can complete the deposit transaction by only authenticating themselves without mutual agreement among account holders. However, disposal transactions, including withdrawal transactions, can only be completed by mutual agreement among account holders. Such mutual-sensitive tasks in joint accounts have been challenging in the context of Internet banking service since all the related stakeholders should agree and proceed the withdrawal transaction. This is why this kind of financial service has been processed in face-to-face service channel, which requires account holders' physical visits to manage the joint account.

This paper suggests a financial security protocol that can perform disposal transactions of joint account via non-face-to-face service channels over Internet banking. The security protocol is designed by a variation based on the Diffie–Hellman protocol and station-to-station protocol, in which the Diffie–Hellman protocol uses a symmetric session key generation between both parties and the station-to-station protocol protects against man-in-the-middle-attacks. The proposed security protocol is verified a safe and is able to generate a symmetric session key between multilateral parties. Based on the security protocol design that is suggested in this paper, we suggest two applicable cases of financial security service models: single-bank and multiple banks involved for the joint account.

2 Study on the Diffie–Hellman and Station-to-Station Protocols

2.1 Diffie–Hellman Protocol and the Features

The Diffie–Hellman protocol uses a symmetric session key generation between two parties and presents that the commutative law applies on the parties' session key operation. [1,3] Since multiple stakeholders involve the financial service of joint account, we assure that the associative law is valid in the Diffie–Hellman symmetric session key generation, as in Fig. 1.

commutative law	$(g^x \bmod p)^y \bmod p = (g^y \bmod p)^x \bmod p$		
associative law	$[(g^x \bmod p)^y \bmod p]^z \bmod p = [(g^y \bmod p)^z \bmod p]^x \bmod p = [(g^z \bmod p)^x \bmod p]^y \bmod p$		
g : prime number (primitive root, depend on p)		**p** : 300 digits or more prime number (1024 bit)	
x : range of (1~P-2)	**y** : range of (1~p-2)		**z** : range of (1~p-2)

Fig. 1 Validation of commutative law and associative law of the Diffie–Hellman protocol

2.2 Station-to-Station Protocol

The Station-to-Station Protocol uses a symmetric session key generation method between two parties based on the Diffie–Hellman protocol. [2] The different point compared to Diffie–Hellman protocol is in the use of signature by public key certificate. Fig. 2 shows the process of the station-to-station protocol.

Step 1: Alice calculates R1 using own nonce x and sends to Bob (①, ②).

Step 2: Bob calculates R2 using own nonce y and calculates session key k using R1 that was received by Alice. Bob attaches to the message Alice's ID, R1, and R2 and encrypts the message using session key K after signing using Bob's private key. The encrypted message with its own public key certificate, R2, is transmitted to Alice. (③, ④, ⑤)

Step 3: Alice creates a session key K by using the received message R2 from Bob. Then, decrypt the message by using session key K and verifies Bob's signature. If Bob's signature is verified, Alice attaches to the message Bob's ID, R1, and R2 and encrypts the message using session key K after signing using Alice's private key and sends the message to Bob (⑥,⑦,⑧)

Step 4: If Alice's signature is verified, Bob accepts the session key K (⑨).

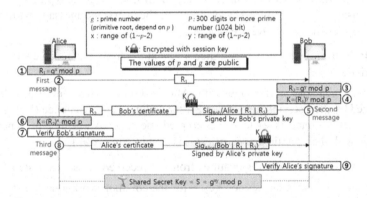

Fig. 2 Station-to-Station Protocol

3 Financial Security Protocol Design for Joint Accounts

3.1 The Proposing Security Protocol

The security protocol suggested in this paper has the following assumptions. First, the certification authority (CA) issues a public key certificate and the public key certificate is trusted. Second, the Key Distributed Center (KDC) distributes a private key and a public key that are used for electronic signature of a message. Fig. 3 presents our financial security protocol that reflects those two assumptions:

Step 1: Alice calculates R1 using her own nonce x and sends it to Bob and Calvin, respectively. (①,②)

Step 2: Both Bob and Calvin calculate R2 an R3 using the R1 received and their own nonce, y and z, and then calculate their session keys, K1 and K2, respectively. Those session keys will use for data encryption between Bob to Alice and Calvin to Alice, respectively. (③)

Bob generates a message with Alice's ID, R1, and R2, which is encrypted by K1 after signing using Bob's private key. Altogether, the encrypted message with Bob's public key certificate and R2 are sent to Alice. As the same way, Calvin generates a message with Alice's ID, R1, and R3, which is encrypted by K2 after signing using Calvin's private key, and sends the encrypted message with Calvin's public key certificate and the R3 to Alice. (④)

Step 3: Alice calculates session keys K1 and K2 using the R2 and R3, received from Bob and Calvin, then verifies signatures of Bob and Calvin. (⑤,⑥)

Alice generates a message with Bob's ID, R1, and R2, Which is encrypted by K1 after signing using Alice's private key. Altogether, the encrypted message with Alice's public key certificate and R3 are sent to Bob. As the same way, Alice generates a message with Calvin's ID, R1, and R3, which is encrypted by K2 after signing using Alice's private key. Altogether, the encrypted message with Alice's public key certificate and R2 are sent to Calvin. (⑦)

Step 4: Bob and Calvin verify Alice's signature from the received messages, respectively. (⑧)

Then they generate a message with Alice's ID, R1, R2, and R3, which is encrypted by their session keys K1, K2 after adding their own nonce (Bob's nonce is y, Calvin's nonce is z) to the encrypted message. Altogether the encrypted message with their public key certificates is sent to Alice. (⑨)

Step 5: Alice verifies users' signatures from the received message on Step 4. (⑩)

Then, Alice generates the session key S for the common work among Alice, Bob, and Calvin. She sends the other nonce to Bob and Calvin. In the view of Bob and Calvin, they need the other nonce required for generating session key S (z is for Bob, y is for Calvin) (⑪,⑫)

Step 6: After Bob and Calvin receive message from Alice, they verify Alice's signature, and generate session key S using the nonce attached in the received message. (Bob uses z, Calvin uses y) (⑬,⑭))

3.2 Safety Verification of Financial Security Protocol

In this section, the security protocol presented in Section 3.1 is assured with the security verification of four information security aspects.

Confidentiality: All the messages transmitted via Internet does not contain the session key itself, but the corresponding necessary information encrypted are transmitted. The arithmetic operating process for making the session key is performed in the individual party. Thus, operating and sharing the session key verifies confidentiality.

Integrity: In the security protocol of Fig. 3, the Sig $_{user}$ (...) means the electronic signature by $_{user}$. Since we assume that the hash function is shared by all the parties in advance, all the parties verify the electronic signature by using the hash function. Message receivers are able to check for forgery of the message.

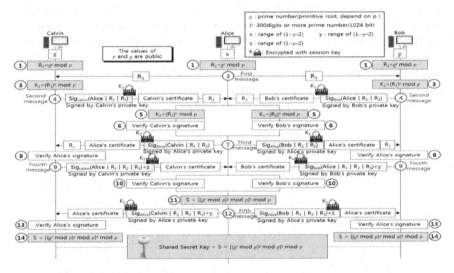

Fig. 3 The Proposing Security Protocol

Authentication: We assume that the KDC distributes a private key and a public key for electronic signature. All the identity of participants are authorized by the KDC, in which the KDC's role is as a Trusted Third Party.

Non-Reputation: Since the KDC distributes a private key and a public key to the authorized participants, the electronic signature generated by them cannot be reputed for legitimacy.

4 The Financial Security Service Model of Joint Account

According to the security protocol design in the Section 3, we suggest two applicable cases of financial security service models for the joint account in this section: single-bank case and multiple banks case.

The first service model is a multilateral agreement procedure through a single bank, which is shown in Fig. 4. Bob, Calvin, Duke are joint account holders of BANK 1 and they have the authentication method. This model is composed of four steps. The first is the **request step,** in which a specific user requests a transaction to other users, and other users receive the agreement request. The second is the **authentication step,** in which users who received the agreement request connect to Internet banking by using their own authentication method. The third is the **agreement step,** in which all users generate a common session key using the financial security protocol introduced in Fig. 3. The fourth is the **processing step,** in which if the session key generation procedure is complete, the transaction process is performed.

The second service model is a multilateral agreement procedure through multiple banks, which is shown in Fig. 5. Duke and Calvin are customers of the Bank1. Edward and Fedora are customers of the Bank2. The joint account is managed by Bank1. The overall structure of this second service model is similar to the first service model in terms of four processing steps and financial security protocol for agreement. The different point of two service models is in bank cooperation. That is, when Bank1 requests Bank2's cooperation, the Bank2 authenticates Edward and Fedora, and join to session key generation process as a representative of customers of Edward and Fedora.

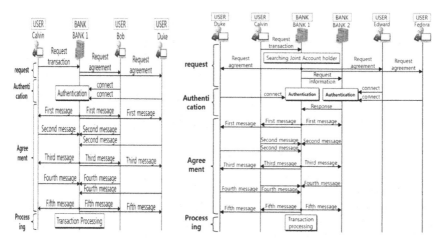

Fig. 4 Multilateral agreement procedure through a single bank

Fig. 5 . Multilateral agreement procedure through multiple banks

5   Conclusion

In this paper, we propose a financial security protocol for joint account based on the Diffie–Hellman protocol and the station-to-station protocol. The financial security protocol we design is considered with threat safety and verified with information security aspects. By using the proposed financial security protocol, we introduce two applicable service models of one or multiple bank cases.

This paper has applied the security protocol to process of joint accounts. Moreover, it is possible to be applicable to other financial services and a variety of fields for the enhanced financial security services.

References

1. Paar, C., Pelzl, J.: Understanding Cryptography: A Textbook for Student and Practitioners, pp. 205–234. Springer (2010)
2. Forozan, B.A.: Cryptography & Network Security, pp. 467–479. McGraw-Hill, Korea (2008)
3. Stamp, M.: Information Security: Principles and Practice, pp. 100–102. Willey (2011)
4. Kim, D.-M., Kim, J.G., Kwon, E.-H.: Bank Teller(First), pp. 153–159. Korea Banking Institute (2004)
5. Sakurai, K., Fukushima, K.: Actual Condition and Issues for Mobile Security System. Journal of Information Processing Systems 3, 54–63 (2007)
6. Tallapally, S., Padmavathy, R.: Cryptanalysis on a Three Party Key Exchange Protocol-STPKE. Journal of Information Processing Systems 6, 43–52 (2010)
7. Kang, C.-G., Choi, Y.-R.: An Algorithm for Secure Key Exchange based on Mutual Entity Authentication. The Transaction of the Korea Information Processing Society 5, 2083–2090 (1998)

Development of a Cursor for Persons with Low Vision

Jong Won Lee, JiSu Park, Jong Il Lee and Jin Gon Shon

Abstract Objects such as windows, icons, menus, and pointers are being smaller and closer on devices display. Small and closely packed object requires more time and efforts in acquisition of target object persons with low vision. Traditional magnifier and cursor technologies are increasing recognition rate with magnification, color contrast, big-sized cursor and separation of the object. But, in the case of low vision, this traditional magnifiers and cursors are not enough to acquire target objects easily. In this paper, we propose a CPLV (cursor for persons with low vision). Basically, CPLV supports magnification, color contrast, and big-sized cursor for persons with low vision. Furthemore, it supports constraint of cursor movement, reduce of hand fatigue, and voice service.

Keywords Cursor · Low vision · HCI · CPLV

1 Introduction

Persons with low vision are who has the corrected visual acuity less than 0.33 and equal to or better than 0.05, and their field of view less than a 10-degree [1]. Graphic user interfaces based on window, icons, menus, and pointer [2].

J.W. Lee · J.G. Shon(✉)
Department of Computer Science, Graduate School,
Korea National Open University, Seoul, South Korea
e-mail: hosori@bohun.or.kr, jgshon@knou.ac.kr

J. Park
Department of Computer Science and Engineering,
Korea University, Seoul, South Korea
e-mail: bluejisu@korea.ac.kr

J.I. Lee
Department of e-Learning, Graduate School,
Korea National Open University, Seoul, South Korea
e-mail: jongilla@naver.com

© Springer Science+Business Media Singapore 2015   529
D.-S. Park et al. (eds.), *Advances in Computer Science and Ubiquitous Computing*,
Lecture Notes in Electrical Engineering 373,
DOI: 10.1007/978-981-10-0281-6_76

Assistive technology for persons with low vision consists of hardware such as assistive device and software such as screen magnifier and cursor. Microsoft magnifier (MSM) is frequently used to screen magnifier. MSM produces a separate rectangular window that displays a magnified portion of screen and MSM supports inverse color and magnifying power range from 1 to 16[3]. MSM has two shortcomings. First, the higher magnification cause lower resolution. Second, it is difficult to differentiate between closely packed objects.

Cursor software supports magnification, color contrast, and big-sized cursor. Pointing magnifier (PM) designed to reduce effort to select small and closely packed target objects for persons with motor impairments[4]. However, PM reduces accuracy to select small and closely packed target objects, if it clicks unintended screen portion in initial positioning, PM needs repositioning of cursors and then reduces performance.

Click-and-Cross cursor and Cross-and-Cross cursor designed to reduce acquisition time and error rate to select small and closely packed target objects for persons with motor impairments using proxy target objects. The proxy increases in size compare to the original target objects. Each proxy is assigned a crossing arc segment that is a large circle appears to area cursor location closely, overlays original space[5]. However, Click and Cross cursor and Cross and Cross cursor are difficult to select target objects because of the small sized proxy and the narrow gaps between objects. And these are losing their cursor position with overshooting of cursor and then reducing performance.

Despite of development of cursor software, the development of cursor application software and web page faces problems such as an increasing number of selectable objects and their closely packed position within finite sized screens. Therefore, these problems increase acquisition time and error rate for persons with low vision. In this paper, the proposed CPLV is a one of kind of cursor software.

2 Design of a Cursor for Persons with Low Vision

CPLV is consist of cogwheel shaped area-cursor and secondary cursor. That is, one is cogwheel shaped area cursor and the other is big-sized cursor moves within area-cursor. Cogwheel shaped area-cursor is consists of circular part for magnifier and saw tooth part for position area of splitting objects. Persons with low vision moves the cogwheel shaped area cursor to magnified area and the objects located under the magnifier itself. Cogwheel shaped area cursor is located over objects, and then target object magnifies and splits. And persons with low vision clicks to fix cogwheel shaped area cursor, the secondary cursor appear to select the magnifying and splitting object. CPLV has basic functions and advance functions. Basic functions consist of magnification, color contrast, and splitting for select the window, icons, menu and pointer easily. Furthermore, it supports the located objects, constraint of cursor movement, reduce of hand fatigue, and voice service as advanced function.

2.1   Basic Functions

Basic functions of CPLV are magnification of objects, splitting of objects, and color contrast. There are explains in detail as following.

First, magnification is essential function to persons with low vision because of who are difficult to recognize small objects. Therefore, cursor for persons with low vision magnifies a part of screen and an object. Expansion of the screen is run by magnify the screen under the area-cursor as like magnifier and the expansion of the object for recognition of them clearly. Magnification of object reduces the resources of consumption by do not magnify of meaningless portion of screen, and Magnification of a recognizable object can easily know the position of the cursor across the screen by reduce hidden spaces due to the screen magnification.

Second, color contrast is essential function to recognize continuous-tone object such as pastel tone for persons with low vision. There are many ways for higher color contrast, inverse color, complementary color, and enhances the color sharpness. Enhancing the color contrast was that elimination the incognizant of higher and lower frequencies which can't recognition by low vision, and manipulating the image spectrum and enhancing the amplitude of recognizable frequency. And Sharpness of color of magnified object was improved with smoothing process and pixel duplication in both horizontal and vertical directions.

Third, splitting of object is necessary function to persons with low vision who are difficult to distinguish intended object from obscure gap and narrow gap between objects. Split method of objects that moves the object under the magnifier itself to rim of area cursor such as cogwheel parts. Magnified objects of cogwheel parts do not interfere with view of magnifier as area cursor, because there is no overlapping each other. If there are too many objects under the magnifier, CPLV gives priority to object which comes in first in range of area cursor when to place rim of area cursor.

2.2   Advanced Functions

Advanced functions of CPLV are placement of splitting objects, limitation of cursor movement, and voice service which are explained in detail as following.

First, placement of splitting objects reduces fatigue of hand. Fatigue of hand happens to persons with low vision when they fail in acquisition of intended target objects. Failure of acquisition cause additional clicks and fatigue of hand. Reducing method of hand fatigue placed splitting objects used on muscle recoil phenomenon. Muscle stores elastic energy in muscle contraction before relaxation of muscle. CPLV uses this elastic energy for placement of splitting objects. While area cursor move to one direction for target objects, if object is under magnifier, splitting objects places in counter direction of cogwheel parts of area cursor.

Second, although size of cursor is bigger than before, persons with low vision are difficult to aware position of cursor when position of cursor gets out of their field of view. Limitation of cursor movement is secondary cursor moves within range of area cursor instead of moving of the area cursor within full screen.

Through limitation of cursor movement, CPLV prevents secondary cursor from overshooting cogwheel parts of area cursor, and it prevents persons with low vision from missing their cursor position.

Third, CPLV provides visual information and auditory information to persons with low vision that is more effective than providing visual information only, because alternative information is very useful for handicapped persons. Generally, Voice information characterizes image using information of objects and transforms each characters of characterized image into phonetic symbol. Phonetic symbol transforms phonemes through prosodic testing then it generates waveform and makes audio clips. Voice information removes the errors which occur in acquisition of target objects through process of reconfirmation.

3   Performance Evaluation

3.1   Experimental Design

Forty five adult Participants has following characteristics: are 25 male (56%) and 20 female (44%) with ages ranging from 25 to 71 (M±SD=33.11±6.73), has visual acuity of low vision in naked eyes. Experiment were run using a 17" LCD monitor (1280 x 1024, 32bits) connected to desktops running Windows XP, Microsoft comparable mice (default cursor speed). Participants start the experiment in sitting position comfortably. The distance between participants and screen is within 40cm. Statistical analysis is performed using SPSS version 15 for Windows. Paired t-test with double blind method is used to examine mean comparison between acquisition times.

3.2   Experimental Result

Cursor for persons with low vision was coded in C# .NET 2.0 and was implemented Hangul(word processor) 2010; one of an applications and Naver homepage; one of a web pages. Figure 1 shows implementation of CPLV.

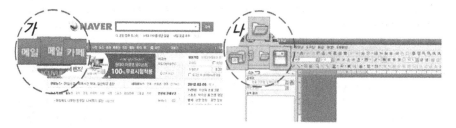

Fig. 1 A cursor for persons with low vision

Figure 2 shows performance evaluation of basic functions. Experimental result of basic functions explained in detail as following. Acquisition time is the main dependent measure, and is the time taken to move and successfully select the target object, and comparisons are significant at p<0.05 unless other mention. Average target acquisition time is 3.18±0.12 second by CPLV, 5.03±0.63 second by PM, and 2.89±0.18 second by MSM. There is significant improvement in performance that average target acquisition time by CPLV is 1.85±3.92 second faster than PM (<0.01). Average Target acquisition time of MSM is 0.28±1.28 second faster than CPLV but there is no significant improvement in performance. There is no error in target acquisition by CPLV but average error rate is 4% by MSM in 2 failure and 17% by PM in 8 failure of target acquisition among both 45 trials.

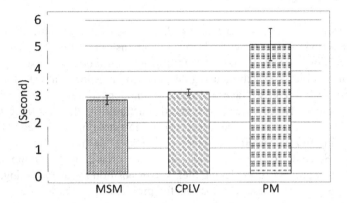

Fig. 2 Result of Performance for basic function

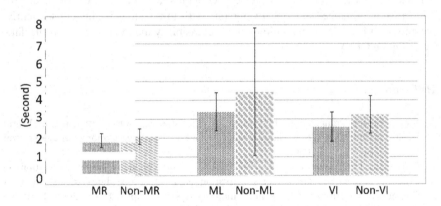

Fig. 3 Result of Performance for advanced function

Figure 3 shows performance evaluation of advanced functions. Experimental result of advanced functions explained in detail as following. Figure 3 shows comparison of target acquisition times between muscle recoil (MR) and non-muscle recoil (Non-MR), movement limitation (ML) and non-movement limitation (Non-ML), voice information (VI) and non-voice information (Non-VI).

First, Average target acquisition time is 2.08±0.41 second on Non-MR and 1.87±0.37 second on MR. There is significant improvement in performance that average target acquisition time of MR is 0.22 second faster than non-MR (<0.01).

Second, Average target acquisition time is 4.45±3.38 second by Non-ML and 3.39±1.01 second by ML. There is significant improvement in performance that average target acquisition time of ML is 0.74 second faster than non-ML (<0.01).

Third, Average target acquisition time is 3.24±0.99 second by Non-VI and 2.59±0.77 second by ML. There is significant improvement in performance that average target acquisition time of VI is 0.66 second faster than non-VI (<0.01).

4   Conclusion

Development of cursor for persons with low vision reduces the difficulties in object selection using basic and advanced functions of traditional cursors function. Basic function of CPLV has two following contributions. First, it increases precision of objects selection using magnification, color contrast, and big-size cursor. Second, it reduces the errors clicks of unintended object using splitting and widening gaps between objects. Advanced function of CPLV has three following contributions. First, reduce the object acquisition time by placement of splitting objects using muscle recoil phenomenon decreases the muscle fatigue of objects selection. Second, reduce the object acquisition time by limitation of cursor moving range prevents users losing their cursor position for overshooting. Third, it provides users with voice information for problem of unknown function in the form of object, and reduces object acquisition time. According to the performance comparison, CPLV can make faster object acquisition time and can reduce error rate in using advanced cursors for persons with low vision. In this consequently, persons with low vision can enjoy better accessibility than before in user interface by using the CPLV.

References

1. World Health Organization: Prevention of Blindness and Visual Impairment. World Health Organization (2012). http://www.who.int/blindness/causes/priority/en/index5.html (accessed March 27, 2012)
2. Grossman, T., Balakrishnan, R.: The bubble cursor: enhancing target acquisition by dynamic resizing of the cursor's activation area. In: Proceedings of the SIGCHI Conference on Human Factors in Computing Systems, April 02–07, 2005, Portland, Oregon, USA, pp. 281–290 (2005)
3. Microsoft: turn-on-magnifier. Microsoft (2015). http://windows.microsoft.com/ko-kr/windows-xp/help/turn-on-magnifier (accessed June 9, 2015)
4. Jansen, A., Findlater, L., Wobbrock, J.O.: From the lab to the world: lessons from extending a pointing technique for real-world use. In: Proceedings of the 2011 Annual Conference Extended Abstracts on Human Factors in Computing Systems, May 07–12, 2011, Vancouver, BC, Canada, pp. 1867–1872 (2011)
5. Findlater, L., et al.: Enhanced area cursors: reducing fine pointing demands for people with motor impairments. In: Proceedings of the 23th Annual ACM Symposium on User Interface Software and Technology (UIST), pp. 153–162 (2010)

Implementation of Unified Gesture Library for Multiple-Devices Environments

EunJi Song, Sunghan Kim and Geun-Hyung Kim

Abstract In recent years, several input devices have changed the way users interact with computers. Gestures can be performed in different ways, such as moving a mouse, pointer, touching a screen, or making hand movements in a three-dimensional (3D) space. With the advent of various smart platforms with different gesture-input devices, application providers have developed specific web applications for each new gesture-input device. In previous years, developers would create new applications whenever new input devices appeared. Therefore, a unified gesture library appears necessary to resolve the inconvenience that a developer must face, as he considers the specific input-device characteristics in order to adapt a web application for various smart device platforms. In this paper, we defined common gestures for mouse, touch, and pointer devices, as well as a 3D hand-movement detector, and proposed a unified gesture library that produces a high-level event when a specific gesture occurs. In addition, we implemented the unified gesture library supporting common gestures, a circle gesture, a two finger drag gesture, and 3D hand- gesture recognition functions. Finally, we verified that the unified gesture library supported mouse or pointer gestures, multi-touch gestures, and hand gestures in 3D space by sample applications.

Keywords Unified gestures · Multiple input devices · 3D hand gesture · 3D hand movement detector

E. Song · G.-H. Kim(✉)
Department of Visual Information Engineering, Dong-Eui University,
177 Eomgwang-no BusanJin-Gu, Busan 614-714, Korea
e-mail: sej1272@naver.com, geunkim@deu.ac.kr

S. Kim
Electronics and Telecommunications Research Institute,
218 Gajeong-Ro Yuseong-Gu, Daejeon 305-700, Korea
e-mail: sh-kim@etri.re.kr

© Springer Science+Business Media Singapore 2015
D.-S. Park et al. (eds.), *Advances in Computer Science and Ubiquitous Computing*,
Lecture Notes in Electrical Engineering 373,
DOI: 10.1007/978-981-10-0281-6_77

1   Introduction

As new various input devices, which interact with computer differently, are released, these new devices exploit gestures performed in different ways. Current user interface (UI) programming frameworks are still bound to the observer pattern where events occur atomically in time and are notified through messages or callbacks [1]. In addition, support for gestures in programming interface frameworks has used in the same paradigm by hiding the gesture recognition logic, which means providing high-level events when a user's gesture is completed and providing intermediate feedback to the handling mechanisms of low-level events that are not correlated with high-level events [1].

A UI is something through which we as users interact with computing systems [2]. The interactions between users and computing systems occur in the UI space, and allow users to effectively operate and control the computer systems. The UI has evolved continuously along with the computing system. A first-generation UI is a command line interface (CLI) that mostly uses a keyboard for input and a screen for output for interactions and feedback.

The graphical user interface (GUI) is considered the second generation UI, in which users are encouraged to look for items on the screen, e.g., icons, pop-ups, dialogs, menus, etc. The mouse was introduced as a way to interact with UI components for GUIs. With the development of GUIs, more sophisticated and direct interaction devices became available, such as pointer devices and touch screens. As the next generation UI, natural user interfaces (NUIs) are introduced. They support a user's cognition and interact with computer systems in a natural manner, without interaction devices such as a mouse or keyboard. Interaction devices for NUIs include eye tracking, sensors, cameras, wearable equipment, etc.

As various device types exist, users can own one or more devices, including smart phones tablet PCs. smart TVs, etc. Therefore, users want to have a similar service experience for the same application using their different devices. With the advances in HTML5 (Hyper Text Markup Language 5) technology, application developers have a chance to provide the same web applications for different devices. However, application developers must consider a variety of UIs to develop web applications for the different devices, because the devices produce different event types. To resolve this problem, several gesture libraries have been developed [3][4]. In this paper, we describe a unified gesture recognition platform, called the UGesture platform, to support the mouse, touch screen, pointer, and 3D hand movement detector [5], one of NUIs consisting of sensors.

The remainder of this paper is organized as follows. In Section 2, we discuss related work. We describe the key operations of the UGesture platform in Section 3. In Section 4, we describe the web applications implemented on the UGesture platform and the gestures that are used for each application. Finally, the paper is concluded in Section 5.

2   Related Works

Because devices are different, their supporting gestures may also differ. Application developers must consider the discrepancies in gestures among devices when they develop an application that runs on different devices. To remove this discrepancy for touch gestures, Craig classified touch gestures into core gestures, basic operations, operations related to an object, and navigation operations. In Craig's classification [6], core gestures include tap, double tap, drag, flick, pinch, spread, press, press and tap, press and drag, and rotate. However, not all devices in real situations support all the core gestures. In addition, the gestures' name may be defined differently on the devices, even though the gestures may be the same.

Gesture libraries allow application developers to develop web applications that work with various devices, regardless of the events that the input devices generate. Users typically want to use web applications with the same gestures on any device.

One of the gesture libraries, the HammerJS library [4] defines a common gesture API for mouse, touch, and pointer interfaces and recognizes the touch, tap, double tap, swipe, drag, transform, pinch, rotate, release, and hold gestures. The library uses a gesture object, a tap object, and a transform object. The HammerJS library creates a Hammer instance, gathers the event data, issues gesture events and calls an event handler.

3   UGesture Platform

Conventional gesture libraries are classified into two groups. One focuses on defining common touch gestures for different mobile devices, and the other focuses on defining common gestures to interact with a mouse or touch screen. With the advances in NUI, motion detection technology, several devices which support 3D motion detection are provided. In this paper, we considered the LeapMotion controller [5] as a 3D movement detector and defined the UGesture platform that supports a mouse, touch interface, and 3D movement detector.

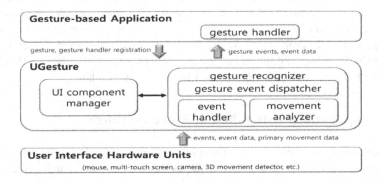

Fig. 1 Functional Architecture of a UGesture Platform

Fig. 1 shows the functional architecture of the UGesture platform and its interactions with other layers. The UGesture platform hides the underlying UI hardware units from gesture-based applications. Therefore, application developers can develop applications on the UGesture platform irrespective of input devices.

Application developers register the required gesture and the gesture handler with the UGesture platform. The platform triggers the gesture handler with data related to the gesture. The UGesture platform includes a UI component manager and a gesture recognizer. The gesture recognizer consists of an event handler, movement analyzer, and gesture event dispatcher. The event handler is used to handle UI input device events and may correspond to part of a gesture. The movement analyzer analyzes the movement data from sensors and checks whether the movement sequence corresponds to a gesture. If the movement analyzer detects a gesture, interacts with the gesture event dispatcher to a generate gesture event for the application.

The LeapMotion controller captures the movement of the hand or fingers on a frame-by-frame basis and only provides a list of movement trajectories. Unlike other gesture input devices, the 3D movement detector does not issue events related to the gestures and does not bind an occurred motion to the UI component on which it occurred. Therefore, when a gesture happens, the UGesture platform recognizes the defined gesture, identifies the UI component on which the motion occurred, and issues the corresponding events.

In order to develop the UGesture platform to support the LeapMotion controller, we defined three events: motion start event, motion move event, and motion end event, corresponding to the start of a gesture, movement, and the end of a gesture, respectively. The start and end of the gesture are based on the origin of the z-axis in the LeapMotion Controller. When a detected motion occurs below the origin of the z-axis, the UGesture platform issues a start event. The platform issues an end event if the detected motion occurs above the origin of the z-axis.

The UGesture platform creates a UGesture instance to store the UI component's location and related information. When an application developer registers the gesture of a specific UI component, the UGesture platform obtains the location information of the component and stores it. The UGesture platform manipulates the gesture registration on specific UI component. When one of the three events occurs, the UGesture platform checks if the position belongs to one of the registered UI components. When the UGesture platform recognizes the UI component on which the motion occurred, it checks whether a registered gesture has occurred, and issues the appropriate gesture events, if it has.

4   Applications Using the UGesture Platform

In order to verify the developed UGesture platform, we implemented two sample applications that work with different input devices; mouse, touch screen, and 3D movement detector. The first is a drawing application shown in Fig. 2.

To implement the drawing application, we used the jCanvaScript library [7] to manage the content of the HTML5 canvas element. In this application, we used the following gestures: tap, double tap, drag, pinch, and rotate.

For this sample application, we defined two shapes: a rectangle and a circle. We also defined three functions: drawing a shape, filling the background with a selected color, and drawing and filling a shape with a selected color. To draw a rectangle, we used the movement of two contact points, which represent the lower left point and the upper right point of the rectangle. To draw a circle, we used the rotate gesture and the distance of two contact points as the radius of a circle. In addition, we used a tap gesture and a double tap gesture to select the color of the shape and the color of the background respectively. Fig. 2 (c) shows the interaction for drawing a rectangle using two fingers.

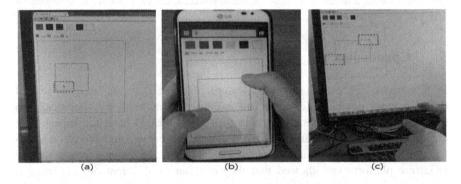

Fig. 2 Screens of the drawing application that interacts between the user and computer systems: (a) using a mouse, (b) using a touch interface, and (c) using a 3D hand movement detector.

The second application was a music player as shown in Fig. 3. This application consisted of an image area, a playback control, and a playlist. The image area was used to capture the gesture. In this application, we used the following gestures: tap, double tap, hold, swipe, two-finger drag, and circle. The tap gesture was used to play the first music item in the playlist, and the double tap gesture was used to pause the playback. The hold gesture was to mute, and the swipe up and swipe down were used to control the volume. The swipe left and swipe right were used to navigate the items in the playlist. Finally, the rotate gesture was used to restart the music playback. The tap and hold gestures were also applied to the playback control panel. The two-finger drag gesture was bound to the playlist object and used to scroll the items in the playlist. Fig. 3 (c) shows the 3 D movement interaction for playing the content and scrolling the playlist.

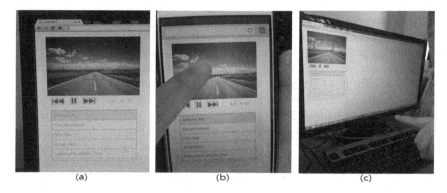

Fig. 3 Screens of the media player application that interacts between the user and computer systems: (a) using a mouse, (b) using a touch interface, and (c) using a 3D hand movement detector.

5   Conclusion

In this paper, we proposed the UGesture platform that can support several different input devices and developed a unified gesture library specifically to support 3D hand movements. In order to support the 3D hand movements, we developed a binding mechanism between a gesture event and an element on a web page. Finally, we implemented web-based sample applications based on the UGesture platform and showed that they can run on different devices using a mouse, touch interface, or 3D movement detector for the same user experience. As future work, we will extend the UGesture Platform to support other UI devices including a Kinect and study a gesture description method that works regardless of input devices.

Acknowledgement This research was supported by Basic Science Research Program through the National Research Found (NRF) funded by Ministry of Education, Science and Technology (NRF-2010-0025069) and Institute for Information & communications Technology Promotion (IITP) grant funded by the Korea government (MSIP)[R0166-15-1002, Convergence-based Web Standardization Development].

References

1. Spano, L.D., Cisternino, A., Paterno, F.: A composition model for gesture definition. In: HCSE 2012. LNCS, vol. 7623 (2012)
2. Heikkinen, K., Porras, J.: UIs – Past, present and future. Wireless World Research Forum White Paper Version 1.0, No 10, December 2013
3. Micro JavaScript Library that simplifies your mobile projects. http://quojs.tapquo.com
4. Add touch gestures to your page. http://hammerjs.github.io
5. Reach into new worlds. http://www.leapmition.com
6. Villamor, C., Willis, D., Wroblewski, L.: Touch Gesture Reference Guide (2010)
7. jCanvaScript library document. http://www.jcscript.com/documentation/

Design of the Real-Time Mobile Push System for Implementation of the Shipboard Smart Working

Joonheung Park, Taehoon Koh, Jun-Ho Huh, Taeyoung Kim, Jeongho Lee, Jaesoon Kang, Donghyun Ju, Jeongdae Kim, Junwon Lee, Taewook Hwang, Youngjoon Park and Kyungryong Seo

Abstract The ships sailing on the sea often have to transmit data through satellites to establish data communications with shore but it is difficult for them to receive real-time data unless they make connection artificially. This is mainly because of high satellite communication charges and its low speed. Currently, there's no real-time push system optimized for the satellite service exits to overcome this problem. For this reason, young sailors tend to avoid working on the ships where they cannot use SNS services to contact their families or friends on land. As a matter of fact, they select the ships where they can communicate with landside people at any time during they are at sea. Meanwhile, the International Labor Organization (ILO) is encouraging the shipowners and shipping companies to have their ships equipped with the system which facilitates an all-day-long communications to promote the welfare of the crews. Therefore, to comply with this guideline by the ILO, it is essential to develop a Push System that enables real-time communications between the sailing ships and land-based people or facilities. Thus, we've designed a real-time push system which is optimized for the satellite communications. The system allows real-time communications while

T. Koh · J.-H. Huh · K. Seo
Department of Computer Engineering, Pukyong National University,
Busan, Republic of Korea
e-mail: krseo@pknu.ac.kr

J. Park · T. Koh · J.-H. Huh · T. Kim · J. Lee · J. Kang · D. Ju · J. Kim ·
J. Lee · T. Hwang · Y. Park · K. Seo(✉)
SUNCOM Co., Busan, Republic of Korea
e-mail: krseo@pknu.ac.kr

This article contains the contents of research which has been conducted with support of the 1st-year Nurimaru R&BD project research fund granted by the Busan IT Industry Promotion Agency of the Republic of Korea in 2015.

© Springer Science+Business Media Singapore 2015
D.-S. Park et al. (eds.), *Advances in Computer Science and Ubiquitous Computing*,
Lecture Notes in Electrical Engineering 373,
DOI: 10.1007/978-981-10-0281-6_78

bearing down excessive communication costs. We consider that this system will bring more benefits in terms of technological, economical, industrial and social aspects compared to existing systems.

Keywords Design · ICT · Push system · Real-Time Mobile Push System

1   Introduction

The introduction of the mobile devices such as Smart phones and tablets people are able to easily and conveniently receive a variety of available service notifications by SMS or the push systems using existing websites or Apps [1 - 4]. The push systems are being provided by Google and Apple independently and used as various means of information/message delivery (e.g., SMS, SNS, news, stock info. broadcasting, etc.) as the applications using these push systems are increasing explosively.

However, those ships on sailing on the sea have no other way but to transmit data through satellites to achieve data communications with the land-based parties. As mentioned above, such data exchanges are limited by the high satellite service charges and its relatively low speeds, forcing the ships to artificially establish connections with the land to get the real-time data. So far, there's no optimal satellite-oriented push systems are available. Currently, the ship-to-shore communications are carried out mostly with the satellite phones and E-mails but in a situation where the ships are being built increasingly larger and the number of domestic crews (Korean crews) is getting less and less, the demands for the real-time two-way communications with portable mobile devices are ever increasing. And accordingly, many young sailors in South Korea deliberately avoid working on a ship where they can't use SNS services offered by the Korean ICT companies (e.g., Daum Kakao Talk, Naver Line, etc.) whenever possible and whoever they want have a chat with. Although such demands and ILO advisement may justify the necessity of developing a better and efficient mobile communication technology which has a sophisticated push system, the problem related to the satellite service charges still remains.

The satellite-oriented transmission and reception systems embedded in the sea-going ships bear the burden of very high service charges. That is, currently, 1MB data and 1 SMS (approx. 140 bytes) transmissions will incur the fees of about 420,000 won and 600~1,000 won, respectively. Also, if the crews wish to use the SNS services mentioned above without appropriate additional equipments, they will have to pay tens of millions of won per month, not to mention the satellite communication equipment, VSAT, which costs around 100 million won. Another concern is the communication security problems. The vital information of the ship could be leaked or hacked when crews contact the non-authorized parties both on and off shore.

When we take the efficiency of communications into consideration, the current systems do not offer much benefit for the ship's operations as there's no standby system to notify data arrivals on real-time basis unless the connection with the satellite(s) remains open. Hence, we've designed the real-time push system which is optimized for the satellite communications, reduces communication costs, enables the real-time communications between ships and shore, and notifies the arrival of messages or data.

2   Related Research

2.1   Push System

The push system is a information or data delivery system where the process begins at the central server when a request is made. It automatically provides news or other information whether the user requests them or not. Following [Fig. 1] is an example of the message service that uses the push system [4 - 8].

[Fig. 1] is divided three sections and the first yellow colored column (Yellow Pay Service) shows the 'log-ins' and a 'deposit request' of 4,000 won. The message in the middle of the second column informs the user that there's been a deposit made to the user's account (IBK bank) at 19:32 and the user should press a 'check' key to see the transaction result. The third message on the right is the 'One-touch Notification Push Service' by Woori Bank.

Fig. 1 An example of the message service using the push system

Information deliveries with the push system are carried out numerous times when one uses smart phone. For instance, when people write comments on someone's Facebook account, they will be transmitted to the smart phone of account holder in real time. That happens because the Facebook application in the user's smart phone normally remains as the "Background State" and notifies the user once an event is triggered or the service state has been changed for some reason. Currently, the APNs (Apple Push Notification Service), GCM (Google Cloud Messaging) and MPNs (Microsoft Push Notification Service) have been introduced for each smart phone operating system available.

The push system is a suitable alternative to the polling service and overcomes the limitations the polling service has. As in [Fig. 2], when the information is exchanged between the parties on the shore and ship with the polling method, as much as 5~8mA of power is consumed on a standby mode but when the polling is activated for about 10 seconds on the network, up to 180~200mA of power is used.

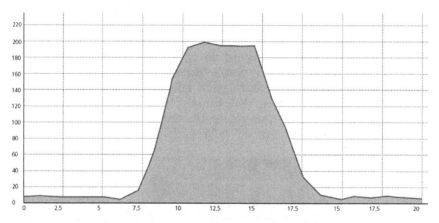

Fig. 2 The battery power consumption during polling

The daily power consumptions (mA/day) with short pollings (power consumption: 0.5mA/polling) are 144mA and 48mA for the intervals of 5 minutes and 15 minutes, respectively. The results may vary depending on the hardware but still the consumption rates are not negligible.

3   Design of the Real-time Mobile Push System for Satellite Communication to Implement Smart Work on the Ship

Despite of the fact that the development and application of the push system are accelerating, there's a problem that the application using the system must be continuously executed and connected to the communication networks all the time.

The push systems like the one in [Fig. 3] are increasingly wide spread on the market as the mobile applications are expanding explosively around the world (e.g., Google, Apple and others).

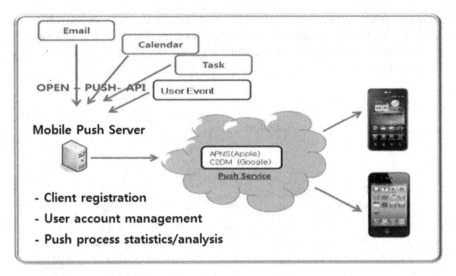

Fig. 3 Current configuration of current push system on land

[Fig. 4] shows current satellite communication system between land and a sailing ship. Here, real-time data delivery is impossible and there's no suitable push system available. This satellite communication method requires artificial and periodic connections to transmit/receive various data.

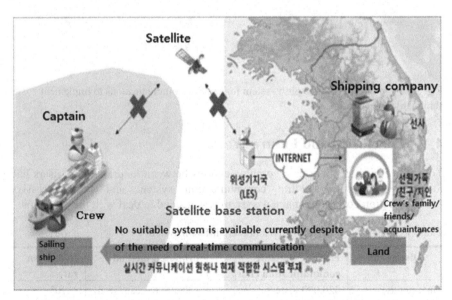

Fig. 4 Current satellite communication system configuration between land and ship

[Fig. 5] describes the communication method which utilizes the real-time push system to implement the Smart Work on a ship. Uninterrupted two-way communications between the parties on both land and shore are possible and therefore the design is expected to contribute to crews' welfare and productivity by enabling the Smart Work on the ship. Also, the design may prove to be very useful foundation technology for implementation of IoT on the ships.

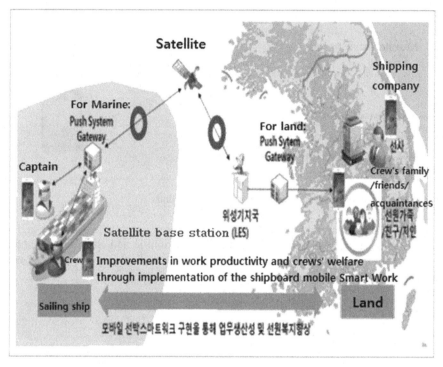

Fig. 5 Design of the real-time push system for satellite communications to implement shipboard Smart Work

4   Conclusion and Future Work

To achieve bi-directional data communications between land and ship, users had to connect to the satellite communication system and bear excessive communication costs so that a new system was needed to deal with the problems.

Table 1 Comparison of costs involved in each different system

Distinction	Existing Satellite communication charges	Proposed method	Reduction rate
Internet use (1Mb)	420,000 Korean (won) / 1MB	130,000 Korean (won) / 1MB	70%
SMS	600 Korean (won) / each	100 Korean (won) / each	84%

For the expected effects, we need to consider them in technological, economical, industrial and finally, social domains.

First, the technological domain, South Korea has the optimal infrastructures for both IT and ship-building businesses so that it is possible to take the initiative in setting the world standard for the through early and extensive technological development. By doing so, we can expect our design to be accepted as a standard maritime communication support technology for the e-Navigation scheme which will be in effect in 2018. Also, the design can support the land-based monitoring system by adopting the IoT application technology (i.e., Machin-to-Machine communications). Other possibility is that the design provide a basic platform to create the contents market on ships targeting the crews or the passengers.

Second, the economical and industrial aspects. The crews can use the satellite communication conveniently with their smart phones and pay for the service individually so that a variety of additional markets can be developed. The reduced communication costs will increase the use of satellite communication and foster the satellite communication industry in return. Next, the increase in productivity by implementing the Smart Work can be expected and remote collaboration between involved parties will be more amicable.

Finally, the social aspect. Convenient real-time communications will contribute to the maritime safety. That is, all the check points and precautions will be monitored throughout the navigation period. Other benefits can be emotional stability of the crew. They can get the information at anytime from their families, friends and acquaintances so that they do not feel isolated.

We will introduce the relevant algorithm, implemented systems and results in our extended thesis after the patent and program registrations.

Acknowledgements This article contains the contents of research which has been conducted with support of the 1st-year Nurimaru R&BD project research fund granted by the Busan IT Industry Promotion Agency of the Republic of Korea in 2015.

References

1. Fan, Z.-P., Feng, Y., Sun, Y.-H., Feng, B., You,T.-H.: A framework on compound knowledge push systemoriented to organizational employees. In: LNCS, pp. 622–630. Springer (2005)
2. Podnar, I., Hauswirth, M., Jazayeri, M.: Mobile push: delivering content to mobile users. In: IEEE 22nd International Conference on Distributed Computing Systems Workshops, pp. 1–6 (2002)
3. Lin, Y.-W., Lin, C.-W.: An Intelligent Push System for Mobile Clients with Wireless Information Appliances. IEEE Transactions on Consumer Electronics **50**(3), 952–961 (2004)
4. Kakali, V.L., Papadimitriou, G.I., Nicopolitidis, P., Pomportsis, A.S.: A New Class of Wireless Push Systems. IEEE Transactions on Vehicular Technology **58**(8), 4529–4539 (2009)

5. Kim, Y., Lee, J., Park, S., Choi, B.: Mobile advertisement system using data push scheduling based on user preference. In: 2009 IEEE Wireless Telecommunications Symposium, pp. 1–5 (2009)
6. Ghrayeb, O., Phojanamongkolkij, N., Tan, B.A.: A hybrid push/pull system in assemble-to-order manufacturing environment. Springer Science+ Business Media, 379–387 (2008)
7. Kakali, V.L., Sarigiannidis, P.G., Papadimitriou, G.I., Pomportsis, A.S.: A Novel Adaptive Framework for Wireless Push Systems Based on Distributed Learning Automata. Springer Science+ Business Media, 591–606 (2009)
8. Nicopolitidis, P., Kakali, V., Papadimitriou, G.: Andreas Pomportsis.: On Performance Improvement of Wireless Push Systems via Smart Antennas. IEEE Transactions on Communications **60**(2), 312–316 (2012)
9. Huh, J.-H., Seo, K.: RUDP design and implementation using OPNET simulation. In: Computer Science and its Applications. LNEE. Springer, Heidelberg, vol. 330, pp. 913–919 (2015)

A Shipboard Secret Ballot System for the ICT-Isolated Ocean Crews

Jun-Ho Huh, Taehoon Koh and Kyungryong Seo

Abstract Due to the cost and technical difficulties, the crews on a ship where they cannot communicate with the land-based families, friends or associates often encounter a number of difficulties in adapting themselves to the outside rapid social change as they are being left out of the ICT-available environment. Thus, in this paper, we've implemented and proposed a Shipboard Secret Ballot System that supports on-board electronic secret voting while ensuring the publicness to let the crews to exercise their rights based on the information acquired from the outside world, and to let them have a sense of belonging.

Keywords Shipboard secret ballot system · ICT · Ballot system · Voting system

1 Introduction

It is evident that the citizens of the Republic of Korea (ROK) are living in a deluge of information supplied by the cutting-edge infrastructures. Despite of the highly developed technologies, many of ocean cruise ships and their crews are often isolated from the benefits offered by these technologies. Depending on the type of ships, crews spend their time on a ship as long as a week to two years and in most of cases, they are completely isolated from the information from the outside world.

In addition to the ocean crews, the servicemen and women in the navy, coastguard, military expeditions and overseas workers can be subjected to the isolation as the land-based communication infrastructures will not be available to them. This situation well justifies the necessity of developing an IT-based technology that can resolve such isolations [1 - 8].

J.-H. Huh · T. Koh · K. Seo(✉)
Department of Computer Engineering, Pukyong National University,
Busan, Republic of Korea
e-mail: krseo@pknu.ac.kr

J.-H. Huh · T. Koh
SUNCOM Co., Busan, Republic of Korea

© Springer Science+Business Media Singapore 2015
D.-S. Park et al. (eds.), *Advances in Computer Science and Ubiquitous Computing,*
Lecture Notes in Electrical Engineering 373,
DOI: 10.1007/978-981-10-0281-6_79

549

Being away from land, ocean cruising ships are limited in their communication capabilities with outside as they can communicate with land only with the satellite communication service provided by 'INMARSAT', which covers most part of the entire earth. Although it is true that the satellite-oriented communication paradigm has changed and improved after the launching of the 4th generation INMARSAT-4 satellite in 2009, it still offers a communication speed similar to the one that has been applied around 20 years ago and a high cost of approx. 360 won (USD 0.30) per second. Following the rapid development of information technology, the ships that use E-mails and other communication methods are gradually increasing but the uses are limited to the official or business purposes so that the crews cannot easily communicate (e.g., e-mails, etc.) with their families or acquire various real-time information concerning ever-changing international situations.

Only about 10 or less E-mail solutions are being used in the ocean cruising ships around the world currently. Even though some shipping companies provide domestic news in summarized form through E-mails, such cases are still rare and in reality, they limit the service to a couple of times in a week due to high communication costs. Also, the majority of the long-term deep-sea fishing vessels are not equipped with the computer(s) that can be used by the crews in their free time – the opposition between the crews who crave outside world information and the shipowners who are keen on saving operation costs.

Therefore, although there could be many problems associated with the operation cost, the burden of communication cost must be reduced to solve the problem. To do so, the problems involved in the communication costs and the development of a solution that has the high-degree of validated compression technique and reliable data transmission capability on the basis of IT technology optimized to the ships must be dealt with to enable cost-efficient data transmission and communication through current low-speed and costly satellites. Together with the solution development, the measures for discovering suitable news providers, setting lower communication cost and providing/installing/servicing the adequate hardware systems must be taken.

Thus, in this paper, we propose a Shipboard Secret Ballot System as one of the solutions that will resolve ICT-isolated ocean crews' problem.

2   Related Research

2.1   Data Flow Diagram of the On-board Secret Ballot System

[Fig. 1] is a data compression algorithm-based solution to reduce communication costs. Once a news provider acquires data from the SUNCOM main server, the news will be displayed on a Ship Digital Information Display taking the steps in the order of Data processing, LES, INMARSAT satellite, Ship radio, and Data processing and output.

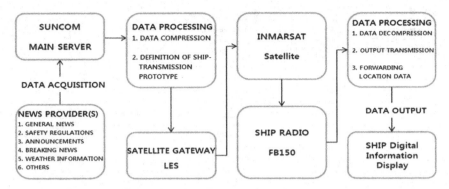

Fig. 1 A solution based on the data compression and transmission algorithm to reduce communication costs

2.2 VMS / e-NOAD

Through the Long-Range Identification and Tracking of Ships (LRIT), the relevant authorities can not only track down the locations of the ships with their own national flag but also monitor other foreign-flagged ships as well at the central control room. This system also enables real-time transmission for the typhoon warnings, expected damages, news flashes and statements released to the public. Also, the English-version domestic news is provided to the foreign crews and other DID information can be supplied to assist the ships to navigate safely.

2.3 e-NOA/D (Electronic Notification of Arrival / Departure) System

After 9.11, the US government issued a mandatory regulation for all those ships entering and leaving the US ports to report their status 48 hours prior to port entry with e-NOAD system to computerize over all movements. Singapore and some of the EU countries are adopting the similar systems but in Republic of Korea, who aims to be a most sophisticated IT-power, such institutional strategy is yet to be realized.

In order to ensure the transparency of on-board secret voting, the ship's captain could be appointed as a presiding officer and the temporary polling box can be set up at the living quarters.

3 Shipboard Secret Ballot Algorithm

[Fig. 2] describes the on-board voting algorithm proposed in this paper and the process can be divided into the 6 major steps.

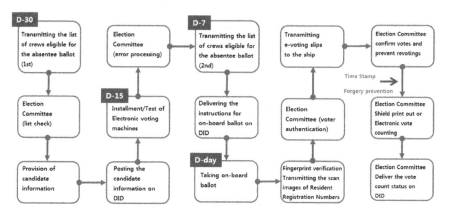

Fig. 2 Proposed shipboard secret ballot algorithm in detail

3.1 1st Step: Confirmation of On-Board Crews Eligible for the Ballot (D-30 days)

For crews of ROK nationality, their eligibility can be confirmed by consulting the Korean Seafarers' Union and their on-duty statuses can be checked through the Immigration Office of the Ministry of Justice.

Since ships keep the crew list themselves legally and secure crews' embarkation/disembarkation schedules routinely, sharing of relevant database will not be the problem. Crews can exercise their voting rights even when there are no polling stations available on the shore or when it is difficult for them to disembark from the ship. After the Election Committee collects the initial crew list across the board, they provide current candidates' promotional materials in accordance with crews' residential areas by checking their resident registration numbers.

Provision/Installation/Utilization of On-board e-voting machines: Since the satellite signals reception mechanisms vary depending on the classes and types of ships and sometimes some of them do not even have an IT infrastructure, the government supplies the voting machines. Each machine is embedded with a compact satellite terminal (FBB150), DID device, fingerprint recognition device and webcam.

The machines should be used for the absentee ballot but it can be used to improve crews' welfare on the ship also by using them to provide daily and breaking news routinely.

3.2 2nd Step: Installation And tests of the Voting Machines (D-15 days)

The installation is simple as it just needs to make USB-connections so that no experts are necessary.

After checking the working condition of the machines by conducting the tests, the Election Committee is to check their readiness for voting procedure. In case of trouble, the committee appoints a certain maintenance company to take care of the problems (D-15 days).

3.3   3rd Step: Updating the Ballot-Eligible Crew List (2nd) (D-7 days)

After 15days of the 1st list, any changes in the list must be reported to the Election Committee to be reflected on their DB for updates, followed by posting of the instructions for the ballot. Under the law, the instructions related to the confidentiality of voting and the voting procedures must be consistently posted to promote participations in the ballot. Meanwhile, during the election period, provision of general news will be suspended to let the crews to concentrate on the ballot.

3.4   4th Step: Commencement of the Ballot (D-day)

After appointing the ship's captain as a presiding officer for the ballot, install polling station(s) at the living quarters in the ship (e.g., mess hall, office or bridge) and make sure that confidentiality will be guaranteed and adequate measures have been taken to prevent frauds.

Election frauds, including the proxy voting, can be prevented by identifying the voters with a fingerprint recognition device, carrying out real-time transmissions of the resident registration number inputs, and the actions of voters to the committee with webcams.

By generating the " * " sign on all of the e-voting slips using a keypad terminal, make sure that no one else but the voter him/herself knows whom he/she has placed a vote.

When the ballot has been completed without an incident, post a phrase on the displays announcing that the ballot is over and prevent any re-voting attempts.

3.5   5th Step: Vote Counting Method

Shield Print Out: As soon as a voter selects a candidate, the Shield printer at the Election Committee outputs the result and it will be shielded and kept in an envelope.

All the voting statuses from the ships will be kept together, counted using the electronic vote counters, and transferred as images.

The votes counted with the electronic vote counters will finally be printed on the plain voting slips in a lump without putting them in envelopes afterwards.

3.6 6th Step: Transmitting the Vote Count Results (ship's DID)

By transmitting the status of the voting tally to the ships' DIDs as a text data, the system will contribute in increasing the level of reliability and participation.

4   Shipboard Secret Ballot System

[Fig. 3] shows the outline of the Shipboard Secret Ballot System. The bottom line is to prevent any hacking attempts by letting the Election Committee to arrange a server themselves directly and using the closed intranet network.

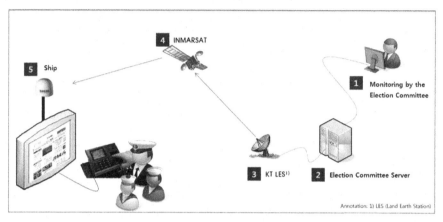

Fig. 3 Outline of Shipboard Secret Ballot System

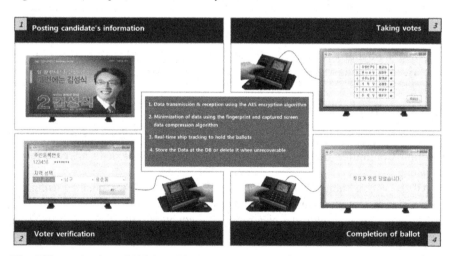

Fig. 4 The mechanism of Shipboard Secret Ballot System

[Fig. 4] is a mechanism of the On-board Secret Ballot System proposed in this paper and the key here is to prevent excessive traffic and costs by minimizing the data exchange volume. As in the picture, the candidate's information is posted first and then a voter is verified. The next step is to place a vote and finally the resulting data will be transmitted/received using an encryption technology adopting the Advanced Encryption Standard (AES). The fingerprints and screen capture data will be minimized using the compression algorithm to perform efficient data transmission and reception. The locations of the participating ships will be tracked on a real-time basis while the crews take the votes, during which each voting result is stored at the database or deleted when it's deemed to be unrecoverable.

This [Fig. 5] shows the process where SMS messages are being sent and received between a user on the land and a crew using crew's personal mobile phone or smart phone. Once a SMS text message is transmitted from the ship, it will then be transmitted to the SMS G/W Server (CDMA) through a satellite(s) and finally delivered to the receiver(s) by the relevant telecommunication company. We will disclose the details in the journal as an extended article after we register the technology for a patent.

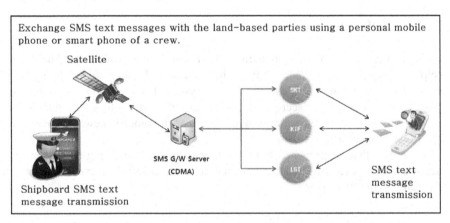

Fig. 5 SMS text message transmission mechanism

[Fig. 6] is a part of GUI (Graphical User Interface) of the Shipboard Secret Ballot System and the method here is to check current location of the ship in real time by acquiring a bit of text message information. The process is that the ship's location is checked with a visual map first and then the internet –based ASP service will be provided, after which a real-time positioning become possible followed by followed by indication of ship's location, speed, direction and time. We are planning to use the materials released by the Google Map and satellite operation companies with their approval. The design and its implementation are described in [Fig. 6]. The details will also be disclosed as an extended journal article after the program has been registered.

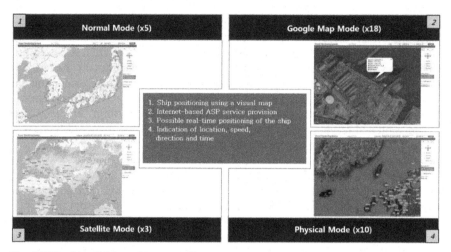

Fig. 6 GUI of the Shipboard Secret Ballot System

5   Conclusion

In this paper, we propose a Shipboard Secret Ballot System which could help can help the ocean crews isolated from the benefits of ICT technology. The system emphasizes the publicness in establishing the system that supports the seafarers to take a transparent and fair secret ballot to exercise their rights by offering them an appropriate means of information acquisition through domestic news, E-mails and other communicational contents.

We expect that the proposed system will be helpful to those long-term deep-sea ocean crews to exercise their fundamental rights in many democratic countries, including Republic of Korea.

References

1. Kohno, T., Stubblefield, A., Rubin, A.D., Wallach, D.S.: Analysis of an electronic voting system. In: 2004 IEEE Symposium on Security and Privacy, pp. 1–14 (2004)
2. Bannet, J., Price, D.W., Rudys, A., Singer, J., Wallach, D.S.: Hack-a-Vote: Security Issues with Electronic Voting Systems. IEEE Security & Privacy, 32–37 (2004)
3. Al-Ameen, A., Talab, S.A.: E-Voting Systems Security Issues. International Journal of Networked Computing and Advanced Information Management **3**(1), 25–34 (2013)
4. Hoque, M.M.: Simplified Electronic Voting Machine System. International Journal of Advanced Science and Technology **62**, 97–102 (2014)
5. Park, H.: Present and Future of the Electronic Election. Korea Information Processing Society Review **19**(5), 21–27 (2012). (In Korean)

6. Lee, K.: Current Status of the Mobile Electronic Election Scheme and its Prospect. Korea Information Processing Society Review **19**(5), 28–33 (2012). (In Korean)
7. Jeon, W., Lee, Y.: Analysis of Research Trend on the Electronic Ballot System. Korea Information Processing Society Review **19**(5), 34–42 (2012). (In Korean)
8. Lee, J.: The Electronic Ballot Software for the Close/Long Range Electronic Election System. Korea Information Processing Society Review **19**(5), 48–52 (2012). (In Korean)
9. Huh, J.-H., Seo, K.: RUDP design and implementation using OPNET simulation. In: Computer Science and Its Applications. LNEE, vol. 330, pp. 913–919. Springer, Heidelberg (2015)

Secure Deletion for Flash-Based Self-Encrypting Drives

Hanwook Lee, Jongho Mun, Jiye Kim, Youngsook Lee and Dongho Won

Abstract Recently, to prevent sensitive data being leaked, self-encrypting drive has been introduced, which encrypts all the data by the drive itself. Although a self-encrypting drive is considered to be secure, all the data stored in the drive might be compromised if the authentication key is exposed. It is, therefore, crucial to provide functions capable of securely deleting individual data in response to a user request or automatically, especially in the flash-based self-encrypting drives. In this paper, we propose a novel design of flash-based self-encrypting drive; in which not only individual data but also data of the previous version left behind during the overwriting can be securely deleted without I/O performance degradation.

Keywords Secure deletion · Flash memory · Self-encrypting drive

1 Introduction

The reason for the recent phenomenon of sensitive data leakage or loss of privacy is not only the loss or theft of computers but also incomplete data erasure on the storage devices to be destroyed. Although degaussing devices for magnetic data storage or repetitive overwriting should be considered for the method of sanitizing the storage devices, the former is too expensive for individuals or small business use, and the latter

H. Lee · J. Mun · J. Kim · D. Won(✉)
Department of Computer Engineering, Sungkyunkwan University, Seoul, Koera
e-mail: {hwlee,jhmoon,jykim,dhwon}@security.re.kr

H. Lee
Korea Financial Telecommunications & Clearings Institute, Seoul, Korea

Y. Lee
Department of Cyber Investigation Police, Howon University, Gunsan-Si, Korea
e-mail: ysooklee@howon.ac.kr

© Springer Science+Business Media Singapore 2015   559
D.-S. Park et al. (eds.), *Advances in Computer Science and Ubiquitous Computing*,
Lecture Notes in Electrical Engineering 373,
DOI: 10.1007/978-981-10-0281-6_80

takes a considerable amount of time. As a result, a disposed storage device containing incompletely deleted data might be obtained by somebody, and the data remaining in the storage device can be retrieved later. As such, considerable attention has recently been paid to storing data in a storage device using data encryption so that the data are secured from the loss or theft, and at the same time, a fast deletion of the entire storage is possible by removing only the encryption key later.

A study on storing data using encryption can be categorized into file system-level encryption, which performs encryption on the file or directory unit, and full disk encryption (FDE), which targets the entire storage for encryption. In particular, a hardware-based FDE is called a self-encryption drive (SED) because it can perform encryption and decryption with a unique key, which is stored inside the storage [1]. As a result, the entire storage can be quickly sanitized because the destruction of only the encryption key is sufficient to delete the entire data stored in the storage device at the end of the service life or for re-purposing.

It is crucial to provide functions capable of securely deleting individual data in response to a user request or automatically. In the case of a hard-disk based SED, it is possible to perform secure deletion through repetitive overwrites on the same physical location. However, it is a different for a flash-based SED. When the flash memory attempts to write data on a block in which data were already written, it should first erase the data on the corresponding block before rewriting the area. This erasure process is slow, causing significant performance degradation. To improve this problem, a flash drive contains a software module called a flash translation layer (FTL). When overwriting is required, the FTL first assigns an empty space rather than the required physical space for writing and maps the physical address and the logical address with each other to maintain consistency for externally accessed addresses, postponing the required erasure. Therefore, it is not guaranteed that the same logical addresses would refer to the same physical location. Moreover, during the overwriting process in a flash memory, data of the previous version remain in memory until it is actually erased. Studies on secure deletion in a flash memory involve a method for reducing the number of erasures via zero-overwriting [2] or file encryption methods [3-6], which requires the support of a file system.

In this paper, a design of flash-based self-encrypting drive is proposed, in which individual data can be securely deleted without performance degradation in a flash memory-based SED.

2   Background

2.1   Self-Encrypting Drive

Each SED has a unique encryption key, which is securely generated from its internal entropy source. This key, called a media encryption key (MEK), is used to perform full disk encryption. MEK is encrypted with an authentication key (AK), which is (derived from) the user's password, and stored within the SED. The reason for using the additional key, the MEK, instead of the AK, to encrypt data is to avoid rewriting all the encrypted data. When a new AK has been set, it is sufficient to encrypt the MEK with the new one. [7]

Each time an SED is powered up, it requires a pre-boot authentication process. The host tries to read the master boot record (MBR) to recognize the file system, but loads the pre-boot OS, because the SED redirects the host to the pre-boot area instead of the MBR. After receiving the AK from the user input, the SED computes an AK hash value. If this value matches the internally saved value, the MEK is decrypted with the AK. Finally, the MEK decrypts the original MBR and transfers it to the host.

Since all the data in the SED have been encrypted with the MEK, once the MEK value has been deleted or changed, no data can be read. This characteristic can be used to sanitize the SED by the quick deletion of all data.

2.2   Phase-Change Memory

The research field of non-charge-based non-volatile memory technologies is one of the most active in terms of improving slow data write and low endurance of non-volatile memory, such as flash memory. Among these technologies, phase-change memory (PCM) is considered a leading candidate for the next byte-addressable non-volatile memory. PCM, like DRAM, can be rewritten without the previous state having to be rewritten. Although PCM has a limited number of writing operations due to the mechanism resulting from the thermal expansion of phase change materials, which degrades its performance, its endurance is several times better than that of flash memory [8]. Unlike flash memory, where access is according to the page unit, it can be randomly accessed. Table 2 shows a comparison of PCM with DRAM and NAND flash memory. It can clearly be seen that PCM is at least 100 times faster than the NAND flash memory in terms of the read and write latencies.

Table 1 Typical distinction among DRAM, PCM, and NAND [9]

	DRAM	PCM	NAND
Non-volatile	No	Yes	Yes
Erase required	Bit	Bit	Block
Read latency	50 ns	50~100 ns	10-25 μs
Write latency	~20-50 ns	~ 1 μs	~ 100 us
Endurance	-	10^8	$10^{4\text{-}5}$

3   Secure Deletion for Self-Encrypting Flash Memory

The SED uses its internally embedded unique key to encrypt all the data, and therefore it is easy to delete the entire drive securely by destroying the encryption key.

However, this rarely happens because the entire drive is deleted only when the life cycle of the drive is finished or the drive is re-purposed. On the other hand, the users can frequently require a secure deletion function to erase individual files. The hard-disk based SED can provide a solution by overwriting the data multiple times. However, in the case of flash-based SED, it would be unreasonable to expect the same results as those of the hard-disk based SED. In this section, we propose a new design of flash-based SED which allocates a different encryption key per page of the flash memory. When a new page of the flash memory is allocated for writing new data, the proposed SED creates a page's encryption key, the page encryption key (PEK). The new data are encrypted with the PEK and saved in the flash memory. The PEK is saved in a separate PEK storage space away from the data storage areas, after being encrypted with a unique key, the key encryption key (KEK). Finally a request for deleting the page is sent out, it is necessary to erase only the PEK.

A. Pre-boot Authentication

Pre-boot authentication is processed only once when the flash-based SED is powered up. Its function is to receive the AK, verify it, and load the KEK for internal use. It also delivers the actual master boot record (MBR) to the host. When our SED is powered up, it first sends the pre-boot information to the system and waits for the AK input.

When the AK is input, the firmware/hardware in our SED computes AK hash value with cryptographic hash function, such as SHA-1, to be compared with the saved value in the hidden storage. If the values are matched, the firmware/hardware decrypts the KEK, which was encrypted with the clear AK. The MBR is encrypted and saved at page 0; as a result, our SED reads the encrypted PEK for physical page address 0, denoted by PEK_0, from the PEK storage, and decrypts it with the KEK. Finally, the MBR is decrypted with PEK_0 and sent to the host. The KEK remains at the firmware/hardware until our SED is powered down, and is used for en/decrypting PEKs.

B. Data Write

An encryption key is created or erased whenever a page write or deletion requested, respectively. Now, let us describe the data write request process in detail. First, when the logical address is received from the file system to write data, the FTL in the SED performs a scan to find whether physical addresses are assigned for the logical address. If no physical addresses are assigned, the FTL assigns a new block and starts the mapping table update. Next, a PEK is generated from a key generation function, called the cryptographically secure pseudo-random number generator (CSPRNG), encrypted with the KEK, and saved in the PEK storage. Finally, the clear text is encrypted with the PEK and saved.

C. Data Read

Reading data involves reading the encrypted data, decrypting and sending them to the file system. This is the reverse of writing data. To read the data, the flash-based SED receives a logical address from the file system, and the FTL converts the logical address to i. Then, the PEK_i in the PEK storage is decrypted with KEK. The page i's encrypted data in the flash memory is finally decrypted with PEK_i to be sent to the file system.

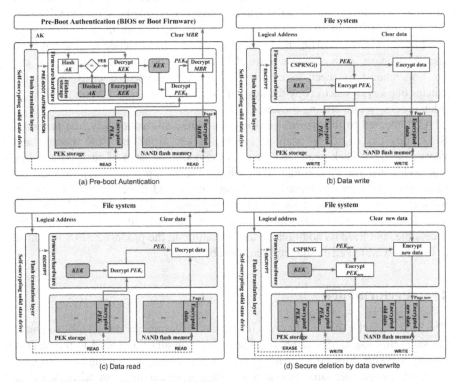

Fig. 1 Block diagram of the proposed SED

D. Secure Deletion

Secure deletion of data should be performed not only at the user's request but also automatically when data have been overwritten. In the first case, when the user requests that specified data be erased, the file system takes logical addresses to the flash-based SED to be deleted. The flash-based SED deletes the PEK associated with a physical address, which is mapped on the logical address. Although the encrypted data can exist in the flash memory before erasing the corresponding block, but since there is no encryption key, they cannot be recovered. In other cases, when overwrite has occurred due to, for example, an update, the flash memory cannot guarantee that the overwriting will occur in the same physical space. This could leave data of the previous version undeleted and retained in the memory. Even in this instance, the problem can be solved by deleting the PEK. In the method described above, even if data remain in the memory, the encryption key is missing, thus making recovery of the data impossible.

E. PEK Storage

The proposed SED, as compared to the typical SED, needs additional extra space for the PEK data saves. If the space in which the PEK data are saved is the previous NAND flash memory, critical inefficiency occurs. The PEK should be saved in the minimal unit of write, but the NAND flash memory's minimal size is 2 KB. However, in the most symmetric encryption algorithms, the key sizes are 16 to 32 bytes, which is inefficient. In addition, when a data deletion request has been issued, deletion of the keys is more than satisfies the request, but when a key deletion request has been issued, the key must be deleted. Even if the system deletes one key, it has to copy the keys of the remaining page in the same block to other blocks, before erasing the key.

As compared to the typical SED using a single encryption key, the proposed SED adds the additional step of reading a key from the PEK storage and decrypting it for each page read, or adds the additional step of encrypting a key and saving it to the PEK storage for each page write. The time taken for the keys to be encrypted and decrypted is relatively short, and therefore, it is crucial to minimize the processing time when reading and writing in the PEK storage. Therefore, the PEK storage must provide low latency for small I/O, be byte-accessible, and need no additional erase operation. In the proposed SED, we chose to use the PCM to satisfy the requirements above.

4 Performance

The SED reads the file and decrypts it, and then transmits the data. This can create the same time lag as descrambling. The encryption algorithm used in the SED is usually the AES. Since it has a high performance design, encrypting or decrypting a key takes negligible time compared to reading a page in the flash memory. Fig. 2 is a diagram in which the times taken by the proposed SED and the typical SED for the reading and writing process are compared. Although the proposed SED includes the additional step of reading from the PEK storage and decrypting the data as compared to the typical SED, this step can be performed in parallel that it takes flash memory to read one page. Let us denote the time it takes to read one page in the NAND flash memory by t_{RP}, the time it takes to read from the PEK storage by t_{RK}, and the time it takes to decrypt the PEK by t_{DK}. The proposed SED does not have any performance issues as compared to typical SEDs if it satisfies $t_{RP} > t_{RK} + t_{DK}$.

For writing data, as compared to typical SEDs, the proposed SED includes the extra step of random-number generation using CSPRNG, encrypting the PEK to save in the PEK storage. Thus, an overhead can be created. We avoid this overhead by generating the new PEK in the vacant space after the PEK has been encrypted instead of performing CSPRING in time. In addition, Encrypting a PEK and saving it to PEK storage can be performed in parallel as writing one page in NAND flash memory.

We denote the time it takes for the NAND flash memory to write a page by t_{WP}, the PEK writing time to the PEK storage by t_{WK}, the time it takes to encrypt the PEK by t_{EK}, and lastly the time it takes to create the PEK by t_{GK}. The performance of the proposed SED is not lacking as compared to that of typical SEDs if it satisfies $t_{WP} > t_{DK} + \text{MAX}(t_{GK}, t_{WK})$.

Now, let us calculate how much PEK storage is needed as compared to the NAND flash memory. We denote the NAND flash memory's entire storage capacity by T, the size of the page by P, and the size of encrypted PEK by K. The required size of PEK storage of the proposed SED, S, is then $S = K \cdot T/P$. If the proposed SED using an 128-bit AES key and applying a large block flash memory, the PEK storage, PCM, needs only 0.8% extra space compared with the NAND flash memory space.

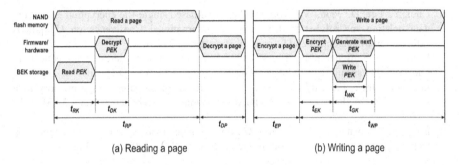

(a) Reading a page   (b) Writing a page

Fig. 2 Timing diagram for (a) reading a page and (b) writing a page from our SED

5   Conclusion

In this paper, we proposed a novel design for the flash-based SED which supports secure deletion for individual data according to the user's request. Besides, data of the previous version due to the file update in flash memory are deleted securely and automatically. Previous research studies for secure deletion have mainly addressed on non-encrypting storage devices. The proposed SED, as compared to the typical SED, includes extra steps to encrypt or decrypt a key and to read or write the encrypted key to PEK storage each time it reads and writes to the flash memory. Even though we included these extra steps in the proposed SED, the performance was comparable with that of typical SEDs.

Acknowledgements. This research was supported by the Basic Science Research Program through the National Research Foundation of Korea (NRF) funded by the Ministry of Science, ICT, and Future Planning (2014R1A1A2002775).

References

1. Müller, T., Latzo, T., Freiling, F.C.: Self-encrypting disks pose self-decrypting risks. In: Annual Computer Security Applications Conference (ACSAC), Orlando, Florida, USA (2011)
2. Sun, K., Choi, J., Lee, D., Noh, S.H.: Models and design of an adaptive hybrid scheme for secure deletion of data in consumer electronics. IEEE Trans. Consumer Electronics **54**, 100–104 (2008)
3. Lee, J., Heo, J., Cho, Y., Hong, J., Shin, S.: Secure deletion for NAND flash file system. In: Proc. ACM SAC, pp. 1710–1714 (2008)
4. Lee, B., Son, K., Won, D., Kim, S.: Secure data deletion for USB flash memory. J. Inf. Sci. Eng. **27**(3), 933–952 (2011)
5. Reardon, J., Capkun, S., Basin, D.: Data node encrypted file system: efficient secure deletion for flash memory. In: 21st USENIX Conference on Security Symposium (2012)
6. Choi, Y., Lee, D., Jeon, W., Won, D.: Password-based single-file encryption and secure data deletion for solid-state drive. In: 8th International Conference on Ubiquitous Information Management and Communication, pp. 1–7 (2014)
7. Willett, M.: Self-encrypting drives (2009). http://www.snia.org/sites/default/education/tutorials/2009/spring/security/MichaelWillett_Self_Encrypting_Drives.pdf
8. Lee, B.C., Ipek, E., Mutlu, O., Burger, D.: Architecting phase change memory as a scalable DRAM alternative. ACM SIGARCH Computer Architecture News **37**(3), 2–13 (2009)
9. Eilert, S., Leinwander, M., Crisenza, G.: Phase change memory (PCM): a new memory technology to enable new memory usage models. In: IMW 2009, pp. 1–2 (2009)

A Distributed Mobility Support in SDN-Based LTE/EPC Architecture

Yong-hwan Kim, Hyun-Kyo Lim, Kyoung-Han Kim, Youn-Hee Han and JooSang Youn

Abstract As smart phone has rapidly proliferated over the past few years, LTE operators endeavor to cope with large mobile data traffic volumes. To solve such problems, we propose a new SDN-based distributed mobility management supporting distributed P-GWs and centralized control plane in LTE/EPC networks. For enhancing network performance more, we also propose a route optimization strategy for internal traffic exchanged between LTE users. The proposed solutions' performance is compared with the conventional LTE/EPC network's scheme in terms of the P-GW data processing volume and the number of valid data sessions. The comparison results show that the proposed solution can be an efficient way to enhance the scalability of LTE/EPC core networks.

Keywords Distributed mobility support · LTE/EPC · SDN

1 Introduction

Due to the increasing popularity of smart-phones and other mobile devices, mobile data traffic is expected to highly increase. Accordingly, LTE operators are in urgent need of means to cope with such increase in data volumes. The evolution of mobile network architectures toward flat architectures is being considered as a key solution to cope with the explosion of mobile data traffic [1]. For this, IETF

Y.-h. Kim · H.-K. Lim · K.-H. Kim · Y.-H. Han
Advanced Technology Research Center,
Korea University of Technology and Education, Cheonan, Republic of Korea
e-mail: {cherish,glenn89,goslim,yhhan}@koreatech.ac.kr

J. Youn(✉)
Department of Multimedia Engineering, Dong-Eui University, Busan, Republic of Korea
e-mail: jsyoun@deu.ac.kr

© Springer Science+Business Media Singapore 2015　　　　　　　　　　　567
D.-S. Park et al. (eds.), *Advances in Computer Science and Ubiquitous Computing*,
Lecture Notes in Electrical Engineering 373,
DOI: 10.1007/978-981-10-0281-6_81

Distributed Mobility Management (DMM) working group is trying to make standardizations of IP mobility anchors at an access network level [2, 3].

As 3GPP evolves to the Evolved Packet Core (EPC) network, the network hierarchy is also being flattened. In the hierarchy, the number of LTE/EPC network elements decreases, where the main elements in the data plane are Packet Data Network Gateway (P-GW), Serving Gateway (S-GW), and Evolved Node B (eNodeB). The number of hierarchy levels is reduced in the EPC network compared with the GPRS/UMTS network. In later deployment, P-GW and S-GW are expected to be co-located in the same physical element, thus the number of different network elements will be further reduced. In this hierarchy, P-GW provides access to Packet Data Network (PDN) by assigning an IP address to User Equipments (UEs) and serves as the mobility anchor point. UEs do not change their mobility anchor points if they remain attached to the same operator's access network. So, all data traffic for an UE is routed over P-GW and thus P-GW is needed to handle it.

As another trends in mobile networks, Software-Defined Networking (SDN) and Network Functions Virtualization (NFV) are being actively adopted by mobile operators. SDN and NFV can enable resource and service orchestration with dynamic provisioning by deploying and utilizing virtually-centralized control plane. In parallel with DMM technology, SDN and NFV will lead to the deployment of more flatter systems in which mobility anchor points can be placed closer to the UEs and the local breakout/traffic offload can be facilitated [4]. Furthermore, a virtually centralized control plane leads to more flexibility by having the S-/P-GWs data-plane deployed in a distributed manner [5]. It is however expected that the change of UE's mobility anchors upon handover will happen far more often due to high number of mobility anchors deployed at access networks. So, DMM should efficiently keep ongoing sessions active in case of handovers that require mobility anchor change.

In this paper, we propose a new distributed LTE/EPC network architecture based on SDN and NFV supporting distributed P-GWs. It is designed based on the following requirements: 1) Distributing P-GWs closer to the UEs, 2) Virtually centralizing control plane, 3) Separation of the control and data planes, and 4) Efficient distributed mobility management. We also propose the route optimization for internal traffic exchanged by LTE/EPC UEs to enhance network performance. We have implemented the proposed solution on NS3-LENA simulation environment [7]. The proposed solution's performance is compared with the conventional LTE/EPC network system's scheme in terms of the P-GW data processing volume and the number of valid data sessions. The comparison results show that the proposed LTE/EPC architecture and DMM solution can be a more efficient way to support the scalability of LTE/EPC core network.

The rest of this paper is organized as follows. Section 2 proposes a SDN-based DMM approach in LTE/EPC. Section 3 investigates the performance of our proposal by simulations, and Section 4 finally concludes this paper.

2   SDN-Based Distributed Mobility Support

2.1   SDN-Based DMM Architecture

Fig. 1 shows the reference model for our SDN-based DMM (SDMM) architecture in LTE/EPC. As shown in the figure, multiple distributed components are deployed in different areas of the network, instead of employing a single centralized infrastructure based on P-GW, A new network element named PDN edge gateway (P-EGW) is similar to the existing P-GW in term of the functionalities and the roles. However, P-EGWs will be deployed near to the radio area network (RAN), so that their location will be somewhere between eNB and S-GW. P-EGWs have also the role of SDN switches to communicate with the SDN controller deployed in virtually and centralized control plane. These distributed data plane allows scalability and flexibility to LTE/EPC networks.

The virtually control plane consists of much functions supported by Mobility Management Entity (MME), Home Subscriber Server (HSS), and Policy and Charging Rules Function (PCRF) which are control parts in conventional LTE/EPC networks. Such functions are deployed in a form of cloud system using NFV technology and can communicate with the entities in data plane via SDN technology. The centralized control plane can provide flexible and dynamic framework to LTE/EPC networks. In order to separate the LTE/EPC network's data and control planes, SDN technology is used to not only EPC core networks but also the edge network. The separation of the control and data planes allows sophisticated traffic management to be driven by software, rather than purely by hardware routers.

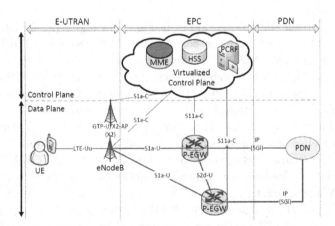

Fig. 1 Reference model for SDN-based DMM architecture in LTE/EPC

2.2   SDN-Based DMM (SDMM)

In the proposed SDMM, there are two types of P-EGWs assigned to a PDN connection of an UE. The one is anchor P-EGW which plays a role of mobility anchor for the PDN connection, and the other is access P-EGW which maintains a normal Evolved Packet System (EPS) bearer for the PDN connection on the current location of the UE.

When an UE initially attaches to an anchor P-EGW, it obtains an IP address called Home Address (HoA). When the UE moves and attaches to a different P-EGW (Access P-EGW), the access P-EGW finds out the address assigned to the UE during the authentication phase via SDN controller in the control plane. And then, a routing update between the anchor P-EGW and the access P-EGW is performed for session continuity by SDN controller in the control plane. In this way, the reachability of the UE is ensured while moving within the LTE/EPC network.

Meanwhile, when the UE in a new P-EGW's area tries to create a new PDN connection, the new P-EGW is in charge of creating the new PDN connection as anchor P-EGW. That is, an UE can have multiple PDN connections through different anchor P-EGWs. Therefore, these IP addresses related with multiple P-EGWs will be managed simultaneously in the UE.

The location information of UE becomes <UE ID, UE IP Address, Anchor P-EGW ID, Access P-EGW ID> per UE's PDN connection. UE IP Address is unique for PDN connection managed by anchor P-EGW. Anchor P-EGW ID and Access P-EGW ID are also needed to find the location of UE in distributed data plane. This location information of UE is managed by the control plane in a centralized manner and obtained by an entity in data plane through the mediation of SDN controller.

Examples of our solution are shown in Fig. 2. In Fig. 2 (a), an UE initially attaches to P-EGW1 and starts to communicate with CN1 located in PDN. As explained previously, P-EGW1 becomes the anchor P-EGW for the PDN connection for this communication. When the UE moves to P-EGW2, which becomes an access P-EGW for the PDN connection, P-EGW2 acquires the UE's information via SDN controller in the control plane during the authentication phase. And then P-EGW2 requests for routing update to the control plane. When SDN controller in the control plane receives this request, it performs a routing update to SDN switches between anchor P-EGW1 and access P-EGW2. When the routing update is finished, the session data anchored P-EGW1 is redirected to P-EGW2. When the UE moves to P-EGW3 as shown in Fig. 2 (b), the session data anchored P-EGW1 is redirected to P-EGW3 though the same procedures. Meanwhile, when the UE creates a new session to CN2 located in PDN, P-EGW3 is the anchor P-EGW for the new session using a new IP address.

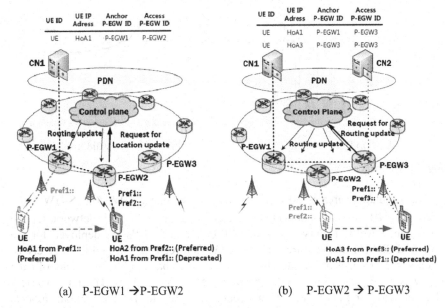

(a)   P-EGW1 →P-EGW2   (b)   P-EGW2 → P-EGW3

Fig. 2 Procedure for UE movement in proposed SDMM

The data transferred after the movement of an UE in the proposed SDMM is routed through sub-optimal path, since the data should be redirected by an anchor P-EGW. However, the proposed SDMM can reduce the amount of signal messages related with routing update and the handover latency compared with the existing routing-based DMM approaches [6] Furthermore, it can remove the cost related with tunneling, while the tunneling cost is imposed on the existing host/network based DMM approaches [3].

An IP address assigned for a PDN connection managed by an LTE/EPC network operator can be made up based on pre-defined prefix information. Therefore, if the IP address has such a pre-defined prefix, the data communication can be judged to be the one of inner traffic exchanged between UEs in the operator's network. In this case, the route for this data communication can be optimized by routing updates to SDN switches between the P-EGWs associated to such UEs.

3   Performance Evaluation

In this section, the proposed SDMM's performance is compared to the mobility management's one in the conventional LTE/EPC networks in terms of the P-GW data processing volume per unit time and the number of valid data sessions. For the analysis, we used the open source simulator NS-3.16 with the LENA LTE module (M4 release) [7].

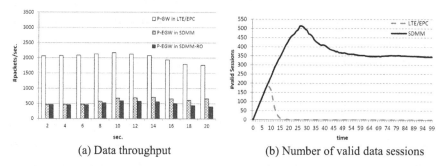

(a) Data throughput   (b) Number of valid data sessions

Fig. 3 Performance evaluation

The simulation area is 30 *km* × *30 km*, where one P-GW, four S-GWs, and thirty-six eNBs are inter-connected in the conventional LTE/SAE network. On the other hand, twelve P-EGWs instead of P-GW and S-GWs are deployed for the proposed SDMM architecture. Each P-EGW is directly connected to three eNBs. Each wired link is a full-duplex link with 10 *Gbps* bandwidth. The delay of each link was configured to be a value between 25 *ms* and 125 *ms*. The LTE wireless channel and modulation/encoding models are set to be the default models provided by NS-3.16-LENA. 100 UEs are deployed randomly in the area and a voice over IP traffic is exchanged between UEs. A half of UEs sends 80 *Bytes* UDP data to the remaining UEs every 10*ms*, so that 64 *Kbps* codec is emulated in the simulation.

Fig. 3 (a) shows the P-GW and the P-EGW's data processing volume per unit time over the given traffic scenario. The data processing volume observed at P-EGWs is apparently quite less than that at P-GW, due to the distributed architecture of P-EGWs. However, it is noted that the total amount of data processing volume observed at all P-EGWs is more than the one at P-GW. It is because more data traffic is redirected between P-EGWs. However, the data processing volume becomes low when the data traffic route is optimized.

Fig. 3 (b) plots the number of valid sessions observed in the conventional and proposed LTE/EPC networks. A session is valid in terms of service quality if its packet drop rate is lower than 10%. In the experiment, we deployed stationary 100 UEs randomly in the area and let each UE create a new session composed of 1*Mbps* UDP data traffic delivered to another UE every 40*ms*. As shown in the figure, the proposed SDMM scheme can create valid sessions twice more than the one used in the conventional LTE/EPC networks. Data traffic are concentrated to the P-GW in the conventional LTE/EPC network, while there is not such a single point in SDMM and all data traffic are directly routed between P-EGWs. That is, the SDMM scheme does not suffer from the bottleneck problem at a specific node in the network, and thus it is more scalable than the one of the conventional LTE/EPC networks.

4 Conclusions

In this paper, a new SDN-based distributed LTE/EPC network architecture and DMM scheme called SDMM are proposed. P-EGWs' distribution close to LTE radio area networks can facilitate the traffic offload and perform local breakout as close as possible to UE, while reducing the load in LTE/EPC core network. Through our performance evaluation, we showed that the proposed ones become more efficient way to enhance the scalability of LTE/EPC core networks. In future, we will do more intensive simulation study in terms of signaling costs and handover latency.

Acknowledgments This research was supported by the MSIP (Ministry of Science, ICT and Future Planning, 2014H1C1A1066391), Korea, and also supported by Basic Science Research Program through the National Research Foundation of Korea (NRF) funded by the Ministry of Education (No. NRF-2013R1A1A2010050).

References

1. Chan, H.A., Yokota, H., Xie, J., Seite, P., Liu, D.: Distributed and Dynamic Mobility Management in Mobile Internet: Current Approaches and Issues. Journal of Communications **6**(1), 4–15 (2011)
2. Liu, D., et al.: Distributed Mobility Management: Current Practices and Gap Analysis. IETF RFC 7429, January 2015
3. Bernardos, C.J.: PMIPv6-based Distributed Anchoring. IETF Internet Draft, draft-bernardos-dmm-distributed-anchoring-05, March 2015
4. Wang, Y., Bi, J.: A solution for IP mobility support in software defined networks. In: Proc. of 23rd ICCCN (2014)
5. Valtulina, L., Karimzadeh, M., Karagiannis, G., Heijenk, G., Pras, A.: Performance Evaluation of a SDN/OpenFlow-Based Distributed Mobility Management (DMM) Approach in Virtualized LTE Systems. In: Proc of Globecom 2014 Workshop - Cloud Computing Systems, Networks, and Applications (2014)
6. Chan, H., Pentikousis, K.: Enhanced Mobility Anchoring. IETF Internet draft, draft-chan-dmm-enhanced-mobility-anchoring-00 (work in progress), July 2014
7. LENA Design Documentation. http://lena.cttc.es/manual/lte-design.htm

Implementation of Wireless Communication-Based Video Control System for Media Façade

Sooyeon Lim and Jaeha Lyu

Abstract The purpose of this research is to implement media façade system using wireless communication without the occurrence of the separate wired facility or data communication cost. It suggested, designed and implemented the principle of video playback device that transmits the control signal using long distance RF communication module. RF control system suggested in this research guarantees the free mobility comparing to the existing wired equipment and can be utilized for video control of media façade within the distance range of maximum 5km. And, it is capable of coping with several environmental conditions actively when establishing system and resolving various wiring treatment.

Keywords Media façade · RF communication · Video control

1 Introduction

Media façade risen as new method of media art recently means new concept of media art form lead to the ground of media communication using the building external wall as large screen as one of media art fields that adopt mass media that is the important means of modern communication to the art. The ripple effect of media façade that is the technology that establishes communication interface using various video art and design to external wall of the buildings is getting increased together with development of information and communications [1][2].

Media façade requires enough test as it uses the external wall of building as display space and the test and edit processes are completed by producing the

S. Lim(✉)
Educational Development Institute, Dongyang University, Yeongju, Korea
e-mail: sylim@dyu.ac.kr

J. Lyu
Department of Fine Art, Kyungpook National University, Daegu, Korea

© Springer Science+Business Media Singapore 2015
D.-S. Park et al. (eds.), *Advances in Computer Science and Ubiquitous Computing*,
Lecture Notes in Electrical Engineering 373,
DOI: 10.1007/978-981-10-0281-6_82

miniature of the building (small sample) sometimes as it takes long time and huge cost as much as such time to test that onto the actual building for error range and edit of video.

But, there is a difficulty to narrow the error range of video as the error rate between the miniature and size of actual building increases and it is necessary to combine a number of computers as the network for operating multiple displays that can cover the large-scaled building. Especially in case of media façade, the cooperation with the city is required such as local transportation information now that trial performance is mainly conducted in metropolitan city considering the number of audience. In case of establishing the wired network in complicated transportation situation or building-concentrated area, huge cost should be paid regarding time and equipment. So, it is required to minimize the work at site through correct preliminary work and simplification of facility.

This research aims to realize media façade system that does not require the fixed communication cost and is capable of video control within 5km by applying long-distance RF communication module and considerate its usability. It is willing to control media façade video in real time by composing RF wireless network in message forwarding method without obstacle or distance limit.

2   Backgrounds

2.1   Technology of Realizing Media Façade

The technical method that can implement media façade be divided into two. They are Projection Mapping and LED [3].

LED needs the control unit definitely to array LED in mixed way and express dynamics and colors of LED, and it uses two communication methods (DMX Control, ETHERNET Control) mainly. LED has a number merits such as high luminance, long lifetime of average 50,000 hours, use of direct current, operation at low voltage, excellent water proof function and expression of various colors with one chip. But, it has the demerits that the price is more expensive than other light emitting devices, luminous efficiency is lower than fluorescent lamp and lifetime gets short in high temperature as it is weak to the change of current and temperature [4]. The examples of large-scaled media façade that uses LED are Ars Electronica Center (www.aec.at) at Linz, Austria and Sky Canopy(www.hok.com) at Suzhou Industrial Park(SIP) in China [5][6].

Projection mapping method is the method that preferred recently as the one that performs high quality of video by projecting the projection video onto whole or part of the building using several beam projectors without installing separate screen. Especially in case of old buildings, it has the merit that it can induce the higher concentration than any other outdoor advertising media as it is capable of stylish change without burden of huge cost. The examples of media façade in projection mapping method are "Wings of Desire" that seeper (www.seeper.com)

of UK showed as the finale of IDFB (International Dance Festival Birmingham), "Sydney Opera" that urban screen (www.urbanscreen. com) of Germany performed at Sydney Opera House in Australia, and "bucur555" that macula(www.themacula.com) of France showed at Bucharest festival in Romania.

2.2 Multi Image Display

As projection mapping method that uses external parts of the buildings has the set expressing range for each projector, it projects the video by deciding the number of projectors according to the size of building and connecting them to the server.
That is, it requires the large size of multimedia video display that is capable of controlling multiple projectors at a time. There are Watchout System and Showlogix as the multimedia video display system that is currently developed.

Watchout System [7] is the system that high quality display system at the excessive level of HD that shows various dynamic effects with easy operation as the multiple video displayer in software method developed in Sweden. In Watchout System, one production server is connected to display server that is capable of endless extension with ethernet network.

Showlogix [8] is the multi-view realization system developed by Israel. It can realize multi-view video, edge blending and warping and express synchronization between players, scheduling function and special effect in free way. It is capable of controlling hundreds of various devices connected to network.

The something in common of previously-mentioned systems is that the devices should be connected with ethernet. There are two ways of establishing ethernet, wired and wireless. Multi-display realization system that is currently available is based on wired protocol in most cases. Wired-network-based system has the demerit that the cost of establishing is very high and there are many cases that the installation is not easy.

In case of wireless LAN to overcome spatial and cost restriction and limit, its installation cost can be saved and the works of extension and reconfiguration are easy because cable is not used. The best cost-saving effect can be achieved in the environment whose system is changed frequently especially. But, if using ZigBee communication [9] or wireless LAN that is near field communication, there is the demerits that it does not exceed maximum 200m even in outdoor case due to the short transceiving range of radio wave and it seldom can pass through the obstacle.

In order to supplement that, there is a way of installing the router that is capable of relay in the middle, however, multi-hop method that communicates through several paths has the demerit that the performance of TCP falls down seriously.

In case of large-scaled media facade, a few dozens of beam projectors should be installed through a few kilometers depending on the size of building to be projected. In this case, its configuration and installation are very complicated and difficult to communicate when using near field communication regardless of wired or wireless. So, it is required to research large-scaled media facade control system that uses long-distance wireless communication.

3   Wireless Video Control System that Uses RF Communication

3.1   RF Communication

Wireless communication means the wireless communication that uses radio wave in general and it is called RF(Radio Frequency) communication. The basic principle of wireless communication that uses radio wave is the same as radio. Wireless frequency is managed as national resource to control by each country and the frequency band that is called as ISM(Industrial, Scientific, Medical) band is the exceptional. It is possible to establish the stable wireless data communication network without communication cost by using RF module that uses ISM frequency band for long-distance wireless data communication of 5km.

The purpose of this research is to use wireless data communication that has maximum 5km of range for video control of media facade. Recently, the commercialization and production launch of low-price and long-distance RF communication module that does not requires separate communication cost and its installation is easy are carried out frequently. We implemented 1:N communication-based video control system by applying ISB band long-distance RF communication model that is being sold. General RF communication module in the market supports RS-232C specifications and provides the interface of H/W and S/W. The completed module in PCB shape without case is provided at the price of less than 100,000 Korean Won regarding the price of RF communication module in respect of cost and it is advantageous because there is not communication cost comparing to Wibro and LTE that is broadband mobile communication service as it uses ISM band.

IRF4020P of RF Techwin [10] was selected as RF model used in this research considering mass production of product, securement of product, miniaturization of product and convenience of technical support. The selected model supports the repeating function by S/W setting as the long-distance communication module that is operated in the specifications such as low power(50mA), low voltage(5V), 447.2750~447.9750 MHz that is ISM bandwidth of Korea and data rate 2400bps.

3.2   Wireless Video Control System

Video control system that uses RF communication is server and client system that consists of three parts as following. Main program design is implemented with C++ language.

- Central processing part: command the order to RF transceiving device (at client)
- Communication part: RF transceiver
- Video controlling part: SW that controls video by the signal of server (transmitting)

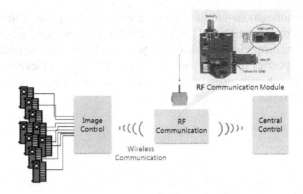

Fig. 1 System configuration with RF communication

Central processing part is connected to video controlling part in wireless using RF transceiving device. Clients connected to server that is central processing part receive the information from the server and check the information of transmitting part by analyzing the received information along the protocol. The check of transmitting ID carried out at RF receiving part is for avoiding the interference to other networks. If the check of transmitting ID is completed, it controls the playing data of video along the received control signal. Fig. 4 shows the operation flow of client according to the control signal from the server.

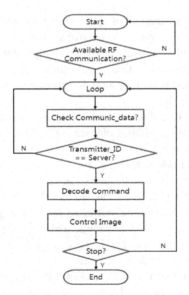

Fig. 2 Client functioning by server control signal

Communication port should be open the application program of this system can be accessed using port number. The individual ID is set in each RF module and, if the protocol of transmitting and receiving parts is checked as the correct one, the receiving part operates the designated operation. The communication flow of this system has the characteristic that is conducted in simplex considering the operating characteristics of media facade.

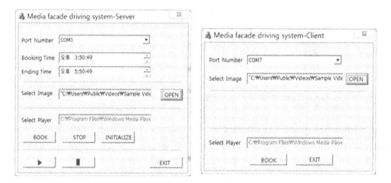

Fig. 3 Wireless Media Facade Driving Program

4   Conclusion

This research suggested the standard of media facade monitoring service in future by offering wireless communication video control system that is capable of managing regardless of location of the users applying long-distance RF communication module. If the user registers ID value of RF module connected to beam projector no matter where he or she is as it mounts RF module in several beam projectors, real-time control is available.

The system suggested in this research composed of RF wireless network in message forwarding method without obstacle or distance limit. So, free mobility was guaranteed without the fixed communication cost and it could be used for video control of media facade within the range of maximum 5km.

As the results of examining the usability the suggested video control system, it was recognized that the signal got weak due to the obstacle and interference at building-concentrated area. It is the matter of how the signal can be transmitted well without the loss within the given range and one of core research themes of wireless communication technology. It is expected to be resolved by using wireless signal amplifier or wireless repeater.

Therefore, we are sure that this study is a useful element in the design of actual stage image for smooth interaction between the audience and the stage. Also, the proposed algorithm will be also used as the development algorithm for effective interaction of physical functional games.

References

1. Brynskov, M., Dalsgaard, P., Ebsen, T., Fritsch, J., Halskov, K., Nielsen, R.: Staging Urban Interactions with Media Facades. Human-Computer Interaction - INTERACT **2009**, 154–167 (2009)
2. Dalsgaard, P., Halskov, K.: Designing urban media façades: cases and challenges. In: SIGCHI Conference on Human Factors in Computing Systems, pp. 2277–2286 (2010)
3. Kim, J.H.: New media: A study on the technical realization of media façade. The Korean Journal of Art and Media **9**(1), 54–66 (2010)
4. Lee, C.J., Ha, D.O., Han, T.W.: Expression method of simulation which is execution of media façade. Journal of the Korea Industrial Information System Society **14**(5), 91–101 (2009)
5. Diniz, N.V., Duarte, C.A., Guimarães, N.M.: Mapping interaction onto media façades. In: 2012 International Symposium on Pervasive Displays, pp. 14:1–14:6 (2012)
6. Fabric Architecture (2010). https://secure.ifai.com/fabarch/articles/0110_nw_a_suzhou.html
7. DATATON. http://www.dataton.com/watchout
8. Showlogix. http://www.showlogix.com
9. Boquete, L., Ascariz, J.M.R., Cantos, J., Barea, R., Miguel, J.M., Ortega, S., Peixoto, N.: A portable wireless biometric multi-channel system. Measurement-Journal of the International Measurement Confederation **45**(6), 1587–1598 (2012)
10. RF RECHWIN. http://www.rftechwin.com

Adaptive Clustering of Merchandise Code for Recommender System with an Immediate Effect

Young Sung Cho and SongChul Moon

Abstract This paper proposes recommending method with an immediate effect through adaptive clustering method based on segmented merchandise code in e-commerce. Using an implicit method without onerous question and answer to the users, it is necessary for us to do the task of clustering using merchandise code in purchase data extracted from whole data, to join customer's data, to keep the analysis of RFM (Recency, Frequency, and Monetary) in order to segment merchandise with an immediate effect by Bayesian suggestion to consider frequently changing customer's preference. We reflect the importance of attribute for merchandise code and then take adaptive clustering of merchandise code customer prefer to forecast frequently changing customer's preference of merchandise code efficiently. We carry out experiments with data set of internet cosmetic shopping mall to measure its performance. We report some of the experimental results.

Keywords BN(Bayesian Network) · Clustering · RFM model

1 Introduction

Along with the advent of ubiquitous computing environment, it is becoming a part of our common life style that the demands for enjoying the wireless internet using intelligent portable device such as smart phone and iPad, are increasing anytime or anyplace without any restriction of time and place [1][3]. In these trends, the

Y.S. Cho
Database and Bioinformatics Laboratory, Computer Science Department in College
of Electrical and Computer Engineering, Chungbuk National University, Cheongju, Korea
e-mail: youngscho@empal.com

S. Moon(✉)
Department of Computer Science, Namseoul University, Cheonan, Korea
e-mail: moon@nsu.ac.kr

© Springer Science+Business Media Singapore 2015   583
D.-S. Park et al. (eds.), *Advances in Computer Science and Ubiquitous Computing*,
Lecture Notes in Electrical Engineering 373,
DOI: 10.1007/978-981-10-0281-6_83

personalization becomes a very important technology which can find exact information to present users. The recommendation system helps customers to find items easily and helps the e-commerce companies to set easily their target customer by automated recommending process. Recently, using an implicit method without onerous question and answer to the users, not used user's profile for rating,it has been actually processed the research of recommendation to improve the accuracy of recommendation, to reduce customers' searching effort to find out the items in e-commerce. Therefore, customers and companies can take some benefit from recommendation system on e-commerce environment. A recommendation system using data mining technique based on RFMto meet the needs of customers changing according to the time, has been actually processed the research [1][3]. And also, collaborative filtering may not provide high quality recommendation because it does not consider user's preference on the attribute, the first rater problem, and the sparsity problem[5]. Now a day, the demands for e-commerce and many different items on e-commerce are increasing. It is crucial to have different value of each item considered the importance of attribute of merchandise or service. In this paper, we suggest recommending method with an immediate effect based on adaptive clustering of merchandise code in e-commerce. We can improve the performance of recommendation through adaptive clustering merchandise code with an immediate effect. The next section briefly reviews the literature related to studies. Section 3 is described a new method for recommender system in detail, such as system architecture with sub modules, the algorithm for proposing system, and the procedure of processing the recommendation. Section 4 describes the evaluation of this system in order to prove the criteria of logicality and efficiency through the implementation and the experiment. In section 5, finally it is described the conclusion of paper and further research direction.

2   Related Works

2.1   RFM Model

RFM model is generally known in database marketing and direct marketing. RFM model is the method which distinguishes the customer quantitatively. It develops the marketing activity with the customer and gives the customer the item's score with categories in order to provide customer with the value of the customer in database marketing. RFM marketing analysis method is used in order to segmentation of customers[2]. It is easy for us to recommend the item with high purchasability using the customer's score and the item's score. The RFM score can be a basis factor how to determine purchasing behavior on the internet shopping mall, is helpful to buy the item which they really want by the personalized recommendation. One well-known commercial approach uses five bins per attributes, which yields 125 cells of segment. The following express in presents RFM score to be able to create an RFM analysis. The RFM score will be shown how to determine the customer as follows, will be used in this paper. The variables (W_1, W_2, W_3) are weights. The categories (R, F, M) have five bins.

$$\text{RFM score} = W_1 \times R + W_2 \times F + W_3 \times M \tag{1}$$

The RFM score is correlated to the interest of e-commerce[3]. It is necessary for us to keep the analysis of RFM to be able to reflect the attributes of the item in order to find the items with high purchasability. The analytical method which brings better rate of the reaction than RFM model is still not discovered, but even then the neural network did not brings better result than RFM models.

2.2 Clustering

Clustering is defined as the process of grouping physical or abstract objects into classes of similar objects. It involves classifying or segmenting the data into groups based on the natural structure of the data. Its techniques[6,7] fall into a group of undirected data mining tools. The principle of clustering is maximizing the similarity inside an object group and minimizing the similarity between the object groups. This algorithm is a kind of customer's segmentation methods commonly used in data mining, can often use to k-means clustering algorithm. K-means is the most well-known and commonly, used partition methods are the simplest clustering algorithm. In the k-means algorithm, cluster similarity is measured in regard to the mean value of the objects in a cluster, which can be viewed as the cluster's center of gravity. This algorithm uses as input a predefined number of clusters that is the k from its name. Mean stands for an average, an average location of all the members of a particular cluster. The euclidean norm is often chosen as a natural distance which customer a between k measure in the k-means algorithm[8]. The a_i means the preference of attribute i for customer a.

$$d_{a,k} = \sqrt{\sum_i (a_i - k_i)^2} \tag{2}$$

There are two part of k-means algorithm. The 1st part is that partition the objects into k clusters. The 2nd part is that iteratively reallocate objects to improve the clustering. The system can use Euclidean distance metric for similarity. In this paper, we can do clustering the customers' data to segment customers and finally forms groups of customers with different features, the extracted purchase data from whole data using RFM score of item score, to join the cluster of customer. Through analyzing different groups of customers' data, we try to do the recommendation for the target customers of internet shopping mall efficiently.

2.3 BN(Bayesian Network)

BN can be used to model the joint probability distribution of multiple random variables. With the BN, we formulate an item preference model in the form of a joint probability distribution. In the case of item recommendation, the problem is finding items that a given user is likely to rate highly. For this purpose, we calculate the conditional probability for the target user U = u, the candidate item C = c and then recommend items in order of probability. Alternatively, we may calculate the conditional probability for the target user and rating to find items that are highly likely to obtain a positive rating. The recommendation system may receive user feedback for final purchase behavior, and periodically, the system updates the parameters of the item preference. BN model using final purchase data by using the Bayesian inference engine as the decision of behavior of buying additional item in order to increase the precision of the recommendation. Although the preference model can be used in many ways, we explain the typical ways for item recommendation. Here, since a recommendation system can use the same item preference BN model can have two type of the calculation of probability, one is prior probability, the other is posterior probability. The users can be commonly used to update the parameters of the model and thus increase the precision of both the recommendation and promotion. Bayesian probability measures a degree of belief. Bayes theorem then links the degree of belief in a proposition before and after accounting for evidence. For proposition C_i and evidence X,

· $P(C_i)$, the prior, is the initial degree of belief in C_i
· $P(C_i|X)$, the posterior, is the degree of belief having counted for X.
· the quotient $P(X|C_i)/P(X)$ represents the support X provides for C_i

Bayes' theorem gives the relationship between the probabilities of C_i and X, $P(C_i)$ and P(X), and the conditional probabilities of C_i given X and X given C_i, $P(C_i|X)$ and $P(X|C_i)$. For example,suppose an experiment is performed many times. $P(C_i)$ is the proportion of outcomes with property C_i, and $P(X)$ that with property X. $P(X|C_i)$ is the proportion of outcomes with property X *out of* outcomes with property C_i, and $P(C_i|X)$ the proportion of those with C_i out of those with X. In Bayesian inference, the posterior distribution is proportional to the product of the likelihood and the prior distribution. For parameters C_i and data X. It is most common form as follows [5]. For events C_i and X, provided that $P(X) \neq 0$,

$$P(C_i|X) = \frac{P(X|c_i)P(c_i)}{P(X)} , 1 \leq i \leq m \qquad (3)$$

The denominator is the marginal likelihood of the data, which is the integral of the likelihood against the prior distribution. In many applications, the event X is fixed in the discussion, and we wish to consider the impact of its having been observed on our belief in various possible events C_i. In such a situation the denominator of the last expression, the probability of the given evidence X, is fixed. For more on the application of Bayes' theorem under the Bayesian interpretation of probability, we can apply it in the application using Bayesian learning. We can apply the algorithms for the preference of merchandise code based RFM using Bayesian theorem in previous paper. For any set of random variables, the probability of any member of a joint distribution can be calculated from conditional probabilities using the chain rule (given a topological ordering of X) as follows:

$$
\begin{aligned}
P(X|C_i) &= P(x_1, x_2, \ldots, x_n, C_i)\, P(C_i) \\
&= P(x_1|C_i)\,P(x_2|C_i)\ldots P(x_n|C_i)\, P(C_i) \\
&= P(C_i) \prod_{k=1}^{n} P(x_k|C_i)
\end{aligned}
\tag{4}
$$

3 Our Proposal for a Recommender System with an Immediate Effect

3.1 Adaptive Clustering of Merchandise Code

In this section, we can describe the adaptive clustering of merchandise code for recommender system with an immediate effect. As you know that, CF(Collaborative Filtering) has been identified with problems of its inability to reflect contents of the items, the complexity of selection in the neighborhood, sparsity and scalability. Despite many efforts to overcome these defects, it still fails to reflects the attributes of the item. To solve this problem, we use an implicit method of using purchase history data and customer's data. It adopts the method of evaluation based on personal profile, which has been identified with difficulties in accurately analyzing the customers' level of interest and tendencies, as well as the problems of cost, consequently leaving customers unsatisfied. Also it is difficult to use the unique recommendation method withhierarchy of each customer who has various characteristics in the existing recommender system. To solve this problem, we have used the analysis of RFM to segment customer and merchandise, to reflect the importance of attribute for them. Nowaday, RFM model accomplishes a foundation of the technique which selects the target customer using user's information. Considering "Pareto Principle (80/20Rule)" in the marketing strategic, analytical technique is intensive marketing of the above 20% best in the respect of the quantity of purchase. As a matter of course, we can use the above 20% most in the respect of the quantity of the count of customer. We can use the rank of RFM score with the quantity of purchase to extract the most frequent purchase data from the whole purchase data. We use 319 customers who have had the experience to buy items in e-shopping mall. As a result of that, we have known 310 customers who have had the purchase from the result of purchase historical data. There are the 6 levels by each rank of RFM score of customer. The following figure 1 show the result with statistics of the output for possession of customers by each rank of RFM score of customer. In case of each level, it is shown that the number of customer is output for possession of customers. In case of the level 1, it is shown that the number is greater equal than 90 points of RFM score for customer, the level 2 is the range of RFM score (RFM score ≥ 80 points and RFM score < 90 points), the level 3 is the range of RFM score (RFM score ≥ 60 points and RFM score < 80 points), the level 4 is the range of RFM score (RFM score ≥ 40 points and RFM score < 60 points), and the level 5 is the range of RFM score (RFM score ≥ 20 points and RFM score < 40 points), the level 6 is the range of RFM score (RFM score < 20 points). Fig. 1. The result of the graphical statistics of the output for possession of customers.

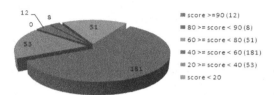

Fig. 1 The result of the graphical statistics of the output for possession of customers

RFM model would be rather to maximize the profit, it is a simple principle of cost-effective mostly. At first, the RFM score of customer is computed to reflect the attributes of the customer. The system can use the extracted purchase data(sale_dat3) with a lot of purchasing counts, between the score is more than 19 points and the score is less than 41 points. As considering of the Pareto Principle, we can use the purchase data extracted from whole data to join the customer information for pre-processing to be possible to recommend merchandise with an immediate effect. The system can create the cluster of purchase data sorted by merchandise code for the preprocessing task on the analytical agent. The system can compute customer's probability of preference of all categories of merchandise code in clustering data which is selected by demographic variable and customer's score. As a result, the system has finished the ready to recommend merchandises with high probability in merchandise code belonged to brand items using the highest RFM score of merchandise code. There are two type of the preprocessing task. One is the adaptive clustering of merchandise code for recommender system with an immediate effect. The other is creating of clusters with neighborhood user-group by customers' the code of classification such as age, gender, occupation, propensity of customer, and skin type via the task of preprocessing for the adaptive clustering of merchandise code for recommender system using Bayesian suggestion by customers' propensity on the purpose of recommending merchandise in e-commerce with high purchasablity. As a matter of course, we use the adaptive clustering of merchandise code for recommender system with an immediate effect via a successful structure based on the purchase data (sale_dat3) extracted from the whole data(sale), to adjust the result of user's probability of preference. The procedural proposing recommending steps for recommender with an immediate effect using Bayesian suggestion are depicted as the following Table 1.

Table 1 The procedural proposing recommending steps for recommener with an immediate effect

---

Step 1: The login user reads user's information as customer's information. And then recognizes the classification code, customer's RFM score in customer's information.

Step 2: The system selects the cluster classified with the code of classification of login user, as the basis of   the social variables as demographic valuables such as age, gender, an occupation, user's   propensity and customer's RFM score.

Step 3: The system scans user's probability of preference of the categories of merchandise in the selected cluster.

Step 4: First, the system recommends the merchandises using prior probability according to the preference in the selected cluster .

Step 5: Then, the system recommends the merchandises according to the information of recommendation which is applied by posterior probability through Bayesian learning, if a customer wanted to buy any items as additional purchase.

Step 6: The system makes user's TOP-4 of merchandises in the list of merchandises to recommend merchandises which is similar to propensity of login user.

Step 7: The system executes the cross comparison with purchased history data in order to avoid the duplicated recommendation which it has ever taken.

This proposed method uses the implicit method without onerous question and answer to the users, unlike the other evaluation techniques. We apply RFM method which can analyze the tendency of the various personalization based on the purchase data extracted the most frequently from whole data, to join customer's data. Then, we used recommending method using the adaptive clustering of merchandise code for recommender with an immediate effect.

4   The Environment of Implementation and Experiment and Evaluation

4.1   Experimental Data for Evaluation

We have 319 users who have had the experience to buy items in e-shopping mall, 580 cosmetic items used in current industry. We have 1600 results of purchased data recommended in order to evaluate the proposing system[2]. For doing that, we do the implementation for prototyping of the internet shopping mall which handles the cosmetics professionally. We carry out experiments with data set of internet cosmetic shopping mall to measure its performance. We report some of the experimental results. We use learning data set for 12 months and testing data set for 3 months[3]. The 1st proposing system is used by adaptive clustering merchandise code with an immediate effect using Bayesian suggestion(ACMC), called by "proposal", the 2nd previous system[4] isused by k-means Clustering of Item Preference (KCIP), the third system is existing system.

4.2   Experiment and Evaluation

We carry out experiments with data set of internet cosmetic shopping mall to measure its performance for recommender system using Bayesian suggestion. We make the evaluation, which is precision, recall and F-measure for proposing system. For performance evaluation of the system, we used metrics most widely used for recommender systems using learning data set and testing data set, precision, recall and F-measure defined as follows: Precision = recommended relevant items / all recommended items, Recall = recommended relevant items / total relevant items, F-measure = 2(Precision * Recall) / (Precision + Recall). The metrics of evaluation for recommender system in our system is used in the field of information retrieval commonly[9].

Table 2 The result of precision, recall and F-measure for recommender system

Cluster	Proposal(ACMCE)			Previous(KCIP)			Existing		
	Precision1	Recall1	F-measure1	Precision2	Recall2	F-measure2	Precision3	Recall3	F-measure3
C1	46.87	81.03	59.39	56.98	91.44	65.90	56.98	50.89	50.21
C2	43.46	81.58	56.71	48.79	55.70	48.41	48.79	31.32	35.64
C3	44.38	84.62	58.22	49.36	52.53	48.09	49.36	29.54	35.06
C4	45.77	77.33	57.50	55.50	23.93	32.19	44.26	21.81	27.65
C5	42.98	65.28	51.84	52.49	38.37	41.95	52.49	34.98	39.75
C6	54.17	78.87	64.23	50.41	47.40	45.24	50.41	43.21	43.10
C7	48.13	81.82	60.61	50.93	37.23	40.03	50.93	36.60	39.64
C8	41.66	75.00	53.57	47.41	27.27	32.60	47.41	26.81	32.26
C9	53.23	79.41	63.74	43.60	37.23	38.17	43.60	36.60	37.82
C10	27.28	75.00	40.01	46.68	28.45	32.62	46.68	25.19	30.28
C11	47.18	85.56	60.82	67.18	20.69	31.17	46.53	18.32	25.10
C12	58.64	93.94	72.20	67.23	62.50	60.94	67.23	55.34	57.10

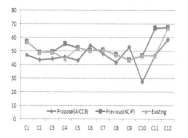

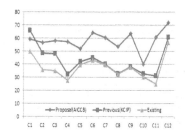

Fig. 2 Precision, the result of recommender system

Fig. 3 F-measure, the result of recommender system

The proposing system's overall performance evaluation presents the result of evaluation for recommender system on the Table 2. The proposedis improved better than the previous systems and existing system. Our proposing system of adaptive clustering merchandise code with an immediate effect using Bayesian suggestion(ACMC)is higher 45.73% in recall than the previous systems and existing system. And then the proposing system's F-measure is higher 20.44 % than both of them even if its precision is lower 4.24% than existing system. And also, our proposing system(ACMC)is higher 36.39% in recall, higher 15.13% in F-measure than both of them even if it is lower 6.90% in precision than previous system(KCIP). As a result, we had the recommender system to be able to recommend the items with an immediate effect. The Fig.5 is shown in the screen of result for recommending itemsof cosmetics on a smart phone. The proposing system is better performance than their performance.

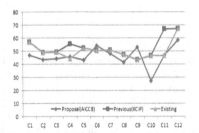
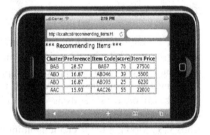

Fig. 4 Recall, the result of recommender system

Fig. 5 The screen of result for recommending items

5 Conclusions

Recently, u-commerce as an application field under ubiquitous computing environment required by real time accessibility and agility,is in the limelight[2]. We proposed recommending method with an immediate effect through adaptive clustering method based on segmented merchandise code in e-commercein order to improve the accuracy of recommender, to reduce customers' searching effort to find out the items. The proposed method was improved in this paper. Then,we mentioned the following into the improvement. At the first, to make data processing promptly, we created purchase data extracted the most frequently from whole data, to join customer's data, then took the task of pre-processing adaptive clustering method based on the variables of the user's information and customer's RFM scores for recommender system. At the second, we constructedthe information based on analyzing the preference throughitem categories of each customer, the information based on the item classification andthe information based on the customer's social variablesm such as age, gender, occupation, customer's propensity, to get the preference of item categories. At the third, for constructing personalized recommender, we applied subdivided techniques to the user history which contains the change of the time and the purchase history and then we used the RFM scoring method of the customers and items to do recommending the process promptly. We had described that the performance of the proposing system of adaptive clustering merchandise code with an immediate effect using Bayesian suggestion was improved better than both of them. We carried out experiments with data set of internet cosmetic shopping mall to measure its performance. We reported some of the experimental results. It was meaningful to present recommending method with an immediate effect through adaptive clustering method based on segmented merchandise code in e-commerce under ubiquitous computing environment.

Acknowledgements This study was supported by Namseoul University.

References

1. Cho, Y.S., Kim, K.A., Moon, S.C., Park, S.H., Ryu, K.H.: Effective purchase pattern mining with weight based on FRAT analysis for recommender in e-commerce. In: Computer Science and its Applications, pp. 443–454 (2015)
2. Khajvand, M., Zolfaghar, K., Ashoori, S., Alizadeh, S.: Estimating customer lifetime value based on RFM analysis of customer purchase behavior: Case study. Procedia Computer Science **3**, 57–63 (2011)
3. Cho, Y.S., Moon, S.C.: Weighted mining frequent itemsets using FP-tree based on RFM for personalized u-commerce recommendation system. In: Mobile, Ubiquitous, and Intelligent Computing, pp. 441–450 (2014)
4. Cho, Y.S., Moon, S.C., Noh, S.C., Ryu, K.H.: Implementation of personalized recommendation system using k-means clustering of item category based on RFM. In: IEEE International Conference on Management of Innovation and Technology (ICMIT), pp. 378–383 (2012)
5. Jung, K.Y.: User preference through learning user profile for ubiquitous recommendation systems. In: Knowledge-Based Intelligent Information and Engineering Systems, pp. 163–170 (2006)
6. Hand, D., Mannila, H., Smyth, P.: Principles of Data Mining. The MIT Press (2001)
7. Collier, K., Carey, B., Grusy, E., Marjaniemi, C., Sautter D.: A Perspective on Data Mining. Northern Arizona University (1998)
8. Hastie, T., Tibshirani, R., Friedman, J.: The Elements of Statistical Learning – Data Mining, Inference, and Prediction. Springer (2001)
9. Herlocker, J.L., Kosran, J.A., Borchers, A., Riedl, J.: An algorithm framework for performing collaborative filtering. In: Proceedings of the Conference on Research and Development in Information Research and Development in Information Retrival (1999)

Bayesian Probability-Based Motion Estimation Method in Ubiquitous Computing Environments

Phil Young Kim, Yunsick Sung and Jonghyuk Park

Abstract Natural User Interface/Natural User Experience (NUI/NUX) is one of the core techniques to control deployed devices in ubiquitous computing environments. User friendly user interfaces can be provided by estimating and utilizing locations and gestures/postures of users. However, the recognition of users' gestures/postures is limited because of the number of the available deployed motion recognition sensors. This paper proposes a Bayesian probability-based motion estimation method to control users' unestimated motions. Given that the whole postures/gestures can be predicted based on the estimated motion, the intention of user can be also predicted. In the experiments, Myos were utilized as motion recognition sensors. The proposed method was validated showing the result through a virtual character. By collecting the data of users' motions in advance, unmeasured motions could be deducted by the proposed method.

Keywords Bayeisan probability · Motion estimation · Motion recognition sensor · Myo · Electromyograms

1 Introduction

Recently the importance of the interaction with users in the field ubiquitous computing environments has emphasized increasingly. The research of Natural

P.Y. Kim · Y. Sung(✉)
Faculty of Computer Engineering, Keimyung University, Daegu 42601, South Korea
e-mail: {kimpy1111,yunsick}@kmu.ac.kr

J. Park
Department of Computer Engineering, Seoul National University of Science and Technology,
Seoul 139-743, South Korea
e-mail: jhpark1@seoultech.ac.kr

© Springer Science+Business Media Singapore 2015
D.-S. Park et al. (eds.), *Advances in Computer Science and Ubiquitous Computing*,
Lecture Notes in Electrical Engineering 373,
DOI: 10.1007/978-981-10-0281-6_84

User Interface/Natural User Experience (NUI/NUX) reflects the importance [1]. Given that the diverse kinds of wearable devices are released, the research to handle deployed devices based the wearable devices has introduced in ubiquitous computing environments. For examples, Leap motion [2] and Myo [3] are utilized. Leap motion recognizes fingers accurately and Myo recognizes the motions based on sensor values and muscle data. However, given that only sensor-attached partial bodies can be estimated, the whole motion of the body cannot be estimated accurately. For an example, when a muscle recognition sensor is attached only to a forearm, it is very difficult to estimate the motion of an upper arm.

This paper proposes a Bayesian probability-based motion estimation and control method for estimating the motions that are not measured by the attached motion recognition sensors. At the first time, users utilize two motion recognition sensors and collect the data of the two motions. The relationship between the two motions is analyzed by applying Bayesian probability. The motions that are not measured are estimated based on the analyzed Bayesian probability with other measured motions.

The rest of this paper is organized as follows: Section 2 introduces motion estimation processes. Section 3 explains the experiment by applying the proposed method. Finally, Section 4 conclusions of this paper.

2   Motion Estimation Processes

This paper defines a motion as the movement of a part of a posture. Motion recognition sensors measure different motions. Measured motions performed not depending on other motions are defined as independent motions. Estimated motions based on independent motions are defined as dependent motions. Measured values of a motion are called by properties in this paper. While estimating dependent motions, motion recognition sensors are not attached to dependent motions but the dependent motions are estimated and controlled depending on multiple independent motions.

First, the partial parts of a body corresponding to independent motions and dependent motions are attached by motion recognition sensors, are and the motion recognition sensors stores the measured values as the properties of motions. The ith motion is denoted as m_i. The kth property of the motions m_i at time t are defined by $x_{i,k,t}$.

Next, the Bayesian probability $p_{i,k,t}$=P($x_{i,k,t} \mid x_{1,k,t}, x_{2,k,t}, \ldots$) of the kth property of the dependent motion m_i at time t is obtained based on the properties of independent motions. In addition, if the properties of motions are expressed by numbers diversely, each Bayesian probability becomes too small to be handled. In that case, Bayesian probabilities are grouped by being divided by the interval δ_k. Therefore, The Bayesian probability $p_{i,k,t}$ is expressed as P($\left\lfloor \frac{x_{i,k,t}}{\delta_k} \right\rfloor \mid \left\lfloor \frac{x_{1,k,t}}{\delta_k} \right\rfloor, \left\lfloor \frac{x_{2,k,t}}{\delta_k} \right\rfloor, \ldots$).

Whenever independent motions are performed, dependent motions are estimated the calculated Bayesian probability and performed. The estimated independent motions are selected by Equation (1). As results, the property corresponding to the maximum probability of each dependent property is utilized to choose the motion corresponding to the property.

$$x_{i,k,t} = \frac{arg}{x_{i^*,k}} \text{Max}(p_{i^*,k,t}) \tag{1}$$

3   Experiment

In the experiments, "Standby", "Place yourself", and "Proceed postures" are utilized as shown in Figure 1. Each posture was performed and measured 30 times. The motions of an upper arm were estimated according to the motions of a forearm based on Bayesian probability. To measure the motion of the two arms in advance, two muscle sensor-based motion recognition device, Myos [3], were utilized. The two Myos were attached to the forearm, the motion m_1, and to the upper arm, the motion m_2, to measure the properties of the forearm and the upper arm in advance as shown in Figure 2(a). The motion m_3 (Figure 2(b)) is the motion of a forearm during estimating the motion of the upper arm.

| (a) Hold/Stand by | (b) Place yourself | (c) Proceed |

Fig. 1 The postures of three hand signals were utilized.

(a) Measurement: Myos were attached   (b) Estimation: Myo was only attached
      to a forearm and upper arm   to a forearm

Fig. 2 Myo-attached arms are shown.

Each Myo measures motions storing the corresponding properties as shown in Table 1. For an example, the values of the elements of an orientation were measured from -1 to +1, transformed to the value from -180 to 180 and utilized to calculate the Bayesian probability of motions.

Table 1 A Myo measures the properties of motions through gyroscope, acceleration, orientation and EMG.

Myo x_i			
Gyroscope	**Acceleration**	**Orientation**	**EMG**
X: $x_{i,gx,t}$ (-500-α ~ +500+α)	X: $x_{i,ax,t}$ (-5 ~ +5)	X: $x_{i,ox,t}$ (-1 ~ +1)	e_1: $x_{i,e_1,t}$ (-127~+127) e_2: $x_{i,e_2,t}$ (-127~+127)
Y: $x_{i,gy,t}$ (-500-α ~ +500+α)	Y: $x_{i,ay,t}$ (-5 ~ +5)	Y: $x_{i,oy,t}$ (-1 ~ +1)	e_3: $x_{i,e_3,t}$ (-127~+127) e_4: $x_{i,e_4,t}$ (-127~+127)
Z: $x_{i,gz,t}$ (-500-α ~ +500+α)	Z: $x_{i,az,t}$ (-5 ~ +5)	Z: $x_{i,oz,t}$ (-1 ~ +1)	e_5: $x_{i,e_5,t}$ (-127~+127) e_6: $x_{i,e_6,t}$ (-127~+127)
		W: $x_{i,ow,t}$ (-1 ~ +1)	e_7: $x_{i,e_7,t}$ (-127~+127) e_8: $x_{i,e_8,t}$ (-127~+127)

In the experiment, all δ were set by 30. Three types of motions were performed 30 times per a motion. 90 data of motions were collected and were utilized to calculate Bayesian probabilities. For an example, the probabilities of four elements in orientation were obtained as shown in Table 2.

Table 2 The Bayesian probabilities of an upper arm according to a forearm were calculated.

(a) $x_{i,ox,t}$ — $x_{i,gx,t}$, Upper Arm m_1 (columns) / Forearm m_2 (rows)

Angle	-180~-150	-150~-120	-120~-90	-90~-60	-60~-30	-30~0	0~30	30~60	60~90	90~120	120~150	150~180
-180~-150	0	0	0	0	0	0	0	0	0	0	0	0
-150~-120	0	0	0	0	0	0	0	0	0	0	0	0
-120~-90	0	0	0	0	0	0	0	0	9	4	0	0
-90~-60	0	0	0	0	0	0	0	0	2	16	2	0
-60~-30	0	0	0	0	0	0	0	0	4	17	3	0
-30~0	0	0	0	0	0	0	0	0	3	0	0	0
0~30	0	0	0	0	0	0	0	0	0	0	0	0
30~60	0	0	0	0	1	1	0	0	0	0	0	0
60~90	0	0	0	0	13	5	0	0	0	0	0	0
90~120	0	0	0	2	8	0	0	0	0	0	0	0
120~150	0	0	0	0	0	0	0	0	0	0	0	0
150~180	0	0	0	0	0	0	0	0	0	0	0	0

(b) $x_{i,oy,t}$ — $x_{i,gy,t}$, Upper Arm m_1 (columns) / Forearm m_2 (rows)

Angle	-180~-150	-150~-120	-120~-90	-90~-60	-60~-30	-30~0	0~30	30~60	60~90	90~120	120~150	150~180
-180~-150	0	0	0	0	0	0	0	0	0	0	0	0
-150~-120	0	0	0	0	0	0	0	0	0	0	0	0
-120~-90	0	0	0	0	0	0	0	0	0	12	2	0
-90~-60	0	0	0	0	0	0	0	0	3	11	0	0
-60~-30	0	0	0	0	0	0	0	1	1	0	0	0
-30~0	0	0	0	0	0	0	2	1	0	0	0	0
0~30	0	0	0	0	0	0	5	2	0	0	0	0
30~60	0	0	0	0	11	7	6	0	0	0	0	0
60~90	0	0	0	8	8	5	0	0	0	0	0	0
90~120	0	0	3	2	0	0	0	0	0	0	0	0
120~150	0	0	0	0	0	0	0	0	0	0	0	0
150~180	0	0	0	0	0	0	0	0	0	0	0	0

(c) $x_{i,oz,t}$ — $x_{i,gz,t}$, Upper Arm m_1 (columns) / Forearm m_2 (rows)

Angle	-180~-150	-150~-120	-120~-90	-90~-60	-60~-30	-30~0	0~30	30~60	60~90	90~120	120~150	150~180
-180~-150	0	0	0	0	0	0	0	0	0	3	26	1
-150~-120	0	0	0	0	0	0	0	0	0	8	14	0
-120~-90	0	0	0	0	0	0	0	0	4	18	11	0
-90~-60	0	0	0	0	0	0	0	0	2	2	0	0
-60~-30	0	0	0	0	0	0	0	0	1	0	0	0
-30~0	0	0	0	0	0	0	0	0	0	0	0	0
0~30	0	0	0	0	0	0	0	0	0	0	0	0
30~60	0	0	0	0	0	0	0	0	0	0	0	0
60~90	0	0	0	0	0	0	0	0	0	0	0	0
90~120	0	0	0	0	0	0	0	0	0	0	0	0
120~150	0	0	0	0	0	0	0	0	0	0	0	0
150~180	0	0	0	0	0	0	0	0	0	0	0	0

(d) $x_{i,ow,t}$ — $x_{i,gw,t}$, Upper Arm m_1 (columns) / Forearm m_2 (rows)

Angle	-180~-150	-150~-120	-120~-90	-90~-60	-60~-30	-30~0	0~30	30~60	60~90	90~120	120~150	150~180
-180~-150	0	0	0	0	0	0	0	0	0	0	0	0
-150~-120	0	0	0	0	0	0	0	0	0	0	0	0
-120~-90	0	0	0	0	0	0	0	7	1	3	0	0
-90~-60	0	0	0	0	0	0	5	6	18	3	0	0
-60~-30	0	0	0	0	0	2	8	3	1	0	0	0
-30~0	0	0	0	0	0	1	15	3	2	0	0	0
0~30	0	0	0	0	0	2	8	0	0	0	0	0
30~60	0	0	0	0	1	1	0	0	0	0	0	0
60~90	0	0	0	0	0	0	0	0	0	0	0	0
90~120	0	0	0	0	0	0	0	0	0	0	0	0
120~150	0	0	0	0	0	0	0	0	0	0	0	0
150~180	0	0	0	0	0	0	0	0	0	0	0	0

In the first experiment, Myos were attached to a forearm and an upper arm and collected the properties of the two arms. Figure 3 shows that when the measured properties are expressed by a virtual character, the forearm and the upper arm moved as the movements of a subject as shown in Figure 1.

(a) Hold/Stend by   (b) Place yourself   (c) Proceed

Fig. 3 The right arm of a virtual character was controlled based on the properties of a forearm and an upper arm with Myos.

The control of the forearm is unnatural as shown in Figure 4 if an upper arm is fixed and the forearm was measured by a Myo, because of the fixed upper arm.

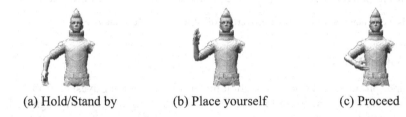

(a) Hold/Stand by   (b) Place yourself   (c) Proceed

Fig. 4 Only a forearm of a virtual character is control by a Myo. An upper arm is fixed.

Figure 5 shows the result when the elbow between a forearm and an upper arm was fixed, only a single Myo was utilized on the forearm and the only shoulder was controlled according to the movements of the forearm. When the directions of the forearm and the upper arm were same in the case of Figure 6(a) and Figure 6(b), the motions looked natural. However, when the two directions were difference as shown in Figure 6(c), the inaccurate motion was performed.

(a) Hold/Stand by   (b) Place yourself   (c) Proceed

Fig. 5 Only a forearm of a right arm attached a Myo and then the right shoulder was controlled.

Table 3 shows the properties of a forearm and an estimated upper arm when the proposed method was applied to a forearm. Therefore, given that only the properties of the forearm were measured, the upper arm could be controlled as shown in Figure 6.

Table 3 The revised properties of an upper arm are expressed.

Properties	Forearm				Estimated Upper Arm			
	$x_{i,ox,t}$	$x_{i,oy,t}$	$x_{i,oz,t}$	$x_{i,ow,t}$	$x_{i,ox,t}$	$x_{i,oy,t}$	$x_{i,oz,t}$	$x_{i,ow,t}$
Hold/Stand by	87.35	110.59	58.73	-95.31	-30	-90	90	90
Place yourself	14.31	-92.40	-2.18	-153.79	-30	120	90	90
Proceed	-82.17	-22.87	60.86	146.34	120	30	90	30

(a) Hold / Stand by   (b) Place yourself   (c) Proceed

Fig. 6 The estimated and controlled properties of an upper are described.

4   Conclusion

This paper proposed a motion estimation method based on the Bayesian probabilities of the previously measured properties of multiple motions. Even though each motion requires a motion recognition device to estimate own movements, an unattached motion can be estimated by utilizing the proposed method. In the experiments, three Myos were utilized: two Myos for collecting the properties of two motions and the other for estimating the motions of an upper arm.

Acknowledgement Following are results of a study on the "LINC" Project, supported by the Ministry of Education(MOE) and the National Research Foundation of Korea(NRF).

References

1. Kwak, J., Sung., Y.: Indoor location-based natural user interface for ubiquitous computing environment. In: The 4th International Conference on Ubiquitous Computing Application and Wireless Sensor Network. Jeju, July 8–10 2015
2. Leap motion. https://www.leapmotion.com/
3. Myo. http://myo.com/
4. Kim, P.Y., Kim, J.W., Sung, Y.: Bayesian probability-based hand property control method. In: International Conference on Intelligent Technologies and Engineering Systems. ICCK, Kaohsiung, Taiwan, December 19–21 (2014)
5. Son, J., Sung, Y.: Bayesian probability and user experience-based smart UI design method. In: International Conference on Intelligent Technologies and Engineering Systems. ICCK, Kaohsiung, Taiwan, December 19–21 (2014)

Occlusion Detection Using Multi-mode Mean-shift Tracking

Eun-Sub Kim, Min Hong and Yoo-Joo Choi

Abstract In this paper, we propose an advanced mean-shift tracking based on multi-mode kernel considering the background weight in which the kernel is divided into multiple sub-kernels in order to detect the partial or full occlusion. The proposed method includes occlusion detection based on coefficient of variance of Bhattacharya coefficients for multi-mode kernels. Experimental results show that the proposed method is able to robustly track a target object with partial and long term full occlusion in moving camera environment.

Keywords Visual object tracking · Kernel-based tracking · Multi-model kernel tracking · Occlusion handling

1 Introduction

Recently, in accordance with the high demands of real-time applications for object tacking, the research of kernel-based mean-shift tracking technique known as the real-time object tacking method has been studying briskly using video images in various fields such as surveillance systems, games and movies. One of major issues to solve in kernel-based mean-shift tracking is detecting the partial or full occlusion of target objects and handing when occlusion is occurred.

E.-S. Kim
Korea Electronics Technology Institute, Sungnam, Korea
e-mail: maycos@naver.com

M. Hong
Department Computer Software Engineering, Soonchunhyang University, Asan, Korea
e-mail: mhong@sch.ac.kr

Y.-J. Choi(✉)
Department of Nemedia Content, Korean German Institute of Technology, Seoul, Korea
e-mail: yjchoi@kgit.ac.kr

© Springer Science+Business Media Singapore 2015   599
D.-S. Park et al. (eds.), *Advances in Computer Science and Ubiquitous Computing*,
Lecture Notes in Electrical Engineering 373,
DOI: 10.1007/978-981-10-0281-6_85

The real environments which require real-time object tracking can use a fixed camera or a moving camera such as PTZ(Pan-Tilt-Zoom) camera to take a sequence of video frames. In general, when a moving object is detected by a fixed camera, moving object region could be tracked by using the background subtraction algorithm [10]. In this case, adaptive background modeling techniques have been studied to reflect the background changes [9]. In case of object tracking using the background subtraction algorithm, all regions of moving objects are extracted. In addition, each color, texture features, and moving direction and speed of moving objects should be analyzed to resolve the partial or full occlusion states [1][4]. With a fixed camera, the speed of moving objects and the location of object in next frame image can be readily estimated in a sequential frame images. But it is not easy to estimate the speed and location of moving objects in next frame image which includes multiple moving objects with a moving camera. Also, the moving object region could not be tracked with a simple background subtraction because the background information is changed continuously. To solve this problem, the kernel-based tacking method has been widely applied to various real environments. In kernel-based tracking, the target object is manually selected in the initial frame and the target model is built based on color information of the selected object region. Then the algorithm moves the kernel, that is, the selected region from the initial position to the target candidate position similar to the target model in the continuous frames. In this case, when the target object is partially or fully occluded by other objects, the object tacking is failed because the special properties of target object in kernel are significantly changed. Some new approaches, such as particle tacking method, which show the robust tracking results, however require high computational burden. The hybrid method to combine different kinds of tracking methods have been tried to resolve this problem [6]. As another approch, a single kernel region can be divided into some sub-regions [6] or fragments [5] and then the similarity analysis between target models in each sub-region for similarity distinction in the multiple local mode leads the advantages for partial occlusion handling. These similarity analysis methods that are available for multiple local mode can provide relatively robust tracking results against the traditional single kernel method. Typical hybrid method [6] that combines the mean-shift tracking approach and particle tracking approach sacrifices the tracking performance with unnecessary particle tracking process when there is no occlusions in video frames. The basic method which utilizes the multi-mode tracking techniques are not concretely specified the working process when the whole target object is fully covered by other objects [5][6].

In this paper, we propose the multi-mode mean-shift tracking algorithm to detect partial and full occlusion automatically for robust tracking of the target object, when partial and full occlusion are happened in video images taken from moving cameras.

2   Mean-shift Tracking Using Multi-mode Kernel

2.1   Partitioning the Target Region

In our method, the topology for partitioning a target region into sub-regions proposed in [6] was applied. The bounding box for a target object is partitioned into non-overlapping sub-regions of a same aspect ratio, i.e., let the bounding box of a candidate object be partitioned into M sub-regions, Ri, i=1,...,M. Any two partitioned sub-regions satisfy $Ri \cap Rj = \emptyset$ if i≠j and $\cup i\, Ri = R$. We assume that all disjoint sub-regions have the same aspect ratio and partitioned sub-regions are non-symmetric. This constraint is to simplify the kernel bandwidth estimate and to automatically divide a target region without the manual operation. Partitioning a target region into smaller sub-regions would allow the mean shift to search several modes.

2.2   Target Model Using Multi-mode Global Kernel Weight

In this subsection, we describe the object similarity metric based on the Bhattacharyya coefficient that is applied in the partitioned sub-regions, using spatial kernel-weighted color histograms and multi-mode local kernel weight.

In our method, we define nine non-overlapping sub-regions of a same aspect ratio which are rectangular in shape. A target model $\therefore = \dot{\therefore}$ $\blacksquare$ $\therefore$ is defined by the probability of the feature u=1..m for sub-region $\therefore = \therefore\dot{\therefore}\blacksquare$ (with $\bar{\therefore}\blacksquare\therefore = \ldots$). The probability q_u^i for sub-region R^i is defined by Eq. (1).

$$q_u^i = C^i v_u \sum_{q \in r} k(\left\| \frac{y_c - x_j}{h} \right\|^2) \delta[b_u(x_j) - u]$$

(1)

where v_u is a background weight for background interference reduction. v_u is computed using background histogram $\therefore\dot{\therefore}\blacksquare$ (with $\bar{\therefore}\blacksquare\therefore = \ldots$ and its smallest nonzero entry o^*. This representation is computed in a region around the target. We used a background area equal to three times the target area. The weights v_u is defined by

$$\ldots = \ldots \frac{\therefore}{\therefore} \ldots \blacksquare \ldots$$

(2)

C^i is the normalizing constant for the color histogram q_u^i. y_c is a center pixel location of the kernel. $k(x)$ is a kernel with Epanechnikov profile, $\therefore = \frac{\therefore}{\therefore} \ldots \therefore \therefore \therefore \therefore$ h is bandwidth of a kernel profile for the kernel. In the

Khan's method [6], q_u^i did not include v_u and kernel weights were computed based on the center of each sub-kernel, not a center of the global kernel. Fig. 1 represents the difference between sub-kernels in Khan's method and ours.

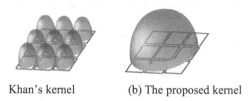

(a)   Khan's kernel   (b) The proposed kernel

Fig. 1 Comparison of multi-mode sub-kernels

2.3   *Target Candidate Using Multi-mode Global Kernel Weight*

A target candidate $. = \vdots \cdot \ \blacksquare \ :$ is defined by the probability of the feature
$u=1..m$ for sub-region $. = \vdots \cdot \vdots \blacksquare \cdot$ (with $\bar{\cdot} \cdot \blacksquare \_\cdot \cdot = \_$). The probability p_u^i
for sub-region R^i is computed as

$$p_u^i(y_0) = C^i \sum_{x_i \in R^i} k \left(\left\| \frac{y_0 - x_j}{h} \right\|^2 \right) \delta[b_u(y_j) - u] \tag{3}$$

where C^i is the normalizing constant for the color histogram p_u^i. y_0 and h are a
center pixel location of the global kernel and kernel bandwidth, respectively.
Note that background weight v_u is not included in computing probability p_u^i.

2.4   *Mean-shift Tracking*

The search for the new target location y_l in the current frame is repeatedly
computed as Eq. (4) from the initial estimated location y_0.

$$\hat{y}_1 = \frac{\sum\limits_{i=1}^{M} \sum\limits_{x_j \in R^i} x_j w_j^i g(D)}{\sum\limits_{i=1}^{M} \sum\limits_{x_j \in R^i} w_j^i g(D)} \tag{4}$$

where $\ \vdots \ = -\vdots \ \cdot \ \vdots\vdots\cdot$

$$\cdot = \bar{\cdot} \blacksquare \cdot \vdots \cdot \cdot \ \ - \cdot \cdot \ \frac{\bar{\ \ }}{\cdot \cdot \ \cdot}, \tag{5}$$

$$D = \left\| \frac{y_o - x_j}{h} \right\|^2 \tag{6}$$

If $\ . - \ ; \ \text{\it v; !}$ or the number of mean-shift iterations comes to N_{max},
typically taken equal to 20, y_l is determined as the center of the most similar

region to the target object. Otherwise, y_l is assigned to y_0, and nine non-overlapping sub-regions and centers of sub-regions are redefined. Then y_l is recomputed using Eq. (4).

3   Occlusion Detection Using Coefficient Variation of Bhattachryya Coefficient

The proposed occlusion detection method calculates the similarity distribution between target and candidate model to determine the occlusion. The proposed similarity function utilizes the Bhattacharyya coefficient to compare the probability distribution between models. Equation (5) shows the similarity of partitioned region between target model (q) and target candidate (p).

$$\rho^i(y) = \rho^i[p^i(y),q^i] = \sum_{u=1}^{m} \sqrt{p^i_u(y)q^i_u} \quad , i = 1,...M \tag{5}$$

The similarity function $\rho^i(y)$ is the similarity value for i which the ROI(Region Of Interest) is partitioned into M and total M numbers of similarity value are computed by equation (5). The $\rho^i(y)$ value means the similarity of each partitioned regions in ROI with the corresponding regions in target model based on y axis. When the partial occlusion is occurred, the similarity value of covered region is relatively lower than other regions. The proposed method uses the coefficient of variation as a decision function of occlusion to reduce the error range of occlusion detection according to the range of observed values. The coefficient of variation can be computed by equation (6) and means the proportional value for the relative difference of distribution.

$$CV(y) = \frac{\sigma}{\bar{x}} \tag{6}$$

where the arithmetic mean can by calculated by equation (7).

$$\bar{x} = \frac{\sum_{i=1}^{M} \rho^i(y)}{M}, \quad \sigma = \sqrt{\frac{\sum_{i=1}^{M}[\rho^i(y)]^2}{M} - (\bar{x})^2} \tag{7}$$

The partial occlusion can be detected, when the value $CV(y) > th_{oc}$ is over the specific threshold. When the number of frames which are determined as the partial occlusion is over the specific threshold, the proposed method determines it as the full occlusion

4   Experimental Results

In order to show the improvement of stability of the proposed method, we implemented the previous multi-mode mean-shift tracking which was proposed in

[6] for occlusion handling. We investigated the tracking errors of [6]'s method and ours. The tracking error is defined by the Euclidean distance from the center of ground truth object to the tracked kernel center for each video frame. To conduct the comparative tests, we selected four video sequences which have characteristic features in camera moving condition, background clutter and the occlusion of a target object.

4.1 Tracking Errors

Table 1 compares tracking errors of the previous multi-mode method[3] and ours. Moreover, the accuracy improvement ratios after applying our method are shown. The accuracy improvement(AI) ratio was computed by

$$\cdot \cdot = \frac{E_{\cdots} - E_{\cdots}}{E_{\cdots}} \qquad \cdot \cdot \qquad (8)$$

where E_{prev} and E_{ours} are the average tracking errors of the previous multi-mode method[3] and ours, respectively. In the comparative test using the selected four test sequences, the accuracy of the proposed method was improved by the average 606.8 % . Fig. 2. shows the tracking results for the visor1 test sequences.

Table 1 Tracking errors and Accuracy Improvement

	Egtest05	Tiger1	Visor1	Woman
Previous Method	71.28 [fail]	34.89	11.90	139.0 [fail]
Ours	8.01	21.29	9.40	8.44
Accuracy Improvement(%)	789.9	63.9	26.6	1546.9

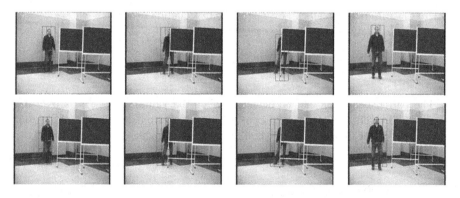

Fig. 2 Tracking results for the visor1 test sequences. (top row) results of the previous method and (bottom row) results of the proposed method.

4.2   Occlusion Detection Results

The proposed occlusion detection method with the coefficient of variation, the standard deviation and arithmetic mean of each index frame, and major frame image are shown in Fig 3. The value of coefficient of variation is increasing from #232 frame when the partial occlusion is happened and then it is sharply increasing from #238 frame when the full occlusion is just happened. However, the result graph of standard deviation and arithmetic mean shows that it is inappropriate to distinguish the starting point of normal detection section, partial occlusion section, and full occlusion section. In this paper, we concluded that the coefficient of variation is appropriate to robustly detect the occlusion.

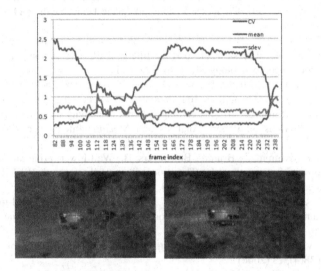

Fig. 3 Comparative test of three occlusion handling functions using coefficient variance, mean, and standard deviation of Battacharyya coefficient.

5   Conclusions

In this paper, we propose the multi-mode global kernel weight method to track the object stably even if partial and full occlusion happens in videos taken from the moving camera. Also, we propose the occlusion detection method using coefficient variance of Battacharyya coefficient that is the multi-mode sub kernel unit. According to our experimental results, the proposed method improves the performance of occlusion detection than other traditional methods.

In the future work, we would like to analyze the stable occlusion heading methods according to the result of occlusion detection in order to design the appropriate algorithm to handle the real-time situation.

References

1. Collins, R.T., Lipton, A.J., Fujiyoshi, H., Kanade, T.: Algorithms for cooperative multisensor surveillance. Proceedings of the IEEE **89**(10), 1456–1477 (2001)
2. Meer, P.: Kernel-based object tracking. IEEE Transactions on pattern analysis and machine intelligence **25**(5) (2003)
3. Elgammal, A., Harwood, D., Davis, L.: Non-parametric model for background subtraction. In: Computer Vision—ECCV 2000, pp. 751–767 (2000)
4. Haritaoglu, I., Harwood, D., Davis, L.S.: W4: real-time surveillance of people and their activities. IEEE Transactions on Pattern Analysis and Machine Intelligence **22**(8), 809–830 (2000)
5. Jeyakar, J., Babu, R.V., Ramakrishnan, K.R.: Robust object tracking with background-weighted local kernels. Computer Vision and Image Understanding **112**(3), 296–309 (2008)
6. Khan, Z.H., Gu, I.Y., Backhouse, A.G.: Robust visual object tracking using multi-mode anisotropic mean shift and particle filters. IEEE Transactions on Circuits and Systems for Video Technology **21**(1), 74–87 (2011)
7. Ning, J., Zhang, L., Zhang, D., Wu, C.: Robust mean-shift tracking with corrected background-weighted histogram. IET Computer Vision **6**(1), 62–69 (2012)
8. Stauffer, C., Grimson, W.E.L.: Learning patterns of activity using real-time tracking. IEEE Transactions on Pattern Analysis and Machine Intelligence **22**(8), 747–757 (2000)
9. Toyama, K., Krumm, J., Brumitt, B., Meyers, B.: Wallflower: principles and practice of background maintenance. In: The Proceedings of the Seventh IEEE International Conference on In Computer Vision, 1999, vol. 1, pp. 255–261 (1999)
10. Wren, C.R., Azarbayejani, A., Darrell, T., Pentland, A.P.: Pfinder: Real-time tracking of the human body. IEEE Transactions on Pattern Analysis and Machine Intelligence **19**(7), 780–785 (1997)
11. Yilmaz, A., Javed, O., Shah, M.: Object tracking: A survey. ACM Computing Surveys (CSUR) **38**(4), 13 (2006)
12. Comaniciu, D., Meer, P.: Mean shift: A robust approach toward feature space analysis. IEEE Transactions on Pattern Analysis and Machine Intelligence **24**(5), 603–619 (2002)

OCSP Modification for Supporting Anonymity and High-Speed Processing in Vehicle Communication System

Jaewon Lee, Beom-Jin Choi, Seol-Hee Seon and Eun-Gi Kim

Abstract In inter-vehicular communication, the message is composed of user data and sender's certificate to verify message forgery. A Vehicle receiving message validates the certificate validity before verifying message integrity. If the certificate is valid, the vehicle verifies message integrity by using the certificate. However, this traditional communication method has disadvantage such as personal information leakage, traffic increase, and long processing time. This study proposes the method which modifies Online Certificate Status Protocol (OCSP) for supporting the high-speed certificate verification and the anonymity of vehicle.

Keywords Certificate · OCSP · Digital signature · CRL · Public key

1 Introduction

A Vehicle has developed a means of transportation with various additional functions through a combination of information and communication technology. Lately, vehicle communication system is composed of Vehicle-to-Vehicle (V2V) network and Vehicle-to-Infrastructure (V2I) network. Driver can safely drive by using the system. In the vehicle communication system, if security is not provided, driver is exposed to critical risk.

In V2V or V2I communication network, the traditional message transmission method using the certificate goes through procedure which validates the certificate. There are two cases for method checking the certificate.

J. Lee(✉) · B.-J. Choi · S.-H. Seon · E.-G. Kim
Department of Information and Communication Engineering,
Hanbat National University, Daejeon, Korea
e-mail: ljw198512@gmail.com

© Springer Science+Business Media Singapore 2015   607
D.-S. Park et al. (eds.), *Advances in Computer Science and Ubiquitous Computing*,
Lecture Notes in Electrical Engineering 373,
DOI: 10.1007/978-981-10-0281-6_86

The first case is to use Certificate Revocation List (CRL) [1]. A vehicle periodically receives the CRL from the Certification Authority (CA) and checks whether the certificate was revoked. The vehicle simply checks CRL, but can't identify in real time whether the certificate was revoked.

The second case is to use Online Certificate Status Protocol (OCSP) [3]. A vehicle sends OCSP Request and receives OCSP Response in real time, and checks whether the certificate was revoked. On the other hand, network traffic and vehicle's message processing time are increased by checking the certificate validity.

In this study, we assume that the certificate is validated by OCSP.

The traditional message transmission method using the certificate goes through with query process at OCSP Responder. In this case, a receiver's message processing time is delayed as much as the query process time. Also, sender's personal information may be exposed to an attacker because sender uses the certificate.

In this paper, to complement the above shortcoming, we propose the method using OCSP Response which contains the vehicle's public key instead of the certificate. The sender maintains anonymity by using OCSP Response instead of the certificate. The message is signed by using sender's private key, and message integrity is guaranteed.

The rest of the paper is organized as follows. In section 2, we explain the traditional message transmission method that uses the certificate. In section 3, we present message transmission method that uses proposed OCSP Response. Finally, we conclude this paper in Section 4.

2   Traditional Message Transmission Method Using Certificate

In inter-vehicular communication, the message is composed of user data, sender's certificate and sender's signature. The user data includes accident time, place and etc. Sender's certificate and sender's signature is used for verifying message forgery. In this case, all vehicles receiving message need to validate the certificate validity. If all vehicles validate the certificate validity, network traffic and vehicle's message processing time are increased by checking the certificate validity. Also, personal information leakage occurs by using the certificate.

2.1 Message Format

Fig. 1 shows the traditional message format using in inter-vehicular communication system. The user data and vehicle's certificate generate vehicle's signature using vehicle's private key.

User data (Accident Time, Place, etc.)	Vehicle's Certificate	Vehicle's Signature

Fig. 1 Message format using certificate

2.2 Operations

Fig. 2 shows the message transmission method using the certificate in the vehicle communication system.

1) The sender generates the signature by using sender's private key, user data and sender's certificate.
2) And the sender transmits message to receiver (vehicle or RSU).
3) The receiver transmits OCSP Request to OCSP Responder in order to validate the certificate validity in message.
4) OCSP Responder validates the certificate validity and transmits OCSP Response to receiver.
5) The receiver checks certificate status field of OCSP Response.
6) If the certificate is valid, the receiver checks message integrity by validating vehicle's signature using sender's public key in the certificate.
7) The receiver repeats process of 1 to 5, when receiving message each time.

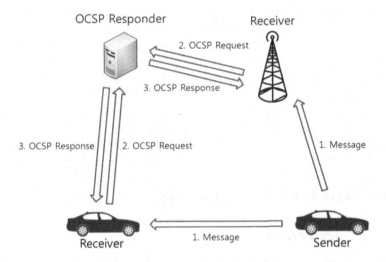

Fig. 2 Message transmission method using certificate

3   Message Transmission Method Using Proposed OCSP Response

In this paper, we assume that proposed method is used in certain section such as highway and all vehicles have OCSP Responder's public key. The sender uses a message containing OCSP Response which has the vehicle's public key instead of vehicle's certificate in order that a procedure for the certificate validation is minimized. Also, the anonymity of message is guaranteed because the message excludes the certificate which includes personal information.

3.1 Proposed OCSP Response Format

When entering the highway, the vehicle transmits OCSP Request to OCSP Responder. OCSP Responder validates the certificate validity through certificate's serial number in the OCSP Response. If the certificate is valid, OCSP Responder transmits OCSP Response including "status field in traditional OCSP Response", "vehicle's public key in the certificate" and etc., to the vehicle.

Fig. 3 shows OCSP Response format defined by RFC 6960. We propose that OCSP Response includes "(1) vehiclePublicKey".

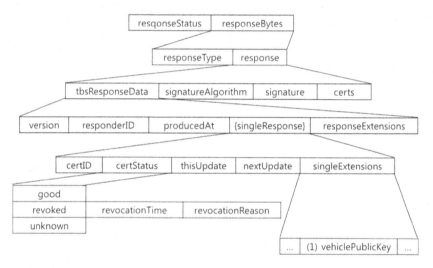

Fig. 3 Proposed OCSP Response including public key

3.2 Message Format

Fig. 4 shows the message format using proposed OCSP Response. The user data and OCSP Response including public key generate vehicle's signature using vehicle's private key.

User data (Accident Time, Place, etc.)	OCSP Response including Public Key	Vehicle's Signature

Fig. 4 Message format using proposed OCSP Response

3.3   Operations

Fig. 5 is the message transmission method using proposed OCSP Response in the vehicle communication system.

1) When entering the highway, all vehicles transmit OCSP Request to OCSP Responder.
2) OCSP Responder which receives the OCSP Request verifies the certificate status.
3) If certificate status is valid, OCSP Responder transmits proposed OCSP Response which is signed by OCSP Responder's private key to the Sender.
4) When sending message, the sender generates the signature by using vehicle's private key, user data, and proposed OCSP Response
5) And the sender transmits proposed message to the receiver (vehicle or RSU).
6) The receiver validates proposed OCSP Response using OCSP Responder's public key.
7) If OCSP Response is valid, the receiver verifies message integrity through the signature validation by using public key in proposed OCSP Response.

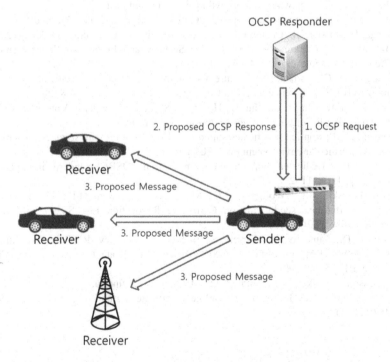

Fig. 5 Message transmission method using proposed OCSP Response

4 Conclusions

When a message is sent with a certificate, a procedure for the certificate validation occurs frequently and the sending time of message is delayed. Also, this method has disadvantage such as personal information leakage because of message including certificate.

In this paper, all vehicles validate their certificate at the beginning of entering the highway and obtain the OCSP Response including their public key. Accordingly, vehicle sending the message can guarantee anonymity because the certificate is not included in the message. Vehicle receiving the message can check that whether the message was maliciously forged and tampered through the signature in the message.

Acknowledgments This research was financially supported by the Ministry of Education (MOE) and National Research Foundation of Korea (NRF) through the Human Resource Training Project for Regional Innovation (No. 2013H1B8A2032154).

References

1. Wikipedia. https://en.wikipedia.org/wiki/Revocation_list
2. Wikipedia. https://en.wikipedia.org/wiki/Certificate_authority
3. Wikipedia. https://en.wikipedia.org/wiki/Online_Certificate_Status_Protocol
4. Viega, J., Messier, M., Chandra, P.: Network Security with OpenSSL. O'Reilly (2002)
5. Hubaux, J.-P., Capkun, S., Luo, J.: The Security and Privacy of Smart Vehicles. The IEEE Computer Society 49–55 (2004)
6. Laurendeau, C., Barbeau, M.: Secure anonymous broadcasting in vehicular networks. In: 32nd IEEE Conference on Local Computer Networks, pp. 661–668 (2007)
7. Lin, X., Lu, R., Zhang, C., Zhu, H., Ho, P.-H., Shen, X.: Security in Vehicular Ad Hoc Networks. IEEE Communications Magazine 88–95 (2008)
8. Rabadi, N.M.: Implicit certificates support in IEEE 1609 security services for wireless access in vehicular environment (WAVE), pp. 531–537. IEEE (2010)
9. Kenney, J.B.: Dedicated short-range communications (DSRC) standards in the United States, pp. 1162–1182. IEEE (2011)
10. Nowatkowski, M.E., Wolfgang, J.E., McManus, C., Owen, H.L.: The Effects of Limited Lifetime Pseudonyms on Certificate Revocation List Size in VANETS, pp. 380–383. IEEE (2010)
11. Cooper, D., Santesson, S., Farrell, S., Boeyen, S., Housley, R., Polk, W.: Internet X.509 Public Key Infrastructure Certificate and Certificate Revocation List (CRL) Profile, RFC 5280 (2008)
12. Santesson, S., Myers, M., Ankney, R., Malpani, A., Galperin, S., Adams, C., X.509 Internet Public Key Infrastructure Online Certificate Status Protocol – OCSP, RFC 6960 (2013)

A Study on the New Ethernet Communication Method Using Virtual MAC Address

Jae-Won Ahn, Seung-Peom Park, Kwon-Jeong Yoo and Eun-Gi Kim

Abstract A media access control address (MAC address) is a unique identifier assigned to network interfaces for communications on the physical network segment. MAC addresses are most often assigned by the manufacturer of a network interface controller (NIC) [1]. In traditional Ethernet network, there is a security vulnerability that allows attackers to easily generate forged frame because they can know MAC address of hosts in the subnet. In this study, we have designed new communication method using a virtual MAC address in order to complement a disadvantage of the traditional Ethernet communication. Hosts generate the frame hiding the MAC address of the header field and encrypting the data field to prevent attacks using forged frame in the subnet.

Keywords MAC address · Virtual MAC address · Ethernet · ECDH · AES · Network security

1 Introduction

People get a lot of benefits through the network communication systems. Its side effects such as data spill, malicious software, hacking, and abuse of security vulnerabilities are becoming serious problems, however [2]. In particular, it is becoming a serious problem that attacks using MAC address can sniff or modify a victim's packets in the subnet. The fundamental ways to prepare for these problems have been desired, but this solution has not been proposed. A MAC address is a unique identifier assigned to network interfaces for communications on the physical network segment. MAC addresses are used as a network address for most IEEE 802 network technologies, including Ethernet and Wi-Fi. MAC

J.-W. Ahn(✉) · S.-P. Park · K.-J. Yoo · E.-G. Kim
Department of Information and Communication Engineering,
Hanbat National University, Daejeon, Korea
e-mail: anjaewon7@gmail.com

© Springer Science+Business Media Singapore 2015
D.-S. Park et al. (eds.), *Advances in Computer Science and Ubiquitous Computing*,
Lecture Notes in Electrical Engineering 373,
DOI: 10.1007/978-981-10-0281-6_87

addresses are most often assigned by the manufacturer of a network interface controller (NIC) [1].

Hosts generate the frame including MAC address in the header to communicate with other hosts. The header of the frame is a field to indicate the source and destination of the frame, and generally is not encrypted. In other words, attackers can easily generate forged frame because they can know MAC address of hosts in the subnet. In traditional Ethernet network, there is a security vulnerability that allows attackers to easily sniff or modify victim's packets using forged frame in the subnet.

In this study, we have designed new communication method using a virtual MAC address in order to complement a disadvantage of the traditional Ethernet communication. By using ARP (Address Resolution Protocol), two hosts share a public key derived from ECDH (Elliptic Curve Diffie-Hellman) algorithm and do not share MAC address. Two hosts generate a secret key using the shared public key and ECDH algorithm.

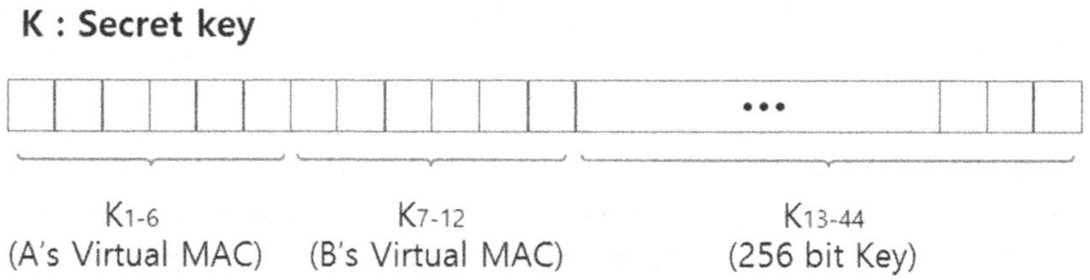

K : Secret key

K_{1-6} K_{7-12} K_{13-44}
(A's Virtual MAC)   (B's Virtual MAC)   (256 bit Key)

Fig. 1 Secret key format

Fig. 1 shows secret key of Host-A and Host-B. Most significant 6 bytes (K_{1-6}) of the generated secret key are used as a virtual MAC address of Host-A and next 6 bytes (K_{7-12}) are used as a virtual MAC address of Host-B. And next 32 bytes (K_{13-44}) are used as key of AES encryption algorithm to encrypt data field of the frame.

This paper is organized as follows: Section 2 describes the design of the proposed communication method and Section 3 describes the operation of the proposed communication method. Finally in Section 4 conclusions are made.

2   Designs

2.1   Frame Format

In this study, we have designed new communication method using a virtual MAC address in order to complement a disadvantage of the traditional Ethernet communication. Fig. 2 shows original Ethernet frame format and Fig. 3 shows proposed Ethernet frame format.

dest NIC's MAC addr	src NIC's MAC addr	type	data

Fig. 2 Ethernet frame format

dest virtual MAC addr	src virtual MAC addr	type	encrypted data

Fig. 3 Proposed Ethernet frame format

Original Ethernet frame includes target's MAC address in destination MAC address filed, and sender's MAC address in source MAC address filed. And data field of the frame is not encrypted. Attackers can easily generate forged frame because they can know MAC address of hosts in the subnet. If attackers successfully intercept the frame, they can see contents of data filed of frame and detect source and destination of the frame through MAC address in header of frame.

Proposed Ethernet frame includes most significant 12 bytes of the generated secret key in destination MAC address filed and source MAC address filed. The data field of the frame is encrypted by using next 32 bytes of the generated secret key as key of AES encryption algorithm. Attackers can't generate forged frame, because they can't know secret key. If attackers successfully intercept the frame, they can't decrypt data field of the frame because they can't know secret key. And they can't detect the source and destination of the frame because there is virtual MAC address in the header of the frame.

2.2 VARP

Logical IP address, as well as physical MAC address, is necessary in order to send and receive data between the hosts in the Internet. ARP is used to convert an IP address to a physical address such as an Ethernet address. There are two kinds of ARP packets called ARP_Request and ARP_Reply. ARP_Request is used to request the MAC address to all hosts in the subnet. ARP_Reply is used to notify the MAC address to a host which sends the ARP_Request. The MAC Address is saved in an ARP cache.

In this study, the ARP is used to share a public key derived from ECDH algorithm and do not share MAC address. This ARP is named VARP (Virtual Address Resolution Protocol). Fig. 4 shows original ARP packet format and Fig. 5 shows proposed VARP packet format.

Hardware type		Protocol type
(1) Hardware Length	Protocol Length	Operation
(2) Sender Hardware Address		
Sender Protocol (IP) Address		
(3) Target Hardware Address		
Target Protocol (IP) Address		

Fig. 4 ARP packet format

Hardware type		Protocol type
(1) Public Key Length	Protocol Length	Operation
(2) Sender Public Key		
Sender Protocol (IP) Address		
(3) Target Public Key		
Target Protocol (IP) Address		

Fig. 5 VARP packet format

Original ARP format includes (1) hardware length field, (2) sender hardware address filed and (3) target hardware address field. ARP is used to share MAC address with other hosts and saves a pair of MAC address and IP address in ARP cache.

Proposed VARP format include (1) public key length filed, (2) sender public key filed and (3) target public key filed. VARP is used to share public key with other hosts and generate a secret key using ECDH algorithm. Most significant 12 bytes of the generated secret key are used as a virtual MAC address of each host. Two hosts save a pair of virtual MAC address and IP address in VARP cache.

3   Operations

In this study, hosts generate the frame hiding the MAC address of the header field and encrypting the data field to prevent attacks using forged frame in the subnet. VARP is used to share public key with other hosts and generate a secret key using the shared public key and ECDH algorithm. Most significant 12 bytes of the generated secret key are used as a virtual MAC address of each host. And next 32 bytes are used as key of AES encryption algorithm to encrypt data field of the frame.

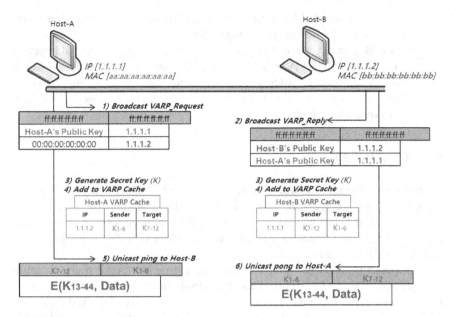

Fig. 6 Proposed Communication Method

Fig. 6 is an example of the operation of Host-A to communication with Host-B.

1) Host-A sends VARP_Request to the all hosts in the subnet to find out Host-B's Public Key. VARP_Request includes Host-A's public key in sender public key filed, and Host-A's IP address (1.1.1.1) in sender IP address filed and Host-B's IP address (2.2.2.2) in target IP address filed. Target public key filed is empty.

2) Host-B confirms that the target's IP address (2.2.2.2) of receiving VARP_Request is the same as its own IP address. Afterwards, Host-B sends VARP_Reply to the all hosts in the subnet. VARP_Reply includes Host-B's public key in sender public key filed and Host-B's IP address (2.2.2.2) in sender IP address filed and Host-A's public key in target public key filed and Host-A's IP address (1.1.1.1) in target IP address filed. Other hosts, except Host-B, drop the VARP_Request because the target IP address (2.2.2.2) is not the same as its own IP address.

3) Two hosts have exchanged public keys. They generate secret key using shared public key and ECDH algorithm.

4) Most significant 6 bytes (K_{1-6}) of the generated secret key are used as a virtual MAC address of Host-A and next 6 bytes (K_{7-12}) are used as a virtual MAC address of Host-B. Two hosts save a pair of virtual MAC address and IP address in VARP cache.

5) Afterward, Host-A creates packets on reference to a pair of virtual MAC address and IP address of Host-B in VARP cache. Destination MAC address is K_{7-12} and source MAC Address is K_{1-12}. The data field of the

frame is encrypted by using a low byte (K13-44) of the generated secret key as key of AES encryption algorithm.

6) Host-B creates packets on reference to a pair of virtual MAC address and IP address of Host-A in VARP cache. Destination MAC address is K1-6 and source MAC Address is K7-12. The data field of the frame is encrypted by using K13-44 of the generated secret key as key of AES encryption algorithm.

4   Conclusions

In traditional Ethernet network, there is a security vulnerability that allows attackers to easily generate forged frame because they can know MAC address of hosts in the subnet. In this study, we have designed new communication method using a virtual MAC address in order to complement a disadvantage of the traditional Ethernet communication.

In proposed Ethernet network, attackers can't generate forged frame, because they can't know secret key. If attackers successfully intercept the frame, they can't decrypt data field of the frame because they can't know secret key. And they can't detect the source and destination of the frame because there is virtual MAC address in the header of the frame.

Acknowledgments This research was financially supported by the Ministry of Education (MOE) and National Research Foundation of Korea (NRF) through the Human Resource Training Project for Regional Innovation (No. 2013H1B8A2032154).

References

1. Wikipedia. https://en.wikipedia.org/wiki/MAC_address
2. Lockhart, A.: Network Security Hacks, 2nd Edition. O'Reilly (2006)
3. Garfinkel, S.: Web Security and Commerce. O'Reilly (1997)
4. Clarke, J., Dhanjani, N.: Network Security Tools. O'Reilly (2005)
5. Hall, E.: Internet Core Protocols. O'Reilly (2000)
6. Forouzan, B.A.: Data Communication and Networking, 3rd edn. McGraw-Hill (2003)
7. Forouzan, B.A.: TCP/IP Protocol Suite, 4th edn. McGraw-Hill (2010)
8. Stevens, W.R.: TCP/IP Illustrated, Volume 1: The Protocols. Addison Wesley (1993)
9. Plummer, D.C.: An Ethernet Address Resolution Protocol, RFC 826 (1982)
10. Black, I.F., Seroussi, G., Smart, N.P.: Advances in Elliptic Curve Cryptography. Cambridge University Press (2005)
11. Hankerson, D., Menezes, A., Vanstone, S.: Guide to Elliptic Curve Cryptography. Springer (2003)
12. Chandra, P., Messier, M., Viega, J.: Network Security with OpenSSL. O'Reilly (2002)
13. Doraswamy, N., Harkins, D.: IPSec: The New Security Standard for the Internet, Intranets, and Virtual Private Networks, 2nd. edn. Prentice Hall PTR (2003)
14. Rosen, R.: Linux Kernel Networking: Implementation and Theory. Apress (2014)

A Hybrid Prediction Model Integrating FCM Clustering Algorithm with Supervised Learning

Seokhwan Yang, Jieun Choi, Sanghoon Bae and Mokdong Chung

Abstract Since most prediction models are still using the algorithm based on supervised learning, they are flawed by various weaknesses, such as the issue of preprocessing the training data, and difficulties in applying them to new patterns other than the trained data. On the other hand, the prediction model that uses only unsupervised learning is flawed by its difficulty in analyzing the result of prediction because no information about the data is given as to when learning is conducted. In this paper, we propose a hybrid prediction model which integrates the FCM clustering algorithm belonging to unsupervised learning with the features of supervised learning that lead to collection of target values. The proposed hybrid prediction model conducts automatic classification without external interference, detects target values inside the data alone, and applies them to deriving numerical prediction results. Thus the proposed model possesses the strong features of both supervised learning and unsupervised learning. We performed a prediction using the actual measurement data in the ITS, and confirmed the accuracy of the result. We expect that the proposed hybrid prediction model may contribute to enhancement of automation standards in various intelligent systems.

Keywords FCM clustering · Regression analysis · Prediction model · ITS (Intelligent Transportation Systems)

1 Introduction

Data mining is a method which analyzes the collected data and detects hidden patterns. The pattern information detected via data mining is widely applied for

S. Yang · M. Chung(✉)
Department of Computer Engineering, Pukyong National University, Busan, Korea
e-mail: {seokhwan,mdchung}@pknu.ac.kr

J. Choi · S. Bae
Department of Spatial Information Engineering, Pukyong National University, Busan, Korea
e-mail: {620jieun,sbae}@pknu.ac.kr

© Springer Science+Business Media Singapore 2015
D.-S. Park et al. (eds.), *Advances in Computer Science and Ubiquitous Computing*,
Lecture Notes in Electrical Engineering 373,
DOI: 10.1007/978-981-10-0281-6_88

making predictions in various fields. In particular, the information and the patterns hidden within the data have important meanings because they may have an effectuality which has not yet been identified by the existing methods of analysis. Research on prediction models have mobilized a variety of methods including statistics, artificial intelligence, and machine learning. However, the majority of prediction models are still relying on algorithms based on supervised learning.

Although prediction models based on supervised learning have high accuracy due to the characteristic features of supervised learning that repeat correcting weighted values until the target value is reached, at the same time they have several problems. First, since a vast amount of classified data is necessary for learning, the prediction models based on supervised learning require the preprocessing of the collected data and the analysis of the manager. Depending on the reliability of the training data, the functionality of the prediction model varies greatly, and only the trained data have to be used, making it difficult to use the data with new patterns. Another weakness of prediction models based on supervised learning is the difficulty in gradual learning for the data that are input in real time [1].

For these reasons, prediction models applying unsupervised learning have been studied under conditions where learning is conducted without fixing the target values. However, the prediction model that uses only unsupervised learning is flawed by its difficulty in analyzing the result of prediction because no information about the data is given as to when learning is conducted [2].

This paper proposes a model which makes predictions by combining the FCM (Fuzzy C-Means) clustering algorithm that belongs to unsupervised learning with the concept of target values that is the characteristic feature of supervised learning. The target value used in the proposed model is detected from inside the input data to maintain the advantage of unsupervised learning for which the manager's preprocessing is unnecessary. In the proposed prediction model, the time interval to be predicted is input as the initial value, and then the learning data are classified via FCM clustering algorithm. Finally, we measure the error pattern between the classification result and the target value detected from input data, and the prediction results are deduced.

The remainder of this paper is organized as follows. Section 2 surveys the related work and the theoretical backgrounds. Section 3 introduces the proposed Hybrid FCM clustering model. In section 4, we apply the proposed model to a comparative analysis of the experimental result of predicted travel time in variance in precipitations and the experimental result of the Back-Propagation (BP) model. Finally, section 5 presents our conclusion and the direction of our future research.

2 Related Work

2.1 FCM (Fuzzy C-Means) Clustering Algorithm

Clustering is a method of classifying given data by comparing them to the pre-fixed class until a class closest to the fixed class is found. The FCM clustering algorithm is a data-classifying algorithm that uses the Fuzzy division technique which classifies data points according to the membership degrees. The membership function U of the FCM clustering algorithm has elements that have values ranging between 0 and 1, and the sum of membership values for the data set is always 1 [3,4]. The cost function of the FCM algorithm has the following form:

$$J(u_{ik}, v_i) = \sum_{i=1}^{c} \sum_{k=1}^{n} u_{ik}^m (d_{ik})^2, \quad (1 \le m < \infty, v_i = \{v_{i1}, v_{i2}, \cdots, v_{ij}, \cdots, v_{iL}\})$$

$$d_{ik} = d(x_k - v_i) = \left[\sum_{j=1}^{L} (x_{kj} - v_{ij})^2 \right]^{\frac{1}{2}}, \quad v_{ij} = \frac{\sum_{k=1}^{n} (u_{ik})^m x_{kj}}{\sum_{k=1}^{n} (u_{ik})^m}, \quad u_{ik} = \frac{1}{\sum_{j=1}^{c} \left(\frac{d_{ik}}{d_{jk}} \right)^{\frac{2}{m}-1}}$$

u_{ik}: the membership degree of k-th data of x_k that belongs to i-th cluster

v_i: the centroid vector of i-th cluster

m : the parameter that controls the degree of fuzziness in the classifying process. In general $m = 2$

d_{ik}: distance between k-th data x_k that belongs to the cluster and the central vector v_i of the i-th cluster

$J(u_{ik}, v_i)$: cost function of FCM clustering algorithm

2.2 Numerical Prediction Using the FCM Clustering Algorithm

Since the purpose of the FCM clustering algorithm is classification in general, it is not used for numerical predictions. However, the FCM clustering algorithm has an advantage because it is capable of classifying various situations that have not been defined in advance, given that it makes use of the similarity of each piece of data.

Won Shik Park et al., (2009) used the FCM clustering algorithm to develop the method of classifying traveling characteristics of traffic flow of the TCS (Toll Collection System) data [5]. Driving patterns are classified and representative values are selected first by classifying the TCS data using the FCM clustering algorithm, and then calculating the course transit time considering the hourly driving patterns on highways. The course transit time is calculated by running a PIFAB (Progressive Iterative Forward and Backward) [6] search on the representative values of the selected sections and also by a comparative evaluation of the representative values and errors.

However, the experimental results of this method include an error ratio of 10% and an error time of more than 5 minutes, which may be attributed to the fluctuations of error ratio owing to the size of each cluster and the distribution of data points when only the FCM clustering algorithm is used alone.

3   Design of A Hybrid Prediction Model

3.1   Process of the Proposed Prediction Model

The FCM clustering algorithm is a simple and effective algorithm that repeats arithmetic operations to minimize the objective function, and thus it seeks out the desired membership function. Since the FCM is an algorithm designed for automatic classification, its advantage is that it does not require preprocessing of the collected data and the manager's analysis. However, it is unable to make precise predictions on particular numerical values because classification results are represented in the forms of the clusters that have been defined to have section scopes [2].

To solve these problems, this paper makes numerical predictions by using the centroid values of the clusters deduced by the result of the FCM clustering classification. To correct the errors generated by the use of the cluster's centroid values, the concept of target value in supervised learning is applied to the FCM clustering algorithm. In the Markov Process, the status of the system at the present point t is influenced only by the past point $t-1$.

Therefore, various prediction models predict the status of the $t+1$ point using the data of the point t. Unlike the pure supervised learning, the proposed model does not receive the input of target values. Therefore, it compares the data at point $t-1$ with the data at point t in order to compare the accuracy of the predicted results. The determination coefficients are calculated by conducting a regression analysis on the correlations of the data contained in each cluster (Step 6), and the interaction formula thus derived is applied to the data at point t and the status at point $t+1$ is predicted. Fig. 1 shows the entire process of the proposed Hybrid FCM clustering.

[Step 1~4] Perform FCM clustering and also set time intervals to predict.

[Step 5] Select the centroid vector of the classified clusters as the default to predict. Set all data at point $t-1$ and t as the criteria. Here t represents the present time.

[Step 6] Perform the regression analysis using the data point of each classified cluster, and calculate the regression determination coefficients. We use the input vector at $t-1$ point and the traffic time data at t point in the regression analysis.

[Step 7] Apply input data to the regression equation derived from Step 6.

[Step 8] Finally, the proposed algorithm obtains the corrected prediction value.

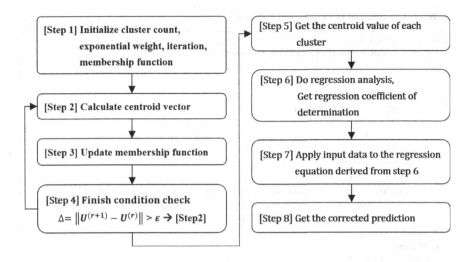

Fig. 1 Process of Hybrid FCM clustering algorithm

3.2   Algorithm of the Proposed Model and its Verification

To verify the accuracy of the proposed model, we use the indexes of statistics-based RMSE (Root Mean Square Error) and MAPE (Mean Absolute Percentage Error). RMSE is a method of measuring the difference between the estimated value or the value predicted by the model and the value observed in the actual environment. It is a significant intuitive method of evaluating the accuracy of the approximate model [7].

Calculation formula of RMSE is as follows:

$$RMSE = \sqrt{\frac{1}{n}\sum_{i=1}^{n}[y(x_i) - \hat{y}(x_i)]^2} \tag{1}$$

where $y(x_i)$ is the actual function value at x_i, $\hat{y}(x_i)$ is the approximate function value at x_i, and n means the number of experimental data to evaluate the approximate model.

MAPE is the method of measuring the accuracy of the method in calculating the predicted value for time series values. The calculation formula is as follows:

$$MAPE\ (\%) = \left(\frac{1}{n}\sum_{i=1}^{n}\left|\frac{y(x_i) - \hat{y}(x_i)}{y(x_i)}\right| \times 100\right) \tag{2}$$

where $y(x_i)$ is the actual function value at x_i, $\hat{y}(x_i)$ is the predicted value at x_i, n is the number of experimental data. In general, the smaller the values of RMSE and MAPE, the accuracy of the model is considered to be greater. Table 1 shows the proposed algorithm in detail.

Table 1 Algorithm of Hybrid FCM clustering

```
set time_unit_length
initialize cluster_count, exp_weight, iteration, membership_ft

Loop
    get input_data
    // Stage 1: FCM Clustering
    While (eps > (membership_ft(r) - membership_ft(r-1)))
        calculate centroid_vector; update membership_ft; set result_fcm
    End While
    // Stage 2: Regression Analysis
    Loop (i = 1 to cluster_count)
        dataset_list[i] = getlist(result_fcm, i); check correlation_coefficient;
        set independent_variable; execute regression_analysis; set regression_equ;
        re[i] = regression_eq(dataset_list);
    End Loop
End Loop
// Stage 3: Prediction
index = get cluster_number(input_data)
predicted_value = revise(index, input_data)
Function revise(index, input_data)
    Loop (i = 1 to count(input_data))
        result = multiple(index, input_data, centroid_vector[index], re[index])
    End Loop
    return result
End function
```

4   Implementation and Evaluation

This paper proposes a model which makes numerical prediction by correcting the error between the centroid values of the clusters and the target values detected from the input data themselves. To test its features, the proposed model is applied to the traffic condition classification function of an intelligent transportation system.

For the experimental data, Hanbat Street of Daejeon City, Korea was set for the scope, and data were collected on travel time, traffic volume, and precipitation for the five links of the Hanbat Street. The travel time collected by RSE (Road-Side Equipment), the traffic volume collected by VDS (Vehicle Detection System) and KMA's (Korea Meteorological Administration) precipitation data from June 20, 2011, to July 18, 2011, were used as the training data for the traffic time prediction model. The data collected from May 18 to May 20, 2011, were used as the test data to test the prediction model. In addition, a comparative analysis was conducted on the result obtained by the proposed model and the prediction result offered by the widely used conventional BP algorithm. The experiment was conducted on a PC equipped with Intel i5-2400 3.1GHz CPU, 4GB RAM and the operating system was Microsoft Windows 7(64bit).

4.1 Results Predicted by Using Back-Propagation Algorithm

In this experiment, first, the prediction was made by using the BP algorithm which is widely applied in the conventional studies. Fig. 2 and Fig. 3 show the result of the experiment conducted using the BP algorithm. The prediction value is similar to the actual measurement value, and the accuracy is especially high on sunny days. The result of accuracy analysis shows that RMSE was 0.3731 and MAPE was 5.3735% on the rainy day (May 20, 2011), whereas on the sunny day (May 18, 2011), RMSE was 0.2102 and MAPE was 3.9282%. The BP method yielded a high prediction accuracy on the whole, and the error rate was big for the peak time.

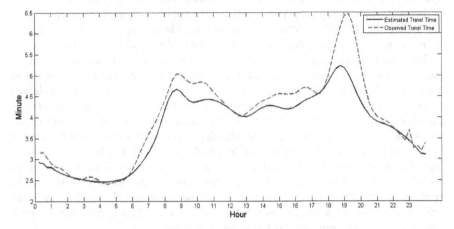

Fig. 2 Estimated travel time and observed travel time on a rainy day compared.

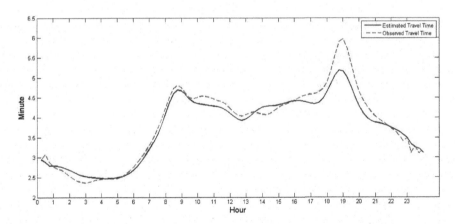

Fig. 3 Predicted travel time and observed travel time on a sunny day compared.

However, the BP method is flawed by the weakness that gradual improvement of learning is difficult, and requires time consuming preparatory work for preprocessing. Another weakness is that if a new pattern of data, which has not been included in the training data at the learning stage, is input, prediction becomes difficult [6].

4.2   Hybrid Fuzzy C-Means (FCM) Clustering

4.2.1   Stage 1 of Hybrid FCM: Implementation of FCM Clustering

To improve on the weakness of BP described above, the FCM clustering algorithm was applied in this experiment. The BP method is liable to fail to approximate to the correct result in the learning process [6], the FCM clustering is a stable algorithm whose approximation is always confirmable. Furthermore, even in the case where new patterns of data that have not been input at the learning state are input, the FCM clustering is capable of classifying them by using the information of the cluster that is situated nearest to the input data point.

The data used in this research is the 15-minute time series data, and one day is composed of 96 data. Since the time points to be predicted are the same time points a day later, the time units to be predicted by the Hybrid FCM clustering model are set at 96. The centroid value of each cluster is the data at $t-1$ point, and the error is calculated by comparing it with the data at t point because it is the result of the renewal of the centroid vector and membership function that have been carried out to the final $t-1$ point. When the section to which the related data belong is deduced instead of using its exact value and its centroid value for prediction, there is a high probability of an error smaller than the radius of the cluster. Fig. 4 shows the comparison between the centroid value of the result of the FCM clustering classification and the actual measurement value. The prediction value for the peak time differs greatly from the actual measurement value.

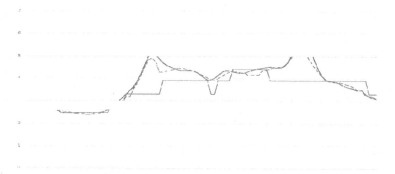

Fig. 4 The comparison between the centroid value of the result of the FCM clustering classification and the actual measurement value

4.2.2   Stage 2 of Hybrid FCM: Correction of the FCM Clustering Result

To correct the result, regression analysis is performed on the data contained in each cluster as a result of FCM clustering, and the prediction value is deduced by applying the regression coefficient of determination.

Table 2 Results of regression analysis on seven clusters

Cluster Variables	1	2	3	4	5	6	7
y	0.981	0.288	0.703	1.119	1.911	-1.075	1.997
x1(peak)	0	-0.001	0.000	0.042	0	-0.134	0
x2(time1)	0.588	0.726	0.744	0.628	0.189	0.582	0.229
x3(vol.1)	0.001	0.001	0.000	0.001	0.001	0.006	0.000
x4(rain1)	-0.016	-0.017	-0.011	-0.007	-0.001	-0.022	0.009
x5(rain2)	0.000	0.010	0.021	0.011	0.006	-0.003	0.002

$$\hat{y} = 0.981267207 + 0.588580577\,x_2 + 0.001065903\,x_3 - 0.01620481\,x_4 + 0.000367157\,x_5 \quad (3)$$

Table 2 shows clusters 1, 5, and 7 indicate that the peak time is an independent value in which the regression relations with the value to be predicted do not apply. The regression equation applies to all the other values except for the peak time. Equation 3 shows the regression equation for the first cluster.

4.2.3   Stage 3 of Hybrid FCM: The Predicted Results

Table 3 compares the results predicted in the number of clusters and BP's prediction results. The results predicted by the proposed Hybrid FCM clustering model are similar to the BP's results. It is worth noting that the predictions made by the proposed one for the seven clusters for the rainy day were even more accurate than the BPs. The experimental result proved that the proposed model not only has the advantages of unsupervised learning but is also suitable for numerical prediction.

Table 3 The comparison between prediction results for the number of clusters

Sunny Day

Clusters	Error Average		RMSE		MAPE	
	Hybrid FCM	BP	Hybrid FCM	BP	Hybrid FCM	BP
5	0.1794	0.1595	0.2312	0.2102	4.4208	3.9282
6	0.1759	0.1595	0.2239	0.2102	4.3463	3.9282
7	0.1619	0.1595	**0.2100**	**0.2102**	4.0360	3.9282
8	0.1695	0.1595	0.2240	0.2102	4.1757	3.9282
9	0.1685	0.1595	0.2215	0.2102	4.1592	3.9282

Table 3 (continued)

Rainy Day

Clusters	Error Average		RMSE		MAPE	
	Hybrid FCM	BP	Hybrid FCM	BP	Hybrid FCM	BP
5	0.2570	0.2429	0.3837	0.3731	5.6592	5.3735
6	0.2445	0.2429	0.3732	0.3731	5.3884	5.3735
7	**0.2385**	**0.2429**	**0.3567**	**0.3731**	**5.3663**	**5.3735**
8	**0.2424**	**0.2429**	**0.3696**	**0.3731**	5.3871	5.3735
9	0.2546	0.2429	0.3852	0.3731	5.6344	5.3735

4.3   Analysis of the Function of the Proposed Model

Table 4 shows the comparisons between the proposed one and other methods.

Table 4 Comparison of the BP, FCM and Hybrid FCM

Compare Item	BP	FCM	Proposed Hybrid FCM
Error range(min.)	0.16~0.24	0.24~5.39	0.16~0.26
Preprocessing	Required	Not needed	Not needed
Preliminary data analysis	Required	Not needed	Not needed
Incremental learning	Impossible	Possible	Possible
Automatic classification	additional rules	Possible	Possible
Numerical prediction	High accuracy	Low accuracy or difficulty	High accuracy

The proposed model has four major advantages over the conventional method. First, the model performs learning and prediction through automatic classification that utilizes the FCM clustering algorithm. Second, because it uses the clustering algorithm, it is capable of gradual learning by making use of input data. Third, it is capable of broadening the application scope of the FCM algorithm by adding a numerical prediction function to the conventional FCM. Fourth, it does not require the user's data control because it performs automatic calculations using only input data.

5   Conclusion

In this paper, we proposed a Hybrid FCM model which combines the high accuracy rates of supervised learning and unsupervised learning. It is capable of automatic classification by deducing the learning target value, which is the characteristic feature of supervised learning, and then by applying it to the FCM clustering algorithm of unsupervised learning.

We applied the proposed Hybrid FCM clustering model to an experiment in an intelligent transportation system. The experimental result indicated that the overall results obtained by the proposed model are similar to those of the BP algorithm, and under certain conditions, it produced even higher prediction accuracies.

The application of the proposed model verified that its accuracy is similar to that of the BP algorithm that has been widely used. At the same time, it also demonstrated that the proposed model can solve the problem of the conventional prediction model based on supervised learning which requires a time consuming pretreatment process that accompanies heavy work load. Moreover, the proposed model is capable of taking advantage of the prediction model based on unsupervised learning, which can automatically classify the input data.

We expect that the proposed model may also be applied to other fields as well for two reasons. First, the information required for learning and prediction is extracted from the data sets which are continuously input at $t-1$ and t points without the user's data control and pretreatment process. Secondly, the proposed model has a data-oriented process that performs error correction and prediction through regression analysis at the classified result data sets.

Although the aim of the experiment was the real time learning and prediction, the consumption of the system resources increased when the regression analysis was performed on each cluster in real time. Therefore, our future research will focus on an improved Hybrid FCM clustering model which minimizes the consumption of system resources and applies the real time data which is collected continuously to learning.

References

1. Shmueli, G., Patel, N.R., Bruce, P.C.: Data Mining For Business Intelligence: Concepts, Techniques, and Applications in Microsoft Office Excel® with XLMiner® 2nd. John Wiley & Sons, October 26, 2010
2. Oh, S.-K.: Computational Intelligence by Programming focused on Fuzzy, Neural Net-works, and Genetic Algorithms. Naeha Publishing Co. (2002)
3. Park, W.-S., Kim, D.-K., Yang, Y.-K.: Driving Characteristics Classification of TCS Data Based on Fuzzy c-means Clustering Algorithms. In: The 31st Conf. of the Korea Information Processing Society (KIPS 2009), April 23-24, pp.1021–1024 (2009)
4. Bezdek, J.C., Ehrlich, R., Full, W.: FCM: The Fuzzy c-Means Clustering Algorithm. Computers & Geosciences **10**(2-3), 191–203 (1984)
5. Namkoong, S.: Progressive Iterative Forward and Backward (PIFAB) Search Method to Estimate Path-Travel Time on Freeways Using Toll Collection System Data. Journal of Korean Society of Transportation **23**(5), 147–155 (2005)
6. Kim, D.S.: Neural Networks Theory and Applications, 17th edn. Jinhan M&B (2006)

Environmental Monitoring Over Wide Area in Internet of Things

Sun Ok Yang, Joon-Min Gil and Sungsuk Kim

Abstract In this paper, we have interests on one application for IoT: Environmental Monitoring. Lots of smart objects deploys at some position, gathers data periodically. Several kinds of techniques should support for the application to do the job. In this paper, we focus on two among them: local storage management algorithm and periodic data gathering by movable data collector.

Keywords Internet of Things (IoT) · Smart objects · Data gathering

1 Introduction

The basic idea of *Internet of Things (IoT)* is the pervasive presence around us of a variety of things or objects. They can be Radio-Frequency Identification(RFID) tags, sensors, actuators, mobile phones and so on, which, through unique addressing schemes, are able to interact with each other and cooperate with their neighbors to reach common goals [1].

This innovation will be enabled by the embedding of electronics into everyday physical objects, making them "smart" and letting them seamlessly integrate within the resulting infrastructure. Generally IoT builds on three pillars, related to the ability of smart objects to: (i) be identifiable (anything identifies itself), (ii) to communicate (anything communicates) and (iii) to interact (anything interacts) – either among themselves, building networks of interconnected objects, or with end-users or either entities in the network [2].

S.O. Yang(✉) · S. Kim
Department of Computer Science, Seokyeong University, Seongbuk-gu, South Korea
e-mail: soyang9149@gmail.com, sskim03@skuniv.ac.kr

J.-M. Gil
School of Information Technology Engineering, Catholic University of Daegu,
Gyeongsan-si, South Korea
e-mail: jmgil@cu.ac.kr

© Springer Science+Business Media Singapore 2015
D.-S. Park et al. (eds.), *Advances in Computer Science and Ubiquitous Computing*,
Lecture Notes in Electrical Engineering 373,
DOI: 10.1007/978-981-10-0281-6_89

IoT related technologies can open up new business opportunities; there are several kinds of application fields and market sectors where IoT solutions can provide competitive advantages over current solutions. One of them is environmental monitoring which is our main consideration in this paper. In this case, a key role is played by the ability of sensing, in a distributed and self-managing fashion, natural phenomena and processes (e.g., temperature, wind, rainfall, river height), as well as to seamlessly integrate such heterogeneous data into global applications (see Fig.1). The vast deployment of miniaturized devices may enable access to critical areas, whereby the presence of human operators might not represent a viable option.

Fig. 1 Environmental monitoring by using small size devices

In this kind of application, each object may deploy in a position, gather interesting data periodically, store them locally, and deliver them when needed. Thus, several kinds of techniques should support for the application to do the job. In this paper, we focus on two among them: local storage management algorithm and periodic data gathering by movable data collector. We name our system *EMoMC* (*Environmental MOnitoring by Movable data Collector*).

2 EMoMC

2.1 Components in System

In the application, there are lots of smart objects. They have limited size of local storage and can connect directly with some nodes within communication range. Among them, some objects are selected as *Connecting Node (CN)* based on their position. *CN*s have role of representing node for some objects N_i ($0 \leq i \leq NU$(all sensors)) within their region, and thus get messages from N_i.

The process to select *CN*s is simple. Let us think that there is a minimum bounding rectangle containing all sensor nodes. All sensor nodes are divided into two parts according to their X-coordinate and then make two the minimum bounding rectangles. Each rectangle contains half the number of objects. And each object group is again divided according to Y-coordinate and two minimum bounding rectangles are generated (see Fig. 2). This partitioning will be ended when one of both conditions is satisfied:

- in a region, there is only one or zero object node;
- the width or height of each region is smaller than object's communication range ×2.

When each object gets environmental data, it just stores them locally instead of delivering to the server or operator. To gather all data in the network, a *Mobile Collector (MC)* will move along the predetermined path to a CN_j and collects data from objects via CN_j.

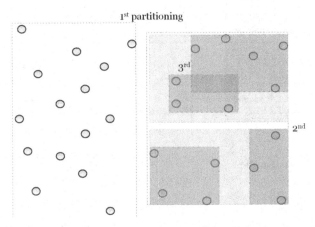

Fig. 2 Selection of CNs from objects

2.2   *Local Storage Management Algorithm*

Generally, an object node has limited size of storage (1-100Kbytes) and thus the free space becomes insufficient over time although it gathers data periodically. In some research approaches, authors tried to reduce the data size by making those data inaccurate (approximation, delayed prediction, in-network aggregation, etc). However, we will devise an algorithm in storage-efficient way, preserving data accuracy.

When an object measures and gets an environmental datum, it has to store it as a pair of (*Time, Value*). To reduce storage usage, at first the storage is assumed to be divided into two parts: *Correct Value Area (CV)* and *Deviation Value Area (DV)*. In *CV*, the original measured data are stored in form of (*Time, Value, Pointer*) where *Value* is the correct value and *Pointer* indicates the related *DV* row; the detailed explanations about it will be shown in next subsection. *DV* contains the deviation values between the current data and the previous data. One *DV* row is N bits and N' bits are allocated for one deviation value ($N > N'$). Thus a *DV* row can contain $\lfloor \frac{N}{N'} \rfloor$ deviation values (see Fig. 3).

In the figure, N' bits play a role of threshold to determine whether the current gathered value (v_c) at period T_i is stored into CV or DV. That is, a deviation value (d) is calculated by subtracting the previous period value (v_p) from v_c and check whether d can be expressed by N' bits; if not possible, the current value(v_c) is added into CV in form of (T_i, v_c, null); otherwise, only the deviation value d is inserted into a DV row. The detailed explanations will be shown in next subsection.

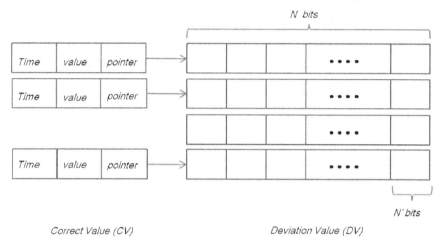

Fig. 3 local storage

2.3   MC's Data Collecting Algorithm

After initial setup, all objects start on their roles. When $A\%$ of local storage ($0 \leq A \leq 100$) in a object node is full, the node thinks that it has to notify the MC of the fact before the storage is completely full. To do so, it will send an urgent message U_{msg} to its CN containing the information {src, $dest$, $TYPE$, $Time$, TTL}, which means the source ID, the destination ID, the message Type (A or B), the current period, and $Time To Live$. B Type field is more urgent than A; that is, B Type means that $B\%$ is full and $B > A$.

When a CN gets the message, it first stores the message locally. And it also changes some field values in the message, and then propagates it to its neighbor CNs. While TTL is positive, the object nodes receiving the message decrease TTL by 1 and also deliver it to the destination node.

Thus the message can be sent from the general object nodes or a CN, and so the data contained in the message are different according to the sender. In case of a general sensor node, src is the ID of itself, $dest$ is its CN, and TTL is the hop number between itself and its CN. Otherwise, in case of CN, src is also the ID of CN itself, $dest$ are the ID set of CNs who are neighbor CNs and have to get this message. Therefore, TTL is the maximum value of the hop number between itself and them.

MC initially starts to move to the first CN_0, CN_1, and CN_2... When it meets a CN_i, it collects data from both CN_i itself and all sensors who knows CN_i as their *CN*. And the *MC* has to determine next target *CN*. In this case that CN_i already got one more U_{msg} messages from the other CN_k ($k \geq 1$), *MC* calculated the available time (T_r) from the messages as following:

$$T_r = \frac{100 - \text{UsageA}}{S} - \text{Time} \qquad (1)$$

where *S* is the average *CV* row size, *Time* is obtained from the U_{msg}, and *UsageA* means the used buffer volume (%) when a sensor sends *A*-type message. Of course, in case of *B*-Type message, *UsageB* is used instead of *UsageA*. The fraction in (1) is to calculate the estimated remaining time until the buffer is completely full.

After calculation, *MC* stores the messages locally and moves to the next *CN*. At there, it also considers the stored messages in determining next target.

3   Simulation Works

In this paper, we have purchased the weather data such as temperature, humidity, wind direction/speed, sunshine and rain for Seoul, Korea every 30 minutes from 2005 to 2009 from Korea Meteorological Administration (KMA) [3]. Among those data, temperature, humidity and daylight exhibited are comparatively regular data but wind direction/speed is not. We make a simulator by using C# on Windows 7. The simulation area is 500 × 500, the communication range of a sensor is 3, default number of sensors is 300.

Due to the limitation of paper, we just mention only one result.

3.1   Local Storage Management Algorithm

In the first experiments, *MC* moves statically; that is, it just visits CN_0, CN_1 and CN_2, and there is no route change. Fig. 4 shows the effects of both *N'* size and data regularity on the data size generated during a day. In the figure, 32 bits and 64 bits means the size of one *DV* row. And X-axis shows the number of bits assigned for storing one deviation value (i.e., *N'* bits) and Y-axis means the average data size gathered and stored during a day.

Our proposed algorithm is to enhance local storage usage by storing deviation values, and thus we first show how much our algorithm decreases the size of local data. If all data are stored into *CV*, total data is 10752 bytes where `sizeof(int)` is 32 bits. In the figure, in case of regular data (a) such as temperature, the average number of daily generated *CV* was 7.6 rows when $N' = 3$, implying that most gathered data were stored in *DV* instead of *CV*. If the data volume obtained from the experiments is compared with the original data size (= 10752 bytes) per a day, it is only 29-44%, and it is evident that a significant amount of buffer was saved. Meanwhile, in case of $N = 32$ bits, the performance decreased along with increasing N'. This is because many deviation values had to be stored in one CV, and DV rows had to be frequently created due to small N value. In case of weak regular data (b), its buffer usage is 40-59% compared with the original one.

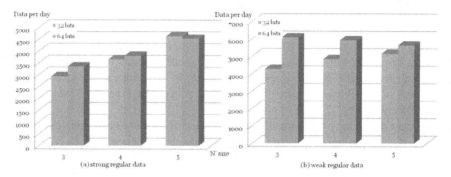

Fig. 4 Effects on threshold

4 Conclusion

In this paper, we have interests on a mobile data collector (MC), which moves around and collects data from nearby sensors. Unlike normal routing techniques, each sensor has to store the gathered data locally, and thus we first devise local buffer management algorithm in a storage-efficient way. To do so, local buffer is divided into two parts and the deviation value between data during the two consecutive periods is stored. And then the dynamic route determination algorithm for MC is also proposed. In the algorithm, connecting nodes (CNs) are selected at initial phase, and the urgent messages from sensors or CNs are used to determine the next visiting CN. Through the extensive simulation works, we show the effects of the proposed algorithms.

References

1. Atzori, L., Iera, A., Morabito, G.: The Internet of Things: A survey. Journal of Computer Networks **54**, 2787–2805 (2010)
2. Miorandi, D., Sicari, S., Pellegrini, F.D., Chlamtac, I.: Internet of things: Vision, applications and research challenges. Ad Hoc Networks **10**, 1497–1516 (2012)
3. Korea Meteorological Administration. http://www.kma.go.kr

Speech-to-Text-Based Life Log System for Smartphones

Dongmahn Seo and Joon-Min Gil

Abstract In this paper, we propose a life log system which provides a real-time voice recording with a speech-to-text feature over smartphone environments. The proposed system records data of user life using a microphone of smartphone. Recorded data are sent to a server, analyzed, and stored. Recorded data are dictated using a speech-to-text service, and saved as text files. The proposed system is implemented as a prototype system and evaluated. Users of our system are able to search their life log sound files using text.

Keywords Life log · Cloud · Smartphone · Speech to text

1 Introduction

A life log is a digital data logging for human daily life. Many life log projects have been proposed [1~7]. In life log projects, digital data are collected using cameras [1, 4, 5], wearable devices [2, 3], and a remote controller [6]. However, legacy projects are using wearable devices or specific purpose devices, which are not commonly used, or only photographs and videos are focused to research.

Speech recognition technologies based on word dictionary have enough accuracy. Current speech-to-text technologies for a sentence and a paragraph have less accuracy than word dictation. However, we assume that searching voice record files is available if accuracy of sentence or paragraph dictation is more than half. Because users use a few keywords when they search recorded voice files.

D. Seo(✉) · J.-M. Gil
Catholic University of Daegu, Gyeongsan-si, South Korea
e-mail: {sarum,jmgil}@cu.ac.kr

This research was supported by Basic Science Research Program through the National Research Foundation of Korea(NRF) funded by the Ministry of Science, ICT & Future Planning (NRF-2015R1C1A1A02036686).

© Springer Science+Business Media Singapore 2015
D.-S. Park et al. (eds.), *Advances in Computer Science and Ubiquitous Computing*,
Lecture Notes in Electrical Engineering 373,
DOI: 10.1007/978-981-10-0281-6_90

These keywords probably appear several times in a voice file because the topic of a speech or a dialog is related on these keywords.

We assume that a smartphone is a suitable device for life log, because smartphones belong in human life on 24 hours 7 days a week and have various sensors, a camera, a microphone, computing powers and network connectivity.

In this paper, we firstly propose a new life log system using user voice recording with dictated texts as far as we know. The proposed system records user's voice and phone calls using smartphones whenever a user wants. Recorded sound files are dictated for storing text files by a speech-to-text service. Recorded sound files and dictated text files are stored in a life log server. The server provides life log file lists and searching features for life log users.

2   Proposed System Overview

Fig. 1 shows a concept of the proposed system. A smartphone records user's voices and phone calls using a microphone of the smartphone. Recorded sound files are sent to a server, and the server converts sound file format and requests texts to a speech-to-text server. Recorded sound files and dictated text files are coupled a pair of life log data for management. Pairs of life log data are provided to smartphones and web browsers via a web service. Users of the proposed system are able to search and consume their life log data.

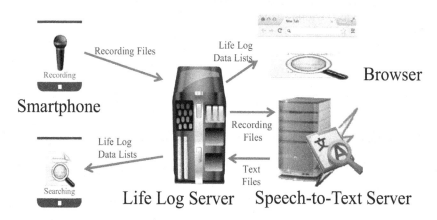

Fig. 1 Concept of the proposed system.

Fig. 2 describes a software architecture for a prototype implementation of the proposed system. The prototype system consists of a life log smartphone application, a life log server and a speech-to-text server. The life log smartphone application is a life log data collector and a client program. In a general situation, a MIC recording module is working to collect life log data using a microphone of smartphone. When a user makes or receives a phone call, a call recording module records a phone call. These two modules are controlled by a background processing module which is connected with a life log server. Recorded sound files are one-minute-length.WAV format files. The background processing module send the files to the life log server, and then delete them in the smartphone. A life log web client provides a web interface for life log data consumption and searching.

The life log server manages life log data files, which are pairs of a sound file and a text file and provides a web service for a user interface. To dictate sound files, a speech-to-text (STT) module controls a flac encoding module and a STT request Module. The flac encoding module converts .WAV format files which are received from the life log application into .FLAC format files. The STT request module requests dictated text files from sound files converted by the flac encoding module to the speech-to-text server. A life log web server provides life log data stored in the life log server to the life log web client. The speech-to-text server dictates recorded files and provides corresponding texts.

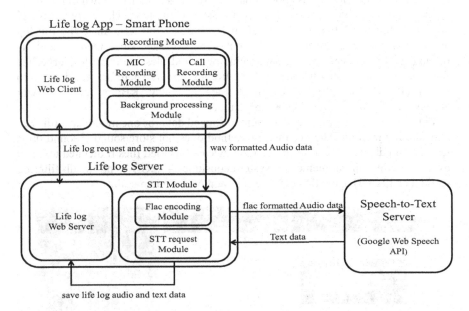

Fig. 2 Software architecture of the proposed system.

To evaluate the proposed system, we implement a prototype system. The life log application is implemented with Java language over Android environments. In the life log server, the life log web server is built with Apache, PHP and Mysql (APM) and the STT module is programmed in Java language. For a speech-to-text server, we apply Google Web Speech API. (http://www.google.com/intl/en/chrome/demos/speech.html).

3   Experimental Results

Fig. 3 shows snapshots of the life log application on a smartphone. The first picture is a life log data search interface, the second picture shows an interface for life log data lists, and the last picture is an interface for listening a selected sound file. For life log data searching, the prototype system provides three options. The first option is keyword-based search using dictated texts, the second option is time of date-based search using time of date at recording and the other option is location-based search using GPS coordinate of recording location.

In this paper, we experiment with the implement prototype system in order to obtain optimal sampling rate of recording sound files. Table I and Fig. 4 explain average dictation accuracies between sampling rates of sound files and buffer sizes to send to the speech-to-text server. As shown in the table, the maximum dictation accuracy is little proportional to the sampling rates, but the buffer sizes have no relation to dictation accuracies. However, we find out that the maximum accuracy is obtained when a buffer size is approximately 9 seconds length as shown in Fig. 4.

In the prototype system, every sound file is 1 minute long. Therefore, an 11025 Hz sound file is 1320 KB, and a 48000 Hz file is 5760 KB. Since the life log application runs on smartphones with mobile network and small memory and the life log server stores all life log data including sound files and text files, large size files are overload to smartphones and the life log server. Therefore, we decided to use 16000 Hz sampling rate sound files and 256 KB buffer size with approximately 68% average dictation accuracy.

Fig. 5 shows an average total amount of logged data for one month according to two filtering method. The db filtering method does not store sound if sound db is less than 45 db. The text filtering method deletes recorded files if dictated text is 0 byte. Therefore, the implemented system consumes 43.2GB using the db filtering method or 12.6GB using the text filtering method for one month.

Fig. 3 Snapshots of the life log application. (From the left, Search page, Result page, and Play page of a recorded file)

Table 1 Average Accuracy between Sampling Rate and Buffer Size

Hz \ KB	64	96	128	256	512	768	1024
11025	58.0%	44.4%	64.8%	**66.4%**	n/a	n/a	n/a
16000	56.1%	60.7%	62.4%	**67.9%**	53.3%	n/a	n/a
22050	53.5%	54.4%	61.7%	65.8%	**67.4%**	n/a	n/a
32000	45.3%	50.3%	55.8%	66.3%	**67.4%**	66.9%	n/a
44100	41.6%	53.6%	53.6%	59.7%	66.8%	**70.6%**	53.6%
48000	43.8%	55.7%	55.7%	64.4%	69.0%	**70.0%**	67.0%

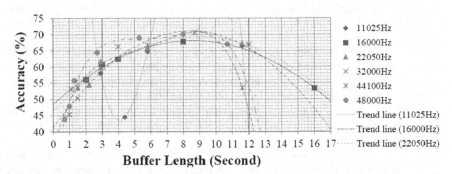

Fig. 4 Average Accuracy with Buffer Length.

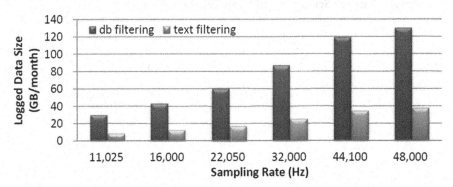

Fig. 5 Logged data size.

4   Conclusion and Future Works

In this paper, a life log system based on Speech-to-Text over Smartphone was proposed. The proposed system recorded information of user life using a microphone and GPS of smartphone. We implemented a prototype system to evaluate the proposed system. With the implemented system, the optimal sampling rate and the optimal buffer size were estimated. Users of our system were able to search their life log sound files using text using the implemented system.

In future works, we are planning to extend our life log system with videos, photographs, emails, call logs, and text messages.

References

1. Tancharoen, D., Yamasaki, T., Aizawa, K.: Practical experience recording and indexing of life log video. In: Proc. of the 2nd ACM Workshop on Continuous Archival and Retrieval of Personal Experiences, pp. 61–66 (2005)
2. Aizawa, K.: Digitizing personal experiences: capture and retrieval of life log. In: Proc. of the 11th international Multimedia Modeling Conference 2005, pp. 10–15, January 2005
3. Hori, T., Aizawa, K.: Capturing life-log and retrieval based on contexts. In: IEEE Intl. Conf. on Multimedia and Expo, vol. 1, pp. 27–30, June 2004
4. Tancharoen, D., Aizawa, K.: Novel concept for video retrieval in life log application. In: Lecture Notes in Computer Science, vol. 3332, pp. 915–923 (2005)
5. Gomi, A., Itoh, T.: A personal photograph browser for life log analysis based on location, time, and person. In: Proc. of the 2011 ACM Symposium on Applied Computing, pp. 1245–1251 (2011)
6. Abe, M., Morinishi, Y., Maeda, A., Aoki, M., Inagaki, H.: A life log collector integrated with a remote-controller for enabling user centric services. IEEE Trans. on Consum. Electron. **55**(1), 295–302 (2009)
7. Makino, Y., Murao, M., Maeno, T.: Life log system based on tactile sound. In: Lecture Notes in Computer Science, vol. 6191, pp. 292–297 (2010)

The Packet Filtering Method with Packet Delay Distribution Forecasting for Stability and Synchronization in a Heterogeneous Network

Kyeong-Rae Cho, You-Boo Jeon, Keun-Ho Lee and Jin-Soo Park

Abstract This paper presents the packet filtering method using a packet delay distribution in order to improve the stability of the time synchronization between devices in a heterogeneous network environment. Also, the paper uses the tracing packet for the estimation of the packet delay distribution and studied for applying to the secondary nodes between devices in which a heterogeneous network environment. We confirmed that this method is effective in a heterogeneous network environment through the congested results of the numerical simulation.

Keywords Heterogeneous network · Packet Filtering Method (PFM) · Delay jitter

1 Introduction

In this paper, we studied the two packet filtering methods(PFM) and two methods of exchange schemes. PFM is a method is a method using a first estimated packet delay distribution is proposed. The first method, packet filtering with estimation of

K.-R. Cho
Department of Computer Software, Seoil University, Seoul 131-702, Republic of Korea
e-mail: krcho@seoil.ac.kr

Y.-B. Jeon(✉) · J.-S. Park
Industry Academic Cooperation Foundation, Soonchunhyang University,
Asan-si, Chungcheongnam-do 336-745, Republic of Korea
e-mail: jeonyb@sch.ac.kr, vtjinsoo@gmail.com

K.-H. Lee
Division of Information and Communication, Baekseok University,
Cheonan-si, Chungcheongnam-do 330-704, Republic of Korea
e-mail: root1004@bu.ac.kr

© Springer Science+Business Media Singapore 2015   643
D.-S. Park et al. (eds.), *Advances in Computer Science and Ubiquitous Computing*,
Lecture Notes in Electrical Engineering 373,
DOI: 10.1007/978-981-10-0281-6_91

packet queuing delay, estimation method using a tracing packet queuing are described in Figure 1. This method is usually used to filter the noise delay [1]. Figure 1 shows the sequence of messages modified between the primary node and secondary node. This sequence is a method based on delay-request, response method of tracing packets between devices in a heterogeneous network. Synchronization or delay request, response packets are sent with n_p ($n_p \geq 2$) of tracing packets [2]. All tracing packets are all the same size and have a synchronization or delay-request, response packets. And tracing all packets are all treated as an event message. Primary and secondary node in the node, both the time stamp at the time of sending and receiving and recording the time interval from the receiving node (inter-departure time, $\tau_{dept(k)}$ ($k = 1,2,3 \dots n_p$)) and the arrival time interval (inter-arrival time, $\tau_{arrv(k)}$ ($k=1,2,3\dots n_p$)) is calculated by the time stamp. The occurrence of noise delay (delay jitter) can be estimated by the difference of arrival time intervals [3].

i) IF($\tau_{arrv\,(j-1)}$ - $\tau_{dept\,(j-1)}$) > $\Delta\tau$ or ($\tau_{arrv\,(j-1)}$ - $\tau_{dept\,(j-1)}$)< - $\Delta\tau$ when the noise occurs,
ii) Unless the above (not the case) is considered to be a noise is not occurred.

Where $\Delta\tau$ should be determined carefully to suit the specifications of the secondary nodes.

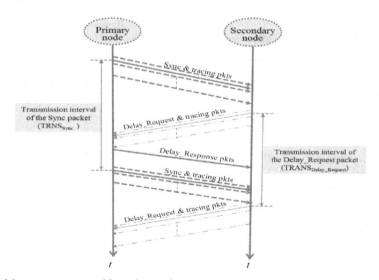

Fig. 1 Message sequence with tracing packet

The first packet is delayed noise does not exist in the estimation step is to be determined in living packet (surviving packet) from the alive packet time interval (inter-departure time, $\tau_{dept(m)}$) and the arrival time interval (inter-arrival time, the $\tau_{arrv\,(m)}$) is determined. The first and the last packet of the group is determined to be alive packet. This repetition is effective to enhance the accuracy of estimation of a delay in an environment where noise is noise occurred delay. It can be confirmed that the queuing delay is effectively removed by this method. These estimates per group delay by the noise can be one of the smallest packet selection. A second method, Packet Filtering with Estimation of Packet Transmission delay distribution. Estimation of the distribution of the packet transmission delay is aimed to determine the interchangeability of the monitoring method of the estimation queuing. We can obtain only the calculated transmission delay. This is the value to be stored and the time difference between transmission delay times, as well as the sender and the receiver. Nevertheless, the distribution of the calculated transmission delay time has a value sufficiently. If able to extract the mode the value of the transmission delay timing packets received by the secondary distribution node may be controlled with the clock of the secondary compared to the sync packet to maintain a small difference [4]. As a result the slave node can be kept small and the clock offset. Filtering the K-th group of the timing packet and the tracing packet is as follows. First validly delay the distribution of the first group of current and W-th(delay distribution) is collected and summarized. 'W' is taken as the window size for the distribution estimation delay. Only to get in front of the transmission delay, calculated as if the story here and there and that includes the transmission delay time, as well as to the time difference between the Primary and secondary node. Variation of the calculated transmission delay may have to be modified in accordance with the adjustment of the secondary clock, the transmission delay of the n-th packet of the k-th group $d^k_{ps(n)}$, expression and correct it $d^k_{ps(i)}$ the j-th group of the modified delay time of the i-th packet $d^{ij}_{ps(i)}$, $d'^{j}_{ps(i)}$ $(0 < j \leq k - w)$ * $d'^{j}_{ps(i)} = d^{j}_{ps(n)} + O^{k}_{j\,accur}$, can be expressed as * $d^{ij}_{ps(i)} = d^{j}_{ps(n)} + O^{k}_{j\,accur}$, Which is the cumulative clock offset $O^{k}_{j\,accur}$. In general, match the mode of the estimated value of the delay profile is the target bin, but the network load suddenly changes in a short time, can be changed in a short period of time the target bin. This case may destabilize the system. The value having a larger value than the value in the previous mode, even if the detected target bin, not just the changes. Only beyond the critical of a range(τ_f) to change the value of the target bin. The width of the bin(τ_b) also, for example, clock quality, considering the configuration of the secondary clock, such as robustness to noise of the clock system must be determined. Estimated clock drift of the node of the secondary affecting the accuracy of considered to be affected also. A frequency drift to create an estimate of the delay to obscure it is impossible to accurately estimate the mode value of the distribution result. One solution is to create a larger width τ_b of the bin, but this can lead to large variations in the secondary clock.

2   Applications of Packet Filtering Method

2.1   Packet Filtering Method(PFM)

This paper, we discuss the filtering method used in the clock control method of the system between devices in a heterogeneous network environment. Said filter mechanism with the availability of queued estimation in the previous study. Figure 2 shows how to use a mix of the two packet filtering method.

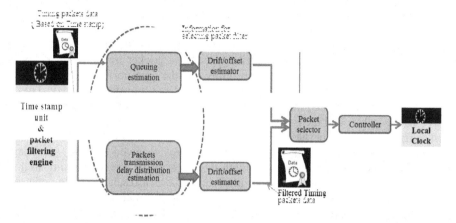

Fig. 2 Example of use a mix of the two packet filtering method (PFM) schemes

The more useful of the two is selected by the offset and drift estimates both the selector in the packet filter. Using the selected offset and drift PI controller adjusts the local clock of the secondary.

There are two variables for selecting the filtering method. The first variable is the extraction ratio of the R_d and timing packets in the tracing packet and the W group R_q. The second variable is the change in the target bin (Fluctuation). If extraction of a timing packet rate R_q in the process of selection is greater than a predetermined size τ_r select the method by queuing delay estimation. In this case, this method in case of a common network without congestion is most effective. If extraction rate R_q of timing packets is less than a certain size τ_r the ratio R_d of tracing packets and compares greater than a certain size τ_r. If the larger, Compares the variation of the target bin (Fluctuation) δ_d is less than a predetermined size d_d. If the less, it is effective to use the delay spread estimation method. This means that the network congestion situation, but the situation is adjustable through estimation. Otherwise, in case the adjustment cannot be estimated through the case, the more good way to maintain a local clock until the network conditions stabilize. In general, if the secondary transfer to the Primary and secondary rate and the delay is not equal to the asymmetry of the Primary transfer the consideration of non-symmetric transmission environment because the process of estimation is made large. Asymmetric delay($\text{Delay}_{\text{Async}}$) is to be estimated represented by the following formula.

$$\text{Delay}_{\text{Async}} = ((d_{ps} - d_{0ps}) - (d_{sp} - d_{0sp}) / 2$$

When calculating the offset on the basis of this, to obtain the following results:

$$\text{Offset}(O_s) = (t_{\text{Sync\_Rx}} - t_{\text{Sync\_Tx}}) - (\text{meanPathDelay}) - \text{Delay}_{\text{Async}}$$

2.2 Numerical Simulation

In this paper, the configuration of the simulation using a network simulation tool NS-2 to obtain a result. Using the modified message sequence shown in Figure 1, the simulation was carried out for a list of the variables used in the simulation shown in Table 1.

Table 1 Argument values of simulation

Criteria	Values
Mean transmission interval of the Delay_Request packet(TRANS$_{Delay\_Request}$)	0.5 sec
Transmission interval of the Sync packet (TRNS$_{Sync}$)	0.5 sec
Number of tracing packets Tx with a Sync/Delay_Request packet(n_p)	20
Inter departure time of timing packets($\tau_{dept(k)}$) ($k = 1, 2, 3, \ldots, n_p$)	135μs
Criteria for the queuing estimation($\Delta\tau$)	150 μs
Width of the bin of the estimated packet delay distribution(τ_b)	3 μs
Window for the estimation packet delay distribution (w)	50 bursts
Criteria of the packet extraction rate for selecting the filtering scheme(τ_r)	5 %
Criteria of the updating packet extraction/filtering with packet delay estimation(τ_f)	40
Criteria of the fluctuation in the packet extraction/packet delay estimation(δ_d)	10 μs
Band rate	100 Mbps

Shown in Figure 3 is a graph comparing how much matching estimated from the actual packet delay in packet delay distribution at 80% and 50% of traffic load.

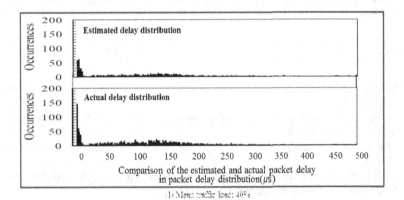

Fig. 3 Example of traffic load distribution delay in the 40% and 70%.

The width of the bin is $3\,\mu s$ can be seen the result of the estimation is as seen from the above results confirm the extract, based on the 1550 samples with a repeating 50 times that estimated to be almost the same distribution as the actual traffic.

3 Conclusion

In this paper, we were queuing delay and packet delay distribution estimation method was estimated to correct the offset in congested network conditions using appropriate methods through accurate estimation at $+$ -$5\mu s$ level. To this end, consider the asymmetric environment to estimate the correct offset of the asymmetric environment through the expression for correcting this. However $\Delta\tau$, τ_f, w, to a practical implementation by a number of variables, such as crude δd has a limit in the hard consideration.

Acknowledgments This research was supported by the MSIP(Ministry of Science, ICT and Future Planning), Korea, under the C-ITRC(Convergence Information Technology Research Center) (IITP-2015-IITP-2015-H8601-15-1009) supervised by the IITP(Institute for Information & communications Technology Promotion).

References

1. Software-only implementations of slave clocks with sub-microsecond accuracy. In: ISPCS (2008)
2. Packet network timing measurement and analysis using an IEEE 1588 probe and new metrics. In: ISPCS (2009)
3. Cluster TDEV a new performance metric for timing over packet networks. In: ISPCS (2010)
4. Cho, K.-R., et al.: Smart mobile Changes the future, pp. 112–128, 220–221, 295–329. Maeil Business Newspaper & MK Inc. (2011)

An Automatic Feedback System Based on Confidence Deviations of Prediction and Detection Models for English Phrase Break

Byeongchang Kim and Gary Geunbae Lee

Abstract This paper presents a method to construct a feedback provision model for English phrase breaks by utilizing confidence deviations of prediction and detection models. The proposed method consists of prediction, detection and feedback provision models. The prediction and detection models adopted conditional random fields classifiers performed on the Boston University radio news corpus, and achieved accuracies of 90.15% and 91.62%, respectively. The feedback provision model determines three types of feedbacks for each disjunction using the differences between the prediction and detection confidences. In a validation experiment for the feedback provision, the proposed method demonstrated a Pearson's correlation coefficient of 0.74 between the feedback provision model's scores and human fluency assessments.

Keywords Computer-assisted language learning · English phrase break · Automatic feedback system

1 Introduction

To learn a foreign language efficiently and rapidly, learners should frequently practice the target language. Computer-assisted language learning (CALL) can provide learners with immediate and easy methods for practicing the target language. In addition to pronunciation, CALL research must focus on prosody, which conveys higher-level information that is not contained at the segmental level [1].

B. Kim(✉)
Catholic University of Daegu, Gyeongsan, Gyeongbuk, South Korea
e-mail: bckim@cu.ac.kr

G.G. Lee
Pohang University of Science and Technology, Pohang, Gyeongbuk, South Korea
e-mail: gblee@postech.ac.kr

© Springer Science+Business Media Singapore 2015   649
D.-S. Park et al. (eds.), *Advances in Computer Science and Ubiquitous Computing*,
Lecture Notes in Electrical Engineering 373,
DOI: 10.1007/978-981-10-0281-6_92

To extract prosodic information from a given utterance or text transcription, prosodic units require phrase chunking, which is directly related to phrase breaks. Thus, to practice prosody in language learning, determining the location of phrase breaks is the first step to be mastered.

This paper proposes a method to provide non-native English learners with intuitive feedbacks for phrase breaks. The method consists of three models: phrase break prediction, detection and feedback provision. The goal for the prediction and detection models is to achieve accuracies comparable with previous work, whereas the feedback provision model represents a novel attempt to provide intuitive feedbacks that easily help learners improve their skills and habits regarding prosodic phrasing and phrase breaks in their utterances.

2   Material

2.1   A Corpus for Phrase Break Models

Previous work related to prosody has used various corpora including the BU corpus[2], the Boston Directions Corpus and the Machine Readable Spoken English Corpus [3]; those corpora were annotated using prosodic labels. To construct prediction and detection models for phrase breaks, we chose the BU corpus annotated with Tones and Break Indices (ToBI) [4].

The BU corpus consists of seven hours of speech recorded in the radio news style which is distinct from normal conversation. The accurate and intelligible speech of this news style contains richer information regarding phrase breaks com-pared with normal conversation.

2.2   Preprocessing the BU Corpus

The BU corpus is partially marked with prosodic labeling, which includes the representation of break indices. Before utilizing the BU corpus, a preprocessing step is required to exclude the incomplete subsets. These incomplete subsets can reduce the accuracies of the prediction and detection models and cause incorrect feedback to be provided to learners.

The BU corpus was partially and independently annotated by different labelers. The inter-labeler agreement for the ToBI's break indices was 95% of 989 words within the chosen paragraphs [2], which can be considered to be the upper-bound on the accuracies of the prediction and detection models for phrase breaks.

Though the BU corpus provides various resources, the proposed method utilized only text transcriptions, break labels and speech files to train the prediction and detection models. The other necessary resources were produced by the proposed method itself; for example, POS tags were obtained using a POS tagger with an accuracy of 96.3%.

3   Methods

To provide intuitive feedbacks for phrase breaks given a sentence and the corresponding speech, the proposed method consists of prediction, detection and feedback provision models (Fig. 1), which were performed at the word level. The proposed method focused on a feedback provision model that determines three feedback types at each disjunction between words based on the confidence deviations of the prediction and detection models. In the ToBI's break indices of 0-4, the proposed method classified break indices 3 and 4 as indicative of the presence of breaks and the other indices as indicative of the absence in a coarse grouping manner to avoid the data sparseness problem.

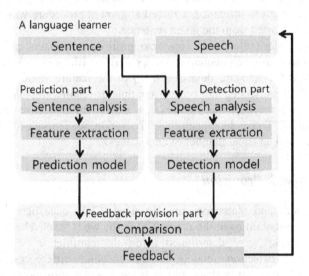

Fig. 1 Architecture and data flow

3.1   Prediction Model

To achieve accurate phrase break predictions based on given sentences without using any additional information, we adopted a conditional random fields (CRF) classifier [5]. Based on related studies, such as [6-8], we chose the following syntactic and lexical features to train the prediction model: word identity, POS tag, word class (function word or content word), number of syllables and punctuation mark types.

Among those features, punctuation marks are considered to be important because they split a long sentence into smaller components. We utilized a contextual window that contained the previous two and next three words.

3.2   Detection Model

As in the prediction model, we trained the detection model using a CRF classifier. In addition to acoustic features directly extracted from utterances, we needed to utilize syntactic and lexical features to achieve a high detection accuracy for phrase breaks, as found by [6] and [8]. We utilized the following features to build the detection model: duration of the syllable, duration of the vowel in the syllable, normalized pitch mean of the syllable, normalized intensity mean of the syllable, silence after the word and the syntactic and lexical features used in the prediction model.

Acoustic feature extraction was conducted at the last syllable of each word because phrase breaks are detected at word boundaries. To determine the timeline of words for the given utterance, a forced alignment procedure was performed on the word list imported from the given sentence.

To reduce undesirable fluctuations among different speakers, acoustic features were normalized using the equation $z = (x - \mu) / \sigma$ (the z-score), where μ is the mean and σ is the standard deviation (s.d.) of the feature value x. Additionally, because CRF classifiers cannot process continuous feature variables, which is the natural form of the acoustic features, we needed to discretize the continuous feature variables. To conduct quantile discretization, we introduced 10 bins, where each bin received an equal number of data values.

3.3   Feedback Provision

The prediction and detection models used CRF classifiers to determine confidences (probabilities in the range 0-1) in the word level. The feedback types for phrase breaks were determined using the confidence deviations of the prediction and detection models. Small absolute differences between the prediction and detection probabilities indicate that the corresponding disjunction of speech is similar to the predicted disjunction. The feedback types are the following: "positive", "negative" and "ambiguous". Positive feedback indicates that learners correctly paused or progressed at the corresponding disjunctions. Negative feedback helps learners correct mistaken phrase breaks. If the confidence deviation of the prediction and detection models cannot be used to determine whether the feedback should be positive or negative, the corresponding disjunction is considered to be ambiguous or indeterminate. The feedback is provided as follows:

$$Feedback = \begin{cases} Positive, & if \ |\pi_{pre} - \pi_{det}| < \theta_1 \\ Indeterminate, & if \ \theta_1 \le |\pi_{pre} - \pi_{det}| \le \theta_2 \\ Negative, & if \ |\pi_{pre} - \pi_{det}| > \theta_2 \end{cases} \qquad (1)$$

where π_{pre} and π_{det} are the confidence output by the prediction and detection models, respectively, and θ_1 and θ_2 are the decision boundaries for the feedback signs.

4   Results

To validate the phrase break prediction and detection models, we compared the accuracies of the models with those of previous works. In addition, to verify that the proposed method effectively provides non-native learners with intuitive feedbacks for phrase breaks, we calculated the correlations between the feedback scores determined using the proposed method and phonetic experts' fluency assessments of non-native learners' speeches. Furthermore, by having non-native learners use the system that adopted the proposed method, we collected assessments of the usability of the system.

4.1   Validation of the Prediction and Detection Models

To assess the phrase break prediction and detection models, we conducted five-fold cross validations in which the utterances in the BU corpus were split with a proportion of approximately 4:1 into the training set and test set. For training the models, we adopted our own CRF tool, which is based on [5].

Table 1 Comparison of the phrase break prediction and detection models with previous work

Models	Accuracy		
	[6]	[9]	Ours
Prediction	89.76	91.18	**90.15**
Detection	91.06	92.11	**91.62**

For comparative evaluation of the prediction and detection models, we refer to the recently reported work of [6] and [9], which used the BU corpus to build prediction and detection models for phrase breaks. The accuracies of the proposed prediction and detection models (Table 1) are slightly higher than those of [6] and slightly lower than those of [9]. However, because the differences between the accuracies of their models and ours are not significant, the proposed work is considered to achieve accuracies sufficient for real CALL activities.

Table 2 Accuracies, precisions, recalls and F-measures of the prediction and detection models

Models	Accuracy	Precision	Recall	F-measure
Prediction	90.15	84.02	80.08	82.00
Detection	91.62	86.14	83.67	84.88

In addition to the accuracy, it is necessary to check the precision, recall and F-measure values of the models (Table 2) because the proposed method performed binary classification tasks, in which accuracies above 0.5 can be achieved by identifying a majority of breaks and non-breaks. In the BU corpus used to train the models, the accuracy derived by selecting non-breaks as the majority is 71.88%. The precision, recall and F-measure values (Table 2) are considered to be sufficient to accurately predict and detect phrase breaks in our classification tasks.

4.2 Validation of the Feedback Provision Model

In an experiment to validate the proposed feedback provision model, we checked 5500 sentences (speeches) uttered by 75 Koreans whose fluency was rated by seven phonetic experts. We calculated Pearson's correlation coefficients between feedback scores (Eq. 2) yielded by the proposed method and the fluency assessments provided by the seven experts for varying $\theta1$ and $\theta2$ (Eq. 1) from 0 to 1 using a step size of 0.1.

$$score = \frac{Feedback\ count\ of\ "positive"\ type}{Entire\ feedback\ count\ except\ "ambiguous"\ type} \tag{2}$$

We found a maximum correlation coefficient of 0.74 at 0.2 for $\theta1$ and 0.7 for $\theta2$, for which the proposed method can be considered to effectively provide phrase break feedbacks that are highly correlated with the learners' actual fluency.

5   Conclusion

This paper proposed a method to provide non-native English learners with intuitive feedbacks for phrase breaks. The proposed method utilized the BU corpus to train prediction and detection models for phrase breaks using CRF classifiers. The prediction and detection models were validated to achieve accuracies of 90.15% and 91.62%, respectively, which are comparable with those of previous work and sufficient to be utilized in real CALL systems. In the validation for the feedback provision, the Pearson's correlation coefficient between the feedback scores yielded by the proposed method and fluency assessments provided by phonetic experts was 0.74; this result verifies that the proposed method provided effective feedbacks that accurately reflect the learners' fluency. Consequently, the proposed method can be adopted in real CALL systems for phrase break learning to provide effective feedbacks with sufficient usability.

References

1. Mareuil, P., Vieru-Dimulescu, B.: The contribution of prosody to the perception of foreign accent, Phonetica, vol. 63, pp. 247–267 (2006)
2. Ostendorf, M., Price, P., Shattuck-Hufnagel, S.: The Boston University radio news corpus, Linguistic Data Consortium (1995)

3. Roach, P., Knowles, G., Varadi, T., Arnfield, S.: Marsec: A machine-readable spoken English corpus. Journal of the International Phonetic Association **23**, 47–53 (1993)
4. Silverman, K., Beckman, M., Pitrelli, J., Ostendorf, M., Wightman, C., Price, P., Pierrehumbert, J., Hirschberg, J.: ToBI: a standard for labeling prosody. In: Proc. of International Conference on Spoken Language Processing (ICSLP), pp. 867–870 (1992)
5. Lafferty, J., McCallum, A., Pereira, F. C.N.: Conditional random fields: probabilistic models for segmenting and labeling sequence data. In: Proc. of ICML 2001, pp. 282–289 (2001)
6. Jeon, J.H., Liu, Y.: Automatic prosodic events detection using syllable-based acoustic and syntactic features. In: Proc. of ICASSP, pp. 4565–4568 (2009)
7. Taylor, P., Black, A.W.: Assigning phrase breaks from part-of-speech sequences. Computer speech and language **12**, 99–117 (1998)
8. Rangarajan Sridhar, V., Bangalore, S., Narayanan, S.S.: Exploiting acoustic and syntactic features for automatic prosody labeling in a maximum entropy framework. IEEE Transactions on Audio, Speech, and Language processing, pp. 797–811 (2008)
9. Qian, Y., Wu, Z., Ma, X., Soong, F.: Automatic prosody prediction and detection with Conditional Random Field (CRF) models. 2010 7th International Symposium on Chinese Spoken Language Processing (ISCSLP), pp. 135–138 (2010)

Conceptual Design of a Network-Based Personalized Research Environment

Heeseok Choi, Jiyoung Park, Hyungseop Shim and Beomjong You

Abstract With advancement in information technologies and a better mobile environment, virtual and personal research environments can be made available for researchers. On the other hand, as more investment is being made in R&D, the efforts to enhance R&D productivity are becoming important. This study proposes a conceptual design of network-based personalized research environment that is called NePRE for developing the tool to assist researchers in their R&D efforts. It can be more easily utilized by researchers in their R&D information activities. To do this, we analyze existing research support environments in terms of R&D information activities. And we also analyze changes of information environment in terms of personalization. Subsequently we define three design principles of NePRE for personalization. Finally we define its key functions to assist researchers with respect to six R&D information activities.

Keywords Research environment · Personalization · Community curation · Virtual collaboration

1 Introduction

With advancement in information technologies and a better mobile environment, virtual computing power can be made available for researchers independently from their location. On the other hand, as more investment is being made in R&D, efforts to enhance R&D productivity of researchers are becoming more important. Light-weight applications are widely being developed and disseminated to assist researchers in their R&D efforts. Furthermore, it is not still easy to perform information-aid R&D since users have to individually access many services for different purposes. In addition, recent changes to information environment makes personalized service more important for convenient information usage by researchers.

H. Choi(✉) · J. Park · H. Shim · B. You
Korea Institute of Science and Technology Information, Daejeon, South Korea
e-mail: {choihs,julia.park,hsshim,ybj}@kisti.re.kr

© Springer Science+Business Media Singapore 2015   657
D.-S. Park et al. (eds.), *Advances in Computer Science and Ubiquitous Computing*,
Lecture Notes in Electrical Engineering 373,
DOI: 10.1007/978-981-10-0281-6_93

This study proposes a conceptual design of network-based personalized research environment for developing the tool to assist researchers in their R&D efforts. It can be easily and conveniently utilized by researchers in their R&D information activities. To do this, we first look into research support environments in terms of information activities in R&D. And we also analyze changes of information environment in terms of personalization. Subsequently we define three design principles of NePRE. Finally we define its key functions to assist researchers with respect to six R&D information activities.

2   Related Works

2.1   Research Support Environments

Research support environments were widely developed and being made good use of assisting researchers' R&D activities. We divided them into virtual research environments, social networking services, bibliometric tools, and analytic tools.

- Virtual research environment (VRE) for researchers: this is a very important aspect for the success of scientific projects [1,2]. Its purpose is to provide researchers with the tools and services necessary to research as efficiently and effectively as possible. For example, RIC [3] supports researchers in managing the increasingly complex range of tasks involved in carrying out research. Specifically, it provides structure to the research process, easy access to resources, guidance and tools to manage information assets, along with integrated collaboration services. And myExperiment [4] supports the sharing of research objects used by scientists, such as scientific workflows ranging from data and methods to scholarly publications.
- Social networking service (SNS) for researchers: this is widely used in research communities for communicating and collaborating among researchers. For example, ResearchGate [5] allows researchers around the world to collaborate more easily. It discovers scientific knowledge, and makes your research visible. For a common purpose of advancing scientific research, it links researchers from around the world. It is changing how scientists share and advance research in digital age. And Academia [6] is a platform for academics to share research papers, and it is a website where researchers can post their articles and discover and read articles posted by others. It combines the archival role of repositories with social networking features, such as profiles, news feeds, recommendations, and the ability to follow individuals and topics [6].
- Bibliometric tool for researchers: this basically uses a bibliometric information, and it helps researchers write references in their papers by searching papers and store them. For example, Mendeley [7] is a reference manager and academic social network. It makes your own fully-searchable library in seconds, cite as you write, and read and annotate your PDFs on any device. It manages bibliographic information, and helps researchers generate references when they write them in a paper. In addition it helps

find collaborative researchers of the world, and supports composing community with them. And RefWorks [8] is an online research management, writing and collaboration tool. It is designed to help researchers easily gather, manage, store and share all types of information, as well as generate citations and bibliographies. If researchers need to manage information for any reason whether it is for writing, research or collaboration.

- Analytic tool for researchers: this provides some insights for a research. For example, SciVal suite [9] provides a critical information about performance and expertise to help enable informed decision-making and drive successful outcomes. It is composed of SciVal Spotlight, SciVal Strata, and SciVal Experts. It helps decision makers responsible for research management to assess institutional strengths and demonstrable competencies within a global, scientific landscape of disciplines and competitor. And it helps decision makers to identify researcher expertise and enable collaboration within the organization and across institutions. And it also helps them to measure individual or team performance across a flexible spectrum of benchmarks and measures.

2.2   Changes in Information Service Environment

Information service environment was largely changed with advancement in information technologies and personal needs. We looked into changes in information service environment in the view of service personalization as follows.

- Open expansion of information and data: the demand for publicly available information and data is increasing due to government's 3.0 and activation policies of creative economy in Korea. Furthermore, there are more projects for publishing and sharing public data. In addition, there are an increasing number of data standardization as LOD (Linked Open Data) and LOD construction.
- Enhancement of personal information: people are more aware of protecting personal information and leakages of personal information in terms of the society and technology. Therefore, regulations and institutions are improved for enhancing personal information protection, and people involved therein have studied how to further enhance the technology. Collecting, using, providing, processing and managing personal information is strictly regulated to minimize personal information leakage. As data contributing to identify personal identity are gradually opened and shared, more efforts are required to protect personal information.
- Popularization of social networking service: Various social networking services and platforms, for example, facebook, youtube, twitter, Kakaotalk, LinkedIn, and ResearchGate are widely used. In the future social networking services will be connected with web portals or mobile services to enhance their functionalities.
- Very big contents: big data and IoT (Internet of Things) technology is developing fast, and services using them are appearing. Data are now more

abundant and diversified than before, and non-literature data as well as literature-centered data will be handled more importantly.

- Advances in web platform technology: as web technology develops, services in various formats have been developed and distributed, for example, mobile apps and web apps. In particular, web-based application S/W based on web standard HTML 5 is even more valued, that can be installed and used in all devices from smartphones to smart TVs where web browsers operate.

3   Design of a Personalized Research Environment

We first defined the personalization [10] with views of service, content, and platform. Three points of personalization are closely related to each other in realizing them. Directions toward personalization for supporting research are as follows.

- **Personalization for services:** the change is necessary from service user's access to each web portal to their access to information provided by each web portal through one tool (application S/W). Although web portals change and develop to be user-customized and personalized web portals, they are too personalized, repeat and are one of a plurality of personalized services. A standard based on XML like RSS for summarizing, sharing, sending and receiving contents among various web sites is already available. Therefore, it should be a shift from the web portals that is shared service channels to a single tool that is simply accessed and utilized by an individual.
- **Personalization for contents:** it is very important to provide and share information scattered on the Internet in line with information publication and release on the basis of subjects or various correlations. To this end, it is essential to provide contents as more condensed and reliable data by using technologies including contents curation [11], clipping, cataloging, clustering, and mash-up.
- **Personalization for platforms:** since web apps are based on the web, they can be operated just by web browsers. That is, they can be driven independently of terminal and OS types. For a conventional application (usually, we call it a native app), its application software adapted to each OS and device should be developed and installed in terminals to drive it. It is necessary to optimize the app adapted whenever a new version of device is launched including iPhone, iPad, Galaxy S, Galaxy Tab, Motorola and Blackberry. In addition, fine tuning is required depending on OS versions or screen size. In particular, the term 'N-screen' (multi-screen) is currently used in the IT industry, which signifies user's diversified physical environment for access to information. Therefore, web browser-based web app types can contribute to independent platforms.

Subsequently, we present a conceptual design on a personal research environment (we call it NePRE) that can be easily used with installation on various device environment of each individual researcher. The NePRE will provide a collection of functions to support the R&D life-cycle of researchers as follows.

- **Search.** The NePRE will provide researchers with an easy access to better resources based on statistical analysis of citation. The search operation will execute contents curation over simple search via participation of other researchers. The contents curation finds high quality of resources that can be more reliable using similar researchers' experience while simple search just finds the resources closely matched to input keyword.
- **Collect.** A number of information resources have been already identified due to national and international policy of opening data. The NePRE will provide means to gain access to and leverage contents. The user can automatically receive updates from specific resources sites, using e-mail and RSS feeds. Good articles can be recommended or identified by some researchers involved to similar topics. The NePRE will provide a link to the content supplier's site for share.
- **Analyze**. The NePRE will provide an insight about technologic trends, competition relationship, and promising technologies. The NePRE will support analysis of technology's ripple effect, researcher network, convergence relationships between technology groups. And the NePRE will help us find research topics, identify key patents, and understand status of technology development based on technology keywords.
- **Collaborate.** The NePRE will be designed to make it easy to identify potential collaborators, create community, and share the results of other researcher like other tools [4,5]. Both national and international information resources will be outlined and summarized via the participation of other researchers. The NePRE will encourage researchers to work together to develop and populate their research results like RIC and ResearchGate. For collaborative research, community participation and enhancements are very important. Therefore the NePRE will provide means to encourage collaboration between communities.
- **Store.** The NePRE will offer an enhanced function to allow the researcher to save searches on cloud environment for seamless usage from any device. The NePRE will support clipping or scrapping searched results. The cloud storage will store metadata associated with the documents, together with links to the full-text where permitted. When the resources are stored on a cloud environment, they are automatically related other resources each other and identified to be unique. The user can also easily access to all resources across various kinds of devices. Each user will access to individual storage on cloud environment via an online account. Manage information resources. In addition, each user will access to the resources of third party or local databases. It will categorize resources according to personal criteria.
- **Publish.** The NePRE will support publication life-cycle, from literature search and retrieval, papers annotation, and bibliography management to self – archiving like RIC. The NePRE will support making template-based document such as papers, patents, and research reports. The NePRE will support automatically manage and list up references, and help researchers find sources of resources. The NePRE will support writing short memos within documents and it will help share them with colleagues or in community.

4   Conclusion

This study suggests a conceptual design of network-based personalized research environment for developing personal knowledge tools researchers can use easily in their R&D. To do this we first analyzed services and tools in terms of information activities in R&D. And we also analyzed changes of information environment in terms of personalization. Subsequently we presented key functions of NePRE in six R&D information activities, and presented three design principles in order to implement the NePRE.

Future studies will focus on establishing a method of efficient connection and use of science and technology information resources by means of personal knowledge tools. It is necessary to study how to facilitate efficient classification and storage of individual researchers' information resources, and integration with connected data. It is necessary to study how to design light-weight personal knowledge tools.

References

1. Carusi, A., Reimer, T.: Virtual Research Environment Collaborative Landscape Study (2010)
2. Roth, B., et al.: Towards a generic cloud-based virtual research environment. In: 35th IEEE Annual Computer Software and Applications Conference Workshops, pp. 267–272 (2011)
3. Barga, R.S.: A virtual research environment (VRE) for bioscience researchers. In: International Conference on Advanced Engineering Computing and Applications in Sciences, pp. 31–38 (2007)
4. De Roure, D.: myExperiment: defining the social virtual research environment. In: 4th IEEE International Conference on Science, pp. 182–189 (2008)
5. ResearchGate: www.researchgate.net
6. Academia: Open Access Meets Discoverability: Citations to Articles Posted to Academia.edu. www.academia.edu
7. Mendeley: www.mendeley.com
8. RefWorks: www.refworks.com
9. Elsevier SchVal Suite: www.info.scival.com
10. Sunikka, A., Bragge, J.: What who and where: insights into personalization. In: Proceedings of the 41st Hawaii International Conference on System Sciences (2008)
11. Forbes: 5 Ways to Use Content Curation for Marketing and Tools to do it (2012)
12. Yao, Y.Y.: A framework for web-based research support systems. In: Proceedings of the 27th International Computer Software and Applications Conference (COMPSAC) (2003)
13. Kim, S., Yao, J.: Mobile research support systems. In: 26th IEEE Canadian Conference of Electrical and Computer Engineering (CCECE) (2013)
14. JISC briefing paper: Digital Information Seekers (2010). www.jisc.ac.kr/publications/reports/2010

Design and Implementation of the Basic Technology for Realtime Smart Metering System Using Power Line Communication for Smart Grid

Jun-Ho Huh, Dong-Geun Lee and Kyungryong Seo

Abstract Recently, in Republic of Korea, the PLC (Power Line Communication) technology is getting a limelight as a foundation technology for the Smart Metering field, and for that reason, we've implemented a similar foundation technology applicable to the Realtime Smart Metering System which enables remote monitoring and management of power usages using the PLC. In this paper, the PLC-utilized remote power metering and monitoring and control of power usages of respective power supplies have been implemented for the system. In addition, In-home headcount monitoring algorithm is proposed in an attempt to achieve efficient power usage management. It is expected that the proposed foundation technology will eventually become a foundation technology for the Internet of Things (IoT). We also anticipate that the technology could be useful for the establishment of an adequate power saving scheme with continuous power usage monitoring and will provide better user-convenience by controlling the power remotely and efficiently. Furthermore, since it provides relatively easy addition or removal of devices within the household, the system will have wider scalability.

Keywords PLC · Smart grid · Realtime Smart Metering System · IoT

J.-H. Huh · K. Seo(✉)
Department of Computer Engineering, Pukyong National University at Daeyeon,
Busan, South Korea
e-mail: krseo@pknu.ac.kr

D.-G. Lee
Department of Mechanical Engineering, Pukyong National University at Yongdang,
Busan, Republic of Korea

J.-H. Huh—Senior Research Engineer of SUNCOM Co., Republic of Korea

© Springer Science+Business Media Singapore 2015
D.-S. Park et al. (eds.), *Advances in Computer Science and Ubiquitous Computing*,
Lecture Notes in Electrical Engineering 373,
DOI: 10.1007/978-981-10-0281-6_94

1   Introduction

The communication through the power line is referred as the PLC (Power Line Communications) and it is a common name for the communication modes for the information delivery. This technology transmits data through the power lines on which the information is stored with the form of high frequency signal. The frequency signals can be separated from the power line by using an exclusive power line modem and be transmitted to the terminal devices [1, 2]. With the PLC technology, voices and data will be transmitted on the frequency signal through the power line-related mediums which have originally been designed for power transmission purpose. Again, the power line communication mediums are exclusively designed for power transmission, not for data transmission. Its channel characteristics are neither ideal nor suitable for data transmission. Considerable disadvantages can arise from the line noises and serious signal attenuation [1, 3, 4]. Improvement of the PLC's reliability can be attained considering the physical layers. For example, channel estimation and selection, filtering design, power distribution and so on [1, 4, 5]. Currently the PLC technology is being used for the Home-Network construction or factory automation. Resulting from the improvements in both the stability and the transmission speed of more sophisticated digital PLC method, it's field of application is growing [1] and well-adapted [1 - 12].

2   Related Research

2.1   PLC between Electronic Ammeter and Main Server

[Fig. 1] shows the PLC between the electronic ammeter and the main server. To use the PLC module, a DCU (Date Concentration Unit) has been created by combining it with the MCU which is to control the module. This DCU stores the usage measured by the ammeter and assumes the role of delivering it to the main server on a regular cycle. Then, the measured usage will be stored on the server after being linked to the PLC module of the main server, subsequently outputting it on the Smart Phone or the external display as needed by the user. RS-232 communication is performed among the PLC module, MCU and main server. This normally is a serial-type interface that accesses modems or others and usually called a "serial port". To achieve this communication method, RS-232 communication must be made possible, and also, a communication protocol will be required.

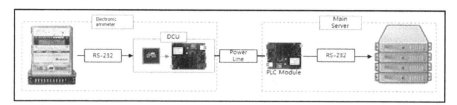

Fig. 1 PLC between electronic ammeter and main server

2.2   Connection System Diagram

[Fig. 2] is a connection system diagram between watt-hour meter, field terminal and DCU, where the field connection unit is connected using a modem communication line projecting out of digital watt-hour meter. Then, the PLC module is connected with RS-232 communication and field connection units, finally establishing the communication with the server through power line. The connection between the digital watt-hour meter and field connection unit is made using the optical communication medium. This part is supplied by the manufacturer of the digital watt-hour meter so that we've used the field connection unit in its original form - judging that it'd be a better choice.

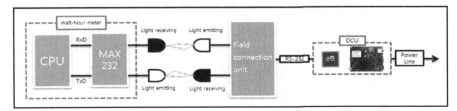

Fig. 2 Connection system diagram between watt-hour meter, field terminal and DCU

3   Basic Technology for Realtime Smart Metering System

3.1   Relationship Diagram of the Power System Control
         Programs Using PLC

The implementation diagram of the usage information transmission is shown in [Fig. 3] The power usage information on the watt-hour meter is recorded through MCU (yellow quadrangle). It transmits the information through the PLC module (red quadrangle).

Fig. 3 Implementation diagram   **Fig. 4** Server implementation diagram

Server implementation diagram ① and ② in [Fig. 4] is the parts that are used to control communication of each device. ③ is used for the communication support between the server and the watt-hour meter. These have been implemented with C and C++ languages and the flow chart is presented in [Formula 1~23]. Indicators are indicated as *Ap* and *An*, where *Ap* is the total numbers of past access address and *An* is the total numbers of present access address.

$$f(x) = (\sum Number(A_p) - \sum Number(A_n)) \tag{1}$$

$$\textbf{If } f(x) = 0 \tag{2}$$

$$\textbf{If } \sum Time(A_n) \geq 7 Days \tag{3}$$

$$\textbf{Then } \text{Warning Message transmission to the User's family} \tag{4}$$

$$\textbf{Not } Return(f(x)) \tag{5}$$

$$\textbf{Not } (=f(x) \neq 0) \tag{6}$$

$$\textbf{If } \sum Number(A_n) \geq 2 (= g(x)) \tag{7}$$

$$\textbf{Then } Return(g(x)) \tag{8}$$

$$\textbf{Not} \tag{9}$$

$$\textbf{If } \sum Time(A_n) \geq 3 Days \tag{10}$$

$$\textbf{Then } \text{transmit visitation request message transmission to the volunteer center} \tag{11}$$

$$\textbf{Not } Return(f(x)) \tag{12}$$

Formula 1 compares the current access address with the past access address; Formula 2 is the case when the current access address and past access address are identical; Formula 3 is when the current access address's access time remains the same for more than a week; then, Formula 4, Warning Message is transmitted to the User's family; and Formula 5 returns to $f(x)$ comparison. Formula 6 is the case when current access address and past access address are not identical; Formula 7 is for the case when assumption is made that the number of current access address is more than $2(g(x))$; returns to $g(x)$ comparison in Formula 8; for Formula 10, if the access address's access time remains the same for 3 or more days; then transmit visitation request to the volunteer center in Formula 11; and finally returning to $f(x)$ comparison in Formula 12.

$$P(x) = \text{Wi-Fi determines the number of connecting users} \tag{13}$$

$$H_{IP} = \text{Home IP address} \tag{14}$$

$$P(x) = \text{① Wi-Fi connection} \tag{15}$$
$$\text{② SSID search}$$
$$\text{③ } N_{IP} = \text{IP address of accessing Wi-Fi router}$$

$$\textbf{If } H_{IP} = N_{IP} \tag{16}$$

$$\textbf{Then } P(x) = C(x) + 1 \tag{17}$$

$$\text{Wi-Fi disconnection} \tag{18}$$

$$\textbf{if } H_{IP} = N_{IP} \tag{19}$$

$$\textbf{Then } P(x) = C(x) - 1 \tag{20}$$

$$\text{end} \tag{21}$$

$$\textbf{Not } \text{end} \tag{22}$$

Formula 13, Wi-Fi determines the number of connecting users; Formula 14 allocates home IP address; in Formula 15, ① checks Wi-Fi connection; ② searches SSID; ③ allocates current access address; Formula 16 checks whether the home address and the current access address are identical; Formula 17 checks increase in the number of connecting users; Formula 18 disconnects Wi-Fi; Formula 19 checks whether the home address and current access address are identical; Formula 20 reduces the number of connecting users; Formula 22 ends the function; Formula 23 returns to P(x).

3.2 RS232-Based Protocol Design

A certain protocol is necessary to link a server to respective devices using the PLC. For the reason that PLC module assumes the Broadcasting type form basically, each device needs a distinct unique number, and for later usage purposes, 'Unused' is included. The protocol has been created referring to the TCP/IP model and the '+' sign, which means that the message transmission has ended, must be included at the last part. Also, the '-' signs must be included in between each word. The diagram showing the execution process for the commands between the devices is shown in [Fig. 5].

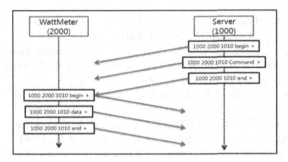

Fig. 5 RS232-based protocol design

Basically, the server is a main subject so that there's no synchronization process and all of the devices are on standby until the communications are directed to them. Additionally, in order to control each device group, respective group broadcasting numbers have been designated. This is to effectively control entire power sockets in home simultaneously, and the foremost number of each device group is the broadcasting number (e.g., in the case of general device group, the number is 3000).

4   Conclusion

The problems related to the proposed system can be classified as following three kinds. First, the time needed for a network connection. A few seconds delay takes place for the present implemented system to connect with WiFi, check In-home headcount, and then control the home devices through the PLC. Such a delay time could lead to user-inconvenience. Second, because of the nature of the PLC, vulnerability to the noises and its instability limit system operation, exposing its environmental vulnerability. However, such vulnerabilities can be seen as the problems which have originated from the limitations of present PLC and network technologies, rather than the problems arisen from the system design and implementation stages. Therefore, it is expected that the problems will be surmounted. Finally, as an external environmental problem, the users might not carry smartphones. Since many senior citizens and children often do not carry this device, the generality of the system could be reduced. However, should the current trend of increasing generalization of wearable devices such as smart watches and the proposed system are combined together, we expect that there will be boundless possibilities for the system even if the users do not use smartphones or tablet PCs. They can be replaced with remote controller devices like key holders or other accessory-type devices. In future studies, the models which could overcome such problems will be dealt with.

References

1. Hrasnica, H., Haidine, A., Lehnert, R.: Broadband Powerline Communications Networks: Network Design. Wiley, New York (2004)
2. Xiao-sheng, L., Liang, Z., Yan, Z., Dian-guo, X.: Performance analysis of power line communication network model based on spider web. In: 8th International Conference on Power Electronics ECCE Asia, pp. 953–959 (2011)
3. Sabolic, D.: Influence of the transmission medium quality on the automatic meter reading system capacity. IEEE Trans. Power Del. **18**, 725–728 (2003)
4. Rus, C., Kontola, K., Igor, D.D.C., Defee, I.: Mobile TV content to home WLAN. IEEE Transactions on Consumer Electronics **54**, 1038–1041 (2008)
5. Peter, M.C., Desbonnet, J., Bigioi, P., Lupu, L.: Home network infrastructure for handheld/wearable appliances. IEEE Transactions on Consumer Electronics **48**, 490–495 (2002)
6. Hwang, T., Park, H., Paik, E.: Location-aware UPnP AV session manager for smart home. In: IEEE First International Conference on Networked Digital Technologies, pp. 106–109 (2009)
7. Helal, S., Mann, W., El-Zabadani, H., King, J., Kaddoura, Y., Jansen, E.: The Gator Tech Smart house: A programmable pervasive space. IEEE Computer Society, 50–59 (2005)
8. Liau, W.-H., Chao-Lin, W.: Li-Chen Fu.: Inhabitants tracking system in a cluttered home Environment Via Floor Load sensors. IEEE Transactions on Automation Science and Engineering **5**, 10–20 (2008)
9. Ou, C.-Z., Lin, B.-S., Chang, C.-J., Lin, C.-T.: Brain computer interface-based smart environmental control system. In: 2012 Eighth International Conference on Intelligent Information Hiding and Multimedia Signal Processing IEEE Computer Society, pp. 281–284 (2012)

10. Huh, J.-H., Seo, K.: Design and implementation of the basic technology for solitary senior citizen's lonely death monitoring system using PLC. Journal of Korea Multimedia Society **18**(6), 742–752 (2015)
11. Huh, J.-H., Seo, K.: AMI design and testbed using intelligent agents on the PLC network. In: The 2015 World Congress on Information Technology Applications and Services "Advanced Mobile, Communications, Security, Multimedia, Vehicular, Cloud, IoT, and Computing" (World-IT 2015) (2015)
12. Huh, J.-H., Lee, D.-G., Seo, K.: Design and implementation of the base technology for realtime smart metering system using PLC. In: The 2015 World Congress on Information Technology Applications and Services Proceedings of the "Advanced Mobile, Communications, Security, Multimedia, Vehicular, Cloud, IoT, and Computing" (World-IT 2015) (2015)

Parallel Balanced Team Formation Clustering Based on MapReduce

Byoung Wook Kim, Ja Mee Kim, Won Gyu Lee and Jin Gon Shon

Abstract For effective cooperative learning grouping student is important. Grouping students can be generalized to the problem that clustering objects into some clusters from a computer science point of view. The large datasets, expensive task of clustering computationally and high dimensionality makes clustering of very large scale of data a challenging task. To effectively process very large datasets for clustering, parallel and distributed architectures have developed. MapReduce is a programming model that is used for handling large volumes of data over a distributed computing environment in parallel. In this paper, we present a Parallel Balanced Team Formation (PBTF) clustering algorithm for the MapReduce framework. The purpose of PBTF is to find partitions with high homogeneity in a group and high heterogeneity between groups in parallel.

Keywords Grouping student · Parallel · Mapreduce · Team formation

B.W. Kim
Department of Computer Science Educationa, Korea University, Seoul, Korea
e-mail: byoungwook.kim@inc.korea.ac.kr

J.M. Kim
Major of Computer Science Education, Graduate School of Education, Korea University, Seoul, Korea
e-mail: jamee.kim@inc.korea.ac.kr

W.G. Lee
Department of Computer Science and Engineering, Korea University, Seoul, Korea
e-mail: lee@inc.korea.ac.kr

J.G. Shon(✉)
Department of Computer Science, Korea National Open University, Seoul, Korea
e-mail: jgshon@knou.ac.kr

© Springer Science+Business Media Singapore 2015 671
D.-S. Park et al. (eds.), *Advances in Computer Science and Ubiquitous Computing*,
Lecture Notes in Electrical Engineering 373,
DOI: 10.1007/978-981-10-0281-6_95

1   Introduction

Recently, some studies are underway to divide the students into some group in the computational aspects [2][3]. To have the desired effect, it is important to group with considering group member's skills needed by the task and communication ability to communicate each other member. Grouping students can be generalized to the problem that clustering objects into some clusters from a computer science point of view.

Cluster plays an important role in many respects such as knowledge discovery, data mining, information retrieval and pattern classification. The large datasets emerging by the technical progress, makes clustering of very large scale of data a challenging task since clustering is computationally expensive task and most of real world data is high dimensional. To effectively process these very large datasets, parallel and distributed architectures have become popular [4], and various parallel clustering algorithms have been developed and applied to many applications [5].

MapReduce is a distributed computing framework for large datasets as a programming model. MapReduce is consisted of two computation stages – map and reduce. In the map stage, an input dataset is partitioned into disjoint parts and distributed to workers called mappers. A map function processes a <key, value> pair to generate a set of intermediate <key, value> pairs. The mappers execute given tasks on local data. A reduce function integrates all intermediate values associated with the same intermediate key. The advantage of MapReduce is the fact that many map tasks can run in parallel.

In this paper, we adapt Balanced Team Formation algorithm using MapReduce framework to make the clustering method applicable to large scale data. By defining map, combine and reduce function, the proposed algorithm can be parallel executed effectively.

2   Parallel Balanced Team Formation Algorithm

In this section we present the main design for Parallel Balanced Team Formation (PBTF) algorithm based on MapReduce. We give a brief overview of the BTF algorithm presented by our previous research and define the parallel parts and serial parts in the algorithms.

2.1   Balanced Team Formation Algorithm

BTF algorithm is a clustering method which finds partitions with high homogeneity in a group and high heterogeneity between groups [1]. It takes the input parameter, k, and partitions a set of n students $S = \{s_1, s_2, \cdots, s_n\}$ into k clusters so that the resulting intra-cluster heterogeneity is high and the inter-cluster homogeneity is also high. The formal definitions of this problem is given below.

Problem (Team Formation for Cooperation Learning). *Given an integer k and a set n students $S = \{s_1, s_2, \cdots, s_n\}$, and task T, find a partition of S into groups $G_1, \cdots, G_l$, where the size of each group is k and $H_e(G)$ and $H_o(G)$ are maximized.*

Cluster homogeneity and heterogeneity are measured according to [1]. The BTF algorithm is shown in algorithm 1.

BTF algorithm has a similarity with k-means algorithm in term of clustering given a set of n student into a given k clusters based on similarity measure and comparing each data object to all cluster center to measure similarity between an data object and a cluster center. However, the differences of BTF algorithm and k-means algorithm are that BTF algorithm does not need iteration operation and have to yield clusters with having the same number of data object included in a cluster equally.

Algorithm 1. Balanced Team Formation Algorithm

Input: Set of students $S = \{s_1, s_2, \cdots, s_n\}$ with sorted total abilities $\theta_1 > \theta_2 > \cdots > \theta_n$, group size k, and number of groups r *($2 \leq r \leq n/2$)*.
Output: Group G.
1: $G = \emptyset$
2: Randomly choose r students as s_i $(i = 1, \cdots, r)$
3: $S \leftarrow S \setminus \{s_i\}$
4: **for** i=1 to r **do**   /*assign r student to each group */
5:     $G_i \leftarrow s_i$
6: **for** j=1 to $(n\text{-}r)$ **do**
7:     **for** i=1 to r **do**
8:         $G_i \leftarrow s_j$ /*assign a student to a group temporary */
9:         $H_i \leftarrow H_e(G) + H_o(G)$ /*calculate homogeneity and heterogeneity*/
10:        $G_i \leftarrow G_i \setminus \{s_j\}$ /*remove a student from G_i*/
11:    Assign $G_i \leftarrow s_j$ with the highest H_i among G
12: **return** G

2.2   Parallel Balanced Team Formation Algorithm

In this paper, we present PBTF algorithm based on MapReduce. First, PBTF splits the data set into several smaller partitions on the mapper. The map function assigns each student vector to the closet cluster. The combine function is executed on each machine to combine each <key, value> pair with same key. In order to reduce the cost of network communication, a combiner function deals with partial combination of the intermediate values with the same key within the same map. The reduce function performs the procedure of updating the new centers.

2.2.1   Map-Function

The input data containing student information is stored as a sequence file of <key, value> pairs. The key is the the number of final cluster, k, and the value is a vector of the student information of this record. The dataset is split and broadcast to all mappers. The cluster homogeneity and heterogeneity's computations are executed in parallel.

For each map, BTF is executed on given values which are student vector assigned each map function. Given the student vector, a mapper can partition each student vector to each cluster. The intermediate values are then consisted of two parts: the number of cluster and the student vector. The pseudocode of map function is shown in Algorithm 2.

Algorithm 2. map (*key, value*)

Input: The *key* is the number of total cluster and the *value* is the student vectors

Output: *<key', value'>* pair where the *key'* is the number of cluster and *value'* is the student id

1: Execute BTF function (*value*)

2: Store the results of BTF function into *<key', value'>*

3: Output *<key', value'>*

2.2.2   Combine-Function

After each map, we adopt a combiner to combine the intermediate data of the same map task. As using combiner, we can reduce the communication cost, because the intermediate data is stored in local disk of each machine.

In the combine function, we integrate the values of the each student vector assigned to the same cluster. The pseudocode for combine function is shown in Algorithm 3.

Algorithm 3. combine (*key, V*)

Input: The *key* is the number of cluster, the *V* is the list of the student vector

Output: *<key', value'>* pair where the *key'* is the number of cluster and *value'* is the list of student id

1: Assign *key' = key*

2: while (V.IsNotEmpty())

3:    *value' = value' +* "," *+* V.getID()   /* Construct *value'* as a string consisted of the list of student id comma-seperated */

4: Output *<key', value'>*

2.2.3   Reduce-Function

The input data of the reduce function is obtained from the combine function of each machine. As described in the combine function, the input data includes partial sum of the student vectors in the same cluster. In reduce function, we integrate all the student vectors assigned to the same cluster. The pseudocode for reduce function is shown in Algorithm 4.

Algorithm 4. reduce (*key, V*)

Input: The *key* is the number of cluster and the *V* is the list of the student id comma-seperated

Output: <*key', value'*> pair where the *key'* is the number of cluster and *value'* is the list of student id

1: Assign *key' = key*

2: while (V.IsNotEmpty())

3:    *value' = value'* + "," + V.getID()   /* Construct *value'* as a string consisted of the list of student id comma-seperated */

4: Output <*key', value'*>

---

3 Conclusion

The increase of data and the need to extract useful information from the large datasets motivates the development of scalable algorithms for data mining. In this paper, we proposed PBTF algorithm for finding partitions with high homogeneity in a group and high heterogeneity between groups in parallel using MapReduce. We developed three functions; map function, combine function and reduce function. PBTF algorithm extended BTF algorithm to be executed in parallel. BTF algorithm is executed for a given partition data in each mapper.

In the future, we plan to evaluate the performance of our algorithm with respect to scaleup, speedup and sizeup and to implement our algorithm on in-memory distributed computing environment.

References

1. Kim, B.W., Chun, S.K., Lee, W.G., Shon, J.G.: The greedy approach to group students for cooperative learning. In: The 16th International Conference on Parallel and Distributed Computing, Applications and Technologies (2015)
2. Majumder, A., Datta, S., Naidu, K.V.M.: Capacitated team formation problem on social network. In: KDD, pp. 1005–1013 (2012)
3. Agrawal, R., Golshan, B., Terzi, E.: Grouping students in educational settings. In: Proceedings of the 20th ACM SIGKDD International Conference on Knowledge Discovery and Data Mining (KDD 2014), pp. 1017–1026. ACM, New York (2014)
4. Abouzeid, A., Bajda-Pawlikowski, K., Abadi, D.J., Silberschatz, A., Rasin, A.: HadoopDB: an architectural hybrid of MapReduce and DBMS technologies for analytical workloads. In: Pro-ceedings of the Conference on Very Large Databases (2009)
5. Rasmussen, E.M., Willett, P.: Efficiency of Hierarchical Agglomerative Clustering Using the ICL Distributed Array Processor. Journal of Documentation **45**(1), 1–24 (1989)

2D Barcode Localization Using Multiple Features Mixture Model

Myeongsuk Pak and Sanghoon Kim

Abstract A 2D barcode region localization system for the automatic inspection of logistics objects has been developed. For the successful 2D barcode localization, variance and frequency of the pixel distribution within average 2D barcodes is modeled and the average model of 2D barcode is combined with the corner features detection to localize the final 2D barcode candidates. An automatic 2D barcode localization software was developed with the multiple features mixture method and we tested our system on real camera images of several popular 2D barcode symbologies.

Keywords 2D barcode · Barcode localization

1   Introduction

1D or 2D Barcode systems are very important in logistics, product packaging and other various commercial applications. Due to the limited capacity of 1D barcodes, 2D barcode systems have been widely used in recent years. In the Two-dimensional barcode symbols, data are encoded in both the height and width of the symbol, and the amount of data that can be contained in a single symbol is significantly greater than that stored in a one dimensional symbol. Obviously, the main advantage of using 2D bar codes is that possibly a large amount of easily- and accurately-read data can "ride" with the item to which it is attached. There are new applications being created for 2D bar code technology every day. More than thirty different 2D-barcodes are currently in use. Common examples of 2D symbologies include DataMatrix, PDF417, Aztec Code, Codeblock, MaxiCode, QR Code. Whereas 1D barcodes are traditionally scanned with rotating laser illumination and linear sensor arrays, 2D barcode systems require imaging sensors for scanning and acquisition. Image-based 2D barcode scanning technologies also have many challenges, but have many efficient functions like very high data capacity and an ability of error recovery.

Many image-based methods for barcode identification have been developed. Ouaviani *et al.* adopted some image processing techniques to segment some of the

M. Pak · S. Kim(✉)
Department of Electrical, Electronic and Control, Hankyong National University,
327, Jungang-ro, Ansung-city, Kyonggi-do, Korea
e-mail: kimsh@hknu.ac.kr

© Springer Science+Business Media Singapore 2015   677
D.-S. Park et al. (eds.), *Advances in Computer Science and Ubiquitous Computing*,
Lecture Notes in Electrical Engineering 373,
DOI: 10.1007/978-981-10-0281-6_96

most common 2D barcodes, including the QR code, Maxicode, Data Matrix, and PDF417 [1]. Jancke et al. [2] introduced a new technique in 2d barcode localization and segmentation, by using a process of thresholding, orientation prediction and then corner localization. Tan et al. [3] develop a recognition algorithm which first finds the location of the finder patterns. Then search for the L-shape guide bar(part of the 2d barcode), and Performing projective mapping to correct symbol distortion to localize the barcode region. Huaqiao et al. [4] localize the barcode region using texture direction analysis and hough transform. Xu and McCloskey focused on solving the localization and de-blurring problem of motion-blurred 2D bar codes using corner feature and motion direction estimation [5].

Most other techniques in literature usually work for single 2D barcode, and rely on finding the unique pattern, or are based on the assumptions that 2D barcodes are close to the scanning cameras. But our researches are focused on to increase the detection rate of 2D barcode area candidate even when the scanning cameras are in distant from the printed barcodes as in practical barcode inspection application systems.

In this research, A 2D barcode region localization system for the automatic inspection of logistics objects has been developed. For the successful 2D barcode localization, variance and frequency of the pixel distribution within average 2D barcodes is modeled and the average model of 2D barcode is combined with the corner features to localize the final 2D barcode candidates. An automatic 2D barcode localization software was developed with the multiple features mixture method and we tested our system on real camera images of several popular 2D barcodes.

2   2D Barcode Detection System in Our Approach

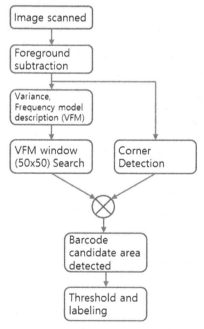

Fig. 1 Overall flow diagram of the research

2.1   2D Barcode Localization

This paper proposes a novel method for localization of 2D QR Barcode images. Firstly, pre-processing like foreground subtraction technique is performed on input image to get the packaged objects moving through the conveyer system in logistics environment. Then variance-frequency model(VFM) of the average 2D QR barcodes is calculated to characterize the normal QR codes based on the variance and frequency of the distribution of the pixels included in 2D QR barcode. And then using the model of VFM features, we scan the whole input image with the 50x50 window to verify the similarity between the VFM distribution and the corresponding scanning area. The detected candidate areas are finally combined with the corner features detected results[5] to increase the detection accuracy of 2D barcode.

2.2   Corner Features for 2D Barcode Localization

We employed corners as low-level features for localizing 2D barcode. A 2D barcode is a pattern comprised of small, rectangular black patches on a white background. As a result, its gradient orientation histogram has two strong peaks at orthogonal orientations. Corner features, whose own localization is determined by orthogonality in local gradient orientation distribution, provides a natural tool for us to localize barcode area from this gradient prior[5].

2.3   Variance-Frequency Distribution Model

Compared to the other patterns in image, all pixel values within 2D QR codes have obvious gray level distribution of black and white group(they keep uniform variance of the distribution) and they are also keeping very high and uniform frequency in gray level changes). So we describe the Variance-Frequency Model(VFM) of the average QR codes to show the general features of the QR codes.

Given our computed density map of input QR code group, we estimate the barcode region with the above model. First, since a barcode has strong black/white contrast, its appearance will still have relatively high variance. We measure a score S_1 as:

$$S_1 = \mathrm{Var}(P) \tag{1}$$

where Var is the variance. Second, a barcode region has a very high frequency of black/white rectangles, thus we define S_2 as:

$$S_2 = \mathrm{Freq}(P) \tag{2}$$

Where Freq is the frequency of the pixels value distribution.

Third, a barcode region must be a concentration of corners, thus we define S_3 as:

$$S_3 = \sum_{(x,y \in C(P)} M(x,y) \tag{3}$$

where C(P) is the set of all detected corners in the patch P, M(x, y) is the corner strength (magnitude) map. In our research, the size of QR code basic cell captured has the range of 3~5 pixels because QR code is usually being captured badly from a long distance camera in the logistics environment, therefore we suppose that the size of QR code is more than 60x60 and VFM window is more than 50x50.

2.4   Estimation of Final QR Code Region

Based on the modeling and score description of QR code candidates region above, we finally estimated the region by simply adding the three scores S1,S2,S3 numerically and we give a candidate definition window only to the region with the all 3 score values

3   Experiments

We tested the proposed algorithm on several test images captured in practical packaged objects printed with QR codes and two examples of them are shown in Fig. 2 and Fig. 3 and the results shows the algorithm is working well even when there are many similar patterns are existing. The proposed algorithm is implemented using Visual C++ programming environments and the test image size is 640x480 and 1280x720 with 24bits color.

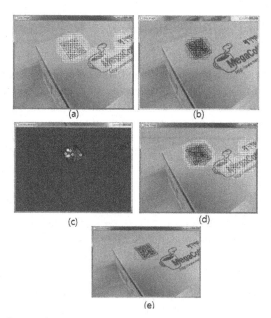

Fig. 2 Experiment for test image 1 (a) detected region with target variance (b) detected region with target frequency (c) detected corner features (d) combined region with VFM and corner features (e) final detected candidate window

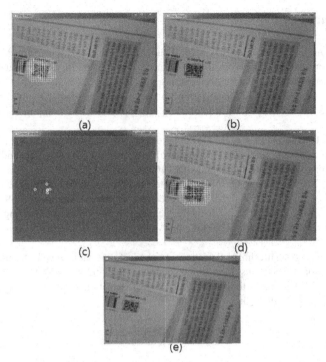

Fig. 3 Experiment for test image 2 (a) detected region with target variance (b) detected region with target frequency (c) detected corner features (d) combined region with VFM and corner features (e) final detected candidate window

4   Conclusions

In this research, A 2D barcode region localization system for the automatic inspection of logistics objects has been developed. For the successful 2D barcode localization, variance and frequency of the pixel distribution within average 2D barcodes is modeled and the average model of 2D barcode is combined with the corner features to localize the final 2D barcode candidates. An automatic 2D barcode localization software was developed with the multiple features mixture method and we tested our system on real camera images of several popular 2D barcodes.

Acknowledgments This work was supported by a research grant from Gyunggi-do(GRRC) in 2013~2015 [(GRRC Hankyong 2012-B02).

References

1. Ouaviani, E., Pavan, A., Bottazzi, M., Brunelli, E., Caselli, F., Guerrero, M.: A common image processing framework for 2D barcode reading. In: 7th International Conference on Image Processing and Its Applications, vol. 2, pp. 652–655 (1999)
2. Jancke, G., Parikh, D.: Localization and segmentation of a 2D high capacity color barcode. In: Proc. 2008 IEEE Workshop on Applications of Computer Vision, pp. 1–6 (2008)
3. Tan, K.T., Chai, D., Kato, H.: Development of a novel finder pattern for effective color 2D-barcode detection. In: International Symposium on Parallel and Distributed Processing with Applications, pp. 1006–1013 (2008)
4. Hu, H., Xu, W., Huang, Q.: A 2D barcode extraction method based on texture direction analysis. In: Fifth International Conference on Image and Graphic, pp. 759–762 (2009)
5. Xu, W., McCloskey, S.: 2D barcode localization and motion deblurring using flutter shutter camera. In: IEEE Workshop on Applications of Computer Vision, pp. 159–165 (2011)
6. Otsu, N.: A Threshold Selection Method from Gray-Level Histogram. IEEE Transactions on Systems, Man, and Cybermetics 9(1), 62–66 (1979)
7. Ng, H.-F.: Automatic thresholding for defect detection. Pattern Recognition Letters 27(14), 1644–1649 (2006)

LPLB: A Novel Approach for Loop Prevention and Load Balancing in Ethernet Ring Networks

Saad Allawi Nsaif, Nguyen Xuan Tien and Jong Myung Rhee

Abstract Ethernet networks using rapid spanning tree protocol (RSTP) to ensure a loop-free topology and provide redundant links as backup paths in case an active link has failed. However, when a failure occurs, RSTP requires a significant amount of reconfiguration time in order to find an alternative path. RSTP is also limited by the number of nodes in a ring network, and its performance degrades when the number of nodes increases. In this paper, we introduce a new approach, called loop prevention and load balancing (LPLB), which can be applied to Ethernet ring networks. If failure occurs, LPLB only requires a very short amount of time to switch to an alternative path and in most cases; LPLB needs zero recovery time for that switching. In addition, under most situations, no data frames are lost when a node switches to an alternative path. Unlike RSTP and the media redundancy protocol (MRP), LPLB also provides load balancing among network links that in turn improves the network performance and reduces the probability of bottleneck occurrence.

Keywords: LPLB · Industrial ethernet · Path redundancy · Loop prevention

1 Introduction

The Ethernet, standardized by the Institute of Electrical and Electronics Engineers (IEEE) in IEEE 802.3 [1], is not capable of supporting a fault-tolerant network. The avoidance of any loops is a basic requirement for every Ethernet network. Each loop would result in data frames circulating forever, thus flooding the network. Since the standard Ethernet does not support fault tolerance, various Ethernet redundancy protocols have been developed and standardized, such as the

S.A. Nsaif · N.X. Tien · J.M. Rhee(✉)
Information and Communications Engineering Department,
Myongji University, Seoul, South Korea
e-mail: {saad.allawi1,nxtien}@gmail.com, jmr77@mju.ac.kr

© Springer Science+Business Media Singapore 2015
D.-S. Park et al. (eds.), *Advances in Computer Science and Ubiquitous Computing,*
Lecture Notes in Electrical Engineering 373,
DOI: 10.1007/978-981-10-0281-6_97

rapid spanning tree protocol (RSTP) [2], the media redundancy protocol (MRP) [3], the parallel redundancy protocol (PRP) [4], and the high-availability seamless redundancy (HSR) [4]. These four protocols are suitable for a ring topology. RSTP can be applied to arbitrary mesh topologies. It implements a distributed computation of a tree based on path costs and priorities. This tree is the active topology, which is established by blocking the switch ports. If a failure occurs, RSTP usually requires an upper bound of 2 seconds per switch [5], because once a link or a switch fails, the network undergoes reconfiguration to rebuild the logical paths. This makes RSTP unsuitable for some message timeout requirements in time-critical applications. In MRP, a dedicated node, the ring manager blocks one of its ring ports in order to establish a line as the active topology. If a failure occurs, this line breaks into two isolated lines that are reconnected by de-blocking the previous blocked port. The reconfiguration time is in the 100 ms range, and it can be guaranteed. In contrast to the two previously mentioned methods, PRP does not change the active topology. It operates on two independent networks. Each frame is replicated on the sending node and transmitted over both networks. The receiving node processes the frame that arrives first and discards the second copy. HSR is a special version of PRP, and it is usually applied to a ring topology. Compared with other approaches, in HSR, the available network bandwidth is halved because two frames of every data frame are transmitted over the ring. The drawback of using RSTP and MRP is that both methods have a switchover delay. PRP and HSR both provide zero recovery time, but PRP requires a duplicated network infrastructure and HSR excessively generates unnecessary redundant frames. In this paper, we propose a novel approach, called looping prevention and load balancing (LPLB), to solve the issues related to using RSTP and MRP protocols in a ring topology. These issues are: 1) it takes a long time for these protocols to find an alternative path during failures and 2) these protocols always disable one of the ring ports to avoid the looping issue, which causes some data frames to take the longest path, even if the destination node is only one hop away from their source node. This may cause some congestion or bottleneck in the ring because not all of the links are fully utilized.

The rest of the paper is organized as follows. In Section 2, we briefly introduce the concept and setup procedure for our proposed LPLB approach. We also describe the monitoring and recovering procedure for that approach. Section 3 presents the results of the LPLB performance analysis. Finally, in Section 4, we present our conclusions and make recommendations for future work.

2   LPLB Approach

2.1   LPLB Concept and Setup Procedure

Our LPLB approach is used in an Ethernet ring to ensure a loop-free network in addition to providing path redundancy with faster reconfiguration and recovery time, in comparison to RSTP or MRP. The LPLB approach is suitable for Ethernet nodes, such as an intelligent electronic device (IED), which has two ports that share the same media access control (MAC) and Internet protocol (IP) addresses, which use the standard Ethernet protocol. The LPLB approach also provides better load balancing among the ring's links. The LPLB approach uses the standard Ethernet frame layout to send and receive data frames. However, it will use only two special frames with private code of Ethertype field, one during the setup process and the second during monitoring and recovery of the network procedure. Since the LPLB approach uses the same layout of the Ethernet frame format, all of the terminal off-shelf Ethernet devices have the ability to work under both the LPLB approach and the standard Ethernet protocol, at a lower cost and with fewer implementation complexities.

The setup procedure for the LPLB approach is described below.

- At the beginning, each Ethernet node broadcasts a special frame, called a port selection (PS) frame, to select the proper port that quickly leads to each destination. This port is called the primary port (PP). The PS frame is sent from the nodes of both ports.
- Each node reads all the PS frame copies that will pass through it. The node will read the source MAC address, the sequence number, and the port number from which the fastest PS frame per each sending node is delivered. This information will be tabulated into a table called a primary port (PP) table. In this table, each node identifies which port has first delivered each PS frame copy per each sending node. Consequently, each node will know which of its ports is fast in terms of reaching each neighbor node.
- Each node will ignore the second PS frame copy per each sending node that will later be passed through it from the second port. However, after each PS frame copy is read, it will be forwarded to the opposite direction until it reaches the source node that is going to delete it.
- Any additional Ethernet node that might be added to the ring during network running will broadcast a PS frame, and, in turn, the other nodes will understand that there is a new node because there is a new PS frame with a new source MAC address that is not listed in the PP table. Therefore, the nodes will update their PP tables and then broadcast their PS frames in order that the new node builds its PP table.

Now, it can be said that the setup process for the LPLB approach is completed and all the nodes are ready to send their data frames to neighbor nodes, according to their PP tables. For example, as shown in Fig. 1, node A will send its data frames to node B through port 2, but it will send its data frames to node F through port 1. This is a type of load balancing that occurs among ring links. Table 1 shows the PP table of node A for the network shown in Fig. 1.

Table 1 Primary port (PP) table for node A, which is the ring network shown in Fig. 1.

Source MAC address	Sequence No.	Primary port No.
Node B	XXXX	2
Node C	XXXX	2
Node D	XXXX	1
Node E	XXXX	1
Node F	XXXX	1

xxxx: Any sequence number

Under multi/broadcast traffic, the LPLB approach will work as follows:

- As shown in Fig. 1, if node A broadcasts a frame, then the first copy will travel through port 1 towards nodes F, E, and D, and each of these nodes will take a copy from the sent frame and then forward it from the opposite direction to the next node. Finally, node D will forward it to node C. The second copy will travel through port 2 towards nodes B and C. and both of those nodes will take a copy and forward it on. Finally, node C will forward the copy to node D.
- Node D will receive two frame copies, one from each direction; however, according to node D's PP table, it is assumed that node D will receive the fastest copy from node A through port 1. Therefore, node D will take a copy from that frame and then forward it to node C through port 2, and node D will delete the second frame copy that will be delivered from port 2 because the PP table shows that node D only receives the fastest copy from node A through port 1, not port 2.
- Suppose that node C's PP table shows that node C receives the fastest frame copy from node A through port 2; in that case, it will take a copy and then forward it to node D. As shown in Fig. 1, node C will delete the second copy that is delivered to it through port 1 as long as its source MAC address is node A.

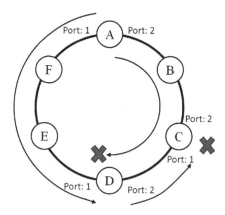

Fig. 1 LPLB behavior under multi/broadcast traffic.

LPLB will provide looping-free delivery and will distribute the data frames from both sides of the nodes, which will ensure a type of load balancing among the ring's links in comparison to the RSTP and MRP protocols.

2.2 Monitoring and Recovery Procedure

The LPLB approach will periodically monitor the network's status and recover the network path if a failure occurs. Moreover, the LPLB will reselect the primary ports of each node to make sure that each node connects with the others through the fastest, failure-free path. However, since the PP table contains the MAC addresses for all the nodes, each node will know which node has the lowest MAC address value. That node will be called the LPLB root node. The root node will manage the monitoring and recovery procedure as described below:

- The root node will periodically broadcast a frame, called the network monitoring status (NMS) frame, to check whether the network has a failure. This frame will be sent through both ports of the root node, and it will be received by the opposite ports. The NMS frame will be sent at one-second intervals, as we assumed. However, within 500 ms, if the root node did not get both of the NMS frame copies it will know that there is a failure in the network. In that case, the root node will broadcast a PS frame, and, in turn, all the nodes will do the same thing in order to reselect the primary ports for each destination node, before updating their PP tables. In other words, LPLB will be restarted.

- However, if the network nodes did not receive any NMS frame within one second from the last time that they saw the NMS frame, then they will broadcast their PS frames to restart LPLB. However, if a failure occurs, the root node will continue to periodically send the NMS frame, without sending the PS frame, because the node has sent that frame earlier after the failure was discovered.

- If any of the network nodes detects a physical-link failure, then that node will send all its data frames from the failure-free direction until the other failed direction is repaired. This is will be done through changes the primary port of the required destination and updates the node's PP table. That node will also broadcast a PS frame to acknowledge the network nodes to reselect the primary ports for each destination by broadcasting their PS frames.

The following section will discuss how the LPLB approach can be used to recover information if a failure occurs, assuming that the data frames that are being transmitted are either unicast traffic or a multi/broadcast traffic.

2.2.1   Unicast Data Frame Traffic

As shown in Fig. 2, if node E has discovered a physical-link failure, then it will broadcast a PS frame through port 2 and update its PP table by changing all the primary ports for all the destination nodes to port 2, and during that time, node E will receive a unicast data frame from node C through port 2 going to node F. In that case, and according to the PP table of node E, node E will send that frame back to node C through port 2 because there is no path to the required destination. This will be done by switching the received data frame from the input queue into the output queue, then sending it back to its source node, which is node C. Consequently, node C will understand that the path of the required destination no longer exists; therefore, node C will redirect that data frame to the opposite port. Node C will also redirect all future data frames going to node F to that opposite port, update its PP table, and then broadcast its PS frame to acknowledge other nodes. This process will occur if node C has not yet received the PS frame of node E and its PP table has not been updated yet.

In general, any node that receives a data frame that has been sent from the same received port will update its PP table by changing the primary port of that destination into the opposite port, and then it will broadcast its PS frame.

In this way, we avoid deleting the sent data frames that have no path to their required destination. Instead, the LPLB approach redirects those data frames back to their source node to let that node redirect the frames from the opposite direction and send them to the required destination. This process enables all the nodes to have the ability to send their data frames at any time, even during the port reselection process. Therefore, most of the data frames will not be lost during a failure. In other words, most of the time, we can say that LPLB provides a redundancy with zero-recovery time, except in the case in which a node has sent a frame through its primary port and the link of that port is down.

2.2.2 Multi/broadcast Data Frame Traffic

As shown in Fig. 2, during the failure of the link between nodes E and F, if node A broadcasts data frames, node E will lose the fastest copies through port 1, but it will keep receiving the second copies that are coming through node D as long as

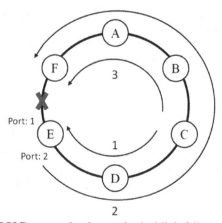

Fig. 2 Behavior of the LPLB approach when a physical-link failure occurs. Node C needs three steps to redirect its traffic from the opposite direction.

node E has detected the physical-link failure, broadcasted its PS frame, and changed all the primary ports of all its destination nodes. In other words, node E will not delete the second copy of the sent frames because the first copy that was supposed to be delivered through node F has stopped and node E has detected that event.

Finally, all the nodes will detect that the failed link is recovered as soon as the root node sends an NMS frame. If both of the NMS frame copies are received, both directions are working fine and the root node will broadcast a PS frame, and, in turn, all the other nodes will do the same to reselect the new primary ports for each destination node.

3   LPLB Performance Analysis

This section will demonstrate that the proposed LPLB approach offers two advantages in addition to avoiding looping storms. The first advantage is that it offers better distribution for the sent data frames among the network links. This is because the LPLB approach uses all the network links during each sending process. As we explained earlier, and as shown in Fig. 1, node A sends frames to node B through port 2, and it sends frames to node F through port 1; thus, the sending node has selected the fastest path to each destination, according to the selection of PS frames or according to the PP table. This behavior is done by each node. Consequently, LPLB provides a type of load balancing among the network links that is not provided by RSTP or other protocols, such as MRP, because those two protocols depend on disabling a link or cutting the ring to avoid the looping issue. However, to convert this advantage to a visible object, we applied LPLB and RSTP, separately, to the network, (Fig. 1), and let each of the nodes send one frame to the other. We then counted the number of frames that passed through

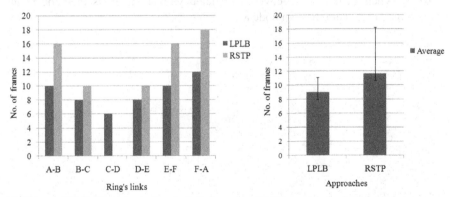

(a) Frames distribution among the ring's link   (b) Mean and standard deviation

Fig. 3 Frame distribution and mean and standard deviation for the LPLB and RSTP approaches.

each network link. The result is shown in Fig. 3. Fig. 3 (a) shows the number of frames per each ring's link and Fig. 3 (b) shows the mean values for the network traffic and the standard deviation for the mentioned scenario under both the LPLB and RSTP approaches. For the LPLB approach, the mean value was 9 frames per link; for the RSTP approach the mean value was 11.6 frames per link. Moreover, the standard deviation was 2 frames for LPLB and 6.6 frames for RSTP, which shows that the LPLB approach offers better load balancing among the ring's links compared to RSTP. The standard deviation values also show that LPLB offers the highest stability for traffic distribution.

The second advantage of using the LPLB approach is that, during the failure period, no data frames will be lost as long as the failure is detected. We simulated this advantage using OMNeT++ simulator version 4.6 [6]. Fig. 4 (a) shows a ring network consisting of 16 nodes, and Fig. 4 (b) shows a graph for the latency of each sent frame using the LPLB approach. In the simulation scenario, node 5 sends 50 frames to node 10; during that time, a link failure occurs between node 8 and node 9, which causes node 8 to redirect the received data frames that were sent from node 5 and send them back to node 5. In addition, all future frames that supposed to be forwarded to node 10 are sent to the other direction of node 5 through node 4. Later, the link failure is recovered and the LPLB approach set node 5 to reuse the fastest path through node 8 towards node 10. As seen from the graph of Fig. 4 (b), the sending process has not stopped in the LPLB protocol as it does in the RSTP approach, in which an alternative path must be found before the sending process can continue. The RSTP approach will also cause the loss of all the data frames that have been sent because the frames could not find a path to the destination node. The graph also shows a spike of about 170 μs, which represents the first period from the failure when node 8 sends he data frames back to node 5 to redirect them from the opposite direction. After that, when node 5 changes its primary port for the destination of node 10, the latency will revert to the value of 107 μs. When the failure is recovered, the latency reverts to 46 μs and node 5 back to use the fastest path through node 8.

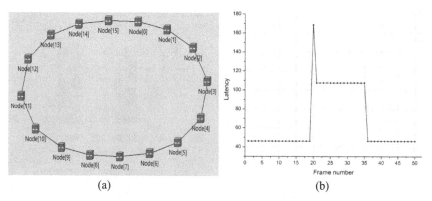

(a)   (b)

Fig. 4 A simulated ring and its latency performance. a) Ring with 16 Ethernet nodes. b) Latency per each data frame sent from node 5 to node 10.

4   Conclusions

This paper proposed a novel approach, called LPLB, to provide path redundancy in Ethernet ring networks without looping issues; LPLB also offers load balancing among the network links and provides zero-recovery time or seamless redundancy if the node has detected the failure before received the data frames. However, if the node is unable to detect the failure, the received frame may be lost if it is forwarded to the failed direction, but the other frames will not be lost because that node will send them back to their source node to redirect them from the opposite direction. LPLB works under the standard Ethernet protocol; therefore, the frame layout does not require any modification. Thus, all the off-shelf Ethernet devices have the ability to work with Ethernet and LPLB without any modifications. LPLB will be suitable for smart grid, substation, and industrial applications that need a very short time-out. In the future, we plan to work on extending the LPLB approach so it can be applied to any type of network topology.

Acknowledgement This work was supported by basic science research program through the National Research Foundation of Korea (NRF) funded by the Ministry of Science, ICT, and Future Planning (No. 2013R1A1A2008406).

References

1. IEEE 802.3: IEEE Standard for Ethernet. http://www.ieee802.org/3/
2. IEEE Standard for Rapid Spanning Tree Protocol (RSTP), IEEE 802.1w (2001)
3. IEC Standard for Industrial communication networks-High availability automation networks-Part 2: Media Redundancy Protocol (MRP), IEC 62439-2, (2010)
4. IEC 62439-3: Industrial Communications Networks: High-Availability Automation Networks, Part 3: Parallel Redundancy Protocol (PRP) and High-Availability Seamless Redundancy (HSR). In: The International Electrotechnical Commission, Geneva, Switzerland (2012)
5. IEC Standard for Communication networks and systems for power utility automation-Part 90-4: Network engineering guidelines, IEC 61850-90-4 (2013)
6. OMNeT++ Version 4.6 Simulator. http://www.omnetpp.org/

Design & Implementation for Emergency Broadcasting Using Agencies' Disaster Information

Geum-Young Min and Hyoung-Seop Shim

Abstract Disaster information is the main element in the disaster management and is the basis for coordination and decision making in emergency situations. We need disaster information for situation management and decision making. In this paper we present an approach to the implement of situation dashboard for disaster management.

Keywords Emergency management · Disaster broadcasting system · Disaster information · Disaster content

1 Introduction

Korea is at risk from a variety of natural and man-made disasters—including landslides, severe winter storms, fire, and transportation incidents—but typhoons and their accompanying floods stand apart; they are both the most damaging and the most frequent of the natural perils facing.

Therefore, Korean government to implements integrated disaster management systems, which would cover the entire country simultaneously. However, it is assessed that its current systems lacks fundamental scientific control and analysis technologies limiting Korea's effective response [1].

As Korea's main public broadcasting network, KBS has been designated by the government as the host broadcaster for disaster broadcasting and, thus, swiftly

G.-Y. Min
Department of MIS, Dongguk University, Jung-gu, Seoul, Korea
e-mail: William1540@naver.com

H.-S. Shim(✉)
Division of Advanced Information Convergence, KISTI, Yuseong-gu, Deajen, Korea
e-mail: hsshim@kisti.re.kr

© Springer Science+Business Media Singapore 2015   693
D.-S. Park et al. (eds.), *Advances in Computer Science and Ubiquitous Computing*,
Lecture Notes in Electrical Engineering 373,
DOI: 10.1007/978-981-10-0281-6_98

delivers accurate disaster status reports to the general public in accordance with the Natural Disaster Protection Act and the Disaster Management Act in order to minimize human and property damage [2]. KBS, the main emergency broadcasting station, opened Disaster Information Center, on 27th June 2011, following the construction of integrated digital disaster broadcasting system.

The center processes information on disaster from Ministry of Public Admini-stration and Security, NEMA (National Emergency Management Agency) and KMA(Korea Meteorological Administration), and broadcasts it through TV, Radio, DMB, Smartphone, Internet and SNS [3]. NEMA, KMA and the Korea Communications Commission are jointly striving for securing consistency and efficiency on disaster management while maintaining a closer cooperation between agencies.

The purpose of this paper is implementation of dashboard system for Disaster broadcasting that can prevent or reduce possible damage caused by disasters.

2 Disaster Broadcasting in Korea

The broadcasting can prevent or reduce possible damage caused by disasters. Disaster broadcasting is urgent in the disaster situation to prepare for disasters, emergency disaster information transfer to emergency broadcasting.

In Korea, by Broadcasting Telecommunication Development Act (Article 40) or in the event of a disaster occur when there is an anticipated broadcaster to prevent the occurrence of a disaster or emergency that has been broadcast [4].

In addition, the KBS specified Primary Disaster Broadcaster, in accordance with the Broadcasting Act of the competent authorities designated as disaster broadcasting. Disaster broadcasting is in charge KCC, NEMA and KMA as shown in Figure 1.

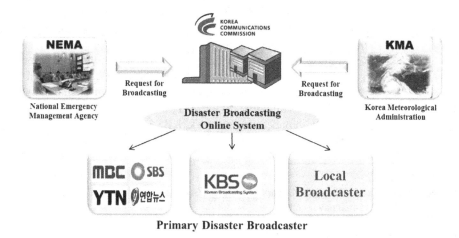

Fig. 1 Disaster Broadcasting Systems related Agencies

3 Disaster Information Contents Analysis

We have analyzed disaster broadcasting contents 10 disaster types, 17 related organizations, 19 disaster information system related agencies, analyzed the disaster information system through site visits and agency cooperation as shown in Table 1. Disaster information data flow as shown in Figure 2.

Table 1 Disaster Information of Disaster Related Agencies

Types	Agencies	Systems	Information
Typhoon	NEMA	Local Disaster Management System	Observation System - Water level etc.
	Meteorological Agency	Disaster Meteorological Information System	Meteorological Information Weather Map etc.
:	:	:	:
Electronic Power	KEPCO	Emergency Situation Management System	Blackout Information etc.

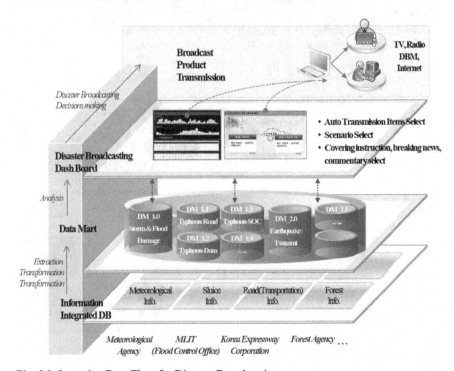

Fig. 2 Information Data Flow for Disaster Broadcasting

4   Implementation of Disaster Situation Dashboard

Disaster information management system (DMIS) supported disaster situation information and relevant information on the type of disaster. And, disaster situation analyzed through satellite image and CCTV as shown in Figure 3.

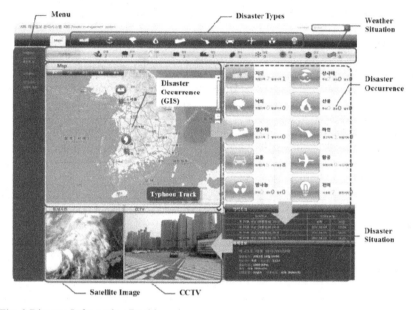

Fig. 3 Disaster Information Dashboard

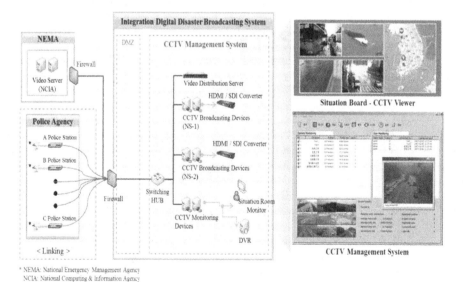

Fig. 4 CCTV Connection

DMIS connected CCTV, browse surveillance camera footage by region and by type (rivers, roads, forests etc.) as shown in Figure 4. An integrated monitoring for all surveillance cameras, about 6,000, installed across the country has been in place.

5   Conclusion

This study is implementation of disaster information dashboard system for disaster broadcasting. Disaster broadcasting is issued to the public to encourage them to take a specific action in response to a disaster event or a threat of a disaster. To prepare for these disasters, we had design of Effective disaster broadcasting System.

Disaster broadcasting system were established monitoring environment for integrated management. Provide disaster situation information and relevant information on the type of disaster.

References

1.   Lee, Y., Song, J.H.: Problems of disaster broadcasting system in Korea. In: Korea-Japan International Symposium, pp. 99–112 (2011)
2.   Kim, Y.S.: Role of media in disseminating information on disaster risk. In: The International Conference on Total Disaster Risk Management, December 2–4, 2003
3.   Choi, S.J.: A study on public alert and warning system using new media. In: Korea Communication Commission (2011)
4.   Young, M.G., Shim, H.S., Hoon, J.D.: Implementation of disaster broadcasting system for efficient situation management using mobile devices. Lecture Notes in Electrical Engineering, vol. 214, pp. 689–697 (2013)

Application and Development of Service Integration Platform for Agricultural Products

Geum-Young Min, Min-Ho Jung and Hyoung-Seop Shim

Abstract The purpose of this study is to developed mobile service integration platform that the customers can purchase agricultural products using mobile devices. Along with the development in agricultural products, it became necessary to develop an e-commerce that allows customers to search information and purchase decision. To development of mobile based virtual store with agricultural products ICT convergence - integration 1)implement of local based service and construction of code interface for smartphone and connected products information using barcode, QR code and NFC, 2)development of DID based V-store, 3)development of mobile integration platform for shopping mall.

Keywords Agricultural products · ICT convergence · Integration platform · Virtual sore

1 Introduction

There are differences depending upon items, the distribution margins for various distribution stages are 40~70% in general agricultural products [1]. The reason is high distribution costs and inefficiency in each distribution stage [2]. Therefore, it is necessary to improve distribution system for agricultural products.

The government establish the Plan for Promotion of Agri-food and ICT Convergence - Integration. this plan is integrating ICT into the agricultural products value chain, such as production, distribution, consumption, and etc [3].

G.-Y. Min M.-H. Jung
Department of MIS, Dongguk University, Jung-gu, Seoul, Korea
e-mail: willam1540@naver.com, minoa4@hanmail.net

H.-S. Shim(✉)
Division of Advanced Information Convergence, KISTI, Yuseong-gu, Deajen, Korea
e-mail: hsshim@kisti.re.kr

© Springer Science+Business Media Singapore 2015

699

D.-S. Park et al. (eds.), *Advances in Computer Science and Ubiquitous Computing*,
Lecture Notes in Electrical Engineering 373,
DOI: 10.1007/978-981-10-0281-6_99

We had developed a mobile store based on the ICT integration model to make a new distribution channel; and an open market type mobile service integration platform that comprehensively connecting the agricultural products shopping malls and relevant sites operated by farmers and the local governments.

2   Agricultural Product's ICT Convergence

2.1   ICT Convergence Strategy for Agricultural Products

The MAFRA(Ministry of Agriculture, Food and Rural Affairs in Korea) has establish the ⌜ Plan for Promotion of Agri-food and ICT Convergence - Integration ⌝ in order to apply ICT technologies to the production, distribution, and consumption of agricultural products for scientific farming [4].

This plan has significant meaning in that it seeks to establish a foothold for agri-food industry growth as a growing future business area and ecosystem construction for ICT convergence – integration, as other major industries create new added-values and improve their competitiveness using ICT [5]. Agricultural products ICT convergence model as shown in the Fig. 1.

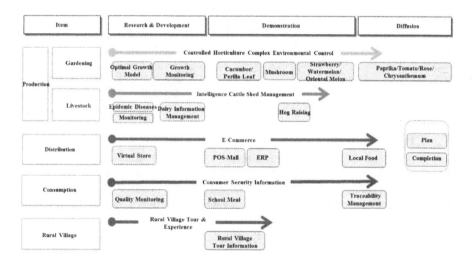

Fig. 1 Agricultural Products ICT Convergence Model

2.2   Agricultural Products ICT Convergence Ecosystem

Agricultural products ICT convergence ecosystem is now being reorganized into a new ecosystem due to the impact of the agri-food market, which is expanding domestic and overseas markets through the government's strong ICT consumption policies [6].

Agricultural products ICT convergence models will be promoted centered on the facility horticulture, growing fruit and livestock industry. ICT convergence ecosystem is focusing on establishing an industrial ecosystem for a wide range of ICT technologies to be applied to the agricultural business. MAFRA will make efforts to establish systemic virtuous circulation from the R&D stage to the model development and promotion stage as shown in the Fig. 2 [7].

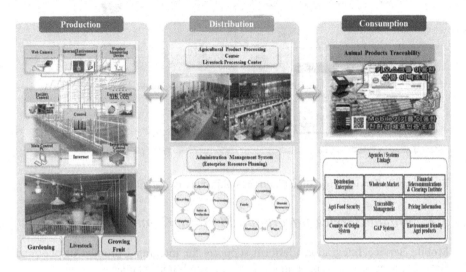

Fig. 2 Agri- food's ICT Convergence Ecosystem

3   Agricultural Products Mobile Service Integration Platform

3.1   Mobile Service Integration Platform

It is designed as a integration platform for conducting effective agricultural products display by interconnecting DID and smart device as shown in the Fig. 3.

Agricultural products information is arranged and distributed through the shopping mall, and it makes efficient management in the display. In addition, integration platform provides delivery system interconnected with shopping malls when selling through V-Store.

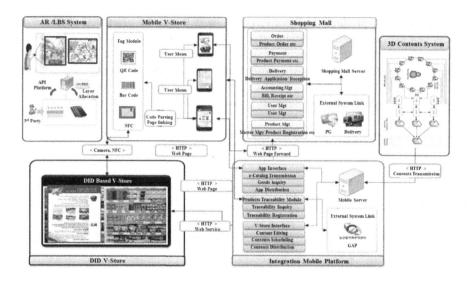

Fig. 3 Mobile Service Integration Platform Architecture in Agricultural Products

3.2 Mobile Service Integration Platform Connection

Mobile Service Integration Platform supports for integrated product display and advertisement by interconnecting DID V-store and Smartphones.

This platform is developed to make efficient display management by making arrangement and distribution of the product (contents) information through the management system as shown in the Fig. 4.

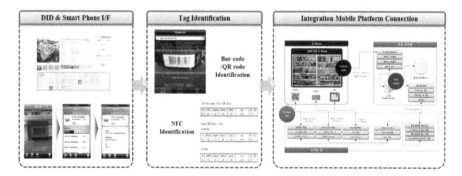

Fig. 4 Mobile Integration Platform Architecture

4   Conclusion

This study developed the mobile service integration platform based on the agricultural products ICT convergence as described below.

First, this study developed Agricultural products display to provide product information by utilizing DID and e-catalog, and provide detailed information by using barcode, QR code and NFC to induce consumer purchase.

Second, this study developed the interface to enable consumers to obtain detailed information about products through barcode, QR code and NFC of DID and e-catalog by using Smartphones.

Third, Mobile Integration Platform supports for integrated product display and advertisement by interconnecting DID and Smartphones.

Future studies will aim to build a Rural Village. The Rural Village will provide information display during the rural theme tour. the information display can be used for information gathering and product purchase.

Acknowledgement This research was supported by the Advanced Technology Development Program (Project Number: 312068-03) of the Ministry of Agriculture, Food and Rural Affairs, Korea.

References

1. Jeon, C.G., Kim, D.H.: Improvement Measures for Agricultural Products Distribution System of Traditional Markets at Consumption Sites. Research Report, Korea Rural Economic Institute (2013)
2. Prime Minister's Office, Civil Economic Agricultural products cultural Marketing Structure Improvement and Income Increase of the Farm-Fishing Households (2010)
3. Ministry of Agriculture, Food & Rural Affairs, The Construction of Ecosystem for ICT Convergence in Agricultural products (2013)
4. Im, J.B.: Plan for Applying Information & Communication Technology (ICT) into Agro-food industry. Agricultural marketing policy (2013)
5. Foodpolis: MAFRA promotes ICT convergence – integration, no. 6 (2013)
6. NIA, National Information White Paper (2013)
7. MAFRA, The Construction of Ecosystem for ICT Convergence in Agricultural products (2013)

Implementation of Kegel Exercises for Prevention of Urinary Incontinence and Treatment Thereof

Jea-Hui Cha and Jong-Wook Jang

Abstract Today, much attention has been paid to the modern healthcare. Such attention has naturally led to the improvement of bio technology. Accordingly, there has been a trend toward growing IT convergence industries. Based on the sharp growth of IT and the development of BT as well as the healthcare applicable to the trend, men and women would be able to take pelvic floor muscle exercises effectively and correctly by themselves.

The Kegel medical equipments which are currently sold in the market make users exercise by giving electrical stimulations compulsively. Users need to take off their bottoms and take the Femcon therapy in a closed room. This causes various restrictions of time, space and hygiene.

This thesis designs a Kegel medical equipment which combines BT and IT, free from restraint in regard to space and hygiene, without the need to take off bottoms.

Keywords Kegel · Embedded system · Medical equipment

1 Introduction

Urinary incontinence or scatacratia defined as a symptom of abnormal urination or bowel movement which causes problems in relation to social activities or hygiene due to involuntary loss of urine or feces (Hwang et al., 1998) does not threaten life but has negative influences on various aspects socially and psychologically.

Along with enlarged uterus, when a fetus's head presses the pudendal nerve of a pregnant woman, or when the fascia, ligaments, nerves or muscles of the pelvic floor is damaged at the time of delivery, the innervation for the musculus sphincter

J.-H. Cha(✉) · J.-W. Jang
Department Computer Engineering, Dong-Eui University,
176 Eomgwangro, Busan Jin-gu, Busan 614-714, Korea
e-mail: ckwpgml5507@naver.com, jwjang@deu.ac.kr

© Springer Science+Business Media Singapore 2015
D.-S. Park et al. (eds.), *Advances in Computer Science and Ubiquitous Computing*,
Lecture Notes in Electrical Engineering 373,
DOI: 10.1007/978-981-10-0281-6_100

ani externus and the pelvic floor muscle may be damaged and accordingly, urinary incontinence or scatacratia may occur[1]. In Korea, there have been a number of prevalence surveys. According to the survey for 13,484 women over the age of 19 in 2005, the overall prevalence appeared to be 24.4%.

The foregoing survey showed an interesting result that the prevalence of the married, unemployed or uneducated women was high. It was estimated that there would be a total of 4.2 million patients suffering urinary incontinence in Korea[2].

The research team of the University of Alabama(UA) in Bermingham, US presented in the latest version of the Journal of the American Medical Association(JAMA) that as a result of 200 American women patients of urinary incontinence taking Kegel exercises, the number of urinary incontinence decreased by approximately 80% or more after 8 weeks. On the contrary, in case of the patients who took drug treatment, the number of urinary incontinence decreased by 70% and in case of the patients who took a placebo, the number of urinary incontinence decreased only by 40%. The research team discovered that only 25% of the patients who took drug treatment showed satisfaction while 86% of the women who took pelvic floor muscle exercises, *i.d.*, Kegel exercises, answered they did not want to change the treatment into another. Accordingly, the pelvic floor muscle exercises were proved to be more effective than drug treatment[3].

Therefore, this thesis has implemented a system that conducts games on mobile or PC through Bluetooth and wireless fidelity by using data of pressure sensors so that Kegel exercises which are easy and simple may be taken.

2   Relevant Studies

The existing equipments for Kegel exercises are largely classified into insertable equipments and extracorporeal equipments. Insertable equipments are exclusively used for women and inserted into a vagina to stimulate and strengthen pelvic floor muscles. Extracorporeal equipments constrict muscles with low- frequency stimulation while a user takes off her bottoms to take exercise.

2.1   Insertable Exercise Equipment

An insertable exercise equipment, as a device which can be directly inserted into a woman's vagina to induce her to take exercise, checks the intensity of exercise through pressure sensors and uses an Android application and the Bluetooth communication so that a user may easily check the quantity, intensity, etc. of exercise. The equipment's advantages are that the pelvic muscle exercise program which repeats contraction, atony and relaxation may be provided by the application and that while inserted, a user may move to some extent.

Fig. 1 Insertable Exercise Equipment (Former Airbee Kegel)

The insertable exercise equipment should be directly inserted into a body for use and accordingly, lubricants such as gel are needed. The equipment is not used by batteries but by recharging[4].

2.2 Extracorporeal Exercise Equipment

An extracorporeal equipment is a sitting cushion type of equipment that directly touches a user's skin and constricts or relaxes pelvic muscles with low-frequency stimulation while a user takes off her bottoms and sits on it[5].

Fig. 2 Extracorporeal Exercise Equipment (Smart Kegel)

3   System Design and Main Functions

This system perceives the exercise intensity of pelvic muscles through pressure sensors and sends the measured values to a computer through an Arduino board. If the value above the upper limit is measured, the value is put into the relevant game through program interworking and then, the game is conducted.

3.1   Structure of Pressure Sensors and Arduino Board System

Figure 3 shows the connection of pressure sensors and the Arduino board. The Arduino board functions as a repeater to deliver the input value to computer. Pressure sensors are connected to the Arduino through jump line. In order to increase the degree of freedom in relation to installation of pressure sensors, the connection is made except for the breadboard and sensing is proceeded with.

The Arduino board is set up with a Wi-Fi module and a Bluetooth module in order to interwork with a PC or mobile and to prepare for communication base.

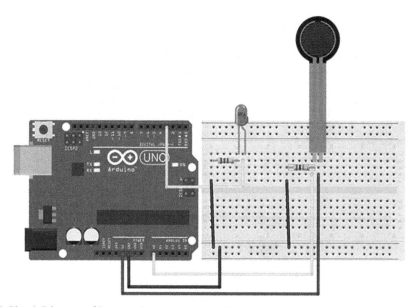

Fig. 3 Circuit Diagram of Pressure Sensors and Arduino Board

3.2   Structure of Communication System

Figure 4 shows the system that the Arduino sends values to a PC and a mobile device through Wi-Fi or Bluetooth.

First, the Wi-Fi communication interworks with the web server of PC and sends data through the Wi-Fi module attached to the Arduino board.

Second, the Bluetooth communication connects the Arduino board and a mobile device through Bluetooth and has data received on the mobile device. By using the sensing value received in that way, a game is conducted and accordingly, Kegel exercises are naturally taken and bio feedback is maximized. The quantity of exercise may be checked daily, weekly and monthly based on the saved database so that systematic exercise effects are expected.

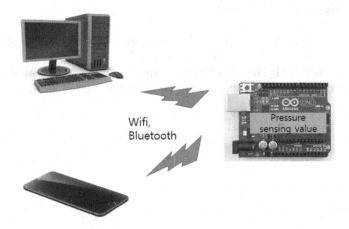

Fig. 4 Arduino Board Communication System

3.3 Construction of a Combined Server

Figure 5 shows a combined server for web and app.

The system collects and processes the signals transmitted from web and mobile, and saves and manages the data collected through the interworking with the combined server. The system makes it possible to provide data that a user wants. By combining web and mobile servers into one, a PC which plays an intermediate role is eliminated and as a result, costs may be also reduced.

Fig. 5 Data Server for Web and App

The main server PC shows a user the data saved on the database through web or a mobile phone. A function that shows the rankings in terms of the quantity of exercise and time through data is included.

Moreover, when a user determines the target baseline and accomplishes it, services applicable thereto will be provided.

3.4 Diagram of Program

Figure 6 is a diagram of program. If the program is carried out, whether there is a connectable device is detected. A connectable device discerns whether the connection is made by Wi-Fi or Bluetooth and after interworking, pressure sensors collect and classify data. If the data are saved onto the server, the program ends.

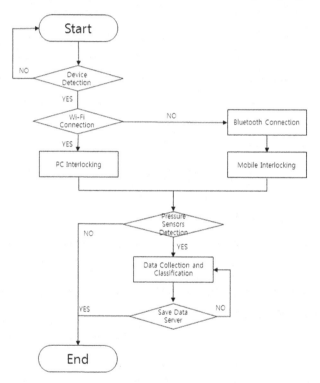

Fig. 6 Diagram of Program

4   Conclusion

Today, much attention has been paid to the modern healthcare. Such attention has naturally led to the improvement of bio technology. Accordingly, there has been a trend toward growing IT convergence industries.

It is presumed that approximately 3 million people suffer urinary incontinence in Korea. It is reported that approximately 200 million or more people suffer the same and about 85% thereof are women.

Most of the existing treatments for urinary incontinence stand for the Femcon therapy and form 60% of the world market. Kegel exercises usually use the Femcon therapy in a closed space, requiring women to be undressed. This causes various restrictions in relation to time, space and hygiene.

This thesis has maximized bio feedbacks by way of making a user naturally do Kegel exercises with putting on her clothes through a game via her computer or smart phone. Moreover, the quantity of exercise measured by a computer and devices through the Bluetooth function and the exercise records kept by a user are inserted into database so that systematic exercise effects may be expected and comparative advantages may be obtained based on inexpensive costs.

In the future, an ergonomic product that can be used more conveniently will be manufactured by applying the system designed by this thesis. Games which may arouse users' interests will be implemented.

Acknowledgments This research was financially supported by the Ministry of Education Science Technology (MEST) and This work was supported by the Human Resource Training Program for Regional Innovation and Creativity through the Ministry of Education and National Research Foundation of Korea(NRF-2015H1C1A1035898).

References

1. Park, S.-H.: Effects of Kegel Exercises for Prevention of Urinary Incontinence and Scatacratia for Women who are Pregnant or Have Given Birth: Systematic Review of Documents. Journal of Korean Academy of Nursing **43**, Item 3, June 2013
2. Urology Laboratory, Ajou University College of Medicine: Urinary Incontinence in Women. Korean J. Fam. Med. **31**, 661–671 (2010)
3. UPI Communications: http://www.kegelclinic.kr/new01/Community/G_sub01_view.html?Idx=260&page=2
4. http://www.air-bee.com/default/00/01.php
5. http://www.kegel365.co.kr/intro/member.html?returnUrl=%2Findex.html

A Functional Relationship Based Attestation Scheme for Detecting Compromised Nodes in Large IoT Networks

Yong-Hyuk Moon, Yong-Sung Jeon and Chan-Hyun Youn

Abstract Despite memory traverse is commonly used for attestation, this approach could not feasibly work for an IoT network that requires scalable and sustainable operations. To overcome this limitation, we propose a functional relationship based attestation scheme, which verifies the integrity of battery-powered devices by analyzing the consistency among neighbors, where a consistent edge between two nodes is given if outputs of the same functions at both nodes are equal to each other. Efficiency of the proposed method is demonstrated in terms of attestation termination and detection speed.

Keywords Attestation · Integrity measurement · Device security · Threat detection · Internet of Things

1 Background

Large-scale connectivity with Internet of Things (IoT) provides a power of cooperation for building a contextual and collective intelligence. Although the large number of end-points participate and interact with each other in this environment, distributed systems are built upon the assumption that each element (computing and communication) is legitimate and trustable. This unreasonable belief makes difficulties in establishing and sustaining reliable networking among

Y.-H. Moon(✉) · Y.-S. Jeon
Electronics and Telecommunications Research Institute (ETRI), Daejeon,
Republic of Korea
e-mail: {yhmoon,ysjeon}@etri.re.kr

C.-H. Youn
Korea Advanced Institute of Science (KAIST), Daejeon, Republic of Korea
e-mail: chyoun@kaist.ac.kr

© Springer Science+Business Media Singapore 2015
D.-S. Park et al. (eds.), *Advances in Computer Science and Ubiquitous Computing*,
Lecture Notes in Electrical Engineering 373,
DOI: 10.1007/978-981-10-0281-6_101

devices, so that devices could be revealed to serious vulnerabilities as evidenced in [1] and [2]. For example, a compromised node could be an active forwarder to propagate malicious codes or perform replay attacks. For this reason, a role of solution that detects exploited devices is crucial. We argue that the existing attestation schemes [3-8] are not feasible for the IoT since battery-powered devices have different requirements (i.e., scalability and sustainability) compared to conventional desktop-class machines.

We briefly review three representative approaches for integrity verification. It is straightforward to use periodic monitors of integrity [3] [4]; however, this technique needs to increase the frequency of monitoring in order to precede unpredictable changes in targets. Continual observation leads to considerable performance overhead and is very weak to transient attacks. To avoid this problem, load-time measurement methods [5] [6] for kernel protection have been studied. One critical drawback of this approach is that a verification routine (e.g., code) can be compromised if an untrusted input is allowed even though the routine is of high integrity. Besides, a process that gets compromised at run time cannot be detected by this technique. In another line of work, temper-resistant hardware based root of trust has been deployed in a device to improve the verification reliability [7] [8]. Due to physical isolation property, these types of hardware generally accommodate a single and static configuration, which takes the role of judgement criterion, so that updates of relevant policy impose additional costs and expose a device to new vulnerabilities.

Particularly, most of existing studies require memory traverse for measuring bytes, resulting in that redundant memory access is unavoidable. Although memory snapshot is somewhat essential in order to verify the security state of target node, the order of memory access could be predictable with a high probability. One more disadvantage is that a memory traverse strongly depends on the memory architecture employed in an IoT device. Despite, the key question of how to effectively attest devices in large-scale networks without memory traverse remains unanswered.

In this paper, we confine our focus to the following two requirements in order to propose a feasible attestation scheme, which does not take a snapshot of memory.

Scalability. Ensuring the integrity of a particular node should be possible by pinpointing malicious or suspicious at least nodes based on strong evidence even in the large-scale device network.

Sustainability. Attestation should not obstruct the management functionalities for an IoT network; and its burden should be restricted to the acceptable notion in terms of service quality.

To satisfy the above requirements, we design a new attestation scheme, which assesses outputs of relevant functions of nodes on a pre-defined path, instead of performing attestation for a single node (e.g., memory fidelity). The proposed scheme also verifies the integrity of suspicious nodes by analyzing the functional relationships (i.e., consistency). The remainder of this paper is organized as follows. Section 2 proposes a functional relationship based attestation scheme.

In Section 3, we discuss related strength and additional issues of the proposed attestation scheme. Simulation results are demonstrated and discussed in terms of termination convergence and detection speed in Section 4. Finally, we conclude the paper and outline issues for future work in Section 5.

2 Proposed Attestation Scheme

This section presents a novel methodology for node attestation based on a concept of integrity verification function (IVF) in a large-scale IoT network. The proposed attestation scheme consists of six phases, such as neighbor discovery, verifier election, path setup, function assessment, path integration and consistency analysis.

2.1 Neighbor Discovery and Verifier Election

IoT devices are connected with many redundant interconnections among network nodes, resulting in that they could form a mesh network over a period of time to relay data through neighbors. Unlike a dedicated verifier based attestation, security status of each node is measured by a random verifier that is elected at every round. This approach maintains sustainability well and is also suitable for scalability.

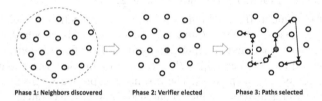

Phase 1: Neighbors discovered   Phase 2: Verifier elected   Phase 3: Paths selected

Fig. 1 A schematic view of the proposed attestation scheme based on per-function evaluation using randomly selected subset of neighbors.

Neighbor Discovery. If a node is newly deployed in an IoT network, it initially attempts to discover neighbors by sending a hello message to other nodes in the near distance [9]. After discovery, the node resets or synchronizes its own timer and then establishes a pair-wise key with every neighbors.

Verifier Election. All nodes have an asynchronous timer of detecting a new event for election within a neighbor group. When an interval of time runs out, one node should be elected for taking a role of verifier. To this end, we adopt a similar approach of a cluster-head selection algorithm, such as (LEACH) [10]. Node i whose time is up sends a freeze message to neighbors. Then, neighbors notify that the round for attestation is updated. Node i computes the following threshold value, T_i, and produces a random number, $\ddot{e}$, between 0 and 1. By comparing each other, node i reaches a conclusion on whether to be a verifier. If $\ddot{e} < T_i$, node i becomes a new verifier for attestation in the current round. Next, the verifier broadcasts an advertisement message to the rest of nodes involved in the same neighborhood; thus, neighbors' timers can be unfrozen.

$$T_i = \frac{p}{1 - p\{r \bmod (p-1)\}} \text{ if } i \in D, \tag{1}$$

where i denotes an identification of node, i.e., node i and D stands for a domain of nodes that have not been a verifier in the last $1/p$ rounds. p is a percentage of verifiers (e.g., $p = 0.02$) and r is a current round. If node i does not belong to D, $T_i = 0$.

Once the aforementioned election algorithm is operated at node i, a random interval of time is set. On the other hand, node i broadcasts an election request to the rest of nodes and then each neighbor executes the same task for election.

2.2 Path Establishment and Function Assessment

We suppose that all devices support network-level encryption, so that additional authentication protocol is not required and a message is transmitted through a secure session established between nodes. Since they contain the same group of IVFs, each one is capable of offering an output for a requested function.

Path Setup. In order to construct a subset of neighbors, two attestation paths are established according to the principle of a random walk algorithm [11]. Moreover, the length of both paths are bounded and equal to each other. For example, path 1, $P_1 = \{8, 3, 2, 6, 5, 4, 1, 7\}$, and path 2, $P_2 = \{6, 4, 1, 3, 2, 8, 7, 5\}$, are constructed with eight nodes at round $i + 1$ as shown in Fig. 2.

Function Assessment. If attestation paths are given, a function of each node is also determined. In case of P_1, IVFs f_1 to f_8 are used to result in an output in numerical order as a response to a random input generated by a verifier. Likewise, functions f_1 to f_8 are selected for P_2, where $|P_1| = |P_2| = L$. A verifier node v sends an input data ε_v to the first node on a given path, P_1. The first node, 8, produces a result using f_1 with ε_v and then forwards its result to the second node, 3, on P_1. Repeatedly, this process is executed to the last node, 7, on the path. Also, a verifier performs the same task again in order to aggregate immediate results to the final node on P_2. Here, we assume that by comparing individual results sequentially, it is found that there are two functionally inconsistent relationships between node 6 and node 3, and node 1 and node 7 as shown in Fig. 2. The measured functional relationships are then translated to a topology as depicted below.

Fig. 2 An example of path setup and intermediate node topology (phase 4). A solid line is an edge with functional consistency between two nodes and a dotted line is for inconsistent relationship.

Remark 1. Although the proposed attestation scheme does not measure and verify all neighbors within a small number of rounds if the size of neighbor group θ is large, all neighbors will be covered by the probabilistic property of random walk over a period of time.

2.3 Path Integration and Consistency Analysis

After phases 3 and 4, a verifier sends topologies as an intermediate result to a gateway. Then, this output is integrated with the last obtained node topology in a gateway for the purpose of topology update. Fig. 3 gives a good example of path integration. Then, a gateway performs a consistency analysis for detecting suspicious nodes.

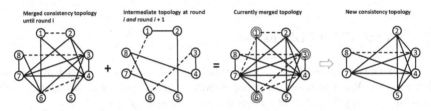

Fig. 3 An illustration of merging a topology with the existing one (phase 5). A circle with two outer lines denotes a node that has more than two inconsistency edges.

Path Integration. As depicted in Fig. 3, a verifier stores a consistency topology integrated until round i and an intermediate topology is built during attestation at round $i + 1$. These two consistency topologies are merged with each other; so, a gateway has a current one as depicted in the third figure of Fig. 3. After this phase, a gateway can point out which node is suspicious according to a predefined inconsistency degree d. For ease of demonstration, we set d to 2 for all nodes; thus, nodes 1, 3, and 6 come under suspicion.

Consistency Analysis. A gateway starts attestation when it is recognized that the following Condition 1 is satisfied in the currently merged graph.

Condition 1. If node i has more than two tolerance degree (i.e., $d_i \geq 2$, $1 \geq i \geq N$), it is said that node i is suspicious. Otherwise, node i turns out to be legitimate. A gateway confirms whether a particular node provides a reliable service or not.

More specifically, the gateway attempts to find functional consistency groups (also known as cliques in graph theory) at every round by using the Bron-Kerbosch (hereinafter referred to as BK) algorithm [12], which compute all cliques in linear time relative to the number of cliques. If the size of IoT network, N, is sufficiently large enough, assuming that the number of legitimate nodes is larger than that of compromised nodes is reasonable. Hence, we establish the second condition in order to verify suspicious nodes.

Condition 2. If node *i* belongs to a complete sub-group that consists of more than half neighbors, it is said that node *i* is consistently trustable. Otherwise, node *i* is treated as a malicious or compromised one, so that node *i* will be isolated.

In the currently merged topology, node 1 does not form any complete sub-group. Also, there is no clique that includes node 6. In case of node 3, there are five different complete sub-groups, such as {2, 3, 5}, {2, 3, 7}, {3, 5, 7}, {2, 3, 4, 7}, {2, 3, 5, 7}, and {2, 3, 4, 5, 7}. We suppose that a size of complete sub-group, *s*, is $\theta / 2$, i.e., $s \geq 4$ according to the aforementioned Condition 2. Therefore, nodes 1 and 6 are isolated and node 3 is maintained in the IoT network as shown in the rightmost figure of Fig. 3.

Remark 2. By setting d_i, a gateway pinpoints suspicious nodes, so that the effort of finding complete sub-groups is reduced. In addition to that, *L* is a fixed constant, so that the BK algorithm can be performed within acceptable time complexity.

3 Analysis and Discussion

So far, we have proposed a functional relationship based attestation scheme for an IoT network. We review the proposed attestation scheme in the three aspects, such as security strength, overhead reduction, and remaining issues.

Strength. In the purely distributed attestation, conventional secret dissemination and its recovery could be vulnerable points against an attacker. However, the proposed attestation scheme does not utilize and share the secret since the same set of functions are loaded on all devices during a manufacturing process. Further, each node has its own interval of time for election (i.e., asynchronous) and its timer is reset with a random interval once the election algorithm is executed. Thus, an attacker is hard to predict the next election time and which node will be elected.

Performance. Our election algorithm allows multiple verifiers; thus, frequent election would occur. However, we can set the number of verifiers that are elected at the similar time zone by adjusting *p* value in Eq. (1). This means that the proposed scheme decides an attestation event in a probabilistic manner and nodes, which have been selected as a verifier in the last $1 / p$ rounds is excluded for a current election.

Improvement. While the proposed election algorithm works effectively by simply comparing a randomly-generated number with a calculated threshold, if we assume that an elected verifier is not be trustable, an additional scoring factor should be added into Eq. (1) for assessing statistical or historical reputation of election candidate nodes.

4   Performance Evaluation

In this section, we aim to evaluate the effectiveness (i.e., detection convergence) of proposed attestation scheme in terms of two indices: attestation termination, which is measured as attestation rounds spent until the number of remaining neighbors is less than the minimum value of L, L_{min}; and detection speed, which describes the ability of how quickly all compromised nodes are confirmed in order to prevent malfunction or error propagation. To construct an IoT network for simulation study, we use a complete graph that has N vertices and each vertex has $N - 1$ connections, so that all vertices could be connected to the others when a path is established. In other words, the possible number of edges, M, is equal to $N(N - 1)/2$. To this end, we use a random graph, where the probability δ controls the density of network topology [11].

If the probability of detecting compromised nodes λ is randomly given and the number of compromised nodes μ are known at every round, the expected number of compromised nodes can be calculated as $\lambda \cdot \mu$. In order to avoid unrealistic value of $\lambda \cdot \mu$, we then set its upper and lower bounds as follows:

$$-\log \alpha \leq \lambda \cdot \mu \leq -\log \beta, \tag{2}$$

where α and β are random numbers and vary over rounds ($0 < \alpha, \beta \leq 1$). This implies that we can identify how many infections are occurred among nodes at a particular attestation round.

If $\lambda \cdot \mu$ value is bounded as described in Eq. (2), we randomly decide which function outputs a wrong result. Otherwise, we assume that no function is infected by malicious code. With the probabilistic infection event based on exponential random numbers, we can emulate malicious code injection to a node. Moreover, the BK algorithm is adopted in our simulation to find a complete sub-group of neighbors when the integrated functional relationship is given to a gateway. This simulation is terminated when all nodes are infected by malicious code or when the number of neighbors remaining is less than L_{min}. Furthermore, we use average values that are obtained by conducting the simulation 100 times for five different neighbor groups. Parameters mainly used in the simulation are summarized in Table 1.

Table 1 Parameter configuration used in the simulation

Parameter	Description	Value
N	The number of nodes	100
θ	The size of neighbor group	20
p	Percentage of verifier	0.02
δ	Density of network topology	0.01
L	A length of attestation path	5, 8
d	A tolerance degree	2, 4
S	A size of complete sub-group	Larger than θ / 2

Attestation Termination. As shown in Fig. 4(a), we measure attestation rounds spent until the number of remaining neighbors is less than L_{min} with four different cases of d and L values. If setting a value of d is low, that is 2, it requires performing analysis more frequently, while a gateway find suspicious ones more rapidly than a case of $d = 4$. Otherwise, IoT network needs to tolerate many suspicious nodes, if a value of d is set too high. Moreover, when L is increased, overhead due to function assessment via communication is possibly expected. On the other hand, a detection rate will be slowly converged to 1 because a gateway can obtain more intermediate topologies for consistency analysis. Due to this reason, two cases of $L = 8$ have less remaining neighbors at the same round, compared to those of $L = 5$, regardless of a value of d. Convergence that is measured by attestation termination time occurs at 36, 23, 46, and 48 rounds, respectively for four cases from top to bottom.

Detection Speed. Next, we only change to fix $d = 2$ and adjust L to floor of $\theta \times 0.4$ under the same simulation configuration in order to investigate different aspect of detection convergence. Fig. 4(b) shows average rounds to find a first suspicious node (5 rounds) as well as to confirm the all compromised nodes (12 rounds) when malicious codes are injected by Eq. (2). This result implies that by adjusting a value of L the proposed scheme can detect all compromised nodes within 7 rounds averagely after a first suspicious node is found. In a real system, the absolute time depends on two factors: an interval of election event and elapsed time for a round.

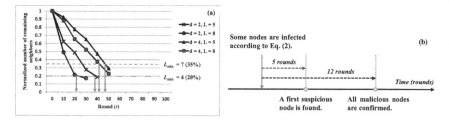

Fig. 4 Detection convergence with adjusting values of d and L.

5 Conclusion and Future Work

We have proposed a functional relationship based attestation scheme, which feeds the same input into multiple neighbor nodes and then compare their outputs in order to create a topology including consistent or inconsistent edges among nodes. The proposed scheme pinpoints suspicious nodes and then verifies their integrity by analyzing consistency without traversing memory. With this approach, attestation is applied to a group of neighbors, so that scalability is guaranteed. Further, our election algorithm only selects one verifier at an interval time and verification is performed by not an end-point node but a gateway. It implies that the sustainability is ensured in the proposed scheme. Our immediate work is to deploy the proposed scheme in real IoT networks and heterogeneity of IVFs will be considered for more constrained environment. We also will explore probabilistic aspects of the propose scheme in future work.

Acknowledgments This research was supported by Institute for Information and Communications Technology Promotion (IITP) grant funded by the Korea government (MSIP) [R-20150518-001267, Development of Operating System Security Core Technology for the Smart Lightweight IoT Devices].

References

1. Atzori, L., Iera, A., Morabito, G.: The Internet of Things: A survey. Computer Networks **54**, 2787–2805 (2010). doi:10.1016/j.comnet.2010.05.010
2. Gubbi, J., Buyya, R., Marusic, S., Palaniswami, M.: Internet of Things (IoT): A vision, architectural elements, and future directions. Future Generation Computer Systems **29**, 1645–1660 (2013)
3. Loscocco, P.A., Wilson, P.W., Pendergrass, J.A., McDonell, C.D.: Linux kernel integrity measurement using contextual inspection. In: Proceedings of the 2nd ACM Workshop on Scalable Trusted Computing (STC 2007), pp. 21–29. ACM (2007)
4. Petroni Jr., N.L., Hicks, M.: Automated detection of persistent kernel control-flow attacks. In: CCS 2007: Proceedings of the 14th ACM Conference on Computer and Communications Security, pp. 103–115. ACM (2007)
5. Jaeger, T., Sailer, R., Shankar, U.: PRIMA: policy-reduced integrity measurement architecture. In: Proceeding of the Eleventh ACM Symposium on Access Control Models and Technologies, pp. 19–28 (2006)
6. R. Macdonald, S. Smith, J. Marchesini, and O. Wild. Bear: An open-source virtual secure coprocessor based on TCPA. Technical Report TR2003-471, Department of Computer Science, Dartmouth College, 2003
7. Trusted Computing Group (TCG). TPM Main Specifications. Version 1.2 rev 116, March 1, 2011. http://www.trustedcomputinggroup.org/resources/tpm_main_specification
8. Sailer, R., Zhang, X., Jaeger, T., van Doorn, L.: Design and implementation of a tcg-based integrity measurement architecture. In: Proceedings of the 13th USENIX Security Symposium, August 9–13, 2004, San Diego, CA, USA, pp. 223–238 (2004)
9. Ganesh, A.J., Kermarrec, A.M., Massoulié, L.: Peer-to-peer membership management for gossip-based protocols. IEEE Trans. Comput. **52**(2), 139–149 (2003)
10. Heinzelman, W.R., Chandrakasan, A., Balakrishnan, H.: Energy-efficient communication protocol for wireless microsensor networks. In: Proceedings of the 33rd Hawaii International Conference on System Sciences, vol. 8 (2000)
11. Erdős, P., Rényi, A.: On Random Graphs. Publicationes Mathematicae **6**, 290–297 (1959)
12. Bron, C., Kerbosch, J.: Algorithm 457: finding all cliques of an undirected graph. Commun. ACM **16**(9), 575–577 (1973). doi:10.1145/362342.362367

Multiple Service Robot Synchronization and Control with Surveillance System Assistance for Confined Indoor Area Applications

Doug Kim, Sangjin Hong and Nammee Moon

Abstract This paper proposes an efficient navigation control and synchronization mechanism of multiple networked robots for operation in large confined areas. An adaptable grid based navigation and control strategy is adopted to eliminate potential collisions among robots. Unexpected obstacles are handled and the speed of individual robot is maintained using the node-ordering technique. The proposed navigation control and synchronization mechanism is scalable and can be easily extended to multi-cell large environment. The obstacles information is gathered through local information by the robots for better planning of the navigation. The system collaborates with the existing surveillance systems in case of additional visual information is necessary. The interaction with the surveillance system is minimized to reduce potential overhead.

Keywords Multiple robot synchronization · Surveillance system coexistence · Service robots · Confined area navigation

D. Kim
Farmingdale State University, New York, USA

S. Hong
Stony Brook University, New York, USA

N. Moon(✉)
Hoseo University, Asan, South Korea
e-mail: mnm@hoseo.edu

S. Hong–This research is supported by the International Collaborative R&D Program of the Ministry of Knowledge Economy (MKE), the Korean government, as a result of Development of Security Threat Control System with Multi-Sensor Integration and Image Analysis Project, 2010-TD-300802-002.

D.-S. Park et al. (eds.), *Advances in Computer Science and Ubiquitous Computing*,
Lecture Notes in Electrical Engineering 373,
DOI: 10.1007/978-981-10-0281-6_102

1   Introduction

There have been numerous studies on the multi-robot navigation and control systems for decades for various health service applications [1][2][3][4] as well as for the warehouse applications. Multiple robots navigation and control in large confined area are particularly interesting since the robots can be used to support various human activities. In these applications, the robots navigate within the area for delivering the objects to the destinations. Within the area, these robots use multiple sensors to navigate through the complex environment [5][6][7].

One of the key objectives for the navigation system is to control the multiple robots simultaneously without interfering the human traffics. Moreover, the robots must be able to handle static obstacles as well as unexpected dynamic obstacles. As the number of operating robots increases, the navigation system must be designed to minimize performance degradation and hence the system should be scalable. The navigation system must be able to interact with existing surveillance systems with minimum intervention of normal surveillance operation. This is very critical such that the navigation system does not have any visual information and whenever the system requires assistance from the surveillance system, the navigation system should be able to collaborate. The interaction overview is illustrated in Fig. 1(a). While minimizing the interaction with the surveillance system, the navigation system fully utilizes the sensor information from the robots for constructing the obstacles map dynamically to be used for better controlling the flow of the robots. With the minimum amount of interaction, the proposed system is designed to monitor the obstacles placement information in addition to the local gathering through the robots. In order to handle the dynamic situational changes, the navigation system closely interacts with the path planning system where the navigation paths for the robots are created. To handle a large number of concurrent navigation of robots without potential collisions and deadlocks, the speed control management mechanism using the node-ordering is proposed where the node-ordering guarantees the effective flow of robot navigation. Multi-cell relay structure shown in Fig 1(b) is used in the proposed navigation system to be scalable to cover the large areas. While there are distributed and centralized control mechanisms, we use the centralized communication model where the system controls individual robots to navigate.

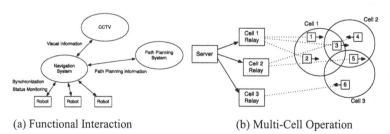

(a) Functional Interaction   (b) Multi-Cell Operation

Fig. 1 (a) The navigation system communicates with multiple robots and existing surveillance system. (b) Illustration of the robot synchronization through multiple relays.

725 Multiple Service Robot Synchronization and Control with Surveillance System

2   Navigation and Synchronization Strategy

2.1   Grid Based Concurrent Navigation

Fig. 2(a) illustrates the overview of the multiple robot concurrent navigation system functions. The navigation server controls the robots through local communications and communicates with the existing surveillance system for visual assistance. The navigation server, the robots, and the surveillance system share the common map for coordination where the map is represented with grids. The size of the grid is adaptable depending on the traffic condition. The path of the robot is defined as a sequence of grids. Hence, the navigating server can determine the position of the robot by its grid location and the navigating server maintains such that no two robots can reside within a grid at any time. The robots are capable of localizing itself within the confined area and respond to the unexpected obstacles with its own proximity sensors. In case the robots fail to localize themselves, the robot requests visual localization assistance from the surveillance systems through the server.

Fig. 2(b) illustrates basic functional mechanism of the robots communicating with server for synchronization and control. The server maintains the robot status information where the status for each robot includes the current speed, the navigation status whether the robot is currently in idle, stop, or moving. The sensor range and the server status are also maintained. The server status indicates the outstanding responses that the server must perform.

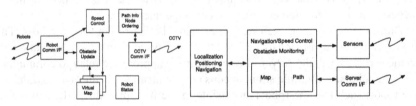

(a) Server Functional Interaction   (b) Robot Functional Interaction

Fig. 2 (a) The server communicates with the visual sensor network for visual service and controls the robot using the map and obstacle information. (b) The operation of the robots consists position computation, navigation and speed control, and interface with the server.

Each time the robot arrives the grid, the robot sends the arrival notification to the server and continues to navigate. Upon receiving the message, the server responds with acknowledgement to the robot. If the server's acknowledgement is not received before getting out of the current grid, the robot stops. The reason for the stop is to avoid potential collision by violating multiple robots in the same grid condition. This interaction is illustrated in Fig. 3(a). As shown in Fig. 3(b), the frequency of the robot communication depends on the size of grids. If the grid size is small, the communication traffic increases. Moreover, the communication traffics increase as the number of the robots increases. Hence, in the normal operation, the system prefers to use the large grid size as possible. Because the server receives all the status information from the robots, the server is able to maintain the location of all robots. The server makes sure that no two or more robots go to the same grid to avoid collision. As we will discuss the speed control in the later section, the speeds of the robots are properly maintained.

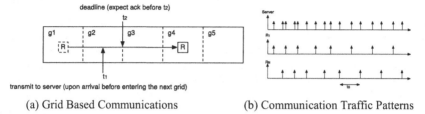

(a) Grid Based Communications   (b) Communication Traffic Patterns

Fig. 3 (a) The robots sends arrival message to the server at t_1 and expects the server acknowledgement by t_2. (b) The server and robots communication frequencies.

2.2 Extension to Multi-Cell Environments

In order to support the large-scale environment, multiple communication relays are utilized where the server and the robots communicate through the relays. Each robot is assumed to have the capability to communicate through all channels. Since the communication coverage is affected by the environment factors, it is not certain to identify the place where the handoff from a cell to another cell is possible. Therefore, the system requires the effective handoff protocol, which can switch from a cell to another under the coverage uncertainty.

During the operation, each relay maintains the list of connected robots. These list of robots are continuously updated by the relays and send the information back to the server. Thus, the server always select the appropriate relay to communicate with the robots. If multiple simultaneous communication channels are available by the robots such that the robots are visible in the multiple relays, the server sends the message to these multiple relays.

2.3 Role of Surveillance Network

Typical robot navigation systems do not have the capability of getting visual information. The proposed navigation system interfaces with the existing surveillance network for visual assistance. Whenever the server needs visual assistance for the robot localization and monitoring the traffics, the server requests the surveillance system. However, the surveillance system usually has its own operation and may not have computation resources to support the navigation system. In order to minimize the overhead by the surveillance systems, two service parameters are defined. $T_{service}$ is the minimum time between the services and $N_{service}$ is the number of requests that the surveillance systems may provide to the server.

3   Environment Adaptations

3.1   Dynamic Obstacle Management

From the navigating robot perspectives, there are three types of obstacles. The first type is when the robots are navigating in the same corridor. This type is not critical since the navigating server knows the directions of navigation and the positions of grid of all robots. Hence, when there are multiple robots navigating within the same corridor, the sensor ranges are adjusted accordingly to eliminate unnecessary stoppage by the robots. The second type is when the obstacles are due to the moving objects such as people. When the robot faces with the moving obstacles, before sending the message to the server, the robot monitors the changes in distance with the robots. If the distance is changing, the robot considers that this is moving obstacles and waits until the obstacles are cleared from the sensor range. Once it is cleared, the robot navigates. The third type is the static obstacle that is semi-permanently placed within the corridor. When the robot is faced with the static obstacles, the robot needs additional path information to navigate away from the obstacles.

The static obstacles can disturb the other robots in the later time. So in order to minimize unnecessary stopping time, the obstacles map is generated. Then the navigation server uses the obstacles map information to send the grid types and sensor range of robots entering the corridor to avoid the stoppage. When the obstacles are removed and the next robot does not sense the obstacles, the server removes the obstacles from the obstacles map. Thus, the navigating robots constantly update the obstacles map. When the navigating server is connected to the surveillance system, the obstacles map can be generated from the visual information as well. Note that when the static obstacles are presented which were not considered in the initial path planning, the presence of the obstacles modifies the capacity resources of the map.

3.2   Virtual Grid Adaptation for Fine Grain Navigation

Grid adaptation during the navigation is very critical for maximizing the robot utilization. The grid changes to the finer grid whenever the obstacles are detected or there are other robots navigating in the same corridor. In the normal navigation, largest possible grid size is used since the usage of smaller grid sizes increases the communication overhead. In this situation, the size of grid must be changed to provide finer navigation.

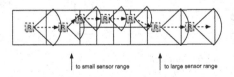

Fig. 4 The robot has a front distance sensor to detect any obstacles. The sensor range changes depending on the grid size used.

The distance range of the sensor is set by the navigation server depending on the size of grid used and/or depending on the traffic condition. This is illustrated in Fig. 4. When the obstacles are detected, the robot stops and sends message to the server and waits for next command. If the obstacles are cleared before the navigation server action, the robot sends message to the navigation server. If the obstacles are not cleared for certain amount of time, the navigation server may regenerate the path with finer grid sizes.

3.3   Speed Control with Node-Ordering

When the paths are generated for the robots, the node-ordering data structure is also updated. The entries maintained in the internal data structure for each node indicate the robots index, earliest arrival time, latest arrival time, and expected arrival time. The earliest arrival time and the latest arrival time are determined during the path generation suggesting that the robot should not arrive before the earliest arrival time and should arrive before the latest arrival time. The expected arrival time is estimated by the server and should be between the two timing parameters. Based on the timing parameters specified in the node-ordering data structure, the speeds of the robots are computed. If any of the timing is violated, paths must be rescheduled.

4   Evaluations

A large-scale map with complex confined corridors with the dimension as illustrated in Fig. 5(a) is considered in the simulation. To cover the entire map, 4 cells are assumed as shown in the figure. The visual sensor coverage sectors are divided into the corridors. The dimension of the coarse grid is the maximum width of the corridor. The maximum speed of the robot is 4m/s and average object speed is 2m/s.

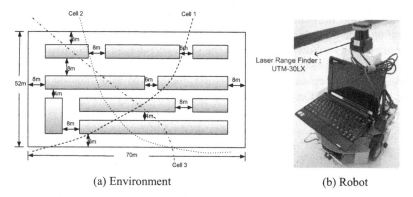

(a) Environment   (b) Robot

Fig. 5 (a) Illustration of the map used in the simulation. (b) The illustration of a mobile robot

Fig. 6(a) illustrates the speed variation of the robots when all the robots are navigating at the same time. Due to the node ordering, the speed variations and service duration are visible. Fig. 6(b) illustrates the communication traffic pattern between the robots and the server when the robots are navigating simultaneously.

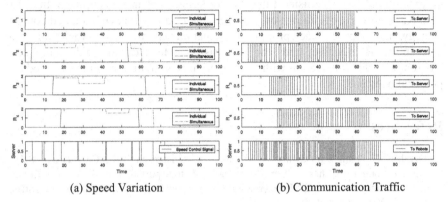

(a) Speed Variation   (b) Communication Traffic

Fig. 6 (a) Speed variations as a function of the time when all robots are simultaneously navigating. Solid lines represent the speed of each robots when the robot is navigating individually. (b) Communication traffics between the robots and the server in the ideal situations.

Interaction between the server and the surveillance network is also evaluated with different values of $N_{service}$ and $T_{service}$. In the simulation, the robots randomly generate localization errors and request the server for the assistance. If the requests are from the robots in the same corridor, the server does not send additional request to the surveillance network. The communication message generated by the server for the different service parameters is illustrated in Fig. 7. In this evaluation, random position requests are generated, which requires the assistance from the visual CCTV network.

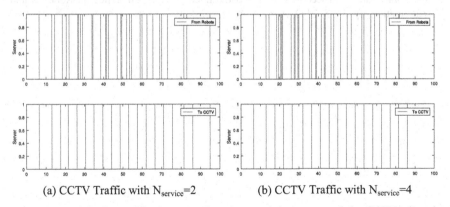

(a) CCTV Traffic with $N_{service}$=2   (b) CCTV Traffic with $N_{service}$=4

Fig. 7 Communication traffics between the robots and the server and the CCTV when the visual assistance is required. (a) $N_{service}$=2 (b) $N_{service}$=4. In both case, $T_{service}$ = 5sec.

5   Conclusions

This paper proposes an efficient navigation control and synchronization mechanism of multiple networked robots for operation in large confined areas. In order to eliminate the collisions, the node-ordering mechanism is utilized. In order to maximize the robots utilization, an adaptable grid based navigation and control strategy is adopted. The proposed navigation control and synchronization mechanism is scalable and can be easily extended for multi-cell large environment. The interaction with the surveillance system is minimized to reduce potential overhead.

References

1. Ozkil, A.G., Fan, Z., Dawids, S., Aanaes, H., Kristensen, J.K., Christensen, K.H.: Service robots for hospital: a case study of transportstion tasks in a hospital. In: Proceedings of the IEEE International Conference on Automation and Logistics, pp. 289–294, August 2009
2. Carreira, F., Canas, T., Silva, A., Cardeira, C.: i-merc: A mobile robot to deliver meals inside health services. In: Proceedings of IEEE/RAM (2006)
3. Prassler, E., Scholz, J., Fiorini, P.: A robotic wheelchair for crowded public environments. In: IEEE Robotics & Automation Magazine, pp. 38–45, March 2001
4. Choset, H.: Coverage for robotics--A survey of recent results. Ann. Math. and AI **31**, 113–126 (2001)
5. Hada, Y., Takase, K., Ohgaki, K.: Indoor navigation of multiple mobile robots in dynamic environment using iGPS. In: Proceedings of the IEEE International Conference on Robotics & Automation, pp. 2682–2688, May 2002
6. Rossetti, M.D., Felder, R.A., Kumar, A.: Simulation of robotic courier deliveries in hospital distribution services. Health Care Management Science **3**, 201–213 (2000)
7. Dias, M.B., Zinck, M., Zlot, R., Stentz, A.: Robust multirobot coordination in dynamic environments. In: Proceedings of the IEEE International Conference on Robotics and Automation, pp. 3435–3442, April 2004

Author Identification and Analysis for Papers, Reports and Patents

Kwang-Young Kim, Seok-Hyoung Lee, Jung-Sun Yoon and Beom-Jong You

Abstract In recent years, the demand for improved transparency and sharing of data from government organizations, private companies, and data information centers has been increasing. However, the public release of metadata alone is of limited impact. In order to ensure the integrity of such reciprocal interactions, secure individual data identification is required. The present study attempts to analyze methods for reciprocal identification of metadata comprising scientific or technical content that may have diverse origins, such as domestic and foreign dissertations, domestic patents, and domestic research reports.

Keywords Author control · Identification · Disambiguation · Linked data

1 Introduction

The effective sharing of research content with the wider research community has become a focus of recent interest, allowing for the reciprocal exchange of results and opinions between authors and readers alike. It is generally accepted that the tendency toward greater openness in information and knowledge sharing has been spearheaded by the acceleration of web-based technology and has had a considerable impact on diverse fields of research. Accordingly, the EU Commission has begun public discussions on the concept of Science 2.0 with a tendency similar to that which occurred for Web 2.0 [1], while the Korean Ministry of Science, ICT (Information/Communication Technology) and Future Planning has provided a range of pilot services including the launch of the National Science Technology Data Portal [2]. Maire Geoghegan-Quinn, the European Research, Innovation and Science Commissioner, stated, 'Science 2.0 is

K.-Y. Kim(✉) · S.-H. Lee · J.-S. Yoon · B.-J. You
Department of Information Convergence Research,
Korea Institute of Science and Technology Information, Daejeon, South Korea
e-mail: {Glorykim,skyi,jsyoon,ybj}@kisti.re.kr

© Springer Science+Business Media Singapore 2015
D.-S. Park et al. (eds.), *Advances in Computer Science and Ubiquitous Computing*,
Lecture Notes in Electrical Engineering 373,
DOI: 10.1007/978-981-10-0281-6_103

revolutionizing the way we do science - from analyzing and sharing data and publications to cooperating across the globe. It is also allowing citizens to join in the search for new knowledge.' [1]

Secure data identification is a necessary prerequisite to move beyond simply sharing data or metadata, and progressing toward a system of meaningful connections between shared data sources. The JISC (Joint Information System Committee, United Kingdom) has recently implemented a project called Names for efficient author identification. The precise identification of publications written by researchers is a significantly important factor for the successful transition to E-Science. Once accurate author identification is achieved, research communities (not-for-profit organizations, governmental organizations, universities, etc.) consisting of users participating in academic communication can more efficiently develop new services, from which diverse benefits can be gained [3].

Diverse fields of expertise require author identification in order to improve levels of data utility by connecting content in meaningful ways. This can also prove useful for various analytical methods by applying identified data to various academic search engines. Insoo Kang (2008) compared the performance, errors, and effectiveness of author identification upon which diverse machine learning methods were influenced, and analyzed diverse methods to draw conclusive values and calculate qualifying similarities regarding co-authors and dissertation titles. The most accurate levels of author identification reported for author appearances in dissertations published in Korean was 83.8% when only co-author information was used, compared to 95.7% when both co-author and email addresses were used. The best approach tested was SVM with 82.7%, among the three test sets of F1, with an under/over-population group error used in the experiment [6][7].

Taehong Kim (2012) also reported that the process for identifying whether an entity is the same or not based on discriminating relationships is necessary for data linkage and consolidation. His research outline set a goal for entity identification within publications, and proceeded with domain accounting for the most prevalent proportions of the linked data (87 out of 295 datasets) to complement weaknesses within existing entity identification services to gradually identify multi-ontological entities. He then proposed and assessed methods of entity identification based on characteristics to identity and utilize multi-ontological entities based on authorship [8].

Since problems exist regarding a single author using diverse names or different authors with an identical name as described earlier, the author identification process is necessary for accurate bibliographic searches. In order to resolve such ambiguity, Jian Huang et al., (2006) proposed an effective consolidated machine learning framework: by eliminating other authors with a similar name from the search results, using DBSCAN, an author clustering process. The machine learning method proposed in the dissertation was the result of research on over 700,000 dissertations in Citeseer and offered improvements in terms of speed and accuracy in comparison with the traditional SVM method. It also demonstrated flexibility when processing data from additional dissertations [9].

2   Author Control Strategy

We sought to create an automatic author identification program applicable for diverse content such as domestic and foreign theses, domestic patents and research reports, and cross-identify authors within this content. Cross-identified authors contributing to various content types were to be identified and linked.

The content used in our approach consists of metadata which does not contain emails or private phone numbers which would more easily enable author identification. We deliberately chose to focus on the identification of authors using the characteristics of the content in the source material.

Listed below Table 1 are cross-author identification rules for various types of content such as domestic and foreign theses, patents, and research reports. For the crossover authors, the same names are clustered for each content type as for Rule R-1, and crossover authors are identified from the names of the institutions, addresses, co-authors, collaborating researchers and associates. If less than 2 documents are attributable to the authors, Rule R-2 will be applied. If there is no name for the institution in Korean, it will be associated with a nominal English name of the institution.

Table 1 Cross-author identification Rule for domestic and foreign theses, patents, and research reports

Cross Content Rule Number	Rule Content	Remark
R1	(Author | Co-author | Chief researcher| Participating researcher | Inventor | Applicant) & (Name of the institution | Address) | (Co-author | Participating researcher | Applicant)	
R2	(Author | Chief researcher | Inventor)	Applicable if there are less than 2 attributable documents

3   Analysis of Author Identification Results

The experimental study was conducted with object identification for domestic and foreign theses, domestic patents and domestic research reports. The experimental program MetaData conducted the experiment on 90 domestic theses, 27,000,000 international theses, 4,000,000 domestic patents, and 50,000 domestic research reports.

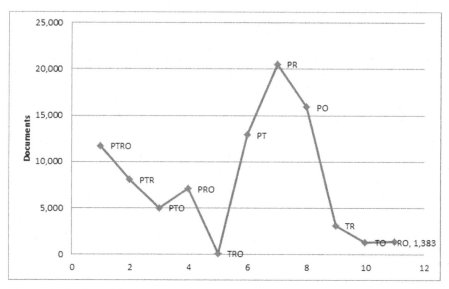

Fig. 1 Distribution of individuals publishing domestic and foreign theses, patents, and reports.

(* P = Domestic Thesis, T = Patent, R = Research Report, O = Foreign Thesis)

The results detailing the researchers who have published in more than 2 types of content are shown in Figure 1. The number of researchers who have published domestic and foreign theses, reports, and patents is 11,703 but the total number of separate publications produced is 1,621,570, a relatively large number considering the small number of the researchers. The number of researchers who have published domestic theses and reports is 20,540 but the total number of separate publications produced is 183,961, a lower total number than the number of researchers. The number of researchers who have published in each category (patents, reports and foreign theses) is relatively low at 111.

4   Conclusions and Future Research

In today's age of greater openness and sharing of data, the release of simple metadata needs to be developed toward more interactive data correspondence between government, industry and data information centers. In order for this to occur, accurate identification tools and the ability to draw connections between data entries is critical. In order to overcome the present limitations inherent in keyword-based information services, as well as to acquire high quality information and maximize search usability, and for higher value data analysis and data amalgamation, system development based on individual data identification is critically needed.

Our results for author identification across various content types revealed that the number of domestic scientists actively producing research results in scientific and technology is 87,257, and of these, 11,703 are working with 4 types of content, 20,310 with 3 types, and 55,244 with 2 types. Furthermore, the 11,703 crossover authors publishing 4 content types are the most productive in terms of active research results through their domestic and foreign theses, patents and reports. Although only a relatively small number of these researchers exist, their output is considerable. Our results also show that there are many authors working on domestic theses and reports, but the overall total productivity in terms of research results in these categories is relatively low. Of particular note, the number of authors who publish their research results in reports or foreign theses are quite low.

References

1. http://europa.eu/rapid/press-release_IP-14-761_en.htm (accessed August 19, 2015)
2. http://www.msip.go.kr/www/brd/m_211/view.do?seq=2068 (accessed August 19, 2015)
3. http://icon.ndsl.kr/i_trend/icon_trendDetail.jsp?record_no=968&trendType=O (accessed August 25, 2015)
4. Lee, S.H., Kwak, S.J.: A Study on the based on FRAD Conceptual Model based Authority Data Scheme for Academic Papers. Journal of the Korean Society for Library and Information Science 45(3), 235–257 (2011). doi:10.4275/KSLIS.2011.45.3.235
5. Cho, J.I.: A Study on the Construction Methods for Author Identification System of Research Outcome based on ORCID. Journal of the Korean BIBLIA Society for Library and Information Science 24(1), 45–62 (2013)
6. Kang, I.S.: Application of Machine Learning Techniques for Resolving Korean Author Names. Journal of the Korean Society for Information Management 25(3), 29–39 (2008). doi:10.3743/KOSIM.2008.25.3.027
7. Kang, I.S., Kim, P., Lee, S.W., Jung, H.M., You, B.J.: A Large-scale Test Set for Author Disambiguation. International Journal of Contents 9(11), 455–464 (2009)
8. Kim, T.H., Jung, H.M., Sung, W.K., Kim, P.: Author Entity Identification using Representative Properties in Linked Data. International Journal of Contents 12(1), 17–29 (2012). doi:10.5392/JKCA.2012.12.01.017
9. Huang, J., Ertekin, S., Giles, C.L.: Fast Author Name Disambiguation in CiteSeer. ISI Technical Report, pp. 1–13 (2006)
10. Pitts, M., Savvana, S., Roy, S.B., Mandava, V., Prasath, D.: ALIAS: author disambiguation in microsoft academic search engine dataset. In: Open Proceedings, pp. 648–651 (2014)

Resource-Aware Job Scheduling and Management System for Semiconductor Application

Hyesuk Park, Seungyun Kim, Taehyun Kim, Junghoon Kim, Youngkwan Park and Young Ik Eom

Abstract The increased number of transistors and complexity of chips has led the rapid increase of the semiconductor design data size. It is crucial for turn-around time (TAT) of mask data preparation (MDP) procedures which are translated from designed data into a set of polygons for mask-writing equipment. MDP procedures require huge computational resources and software to maintain desired TAT. One of our concerns is how to decide priorities of all jobs and effectively manage target TAT using limited resources with many users. In this paper, we introduce a job scheduling system with a newly developed algorithm to control job priorities considering planned date and predict a number of required resources in advance. Furthermore, we propose a resource management technique to maximally utilize limited resources and dramatically reduced TAT, simultaneously. As a practical example, TAT and resource utilization of advanced technology node were improved by 58% and 30%, respectively.

Keywords Resource management · Job Scheduler · MDP

1 Introduction

The process of semiconductor manufacturing involves several steps. The first step is the design of integrated circuits (ICs). The second one is the translation of a file

H. Park(✉) · S. Kim · T. Kim · Y. Park
Semiconductor R&D Center, Samsung Electronics Co. Ltd., Suwon, Gyeonggi-do, Korea
e-mail: {hsraon.park,sweeney.kim,th36.kim,hsraon}@samsung.com

H. Park · J. Kim · Y.I. Eom
College of Information and Communication Engineering,
Sungkyunkwan University, Seoul, Gyeonggi-do, Korea
e-mail: {hsraon,myhuni20,yieom}@skku.edu

© Springer Science+Business Media Singapore 2015   737
D.-S. Park et al. (eds.), *Advances in Computer Science and Ubiquitous Computing*,
Lecture Notes in Electrical Engineering 373,
DOI: 10.1007/978-981-10-0281-6_104

from designed data into a set of polygons for mask-writing equipment, which is called mask data preparation (MDP). The third step is the process of writing the circuit patterns onto a mask by using an electron beam. The final step is the printing of mask's circuit design onto a wafer which is a thin slice of semiconductor material. In these steps, MDP occupies about 50% of the resources for semiconductor manufacturing. For an advanced semiconductor technology that integrates more than 100 million transistors, design complexity and data size increase dramatically. This leads to a tremendous MDP TAT and requires drastic amount of resources as shown in Figure 1.

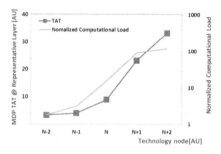

Fig. 1 MDP TAT and normalized computational load grows almost exponentially. (N is the current technology, N+1and 2 are toward technology nodes. N-1and N-2 are previous technology nodes.)

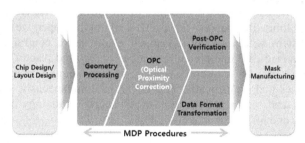

Fig. 2 MDP procedures which are translated an integrated circuit layout into set of instruction that photomask equipment understands.

Figure 2 shows MDP procedures which are series of geometry processing, optical process correction (OPC), verification and data format transformation [1,2]. Several MDP procedures are issued at the same time, and a traditional scheduling system selects one job of which required resources are smaller than available capacity, regardless of planned completion date. After starting, the job has no chance to change the number of resources whether there is a lack of resources or not. In addition, when high job priority is issued in case of fully charged CPU, we have to decide whether to kill another job or to wait. This may cause job rework, decreased efficiency, and waste of resources by human's decision [3]. Although commercial software products provide traditional solutions for large scale distributed systems, they have two critical issues: Job scheduling and Resource management.

1.1   Job Scheduling

First-come first-serve (FCFS)[4-7] and fair-share are typical basic scheduling strategies. In the FCFS scheduling, the dispatch order depends on the order of jobs in the queue, where the jobs are placed in the order of arrival. In the fair-share scheduling, the dispatch order depends on the dynamic share priority of jobs in the queue[7,8]. Figure 3 and Table 1 show examples of FCFS and fair-share.

FCFS: Job1 is starting first because job1 submitted firstly. Then job2 started.

Fair-share: Job4 started next because job4 have smaller execution time than historical data of others, although job3 and job4 are submitted simultaneously. In fact, fair-share scheduling is running in equal proportion of all job based on predefined priority factor.

Urgency and Order: Job5 was submitted lastly, it has top priority due to target date. Moreover, job5-4-2-3-1 order is right schedule for MDP. However, FCFS and fair-share make an order job1-2-4-3-5 following by their unique policies and those will lead to business failure.

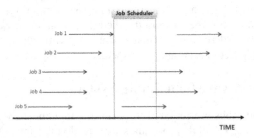

Fig. 3 Traditional job scheduling without considering job priority.

Table 1 Job Information and scheduling results

Job	Submission	Starting	Policy	Target Date	Priority
Job1	1	1	FCFS	6/20	normal
Job2	2	2	FCFS	6/16	normal
Job3	3	4	fair-share	6/17	normal
Job4	3	3	fair-share	6/15	normal
Job5	4	5	FCFS	6/10	urgent

1.2   Resource Management

Resource manager collects load information through the communication with the server, and assigns the available resources to the job according to the requested amount. When the operation is completed, the resources are reclaimed to maintain the status information of the resources.

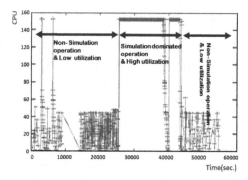

Fig. 4 CPU utilization in one job of MDP procedures which have several operation.

For example, Figure 4 shows CPU utilization of OPC that accounts for the most of the TAT in the MDP procedures. During time period 0~25000 and 45000~60000, only 35% of CPUs were used for simulation. In general, OPC job consists of multiple operations. However, we can utilize and redistribute resources for another job during non-simulation operation[10].

In this paper, we propose a resource-aware scheduling system and a dynamic resource management mechanism to overcome two limitations of current solutions. Section II presents the proposed scheduling system in detail. Section III shows the operations of resource management. Experimental results and conclusion are presented in Section IV and V, respectively.

2   Resource-Aware Scheduling System

More than 100-1000 jobs are required to perform simultaneously for the manufacturing of s single mask. MDP procedures can be divided into OPC-Verification, which requires sequential processing, and Verification-Data Format Transformation, which requires independent and simultaneous processing. Considering these characteristics, job scheduling system is implemented as shown in Figure 5. The proposed system consists of *User Interface* and *Job Scheduler*. After the *Job Scheduler* decides to run a job, it requests necessary resources to the *Resource Manager*.

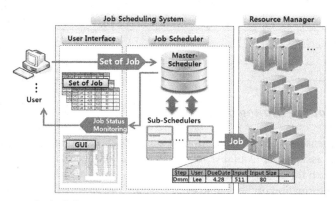

Fig. 5 Structure of scheduling system

We designed target oriented scheduling, which considered the dependency between jobs and due-date, and configured the scheduling unit to set instead of job. For the scheduling of sets, we added *wait* state to the job life cycle, resulting in the states including *wait, pend, run, done, and exit* in the job life cycle.

Job Scheduler is configured as a master-scheduler and several sub-schedulers, and it operates as follows:

1) User generates the set of job using a standard flow and pre-defined due date in project management system. It has a lot of information: input gds, MDP date, the person in charge, customer, and etc.

2) To determine the priority of each job, master-scheduler calculates start date based on due date and estimates how many resources are required for the set.

3) Sub-scheduler submits a job to *the Resource Manager* according to the determined priorities assigned by maser-scheduler.

4) Sub-scheduler monitors the status of the job and delivers the job to the master-scheduler.

5) Repeat 1~4.

A job goes through a series of state transitions until it eventually completes its task as failure or normal termination. The possible states of a job during its life cycle are shown in Figure 6. In general, a job has four states. A job remains in pending state until all conditions for its execution are met. Some of the conditions are availability of the specified resources, job dependency and pre-execution conditions, and so on. For example, if the highest priority job requires 300 CPUs, the job will be in pending state until it gets the availability of 300 CPUs.

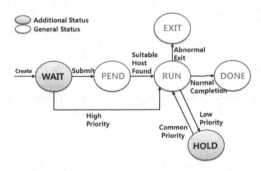

Fig. 6 Life cycle of job in job scheduling system.

Resource-Aware scheduling system has six states such as *wait, pend, run, hold, exit,* and *done,* where wait state and hold state are added into the general model. In our system, we implemented a hotline, to support urgent project, with which a job in *wait* state can go to *run* state bypassing the *pend* state. In the meantime, *Resource Manager* may make a decision to assign the job more resources than other jobs, which results in some jobs with low priorities losing their resources. We added *hold* state for the jobs losing their resources as in the above mentioned situation.

3   Dynamic Resource Management

Resource Manager communicates with each one of the running jobs in order to dynamically determine how many processors can be used for each job without sacrificing the performance of the system [9]. It is a centralized service that collects the necessary information not only on the active jobs but also on the clusters and networks on which these jobs are running. It can use the collected information to dynamically determine how many resources should be allocated to each job and which resources should be preempted and reassigned among the jobs, to improve the overall system performance. By dynamically sharing resources among the running jobs, the *Resource Manager* automatically controls the number of resources assigned to each job. Traditional resource managers use fixed amount of resources during their execution by allocating the resources statically when a job is submitted. In contrast, dynamic resource management estimates the resource requirements of the jobs and operations by communicating with the jobs in execution, and redistributes the resources accordingly. Figure 7(a) and Figure 7(b) shows the results of the experiments that execute the same job using the static and dynamic resource management schemes, respectively: the gray area shows the amount of CPU resources allocated and the back-slashed   area shows the amount of CPU resources actually used by the job.

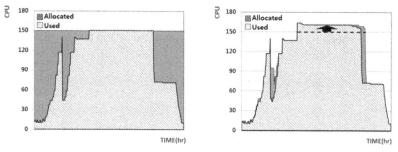

Fig. 7 (a) Traditional resource allocation for running jobs. (b) Dynamic resource allocation for running jobs

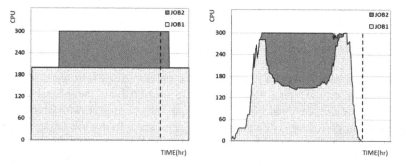

Fig. 8 (a) Traditional resource allocation in 2jobs. (b) Dynamic resource allocation in 2jobs.

As another experiment, we executed two jobs concurrently in 300 CPU systems, with the traditional and dynamic resource management schemes. The results are shown in Figure 8(a). As shown in Figure 8(a), the resource requirement of JOB1 is 200 CPUs and the resource requirement of JOB2 is 100 CPUs. In this case, which uses static resource management, fixed number of resources is allocated to each job during their execution.

On the other hand, Figure 8(b) shows the experimental result of dynamic resource management. In the figure, we can see that more resources are allocated to JOB1 until JOB2 starts. The redistribution of resources are performed after job2 starts, and, as time goes, approximately the same amount of resources are allocated to JOB2 as JOB1. Also, the resources are allocated to JOB1 again as soon as JOB2 finishes.

Despite that we performed our experiments with the same number of jobs on the same 300 CPUs, dynamic resource management scheme finished the jobs earlier than the static scheme. In this experiment, we can see that dynamic resource management scheme achieves higher CPU utilization and shorter TAT. In Figure 8(b), we see that approximately the same amount of resources is allocated to the two jobs when they run concurrently in the system. But, it is possible for the dynamic resource management scheme to use group-specific rate for each group of jobs according to their priorities and allocate resources according the rates of the groups. With this scheduling system, we can give higher-priority jobs with more resources and finish the jobs earlier than others[10].

4   Result

Proposed solutions were applied to the MDP procedures of logic products, and improved the TAT by 58% at most as shown in Figure 9. In addition, the CPU utilization of total system was improved by 30%.

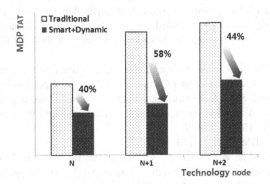

Fig. 9 Comparison MDP TAT between traditional and proposed solutions.

5   Conclusion

In the computing facilities of the semiconductor manufacturing system, the MDP procedures occupy about 50% of the resources of the computing facilities. Moreover, MDP TAT of advanced technology is increased consistently. Both the resource utilization and the TAT can be improved by implementing the job scheduling scheme and resource management technique in a dynamic. In this paper, we introduced a scheduling system with the newly developed algorithm to control job priorities and predict the number of required resources in advance. Also, we proposed a resource management technique to maximally utilize limited resources and dramatically reduced TAT, simultaneously.

In the future, a research on advanced resource management techniques may be necessary to enhance the flexibility of all the system components and to efficient management of resources used in semiconductor manufacturing.

Acknowledgment This research was supported by the MSIP(Ministry of Science, ICT and Future Planning), Korea, under the ITRC(Information Technology Research Center) support program (IITP-2015-(H8501-15-1015)) supervised by the IITP(Institute for Information & communications Technology Promotion). Young Ik Eom is the corresponding author of this paper.

References

1. Leng, N.C., Har, A.J.C., Tat, O.Y.W.: System management with relational database for mask tape-out. In: Quality Electronic Design (ASQED), pp. 128–132 (2012)
2. Schulze, S.F., Bailey, G.E.: Distributed processing in integrated data preparation flow. In: BACUS Symposium on Photomask Technology, pp. 1–12 (2004)
3. Ahn, B.-S., Bang, J.-M., Ji, M.-K., Kang, S., Jang, S.-H., Choi, Y.-H, Ki, W.-T., Choi, S.-W., Han, W.-S.: Optimized distributed computing environment for mask data preparation. In: BACUS Symposium on Photomask Technology, pp. 1–8 (2005)
4. Zhou, S., Zheng, X., Wang, J., Delisle, P.: Utopia: a load sharing facility for large heterogenous distributed computer systems. Technical Report CSRI-257, pp. 1–37 (1992)
5. Gentzsch, W.: Sun grid engine: towards creating a compute power grid. In: Proc. of International Symposium on Cluster Computing and the Grid, pp. 35–36 (2001)
6. Zhao, W., Stankovic, J.A.: Performance analysis of FCFS and improved FCFS scheduling algorithms for dynamic real-time computer systems. In: Real Time Systems Symposium, pp. 156–165(1989)
7. Administering IBM Platform LSF
8. Hasija, M., Kaushik, A., Kumar, P.: OM algorithm for multi-level queue scheduling. In: Machine Intelligence and Research Advancement (ICMIRA), pp. 564–568 (2013)
9. Nouh, A., Jantzen, K., Park, M., Vu, H.T.: Compute resource management and TAT control in mask data prep. In: Metrology, Inspection, and Process Control for Microlithography XXIII, pp. 1–8 (2009)
10. Endo, T., Park, M., Ghosh, P.: Predictable turn-around time for post tape-out flow. In: Proc. of Optical Microlithography XXV, pp. 1–14 (2012)

Live Mobile Learning System with Enhanced User Interaction

Jang Ho Lee and Haedong Hwang

Abstract We have been building a live mobile learning system that provides real-time distance learning for students using tablets. An instructor gives a lecture in front of a tablet by bringing up slides and making annotations on them. Students watch the instructor as well as the slides with annotation being made by him, on their tablets. They can also ask him questions by typing texts in the text chat. Typing texts, however, turned out to be not efficient user interaction method in terms of convenience and speed. Therefore, we enhanced the user interaction by allowing students to talk to the chat which, in turn, handles speech-to-text processing and send the text to the server.

Keywords Live mobile learning · Speech-to-text · Tablet · Android

1 Introduction

Over the last decade, there has been tremendous advancement in mobile device and networking technologies which prompted people's popular use of smartphones and tablets. In this period, researchers in the field of distance learning have become interested in mobile learning than in conventional desktop PC-based learning, since it provides remote students to participate in the distance learning with their mobile devices from anywhere [1].

Classroom Presenter [2] allows students to share slide and annotation with a teacher on their tablets. But it requires students and a teacher to be physically present in the same classroom since it doesn't support video nor method for interaction between students and a teacher such as chat. MLVLS [3] is a live mobile learning system that allows students to watch an instructor as well as slides with

J.H. Lee(✉) · H. Hwang
Department of Computer Engineering, Hongik University,
72-1 Sangsu, Mapo, Seoul 121-791, Korea
e-mail: janghol@cs.hongik.ac.kr, haenara90@naver.com

© Springer Science+Business Media Singapore 2015
D.-S. Park et al. (eds.), *Advances in Computer Science and Ubiquitous Computing*,
Lecture Notes in Electrical Engineering 373,
DOI: 10.1007/978-981-10-0281-6_105

annotation on their mobile devices. However, it only supports smartphone which can give students hard time recognizing the slide with annotation on a small display of smartphone. It also lacks support for interaction between students and an instructor. We had also previously developed a real-time mobile learning system that enables student to watch an instructor and slides with annotation on their tablets [4]. It also supports interaction between students and an instructor via text chat. However, the problem with this system is that students find it inconvenient as well as slow to type texts on a chat for interaction with the instructor.

Thus, we present a live tablet-based learning system with enhanced user interaction. The system basically supports video, audio, slide with annotation from an instructor as well as text chat on student's tablet in real time. Furthermore, in order to overcome the inconvenience and slow speed of text-typing on a chat for interaction, we adopted the speech-to-text technology in the chat so that when a student can not only type texts but also talks, in which case the speech is converted into text and sent to the broadcasting server.

2 Live Mobile Learning System with Enhanced user Interaction

Fig.1. shows the basic concept of the proposed live mobile learning system. The instructor gives a lecture in front of the instructor's tablet. The instructor can make facial expression and use voice while he shows slides and makes annotations on them. Students watch the video of the instructor and the slide with annotation made by him. Students can also ask questions using text chat. The text chat also provides students with speech-to-text feature to enhance interaction.

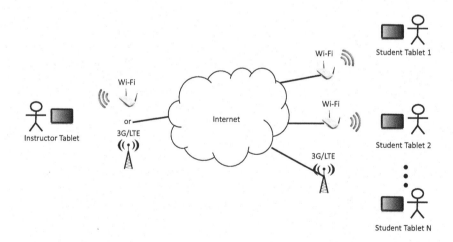

Fig. 1 Live mobile learning system

Fig.2. shows the software architecture of the proposed live mobile learning system with enhanced user interaction. It consists of servers (broadcasting server and speech-to-text server) and tablet clients (one instructor's tablet client and multiple student's tablet clients).

On the instructor's client side, the encoding is done as follows. The video is encoded in H.263 [5], the audio is encoded in G.723.1 [6]. The slide is encoded in JPEG. The annotation actions are packed into groups of input events. The instructor's tablet client sends the encoded data of video, audio, slide and annotation to the broadcasting server, which, in turn, broadcasts them to the multiple students' tablet clients. A client decodes the encoded video, audio and slide and renders the video and slide as well as plays the audio. It also unpacks groups of annotation input events and replays the events on the slide.

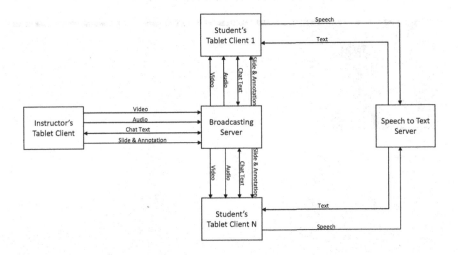

Fig. 2 Architecture of the live mobile learning system with enhanced user interaction

Chat texts between tablet clients are exchanged through the broadcasting server. When a student speaks to the chat, the voice is converted to the text by using the speech-to-text server and then sent to the broadcasting server. For a speech-to-text server, we used Newtone[7], which is a third-party speech recognition framework that supports server-based speech recognition for the Korean language. It should be noted that the system doesn't allow student tablet to send the student's voice to the broadcasting server since it could generate an overload in the broadcasting server that is already busy broadcasting video to student tablet clients. It should also be noted that the session update message (e.g., A joins a class session, B leaves a class session) from any tablet client is sent to the broadcasting server, which, in turn, broadcasts it to all the tablet clients, which is omitted in the Fig.2.

Fig. 3 shows the user interface of the tablet client for the instructor. It consists of 5 panels: video (upper left), chat (lower left), participant list (right side of the chat), slide view (upper right), and slide control (lower right). The slide control panel has 4 slide control buttons (move to the previous slide page, open a slide file, make annotations, and move to the next slide page).

After the instructor opens a slide file, he performs the lecture by showing the slides, explaining them to the students using his voice and making annotations on the slides. He checks his gestures through the video panel. When a student asks a question through chat, he can answer by using slide with annotation, voice and/or chat text.

Fig. 4 shows the user interface of the tablet client for the student. It consists of 4 panels: video (upper left), chat (lower left), participant list (right side of the chat), slide view(upper right), and slide control(lower right).

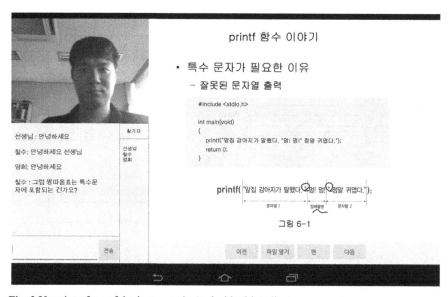

Fig. 3 User interface of the instructor's Android tablet client

During the lecture, students watch the instructor through the video panel, listen to his voice, watch a slide well as annotations being made by him. When a student needs to ask a question or sends a feedback to the instructor, he can do it by using typing texts or speaking on the chat. When an input box of a chat is clicked, a microphone icon is shown along with regular key icons, which is a typical UI for voice recognition apps. Then, when a student speaks to the chat, the voice data is converted to the text.

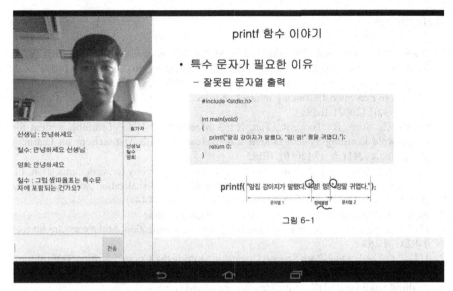

Fig. 4 User Interface of the student's Android tablet client

We are currently building a prototype of the system and its development platform is as follows. The tablet clients for the instructor and students have been implemented in Java using Android Studio 1.2 Integrated Development Environment with Android SDK [8] and Java SE Development Kit 8 [9]. For the client tablet, Samsung Galaxy Note 10.1 with Android 4.4 KitKat has been used. The broadcasting server has also been implemented in Java using Java SE Development Kit 8 on Windows 8.

3   Conclusion

We presented a real-time tablet-based learning system with enhanced user interaction. Students can watch and listen to the instructor as well as watch the slide with annotation being made by him, with their tablets. They can also send a feedback (e.g., asking a question) using text chat. Since typing text in the chat can be slow and inconvenient, we enhanced it by allowing students to talk to the chat, which processes speech-to-text conversion and send the converted text to the server. An Android tablet is used as the mobile device.

We are currently building a prototype of the system. We plan to conduct an empirical study by developing various usage scenarios and find out how effectiveness the system is, when the implementation is completed.

Acknowledgements This work was supported by 2015 Hongik University Research Fund.

References

1. Wains, S.I., Mahmood, W.: Integrating M-learning with E-learning. In: 9th ACM SIGITE Conference on Information Technology Education, pp. 31–38. ACM (2008)
2. Anderson, R., Anderson, R., Davis, P., Linnell, N., Prince, C., Razmov, V., Videon, F.: Classroom Presenter: Enhancing Interactive Education with Digital Ink. IEEE Computer **40**(9), 56–61 (2007). IEEE
3. Ulrich, C., Shen, R., Tong, R., Tan, X.: A Mobile Live Video Learning System for Large-Scale Learning-System Design and Evaluation. IEEE Transactions on Learning Technologies **3**(1), 6–17 (2010). IEEE
4. Lee, J.: Synchronous mobile learning system to cope with slow network connection. In: 11th International Conference on Cooperative Design, Visualization and Engineering (CDVE 2014). LNCS, vol. 8683, pp. 171–174. Springer (2014)
5. Newtone. https://developers.daum.net/services/apis/newtone
6. ITU-T H.263: Video Coding for Low Bit Rate Communication. http://www.itu.int/rec/T-REC-H.263
7. ITU-T G.723.1: Dual-Rate Speech Coder for Multimedia Communication Transmitting at 5.3 and 6.3 kbit/s. http://www.itu.int/rec/T-REC-G.723.1
8. Android Studio and SDK. https://developer.android.com/sdk
9. Java Standard Edition. http://www.oracle.com/technetwork/java/javase/overview

Implementation of Autonomous Navigation Using a Mobile Robot Indoor

Sung Woo Noh, Dong Jin Seo, Tae Gyun Kim, Seong Dae Jeong and Kwang Jin Kim

Abstract This paper describes an implementation of autonomous navigation of a mobile robot indoors. The implementation includes map building, path planning, localization, local path planning and obstacle avoidance. ICP(Iterative closest point) is employed to build grid based map using scanned range data. Dijkstra algorithm plans the shortest distance path from a start position to a goal point. Particle filter estimates the robot position and orientation using the scanned range data. Elastic force is used for local path planning and obstacle avoidance towards a goal position. The algorithms are combined for autonomous navigation in a work area, which comprises indoor environments with different types. The experiments show that the proposed method works well for safe autonomous navigation.

Keywords Autonomous navigation · Localization · Map building · Path planning · Obstacle avoidance

1   Introduction

Autonomous navigation is one of the most fundamental functions for practical use of a mobile robot. There have been abundant researches on autonomous navigation indoors and many of them are implemented for practical use. However, still these algorithms are not easily accessible to users for practical use.

S.W. Noh(✉) · D.J. Seo · T.G. Kim · S.D. Jeong
Sinmyeong Urban Information Co.LTD,
333 Cheomdan Kwagiro Bukgu, Gwangju 500-706, Korea
e-mail: nswking0212@gmail.com

D.J. Seo · T.G. Kim · S.D. Jeong · K.J. Kim
Gwangju Technopark Robot Center,
333 Cheomdan Kwagiro Bukgu, Gwangju 500-706, Korea
e-mail: kjkim@gjtp.or.kr

© Springer Science+Business Media Singapore 2015
D.-S. Park et al. (eds.), *Advances in Computer Science and Ubiquitous Computing*,
Lecture Notes in Electrical Engineering 373,
DOI: 10.1007/978-981-10-0281-6_106

Sometimes, users feel difficulty adapting algorithm components for navigation implementation. As an example, though there are many algorithms for obstacle avoidance, it is not easy to incorporate the algorithm into navigation software for practical implementation. This paper reports one of the successful implementation of the indoor navigation which is practically feasible in the respect that it combines map building, path planning, localization, local path planning and obstacle avoidance, and path tracking in a coherent manner[1]. Section 2 describes each algorithm component for integrated navigation implementation and section 3 shows an example of navigation in a building using the proposed method. Section 4 concludes the research results.

2   Element Technology for Navigation

2.1   *Map Building*

The map required for the navigation is built before the navigation and is provided to the motion algorithms. For the map building, ICP (Iterative closest point) algorithm is used. The ICP provides grid based map using the scanned range data gathered by a Laser range finder (LRF) mounted on a mobile robot which is driven by an operator. Fig. 1 shows a grid map of a building. The map represents an environment of 100m length with 40m width work area of a floor by the grid resolution of 10cm×10cm.

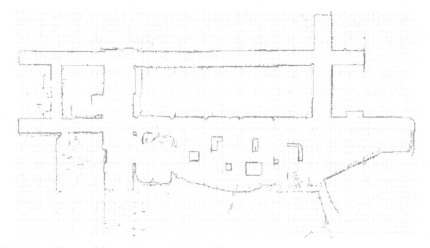

Fig. 1 A grid map of the work area of 100m×40m

2.2   *Localization*

A particle filter based method, called the MCL (Monte Carlo Localization) is used for estimation of robot location[2]. As with usual Bayes filtering estimation methods, the MCL consists of prediction and correction steps. Table 1 shows pseudo code of the MCL algorithm. Every particle of the method represents a possible location of the robot. Line 3 of the Table 1 depicts the prediction step which calculates a possible location of a particle based on dead reckoning. Line 4 assigns belief to each predicted particle using measured sensor data. In our research, the data for belief calculation is the scanned range data to the wall from the LRF on board the robot. Lines 6 to 9 resamples possible locations form the predicted particles to provide corrected possible locations. Therefore, these lines correspond to the correction step and complete an iteration of the MCL.

Table 1 Localization method using particle filter

$$Algorithm\ MCL(X_{t-1}, u_t, z_t, m)$$

$$\{$$

$$\bar{X}_t = X_t = \phi$$

$$for\ m = 1\ to\ M\ do$$

$$x_t^{[m]} = sample\_motion\_model(u_t, x_{t-1}^{[m]})$$

$$\omega_t^{[m]} = measurement\_model(z_t, x_t^{[m]}, m)$$

$$endfor$$

$$for\ m = 1\ to\ M\ do$$

$$draw\ i\ with\ probability \propto \omega_t^{[i]}$$

$$add\ x_t^{[i]}\ to\ X_t$$

$$endfor$$

$$return\ X_t$$

$$\}$$

2.3 Path Planning and Obstacle Avoidance

The global path from a starting location to a goal location is planned using Dijkstra algorithm, which yields the shortest path through way points in the map[3].

Once the way points from a starting location to a goal location are given, local path planning and tracking is needed to move the robot through the way points while avoiding local obstacles detected during the navigation. Applying virtual elastic force and repulsive force to the path segment from the robot to the adjacent way point produces smooth and collision free local path to the way point. The following equation is used for the repulsive force from the obstacles.

$$V_{rep}\left(^j P_i\right) = \begin{cases} \dfrac{1}{2} K_r (d_r - d(^j P_i))^2, & if\ d(^j P_i) < d_r \\ 0, & otherwise \end{cases} \tag{1}$$

$$^r f_i = -\nabla V_{rep}(^j P_i) = K_r (d_r - d(^j P_i)) \frac{\vec{d}}{d} \tag{2}$$

Eq. (1) describes repulsive force field and Eq. (2) calculates the repulsive force $^{r}f_i$ from obstacles to the robot located at $^{j}P_i$. In these equations, $d(^{j}P_i)$ is the shortest distance between $^{j}P_i$ and obstacles, d_r is the virtual distance of repulsive field influence, and K_r is the control parameter of the repulsive force. Elastic force exerted to the point $^{j}P_i$ which is on the local path is given by the Eq. (3) [4].

$$^{e}f_i = K_c \left(\frac{d_i^{j-1}}{d_i^{j-1} + d_i^{j}} (^{j+1}P_i - {}^{j-1}P_i) - (^{j}P_i - {}^{j-1}P_i) \right) \quad (3)$$

d_j^i is the distance between $^{j}P_i$ and $^{j+1}P_i$, and the K_c is the elastic force parameter. The force to the robot at $^{j}P_i$ is the sum of the repulsive force and elastic force. Fig. 2. shows collision avoidance of local path to the way point in a simulation

$$^{j}F_i = {}^{r}f_i + {}^{e}f_i \quad (4)$$

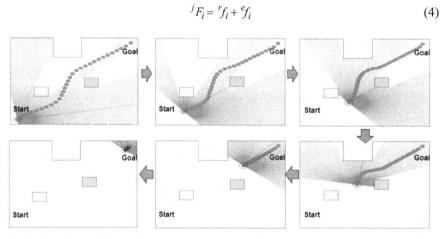

Fig. 2 Obstacle avoidance process using the elastic force

2.4   Integration of the Algorithm Components

The method integrates the algorithm components in a coherent manner[5]. Map building algorithm provides map data to path planning, local path planning and tracking. Localization algorithm provides the robot location to local path planning and tracking. Planned global path is provided to local path planning and tracking. Detected obstacle is fed to local path planning and tracking. There algorithms are interconnected, and finally the local path planning and tracking produces command to the robot, that is, linear speed and angular speed of the robot motion.

3   Experiments

The method is implemented for navigation in a building of work range 100m×40m. The range sensor used for map building, localization, and obstacle detection is the LMS511 of Sick. The experiment uses a differential drive mobile robot. The maximum speed is set to be 1.0m/sec. Fig. 3 shows the navigation path and the path generated by Dijkstra algorithm. Fig. 4 shows an autonomous navigation trajectory.

The overall distance from the initial location to the goal location is 165m, and it takes 255sec for the travel, thus the average speed is 0.65m/sec. One of the corridors has the width of 2.2m and length 9.5m, and the other has the width 2.35m and the length 37m. The robot was able to avoid passers-by in the narrow corridors as well as in the lobby.

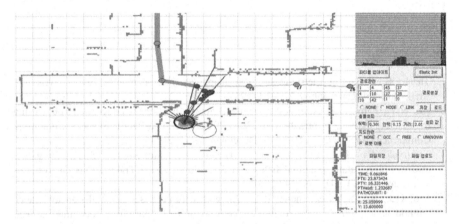

Fig. 3 Trajectory of robot motion in the experiment.

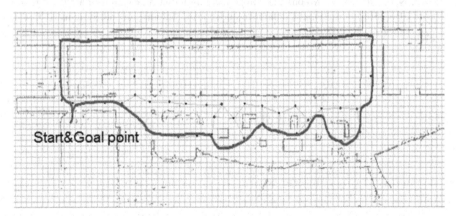

Fig. 4 Trajectory of an autonomous navigation in the experiment.

4 Conclusion

The paper reports an implementation of autonomous navigation in indoor environment. It integrates several algorithms for mapping, planning, localization, and tracking, Through experiments, the method reveals reliable and robust navigation performance. There are other algorithm elements to be integrated to the proposed method such as map update and path re-planning. When the robot detects an obstacle which stationary, then the map should be updated to include the obstacle. Also, if the corridor along which the planned path lies is blocked by obstacles, the path should be re-planned globally.

Acknowledgement The research was supported by 'Software Convergence Technology Development Program', through the Ministry of Science, ICT and Future Planning(S0170-15-1079).

References

1. Thrun, S.: Learning metric-topological maps for indoor mobile robot navigation. Artificial Intelligence **99**(1), 21–71 (1998)
2. Ko, N.Y., Kim, T.G., Noh, S.W.: Monte carlo localization of underwater robot using internal and external information. In: Proc. 2011 IEEE Asia-Pacific Services Computing Conference (APSCC), pp. 410–415, December 12-15, 2011
3. Tsourveloudis, N.C., Valavanis, K.P., Hebert, T.: Autonomous vehicle navigation utilizing electrostatic potential fields and fuzzy logic. IEEE Transactions on Robotics and Automation **17**(4), 490–497 (2001). 8
4. Seo, D.J., Ko, N.Y., Simmons, R.G.: An elastic force based on collision avoidance method and its application to motion coordination of multiple robots. International Journal of Computer Integrated Manufacturing **22**(8), 784–798 (2009)
5. Ko, N.Y., Noh, S.W., Moon, Y.S.: Implementing indoor navigation of a mobile robot. In: 2013 13th International Conference on Control, Automation and Systems (ICCAS), pp. 198–200. IEEE (2013)

Convergence Modeling of Heterogeneous Medical Information for Acute Myocardial Infarction

Meeyeon Lee, Ye-Seul Park, Myung-Hee Kim and Jung-Won Lee

Abstract In recent years, as the big data boom accelerates, the possibility of using personal health and medical data is also growing. However, since most of previous studies have focused on computerizing, storing, and transferring of medical data, it is hard to say that they intelligently use medical data. Particularly, in cases of urgent diseases like acute myocardial infarction (AMI) that prompt diagnosis and treatment is needed, the current hospital information systems are difficult to efficiently provide information. Therefore, in this paper, we propose a convergence modeling method based on semantic relations by analyzing characteristics of medical data for AMI. The proposed method can unify medical data which is separately stored in medical information systems and provide important data as a one record.

Keywords Medical information · Heterogeneous data · Data convergence · Data modeling · AMI (Acute Myocardial Infarction)

1 Introduction

In recent years, the needs for the intelligent management and usage of health and medical data have significantly increased. It was hard to efficiently manage those information in the past since the most of medical institutions used analogue

M. Lee · Y.-S. Park · J.-W. Lee(✉)
Department of Electrical and Computer Engineering, Ajou University,
206 Worldcup-ro, Yeongtong-gu, Suwon 16499, Korea
e-mail: {mylee,jungwony}@ajou.ac.kr, yeseuly777@gmail.com

M.-H. Kim
Department of Computer Science and Engineering,
Ewha Womans University, 52, Ewhayeodae-gil, Seodaemun-gu, Seoul 03760, Korea
e-mail: mhkim@ewha.ac.kr

© Springer Science+Business Media Singapore 2015
D.-S. Park et al. (eds.), *Advances in Computer Science and Ubiquitous Computing*,
Lecture Notes in Electrical Engineering 373,
DOI: 10.1007/978-981-10-0281-6_107

materials like paper charts or films. As efforts to digitalize medical materials have become active, HISs (Hospital Information Systems) including OCS (Order Communication System), PACS (Picture Archiving and Communication Systems), and EMR (Electronic Medical Record) have been developed and widely diffused. Accordingly, many studies for easy storing and efficient transferring of medical data have conducted. Also, some previous works proposed methods to process individual kind of medical data, especially they focused on the analysis, processing, and registration of medical images [1-4]. Recently, the standards such as CCR (Continuity of Care Record) [5], CDA (Clinical Document Architecture) [6], and CCD (Continuity of Care Document) [7] have established for big data in medical domain. They are aiming at the efficient description of EPR (Electronic Patient Record), EHR (Electronic Health Record), PHR (Personal Health Record), and UHR (Universal Health Record). In addition, some systems have been developed in order to infer or determine a patient's condition, aid the diagnosis, and predict the prognosis based on medical data [8, 9].

However, due to the following characteristics of medical domain, the requirements of data handling or utilization are different compared to other general areas.

- Medical data: it is very vast in volume. The more important point is that its types, formats, and attributes are heterogeneous depending on the institutions and diseases. Furthermore, the current HISs generate and store these information in various materials through different systems.
- Medical service: in medical fields, there is a strong concern about the risk of the decision or prediction by the systems. Therefore, we can say that a method for efficient search and provision of medical data is more feasible and helpful.

Therefore, in this paper, we propose a method for a semantic convergence and modeling of medical information. In particular, we focus on acute myocardial infarction (AMI), since it is a critical disease that requires the efficient provision of essential part among vast information for the quick decision in the golden time. The proposed method can extract semantic data which are distributed in various medical materials and unify them into a one record.

2 Analysis of AMI-related Data

In this paper, we collect the medical materials related to AMI from four medical institutions in Korea. Then, we select eight types of materials which are commonly used by institutions and most essential for the expert's decisions.

In general, data can be categorized into three types: structured, semi-structured, and unstructured data.

- Structured data (S): data which has its fixed specification or forms and can be stored in formatted fields.
- Semi-structured data (SS): data which contains kinds of schema, not fixed formats.
- Unstructured data (U): data with irregular format.

Table 1 shows the classification of medical materials related to AMI according to these criteria. The classification of medical data is different from the general way. For example, in normal domain, data of text type is treated as the unstructured type. However, text in medical documents can be analyzed as the semi-structured type, since medical experts write their opinion about the diagnosis or treatments in certain patterns.

Among eight types of materials, we analyze three grayed in Table 1 and unify their data based on semantic relations. The first one is a coronary angiography report. It is a document which contains not only structured data like patient's information (age, gender, etc.) and date, but also semi-structured data. The large portion of this report is text, but they are written in typical patterns to mention patient's medical history, smoking habit, and symptoms. The second material is a coronary angiography which is an image or video type. We can define two groups of data which can be extracted from this material as shown in Fig. 1 (a). A group of metadata is properties of a image/video such as performing date, angles, and so on. Semantic data means information which is supposed to be interpreted by medical experts. In other words, this group includes the location and condition of a lesion in terms of the main coronary artery, or its segments or branches. As the final material, we analyze a coronary arteriogram which can describe a summarization of diagnosis and treatment procedures. Moreover, physicians mark the specific site and severity of lesions on a simplified map of coronary artery related to AMI. This material contains significant data depicted in Fig. 1 (b).

3   Data Convergence Modeling for AMI

The materials of three types analyzed in the previous section contain redundant data. In addition, even though those data are semantically related to each other, they are distributed in different heterogeneous materials. Therefore, rather than the material-oriented data provision of current HISs, a new method is necessary to semantically converge and manage clinical data. Table 2 shows a record specification for AMI derived by converging CAG-related materials of three types.

The proposed record consists of four kinds of clinical information.

- **Patient**: patient's basic information such as personal information and coronary anatomical information, etc.
- **Physical History**: patient's states or habits related to AMI.
- **Vital History**: AMI-related factors among patient's basic medical states.
- **Medical History**: information about the diagnosis or treatments that a patient has been previously received related to AMI. That is, this part includes experts (physicians, institutions, etc.), location and states of lesion, disease names about past diagnosis.

Table 1 AMI-related Data.

Data Type	Clinical Material		Char.
System input	Echocardiography laboratory	• a report generated by echocardiography • including numerical values for each examination item and text comments of experts (physicians)	S SS
Document	Coronary Angiography report	• reports describing information about medical treatments or procedures • text having certain description patterns for AMI-related information (patient history, current states, etc.)	S SS
	PTCA and Stent Deployment report		S SS
Image /Video	EKG (Electrocardiogram)	• an image of EKG examination results • including result images, measurement values, and metadata	S U
	CAG	• videos generated by examinations • including metadata and values measured by examinations	S U
	Echocardiogram		S U
Analogue /Digital chart	Coronary Arteriogram	• A cardiac diagram including major diagnosis and treatment information • generated by handwriting or system input depending on the medical institutions	SS U
	Cardiology Lab Sheet	• a report generated by blood tests related to AMI • generated by handwriting or system input depending on the medical institutions	S SS

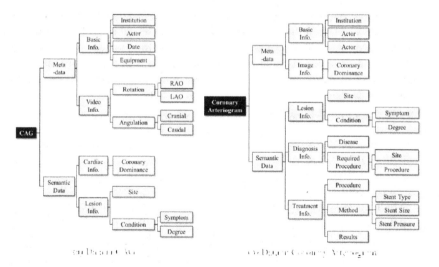

Fig. 1 Data included in two materials implicitly or explicitly.

In the current EMR and PACS, these data associated with each other are represented and stored in heterogeneous materials. On the contrary, the convergence model shown in Table 2 can provide and manage essential information as a single record format of the data level, not the level of materials or documents.

Moreover, each data element is described with following properties:

- **Creator:** medical experts (physicians, inspectors, etc.) or institutions who created and modified the element.
- **Date:** the date on which creators created and changed the value of the element.
- **Importance:** the level of significance of each element. This property can have grades depending on whether the element is essential for the diagnosis and treatment, or on the degree that it influenced AMI.
- **Reference-metadata:** metadata for 'reference' elements of a 'Medical History' group. This property represents metadata of related materials (documents, images/videos, etc.) such as equipment models and angles of examinations.

Table 2 A convergence model of three types of CAG-related data for AMI.

Element		
Patient	Basic Profile	Name
		Age
		Sex
		Address
		Tel Number
	Cardiac Profile	Coronary Dominance
Physical History	Pain	
	Hypertension	
	Diabetes Mellitus	
	Ex-smoke	
	Current-smoke	
	Chest Pain Start/First/Recent	
Vital History	Blood Pressure	
	Pulse Rate	
	Breathing Sound	
	Heart Sound	
	EKG	

Table 2 (*Continued*)

Medical History	Diagnosis	Actor			
		Lesion	Site		
			Condition	Symptom	
				Degree	
		Disease	Disease name		
			Required/Recommended Procedure	Target Site	
				Procedure Name	
		Reference			
	Medical Treatment	Procedure	Actor		
			Procedure Name		
			Target Site		
			Technique		
			Equipment		
			Patient's Condition		
			Result		
		Reference			

The proposed data model shown in Table 2 can be a foundation to promptly and intuitively provide crucial medical information.

4 Conclusion

The medical and clinical information are vast in volume and heterogeneous in terms of type, format, and characteristics. Since the medical institutions manage these information as units of materials and documents, it is difficult to efficiently provide and view the relevant data in the golden time, particularly in the case of AMI which is one of urgent diseases. Therefore, in order to usefully manage these data, a new data model should be defined based on the analysis of characteristics and relations of data included in heterogeneous materials. In this paper, we defined a new data model from materials of three types which are essential for the diagnosis of AMI. In contrast with the current HISs, the proposed model can unify data in various materials as a single record based on their semantic relations. The convergence record will enable medical experts to easily and intuitively search important data for quick decision about diagnosis and treatments.

For the future works, we plan to improve the coverage and completeness of the data model by targeting at multiple institutions. Furthermore, we will implement a medical data provision and management system based on the model.

Acknowledgments This work was supported by Institute for Information & communications Technology Promotion (IITP) grant funded by the Korea government (MSIP) (No.B0101-15-247, Development of Open ICT Healing Platform using Personal Health Data).

References

1. Perera, S., Henson, C., Thirunarayan, K., Sheth, A., Nair, S.: Semantics Driven Approach for knowledge Acquisition From EMRs. IEEE Journal of Biomedical and Health Informatics **18**(2), 515–524 (2014)
2. Na, H.-S., Yun, S.-Y., Park, S.-C.: Design and implementation of mapping system for effective health information data exchange in multi-platform environment. The Journal of Korean Institute of Information Technology **10**(12), 143–150 (2012)
3. Valente, F., Viana-Ferreira, C., Costa, C., Oliveira, J.L.: A RESTful image gateway for multiple medical image repositories. IEEE Transactions on Information Technology in Biomedicine **16**(3), 356–364 (2012)
4. Alvarez, L.R., Vargas Solis, R.C.: DICOM RIS/PACS telemedicine network implementation using free open source software. IEEE Latin America Transactions **11**(1), 168–171 (2013)
5. ASTM International, ASTM E2369-12: standard specification for continuity of care record (CCR) (2012)
6. Health Level 7 International, HL7 implementation guide for CDA release 2. http://www.hl7.org/implement/standards/product_brief.cfm?product_id=7
7. Health Level 7 International, HL7 implementation guide: CDA release 2 – Continuity of Care Document (CCD) (2007)
8. Park, S.H., Park, H.A., Ryu, K.S., Kim, H., Ryu, K.H.: A long-term mortality prediction model for patient with st-segment elevation myocardial infarction using decision tree. In: Proceeding of the Korea Computer Congress, no. 1(C), pp. 139–141 (2012)
9. Martinez-Romero, M., Vazquez-Naya, J.M., Pereira, J., Pereira, M., Pazos, A., Banos, G.: The iOSC3 System - Using Ontologies and SWRL Rules for Intelligent Supervision and Care of Patients with Acute Cardiac Disorders. Computational and Mathematical Methods in Medicine, 1-13 (2013)

Acknowledgments This work was supported by Institute for Information & communications Technology Promotion (IITP) grant funded by the Korea government (MSIP) (No.B0101-15-247, Development of Open ICT Healing Platform using Personal Health Data).

References

1. Perera, S., Henson, C., Thirunarayan, K., Sheth, A., Nair, S.: Semantics Driven Approach for knowledge Acquisition From EMRs. IEEE Journal of Biomedical and Health Informatics **18**(2), 515–524 (2014)
2. Na, H.-S., Yun, S.-Y., Park, S.-C.: Design and implementation of mapping system for effective health information data exchange in multi-platform environment. The Journal of Korean Institute of Information Technology **10**(12), 143–150 (2012)
3. Valente, F., Viana-Ferreira, C., Costa, C., Oliveira, J.L.: A RESTful image gateway for multiple medical image repositories. IEEE Transactions on Information Technology in Biomedicine **16**(3), 356–364 (2012)
4. Alvarez, L.R., Vargas Solis, R.C.: DICOM RIS/PACS telemedicine network implementation using free open source software. IEEE Latin America Transactions **11**(1), 168–171 (2013)
5. ASTM International, ASTM E2369-12: standard specification for continuity of care record (CCR) (2012)
6. Health Level 7 International, HL7 implementation guide for CDA release 2. http://www.hl7.org/implement/standards/product_brief.cfm?product_id=7
7. Health Level 7 International, HL7 implementation guide: CDA release 2 – Continuity of Care Document (CCD) (2007)
8. Park, S.H., Park, H.A., Ryu, K.S., Kim, H., Ryu, K.H.: A long-term mortality prediction model for patient with st-segment elevation myocardial infarction using decision tree. In: Proceeding of the Korea Computer Congress, no. 1(C), pp. 139–141 (2012)
9. Martinez-Romero, M., Vazquez-Naya, J.M., Pereira, J., Pereira, M., Pazos, A., Banos, G.: The iOSC3 System - Using Ontologies and SWRL Rules for Intelligent Supervision and Care of Patients with Acute Cardiac Disorders. Computational and Mathematical Methods in Medicine, 1-13 (2013)

HiL Test Based Fault Localization Method Using Memory Update Frequency

Ki-Yong Choi, Jooyoung Seo, Seung Yeun Jang and Jung-Won Lee

Abstract As the proportion of electronics parts increases, HiL (Hardware in the Loop) testing is performed in order to ensure the quality of automotive software. When the developer is in the process of correcting faults detected by the tester, it would be useful to correct a fault if the tester can provide a fault localization result to the developer. However, the conventional fault localization methods are unsuitable for a tester to test the completed hardware without source code. If the test can provide the internal operating information at the time of fault without the source code and the modification of the hardware logic, it is more effectively for the developers to debug the fault. In this paper, we propose a fault localization method using the statistics of the memory update frequency at the time of fault appearance. The experimental results with the test suites of the actual automobile industry show that the fault candidates were reduced by 47% on average.

Keywords Automotive software · Hil testing · Fault localization · Memory update frequency

1 Introduction

The proportion of electronic parts increases, the automobile software also increases. The quality of automotive software has emerged as an important issue

K.-Y. Choi · J.-W. Lee(✉)
Department of Electrical and Computer Engineering,
Ajou University, Suwon, Republic of Korea
e-mail: {ki815kaisian,jungwony}@ajou.ac.kr

J. Seo
Department of Information and Computer Engineering,
Ajou University, Suwon, Republic of Korea
e-mail: jyseo@ajou.ac.kr

S.Y. Jang
HYUNDAI Motor Group, Electronics Center,
Electronics Control Engineering Development Team, Seoul, South Korea
e-mail: jangdana@hyundai.com

© Springer Science+Business Media Singapore 2015
D.-S. Park et al. (eds.), *Advances in Computer Science and Ubiquitous Computing*,
Lecture Notes in Electrical Engineering 373,
DOI: 10.1007/978-981-10-0281-6_108

so the fulfillment of the test based on the requirement becomes one of the important measures for ensuring quality [1].

The automotive industries have taken the OEM (Original Equipment Manufacturer) system for developing software. Depending on the OEM, the definition of requirements, the development, and the test are performed by different organizations [2]. In this case, the tester of 'A' automobile industry performs the black box test by using the HiL (Hardware in the Loop) simulator upon the requirements to inspect satisfaction of the output for the input and found fault is corrected by the developer of 'B' software company. If the tester provides the operating information generated at the time of that the fault is inspected to the developers, the developer can perform the fault localization more effectively. Then it is possible to start debugging without re-enacting the faulty states. Therefore, in order to provide the information necessary for debugging to the developer, the tester needs to acquire the internal operating information related to the fault during testing.

Using the existing debugger and fault localization method is possible to observe the typical hardware and software operations. However, it is unsuitable for testing the automotive electronic parts by the tester for the following reasons. First, since the tester performs the HiL testing for the completed parts which have been delivered, it is impossible to use the debugger when the debug ports such as JTAG (Joint Test Action Group) and BDM (Background Debug Mode) are not exposed from the prototype. Second, the existing methods for fault localization have used the source code, but the tester performs the black box testing without the source code or the debugger. Therefore, if we can provide the internal operating information related to the found fault without the source code and additional hardware logic for testing, the developer will be able to debug the fault more effectively.

On the other hand, the footprints of the program execution remain in memory. In software testing, there is a method to generate a test case using the DU-pairs. The DU-pairs is defined as "def. and use" in the CFG (Control Flow Graph). At the $A = B + C$, A is defined by 'def.', B and C are defined by 'use' [3]. All these variables defined by the 'def.' and 'use' are stored in the memory by the memory interface instructions such as MOVE, LOAD, and STORE. Therefore, for the test environment without using the source code, it is possible to propose a fault localization method by analyzing the memory dump. In the past, the traditional method using the memory dump was impractical because the enormous size of data needs to be analyzed for localizing the fault [4]. However, if the operating information may be presented at the time of that the fault is inspected, the developers can identify the causes of the fault more easily.

Therefore, we first define a 'frame' by differentiating the memory area to be focused and capturing the slicing area by the unit-time. By tracing the value changes of the same address between the frames, we determine the memory updates and get memory update frequency accumulated by the time and by the address (by the frame). When the fault is detected in testing, we analyze the several frames made at the time before and after of the fault and consider the address with a higher memory update frequency as the fault candidate. And we finally present the set of fault candidates as a result of localization. As an experimental result, our proposed method using the memory update frequency showed that the number of fault candidates was reduced by 47% on average. In addition, we found the test script errors and an initialization error out of the design specification.

2   Related Work

Fault localization is one of the most expensive activities in program debugging [4]. In general, since the developer is corrected by detecting the fault, the existing fault localization method research has been mainly performed based on code.

Program slicing method is a code-based method that is normally used. It is observed by slicing the code based on the location affected by the variables. A code where the fault appears is detected by static or dynamic analysis [5]. Code Coverage method is used to detect the fault in the code executed when a fault appeared with measuring the code coverage at the time [6]. These methods suggested code-based execution flow or data flow for fault localization. On the other hand, Statistical method statistically determines the fault probability of the operations with repetitive executions. And it searches for the code related to the operations where the fault frequently appears [7]. However, this method tried statistical method, since code based, the tester of the black box test cannot apply to HiL testing.

For technique which does not use codes apply to a black box test, there is a method of using a debugging tool and of using logging. Run-stop method using the debuggers, locate for the fault by executing step by step [8]. Since the automotive software tester performs tests on the finished hardware, it is not applicable if the debug port is not equipped at SUT (System Under Test). Fault localization method using the log detects the fault by exploiting the characteristics logged during the program execution. It can find the characteristics of the faults without code or debugging tool, mainly detecting faults by collecting the operation log of the distributed system like Map Reduce [9]. With this method, however, logs collecting method and subject are unsuitable to apply it to fault localization of HiL Testing since this method is in interest of operation state information of the dispersed system over the network.

Thus, the method using a log out of the conventional methods, the tester can apply in HiL Testing without code and debugging tools. Since footprints of the program execution remain in memory, fault localization is possible if analyze the memory dump. Then, it would be possible to find the cause of fault if the memory is analyzed before and after the fault detected when the fault appears as dumping it by slicing into unit time and letting it remain as frames.

3   Finding the Fault Candidates Using Memory Update Frequency

3.1   Problem Definition

The problem to solve in this paper is how the tester provides debug information to the developers for debugging the detected fault by HiL test. Using this frequency of memory updates, fault candidates estimated to be related to the appeared fault are provided to solve it. Before declaring fault candidates, it is defined as follows.

$$\text{Frame}(F_k) \equiv \{V_{A,k} \mid V_{A,k} \text{ is the value of the address A at time } T \times k;$$
$$k \text{ is an index number, T is period of the system main task}\} \quad (1)$$

Eq.(1) shows that k-frame is defined as a k-th dumped memory data at intervals of T from 0-frame, which is a memory data that dumped first, denoted as F_k. As the element, F_k has the value of each address at the time of being dumped.

$$\text{Memory Update}(U_{A,k}) = \begin{cases} 1, & V_{A,k-1} \neq V_{A,k} \\ 0, & V_{A,k-1} = V_{A,k} \end{cases}; k \text{ is a frame number} \quad (2)$$

$$\text{Memory Update Frequency}(UF_{A,R}) = \sum_{k=R\,\text{start}}^{R\,\text{end}} U_{A,k} \quad (3)$$

Eq.(2) shows that by comparing $V_{A,k-1}$ and $V_{A,k}$, when they are difference, memory update ($U_{A,k}$) has a value of 1 and has a 0 if they are same. If the $U_{A,k}$ is 1, it means that the address A is updated at the k-frame. By using this, at Eq.(3) memory update frequency ($UF_{A,R}$) determines how many times update has occurred in the address A in the range R of the frame to be observed. When the value of the address has been updated, as it means that operation related to the address is performed, it is possible to know the action performed in the range R by using the $UF_{A,R}$ of each address.

$$\text{Memory Data}(MD_{A,R}) \equiv \{ A, UF_{A,R}, \{V_{A,k}\} \mid k \in R \} \quad (4)$$
$$\text{Fault Candidates}(FC) \equiv \{ (MD_{A,R}, \text{Sym}_A) \mid$$
$$\text{It is assumed that A and R is related to the fault}\} \quad (5)$$

Memory data ($MD_{A,R}$) defined by Eq.(4) have the number of address A's updates ($UF_{A,R}$) in the range R and the value for each frame ($V_{A,k}$) as elements. And using the $MD_{A,R}$, we define fault candidates by Eq.(5). FC has elements, the MD and the symbol name (Sym_A) in the address. It is assumed that the addresses and the ranges are related to the fault. Therefore, it can be used for debugging, if the FC is provided to the developer.

3.2   Fault Localization Process

Fig.1, it shows the fault localization process using a memory update frequency. The Process is roughly composed of four phases. Each phase is as follows. First, testing phase collects the memory from HiL testing. Then, pre-processing phase is to remove unnecessary data from the memory that were collected using the static information which can be obtained from the execution file. Localization phase locates the fault candidates by using a test specification which has been obtained by analyzing the test script. Finally, post-processing phase searches for the addresses that are non-updated from the addresses in needs of update by using a test specification. Each phase consists of a total of seven stages. The output from each stage is used for fault localization in the subsequent stage. Fig.2 shows running process. (a) displays a Type I, II and (b) is in the process of obtaining the target address in stage4, showing a state the Type I was removed. (c) is an example of finding input driven update range (IDUR) utilizing target address at stage6.

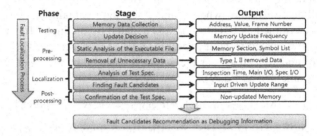

Fig. 1 Fault localization process using memory update frequency

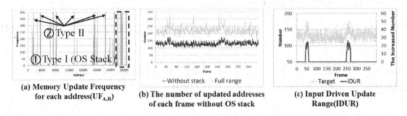

(a) Memory Update Frequency for each address(UF$_{A,R}$)

(b) The number of updated addresses of each frame without OS stack

(c) Input Driven Update Range(IDUR)

Fig. 2 Running the fault localization process

- **Stage1. Memory Data Collection:** Collection of the $V_{A,k}$ in each F_k at intervals of the period T.
- **Stage2. Update Decision:** Determines the $U_{A,k}$ by Eq.(2) from the $U_{A,k}$ collected by Stage1. The $UF_{A,R}$ can be determined by the $U_{A,k}$ according to Eq.(3).
- **Stage3. Static Analysis of the Executable File:** By Binary Utility [10] and the Parser, obtains memory sections and symbol list from the executable file.
- **Stage4. Removal of Unnecessary Data:** To obtain target addresses, removes the addresses of unnecessary Type I, II to fault localization as a result of Stage3. Since Type I is important sections of meaning to the values as OS stacks, the UF is not critical. Type II is the addresses of the symbol that is dynamically used as a buffer or counter. In particular, it has been updated with all the frames. The number of frame and $UF_{A,R}$ are the same value in the addresses.
- **Stage5. Analysis of Test Specification:** The test specification is grasped by analyzing a test script based on Eq.(6), Eq.(7) and Eq.(8).

$$\text{Inspection time} = |\text{ Input time} - \text{Inspect(output) time }| \tag{6}$$

$$\text{Spec. I/O} \equiv \{ (A, \text{Sym}_A) \mid \text{Sym}_A \in (\text{Input signal} \cup \text{Inspection signal})\} \tag{7}$$

$$\text{Main event I/O} \equiv \{ (A, \text{Sym}_A) \mid \text{Sym}_A \in ((\text{Input signal} \cap \text{Inspection signal})$$
$$\cup \text{ Input driven inspection signal})\} \tag{8}$$

- **Stage6. Finding Fault Candidates:** To find the section where input is applied after the input, the number of updated address per each frame ($\sum_A^{\text{Target}} U_{A,k}$; $A \in$ Target) in inspection time is determined, and the tendency is identified by applying the moving average technique. Range of frames where it increases and maintains to above average is determined as IDUR, which is updated by the input. IDUR is assumed to be related to fault appearance, since it is a range where the number of update increases as the output is changed by the input. Thus, $\text{FC}_{\text{stage6}}$ is obtained by Eq.(9).

$$\text{FC}_{\text{stage6}} = \{(\text{MD}_{A,\text{IDUR}}, \text{Sym}_A) \mid \text{UF}_{A,\text{IDUR}} \geq 1; A \in \text{Target} \} \tag{9}$$

- **Stage7. Confirmation of Test Specification:** Finds Address as Eq.(10) which $\text{UF}_{A,\text{IDUR}}$ is 0 from the spec obtained in Stage5. And then it will be presented along with the $\text{FC}_{\text{stage6}}$.

$$\text{FC}_{\text{stage7}} = \{(\text{MD}_{A,\text{IDUR}}, \text{Sym}_A) \mid \text{UF}_{A,\text{IDUR}} = 0; A \in \text{Specification I/O} \} \tag{10}$$
$$\text{FC} = \text{FC}_{\text{stage6}} \cup \text{FC}_{\text{stage7}} \tag{11}$$

Table 1 shows the reason for presenting the address that appears in the test specifications from the address which has not been updated in the last stage7.

Table 1 Relationship memory update and fault

	Related to fault(T)	Not related to fault(F)
Updated memory(P)	Incorrect behavior(**TP**)	Desired behavior(**FP**)
Non-updated memory(N)	Not working the desired behavior(**FN**)	Not interested(**TN**)

TP, if the updated memory is related to the fault, or FP if not. FN, if it is a memory in needs of updated in the non-updated memory, or TN if not. As this paper proposes fault localization method is for the updated memory (TP+FP), a method to find FN is necessary. Meanwhile, if a variable is significant to be updated in the test specification but not updated, it is considered as FN. Thus, the method proposed in this paper removes the memory irrelative to the fault (FP, TN) in the memory, and it will provide the developer with memory update information (TP, FN) of wrong behavior related to the fault.

4   Experimental Result

The HiL testing based fault localization method using memory update frequency proposed was applied to four cases of Table 2. As the result of application, fault candidates were localized 47.8% on average in the updated memory.

Table 2 Experimental result (Size of used memory: 29420 Byte)

Test Script	Number	Case	Update Memory	Fault Candi.	Localization Rate (%)
Warning	#1	Missing signal	1207	477	39.5
Tele-matics	#2	Initialization error	1349	667	49.4
	#3	Fixed signal	1794	821	45.8
	#4	Fixed signal	2067	1168	56.5
Average					47.8

- **#1**: When applying to the case of being decided as pass although the signals appearing in the test script is missed in the transmission process, the missed signal was found in the process of comparing the test specification.
- **#2**: When applied to the case that expected output is not obtained by entering an inappropriate initial value at function test, the fact that the initialization input is wrong, not a fault of function test, has been included in the FC.
- **#3, #4**: When applied to the two fault cases of corrected previously, the state which has been reported as the cause of the fault has been included in the FC.

The result of applying the method provides fault candidates as shown in Fig.3. The non-updated signals of appearing in the specifications are highlighted. Except for 1074(①), addresses which are highlighted in the results of # 1 appeared to the specifications, but they do not change the value. However, the 1074, the value can change in the specification. From this result, the tester can detect a signal missing.

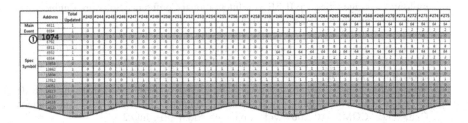

Fig. 3 Fault candidates from test script #1 (without symbol name)

This result was an actual test and fault correction, and each was confirmed from the tester and the developer. The tester was able to check the error that may occur during the production and testing of false test cases. And the developer could obtain information necessary for debugging as it includes the situation when the fault detects.

5   Conclusion

This paper studied the conventional HiL test and proposed a fault localization method using a memory update frequency so that the tester is able to provide the debugging information to the developer when the fault detects. Features and advantage of the proposed method are as follows. First, the source code is unnecessary. Second, application is possible by adding the test agent software to the SUT. Third, it can be applied using the given material even without domain knowledge.

The result of applying the proposed method is confirmed by the developer and the tester respectively. In other words, the objectives of the method, that the tester provide to the developer with a result of fault localization using memory update frequency, is achieved. Furthermore, it can be used to check the error for the test.

When the method is applied afterward, it needs to be verified for the additional data transfer through the repeated experiment assumed at memory data collection, though there was no problem yet. And it is necessary to confirm the effect of the processing load, too. Moreover, the memory update frequency as well, it is expected to advance the study on the method of providing more systematic and associative information clarifying the relationship between the updated addresses with this.

Acknowledgments This research was supported by the HYUNDAI Motor Group, Electronics Center, Korea.

References

1. Conrad, M.: A systematic approach to testing automotive control software. In: Proc. of Convergence (2004)
2. Siegl, S., Hielscher, K.S., German, R.: Model based requirements analysis and testing of automotive systems with timed usage models. In: 2010 18th IEEE International Requirements Engineering Conference (RE), pp. 345–350. IEEE, September 2010
3. Santelices, R., Jones, J., Yu, Y., Harrold, M.J.: Lightweight fault-localization using multiple coverage types. In: IEEE 31st International Conference Software Engineering, ICSE 2009, pp. 56–66. IEEE, May 2009
4. Wong, W.E., Debroy, V.: A survey of software fault localization. Department of Computer Science, University of Texas at Dallas, Tech. Rep. UTDCS-45, 9 (2009)
5. Weiser, M.: Programmers use slices when debugging. Communications of the ACM **25**(7), 446–452 (1982)
6. Wong, W.E., Qi, Y., Zhao, L., Cai, K.Y.: Effective fault localization using code coverage. In: 31st Annual International Computer Software and Applications Conference, COMPSAC 2007, vol. 1, pp. 449–456. IEEE, July 2007
7. Liblit, B., Naik, M., Zheng, A.X., Aiken, A., Jordan, M.I.: Scalable statistical bug isolation. ACM SIGPLAN Notices **40**(6), 15–26 (2005). ACM
8. Vermeulen, B.: Functional debug techniques for embedded systems. IEEE Design & Test of Computers **25**(3), 208–215 (2008)
9. Chun, B.G., Chen, K., Lee, G., Katz, R.H., Shenker, S.: D3: Declarative distributed debugging. Technical report, UC, Berkeley (2008)
10. GNU Binary Utilities. https://sourceware.org/binutils/docs/binutils/index.html

Data Cascading Method for the Large Automotive Data Acquisition Beyond the CAN Bandwidth in HiL Testing

Jeong-Woo Lee, Ki-Yong Choi, Seung Yeun Jang and Jung-Won Lee

Abstract As the proportion of the software in an ECU (Electronic Control Unit) has increased, HiL (Hardware in the Loop) test is performed for verification. However, an extra monitoring technique is needed to provide the debugging information for developers due to the characteristics of the test environment. Thus, this paper suggests a novel method named data cascading algorithm to transfer the large monitoring data generated from the ECU. In this algorithm, the data is divided into segments, considering the communication condition and the operation cycle of the system's main task. Then the segments are transmitted by reiterating the test with the number of segments. As an experimental result, we proved that the massive monitoring data generated from ECU was collected by overcoming the low bandwidth of CAN and the transmitted data was completely the same of the original data without loss.

Keywords Automotive data · Automotive software testing · Controller area network · Hardware-in-the-loop test · Data acquisition

1 Introduction

As the occupancy ratio of the software in automotive ECU (Electronic Control Unit) increases, the software takes responsibility for not only strict real-time

J.-W. Lee · K.-Y. Choi · J.-W. Lee(✉)
Department of Electrical and Computer Engineering,
Ajou University, Suwon, Republic of Korea
e-mail: {ljwoo92,ki815kaisian,jungwony}@ajou.ac.kr

S.Y. Jang
HYUNDAI Motor Group, Eletronics Center,
Electronics Control Engineering Development Team, Seoul, South Korea
e-mail: jangdana@hyundai.com

© Springer Science+Business Media Singapore 2015
D.-S. Park et al. (eds.), *Advances in Computer Science and Ubiquitous Computing*,
Lecture Notes in Electrical Engineering 373,
DOI: 10.1007/978-981-10-0281-6_109

functions such as power-train or chassis control but also functionalities directly connected to the user safety. HiL (Hardware in the loop) test is generally carried out in order to verify the performance of the features as the above [1].

However, due to the characteristic of completed ECU as the target of the automotive HiL test, we cannot use a debugging tool without the available debugging interface. In addition, the modification for attaching an additional hardware interface for testing is prohibited [2]. The automotive industries have taken the OEM (Original Equipment Manufacturer) system for developing software. Depending on the OEM, the definition of requirements, the development, and the test are performed by different organizations. Therefore, the testers provide only the result of fault without the detail testing information, and the developers are required to re-enact the defect to correct for finding the cause of the defect [3]. If the developers can get more internal operating data at the time of the defect, they can debug more effectively. Thus, the special monitoring techniques are needed to provide the debugging information for developers efficiently [4].

Monitoring the large amount of information generated in HiL test has the following restrictions. First, in HiL testing environment, it is difficult to perform the test which is controlled under the commands such as 'start', 'stop', and 'resume'. With an assumption that generating data is controllable by any events is hard. That is, it is impossible to stop or restart the test temporarily by issuing the events since inputs of the test script are composed of data to be inspected and time to monitor. Secondly, adding buffers to the ECU is not permitted. In general, adding certain logic to the test is not recommended to ensure the performance and reliability of ECU. This means that there is no space to hold the data generated during the test.

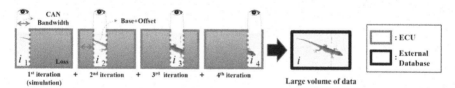

Fig. 1 An Overview of Data Cascading Algorithm

In monitoring the large data, the bandwidth of controller area network (CAN) is a critical limitation in the test environment where unconstrained stopping, restarting, or buffering is not allowed. The CAN protocol, which has the transfer rate up to 1Mbps guaranteed within 40m, is commonly used with 2~3 buses simultaneously at a speed of about 25 to 500Kbps [5]. On the other hand, it is feasible to repeat the test with the same test suite by using the HiL simulator. Therefore, if we consider the limited range of view as CAN bandwidth in fig. 1, and can configure the offset evenly, the entire data can be safely obtained by repeating the simulation.

In this paper, we propose a data cascading method, which enables to collect a large amount of monitoring data beyond the low bandwidth of CAN possible to make the whole picture (i) by assembling the pieces ($i_1 \sim i_4$) in fig. 1. First, we divide the data into segments by considering the minimum time for sending one message packet and the operation cycle of the system's main task. Then, we re-define only the data field of CAN message without the protocol modification. Finally, we present a data cascading algorithm based on the above CAN message protocol. As an experimental result, we verified that the 11Mbytes of data generated in every 10ms for 400 times from ECU's memory were transmitted without any loss or error over CAN.

2 Related Work

The CAN protocol, most commonly used protocol in the ECU, is optimized to transmit short but significant data frequently, such as the value of temperature or RPM [5]. This makes a limit to transfer the large data generated periodically from ECU to the outside. To solve these limitations, methods to transmit the data by compressing and dividing into segments have been proposed.

Kelkar *et al.* proposed an algorithm based on the enhanced data reduction algorithm to provide the better compression range [6]. Wu *et al.* introduced an algorithm by utilizing the features of each signal value, and relocating the values in data field to improve an efficiency [7]. Nevertheless, when the result of compression does not satisfy the required compressibility, the corresponding amount of delay or loss occurs in communication. Furthermore, using extra resources for compressing data may interrupt the operation predefined in the embedded system where limited memory and processing performance are provided.

There is another attempt to transfer over CAN by slicing the data. Shin [8] utilized reserved bit and data field in CAN protocol to define a message type and a sequence information of segments. Even so, this algorithm can cause the loss of data when storage such as a buffer is neither sufficient nor addible. In addition, there is a limit to modify CAN driver to use the area besides the data field in automotive HiL test [9].

For the reasons described above, existing methods for transferring a large volume of data periodically generated from the ECU using CAN protocol are inappropriate to use in the HiL test. Therefore, there is a need to make new approaches to transfer the sizable data using CAN in HiL testing environment where no hardware modification is allowed and using supplementary processor load is restricted not to impede the original behavior.

3 Data Cascading Method

In this section, data cascading method is introduced. It can be materialized by dividing the data into segments, and defining new protocols by utilizing the data field of CAN. At last, detailed algorithms for the sender and receiver are introduced.

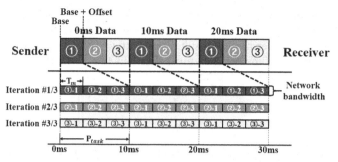

Fig. 2 An Example of Data Slicing Mechanism (P_{task} = 10ms, T_m = 3.3ms)

3.1   Slicing Data into Segments

To employ the data cascading method, the number of segments should be calculated. The number of segments can be determined by using the following information.

- **P_{task}**: The operation cycle of the system's main task (treated as a unit time)
- **T_m**: Minimum time to transfer a message, decided by the protocol in the system
- **C_m**: The number of messages that can be sent in P_{task} ($C_m = P_{task} / T_m$)
- **D_m**: The size of data stored in a message excluding the sequence information
- **$D_{monitor}$**: The area of data to be monitored
- **OCC**: The occupancy of communication bus that existing operation is used

The number of segments is calculated as the equation (1). When the result is not an integer, it is obtained by rounding-up. In this case, it is necessary to guarantee the bus occupancy that can transfer a fixed quantity of data to monitor regularly. The occupancy to be guaranteed can be obtained by subtracting the share of the existing operation from the total amount (1-OCC).

$$\# \, of \, Segments = D_{monitor} / (C_m \times D_m \times (1 - OCC)) \tag{1}$$

Fig. 2 illustrates how data segments are divided and transmitted from sender to receiver by using the data cascading method in an environment where the data is generated in every unit time. The period of a task (P_{task}) is 10ms, and time to transmit a message (T_m, ①-1) is 3.3ms (P_{task} / T_m = 3). An area to monitor is 3 blocks (①②③), which is generated in every 10ms. The size of data delivered per unit time corresponds to one block (①) and it contains 3 messages (①-1~①-3). The test is repeated until the whole data is sent. As each test finishes, the sender transfers the data from the position starting from adding the offset to the previous address. The data transfer is finished when the value obtained by repeatedly adding the offset exceeds the area to be observed.

Table 1 Protocols in Data Cascading Algorithm

Octet	0		1	2	3	4	5	6	7
Bit	0	1:7							
Data Request Protocol	Info.		Reserved	Frame		Start Address		End Address	
Cascading Info. Protocol	Info.		Offset	Frame		Reserved			
Data Transfer Protocol	Info.		Base	Data[0]	Data[1]	Data[2]	Data[3]	Data[4]	Data[5]

3.2   Defining Message Protocol by Reusing CAN Data Field

New protocols used in this algorithm are defined in table 1 by utilizing the data field in CAN. Other fields are inappropriate to utilize since extra modification in CAN driver is needed [7].

• **Data Request Protocol:** Protocol to send the transfer environment information, request the range of required data
  - **Info.:** A flag for indicating the type of message (*Set*: Data Request or Cascading Information Protocol, *Clear*: Data Transfer Protocol)
  - **Frame:** Converted value of the execution time of the test into unit time
  - **Start Address, End Address:** Range of data to receive
• **Cascading Information Protocol:** Protocol to provide the frame and offset information where contemporary data is included
  - **Info.:** Setting the flag to indicate the information is included
  - **Offset:** Sequence of data to transfer out of the entire data segments
  - **Frame:** Converted value of the elapsed time by the unit time in current test
• **Data Transfer Protocol:** Protocol to send the data with sequence information within the on-going segments
  - **Info.:** Clearing the flag to indicate the data is included
  - **Base:** Sequence information within the same segments
  - **Data:** Series of data starts from the base

3.3   Data Cascading Algorithm

Sender and receiver operate with algorithms as shown in fig. 3. Figure (a) is an algorithm of the sender. At first, the data is sliced into segments when the sender received the data request packet (line 3-7). As each execution ends, the sender periodically transmits the segment and CI packet including the current status of the offset and frame. The frame increases whenever the unit time is elapsed (line 11-13). The process is repeated from sending CI packet with increased offset until exceeding the number of segments.

Figure (b) is an algorithm of the receiver. The receiver attains the frame and offset value from the CI packet (line 3-6). Until a new CI packet arrives, receiver parses the CI packet and data transfer packet to acquire the frame, offset, base and data. With the information above, receiver stores the data into an external database (line 8-15).

```
 1: procedure SENDDATASEGMENTS
 2: begin
 3:    if Data request packet received then
 4:       RP ← Data request packet
 5:       F, SA, EA ← RP.Frame, RP.StartAddress, RP.EndAddress
 6:       X, Y ← 0
 7:       S ← Calculate the number of Segments
 8:
 9:       while Y < S do
10:          while X < F do
11:             Send cascading information packet
12:             Transmit Yth data segment
13:             X ← X + 1
14:          end while
15:          X ← 0, Y ← Y+1
16:       end while
17:    else
18:       Waiting for data request packet
19:    end if
20: end
```

```
 1: procedure RECEIVEDATASEGMENTS
 2: begin
 3:    if Cascading information packet received then
 4:       CIP ← Cascading information packet
 5:       F ← CIP.Frame
 6:       Offset ← CIP.Offset
 7:
 8:       while new cascading information packet arrives do
 9:          if Data packet received then
10:             DP ← Data packet
11:             Base ← DP.Base
12:             Data[] ← DP.Data[0] to DP.Data[5]
13:             Save Data[] to external database starts from Base
14:          end if
15:       end while
16:    else
17:       Waiting for cascading information packet
18:    end if
19: end
```

(a) Algorithm of Sending part   (b) Algorithm of Receiving part

Fig. 3 Data Cascading Algorithm

4 Experimental Result

In this section, we implemented the algorithm to acquire the massive data generated from RAM in ECU while HiL testing is in progress, and validated the result. Embedded system, such as the ECU, the data is stored and loaded from RAM in program execution. Therefore, monitoring the RAM for each operating cycle can be utilized in debugging.

4.1 Experiments

Test agent and data collector are appended in general HiL test environment. The data collector is installed on Host PC to send the data request packet, and save the received data. The test agent is implemented as an independent software in the ECU not to corrupt the operation of existing hardware, and to transmit the data to the data collector.

The test was performed with a script consuming 4 seconds at once. The size of RAM to monitor was 29,400Bytes, which started from the address of 0 to 29,399. Other experimental conditions were set as follows.

- P_{task}: 10(ms), T_m: 300(us), D_m: 6(Bytes), $D_{monitor}$: 29,400(Bytes), **OCC:** 55%
- $C_m = (10 \times 10^{-3})/(300 \times 10^{-6}) = 33.33 \approx 33$

   # of Segments $= 29,400/(6 \times 33 \times (1 - 0.55)) = 329.97 \approx 330$

We repeated the script for 330 times to acquire the 29,400 bytes of data generated in the ECU starts from the base address, and stored into the database based on MySQL. The lab id for distinguishing each experiment was set to 74. That is, the total amount of data was consisted of 400 frames, and each frame contained 29,400bytes. As a result shown in fig. 4, 11,760,000 bytes of data were saved with the lab id of 74.

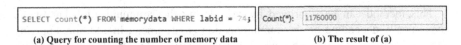

SELECT count(*) FROM memorydata WHERE labid = 74;	Count(*): 11760000
(a) Query for counting the number of memory data	**(b) The result of (a)**

Fig. 4 Total Bytes of Cascaded Data

4.2 Evaluation

The values of output variables which are related to specific function such as strict real-time operations in program execution should be equally arisen from the same inputs. Therefore, we captured the value from the RAM at a randomized time with a debugger to verify the received data using this method.

The result of comparing stacks is shown in fig. 5. Figure (a) is a register view, and its stack part in RAM furnished in Trace32 debugger availed to check internal variables on the general HiL test environment. The address 40006BA4 in R1 indicates a starting address of the stack. The data on the stack is located from the address (R1) with the value of 40006BC4 at intervals of 4bytes.

Figure (b) is a set of collected data starting from address 40006BA4 by applying data cascading algorithm with the same test script used in (a). The whole data set is saved at intervals of 4bytes. Referring to the figure, the data from the same starting address in (a) and (b) are identical. In other words, this algorithm is a reliable to use.

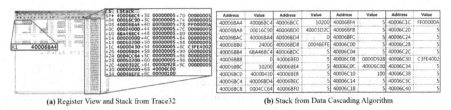

(a) Register View and Stack from Trace32   (b) Stack from Data Cascading Algorithm

Fig. 5 Stack with Trace-32 debugger and Data Cascading Method

5   Conclusion

In order to deliver a large amount of monitoring data in HiL testing environment, there are some limitations to overcome. First, adding buffer is forbidden due to the characteristics of the prototyped ECU. Second, any events cannot control the generation of data. Lastly, using extra resources is prohibited to preserve the original behavior.

In this paper, we proposed a novel method called data cascading algorithm to overcome the low bandwidth of CAN to transmit the large amount of data in HiL testing environment. By using this method, massive monitoring data can be transferred outside of the ECU during the test in spite of the limitations mentioned above. With the proper processing steps, obtained data can be useful for developers where accessing the debugging information is restricted.

This method can be improved by considering the communication bandwidth dynamically depending on the communication environment. In addition, applying the data compression algorithm which has the insignificant effect on the processor can be supportive to enhance the efficiency.

Acknowledgments This research was supported by the HYUNDAI Motor Group, Electronics Center, Korea.

References

1. Bringmann, E., Kramer, A.: Model-based testing of automotive systems. In: 2008 1st International Conference on Software Testing, Verification, and Validation. IEEE (2008)
2. Haufe, J., Fritsch, C., et al.: Real-time debugging of digital integrated circuits. In: Proc. Design, Automation and Test in Europe Conf., User Forum, Paris (2000)
3. Siegl, S., Hielscher, K.S., German, R.: Model based requirements analysis and testing of automotive systems with timed usage models. In: 2010 18th IEEE International Requirements Engineering Conference (RE). IEEE (2010)
4. Watterson, C., Heffernan, D.: Runtime verification and monitoring of embedded systems. IET Software **1**(5), 172–179 (2007)
5. Corriganm, S.: Introduction to the controller area network (CAN). Texas Instrument, Application Report (2008)
6. Kelkar, S.: Control area network based quotient remainder compression-algorithm for automotive applications. In: IECON 2012-38th Annual Conference on IEEE Industrial Electronics Society. IEEE (2012)
7. Wu, Y.J., Chung, J.G.: Efficient controller area network data compression for automobile applications. Journal of Zhejiang University Science **16**, 70–78 (2015)
8. Shin, C.: A framework for fragmenting/reconstituting data frame in Controller Area Network (CAN). In: 2014 16th International Conference on Advanced Communication Technology (ICACT). IEEE (2014)
9. Köhl, S., Jegminat, D.: How to do hardware-in-the-loop simulation right. No. 2005-01-1657. SAE Technical Paper (2005)

Classification Framework for Electropulsegraph Waves

JinSoo Park, Dong Hag Choi, Se Dong Min and Doo-Soon Park

Abstract Electropulsegraphy is a medical device that was invented by an orient medical physician and a few engineers to help the physicians to diagnose patients in more systematic way by analyzing waves generated from the device. With the use of the device, doctors can diagnose patients based on charts rather than by feeling purses manually as in traditional oriental medicine practice. The device uses traditional microphones as sensors attached to particular locations on a body of a patient (e.g., a wrist) and generates a number of waves that reflect the physical states of the patient. The device has been used for several decades by physicians in Korea, and it undergoes functional upgrades both in hardware and software aspects recently. As one of those upgrading efforts, we strive to make the diagnostic process automatically by applying the well-known machine learning algorithm-logistic regression. In this article, we provide a framework of the preliminary study with experimental classification results. Training data sets collected for decades are used to estimate the parameters of the logistic regression. And the parameters are used to classify wave inputs chosen at random.

Keywords Electropulsegraphy · Waveform classification · Logistic regression

1   Introduction

Physicians of oriental medicine have diagnosed the physical states of patients by checking on the pulse of the patients with hands, which often results in irregular and inconsistent diagnostic decisions by the physicians. The diagnostic decision can be different among the physicians even for the same type of symptoms. This phenomenon had led to diagnostic reliability of the physicians for a long period of

J. Park(✉) · D.H. Choi · S.D. Min · D.-S. Park
SoonChunHyang University, RM U1202, 22 Soonchunhyangro,
Shinchang-myeon, Asan-si, ChoongCheongNam-do, South Korea
e-mail: {vtjinsoo,cdh0191,sedongmin,parkds}@sch.ac.kr

© Springer Science+Business Media Singapore 2015
D.-S. Park et al. (eds.), *Advances in Computer Science and Ubiquitous Computing*,
Lecture Notes in Electrical Engineering 373,
DOI: 10.1007/978-981-10-0281-6_110

time, and accordingly the number of patients that rely on oriental medicine is decreasing annually compared to those that trust western medical systems. The physicians has yearned to get dedicated orient medical devices that can help them to diagnose patients in a more reliable and consistent manner and to restore their prides as orient medical practitioners accordingly.

Thus, a group of physicians has teamed up with several electrical engineers to invent a device called 'electropulsegraphy' that uses traditional microphones as sensors to check the pulse of patients. The device listens the sound of radial artery pulse on a patient's right and left wrist and generates waves that reflect the states of the patient. The physicians collected the wave data generated the device during their medical practices for several decades.

Today a group of researchers began to upgrade the machine from both hardware and software perspective. It includes the automatic diagnostic function of the device to help the physicians diagnose patients more reliably. In this article, we provide the results of the related preliminary study of the automatic diagnostic system that employs one of the well-known machine learning algorithms.

Researchers have put a lot of efforts to analyze various waveforms generated from medical devices including the very recent ones [1-4]. Most of these studies are focused on localizing QRS component of the waves to estimate significant medical information of patients, for example, the heartbeat rate rather than the classification of waveforms. In [2], the author made an effort to classify the electrocardiogram based on logistic regression algorithm. However, the study is performed only for ECG waveforms and does not include other type of waveforms.

We strive to apply the logistic regression algorithm for the classification of waveforms generated from our dedicated device, thus the physicians can use the classification results for their diagnostic decisions on the patients.

The logistic regression algorithm is one of the well-known machine learning algorithms, and it has been used for multiclass classification purposes today. One of those application areas can be character recognition systems [5]. The algorithm is used to classify hand-written characters based on derived parameters by the logistic regression algorithm.

This paper is organized as follows. Section 2 describes the regularized logistic regression algorithm and its use for multiclass classification. Section 3 provides explanations on the device and waveforms generated from the device. Also it describes experimental results based on the estimation of the logistic regression. Finally we discuss conclusions of the study and future studies later on.

2   Classifications of Electropulsegraphy Waves Based on Logistic Regression

2.1   Regularized Logistic Regression

In statics, logistic regression is used for prediction of the probability of occurrence of an event by fitting data to a logit function - logistic curve. It is a generalized linear model used for binomial regression. In the logistic regression, hypothesis is represented by a logistic curve as follows.

$$h_\theta(x) = p(y = 1|x; \theta) = g(\theta^T x) = g(\theta_0 + \theta_1 x_1 + \theta_2 x_2 + \theta_3 x_3 + \dots) \qquad (1)$$

And the corresponding cost function in regularized logistic regression can be expressed as

$$J(\theta) = \frac{1}{m} \sum_{i=1}^{m} Cost(h_\theta(x^{(i)}, y^{(i)}))$$
$$= \frac{1}{m} \sum_{i=1}^{m} \left[-y^{(i)} log(h_\theta(x^{(i)})) - (1 - y^{(i)}) log(1 - h_\theta(x^{(i)})) \right] + \frac{\lambda}{2m} \sum_{j=1}^{n} \theta_j^2 \qquad (2)$$

Now we can find parameter θs that minimize the cost function that satisfies $\min_\theta J(\theta)$ by gradient descent algorithm as follows.

$$\theta_j := \theta_j - \alpha \frac{\partial}{\partial \theta_j} J(\theta)$$
$$= \theta_j - \alpha \sum_{i=1}^{m} \left(h_\theta(x^{(i)} - y^{(i)}) \right) x_j^{(i)} \qquad (3)$$

Here, we simultaneously update all θ_j.

First we estimate the regression parameters and then we use those parameters to calculate the probability of the waveforms to distinguish its class. We classify them according to the maximum probability.

2.2   One-vs-all Multiclass Classification

Once the logistic regression classifier is trained for each class i to predict the probability that $y=1$, it can be applied to make prediction for a new input x. On a new input x, to make a prediction, we can pick the class i that maximizes

$$\max_i h_\theta^{(i)}(x) \qquad (4)$$

3   Estimation Results

The device generates 12 waveforms reflecting the states of internal organs respectively, and amplitude for each wave is composed of integer values showing the strength of pulse and has a periodic wave patterns generally as shown in Fig. 1. The device converts an analog signal into a digital signal, and it records the pulse rate, the interval between pulse waves and the waveform of the pulse waves by using a dedicated computer program. We use the data collected for over decades, and only some of the data recently collected are used for the experiment. Through a series of manual works, the data is first transformed and saved into excel format, and it is used later for Matlab simulation.

Among the thirty-three different waveforms obtained from the device, we use only four of them. In other words, thirty-three waveforms can be viewed as generated from twelve different organs according to the theory of oriental medicine. Thus training data sets can be obtained for those waveforms respectively. In this article, we only report the training and estimation results for four different probing waveforms - DaeMac, WanMac, DanMac, and SaeMac.

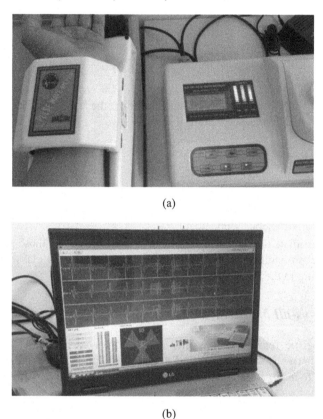

(a)

(b)

Fig. 1 Electropulsegraphy and waves from 12 different probing organs

Fig. 2. illustrates samples of waves from the four different types where each type of wave has periodicity with relatively similar amplitudes and frequencies respectively. Each waveform is composed of 783 integer values as amplitude values. The number of training data is different for each probing types because patients see doctors for different symptoms.

In most medical area literature, the main research interests are focused on detecting QRS parts of waves in order to estimate crucial health indicators such as heart beat rate, and etc. However, the focus of our research lies in the classification of waveforms, so that doctors can determine the symptoms of patients based on the classification results. Accordingly, we emphasize that our research goal is to decide the classes of waveforms for unknown waveforms. During the training process, we first estimate the parameters of the training data set, and the estimated parameters are used to classify the unknown waves using the estimated parameters.

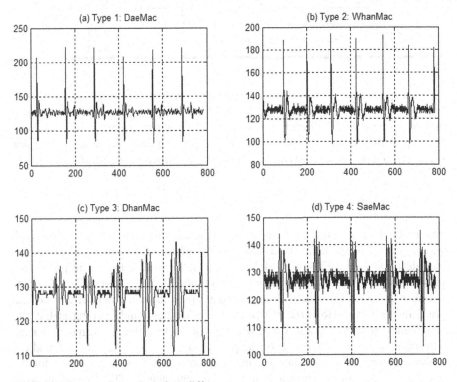

Fig. 2 Sample waveforms from four different types

In the estimation process, we need to determine the learning rate (α) that relates to the entire learning speed. As we increase the learning rate from 0.02 to 0.4, we examine the change in training accuracy as shown in Table 1. It shows that the accuracy is fairly stable with regards to the changing learning rate, so we choose one between 0.08 and 0.3.

Table 1 Training set accuracy depending on learning rate α

	α							
	0.02	0.04	0.06	0.08	0.1	0.2	0.3	0.4
Training set accuracy(%)	94.4	94.87	93.16	95.29	94.44	96.15	96.15	94.44

Table 2 shows the estimation accuracy of training data set measured at four different probe points-DaeMac, WanMac, DanMac and SaeMac. The number of data for each probe points is 22, 30, 87, and 94 respectively. And as shown in table 2, the estimation accuracy is fairly high for each probe points respectively.

Table 2 Estimation accuracy for each training data set

Probe points	Number of data	Estimation accuracy (%)
DaeMac	22	95.45
WanMac	30	96.77
DanMac	87	90.80
SaeMac	94	96.80

The experiment was not performed for large number of data sets due to several reasons. Although the experiment was performed only for four different data sets and each one is small in terms of numbers, the training accuracy is fairly high. We expect that as we increase the number of training data set and execute a few preprocessing procedure, we can achieve better classification results later.

4   Conclusions

This research aims to make a framework for the classification of electropulsegraph waves. The device was invented and led by a group of oriental medical doctors and electrical engineers several decades ago. It measures changes in pulse rate, pulse speed and records the pulse waves through a dedicated computer. The data has been collected for several decades and are used to train young orient medical practitioners today. However, the doctors are having trouble getting stable and consistent diagnostic results so far, which often led to different diagnostic results depending on the level of skills of the practitioners. Thus several doctors and engineers are teamed to renovate the machine so that it generates outputs for the doctors to make the diagnostic results more consistently. The renovation work includes the automatic classification of the waves, so doctors can save the time for waveform classification and lessen the erratic decision rate.

We employ the logistic regression algorithm for this classification work that has been used for various multiclass classification problems. Through the experiment, we demonstrated that the algorithm works well for the medical waveform classification, even though the number of data is small. We expect to achieve better classification results when more data is secured later.

We plan to employ deep learning algorithms and several other machine learning algorithms in order to increase the accuracy of the classification for further studies.

Acknowledgement This research was supported by the MSIP(Ministry of Science, ICT and Future Planning), Korea, and run under the C-ITRC(Convergence Information Technology Research Center) (IITP-2015-IITP-2015-H8601-15-1009) program supervised by the IITP(Institute for Information & communications Technology Promotion).

References

1. Keerthi Priya, P., Umamaheswara Reddy, G.: MATLAB Based GUI for Arrhythmia Detection Using Wavelet Transform. International Journal of Advanced Research in Electrical, Electronics and Instrumentation Engineering 4, 2278–8875 (2015)
2. Escalona-Moran, M.A., Soriano, M.C., Fischer, I., Mirasso, C.R.: Electrocardioram Classification Using Reservior Computing With Logistic Regression. IEEE Journal of Biomedical and Health Informatics 19, 892–898 (2015)
3. Mehta, S.S., Lingayat, N.S.: Support Vector Machine for Cardiac Beat Detection in Single Lead Electrocardiogram. IAENG International Journal of Applied Mathematics 36 (2007)
4. Wulsin, D.F., Gupta, J.R., Mani, R., Blanco, J.A., Litt, B.: Modeling electroencephalography waveforms with semi-supervised deep belief nets: fast classification and anomaly measurement. Journal of Neural Engieering 8, 1–13 (2011)
5. Basu, K., Nangia, R., Pal, U.: Recognition of similar handwritten characters using logistic regression. In: 2012 10th IAPR International Workshop on Document Analysis Systems, pp. 200–204 (2012)

Optimization of LSPL Algorithm for Data Transfer in Sensor Networks Based on LEACH

Wang-Boo Jeong, Dong-Won Park and Young-Ho Sohn

Abstract With the development of ubiquitous computing, wireless sensor networks (WSNs) are now being applied to many areas for the collection of field information. However, the sensor nodes in a WSN have limited energy retention and processing capability. Most notably, battery limitations mean the energy efficiency of the network operation is a crucial issue. Accordingly, this paper presents the optimization of the LSPL algorithm that uses energy more efficiently by reconstructing the data transmission based on LEACH. The capability of the LSPL algorithm is confirmed using an NS2 simulation, and the LSPL 2 algorithm is shown to be the most efficient.

Keywords WSN · LEACH · Sensor nodes · Sensor network · LSPL

1 Introduction

A sensor network is a small, autonomous and independent network formed between sensor nodes randomly allocated in a sensor field for the purpose of obtaining and processing specific information in a network environment without an infrastructure, such as access points[1-3].

W.-B. Jeong
Kyunghwa Girls' High School, Daegu, Korea
e-mail: wbjeong@gmail.com

D.-W. Park
Paichai University, Daejeon, Korea
e-mail: dwpark@pcu.ac.kr

Y.-H. Sohn(✉)
Department of Computer Engineering, Yeungnam University, Gyeongsan, Korea
e-mail: ysohn@ynu.ac.kr

© Springer Science+Business Media Singapore 2015   789
D.-S. Park et al. (eds.), *Advances in Computer Science and Ubiquitous Computing*,
Lecture Notes in Electrical Engineering 373,
DOI: 10.1007/978-981-10-0281-6_111

However, while a sensor node can detect and transfer information, it is restricted in its capability to deal with information and maintain energy. In particular, due to replacement and battery recharging restrictions, efficient energy use is very important for the operation of a sensor network. For example, if each sensor node periodically transmits data to the BS(Base Station), this generates a lot of traffic, resulting in a high energy consumption across the sensor network[4]. Accordingly, this study attempts to improve the energy efficiency by varying the LSPL(LEACH Steady-state Phase Looping) algorithm[5] that constantly collects information by restructuring the data transfer based on LEACH(Low Energy Adaptive Clustering Hierarchy)[6]. The performance is evaluated using NS-2(Network Simulator Version 2)[7,8] and LSPL 2, which repeats the steady-state phase three times, is shown to be the most effective.

2   Relevant Studies

2.1   LEACH Protocol

LEACH[6] receives data from sensor nodes in a cluster, aggregates the data where the cluster heads are repeated, and then sends the aggregated data to the base station. As the efficiency of data transfer in a sensor network is related with energy efficiency[9], LEACH is a typical hierarchy-based routing protocol that stops unnecessary repeated data from being transferred to the base station[10,11]. The LEACH protocol is composed of a unit of time, called a round, and each round consists of a set-up phase and steady-state phase.

During the set-up phase, the determination of whether a sensor node is a cluster head for the present round is based on whether it was a cluster head during the previous round and the ideal number of cluster heads. When a sensor node is determined to be a cluster head for the present round, this fact is made known to the neighboring sensor nodes. The member nodes that receive this information then determine which cluster head to use based on such parameters as the received signal strength, and then send their information to that cluster head to form a cluster.

During the steady-state phase, to save energy, the member nodes in each cluster only transfer data during their own transfer slot and sleep during the other time slots.

2.2   LSPL Algorithm

In the case of LEACH, the cluster heads are connected with member nodes and work separately, plus the set-up phase and steady-state phase are connected. However, separating the two phases and restructuring the data transfer can produce a more effective algorithm.

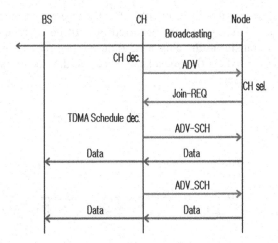

Fig. 1 Outline map of LSPL Algorithm

Therefore, the LSPL algorithm separates the set-up phase from the steady-state phase, and two data transfers are made by the sensor nodes for every one selection of cluster heads, that is there are two steady-state phases for every one set-up phase[5].

As shown in the LSPL algorithm outline map in Figure 1, the member nodes cannot send data repetitively with one ADV-SCH scheduling signal and send data according to a scheduling signal from the cluster head. Therefore, in the LSPL algorithm, ADV-SCH scheduling signals and data are sent repetitively.

3   Optimization of LSPL Algorithm

The LSPL algorithm has already been shown to have a higher capability than LEACH using NS-2[5]. The LEACH algorithm consumes a lot of energy due to the data transfer during the set-up phase. Thus, a sensor network can be operated far more effectively by shortening the length of the set-up phase, while repeating the steady-state phase. Therefore, focusing on the steady-state phase, i.e. sending data between the cluster heads and member nodes, this study investigated the optimization of the operation efficiency of a sensor network using different variations of the LSPL algorithm.

As shown in Figure 2, to optimize the energy efficiency of data transfer based on repeating the steady-state phase, the number of repetitions of the steady-state phase was varied. Thus, the LSPL algorithm was classified to determine the most efficient network operation.

In the case of LSPL 1(=LPSL), the steady-state phase was repeated twice, while LSPL 2 repeated the steady-state phase three times and LSPL 3 repeated it four times. When repeating the steady-state phase five times, the simulation did not work correctly as the data is structured to find the required information from the transferred data and use it efficiently, that is, an error message[Error: Meta size 1422 too large (max = 1000).] occurred due to increased metadata. When repeating the data transfer to the cluster head during one round, this increases the metadata. Therefore, when the metadata exceeded the fixed maximum value (default=1000), errors occurred.

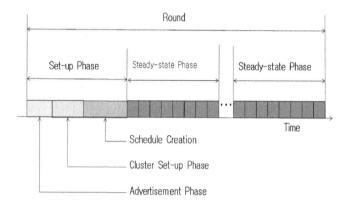

Fig. 2 LSPL Optimization Algorithm for One Round

The simulation measured the integrated energy consumption per unit time, integrated data traffic per unit time, and changes in the survival node numbers per unit time. Furthermore, it compared and analyzed the topology of the survival time, the total energy, and the total data to find the most effective algorithm.

4   Simulation and Analysis

4.1   Experimental Environment

To analyze the performance of the proposed LSPL variations, all sensor nodes were always assumed to have data to send, as with LEACH. To compare the LSPL variations with LEACH, the simulation was conducted using NS-2[7,8] under the same conditions as used for LEACH. And, to operate the NS-2 simulator, this study used a Linux operating system and the following simulation environment:

The simulation mode used a 100m×100m network with 100 sensor nodes randomly arranged and the base station node placed at 50 and 175. The number of cluster heads was 5 per round and the round time changed according to the type of simulation. The amount of data gathered by the sensor nodes was 500Bytes and the size of the transferred message head was 25Bytes. The energy at the BS was 50,000 Joules, which was large to ensure the longest survival in the sensor network. Each sensor node was initially given 2 Joules of energy, and once these 2 Joules were used up, the sensor could no longer join the network.

4.2   Evaluation of LSPL Algorithm Performance

The simulation results were roughly classified into two sections: First, the integrated energy consumption per unit time, integrated data traffic per unit time, and survival node number per unit time were measured over time. Then, at the end of the simulation, the topologies of the survival time, total energy used, and total data sent were compared and analyzed.

1) Integrated Energy Consumption Per Unit Time
The graph in Figure 3 shows the integrated energy consumption over time, where LSPL 1 and LSPL 2 exhibited a more efficient energy use over time when compared with LEACH. However, in the case of LSPL 3 that repeated the steady-state phase four times, a rapid change in energy use was observed, and since the energy use was not efficient, the network quickly stopped working.

Therefore, as shown in Figure 3, LSPL 2, which repeated the steady-state phase three times per round, showed the most efficient energy use and maintained the sensor network the longest.

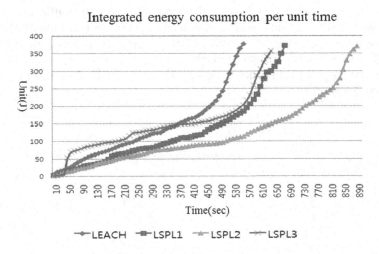

Fig. 3 Integrated energy consumption per unit time for LSPL optimization

2) Integrated Data Traffic Per Unit Time
The graph in Figure 4 shows the integrated data traffic over time, where the amount of data transferred during one round decreased when increasing the number of repetitions of the steady-state phase.

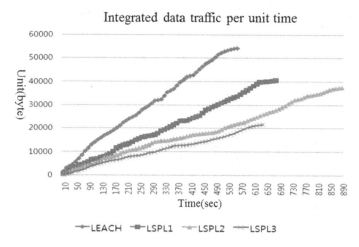

Fig. 4 Integrated data traffic per unit time for LSPL optimization

This also emphasizes the high amount of data sent during the cluster formation process. However, while showing the most efficient data traffic over time in Figure 4, LSPL 3, which repeated the steady-state phase four times, had a lower network survival time. Thus, while LSPL 3 was effective for data traffic, the LSPL 3 sensor nodes consumed more energy due to the increased metadata, plus the number of surviving nodes decreased. Therefore, LSPL 2 was the most overall effective algorithm.

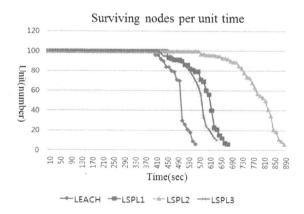

Fig. 5 Surviving nodes per unit time for LSPL optimization

3) Surviving Node Numbers Per Unit Time
The graph in Figure 5 shows the number of surviving nodes per unit time in the sensor network. LSPL 2, which showed the most efficient energy use and longest

surviving network in Figure 3, also had the highest number of surviving sensor nodes. Meanwhile, LSPL 3 had the lowest number of surviving sensor nodes, and the simulation that repeated the stead-state phase five times stopped due to increased metadata and no data appeared. Therefore, it was assumed that the LSPL 3 sensor nodes consumed more energy due to the increased metadata and fewer surviving nodes. Thus, the proposed LSPL variations were shown to have more capability than LEACH, except when the simulation stopped due to increased metadata, plus LSPL 2 had the highest number of surviving nodes over time.

5 Conclusion

Despite their many benefits, wireless sensor nodes have a limited information-processing capacity and energy retention. Thus, the energy efficiency of wireless sensor nodes is very important, especially when human access is difficult and restricts replacement and battery recharging. Therefore, the algorithm variations presented in this paper minimize energy waste, while focusing on energy efficient information transfer by sensor nodes.

Simulations confirmed a superior performance by the LSPL algorithm when compared with LEACH. The highest efficiency was achieved when shortening the length of the set-up phase and sending more data in the steady-state phase. However, the transfer of excessive amounts of data caused one simulation to stop because of increased metadata. As a result, LSPL 2, which repeats the steady-state phase three times, was identified as the most effective algorithm with the longest network survival time and most efficient energy use.

References

1. Corson, S., Macker, J.: Mobile Ad-hoc Networking (MANET), ITEF RFC 2501, January 1999
2. Hac, A.: Wireless Sensor Network Designs. John Wiley & sons, Ltd. (2003)
3. Santi, P.: Topology Control in Wireless Ad Hoc and Sensor Networks. ACM Computing Surveys 37(2), 164–194 (2005)
4. Jeong, W.-B., Bae, J.S., Sohn, Y.-H.: Efficient Architecture for Sensor Data Acquisition Applicable to Wireless Sensor Networks. Sensor Letters 12, 990–993 (2014)
5. Jeong, W.-B., Sohn, Y.-H.: LSPL Algorithm for Efficient Data Transfer in Sensor Network based on LEACH. Journal of Korea Institute of Information Technology 11(10), 105–111 (2013)
6. Heinzelman, D.W., Chandrakasan, A.P., Balakrishnan, H.: An Application-Specific Protocol Architecture for Wireless Microsensor Networks. IEEE Trans. on Wireless Communications 1(4), 660–670 (2002)
7. Fall, K., Varadhan, K. (eds.): Formerly ns Notes and Documentation, http://www.isi.edu/nsnam/ns, University of Southern California (2008)
8. Fall, K., Varadhan, K.: The ns Manual (formerly ns Notes and Documentation), The VINT Project, UC Berkeley, LBL, USC/ISI, and Xerox PARC, April 2006

9. Kim, J.S., Park, C.H., Kim, C.G., Kang, B.W.: An Energy-Efficient Data aggregation using Hierarchical Filtering in Sensor Network. Journal of The Korea Society of Computer and Information 12(1), 79–88 (2007)

10. Liu, X.: A Survey on Clustering Routing Protocols in Wireless Sensor Networks. Sensors 12(8), 11113–11153 (2012)

11. Wei, D., Jin, Y., Vural, S., Moessner, K., Tafazolli, R.: An Energy-Efficient Clustering Solution for Wireless Sensor Networks. IEEE Trans. Wireless Communications 10(11), 3973–3983 (2011)

In-Memory Processing for Nearest User-Specified Group Search

Hong-Jun Jang, Woo-Sung Choi, Kyeong-Seok Hyun, Taehyung Lim, Soon-Young Jung and Jaehwa Chung

Abstract This paper presents a nearest user-specified group (NUG) search which called a clustered NN problem. Given a set of data points P and a query point q, NUG search finds the nearest subset $c \subset P$ ($|c| \geq k$) from q (called user-specified group) that satisfies given conditions. Motivated by the brute-force approach for NUG search requires $O(|P|^2)$ computational cost, we propose a faster algorithm to handle NUG problem with in-memory processing. We first define clustered objects above k as a user-specified group and the NUG search problem. Moreover, the proposed solution converts a NUG search problem to a graph formulation problem, and reduces processing cost with geometric-based heuristics. Our experimental results show that the efficiency and effectiveness of our proposed approach outperforms the conventional one.

Keywords Nearest user-specified group query · k-nearest neighbor query

1 Introduction

This paper proposes a new type of search, called nearest user-specified group (NUG) search, which has many applications in LBSs and spatial data mining. The researches for location-based ad-hoc query processing have focused on retrieving nearest neighbor [1] and its variants [2], [3] (e.g., RNN, GNN, etc.). Unlike nearest neighbor and its variants searches, NUG search find the nearest subset (called user-specified group) that satisfies given condition from a given query point.

H.-J. Jang · W.-S. Choi · K.-S. Hyun · T. Lim · S.-Y. Jung(✉)
Department of Computer Science Education, Korea University, Seoul, Korea
e-mail: {hongjunjang,ws_choi,ks_hyun,th_lim,jsy}@korea.ac.kr

J. Chung
Department of Computer Science, Korea National Open University, Seoul, Korea
e-mail: jaehwachung@knou.ac.kr

© Springer Science+Business Media Singapore 2015
D.-S. Park et al. (eds.), *Advances in Computer Science and Ubiquitous Computing*,
Lecture Notes in Electrical Engineering 373,
DOI: 10.1007/978-981-10-0281-6_112

Fig. 1 shows the results of k-NN search and NUG search, where k is 3 and distance threshold δ is 1km. Although the result of 3NN(q), which is $R_n = \{p_1, p_2, p_3\}$, guarantees that R_n are the closest points set from q, it does not guarantees k points are clustered. On the other hand, NUG search considers both $R_1 = \{p_1, p_6, p_7\}$ and $R_2 = \{p_2, p_4, p_5\}$ which are located within 1km from p_1 and p_2 respectively. If the distances between q and the sets of objects are only affected by p_1, p_2, the result would be R_2. The result of NUG search is determined by the distance from q as well as threshold-based condition which indicates the user specified distance where more than k objects reside. Throughout this paper, the above results (R_1 and R_2) are denoted as user-specified group (u-group).

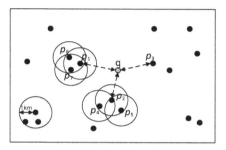

Fig. 1 An example of kNN and NUG queries. (k is 3)

The NUG search can be applied to following examples and many applications in spatial domain.

Example (Location-based services): In a LBS region with a large set of points (such as restaurants or clothing stores etc.), tourists want to visit places that are located within limited distance to have a variety of experiences at a time. In this case, a NUG search would return the nearest set of points with constraints of users.

Recently in spatial databases and mining field, group searching has be received great attention and various researches [4] [5] has been developed in the literature to find the user specified group efficiently. However, the NUG problem proposed in this paper shows lower complexity than the conventional variants. The brute-force approach are obviously too expensive to be of any practical use. Motivated by this issue, we devise new in-memory algorithm to process the NUG search more efficiently.

2   Problem Formulation

We formally define u-group and NUG query in this section. We first define the γ-neighborhood of a point p before define u-group formally.

Definition 1. (γ-*Neighborhood* of p_i) Given a set of data points P, a point $p_i \in P$ and a distance γ, the γ-*neighborhood of p_i*, denoted as γ-$NH(p_i)$, is the subset of P which consists of points (including p_i) which are located less than γ distance from p_i. It is formally defined as follows:

$\gamma\text{-}NH(p_i) = \{p_j \mid p_j \in D,\ dist(p_i, p_j) \leq \gamma\}$,where $dist(x,\ y)$ is the *Euclidean distance* between two points x and y

Let the point set $\gamma\text{-}NH(p_i)$ be S for any $p_i \in P$. Then the S indicates every points p_i from the dataset within the distance γ. That is, it is guaranteed that every points in S has locational locality from p_i. And for a point $p_j \in S - \{p_i\}$, if $\gamma\text{-}NH(p_j)$ have same elements with S, we call the points like p_i and p_j points as the *representatives* of S.

Definition 2. (*User-specified group*) Given a set of data points P, a user-specified distance threshold δ ($\leq \tau$) and cardinality threshold k ($\geq$ 2), a *user-specified group* (*u-group*) is a points set $\delta\text{-}NH(p_i)$ S, such that for any $p_i \in P$, $|S| \geq$ k.

Definition 3. (*Distance from a query point q to a u-group*) Given a query point q and a u-group $G(\delta,\ k)$, the *distance from q to* $G(\delta,\ k)$ is formally defined as follows:

$$\|q, G(\delta, \text{k})\| = \min_{p \in G_{rp}} dist(q, p)$$

, where G_{rp} is the *representatives* of $G(\delta,\ k)$.

Definition 4. (*NUG query*) Let S_G denote a collection of u-groups with both user-specified distance threshold δ and cardinality threshold k ($\geq$ 2). Given a query point q, the *NUG query* finds a u-group $G'(\delta,\ k)$, such that for any $G(\delta,\ k) \in S_G - \{G'(\delta, k)\}$, $\|q, G'(\delta, k)\| \leq \|q, G(\delta, k)\|$.

3   In-Memory NUG Query Processing

In this section, we study for the in-memory algorithm that solves NUG search problem with distance threshold and geometric based heuristics. For a given a NUG query and dataset P, we are to compute $\delta\text{-}NH(p)$ for $p \in P$, then find a set of points which satisfies the cardinality condition among them to search the nearest u-group. To increase the efficiency of $\delta\text{-}NH(p)$ computation and u-group searching, we convert a finding the nearest u-group problem into an adjacent list based graph representation problem. In this graph, each vertex is a data point and edge is neighborhood relationship between two points. This graph is called *the neighborhood graph* of P.

Lemma 1: If $p_j \in \delta\text{-}NH(p_i)$, then $p_i \in \delta\text{-}NH(p_j)$.

The proof of lemma 1 is omitted by the space limitation. According to lemma 1, when $p_j \in \delta\text{-}NH(p_i)$, two points p_i and p_j can be represented as a undirected relationship. Fig. 2 illustrates an example that expresses neighborhood among the eight points in P using undirected graph. In this figure, the circles centered at point p_i cover all elements in $\delta\text{-}NH(p_i)$ of each p_i.

In the adjacency list of this graph, if the number of nodes connected to a first node p_i (including first node) is more than the cardinality threshold k, then node p_i and the connected nodes form a u-group $G(\delta,\ k)$. Moreover, node p_i becomes one of representatives of $G(\delta,\ k)$. The algorithm that we propose through this paper constructs neighborhood graph and applies the plane-sweep based approach to handle NUG search problem effectively. In this algorithm, the primitive idea constructing graph is that sorting every points into the order of coordinate value of an arbitrary axis and connecting neighborhoods of each points to the corresponding node through the sweep line formed by each point. Computing the neighborhood of each point is not able to gain by a single sweep. For each sweep line of p, both forward and backward sweep is necessary in order to compute neighborhood of p correctly. However, according to the lemma 1 and 2, the graph can be constructed by only searching the limited distance sweep line either direction.

Lemma 2: Let l_1, l_2 be two sweep lines in the space ordered by an arbitrary axis Θ. For a given a distance threshold δ, if $|l_1.\text{cord} - l_2.\text{cord}| > \delta$, then for a point p placed on l_1, none of points on l_2 can be an element of δ-$NH(p)$, where $l_1.\text{cord}$ is Θ's coordinate of l_1.

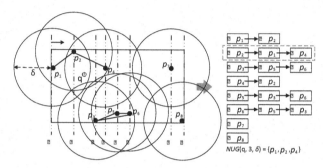

Fig. 2 An example of in-memory query processing using the neighborhood graph of $P = \{p_1, p_2, p_3, p_4, p_5, p_6, p_7, p_8\}$

Algorithm 1 shows in-memory process for NUG query using the adjacent list. In line 2, it stores all of the data points in event queue Q and sort them with an

---

Algorithm 1. In-Memory NUG Processing (MNUG)

Input: a set of points C, a query point q, a distance threshold δ, a cardinality threshold k
Output: a group $R \subseteq C$

1. $R \leftarrow \{\}; i \leftarrow 0$;
2. Initialize the event queue Q with all data points of C;
3. Initialize an empty status structure A; //adjacency-list
4. **while** (C is not empty) **do**
5.   remove the data point p_i with smallest coordinate from Q;
6.   insert p_i into A[i];
7.   **for** ($j \leftarrow i; j > 0; j$--) // by lemma 1
8.   **if** ($|p_i.\text{cord} - A[j-1].\text{cord}| > \delta$) **then** // by lemma 2
9.   **break**;
10.   **else if** ($\delta \geq dist(p_i, A[j-1])$) **then**
11.   A[$j-1$].adj.add(A[i]);
12.   A[i].adj.add(A[$j-1$]);
13. $i \leftarrow i + 1$;
14. $R \leftarrow CalResult(q, k, A)$;
15. **Return** R;

arbitrary axis. Then, lines 5-6 delete a point p with the smallest coordinate value from the Q and creates the first node of adjacent list with p. By lemma 1, in lines 7-12, a previous node searching to backward to compute the neighborhood of p. Moreover, the correct neighborhood of p can be obtained by only extending the search area until the distance between p.cord and node.code becomes larger than δ by lemma 2 When organizing the adjacent list to represent the neighborhood graph is completed, algorithm 1 seeks for the adjacent list (i.e. u-group) that the cardinality of this is bigger than k and has the first node closest to q in lines 14-15. For instance, in Fig. 3, though there are two u-groups, the set $S = \{p_1, p_2, p_4\}$ is returned as a result since p_2 is a representative point that is closest to q.

The algorithm sorts the data points, which takes $O(n \log n)$ time. And the cost for constructing an adjacent list using sweep algorithm is $O(nt)$, where t is a query parameter that shows the average number of sweep lines within the given distance δ. The sequential searching for u-group which is closest to q requires $O(n)$. As a result, the time complexity of the proposed algorithm is $O(n \log n + nt)$.

4 Experimental Results

We performed extensive experiments and evaluated the results to prove the efficiency of the proposed method. Every the experiments are performed on a PC with Intel Core2 Quad Q8200 2.33GHz, 4GB memory. In each setting, we use two synthetic datasets, UN and GA that are distributed uniformly and clustered (i.e., Gaussian distribution) with the Spatial Data Generator [6]. All data objects are located in the space 100K x 100K. The performance of algorithm is measured by CPU time. Table 1 shows the parameters and their settings. Bold font indicates the default values. For each experiment, 1K of queries were executed and average of CPU time is.

Table 1 Defualt Experiment Setting

Parameters	Settings
Number of points	2K, 4K, **6K**, 8K, 10K
Distribution of datasets	Uniform (UN), Gaussian (GA)
Range of δ	S: 150 (75 < δ <= 150), M: **300** (150 < δ <= 300), L: 600 (300 < δ <= 600)
Range of k	s: 10 (5 < k <= 10), m: **20** (10 < k <= 20), l: 40 (20 < k <= 40)

We compare two methods: Plane sweep method (PSM) and Nested Loop Method (NLM). PSM is the proposed MNUG algorithm whereas NLM is straightforward algorithm. NLM searches NUG(q, δ, k) after computing neighborhood of every point in the dataset.

Fig. 3 (a) and (b) show the performances of PSM and NLM when varying the cardinality of datasets from 2K to 10K. These figure show CPU time to finish the searching process for both PSM and NLM in UN and GA datasets. PSM is dozens of faster than NLM. This results in PSM has smaller search space using pruning heuristics than that of NLM which has $O(n^2)$ complexity.

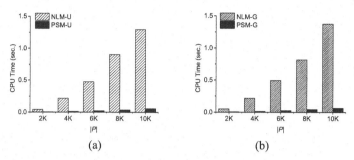

Fig. 3 Time cost vs. cardinality of two datasets (a) UN, (b) GA

Fig. 4 depicts the CPU time of PSM when varying the range of δ and k with UN and GA datasets. Ranges of δ and k affect NUG(q, δ, k) queries which are generated randomly. In this setting, increase of δ results in lower scale of linear increasing of time cost for PSM in both UN and GA datasets, while that of k does not affect the time cost. This causes in having bigger search space for PSM in backward sweep as δ is bigger regardless of k.

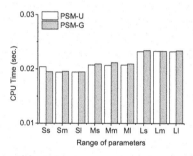

Fig. 4 Time cost vs. range of δ and k

5   Conclusion

This paper studies the nearest user-specified group search that has many real life applications. We introduce plane-sweep based in-memory processing algorithm to find a nearest u-group using geometric heuristic. We have conducted thorough experiments on synthetic datasets having uniform and Gaussian distributions. As a result, our algorithm showed effectiveness in reducing search spaces (while naïve approach can cause full access to datasets). In future, we aim to develop more efficient in-memory and external algorithms for the NUG search, and extend our work for moving object databases.

Acknowledgement This research was supported by Basic Science Research Program through the National Research Foundation of Korea (NRF) funded by the Ministry of Education, Science and Technology (2013-R1A1A2010616).

References

1. Roussopoulos, N., Kelly, S., Vincent, F.: Nearest neighbor queries. In: Proc. ACM SIGMOD (1995)
2. Korn, F., Muthukrishnan, S.: Influence sets based on reverse nearest neighbor queries. In: Proc. ACM SIGMOD (2000)
3. Papadias, D., Shen, Q., Tao, Y., Mouratidis, K.: Group nearest neighbor queries. In: Proc. ICDE (2004)
4. Zhang, D., Chan, C.Y., Tan, K.L.: Nearest group queries. In: SSDBM (2013)
5. Choi, D.W., Chung, C.W.: Nearest neighborhood search in spatial databases. In: ICDE (2015)
6. Spatial Data Generator by Yannis Theodoridis

Choreography Retrieval from the Korean POP Dance Motion Capture Database with Low-Cost Depth Cameras

Dohyung Kim, Minsu Jang, Youngwoo Yoon and Jaehong Kim

Abstract This paper proposes a method for searching a target choreography fraction from a motion capture database of the Korean POP (K-POP) dance. The proposed retrieval system allows users to create their own query sequences by performing dance with low-cost depth cameras. This intuitive search interface is essential for a retrieval of K-POP dance motions that have no official names for unit motions. As a method to describe and measure complex and dynamic dance poses, we utilize a relative angles between joints of interest. For speed up of matching motions, the two-phase approach is proposed which involves fast selection of candidates with key poses and precise comparison between motion segments by using Dynamic Time Warping method. The experimental results on a large database demonstrate that the performance of the system is a sufficiently practical level for real-world applications.

Keywords Choreography retrieval · Korean pop dance · Motion capture · Depth camera · Dynamic Time Warping

1   Introduction

As the amount of the motion capture data increases, retrieval technology for the purpose of motion reuse has attracted considerable attention [1]. Various studies on motion retrieval have been explored also for choreography motions including ballet, classical dance, and traditional dance [2].

The methods for searching target segments of dance motions are threefold. First, methods based on text query like a title of a song, official names for unit dances or choreographer's name are still most popular. However, motion indexing

D. Kim(✉) · M. Jang · Y. Yoon · J. Kim
Electronics and Telecommunications Research Institute, Cheonan-si, Korea
e-mail: {dhkim008,minsu,youngwoo,jhkim504}@etri.re.kr

© Springer Science+Business Media Singapore 2015
D.-S. Park et al. (eds.), *Advances in Computer Science and Ubiquitous Computing*,
Lecture Notes in Electrical Engineering 373,
DOI: 10.1007/978-981-10-0281-6_113

has to be preceded before the retrieval, which requires high-cost operation. Second, target poses created by users' sketches or virtual characters can be utilized, but it is not easy for users to generate a target motion accurately from those poses [3]. Third, query motion based approaches can be considered as a promising solution especially with users' dance motion captured from 3D depth sensors [4]. These methods have the advantage of easy generation of the target motion and an intuitive search interface.

This paper proposes a motion capture retrieval system from a large database consisting of complex and dynamic motions of K-POP dances especially with user-created query sequences.

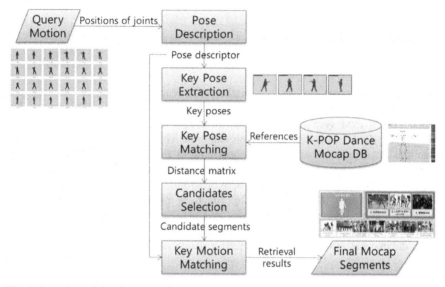

Fig. 1 Overview of the choreography retrieval from the K-POP dance database.

2   Choreography Retrieval System

The flow of the proposed choreography retrieval system from the K-POP dance database is shown in Fig.1.

A user creates a query motion of a sequence of 3D depth maps with low-cost depth cameras like Microsoft Kinect and then the 3D positions of a set of joints in the human skeleton can be estimated by skeleton trackers from the depth maps [5][6].

The sequence of the joints positions are fed into the pose description module that specifies a human pose of each frame. Among the tracked joints shown in Fig. 2 (left), the system takes the 3D positions of the body joints, 1st degree joints, and 2nd degree joints as a pose descriptor. The 1st degree joints means five joints articulated with the body joints and the 2nd degree joints are four joints at the end of the legs and arms.

The system then analyzes the sequence of the pose descriptors and detects several key poses from the query motion consisting of a sequence of poses. The key poses are highly discriminative poses representing a motion segment and can be detected a kind of clustering methods.

The key pose matching module matches each detected key pose with all poses in the K-POP dance database and records distances on the matching distance matrix. Fig. 2 (right) shows how to calculate the matching distance between two poses. A lower part of a right arm of each pose can be represented as a vector with its parent joint (right elbow) located at the origin point O in a three-dimensional space. Parent joints of the 2nd degree joints are the 1st degree joints.

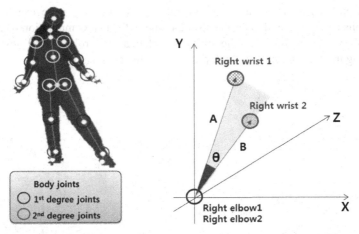

Fig. 2 Joints information used for describing a pose (left) and the distance measure between two body parts (right).

The matching distance between the two body parts can be computed with an equation of $(1-cos\theta)/2$ by finding the angle θ between vector A and B. For example, when the directions of the two vectors are completely equal $(\theta=0)$ the matching distance of the two body segments is 0. For each pose, nine body parts can be generated from the pose descriptor and the distance of each corresponding body parts of two poses is calculated in the same way. The sum of the nine distance values is finally the matching distance between two poses.

By analyzing the matching distance matrix of the key poses, candidate segments where key poses group together within can be selected from the dance database. The role of the candidate selection module is obtaining high scalability of the choreography retrieval system through detecting candidates segments at a high speed.

The matching distance matrix can be illustrated as Fig. 3. The figure shows the matching distance between each of four key poses and all poses in the motion capture database. In order to select candidate segments, the system scans all the dance segments in the database with the time window (red dotted line) and computes the sum of the minimum matching distances of key poses for the distance of the segment on the window. After calculating all of the distances of the segments in the database, the segments are sorted base on the matching distance from lowest to highest and top N segments are chosen for a final set of candidate segments.

The key motion matching module yields final retrieval results by comparing key motions of the query dance motion to all the candidates segments. The key motion means a local motion of about two or three seconds including several previous and next frames of the key pose. In motion retrieval, the discriminating power of pose information taken from single frame is generally lower than that of motion information extracted from multiple frames. In our approach, the local motion embodying information of pose change of the key pose can be highly discriminative. The candidate selection module is for fast computation of the pose dissimilarity by using a simple measure, whereas the role of the key motion matching module is obtaining high precision of the retrieval system.

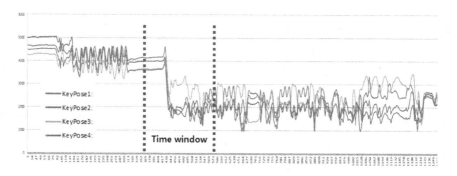

Fig. 3 Matching distance between key poses and all poses in the database.

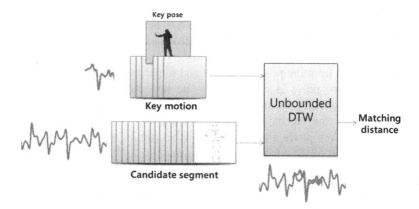

Fig. 4 Subsequence matching distance between single key motion and one candidate segment.

The proposed system makes use of Dynamic Time Warping (DTW) algorithm to get the matching distance between single key motion and one candidate segment. DTW is a popular method for measuring similarity between two sequences of data points which may vary in time or speed. In our case, since the key motion is a fraction of the full query sequence, the matching module needs to solve subsequence matching problem. Therefore we cannot directly use the classical DTW method with boundary restrictions on the warping function. As shown in Fig. 4, instead, we employ an unbounded or open-being-end DTW (OBE-DTW) algorithm which allows the comparison of incomplete input time series with complete references [7]. The subsequence matching distances are computed between all of the key motions and single candidate segment by using the unbounded DTW and the sum of those distances are finally the distance between the query sequence and the candidate segment.

3   Experimental Results

For the evaluation of the proposed method, we built a large motion capture database including 100 dance performances of K-POP songs. The motion data was captured for more than 20 professional dancers in an exclusive studio equipped with 12 infrared cameras. In total, 644,135 frames were recorded for about 358 minutes of performance. For a test set, we captured query sequences by using the Kinect for Windows v2 sensor of Microsoft. Two popular dancing parts from each K-POP song were selected as query segments. In total, 200 segments are gathered from four persons and the average performance time of the query dances was 5.6 seconds. Fig. 5 shows sample depth images and tracked joints of interest.

In the experiment, 200 candidate segments were selected for the input query sequence by the candidate selection module. The precision of the candidate selection, the rate of the true segment existing in the candidate segments, was 97.5%. Fig. 6 shows the precision of the retrieval system by top-X hit rate. The proposed system provides the retrieved segments in ascending order according to the matching distance. The precision of the proposed system at rank 10, the percentage of the target segment existing in the top 10 list of the final retrieval results, was 92%. At rank 20, it was 95%. Considering complexity and degree of variation of the K-POP dance motion, the experimental results show that our system has highly competitive performance in motion retrieval area. Meanwhile, the average response time was 4.7 seconds.

Fig. 5 Sample depth maps and tracked joints captured from Microsoft Kinect

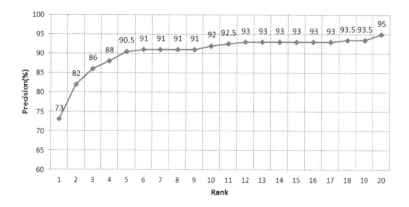

Fig. 6 The precision of the retrieval system by top-X hit rate (rank).

4   Conclusion and Future Work

This paper addressed a method for searching a target choreography fraction from a large motion capture database of K-POP dance. The proposed retrieval system featured with intuitive search interface which allows users to create their own query sequences by performing dance with low-cost depth cameras. The key pose based approach enabled fast retrieval of complex and dynamic motions of K-POP dances. To boost a search speed, the two-phase approach involving fast selection of candidate and precise comparison between motions was proposed.

For improvement of the retrieval accuracy and response time, we plan to develop other pose descriptors which are robust to errors of a skeleton tracker and investigate an algorithm of approximate nearest neighbor search and fast DTW method in the near future.

Acknowledgment This research is supported by Ministry of Culture, Sports and Tourism(MCST) and Korea Creative Content Agency(KOCCA) in the Culture Technology(CT) Research & Development Program 2015.

References

1. Deng, Z., Gu, Q., Li, Q.: Perceptually consistent example-based human motion retrieval. In Proceedings of the 2009 Symposium on Interactive 3D Graphics and Games, pp. 191–198. ACM (2009)
2. Essid, S., Lin, X., Gowing, M., Kordelas, G., Aksay, A., Kelly, P., Fillon, T., et al.: A multimodal dance corpus for research into real-time interaction between humans in online virtual environments. In: ICMI Workshop on Multimodal Corpora For Machine Learning (2011)

3. Chao, M.-W., Lin, C.-H., Assa, J., Lee, T.-Y.: Human motion retrieval from hand-drawn sketch. IEEE Transactions on Visualization and Computer Graphics 18(5), 729–740 (2012)
4. Hu, M.-C., Chen, C.-W., Cheng, W.-H., Chang, C.-H., Lai, J.-H., Ja-Ling, W.: Real-Time Human Movement Retrieval and Assessment With Kinect Sensor. IEEE Transactions on Cybernetics 45(4), 742–753 (2015)
5. Shotton, J., Sharp, T., Kipman, A., Fitzgibbon, A., Finocchio, M., Blake, A., Cook, M., Moore, R.: Real-time human pose recognition in parts from single depth images. Communications of the ACM 56(1), 116–124 (2013)
6. Chang, J.Y., Nam, S.W.: Fast Random-Forest-Based Human Pose Estimation Using a Multi-scale and Cascade Approach. ETRI Journal 35(6), 949–959 (2013)
7. Tormene, P., Giorgino, T., Quaglini, S., Stefanelli, M.: Matching incomplete time series with dynamic time warping: an algorithm and an application to post-stroke rehabilitation. Artificial Intelligence in Medicine 45(1), 11–34 (2009)

Enhancing PIN Input for Preventing Eavesdropping in BLE Legacy Pairing

Jun Kwon Jung and Tai Myong Chung

Abstract IoT services are growing up and up. Therefore many devices for IoT services are needed. One of spotlighted items is beacon with Bluetooth. a beacon can many commercial IoT services by Bluetooth technology. However, Bluetooth pairing model has exploit that the hacker can eavesdrop Bluetooth communication. We focus on Bluetooth pairing on LE legacy. In this paper, we propose the way to defending eavesdropping on Bluetooth LE environment. Our proposal is 3 million times to original pairing.

Keywords Bluetooth · Pairing security · Computational entropy

1 Introduction

The use of services of IoT(Internet of Things) is growing up and up now and many scientists are interested in IoT. IoT is extensive concept of information technology. Thus, IoT services use various devices for IoT services. Recently, most focused item for IoT services is beacon. Beacon is wireless communication device and it communicates other device by Bluetooth technology. Newest beacon is based on Bluetooth 4.0(BLE: Bluetooth Low Energy). Beacon is used for some trial commercial services such as advertising, payment, and so on. As the use of beacon is increasing, interesting of security on beacon is also increasing. Especially, the security of beacon is focused on beacon communication, thus

J.K. Jung(✉)
Department of Electrical and Computer Engineering,
Sungkyunkwan University, Suwon, Korea
e-mail: jkjung@imtl.skku.ac.kr

T.M. Chung
College of Information and Communication Engineering,
Sungkyunkwan University, Suwon, Korea
e-mail: tmchung@skku.edu

© Springer Science+Business Media Singapore 2015   813
D.-S. Park et al. (eds.), *Advances in Computer Science and Ubiquitous Computing*,
Lecture Notes in Electrical Engineering 373,
DOI: 10.1007/978-981-10-0281-6_114

beacon security is concentrated to Bluetooth communication. Bluetooth is against various security attacks. However, Bluetooth cannot perfectly defend all security attacks. In communication, Bluetooth can be sniffed each communication message. Also there are some attack utilities to attack Bluetooth communication. In this paper, we focus eavesdropping on Bluetooth. On Bluetooth specification, it comments that Bluetooth LE legacy has some vulnerabilities of eavesdropping. Thus, Bluetooth has defending technology for eavesdropping at least, we propose how to reinforce defending power of eavesdropping.

Section 2 introduces Bluetooth 4.0 and its vulnerabilities. We show known Bluetooth architecture and its security issues. Section 3 shows Bluetooth eavesdropping possibility and simple solution for eavesdropping. Section 4 calculates that proposed solution is so safe to measure computational power. Then, we mention conclusion and future work.

2 Bluetooth

2.1 Architecture

Bluetooth is a wireless communication standard. It is invented by Ericsson in 1994. It uses various small mobile devices. Bluetooth is suitable for short and simple message communication. Thus, it is used to many human closed devices and services such as earphone, headphone, mouse, keyboard, speaker and so on. Bluetooth is closely smartphone, because the smartphone is relaying and providing service data to Bluetooth devices.

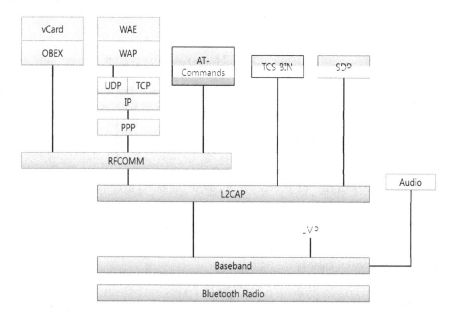

Fig. 1 Bluetooth architecture

Bluetooth has a number of released versions. Recently, Bluetooth 4.0 is released at 2010. Bluetooth 4.0 is called to BLE and this version is focused on utility of mobile environment and consuming low power. Bluetooth is based on basic architecture and has different detail architecture on each version. Figure 1 shows Bluetooth basic architecture.

Bluetooth supports many other standard protocols such as TCP, UDP, SDP and so on. Bluetooth is used various mobile device and it is used to control, I/O, simple data sending. Bluetooth is so useful communication standard. However, it has also some security problems.

2.2 Security Issues

Bluetooth has some security issues. It is a wireless communication standard, thus Bluetooth cannot remove communication exploits perfectly. Well-known security exploits of Bluetooth are bluejacking, bluesnarfing, bluebugging and so on. However, we focus on eavesdropping, not named attack. Of course, Bluetooth provides eavesdropping defending techniques, but it has a loophole of eavesdropping because it provides various association models.

Bluetooth provides four association models: Numeric Comparison, Just Works, Out Of Band, and Passkey Entry. Each association model is designed for own scenario. In these models, Just Works and Passkey Entry model are vulnerable to eavesdropping. However, Just Works model is very simple and it is designed for no more user input, so it is used to headset as simple devices. Thus, Just Works model can be excepted to consider eavesdropping due to insensitive data exchanging. However, Passkey Entry model supports keyboard input for security PIN number. PIN number is a 6-digit number and it uses Bluetooth pairing. If a hacker gets or guesses PIN number, this Bluetooth paired communication is eavesdropped by the hacker. Thus, PIN number's entropy is very important component.

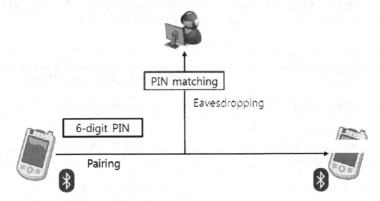

Fig. 2 LE legacy pairing eavesdropping

3   Increasing Bluetooth PIN's Entropy

Bluetooth LE legacy provides 6-digit PIN number. This PIN number's entropy is 1,000,000 because each number can select 0~9. The simple way is increasing PIN digits. However, this way is not clever. Because other association model uses also 6-digit PIN number. To modify PIN digit counts for security on one model inflicts harm to other model's service and gives users the inconvenience. Thus, it is absurd that way is increasing PIN digit count. However, user can use PIN number using user input, the user has various type of keyboard, mostly touch keyboard. Keyboard has not only number input but also alphabet and special characters. Thus, if Bluetooth supports characters on each digit, PIN's entropy is considerably increasing.

Recently, many inputs of Bluetooth service are touch keyboard in smartphone. This keyboard supports alphabet and special characters mandatory. Another type of keyboard is old-fashioned keypad like figure 3. This type may support alphabet or may not support it. However, this type utilizes alphabet input by feature phone keyboard. Each number key is matched some alphabets and special characters. User can input PIN character code push a button repeatedly.

4   Computational Entropy Comparison

Only 6-digit numbers can present 000000~999999. It's entropy is one million. This entropy is not sufficient to prevent eavesdropping Bluetooth pairing. In our proposal, 6-digit PIN can present alphabet and special characters. Alphabet is 26 characters. If PIN can distinguish uppercase and lowercase, each PIN is available 62-kind input. Also, PIN can present special characters, just 6-digit PIN can have more entropy. ASCII characters present all printable number, alphabet, and special characters. ASCII characters are 125 characters except space. Thus, 6-digit PIN will have 3.8 quadrillion entropy value and it is more million times that of default PIN's entropy. This value is close that SSP(Secure Simple Pairing) which is Bluetooth secure default secure mode use 16 numeric digit PIN.

Table 1 Entropy comparison

PIN mode	Computational number of attempts
6 numeric digit	1,000,000
6 ascii digit (number + alphabet + special character)	3,814,697,265,625
16 numeric digit	10,000,000,000,000,000

5 Conclusion

For IoT services, beacon is useful device. Bluetooth is communication standard in beacon, Bluetooth has eavesdropping exploits on LE pairing. In this paper, we propose Bluetooth pairing entropy increasing way by variable keyboard input. We calculate proposed pairing entropy is a few million times to default pairing way. However, it is needed to Bluetooth interface modification to support our proposal although algorithm is simple. In future work, we research how to apply our proposal concept to Bluetooth. We expect Bluetooth LE is more safe to eavesdropping by our proposal.

Acknowledgment Following are results of a study on the "Leaders in INdustry-university Cooperation" Project, supported by the Ministry of Education.

References

1. Spill, D., Bittau, A.: BlueSniff: eve meets alice and bluetooth. In: Proceedings of USENIX Workshop on Offensive Technologies (WOOT), August 2007
2. Jakobsson, M., Wetzel, S.: Security weaknesses in bluetooth. In: Naccache, D. (ed.) CT-RSA 2001. LNCS, vol. 2020, pp. 176–191. Springer, Heidelberg (2001)
3. Bluetooth Technology Website. http://www.bluetooth.com/
4. Aissi, S., Gehrmann, C., Nyberg, K.: Proposal For Enhancing Bluetooth Security Using an Improved Pairing Mechanism. Presented to the Bluetooth Architecture Review Board at the Bluetooth All-Hands Meeting, April 19–23, 2004
5. Legg, G.: The bluejacking, bluesnarfing, bluebugging blues: Bluetooth faces perception of vulnerability, August 2005. http://www.wirelessnetdesignline.com/192200279?printableArticle=true
6. Struman, C.F., Bray, J.: Bluetooth: Connect Without Cables. Prentice Hall, Englewood Cliffs (2001)
7. Scarfone, K., Padgette, J.: Guide to Bluetooth Security: Recommendations of the National Institute of Standards and Technology. Special Publication 800-121, National Institute of Standards and Technology (NIST), U.S. Department of Commerce (2008)
8. Vainio, J.: Bluetooth Security. Helsinki University of Technology (2000). http://www.cse.tkk.fi/fi/opinnot/T-110.5190/2000/bluetoothsecurity/bluesec.html (accessed 16 June, 2011)
9. Bialoglowy, M.: Bluetooth Security Review, Part I. Retrieved from http://www.symantec.com/connect/articles/bluetooth-security-review-part-1 (accessed, 19 June, 2011)
10. Shaked, Y., Wool, A.: Cracking the bluetooth PIN. In: 3rd USENIX/ACM Conf. Mobile Systems, Applications, and Services (MobiSys), pp. 39–50 (2005)
11. Laurie, A., Holtmann, M., HeMrfurt, M.: Bluetooth Hacking: The State of the Art. Retrieved from http://trifinite.org/Downloads/trifinite.presentation22c3berlin.pdf (accessed 10 June, 2011)

Erasure Codes Encoding Performance Enhancing Techniques Using GPGPU Based Non-sparse Coding Vector in Storage Systems

Kwangsoo Lee, Huiseong Heo, Teagun Song and Deok-Hwan Kim

Abstract In this paper, we propose a new GPGPU based erasure codes encoding method using non-sparse coding vector. The proposed method can be applied to various erasure codes and efficiently utilize GPU cores and GPU memory by using non-sparse coding vector instead of large sparse coding matrix, and erasing multiplication operation and reducing unnecessary XOR operations. The experimental results show that the average encoding times of SPC, RDP, EVENODD erasure codes using the proposed method are 39.26%, 37.5%, 37.85% less than traditional GPU based encoding, respectively.

Keywords Storage system · GPGPU · Erasure codes · Non-sparse coding vector

1 Introduction

Erasure Codes is mainly used to generate parity data in RAID system such as RAID-4, 5, 6, and to recover data when the data is lost. Erasure codes uses less storage space than creating replicas of using mirroring (RAID-1). It provides excellent performance in terms of resiliency and I/O. However, it requires a large amount of XOR and multiplication operations so that CPU should be held to encode or decode data for a long time [1].

In order to redeem this disadvantages, recently, GPGPU based erasure coding techniques have been emerged. There are two categories of enhancing erasure coding performance using GPGPU as follows. The former reduces data transfer time between main memory and GPU memory by using hardware such as APU and shared memory [2]. The latter uses GPGPU algorithms to fasten up the

K. Lee · H. Heo · T. Song · D.-H. Kim(✉)
Department of Electronic Engineering, Inha University, Incheon, South Korea
e-mail: {lks89,hhs89,stg90}@iesl.inha.ac.kr, deokhwan@inha.ac

© Springer Science+Business Media Singapore 2015   819
D.-S. Park et al. (eds.), *Advances in Computer Science and Ubiquitous Computing*,
Lecture Notes in Electrical Engineering 373,
DOI: 10.1007/978-981-10-0281-6_115

process of repetitive internal multiplication and XOR operations. But, existing works can only be applied to specific erasure codes [3].

In this paper, we propose a new GPGPU based erasure codes encoding method using non-sparse coding vector to enhance encoding performance in storage systems.

2   Erasure Codes

Erasure Codes is mainly used to generate parity data in RAID system such as RAID-4, 5, 6 to recover data when the data is lost. Erasure codes uses less storage space than creating replicas of using mirroring (RAID-2) and 3x-replication. There are several studies of which its I/O and reliability performances are better than data replication technique [3]. So, erasure codes is widely used in the storage system using multiple disks.

Fig. 1 shows the bit matrix operation and its data structure for generating parity disks, where the number of data disks, k, is 4, the number of parity disks, m, is 1 and word w is 4. At first, it chunks the input data into data chunks with the number of data disks × word size. Then, it performs XOR operation between the data chunk D and identity matrix I and distributes results to k data disks, and performs XOR operation between the data chunk and coding matrix H and then generates m parity disks.

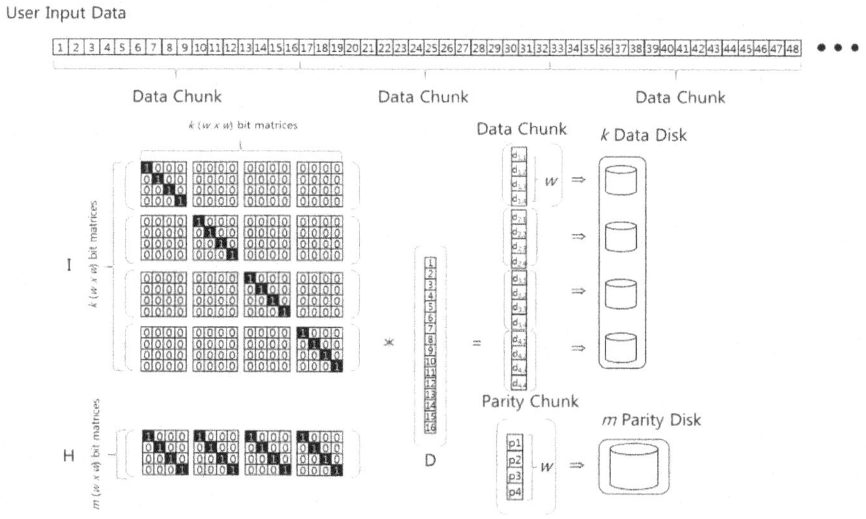

Fig. 1 CPU based bit matrix multiplication where identity matrix I, coding matrix H, data matrix D, the number of data disks k, the number of parity disks m, the number of elements w.

3   GPU Based Erasure Codes Encoding Using Non-sparse Coding Matrix

Fig. 2 shows the Non-Sparse Coding Matrix (NSCM) corresponding to the coding matrix of a given erasure codes. In Fig. 2, there are coding matrix of (a) SPC, (b) RDP, and (c) EVENODD and corresponding non-sparse coding matrix of (d) SPC, (e) RDP, (f) EVENODD.

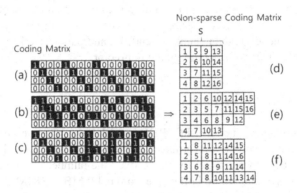

Fig. 2 CM and NSCM for various erasure codes, The coding matrix(CM) of (a) SPC, (b) RDP, (C) EVENODD, corresponding non-sparse coding matrix(NSCM) of (d) SPC, (e) RDP, (f) EVENODD where s is the number of non-zero elements

Here, we can find that most of coding matrix for encoding operation is sparse and multiplication and XOR operations between elements with a value of 0 are useless[4]. First of all, the proposed method is searching for a valid element with a value of 1 for the encoding operation in the original coding matrix, and then configure new non-sparse coding matrix by filling all elements in the matrix with their column index whose value is nonzero. The values in the column of the elements with a value of 1 represent the location values of the data elements for proceeding XOR operations during encoding process.

$$S_i = \text{size of NSCM}_i \text{ row.} \tag{1}$$

$$P_i = \bigoplus_{j=1}^{s_i} d_{\,NSCM\,i,j} \tag{2}$$

A row vector of NSCM represents data elements needed to generate a single parity element. Therefore, as shown in Eq. (2), a parity element is calculated by performing XOR operations between data elements whose location is given by index value of NSCM.

The process of performing GPU-based erasure codes encoding using non-sparse coding matrix is as follows. Initially, elements of non-sparse coding matrix are copied to GPU memory. For GPU based encoding, data must be transferred to the GPU Memory, from Main Memory. Therefore, GPU based encoding algorithm utilizes the stream processing mechanism, when transfer of a data stream (vector d_j) is performed to generate the parity p_i. The vector d_j is allocated into the GPU memory and then the GPU kernel is invoked. Each GPU thread executes XOR operations for a single element of the vector d_{NSCMj}, indexed by non-sparse coding matrix and corresponding elements of coding vectors pi. Notice that it does not need multiplication operations for parity generation whereas existing encoding technique requires multiplication and XOR operation together. Generated parity is transferred back to main memory from GPU memory. Upon completion of encoding, data chunks, metafiles, and parity chunks are written to k data disks and m parity disks through the redirector.

4   Experimental Results

Various erasure codes such as SPC, RDP and EVENODD code are implemented to evaluate the encoding performance of the proposed method on NVIDIA GeForce GTX670. In our experiment, we fixed the number of data disks (k) as 4, word size (w) as 4, and varied buffer size from 0.375MB to 384MB, which is used to send data from main memory to GPU Memory. With respect to various buffer size, we repeated measurement 15 times and averaged the result.

Traditional CPU based encoding spends most of time to perform multiplication and XOR operations by accessing data in main memory whereas the proposed GPU based encoding method needs time to perform the XOR operation on GPU Memory. Additionally, it needs to transfer data from main memory to GPU memory for generating parity and vice versa for writing parity to disk.

$$\text{Gain of encoding time} = (\frac{\text{CPU Encoding time - GPU Encoding time}}{\text{CPU Encoding time}}) * 100 (\%) \qquad (3)$$

Fig. 3 shows gain of encoding time of SPC, RDP and EVENODD erasure codes with respect to various buffer size. When buffer size is small (0.375MB, 0.75MB), it shows negative gain (SPC -60%, RDP -66%, EVENODD -51%) because address translation overhead in GPU is larger than improvement of encoding time between GPU and CPU. When buffer size is larger than or equal to 1.5MB, it shows positive gain and increases gain when buffer size increases up to 48MB. The average gains are 50.9%, 58.8%, 60% in terms of SPC, RDP and EVENODD.

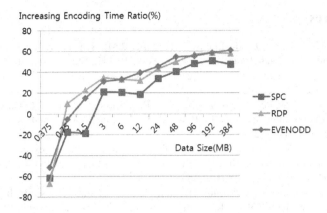

Fig. 3 Gain of Encoding time

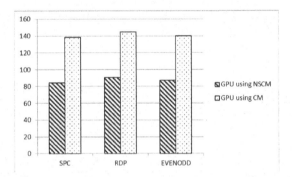

Fig. 4 Encoding time using traditional coding matrix and non-sparse coding matrix

Fig. 4 shows the average encoding time of erasure codes at GPU environment. The encoding time of the proposed method using NSCM is 39.26%, 37.5%, 37.85% better than existing GPU encoding method using CM. It is because there is no multiplication and the number of XOR operations are reduced.

5　Conclusion

This paper presents GPGPU based NSCM encoding technique. It can be applied to various types of erasure codes and can reduce the number of multiplication and XOR operations in GPGPU based encoding process. The Encoding gain of the proposed method increases when buffer size is larger than 1.5MB.

References

1. Pirahandeh, M., Kim, D.-H.: Energy-aware GPU-RAID scheduling for reducing energy consumption in cloud storage systems. In: Park, J.J.J.H., Stojmenovic, I., Jeong, H.Y., Yi, G. (eds.) Computer Science and Its Applications. LNEE, vol. 330, pp. 705–712. Springer, Heidelberg (2015)
2. Roman, W., Woźniak, M., Kuczyński, L.: Efficient Execution of Erasure Codes on AMD APU Architecture. Parallel Processing and Applied Mathematics, pp. 613–621. Springer, Heidelberg (2014)
3. Wyrzykowski, R., Kuczynski, L., Wozniak, M.: Towards efficient execution of erasure codes on multicore architectures. In: Jónasson, K. (ed.) PARA 2010, Part II. LNCS, vol. 7134, pp. 357–367. Springer, Heidelberg (2012)
4. Maheswaran, S., et al.: Xoring elephants: novel erasure codes for big data. In: Proceedings of the VLDB Endowment, vol. 6, no. 5. VLDB Endowment (2013)

FAIR-Based Loss Measurement Model for Enterprise Personal Information Breach

Jang Ho Yun, In Hyun Cho and Kyung Ho Lee

Abstract Loss measurement for personal information breach incidents can be used as a basis for decision making for information security investments. In this vein, reasonable loss measurement is important in determining information security policies. However, the previous research is focused on estimating the amount of loss which is incurred after incidents. In order to be base data for decision making, loss measurement should include incident-causing-factors before incidents occur. In this paper, we propose a loss measurement model based on an improved FAIR (Factor Analysis of Information Risk) risk analysis methodology. Additionally, we verify the effectiveness of the proposed model by applying it to a large scale personal information leakage case.

Keywords Loss measurement model · Personal information breach · Security policy

1 Introduction

Personal information breaches which have occurred recently in South Korea had a critical impact not only on related enterprises, but also nationally. For the last 3 years, about 10 million 4 thousand of personal information have been identified to leak out. Considering the South Korean population of 50 million, it is thought that one Korean's personal information was leaked at least twice.

Breached personal information is being used for spamming, voice phishing, telemarketing, and so forth, which is incurring damages to numerous users.

However, there are few studies and data on the loss of personal information breach in South Korea. It is important to estimate loss reasonably, because precise loss estimation can cover economic and social loss, and consequently decision-making can be facilitated for responsive measures against breaches.

J.H. Yun · I.H. Cho · K.H. Lee(✉)
Korea University Anam Campus, Anam-Dong 5-Ga, Seongbuk-Gu, Seoul, Korea
e-mail: {jang_ho,ihcho13,kevinlee}@korea.ac.kr

© Springer Science+Business Media Singapore 2015
D.-S. Park et al. (eds.), *Advances in Computer Science and Ubiquitous Computing*,
Lecture Notes in Electrical Engineering 373,
DOI: 10.1007/978-981-10-0281-6_116

825

FAIR risk analysis methodology is fit for assessing risk quantitatively. In this methodology, risk factors are specified as loss factors and probability factors, and data of similar past breaches is reflected upon those factors to estimate the magnitude of loss. But this methodology does not consider the timing of loss amount, and enterprise-related information such as the characteristics of an enterprise whose data is breached, the kinds of data, and the past usage of data. In this paper, we propose a model for enterprises to estimate loss of data breach before an incident occurs through an improved FAIR methodology. We verify the effectiveness of the proposed model by applying it to a large scale personal information leakage case in South Korea.

2   Previous Research

2.1   South Korea

In South Korea, research on loss estimation has been done actively since the beginning of 2000. Yoo(2008) categorized damage which can be caused by internet data breaches and calculated loss amount by analyzing factors which consist of damage [1]. Lee(2009) calculated loss amount when information of enterprise technology is leaked out [2]. Although his research is not on loss calculation of personal information leakage, he proposed a loss measurement model for enterprise technology leakage. Han(2011) attempted to calculate economical loss by conducting surveys instead of considering loss amount from users' perspective [3]. Yoo(2014) utilized double-bounded dichotomous choice methodology to pinpoint factors which determine costs to prevent personal information breaches [4].

National organizations also did active research on the related area. Korea Online Privacy Association and Korea Internet & Security Agency estimated social cost of personal information breach by considering social ramifications and various methodologies, and proposed some policy agendas such as the estimation of appropriate amount of investment, in order to minimize social cost [5-7].

2.2   The World

The Ponemon research institute in the US has conducted active research on the loss measurement of personal information breaches. The institute started off with a study which suggests to conceptualize direct and indirect cost of personal information breaches to calculate loss (2009). In 2014, it released a reasonable loss measurement methodology to consider various indirect costs including decrease in customer reputation, and breakaway customers [8].

USENIX, a computing systems organization which was founded in 1975, proposed cost analysis models through I-CAMP I , II (Incident Cost Analysis and Modeling Project). I-CAMP I demonstrated a model for analyzing cost incurred by enterprise insider threats and I-CAMP II added the user-side to the pre-existing model [9], [10].

JNSA(Japan Network Security Association) publishes annual reports on the calculation of loss incurred by personal information breaches. Initially, JNSA calculated loss amount by taking into account direct and indirect cost. Since 2010, it has made a shift toward measuring basic value of breached data and recently published a novel methodology to consider stock prices in loss calculation [11].

2.3   FAIR (Factor Analysis of Information Risk) Methodology

FAIR was proposed by Risk Management Insight in 2005 and has been used in leading corporations. FAIR- ISO/IEC 27005 cookbook expounds on FAIR methodology. Unlike the pre-existing methodologies which mainly calculate risk qualitatively, FAIR methodology develops them by subdividing and analyzing risk scenario factors, considering expected probabilities of risk scenarios, and thereby calculating risk quantitatively. The quantified risk value, or loss can facilitate decision-making [12], [13].

Figure 1 shows the process of the pre-existing FAIR methodology. As a whole, they are divided into frequency section and loss section. At the end, loss is calculated by obtaining expected value.

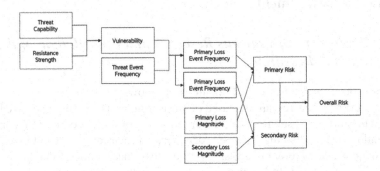

Fig. 1 FAIR process

All components indicated in the process are given a value from level 1 to level 5 through qualitative assessment. The definitions for the components are given as follows.

Threat Event Frequency (TEF) means the number of attempts for an insider to leak personal information. Threat Capability (TCAP) indicates an insider's capability. The higher threat capability is, the more capable an insider is. Resistance strength (RS) is security policies and controls which are set in an organization in order for a malicious insider not to be able to breach data. Vulnerability indicates how vulnerable related information systems are to threats. It is calculated by considering TCAP and RS.

Primary loss event frequency (PLEF) is the number of occurrence of data leakage by an insider and has five rating levels for which different occurrence probability intervals are given. PLEF is used to calculate the maximum and minimum value of loss. Secondary loss event frequency (SLEF) is the number of occurrence of indirect loss when asset loss occurs. Indirect loss does not always occur, but occurs conditionally after direct loss. In this sense, PLEF is considered when SLEF is determined. Primary loss (PL) is loss cost which stakeholders directly bear because of asset loss. Second loss (SL) is loss cost which stakeholders indirectly bear because of asset loss. Primary risk (PR) is risk which stakeholders directly bear because of asset loss. Qualitative PR can be calculated by considering PLM and PLEF and categorized into 5 levels. Quantitative PR can be calculated by considering probability intervals of LEF in PL category.

Secondary risk (SR) is risk which stakeholders indirectly bear because of asset loss. Qualitative SR can be calculated by considering SLM and SLEF and categorized into 5 levels. Quantitative SR can be calculated by considering probability intervals of LEF in SL category. Overall risk (OR) is a total amount of risk. It can be calculated by considering PR and SR, and categorized into 5 levels. Quantified amount of PR and SR is considered to calculate loss caused by data leakage by an insider [14].

3   Loss Measurement Model

3.1   Limitations of Previous Research and Contribution of the Current Research

The previous research has some limitations as follows:

First, the previous research focused on measuring costs which are incurred after incidents such as response cost, labor cost, image recovery cost, recurrence prevention cost, and so on. However, decisions on security policy and investment should be made before incidents. For this, loss measurement should be done before incidents occur. In this paper, we utilize a risk analysis methodology which can be applied to likely future incidents.

Second, the pre-existing FAIR methodology considers limited number of relevant factors in calculating risk. In particular, there is no factor which reflects enterprise environment before incidents. In this paper, we add enterprise-related factors to the FAIR methodology.

Third, timing issue should be handled in order to calculate loss amount by using FAIR methodology. As mentioned above, loss should be measured before incidents. But the pre-existing FAIR methodology uses past data as it is. In this paper, we recalculate past data or future cost from the present point.

3.2 Probability Interval Model for Enterprise Environment

The FAIR model considers the two following aspects to obtain probability intervals of incident occurrence.

The first aspect is threat agents. Depending on the kind of personal information breaches, threat agents differ in terms of purpose, number of attempts, accessibility and so on. So, it is vital to define and analyze threat agents of data breaches reasonably. For this, we follow the pre-existing FAIR methodology. We consider TEF and TCAP to determine threat exposure rating. We use the matrix operation (Notation for this operation is). Detailed criteria for TCAP, TEF are given in Table 3.

The second aspect of consideration is enterprise environment. In order for personal information breaches to occur, personal information which enterprises have is sensitive to individuals and can have economic impact. This aspect will be acronamed as EP. EP can be rated according to (1) and Table 1 [11].

$$\text{Min [EP = Economy Rating + Mental Rating -1]} \qquad (1)$$

Table 1 Privacy Sensitivity classification table (EP MAP)

Economic Distress Level	1	Account number & PIN, Credit card Number expiration date, Financial website ID&PW	Last will and testament	Criminal record, Criminal history, Credit blacklist
	2	Passport information , Purchase records, ISP ID &PW, Credit card number, Financial website ID, Seal certificate	Salary income class, assets, Buildings, land, balance, loans, home income, loan records	-
	3	Name, address, birth date, sex, financial institution name, resident card code, email, health insurance policy number, pension policy number, license number, employee number, member number, telephone number, handle name, pension plan information, company name, school name. job title, occupation, job description, height, weight, blood type, physical information, voice print physical fitness examination	Physical examination, mental health tests, Personality test, pregnancy history, operation history, nursing care record, examination record, physical disability certificate, DNA, sickness history, fingerprint, receipt, dialect, hobbies, work history, grades, test score, mail content, location information	Political party, political opinions, labor union membership, beliefs, creeds, religion, faith, permanent address, symptoms, medical chart, dementia, physical handicaps, learning disability, mental disability, infections, sexual propensities, sex life
		3	2	1
		Metal Distress Level		

Table 2 Table of personal identification (ES)

Rating	Contents
1	Individual may be easily identified ("Name" and "Address" are included)
2	Individual may be identified after certain costs are incurred
3	Difficult to identify the individual Oater than that described above

Also, the more personally-identifiable personal information is, the more impact it has in times of incidents [11]. This aspect will be acronamed as ES. Information classification criteria in Table 1 and 2 are based on those of JNSA which has implemented systematic information classification. Rating levels have been revised according to the proposed model. Finally, security degree of enterprise information system is considered (RS). This is measured based on technical and administrative measures which are given in 'Act on Promotion of Information and Communication Network Utilization and Information Protection, etc.'

After determining ratings for EP, ES, and RS, matrix operation is done according to the process. Values for matrix operation change according to weighted values for EP, ES and RS.

3.3 Loss Measurement Model Where Timing Issue Is Considered

There are two cases where there is a timing issue in loss measurement. First, it is when past data itself is used to calculate loss. Second, it is when future loss is calculated from the present point.

For the first point, suppose that we calculate response cost among direct costs at a time of incident occurrence, t, based on the Ponemon institute's formula. Considering that the institute released its report at a time of t`, we recalculate the response cost with the formula (2).

$$Response^* = \int_r^{r`} \int_t^{t`} Response \times e^{r(t`-t)} \, dt \, dr \quad (2)$$

In (2), 'r' should be a reasonable rate for the cost. It is reasonable to use average wage increase rate for response cost.

3.4 Improved FAIR Methodology Process

Based upon improvements mentioned in Section 3.2 and 3.3, the improved FAIR methodology process can be drawn as in Figure 2.

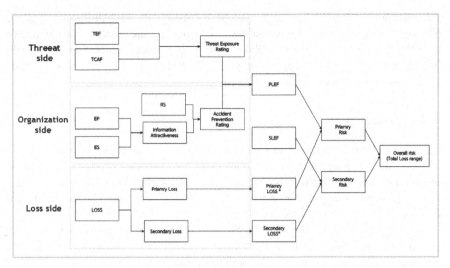

Fig. 2 Processes of damage scale measurement model

In calculating probability intervals, the four factors which an enterprise can control such as sensitivity of information, degree of personal identification, information attractiveness, asset vulnerability. In terms of threat agents, two factors are considered for capability and the number of attack attempts.

In determining loss magnitude, direct and indirect loss are considered, and for each cost present values are assigned.

Overall risk is calculated as a form of a range, based on probability intervals of PLEF and SLEF.

4 Conclusion

4.1 Case Study and Application of the Proposed Model

In 2011, a personal information breach occurred at a South Korean company that an alleged Chinese hacker leaked around 35 million personal information. Breached
-information includes ID, password, resident registration number, name, date of birth, email address, phone number, address, and so forth. Investigation revealed that the hacker had targeted an affiliated company which was prone to hacking, and circumvented via its update server to reach and leak personal information of the victim company [15].

According to the ruling for damages, the accused company had received hundreds of hacking reports annually since a few years before, and vulnerabilities in security systems were pinpointed frequently. Based upon this fact, about 2,882 victims filed a class action suit for damages incurred by the breach and won the case partially. Overall loss is measured by using the probability interval in. Through the application to the past case, we calculate the overall loss magnitude as 3,989 billion won ~ 4,124 billion won. This is much more than about 500 billion won, or an estimated loss of personal information breach in the 3 major card companies in South Korea [16].

4.2 Conclusion

The proposed loss measurement model is significance in several respects.

First, the proposed model calculates loss by considering characteristics of information and assets, the existence of threat agents, the kind of information, and so on. Loss measurement prior to security incidents can provide basic data for determining security investment and policy.

Second, the proposed model is flexible enough to be customized for any relevant organization. FAIR methodology is flexible in considering new factors because it calculates loss magnitude with basic factors.

Third, the proposed model outputs loss magnitude as an interval. Loss measurement is done to estimate future loss and it is hard to estimate the loss as a certain amount. Loss magnitude given at an interval can be used for enterprise decision making process.

The proposed model is a model which reflects enterprise aspects which the previous research failed to reflect. We expect that through the model loss measurement will be more precise and sophisticated, and that security policy and investment will be facilitated.

Acknowledge This research was supported by the MSIP (Ministry of Science, ICT and Future Planning), Korea, under the ITRC (Information Technology Research Center) support program (IITP-2015-R0992-15-1006) supervised by the IITP(Institute for Information & communications Technology Promotion).

References

1. Yoo, J.H., et al.: Estimating Economic Damages from Internet Incidents. Information Society **15**(1) (2008)
2. Lee, K.H.: Study on the Model for Estimation of Financial Loss due to the Industrial Information Leakage. Unpublished doctoral dissertation, Korea University Graduate School of Information Management Engineering (2009)
3. Han, C.H., et al.: A Quantitative Assessment Model of Private Information Breach. Journal of Society for e-Business Studies **15**(4) (2011)
4. You, S.D., et al.: Determinants of Willingness to Pay for Personal Information Protection. Journal of The Korea Institute of Information Security & Cryptology **24**(4) August 2014
5. Korea Online Privacy Association, Analysis of Social Costs in Personal Information Value and Leakage, Research report, September 2013

6. Chai, S.W.: Internet Infringement Accident Damage Costs Calculated Model. SIS 2006-KISA Symposium, June 2006
7. Song, H.I., et al.: Analysis of The Economic Value of Personal Information by the CVM Method. KISA Internet & Security Focus, Focus 2, May 2014
8. Ponemon Institute. http://www.ponemon.org/blog/ponemon-institute-releases-2014-cost-of-data-breach-global-analysis (2015.09.04.)
9. Rezmierski, V., Carroll, A., Hine, J.: A Study on Incident Costs and Frequencies, USENIX Research report, August 2000
10. USENIX, Incident Cost Analysis and Modeling Project I-CAMP II, Report
11. Japan Network Security Association, 2009 Survey Report of Information Security Incident, Report, pp. 48, September 2010
12. FAIRWIKI, september 04, 2015. http://fairwiki.riskmanagementinsight.com
13. The Open Group, ISO/IEC 27005 Cookbook, Technical Guide, pp. 6, October 2010
14. Freund, J., Jones, J.: Measuring and Managing Information Risk A FAIR Approach, pp. 26 (2015)
15. Im Cha. http://navercast.naver.com/contents.nhn?rid=2871&contents_id=81880, Column
16. Company Guide. http://comp.fnguide.com.SVO2/asp/SVD_ijanal.asp?pGB=1&gicode=A066270&cID=&MenuYn=Y&ReportGB=&NewMenuID=110&stkGb=701

Variational Bayesian Inference for Multinomial Dirichlet Gaussian Process Classification Model

Wanhyun Cho, Soonja Kang, Sangkyoon Kim and Soonyoung Park

Abstract In this paper, we propose the variational Bayesian inference algorithm which can drive approximate posterior distributions of both three latent functions and two parameters needed to define the multinomial Dirichlet Gaussian process (GP) classification model. This model consists of three components: a latent function with GP prior, a response function with multiclass, and a link function that relates the latent function and response mean. Here, we consider the variational Bayesian estimation method to estimate the proposed model. This is performed in two parts: one is to derive the variational posterior distribution of auxiliary variables and latent function, another is to derive the variational posterior distribution for the various parameters. Moreover, we have proposed a classification rule that can predict a particular category for a new observation by using the trained model. Finally, we conducted experiment using a well-known Iris data in order to verify the performance of the proposed model. Experimental result reveals that the proposed model shows good performance on this data set.

Keywords Multinomial dirichlet gaussian process classification model · Variational bayesian inference · Approximate posterior distribution · Important sampling method · Iris dataset

1 Introduction

Gaussian processes can conveniently be used to specify prior distributions of latent function for Bayesian inference. In the case of regression with Gaussian

W. Cho(✉) · S. Kang
Department of Statistics and Mathematical Education,
Chonnam National University, Gwangju, South Korea
e-mail: {whcho,sjkang}@chonnam.ac.kr

S. Kim · S. Park
Department of Electronic Engineering, Mokpo National University, Chonnam, South Korea
e-mail: {narciss76,sypark}@mokpo.ac.kr

© Springer Science+Business Media Singapore 2015

D.-S. Park et al. (eds.), *Advances in Computer Science and Ubiquitous Computing*,
Lecture Notes in Electrical Engineering 373,
DOI: 10.1007/978-981-10-0281-6_117

noise, inference can be done simply in closed from, since the likelihood function is Gaussian. But in the case of classification, Gaussian likelihood is inappropriate because the target values are discrete class labels. One prolific line of attack is based on approximating the non-Gaussian likelihood with a tractable Gaussian distribution.

Various solutions have been suggested in recent literature [1-2]. These are Laplace Approximations and Expectation Propagation, Kullback-Leibler divergence minimization comprising variational bounding as a special case, and a factorial approximation. First, Williams et al. [3] proposed the use of a second order Taylor expansion around the posterior mode to a natural way of constructing a Gaussian approximation to the log-posterior distribution. The mode is taken as the mean of the approximate Gaussian. Second, Minka [4] presented a new approximation technique ("Expectation Propagation") in Bayesian networks. This is an iterative method to find approximations based on approximate marginal moments, which can be applied to Gaussian processes. Third, Opper et al. [5] discussed the relationship between the Laplace and the variational approximation and they also considered problem that minimizes the KL-divergence measure between the approximated posterior and the exact posterior. Fourth, Girolami & Rogers [6] introduced the multinomial probit regression model with Gaussian Processes priors, and employed a variational Bayes approximation to drive factored posterior for multinomial probit regression. Fifth Csato et al.[7] presented three simple approximations for the calculation of posterior mean in Gaussian process classification.

In this paper, using the similar idea with Girolami & Rogers, we propose new variaitonal Bayesian inference algorithm that can drive simultaneously posterior distributions of both three latent function and parameters in the multinomial Dirichlet Gaussian process classification model. The proposed algorithm is performed in two steps. The first step is to drive approximate variational posterior distributions for various hidden functions several parameters. The second step is to classify new sample using posterior distribution of the latent function derived based on learning data. Moreover, we have conducted the experiments using the Iris data in order to verify the performance of the proposed algorithm.

2   Multinomial Dirichlet Gaussian Process Classification Model

We consider the multinomial Dirichlet Gaussian process classification model (MDGPCM) with multiclass. The model consists of three components: a latent function with Gaussian Process (GP) prior, a response function with multiclass, and a link function that relates the latent function and response mean.

The graphical representation of the conditional dependency structure for the multinomial Dirichlet classification model with GP priors that we will consider in this paper is shown in Figure 1.

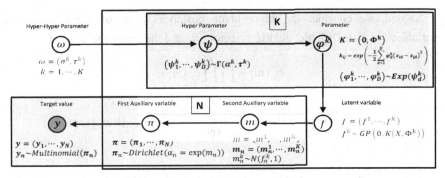

Fig. 1 Multinomial Dirichlet Gaussian Process Classification Model.

And if we define a set of the hidden variables as $\Theta = \{\boldsymbol{\pi}, \mathbf{m}, \mathbf{f}\}$ and the parameters as $\Phi = \{\boldsymbol{\varphi}^{k=1,\cdots,K}, \boldsymbol{\psi}^{k=1,\cdots,K}\}$, the joint likelihood function for all variables and parameters takes the following form:

$$p(\mathbf{Y}, \Theta, \Phi \mid \mathbf{X}, \omega) = \prod_{n=1}^{N} p(\mathbf{y}_n \mid \boldsymbol{\pi}_n) p(\boldsymbol{\pi}_n \mid \boldsymbol{\alpha}_n)$$
$$= \exp(\mathbf{m}_n)) \times \prod_{k=1}^{K} p(\mathbf{m}^k \mid \mathbf{f}^k) p(\mathbf{f}^k \mid K_{(N \times N)}^k (\mathbf{X}, \boldsymbol{\varphi}^k)) p(\boldsymbol{\varphi}^k \mid \boldsymbol{\psi}^k) p(\boldsymbol{\psi}^k \mid \omega^k)$$

where the individual factors are

$$p(\mathbf{Y} \mid \boldsymbol{\pi}) = \prod_{n=1}^{N} \mathrm{Mult}(\mathbf{y}_n \mid 1, \boldsymbol{\pi}_n = (\pi_n^1, \cdots, \pi_n^K)) \quad p(\boldsymbol{\pi} \mid \boldsymbol{\alpha}) = \prod_{n=1}^{N} \mathrm{Dir}(\boldsymbol{\pi}_n \mid \boldsymbol{\alpha}_n = \exp(\mathbf{m}_n))$$

$$p(\mathbf{m} \mid \mathbf{f}) = \prod_{k=1}^{K} \prod_{n=1}^{N} N(m_n^k \mid f_n^k, 1) \quad p(\mathbf{f} \mid \mathbf{X}, \boldsymbol{\varphi}) = \prod_{k=1}^{K} \mathrm{MN}(\mathbf{f}^k \mid 0, K_{(n \times n)}^k (\mathbf{X}, \boldsymbol{\varphi}^k)).$$

$$p(\boldsymbol{\varphi} \mid \boldsymbol{\psi}) = \prod_{k=1}^{K} \prod_{d=1}^{D} Exp(\varphi_d^k \mid \psi_d^k) \quad p(\boldsymbol{\psi} \mid \omega) = \prod_{k=1}^{K} \Gamma(\boldsymbol{\psi}^k \mid \omega^k = (\sigma^k, \tau^k))$$

3 Variantial Bayesian Inference

3.1 Variational Approximate Posterior Distribution

First, we consider the variational approximate posterior $q^*(\boldsymbol{\pi})$ for the parameter vector of classification probabilities $\boldsymbol{\pi}$.

$$q^*(\boldsymbol{\pi}) = \prod_{n=1}^{N} q^*(\boldsymbol{\pi}_n), \, q^*(\boldsymbol{\pi}_n) \sim \mathrm{Dir}(\boldsymbol{\pi}_n \mid \boldsymbol{\beta}_n^*), \, \boldsymbol{\beta}_n^* = (\beta_n^{1*}, \cdots, \beta_n^{K*}),$$

where $\beta_n^{k*} = y_n^k + E_{q(m_n^k)}(\exp(m_n^k))$.

Second, we consider the approximate posterior $q^*(\mathbf{m})$ of normal auxiliary variables $\mathbf{m}$ over Dirichlet auxiliary vector $\boldsymbol{\pi}$ and the latent variables $\mathbf{f}$.

$$q_1^*(\mathbf{m}) = \prod_{n=1}^{N}\prod_{k=1}^{K} q_1^*(m_n^k),$$

where $q_1^*(m_n^k) = \dfrac{r(m_n^k)}{\sum\limits_{l=1}^{K} r(m_n^l)}$, $r(m_n^k) = \dfrac{1}{\Gamma(\exp(m_n^k))}\exp(\mathrm{E}_{q(\pi_n)}[\ln \pi_n^k]\exp(m_n^k))$,

and $\mathrm{E}_{q^*(\pi_n)}[\ln \pi_n^k] = \upsilon(\beta_n^{k*}) - \upsilon(\sum_{l=1}^{K}\beta_n^{l*})$, $\upsilon(x)$ is the digamma function.

Hence, taking exponential both sides of the formula for $q_2^*(\mathbf{m})$ yields

$$q_2^*(\mathbf{m}) = \prod_{n=1}^{N}\prod_{k=1}^{K} q_2^*(m_n^k) ,$$

where the distribution $q_2^*(m_n^k)$ is given by the normal distribution $N(m_n^k \mid \mathrm{E}_{q(f_n^k)}(f_n^k),1)$. Finally, by combine $q_1^*(\mathbf{m})$ and $q_2^*(\mathbf{m})$, we have the approximate posterior $q^*(\mathbf{m})$ of auxiliary latent variables $\mathbf{m}$ given as the following form

$$q^*(\mathbf{m}) = \prod_{n=1}^{N}\prod_{k=1}^{K} q^*(m_n^k),\ q^*(m_n^k) = q_1^*(m_n^k)\times q_2^*(m_n^k) .$$

Therefore, the required posterior expectation $E(m_n^k)$ can be computed as the following manner by using importance sampling method:

$$E(m_n^k) = \int m_n^k q^*(m_n^k)dm_n^k = \int m_n^k q_1^*(m_n^k)\times q_2^*(m_n^k)dm_n^k \approx \sum_{l=1}^{L} m_n^k w(m_n^{k(l)}),$$

where each $m_n^{k(1)},\cdots,m_n^{k(L)}$ are random samples drawn from $N(m_n^k \mid \mathrm{E}_{q(f_n^k)}(f_n^k),1)$, and

$$w(m_n^{k(l)}) = \dfrac{q_1^*(m_n^{k(l)})}{\sum_{s=1}^{L} q_1^*(m_n^{k(s)})} .$$

Third, we consider the approximate posterior $q^*(\mathbf{f})$ of latent variables $\mathbf{f}$ over normal auxiliary variables $\mathbf{m}$ and the parameters $\boldsymbol{\varphi}^{k=1,\cdots,K}$. Finally, we have obtained the approximate posterior $q^*(\mathbf{f})$ for latent function $\mathbf{f}$:

$$q^*(\mathbf{f}) \propto \prod_{k=1}^{K} N\big(\mathbf{f}^k \mid \mathrm{E}(\mathbf{m}^k),\mathbf{I}\big) N\left(\mathbf{f}^k \mid 0,\big(\mathrm{E}_{q(\varphi^k)}[\mathrm{K}_{(N\times N)}(\varphi^k)^{-1}]\big)^{-1}\right) ,$$

and so we have

$$q^*(\mathbf{f}) = \prod_{k=1}^{K} N\left(\mathbf{f}^k \mid \mathrm{E}(\mathbf{f}^k), \Sigma^k\right),$$

where $\mathrm{E}(\mathbf{f}^k) = \Sigma^k E(\mathbf{m}^k)$, and $\Sigma^k = \mathrm{K}_{(N \times N)}^k (\mathrm{E}_{q(\varphi^k)}(\varphi^k))(\mathbf{I} + \mathrm{K}_{(N \times N)}^k (\mathrm{E}_{q(\varphi^k)}(\varphi^k)))^{-1}$.

Fourth, if we also consider the set of hyper-parameters $\Phi = \{\varphi^{k=1,\cdots,K}, \psi^{k=1,\cdots,K}\}$, in this variational treatment, then the expectation of the covariance kernel hyper-parameters $\varphi^{k=1,\cdots,K}$ under the variational posterior distribution $q(\varphi^k)$ can be approximated by drawing S samples such that each $\varphi_s^{kd} \sim Exp(E_{q(\psi^{kd})}(\psi^{kd}))$ and so

$$\mathrm{E}_{q(\varphi^k)}[\varphi^k] \approx \sum_{s=1}^{S} \varphi_s^k w(\varphi_s^k), \quad w(\varphi_s^k) = \frac{N(E_{q^*(\mathbf{f}^k)}(\mathbf{f}^k) \mid 0, \mathbf{K}(\mathbf{X}, \varphi_s^k))}{\sum_{v=1}^{S} N(E_{q^*(\mathbf{f}^k)}(\mathbf{f}^k) \mid 0, \mathbf{K}(\mathbf{X}, \varphi_v^k))}.$$

Finally, the approximate posterior for hyper-parameters ψ_d^k is given as the gamma distribution with parameters $\sigma^k + 1$ and $\tau^k + \mathrm{E}_{q(\varphi_d^k)}(\varphi_d^k)$. And the required posterior mean values are given as follow $E_{q(\psi_d^k)}(\psi_d^k) = \dfrac{\sigma^k + 1}{\tau^k + \mathrm{E}_{q(\varphi_d^k)}(\varphi_d^k)}$.

3.2   Predictive Classification Method for New Sample

We consider the problem of producing a classification probability $p(y_{new} = k \mid \mathbf{x}_{new}, \mathbf{X}, \mathbf{y})$ for a new sample $\mathbf{x}_{new}$. First, the posterior distribution of latent function $\mathbf{f}_{new}$ corresponding with new point $\mathbf{x}_{new}$ is generally given by

$$q(\mathbf{f}_{new} \mid \mathbf{x}_{new}, \mathbf{X}, \Theta) = \int p(\mathbf{f}_{new} \mid \mathbf{x}_{new}, \mathbf{X}, \Theta, \mathbf{f},) q(\mathbf{f} \mid \mathbf{y}, \mathbf{X}, \Theta) d\mathbf{f} .$$

Here, since $p(\mathbf{f}_{new} \mid \mathbf{x}_{new}, \mathbf{X}, \Theta, \mathbf{f},)$ and $q(\mathbf{f} \mid \mathbf{y}, \mathbf{X}, \Theta)$ are both Gaussian, $q(\mathbf{f}_{new} \mid \mathbf{x}_{new}, \mathbf{X}, \Theta)$ will be also be Gaussian. Hence, we need only compute its mean vector and covariance matrix. The predictive mean vector for latent function $\mathbf{f}_{new}$ is given by

$$\begin{aligned} E_q(\mathbf{f}_{new}^k \mid \mathbf{X}, \mathbf{x}_*, \Theta, \mathbf{y}) &= Q(\mathbf{X}, \mathbf{x}_{new} \mid \mathrm{E}(\varphi))^T \mathbf{K}(\mathbf{X}, E(\varphi))^{-1} \boldsymbol{\mu}_\mathbf{f} \\ &= Q(\mathbf{X}, \mathbf{x}_{new} \mid \mathrm{E}(\varphi))^T (\mathbf{I} + \mathbf{K}(\mathbf{X}, E(\varphi)))^{-1} \mathrm{E}(\mathbf{m}) \end{aligned}$$

And the covariance matrix for latent function $\mathbf{f}_{new}$ is given by

$$\begin{aligned} Cov_q(\mathbf{f}_{new} \mid \mathbf{X}, \mathbf{y}, \Theta, \mathbf{x}_{new}) &= \mathrm{diag}(k_{new}^1, \cdots, k_{new}^K) \\ &- Q(\mathbf{X}, \mathbf{x}_{new} \mid \mathrm{E}(\varphi))^T (\mathbf{I} + \mathbf{K}(\mathbf{X}, E(\varphi)))^{-1} Q(\mathbf{X}, \mathbf{x}_{new} \mid \mathrm{E}(\varphi)) \end{aligned}$$

Second, we define the parameters $\alpha_{new}^{s} = (\exp(m_{new}^{1(s)}), \cdots, \exp(m_{new}^{K(s)}))$ of the Dirichlet probability distribution using the extracted samples $(m_{new}^{1(s)}, \cdots, m_{new}^{K(s)})$ and we extract again the classification probabilities vector $\pi_{new}^{(s)} = (\pi_{new}^{1(s)}, \cdots, \pi_{new}^{K(s)})$ from the Dirichlet distribution $\pi_{new}^{(s)} \sim Dir((\pi_{new}^{1(s)}, \cdots, \pi_{new}^{K(s)}) \mid (\exp(m_{new}^{1(s)}), \cdots, \exp(m_{new}^{K(s)})))$. Repeating this procedure S times, we have generated a total S number of classification probabilities vector $\pi_{new}^{(1)}, \cdots, \pi_{new}^{(S)}$. And using them, we calculate the mean of the classification probabilities vectors as follows:

$$\overline{\pi}_{new} = \frac{1}{S}(\pi_{new}^{(1)} + \cdots + \pi_{new}^{(S)})$$

Finally, after we try to find the class C^k that have a maximum of classification probability $\overline{\pi}_{new}^{k}$, we classify new observation x_{new} into this class C^k. That is, $\overline{\pi}_{new}^{k} = \max\{\overline{\pi}_{new}^{1}, \cdots, \overline{\pi}_{new}^{j}, \cdots, \overline{\pi}_{new}^{K}\}$ implies the new sample x_{new} classify into class C^k.

4   Experimental Results

Here, we considered a real data which is called Iris dataset. This dataset consists of 50 samples from each of three species of Iris flowers that are setosa, versicolor and virginica. Four features were measured from each sample, that are the length and the width of sepal and petal, in centimeters. Based on the combination of the four features, we developed a GP classifier model to distinguish the species from each other.

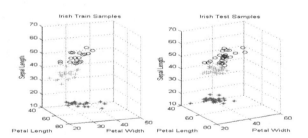

Fig. 2 Graphical Representation of Iris Dataset

Figure 2 shows the Iris dataset at different viewpoints. First, in order to train a model, we have used totally ninety observations that come from three classes. And in order to verify the performance of the model, we have selected sixty samples except ones used in the training step. Next, we want to measure the performance of our proposed model to classify the Iris species.

Table 1 shows the results of the Iris species classification. To calculate the rates, we estimate the number of correctly classified negatives and positives and divide by total number of each species. We have to try many experiments to get meaningful results using randomly selected samples.

Table 1 Classification of Iris species for test samples

	setosa	versicolor	virginica
setosa	1	0	0
versicolor	0	0.64	0.36
virginica	0	0	1

5   Conclusion

This paper proposed a new inference algorithm which can drive simultaneously variaitional approximate posterior distribution of both latent functions and parameters in the multinomial Dirichlet Gaussian process classification model. The proposed algorithm was performed in two steps: each of which is the training step and testing step. First, in the training step, using the variaitonal Bayesian theory, we derived approximate the posterior distribution of the latent function and parameters on the basis of the learning data. Second, in the testing step, using a derived posterior distribution of the latent function, we classify a new sample into the category that have a maximum of classification probabilities. Finally, we have conducted the experiments by using the Iris data in order to verify the performance of the proposed algorithm. Experimental results reveal that the proposed algorithm shows good performance on these datasets. Our future work will extend the proposed method to other video recognition problems such as 3D human action recognition, gesture recognition, and surveillances system.

Acknowledgement This work was jointly supported by the National Research Foundation of Korea Government (2014R1A1A4A0109398) and the research fund of the Chonnam National University (2014-2256).

References

1. Rasmussen, C.E., Williams, C.K.I.: Gaussian Processes for Machine Learning. MIT Press (2006)
2. Nicklisch, H., Rasmussen, C.E.: Approximation for Binary Gaussian process Classification. Journal of Machine Learning Research **9**, 2035–2075 (2008)
3. Williams, C.K.I., Barber, D.: Bayesian Classification with Gaussian Processes. IEEE Tran. On PAMI **12**, 1342–1351 (1998)

4. Minka, T.P.: Expectation Propagation for Approximate Bayesian Inference. In: UAI, pp. 362–369. Morgan Kaufmann (2011)
5. Opper, M., Winther, O.: Gaussian Processes for Classification: Mean Field Algorithms. Neural Computation **12**, 2655–2684 (2000)
6. Girolami, M., Rogers, S.: Variational Bayesian Multinomial Probit Regression with Gaussian Process Priors. Neural Computation **18**, 1790–1817 (2006)
7. Csato, L., Fokoue, E., Opper, M., Schottky, B.: Efficient Approaches to Gaussian Process Classification. Neural Information Processing Systems **12**, 251–257 (2000)

A Log Regression Seasonality Based Approach for Time Series Decomposition Prediction in System Resources

Chul Kim, Sang-Hun Nam and Inwhee Joe

Abstract It has been challenging to predict data in terms of monitoring information technology (IT) resources. In order to obtain the quality and performance of products, changes can be detected and monitored setting up a fixed threshold value based on statistics and operation experiences. Monitoring data by a fixed threshold value may not work properly during none busy hours in exceptional situations whereas a usage change during busy hours can be detected. It is because it cannot reflect the trend of resource usage seasonality as a function of time. The technique based on Time Series Decomposition (TSD) can provide the one with appropriate methodology so that problems can be recognized and diagnosed and the correction can be made ahead of time by detecting a subtle status change of devices in massive IT resources. In this paper, we propose three approaches to predict data such as Intelligent Threshold, Abnormal Pattern Detection, time prediction of reaching target value; the appropriate trend detection of Time Series, optimal seasonality detection and technique using Log Regression Seasonality. The experimental data collected here exhibit that it can reflect the change over time to the prediction data improving its accuracy compared to existing TSD technique.

1 Introduction

Monitoring IT resources has attracted great attention since it is significant to recognize problems in advance and execute system recovery to maintain the

C. Kim(✉) · S.-H. Nam
NKIA Corperation, U space 1B, 10th floor, Daewangpangyoro 660 (Sampyeongdong), Bundang-gu, Seongnam city, Gyeonggi do, South Korea
e-mail: {ki4420,smilegun}@gmail.com

I. Joe
Hanyang Univresity, 222 Wangsimni-ro, Seongdong-gu, Seoul 04763, South Korea
e-mail: iwjoe@hanyang.ac.kr

© Springer Science+Business Media Singapore 2015   843
D.-S. Park et al. (eds.), *Advances in Computer Science and Ubiquitous Computing*,
Lecture Notes in Electrical Engineering 373,
DOI: 10.1007/978-981-10-0281-6_118

quality and availability of particular services. Its general approach is to chart the real-time trend of IT resources as a visual inspection and keep users updated just in case where something exceptional occurs based on threshold values Although the visual inspection can allow the once to analyze a variety of situations it is almost impossible to maintain the inspection performance all the time in reality. In order to compensate this drawback, the monitoring system gives the threshold values to each individual resource. In this paper, we propose that the technique based on Time Series Decomposition (TSD) can provide the one with appropriate methodology so that problems can be recognized and diagnosed and the correction can be made ahead of time by detecting a subtle status change of devices in massive IT resources. It is suggested that the data in the future can be predicted by utilizing a model made through mechanical analysis with resource usages. The data prediction technique using TSD can establish the structure that is easy to understand by analyzing several factors with observed data. We experimentally investigate the prediction accuracy and improvement by making a comparison between the new technique we suggest and existing technique.

In order to evaluate the technique suggested here, we collected the data by using Polestar(TM)[1] For realizing the techniques and visualization analysis, R 3.0.3[2] and Spotfire 7.0 desktop[3] are utilized respectively.

2   Related Work

TDS technique uses the average value of extracted seasonality in order to obtain the prediction data. In general, the trend extraction method in the observed data employs moving average.

2.1   Time Series Decomposition

If the future behavior of the system is not totally different from the past it is completed through three steps, characterization, modeling and forecasting. [5]

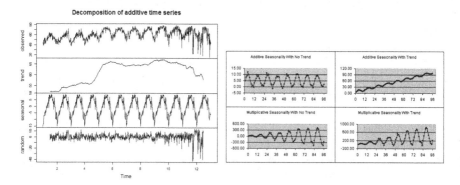

Fig. 1 . a. Time Series Decomposition by R Package {stats}[4] and b. Additive Seasonality and Multiplicative Seasonality[7]

When the data is analyzed by TSD it becomes made out of trend, seasonality and cyclical. Figure 1 shows the seasonality analysis chart of CPU usage using the decompose function of R stats package.[6]

2.2 Seasonality

If some data have a periodic pattern with certain repetition these can be regarded as the seasonality. The seasonality results from weather, vacation and holiday[8] and consists of general regularity and the predictable patterns.[9]

The seasonality is generally repeatable weekly, monthly and quarterly in large within one year.[10] It is oppositely characterized compared to the cyclical which is periodic for more than a year and fluctuates over 2 years.[11][12]

3 Detecting Trend and Seasonality, Forecasting

3.1 Detecting Trend and Seasonality

The first step of TSD modeling is to extract the trend. The trend describes the main stream of observed data and the TSD technique in general takes the trend using moving average. In order to detecting fitted trend, We calculated R^2 values of each model.

In the second step, the seasonality is extracted. Typical TSD sets up the seasonality hourly, daily, weekly, monthly or depending on a particular date. Most of IT resources have such seasonality, but sometimes follow the particular seasonality related to the role of the resource.

3.2 Log Regression Seasonality

The third step is the most important method that reflects to the data prediction by using the extracted seasonality. The existing TSD technique takes the average values of seasonal factors with the same index as the extracted seasonality into account. Utilizing the average value is useful when seasonality keeps a stable pattern over time. However, there are 3 major drawbacks of using the average value in general.

First of all, the phenomena that the amplitude of the prediction data is less than that of the observed data occur.

We suggest that the average seasonality used in the process of producing the prediction data is replaced by the log regression seasonality. This method can not only resolve the problems of the average seasonality mentioned above, but also improve the prediction accuracy through the experiments.

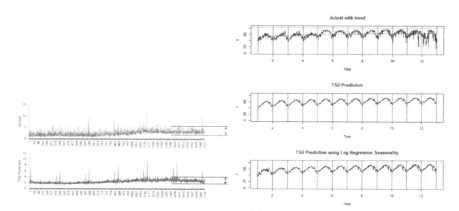

Fig. 2 a. Amplitude difference between of actual and the TSD Prediction and b. Uniformly Seasonality in TSD Prediction and Incremental Crack Appearance in TSD Prediction using Log Regression Seasonality. a shows the amplitude difference between two data sets. The data predicted by TSD has the narrower amplitude that the observed data resulting from the prediction data produced by the average value. b show the crack gets greater as time goes in the observed data while the seasonality with a stable pattern is kept in the prediction data. This occurs because of using average seasonality repeatedly.

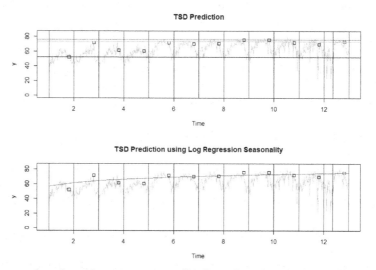

Fig. 3 upper chart shows the average seasonality that reflects the average of the factors with the same seasonal index as the observed data. Lower chart the prediction method using log regression is used.

4   Experiment

In order to assess Log Regression of Time Series Decomposition (LRTSD), the prediction accuracy measure the differences between the predicted and observed values.

The prediction accuracy(R^2) calculate as squared correlation coefficient. R2 equals the squared (absolute) Pearson correlation coefficient of the dependant and explanatory variable in an univariate linear least squares regression. In this case, the value is not directly a measure of how good the modeled values are, but rather a measure of how good a predictor might be constructed from the modeled values (by creating a revised predictor of the form $\alpha + \beta fi$). According to Everitt (p. 78) [16][17][18] [19].

Pearson's correlation coefficient when applied to a population is commonly represented by the Greek letter ρ (rho) and may be referred to as the population correlation coefficient or the population Pearson correlation coefficient. The formula for ρ[20] is:

$$R^2 = \rho^2 \tag{1}$$

4.1 The Prediction Accuracy: TSD vs. LRTSD vs. Simple Linear Regression TSD (SLRTSD)

The three different methods, TSD, LRTSD and SLRTSD, are compared in terms of the prediction accuracy in table 1. The prediction accuracy of LRTSD is higher than that of the existing TSD by 0.045 whilst SLRTSD has the worst accuracy among the three methods studied here.

Table 1 The Prediction Accuracy(R^2)

Observed	TSD	LRTSD	SLRTSD
actual1	0.3613698	0.4553304	0.2303279
actual2	0.5234975	0.5731348	0.4398654
actual3	0.1553385	0.2032027	0.09699809
actual4	0.620493	0.6248148	0.6206843
actual5	0.8693498	0.89282	0.8185788
actual6	0.2479894	0.2928472	0.1635196
actual7	0.2441821	0.2587832	0.2031347
actual8	0.09664329	0.1216329	0.08832122
actual9	0.06361754	0.1249853	0.1855435
actual10	0.1594074	0.2066528	0.13742
actual11	0.174527	0.3839953	0.00294194
actual12	5.99E-05	8.21E-05	0.2368767
actual13	0.1605005	0.2668139	0.06406119
actual14	0.1268307	0.1911757	0.08658126
actual15	0.9303289	0.9949372	0.6049848
actual16	0.3223334	0.5069629	0.0116205
actual17	0.2944435	0.5374753	0.06748486
actual18	0.376608	0.584577	0.1008989
actual19	0.1212291	0.163354	0.07343412
Average	0.307828913	0.388609341	0.222804094

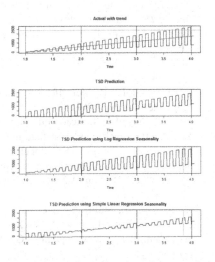

Fig. 4 shows the comparison between the observed data and the data created by using TSD and LRTSD and SLRTSD. Each Prediction Accuracy are 0.9329 and 0.9986 and 0.9437.

5 Conclusions

LRTSD is suggested since it has the high prediction accuracy compared to the existing TSD since it reflects the change of seasonality. Initially, linear regression method is used to predicted, but it is noted that it tends to be biased. As a result, it comes to have the low prediction accuracy. Since the log regression method can get rid of this biased trend it is suggested.

The empirical seasonality is detected based on the measured values by using TSD. It is found that log regression method is useful when the data is predicted.

Finally, This work was supported by Institute for Information & communications Technology Promotion(IITP) grant funded by the Korea government(MSIP) (No.2014-044-081-001, Development of performance failure prediction and analysis system for next-generation data center).

References

1. (주) 엔키아: 엔키아 OmniWorker 소개자료. 한국정보통신설비학회 학술대회, 437–447 (2004)
2. Gentleman, R., Ihaka, R., Bates, D.: The R project for statistical computing (2009). http://www.r-project.org/254
3. Ahlberg, C.: Spotfire: an information exploration environment. ACM SIGMOD Record 25(4), 25–29 (1996)
4. Kendall, M., Stuart, A.: The Advanced Theory of Statistics, vol. 3, pp. 410–414. Griffin (1983)
5. 지원철: 선경망을 이용한 시계열의 분해분석. 대한산업공학회지 25(1), 111–124 (1999)
6. Cowpertwait, P.S.P., Metcalfe, A.V.: Introductory time series with R. Springer Science & Business Media (2009)
7. Oracle® Crystal Ball, Fusion Edition (Version 11.1.1.1.00). Crystal Ball User's Guide, 11.1.2. http://oracle.com/crystalball/index.html (last accessed 16 August 2011)
8. Cancelo, J.R., Espasa, A.: Modelling and forecasting daily semes of electricity demand. Investigaciones economicas 20(3), 359–376 (1996)
9. Barnett, A.G., Dobson, A.J.: Analysing Seasonal Health Data. Springer (2010). ISBN 978-3-642-10747-4
10. Lakonishok, J., Smidt, S.: Are seasonal anomalies real? A ninety-year perspective. Review of Financial Studies 1(4), 403–425 (1988)
11. Hyndman, R.J., Athanasopoulos, G.: Forecasting: principles and practice. OTexts (2014)
12. Montgomery, D.C., Johnson, L.A., Gardiner, J.S.: Forecasting and time series analysis. McGraw-Hill Companies (1990)
13. Hylleberg, S.: Seasonality in regression. Academic Press (2014)
14. Garavaglia, S., Sharma, A.: A smart guide to dummy variables: four applications and a macro. In: Proceedings of the Northeast SAS Users Group Conference (1998)
15. Kruskal, W.H., Tanur, J.M. (ed.): Linear hypotheses. In: International Encyclopedia of Statistics, vol. 1, pp. 523–541. Free Press (1978). Evan J. Williams, I. Regression

16. Steel, R.G.D., Torrie, J.H.: Principles and Procedures of Statistics with Special Reference to the Biological Sciences. McGraw-Hill Companies (1960)
17. Glantz, S.A., Slinker, B.K.: Primer of Applied Regression and Analysis of Variance. McGraw-Hill Companies (1990). ISBN 0-07-023407-8
18. Draper, N.R., Smith, H.: Applied Regression Analysis. Wiley-Interscience (1998). ISBN 0-471-17082-8
19. Everitt, B.S.: Cambridge Dictionary of Statistics, 2nd edn. CUP (2002). ISBN 0-521-81099-X
20. a b c d e Real Statistics Using Excel: Correlation: Basic Concepts, retrieved 2015-02-22

A Study on the Algorithm Design for Improving the Accuracy of Decision Tree

ChangSeup Han, HeeJun Moon and ChangJae Kim

Abstract As the interest in IoT is increasing, various researches using IoT is being carried out. Research using IoT aims to provide appropriate services by recognizing user's condition. These researches utilizing IoT includes the function of supporting user's decision making, but it cannot provide services to user autonomously. In this study, in order to solve existing problems, we improve the accuracy of the decision tree and suggest algorithm offering services to user autonomously. Proposed algorithm consisted of error correction part and limit value processing part. Through the result of experiment, resolve the problems of existing research and show the higher accuracy than the rule-based decision making algorithm.

Keywords Decision tree · IoT · Sensor · Context-aware

1 Introduction

Recently, as the interest in IoT is increasing, IoT is utilized in various fields [1]. IoT based services aim to recognize the user's status through the sensor and provide appropriate services [2]. Also, in the function of the smart home platform includes recognition techniques in accordance with changing conditions. While this smart Home platform includes the ability to support the user's decision-making through sensor-based analysis, this decision making supporting system has a problem that it does not match the goal to provide IoT based services by determining the situation autonomously.

In this study, in order to overcome the problems of the conventional smart home platform, it is used on the basis of the decision tree algorithm to the sensor data. Based on the first sensor data, it becomes possible to compare the data

C. Han · H. Moon · C. Kim(✉)
Graduate School of Software, Soongsil University, Seoul 156-743, Korea
e-mail: suid4916@naver.com, aran0913@gmail.com, winchang@ssu.ac.kr

© Springer Science+Business Media Singapore 2015
D.-S. Park et al. (eds.), *Advances in Computer Science and Ubiquitous Computing*,
Lecture Notes in Electrical Engineering 373,
DOI: 10.1007/978-981-10-0281-6_119

collected later. If the data collected later is different from the reference value, increase the counter value. If the counter value exceeds the limit value, it will execute the appropriate service for the situation.

Decision tree algorithm is composed of error correction part and limit value processing part. When collecting data from the sensors, specific noise is generated. Therefore, in the error correction part, data that contains higher noise will be corrected by the filter. In the limit value processing part, set the limit value for each leaf node and perform specific service when the leaf node data value exceeds the limit value.

This paper is structured as follows. In Chapter 2, we list the domestic related study. In Chapter 3, we propose a decision tree algorithm. In Chapter 4, we perform experiments and analysis. Chapter 5 describes the conclusion of this paper.

2   Related Study

2.1   Decision Tree

Previous studies utilizing decision tree mainly focused on making the prediction of the particular situation [3], [4] and [5]. HyunKyung Kim from Chonbuk National University [3] used the decision tree classifying the investigation group. They looked into the group whose knowledge about a stroke is insufficient. ByongHui Jung from Incheon National University [4] used Rule base 1 utilizing expert rules with Fuzzy function and Rule base 2 utilizing decision tree. It shows higher accuracy than using a single Rule base. YoungJin Kim [5] predicted the probability of fire using the decision tree to classify the weather information.

2.2   IoT Data Processing

IoT, proposed by Auto-ID Center Director Kevin Ashton [6], is a network topology that forms the intelligent relationship mutually such as sensing, networking, information processing without human interference.

Studies for the IoT data processing place emphasis on increasing the reliability and efficiency of the sensor network by processing massive data generated from sensors efficiently [7], [8], [9], [10], and [11]. In the research of SeongYong Joo [7], to solve the previous problem caused by increased data quantity as the size of sensor network extended, suggested improved data collecting middleware to disperse overload and decreased network complexity. However it has the disadvantage of performance evaluation in accordance with the complexity of the network has not been performed. WooJin Joe [8] proposed environmental monitoring technology through smart data logger technique which is necessary for the large scale environmental sensor network for systematic observation of environmental pollution and climate change. But there is a lack of research to implement the generic architecture (general purpose architecture) and the battery efficiency of the sensors needs to be improved. Study of SeokMo Gu [9] has a feature that ensures credibility and immediacy by suggesting distributed data processing technique to process data effectively as the scale of IoT is enlarged. However, it suffers a problem that the test for dynamically changing environment depending on the number of sensors is insufficient. In the research of SoonHyun Kwon [10], they suggested real time and paralleled processing technique in order to solve the problem of real time and massive data processing in accordance with the convergence of semantic web technique and sensor network. As applied to Automated Weather Station (AWS), it ensures the higher efficiency to the larger data. However, nodes have to be dispersed to process a large amount of data. In the research of JeongWon Kim [11], in order to improve the reliability of the data generated by the sensor network, replicate the data that generated by the sensor node to a neighbor node which has higher battery and memory residual value. From that, the network stability is improved over time but the battery efficiency is lower than the copy is not maintained.

3   Improved Decision Tree

Composition of context aware tree presented in this paper is as shown in Fig. 1. Data received from the sensors is transmitted to a management system and appropriately revised through error correction. Sensor data amended by error correction is stored in data warehouse and a new node is created. Calculate the limit value of the new node and compare the counter value with the limit value. If the counter value is bigger than the limit value, the Dictator will give a command to smart device in order to provide services according to the user's situation. In addition, user is able to know their situation through Data Mart.

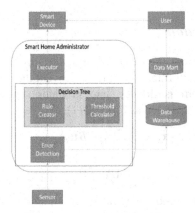

Fig. 1 Configuration diagram of context-aware tree algorithm

3.1 Algorithm Process

In this paper, context aware tree algorithm is constructed as shown in Fig. 2. After operating sensor, the sensor transmits specific data to server. If there is no root node, the first received data will be set as a data value of the root node. After that, if the data received from the sensor is different from the root node data value, compare them and set the data as a new child node and calculate the limit value of the node. If new data is same as the data in the leaf node, increase the counter value. And if the counter value exceeds the limit value, perform a particular service. That is, configure the tree using the data received from the sensors, calculate the limit values of each leaf node and then execute services when the result value exceeds certain counter value.

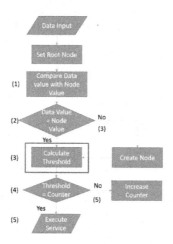

Fig. 2 The process of context-aware tree

3.2 Error Correction

When collecting data from the sensors, particular noise occurs. Therefore in this paper, we applied Kalman filter algorithm to prevent the noise of sensor data [12]. Kalman filter algorithm is divided into prediction step and correction step. Prediction step can be expressed as formula (1), (2).

$$\widehat{x_k^-} = A\widehat{x_{k-1}} + Bu_k \tag{1}$$

$$P_k^- = AP_{k-1}A^T \tag{2}$$

Equation (1) is the predictive value of situation, $\widehat{x_k^-}$ represents expected previous data based on the present point. A and B is state transition model to convert the state and u_k is prediction error noise. In addition, the initial estimated value of the equation (1) is set to the data value of the root node. Equation (2) is a step to obtain a correction value based on the collected data, P_k^- means predictive value of error covariance matrix at this point in time. At correction step, we adjust the actual received data through the filter and it can be expressed as equation (3), (4) and (5).

$$K_k = P_k^- H^T (H P_k^- H^T)^{-1} \tag{3}$$

$$\widehat{x_k} = \widehat{x_k^-} + K_k (z_k - H \widehat{x_k^-}) \tag{4}$$

$$P_k = (I - K_k H) P_k^- \tag{5}$$

Equation (3) is a formula for calculating the Kalman gain, H means state transition model. New data collected through equation (4). By adding noise z from the calculated data from equation (1), corrected data is generated through the filter. Equation (5) is a formula for calculating the error covariance; it corrects the current data when applying the filter.

3.3 Processing of the Limit Value

Limit value calculation part sets the information gain of existing decision tree as a limit value. And it can be expressed as equation (6).

$$IG = -\sum_{i=1}^{x} P(c_x/x) \log_2 P(c_x/x) - \sum_{i=1}^{x} N(c_x/x) \log_2 N(c_x/x) \tag{6}$$

Equation (6) is a formula for calculating the information gain and x is the total sum of all nodes composed of tree. c_x is the increment along the direction. $P(c_x/x)$ increases when they are bigger than the parent node and $N(c_x/x)$ increases when they are smaller than the parent node so that it can determine the limit value. For example, suppose the root node data is 26 and each child node data is 25, 27. When root node data value is 25, it is smaller than the root node value 26. Therefore the value N is increased and the limit value will be 0.5283. Also, in the case of 27, because it is bigger than the root node data value 26, value P will increase and 0.5283 is applied as a limit value. Moreover, the information gain value decreases if the tree becomes skewed tree. It can provide faster service than previous trees.

4 Test and Evaluation

To test decision tree algorithm based situational awareness technique, we focused on the number of service provision time. Also, to test decision tree algorithm, we generated decision tree that predicts only the heat and the cold using temperature sensor. Install a sensor in a room space and collect 1,000 data for experiment per

10 seconds. As shown in Fig. 3, data from the sensor are sent to the server and the server applies error correction to the collected data so that appropriate data can be inserted into the decision tree and stored in data mart. It is designed to notice user if new node is generated or counter value exceeds the limit value.

Fig. 3 Assumed scenario

4.1 Experimental Results and Analysis

In order to evaluate the decision tree algorithm, we carried out comparative evaluation with the existing rule-based decision-making algorithm. Set the temperature range for the operation of smart device using THI value from equation (7).

$$THI = ((9/5)T) - 0.55 * (1 - RH)((9/5)T - 26) + 32 \qquad (7)$$

Equation (7) is a formula for calculating the value of THI, T means the temperature value and RH means relative humidity. Humidity of the experiment site was measured to be 36%. The maximum value of temperature to operate the smart device is 29 °C based on the THI 75 where most people feel discomfort. And set the minimum temperature to stop the smart device at 23 °C based on the THI 68. Thus, rule-based decision-making algorithm is as shown in equation (8).

$$\text{Event} := \text{CurrentTemperature} \geq 29°C \ \text{Action} := \text{SmartDevice} \rightarrow \text{On}$$
$$\text{Event} := \text{CurreentTemperature} \leq 23°C \ \text{Action} := \text{SmartDevice} \rightarrow \text{Off} \qquad (8)$$

First, left side of Fig. 4 is an experimental result graph showing the distribution of temperature change per hour. There were irregular temperature change and temperature was heated 3 times in left side of Fig. 4. Right side of Fig. 4 refers to the number of data collected from the sensors. One data set at 23 degree Celsius, 176 data sets at 24 degree Celsius, 614 data sets at 25 degree Celsius, 168 at 26 degree Celsius, 28 at 27 degree Celsius and 13 data sets at 28 degree Celsius are collected. Totally, 1,000 data sets are collected.

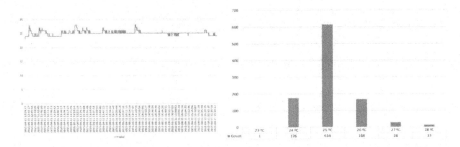

Fig. 4 The collected data through sensor

Fig. 5 shows decision tree generated by the data from the sensors. 23°C when the sensor activated first is generated at the root node, the data of each child node is created as 24°C, 27°C, 26°C, 25°C, and 28°C. The limit value of the leaf node is calculated by 1.05, 0.5. That indicates if the direction of the leaf node goes like skewed tree, the limit value decreases.

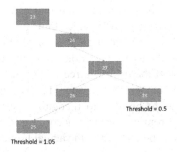

Fig. 5 The result of context-aware tree

Fig. 6 is a graph that compares rule-based decision-making algorithm to decision tree algorithm using the time required and the number of executions of particular services for each node. Because rule-based decision-making algorithm does not reach to the data condition for executing the service, the number of service execution is zero. When the leaf node data is 24, number of times of service execution until a new leaf node is created is 3. This service was to turn off the smart device. Number of times of service execution was 1 when the leaf node data is 27, this service was to turn on the smart device. Service was not performed when the data of leaf node is 26. There was 101 times of service execution at the leaf node data value 25 and this was the condition to turn on the smart device. Moreover, we found that as the decision tree for each threshold became biased, the reaction time for performing the service is reduced.

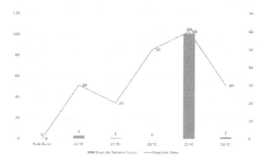

Fig. 6 Comparison in accordance with count of service execution and reaction time following node's threshold

5   Conclusions and Future Challenges

Previous rule-based decision-making algorithm, as it performs the service only if the data matches the rule, there is a problem that cannot perform the service in certain circumstances. Decision tree algorithm presented in this paper, by resolving existing problem and updating the threshold value flexibly, it could execute services that meets user's situation, and improve the user experience. However, the experiment of the decision tree algorithm only provides services by predicting situations with temperature sensor. The direction for future research, by improving accuracy so as to be able to recognize user's status through various sensors and executing appropriate services according to the different situations of the user, it seems that it can be applied to the actual home service.

References

1. Kang, J., Kim, H., Jun, M.: Market and Technical Trends of internet of things. International Journal of Contents **13**, 14–17 (2015)
2. Perer, C., Georgakopoulos, D.: Context aware computing for the internet of things: A survey. IEEE Communications Surveys & Tutorials **16**, 414–454 (2014)
3. Kim, H.K., Jeong, S., Kang, H.: Identification of Subgroups with Lower Level of Stroke Knowledge Using Decision-tree Analysis. Journal of Korean Academy of Nursing **44**, 97–107 (2014)
4. Jeong, B., Park, D., Jung, E., Kim, J., Choi, J.: Dempster-Shafer Theory based Clinical Decision Support System Using Clinical Knowledge and Rule Induction of Decision Tree. Journal of Korean Institute of Information Technology **11**, 85–90 (2013)
5. Kim, Y., Ryu, J., Song, W., Kim, M.: Fire Probability Prediction Based on Weather Information Using Decision Tree. Journal of KIISE **40**, 705–715 (2013)
6. That 'Internet of Things' thing in the real world, things matter more than ideas. http://www.rfidjournal.com/articles/view?4986
7. Joo, S.-Y., Lim, H.-S., Kang, D.-S.: A Middleware with Efficient Memory Management Technique and Advanced Structure for M2M Network. JKIIT **12**, 101–108 (2014)

8. Joe, W., Jiang, M., Jeong, K.: An M2M/IoT based Smart Data Logger for Environmental Sensor Networks. KIISE Transactions on Computing Practices (KTCP) **20**, 1–5 (2014)
9. Gu, S., Kim, Y.: A Distributed Data Processing Scheme Considering Dynamic Network on IoT Environment. The e-Business Studies **15**, 255–271 (2014)
10. Kwon, S., Park, D., Bang, H., Park, Y.: Real-time and Parallel Semantic Translation Technique for Large-Scale Streaming Sensor Data in an IoT Environment. Journal of KIISE **42**, 54–67 (2015)
11. Kim, J.: An Effective Data Distribution Scheme in Sensor Network for Internet of Things. The Journal of The Korea Institute of Electronic Communication Sciences **10**, 769–774 (2015)
12. Faragher, R.: Understanding the Basis of the Kalman Filter via a Simple and Intuitive Derivation. IEEE Signal Processing Magazine **29**, 128–132 (2012)

Design of a Smartphone-Based Driving Habit Monitoring System

Jin-Woo An, Daejin Moon and Dae-Soo Cho

Abstract Cars have supplied rapidly, and most households now own at least one car. With increase of interest in the car, many people also focus on the scheme to saving for maintenance cost and prevention of traffic accident. Vehicle manufacturers provide Telematics and Connected-car services, however, it is possible only the car equipped module for the services. And it is inevitable to pay additional high price cost. By 2014, there are only 7 million cars with module for Connected-Car service in the whole world. Each manufacturer provides independency and exclusive platform service. In this paper, we propose common service of driving habit monitoring. This can provide driving habit monitoring service to anyone who has a smartphone and a car equipped with low price IoT devices. We analysis driving habit using vehicle operating data, user activity data and GPS based on OBD-II and smart phone.

Keywords Connected-car · IoT · OBD-II · Driving habit monitoring

1 Introduction

The IoT(Internet of Things) device market is growing and it makes many examples applied at real life and industry. The IoT is the network of physical objects or "things" embedded with electronics, software, sensors, and connectivity to enable objects to collect and exchange data [1]. The IoT allows objects to be sensed and controlled remotely across existing network infrastructure, creating opportunities for more direct integration between the physical world and

J.-W. An(✉) · D.-S. Cho
Division of Computer Engineering, Dongseo University, Busan, Republic of Korea
e-mail: banhae11@gmail.com, dscho@dongseo.ac.kr

D. Moon
Double-P Corporation, Busan, Republic of Korea
e-mail: wizardyk@gmail.com

© Springer Science+Business Media Singapore 2015   861
D.-S. Park et al. (eds.), *Advances in Computer Science and Ubiquitous Computing*,
Lecture Notes in Electrical Engineering 373,
DOI: 10.1007/978-981-10-0281-6_120

computer-based systems, and resulting in improved efficiency, accuracy and economic benefit. Each thing is uniquely identifiable through its embedded computing system but is able to interoperate within the existing Internet infrastructure. Experts estimate that the IoT will consist of almost 50 billion objects by 2020 [2].

Meanwhile, it is increasing that the perspective of the car as IoT. Recently, Connected-car service has been the focus. It is a car that is equipped with Internet access, and usually also with a wireless local area network. This allows the car to share internet access with other devices both inside as well as outside the vehicle. By Korean Statistical Information Service [3], there are 17 million households and 20 million cars in the country. Most households now own at least one car. With increase of interest in the car, many people also focus on the scheme to saving for maintenance cost and prevention of traffic accident. Vehicle manufacturers provide Telematics and Connected-car service to meet the need for customers.

By 2014, there are only 7 million cars with module for Connected-Car service in the whole world. Hyundai Motors provide service called Bluelink [4] which is technology can connect the car to the internet. User can get the service for diagnostics, consumable management, regular inspection report, and Eco-Driving Coach. AT&T is an American multinational telecommunications corporation. It provides Connected-car service called Autonet [5] in cooperation with GM, Maserati, and so on. Each manufacturer provides independency and exclusive platform service. Customers have to purchase the car equipped connected-car module.

In this paper, we propose common service for driving habit monitoring system. It gathers driving operating raw data and alerts the vehicle driver to eco-driving. Our system has the advantage of that anyone who has a smartphone and a car equipped with low price IoT devices can get the service. We use OBD-II adapter to gather driving operating data and a smartphone. With additional device, like smartwatch, it makes get the service with additional function. User who use our system can get the service such as right driving habit, saving maintenance cost, and prevention traffic accident.

This paper consists of 4 chapters. In chapter 2, we mention related work. In chapter 3, we suggest our recommendation and analysis system for driving habit. And finally, we conclude in chapter 4.

2   Related Work

2.1   OBD-II

OBD (On-board diagnostics)[6] is an automotive term referring to a vehicle's self-diagnostic and reporting capability. OBD systems give the vehicle owner or repair technician access to the status of the various vehicle subsystems. OBD-II is an improvement over OBD-I in both capability and standardization. The OBD-II standard specifies the type of diagnostic connector and its pinout, the electrical signaling protocols available, and the messaging format. It also provides a candidate list of vehicle parameters to monitor along with how to encode the data for each. There is a pin in the connector that provides power for the scan tool from the vehicle battery, which eliminates the need to connect a scan tool to a power source separately. However, some technicians might still connect the scan tool to an auxiliary power source to protect data in the unusual event that a vehicle experiences a loss of electrical power due to a malfunction. Finally, the OBD-II standard provides an extensible list of DTCs. As a result of this standardization, a single device can query the on-board computers in any vehicle.

2.2 Wearable Devices

Wearable devices are clothing and accessories incorporating computer and advanced electronic technologies. The designs often incorporate practical functions and features, but may also have a purely critical or aesthetic agenda. Wearable devices such as activity trackers are a good example of the Internet of Things, since they are part of the network of physical objects or "things" embedded with electronics, software, sensors and connectivity to enable objects to exchange data with a manufacturer, operator and/or other connected devices, without requiring human intervention.

2.3 Driver Monitoring System

There are many research of driver monitoring system. Michimasa et al made a prototype rear-view mirror with a built in image sensor [7]. They use image sensor to analyze driver's awareness by detecting a driver's eye blink and other characteristics of the face. There is driver behavior analysis system through speech emotion understating [8]. The researchers consider emotion as major factor of vehicle accidents. B. Lee proposed driver monitoring system using data fusion [9]. This research uses various data types: eye features, bio-signal variation, in-vehicle temperature, and vehicle speed.

3 System Design

3.1 System Architecture

Most of existing driver monitoring system are uncommon and sometimes use high price tool. In this paper, we suggest common driver monitoring system with low cost. In Fig. 1, we describe our system. When a person rides in own car, and starts to be activity with smart phone or any devices, then our system is working. All devices produce raw data by itself and transmit it to the central server. There are two types of data produced by user. First is OBD-II data, which is vehicle operating data. OBD-II adapter cannot communicate to server by itself. It is necessary to use smart phone for communication. Second is GPS data. Almost of smart phone can receive GPS coordinate. GPS data is also transmitted to the server.

This system can also connect with other IoT device which is available to make data related user activity. Wearable device is good example. It cannot be aware of driving habit, but it can measure biometric data. It is usable to recommend different approach to each driver as case of whose bio-condition. IoT device technology has being constantly developed. The more various devices will make in the future. We have designed our proposed system in consideration of this situation.

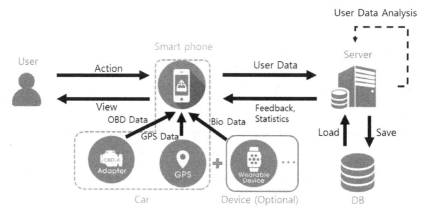

Fig. 1 Proposed System Architecture

3.2 Monitoring Service

TS (Korea Transportation Safety Authority) [9] defines dangerous driving behavior for eco-driving. Table 1 shows the definition. In our system, we define these definitions by using transmitted OBD data. When the event defined dangerous behavior occurs, the server notifies feedback. Some kinds of users want to know their car's maintenance expenditure. We also provide statistics data through smartphone app.

Table 1 Dangerous Driving Behavior Definition by Korea Transportation Safety Authority

Dangerous Driving Behavior		Definition
Speeding Type	Speeding	When you exceeded 20km/h than the road speed limit
	Long-term Speeding	When you exceeded 20km/h than the road speed limit by more than 3 minutes
Rapid Acceleration Type	Rapid Acceleration	When you accelerated 11km/h~25km/h for a second
	Impulsively Start	When you accelerated 11km/h~25km/h for a second from stationary
Abrupt Deceleration Type	Abrupt Deceleration	When you decelerated 7.5km/h~40km/h for a second
	Quick Stop	When you decelerated 7.5km/h~40km/h per a second to stop

All of data produced by user is gathered in the server, and saved the data. Figure 2 shows the data flow between users and server. Individual user data can offer service only current driving condition. By analyzing many user data, it can be aware of surrounding condition. For example, suppose that all of vehicles decelerate around some region. It may lead to something was happened like a car accident. And it is available to recommend that drive slowly to the driver around the region.

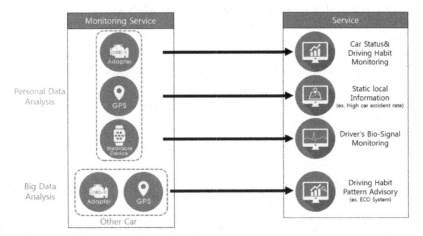

Fig. 2 Data Flow between Users and Server

4   Conclusion

In this paper, we suggest common driving habit monitoring service based on IoT devices. Which can provide the service to anyone who has car equipped low price IoT devices. Proposed system is also designed in consideration to add other devices. In the future, we will implement our proposed system, and try to apply real car and driver.

Acknowledgments This research is results of a study on university for Creative Korea-1 (CK-1) Project and the Leaders in Industry-University Cooperation (LINC) Project, supported by the Ministry of Education.

References

1. Internet of Things Global Standards Initiative, ITU (2015)
2. Evans, D.: The Internet of Things: How the Next Evolution of the Internet Is Changing Everything, Cisco (2015)
3. KOSIS. http://www.kosis.kr/
4. Hyundai Motors. http://www.hyundai.com/
5. Autonet. http://www.theautonet.com

6. On-board diagnostics. https://en.wikipedia.org/wiki/On-board_diagnostics
7. Itoh, M., Mizuno, Y., Yamamoto, S., Mori, S.: Driver's status monitor. In: International Conference on Data Engineering (2005)
8. Kamaruddin, N., Wahab, A.: Driver behavior analysis through speech emotion understating. In: IEEE Intelligent Vehicles Symposium, San Diego, CA, USA, June 21-24, 2010
9. Lee, B.-G., Chung, W.-Y.: A smartphone-based driver safety monitoring system using data fusion. Sensors **12**, 17536–17552 (2012)
10. Korea Transportation Safety Authority. http://www.ts2020.kr/

Analysis of Medical Data Using the Big Data and R

Giyong Choi, Kangwoo Lee, Deokseok Seo, Soonseok Kim, Dongho Kim and Yonghee Lee

Abstract It is medical and health industry that Big Data is most valued. However, Big Data has not been actively introduced in the domestic medical field. In this respect, the present study was aimed to use R, which is an analyzing tool of Big Data, to promote various business models in health and medical field, and to analyze medical data of diseases and genetic information so Big Data can be utilized in the field. In this study, SGA consortium electrocardiogram data was used, which is database of ECG-ViEW provided in http://egcview.org. R was used to analyze the medical Big Data from various angles. RStudio Version 0.98.1103 was used as R tool to perform association rule analysis, outlier diagnosis and simple regression analysis with ECG-ViEW data.

Keywords Big data · ECG · RStudio · Medical data

1 Introduction

Recently, Big Data is drawing a keen attention from the world as a solution to economic and social pending issues. Being interlocked with the rapid spread of smart device infrastructure, the development of Big Data management and technology and the advancement of public awareness, the foundation for using Big Data has already been well established in the world. As smart devices equipped with various sensors spread, it became to collect comprehensive micro data such as individual's life and environment. In addition, as the price of data storage device and cost of communication are rapidly declining, operational capability and dada analysis methods by a computer are also rapidly growing [1, 2].

G. Choi
Graduate School of Information Industry, Halla University, Wonju, Kangwon, Korea

K. Lee · D. Seo · S. Kim · D. Kim · Y. Lee(✉)
Department of Computer Engineering, Halla University, Wonju, Kangwon, Korea
e-mail: yhlee@halla.ac.kr

© Springer Science+Business Media Singapore 2015   867
D.-S. Park et al. (eds.), *Advances in Computer Science and Ubiquitous Computing*,
Lecture Notes in Electrical Engineering 373,
DOI: 10.1007/978-981-10-0281-6_121

In the meantime, Big Data has not been actively introduced in the domestic medical field. As Electronic Medical Record (ERM) system was introduced and expanded, major general hospitals have to enter standard data in Data Warehouse, so it is very limited to analyzing and using bulky data. In addition, the non-standard data that medical staffs input in the electronic chart are hard to search and use for statistical purpose. Furthermore, as healthcare industry shifted its focus on medical service from treatment to prevention and health management, the medical service for forecasting the chance of disease incidence, personalized medical service have been more important. On the other hand, such data as physical checkup, diseases, EMR, and genome analysis, which are collected by bio-sensing and medical imaging technology, have rapidly been cumulated [3,4].

Accordingly, this paper study used R, which is an analyzing tool of Big Data, to promote various business models in healthcare industry, and to analyze medical data of diseases and genetic information so Big Data can be utilized in the field.

2   ECG-ViEW Database

In this study, SGA consortium electrocardiogram (ECG) data was used. They are the database of ECG-ViEW provided in http://egcview.org. R, which is an analyzing tool of Big Data, was used to analyze the medical Big Data from various angles. RStudio Version 0.98.1103 was used as R tool.

2.1   Overall Structure of Tables

Fig. 1 shows ER diagram of the overall tables of ECG-ViEW database.

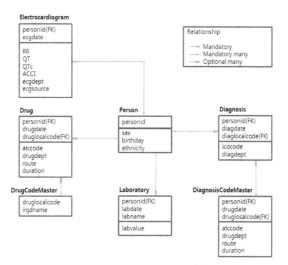

Fig. 1 ER Diagram of Overall Tables

2.2 Explanation of Table

Table 1 explains each table used for data analysis. ECG-ViEW database files consists of 7 excel files in total and the sizes of files are shown in Table 2.

Table 1 Explanation of Table

Table	Explanation
Person	General information of a patient
Electrocardiogram (ECG)	ECG value measured at a certain time point
Drug	Drug information prescribed at a certain time point
Diagnosis	Patient diagnostic information at a certain time point
Laboratory	Lab test results on a patient done at a certain time point
DrugCodeMaster	Prescription code and name of drug
DiagnosisCodeMaster	Corresponding table of diagnosis code and ICD-10 code

Table 2 ECG-ViEW Database Files

File Name	Size
DiagnosisCodeMaster.csv	412KB
DrugCodeMaster.csv	81KB
Electrocardiogram.csv	1,761KB
person.csv	612KB
sample_Diagnosis.csv	14,661KB
sample_drug.csv	111,182KB
sample_laboratory.csv	5,585KB

3   ECG-ViEW Data Analysis

3.1 Association Analysis with ECG-ViEW Database

Association analysis is a non-oriented data mining to seek the association of events in database. R uses "arules" package to conduct association analysis while "arules" uses an apriori algorithm, which is an algorithm of minimum support pruning (MSP) to carry out association analysis. An apriori function, which is used to find association rules, has data, parameter, appearance, and control. The support and reliability of ECG data were set to 0.1, respectively and 'lift' was not designated because its default value is 1. Fig. 2 shows the resultant associations found through the analysis.

Analyzing the results above, we can see that when QTc is 411->475, support rate is 0.111111; reliability is 0.3333333; and life rate is 3.000000. It explains that when QTc is measured to be 411 or 475, the possibility for QTc to become 475 or 411 later is 11% and 25% of patients of 411 can be patients of 475. Plot() function was used to display association rules with support rate and reliability in a scatter diagram.

```
> inspect(rules)
   lhs       rhs       support   confidence  lift
1  {}    => {475}  0.1111111  0.1111111  1.0000
2  {}    => {410}  0.1111111  0.1111111  1.0000
3  {}    => {408}  0.1111111  0.1111111  1.0000
4  {}    => {397}  0.1111111  0.1111111  1.0000
5  {}    => {450}  0.1111111  0.1111111  1.0000
6  {}    => {444}  0.1111111  0.1111111  1.0000
7  {}    => {426}  0.1111111  0.1111111  1.0000
8  {}    => {452}  0.1111111  0.1111111  1.0000
9  {}    => {415}  0.1111111  0.1111111  1.0000
10 {}    => {413}  0.1111111  0.1111111  1.0000
11 {}    => {431}  0.1111111  0.1111111  1.0000
12 {}    => {424}  0.1111111  0.1111111  1.0000
13 {}    => {436}  0.1111111  0.1111111  1.0000
14 {}    => {423}  0.2222222  0.2222222  1.0000
15 {}    => {461}  0.2222222  0.2222222  1.0000
16 {}    => {402}  0.2222222  0.2222222  1.0000
17 {}    => {411}  0.3333333  0.3333333  1.0000
18 {}    => {380}  0.6666667  0.6666667  1.0000
19 {}    => {420}  0.8888889  0.8888889  1.0000
20 {475} => {411}  0.1111111  1.0000000  3.0000
21 {411} => {475}  0.1111111  0.3333333  3.0000
22 {475} => {380}  0.1111111  1.0000000  1.5000
23 {380} => {475}  0.1111111  0.1666667  1.5000
24 {475} => {420}  0.1111111  1.0000000  1.1250
25 {420} => {475}  0.1111111  0.1250000  1.1250
26 {410} => {408}  0.1111111  1.0000000  9.0000
```

Fig. 2 Association Rules

3.2   *Outlier Diagnosis of ECG-ViEW Data*

One of the most important suppositions in regression analysis is that the model using observation values included in data should be appropriate (fit). However, practically, one or two data do not fit the model. Such observation value is called 'outlier'. Fig. 3 shows the graph of outlier analyzed by CAR (companion to

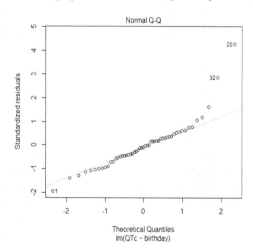

Fig. 3 Outlier Graph

applied regression) package. Fig. 3 let us know that the 20th observation value is an outlier. There are many ways to identify an outlier. What matters is follow-up action after finding the outlier. In some case, outlier can be discarded because they are determined not to be as object for research or used as object for research.

3.3 Simple Regression Analysis of ECG-ViEW Data

Simple regression analysis is a statical technique to predict another variable or factor from one predicting variable. With ECG data, this study built a fit model and analyzed it for QTc by age.

The results are;

Residuals:

Min	1Q	Median	3Q	Max
-44.203	-12.174	-1.876	8.066	100.066

Five values can be obtained from the model's residuals: minimum, first quartile, median, third quartile, and maximum.

Coefficients:

	Estimate	Std. Error	t value	Pr(>|t|)
(Intercept)	157.5823	359.0321	0.439	0.663
birthday	0.1314	0.1833	0.717	0.477

Estimated value means fit regression coefficient; standard error = the standard error of regression coefficient; t = t value to verify the regression coefficient is 0; and Pr(>|t|) = signification probability to verify the regression coefficient is 0. As a result, the fit model is like this.

$$Y = -157.5823 + 0.6087X$$

It demonstrates significance of the model under these conditions: the standard error of Slop b1 is 0.1833, test statistic of null hypothesis H0:b1=0 is 0.717; p = 0.477 (>a=0.001).

Residual standard error: 23.64 on 51 degrees of freedom
Multiple R-squared: 0.009974, Adjusted R-squared: -0.009438
F-statistic: 0.5138 on 1 and 51 DF, p-value: 0.4768

It shows that the square root of estimated sigma mean square error (MSE) of error variance is 23.64; coefficient of determination R square is 0.009974; and test statistics for F is 0.5138. Fig. 4 shows standardized residuals and explanatory power.

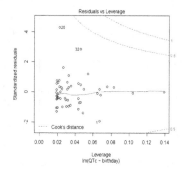

Fig. 4 Graph of Standardized Residuals and Explanatory Power

4   Conclusion

This paper study used R, which is an analyzing tool of Big Data, to promote various business models in healthcare industry, and to analyze medical data of diseases and genetic information so Big Data can be utilized in the field.

In this study, SGA consortium electrocardiogram (ECG) data was used. They are the database of ECG-ViEW provided in http://egcview.org. To use R, which is an analyzing tool of Big Data, this study analyzed the medical Big Data from various angles for association analysis, outlier diagnosis and simple regression analysis. RStudio Version 0.98.1103 was used as R tool.

Association analysis showed that when QTc moved from 411 to 475, it turned out that support rate was 0.111111; reliability was 0.3333333; and 'lift' was 3.000000. This result explained that when QTc is measured to be 411 or 475, the possibility for QTc to become 475 or 411 later is 11% and 25% of patients of 411 can be patients of 475. In addition, outlier diagnosis confirmed that the 20th observation value could be an outlier. Moreover, simple regression analysis found out a fit model. In this paper, medical explanation on the results were put aside, but it is expected that these findings can be useful for doctors to explain medical phenomena of a patient's health conditions.

Acknowledgments

1) This research was supported by the MSIP(Ministry of Science, ICT and Future Planning), Korea, under the C-ITRC(Convergence Information Technology Research Center) (IITP-2015-IITP-2015-H8601-15-1009) supervised by the IITP(Institute for Information & communications Technology Promotion)

2) The research was supported by the 2015 Economic Cooperation Business Project(Business Cooperative R & D) under the Ministry of Trade, Industry & Energy, Korea.

References

1. McSfee, A., Brynjofsson, E.: Big-data: The management revolution. Harvard Business Review, October 2012
2. Chen, C.L.P., Zhang, C.-Y.: Data-intensive applications, challenges, techniques and technologies: A survey on Big Data. Information Sciences (2014)
3. Cosgrove, D.M.: A healthcare model for the 21th centry. Group Practice Journal (2011)
4. Ferrara, E., De Meo, P., Fiumara, G., Baumgartner, R.: Web Data Extraction, Applications and Techniques: A Survey. Knowledge-based systems, June 11, 2014
5. McKinsey Global Institute: The big data revolution in healthcare (2013)
6. Funao, N.: The R Tips (2009)

Homography-Based Motion Detection in Screen Content

Ton Thi Kim Loan, Xuan-Qui Pham, Huu-Quoc Nguyen, Nguyen Dao Tan Tri, Ngo Quang Thai and Eui-Nam Huh

Abstract Screen content has some different characteristics from camera-captured content, such as large motion and repeating patterns which lead to low encoding speed and higher bit-rate. To cope with these problems, homography-based motion detection is proposed to better explore the temporal correlation in screen content. After detecting motion, the motion parameters are forwarded to JPEG encoder for motion compensated predictions. Therefore, it can improve the coding efficiency. Experimental results show the proposed algorithm achieves efficiency in terms of both encoding time and encoding complexity.

Keywords Screen content · JPEG codec · Motion detection

1 Introduction

In recent years, the rapidly development of technology and the increase of user demand, screen contents have become more and more popular. As shown in Fig. 1, applications of screen content may include desktop sharing, remote desktop, remote education, game video, web conferencing. In these applications, the screen content in a remote computer is captured, compressed and transmitted to end users immediately to interact bi-direction. Screen virtualization [1] on the mobile cloud computing architecture is new dimension and is considered as a promising solution to optimize the computing experience for users between the cloud and the client. Using the cloud computing technology, the virtual screen is rendered in the cloud and delivered as images to the client via Internet for play-out. This mechanism not only overcomes software and hardware limitations but also allows users to access the computationally intensive and graphically rich services from the cloud. Hence, the screen encoding method is needed to keep high quality and low bit-rate to achieve the interactive and real-time environment at the remote computer.

T.T.K. Loan(✉) · X.-Q. Pham · H.-Q. Nguyen · N.D.T. Tri · N.Q. Thai · E.-N. Huh
Kyung Hee University, Yongin, South Korea
e-mail: {loanttk,pxuanqui,quoc,tringuyendt,nqthai,johnhuh}@khu.ac.kr

© Springer Science+Business Media Singapore 2015   875
D.-S. Park et al. (eds.), *Advances in Computer Science and Ubiquitous Computing*,
Lecture Notes in Electrical Engineering 373,
DOI: 10.1007/978-981-10-0281-6_122

Basically, screen contents can be divided into two types: text and graphic. Many works have been proposed to separate these two categories. JPEG is the well-known encoding standard, which is employed to exploit the spatial-temporal correlations among frames. Said et al. [2] proposed a block-based approach, which firstly divides an input image into blocks and classifies these blocks into groups with different characteristics as text and picture blocks, and then it compresses different types of blocks with different types of compression algorithms. Queiroz et al. [3] described a layer-based approaches based on mixed raster content (MRC) model, which segments an image into two image layers and a binary mask layers. In [4], a color and index map (BCIM) scheme is proposed to enable efficient screen content coding in H.264/AVC. All these methods [2], [3], [4] prefer to improve the video coding in term of spatial characteristics.

(a)   (b)

Fig. 1 Screen content in different application scenarios. (a) Desktop sharing, (b) Remote education

Considering thin-client devices, it is necessary to preserve enough CPU resources for other working programs. It means that the CPU resources for remote applications are often limited. In some cases, the client-side devices are also simple and weak. Therefore, the different screen contents are processed by single screen compression methods. This paper proposes JPEG encoder and decoder to obtain efficiently in terms of software and hardware usages. Different from the previous approaches, the proposed scheme solves temporal problems rather than spatial ones. That is, the large motion regions, which always happen in screen contents as show Fig. 2, will lower the performance of JPEG. Besides, if the search range of motion estimation (ME) is large, the encoding time could increase. Inversely, if the search range is small, the correlation among frames will be inefficient, which leads to a high bit-rate.

To solve these problems, a motion detection algorithm is proposed to improve the efficiency of JPEG encoding. Particularly, a fast and efficient motion detection using homography with Random Sample Consensus (RANSAC) [5] is used to find global motion with low complexity.

The remainder of this paper is structured as follows. Section 2 describes the characteristic of motion in screen contents. In Section 3, we present the proposed screen content scheme. Section 4 shows our simulation results, which illustrate our improvements compared to other approaches. Finally, conclusions are drawn in Section 5.

Webpage

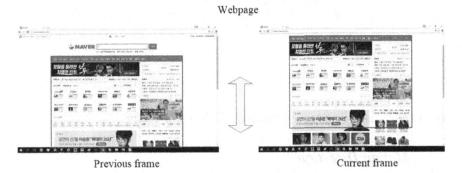

Previous frame   Current frame

Fig. 2 An example about the vertical translation between two consecutive frames from the sequence webpage

2   The Characteristic of Motion in Screen Contents

In screen content applications, the motion is different from the camera-captured one. For example, when a user browses a web page, the motion is generated such as scrolling, rotating, touching, scaling and dragging. Especially, the motion gap between two continuous frames is big when the network condition is bad and some frames have to be dropped. Additionally, the screen content can contain unnatural motions, like as fading in and fading out. Moreover, the screen images are rendered and captured by system, which seems to show no noise compared with the nature video. The motion in screen images is often region-based and arbitrary-sized and the region always has rectangular shape. All above characteristics lead to the fact that the motion search in screen content image can be faster and more efficient than the motion captured by the camera. Therefore, motion search and motion detection are needed to perform for reducing the encoding complexity.

3   Motion Detection

3.1   The Proposed Screen Content Scheme

Fig. 3 shows the scheme of the proposed screen content. To obtain displacement among frames or blocks, we first identify feature points that are no invariant to image scaling, rotation or partially invariant to changes in illumination and viewpoint. Continuously, the motion detection is performed to find out global motion vectors between the current frame and the previous frame. Finally, the motions are passed to JPEG encoder for the inter-prediction of the current frame. Here, the difference detection is not mentioned.

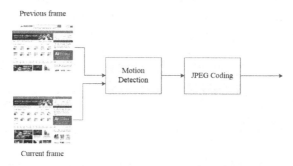

Fig. 3 The flowchart of proposed screen content

3.2   *Motion Detection*

In this paper, we propose an efficient motion detection method which uses the homography between two continuous frames to find the large motion. In specific, feature points are extracted from current frame. The corresponding feature points are then found in the previous frames. If these feature points are found, an affine model or homography is defined, which describes the motion between two consecutive frames based on feature points pairs. This progress is presented in more details as bellows.

3.2.1   Feature Extraction and Matching

The selection of feature points is a very crucial issue in the motion detection. In previous works, the features are region boundaries or image contours and these features move unpredictably with the remains of the image frame. So, when scaling, rotation or change of illumination happens, thus the results can be unreliable. For this reason, we propose the use of Lowe's approach [6] to extract Scale Invariant Feature Transform (SIFT) feature points for every frame in screen sharing system. SIFT is very popular feature extraction method that identifies no variant feature points to image scaling, rotation or partially invariant to changes in illumination and viewpoint.

After the feature points are extracted, the next step is to select the corresponding feature points. Here, the coarse-to-fine approach is proposed for finding the corresponding feature pairs. At the coarse stage, Euclidean distance is used to measure the matching of any two feature points that has the shorter distances the higher matching possibility. Moreover, at the fine stage, we combine RANSAC [5] algorithm to remove the mismatch of feature pairs. Then, top five feature points pairs are selected to form a linear system for deriving the affine transformation between two continuous frames.

Fig. 4 Feature extraction and matching between two successive frames

3.2.2   Motion Detection

Our method is based on the intuition that we can use easily these points to track any form of motion. For example, let us assume that the user's screen contained a big red point. When user scrolls the screen up, the red point also moves up. We might detect the new location easily and compute the displacement vector of this point in the update screen. There are rotation, scale, translation and other motions in vertical and horizontal directions, so the mathematical model in this paper is affine model which effectively describe translation, rotation and scale motion between two consecutive frames. Suppose $P(x, y)$ and $P'(x', y')$ be the pixel location of corresponding feature points in two continuous image frames. The relationship between two pixel locations can be shown by the following affine transformation:

$$P' = H \times P \tag{1}$$

$$\begin{bmatrix} x' \\ y' \\ 1 \end{bmatrix} = \begin{bmatrix} S.\cos(\theta) & -S.\sin(\theta) & T_x \\ S.\sin(\theta) & S.\cos(\theta) & T_y \\ 0 & 0 & 1 \end{bmatrix} \cdot \begin{bmatrix} x \\ y \\ 1 \end{bmatrix} \tag{2}$$

In the affine model, T_x and T_y are the horizontal and vertical translations, θ is the rotation, S is the scale. The transformation matrix is obtained as the results of matching SIFT feature over successive frames. The matching of features is done using Euclidean distance. The matches obtained between two frames are shown in Figure 4. Since the transformation between frames is considered to be affine, the matched feature points can be represented as in

$$Ax = b \tag{3}$$

where the elements of A and b are the matched feature points and x is the unknown affine transformation matrix to be computed. The least squares solution of this equation can be obtained using

$$x = [A^T A]^{-1} A^T b \tag{4}$$

Six parameters for affine motion can be calculated with at least three matched feature point pairs. To robustly estimate this motion, we use RANSAC, which finds the best fit affine model.

4   Experimental results

In this section, we illustrate the compression efficiency by comparing our proposed method with the traditional one, which will send each captured screen to update screen. Three scenarios are tested: compound images (*webpage*), document (*word*), touch UI in Windows 8 (*metro*). In *webpage* and *word*, the motion only includes scrolling up and scrolling down. In *metro*, the motion comes from finger slipping. Table 1 shows the comparison of average encoding time for each frame, which is running in the PC with CPU Core ™ i5- 6470@ 3.4 GHz and 4GB RAM. For *word*, *webpage* and *metro*, 10%-30% encoding time reduction can be obtained. There are also some factors, which affect the encoding time such as capturing, transmission and controlling. Thus, the compression is necessary as fast as possible to guarantee a high frame rate which improves the user experiences significantly.

Table 1 Comparison the encoding time (ms/frame) of different algorithms.

	word	*webpage*	*metro*
JPEG	35	33	30
Proposed	**24**	**27**	**27**
Reduction	32.4%	18.8%	10.7%

5   Conclusions

In this paper, we propose the global motion estimation using affine transformation between two consecutive frames with RANSAC which reduced computational time before passing to the encoder part of JPEG coding. The proposed method can obtain the efficient of the performance in the time and encoding complexity characteristics through experimental results.

Acknowledgments This research was supported by the MSIP (Ministry of Science, ICT and Future Planning), Korea, under the ITRC (Information Technology Research Center) support program (IITP-2015-(H8501-15-1015) supervised by the IITP(Institute for Information & communications Technology Promotion). Institute for Information & communications Technology Promotion (IITP) grant funded by the Korea government (MSIP) (B0101-15-0535, Development of Modularized In-Memory Virtual Desktop System Technology for High Speed Cloud Service). The corresponding author is professor Eui-Nam Huh.

References

1. Lu, Y., Li, S., Shen, H.: Virtualized screen: A third element for cloud-mobile convergence. Multimedia **18**(2), February 2011. IEEE
2. Lan, C., Peng, X., Xu, J., Wu, F.: Intra and inter coding tools for screen contents. In: 5th JCT-VC Meeting, Geneva, Switzerland, Document JCTVC-E145, March 2011
3. de Queiroz, R., Buckley, R., Xu, M.: Mixed raster content (MRC) model for compound image compression. In: Proc. IS&T/SPIE Symp. Electronic Imaging, Visual Communications and Image Processing, vol. 3653, pp. 1106–1117. Citeseer (1999)
4. Ding, W., Lu, Y., Wu, F.: Enable efficient compound image compression in h. 264/AVC intra coding. In: Proc. IEEE Int. Conf. Image Processing, vol. 2, p. II–337 (2007)
5. Fischler, A., Bolles, R.C.: Random Sample Consensus: A Paradigm for Model Fitting with Applications to Image Analysis and Automated Cartography. Comm. of the ACM **24**, 381–395 (1981)
6. Lowe, D.G.: Object recognition from local scale-invariant features. In: The Proceedings of the Seventh IEEE International Conference on Computer Vision, 1999, vol. 2, pp. 1150–1157 (1999)

A Reference Architecture Framework for Orchestration of Participants Systems in IT Ecosystems

Soojin Park, Lee Seungmin and Young B. Park

Abstract Nowadays, IT systems are advancing into our lives in explosive numbers. As new IT systems and devices pervade deeper into all aspects of human routines, our requirements for their system software are diversifying as quickly as the expanding array of functionalities on each system. Such phenomenon has lead up to the proposal of several new paradigms recognizing groups of IT systems as an ecosystem, coining the term IT ecosystem. Each individual system participating in an IT ecosystem has, to a varying degree, autonomy. The ultimate goal of participating systems is to satisfy the global goal of the entire IT ecosystem it is participating in. In order to maintain autonomy and controllability over an IT ecosystem through environmental changes, orchestration strategies must be invented and implemented. In this paper, we propose a reference orchestration architecture framework to support IT ecosystems accomplish the goals of individual participant systems while it simultaneously achieves its system-wide goals. The reference architecture is illustrated in detail using a tangible scenario for collaboration of unmanned vehicles to fight fire.

Keywords Orchestration · Service level management · IT ecosystem

1 Introduction

With fast-paced advent of new software technologies, including recent additions such as cloud computing and Internet of Things (IoT), paradigms in software system

S. Park
Graduate School of Management of Technology, Sogang University, Seoul, South Korea
e-mail: psjdream@sogang.ac.kr

L. Seungmin · Y.B. Park(✉)
Department of Computer Science and Engineering,
Dankook University, Cheonan, South Korea
e-mail: dltmdals1119@gmail.com, ybpark@dankook.ac.kr

© Springer Science+Business Media Singapore 2015
D.-S. Park et al. (eds.), *Advances in Computer Science and Ubiquitous Computing*,
Lecture Notes in Electrical Engineering 373,
DOI: 10.1007/978-981-10-0281-6_123

883

operation has shifted its focus from operation of a single system to operation of a system of systems. In response to these major changes, a number of research groups have introduced new concepts such as software ecosystem or digital ecosystem. Among the latest concepts based on the same idea is the paradigm of IT ecosystem [1][2][5]. IT ecosystem is a complex system consisting of interacting autonomous individual systems, adaptive as a whole based on local adaptiveness [1]. As is the case with biological ecosystems, autonomy among system participants is just as important as controllability across the entire ecosystem.

Recent researches[3][6][7] on self-adaptive software have reported their results obtained from applying MAPE loop to various purposes and industry domains, where MAPE loop is a typical control loop for self-adaptive systems composed of a sequence of four computations - monitor, analyze, plan, and execute - as to autonomously detect context changes and to perform appropriate reconfiguration where deemed necessary. While a single self-adaptive system can ensure adaptiveness by implementing a single MAPE loop, a group of heterogeneous multiple self-adaptive systems such as an IT ecosystem requires additional MAPE loop to implement controls for achieving system-wide goals besides individual MAPE loop implemented for each participant system. The research in [4] proposes patterns for decentralized control in generally large scale, complex, and heterogeneous self-adaptive systems.

In this paper, we propose a reference architecture framework to support IT ecosystems as a composition of MAPE loops embedded in N number of participant systems and an additional MAPE loop to govern interactions among systems and dynamic reconfiguration of participating systems. The rest of this paper is dedicated to illustrating the proposed framework. First, in Section 2, we introduce a metamodel defining the relationships between elements necessary for achieving two different types of goals of IT ecosystems. Section 3 describes a reference architecture defining two different MAPE loops designed to achieve the two different types of goals. Finally, in Section 4, we present our conclusion of this study along with our ongoing works.

2   Meta-model for IT Ecosystem

A metamodel for representing IT ecosystem has been designed. As shown in Fig. 1, an IT ecosystem is composed of collaborations among multiple participant systems (Participants) and its system-wide goals. The goals are divided into two categories: functional service goals (Service Goal) which are functions to be provided to the system use, and quality goals which govern the quality aspects of inter-system collaborations.

Service goals can be regarded as requirements to be fulfilled by implementations of individual MAPE loops in participant systems, whereas goals on quality can be seen as requirements for the MAPE loop controlling the entire IT ecosystem. As system-wide quality does not depend on specific participant's services or collaborations but rather depends on the quality of orchestration over the entire IT ecosystem, we have named the goals on quality aspects Orchestration Goals.

The set of *Capabilities* required for each *Role* in IT ecosystem may differ. In order to obtain the required capabilities per role, *Plans* composed of appropriate *Actions* must be established. Similar to a biological ecosystem, an IT ecosystem is required to adapt to changing environment. Therefore plans are defined not during design phases but rather dynamically in response to context changes.

While the meta model in Fig. 1 does not illustrate specific roles identified for IT ecosystems, we have defined two roles: <<Role>>TeamMember and <<Role>>TeamLeader. The distinction among two roles is according to the aforementioned different goal types. To help understanding the distinction, the relationship between goals and roles will be revisited in Section 3 using a simple fragmentary example depicted by Fig. 3.

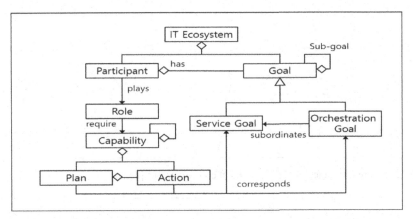

Fig. 1 Meta-model for IT Ecosystem

3   Architecture of IT Ecosystems for Orchestration

We have designed reference architecture for orchestrating inter-system interactions in IT ecosystems using instances of metamodel elements introduced in Section 2. Requirement set defined prior to designing the proposed architecture is as follows:

1. The reference architecture must support two different types of loop interactions, which N number of MAPE loops controlling individual participant systems and an additional MAPE loop controlling the entire IT ecosystem.
2. The objective of adaptation by two different MAPE loops included in the reference architecture is to achieve respective goals specified by the service goal model and the orchestration goal model.
3. The role executing different loops must be explicitly separated. Further, the interaction protocol between different roles must be explicitly defined.

Prior to designing the architecture to satisfy above requirements, we have selected the hierarchical control pattern, among multiple MAPE loop decentralization patterns proposed in [4], as our base pattern. Hierarchical control pattern is applicable where a complex distributed self-adaptive system can be regarded as a single complex self-adaptive system, and the pattern conforms most appropriately to the IT ecosystem concept defined in our research. The pattern depicted in Fig. 2, as proposed in [4], is to achieve identical type of goals assigned to N number of MAPE loops where analysis and planning are performed individually in respective loops but monitoring and execution are performed in an interactive relationship by sharing local information and capabilities of each loop. Our research, however, differs from the original pattern in that we must define the relationship between two heterogeneous MAPE loops striving to achieve two different types of goals, which are service goals and orchestration goals. For this, the original pattern (Fig. 2) had to be modified.

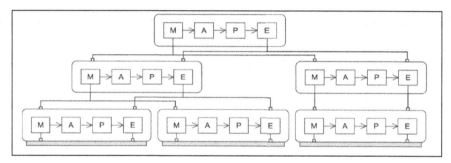

Fig. 2 Concrete instance of the hierarchical control pattern [4]

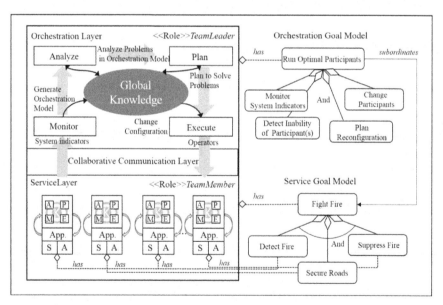

Fig. 3 Reference Architecture of IT Ecosystem for Accomplishing Two Different Goal Model: Service Goal Model and Orchestration Goal Model: M(Monitor), A(Analyze), P(Plan), E(Execute), App.(Application), S(Sensor), A(Actuator).

Fig. 3 depicts the resulting design of the reference architecture. Based on the pattern in Fig. 2, the MAPE loops for achieving orchestration goal are assigned as the higher layer on top of system architecture, while the MAPE loops targeted to achieve service goals (executed by individual systems) are placed on the lower layer. In the hierarchical control pattern, each MAPE loop monitors the identical system indicators and shares acquired information with each other. However, the proposed architecture does not make use of mutual monitoring interaction among components as the MAPE loops in higher and lower layers have different types of goals. While the upper layer MAPE loop monitors indicators which represent the collaboration status or current availability of participant systems within the IT ecosystem, the lower layer MAPE loops monitor the indicators relevant to checking the difference between the service goals assigned to itself and its current activities.

The orchestration goal model and service goal model illustrated on the right-hand side of Fig. 3 describe a portion of the goal model derived from the firefighting scenario using with unmanned vehicles. First of all, the sub-goals of "Fight Fire," which are "Detect Fire," "Secure Roads," and "Suppress Fire," are assigned to individual systems within the service layer on the bottom-left of Fig. 3. Criteria used in decisions for assigning goals to participant systems are the compatibility of sensors (s) and actuator(s) owned by each system. A participant system assigned a service goal is also assigned the <<Role>>TeamMember, thereafter continuously monitoring its own performance and the difference between its goal, and in parallel executing adaptation cycles to satisfy its assigned goal.

Orchestration goals, on the other hand, are assigned to the participant system designated to be the coordinator over the entire IT ecosystem. This particular coordination role is defined as the <<Role>>TeamLeader. Electing the participant system to assume *TeamLeader* role can be performed based on selection strategy such as highest availability or highest capability; the election is performed dynamically based on the designated election strategy. The layer on which *TeamLeader* is executed to follow MAPE adaptation cycle to accomplish orchestration goals is named "*Orchestration Layer*." A monitor in the orchestration layer collects various information such as individual status of systems executing respective MAPE loop or performance index from collaboration configurations, through monitoring indicators, and then generates an orchestration model in global knowledge storage. The generated model is analyzed to check for evidence of risks potentially hindering accomplishment of global goals. If concerns are detected, a new plan for reassigning or removing particular participant systems and replacing them with new participant for collaborations is created. The adaptation plan is executed by driving actuators of *TeamMember* systems. Global knowledge required to execute MAPE loop in orchestration layer includes monitoring data accumulated by participant systems, in addition to information required to maintain controllability across the entire IT ecosystem.

The communication protocol between a *TeamLeader* system and a *TeamMember* system is defined by the "Collaborative Communication Layer" between the two aforementioned layers. Fig. 4 depicts a communication protocol defined to execute "Run Optimal Participants" goal, which is to reconfigure a group of participant systems to propose an optimal set for a given context. The *Calculate CollaborationScore( )* operation shown in the analyzing phase of Fig. 4 is the function evaluating the cost and benefit of each possible participants configuration within the IT ecosystem. The resulting value is the collaboration score, and when the score turns out to be under a defined threshold value then a new participant system configuration is recomposed in an effort to improve the collaboration ROI. With cyclic execution of adaptation loop, the quality of collaboration within an IT ecosystem can be steadily sustained.

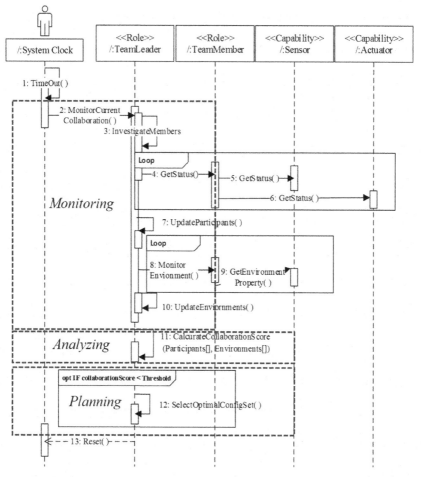

Fig. 4 Communication protocol between *TeamLeader* and *TeamMember* for "Run Optimal Participants".

4   Conclusion

In this paper we proposed a reference architecture framework for supporting the accomplishment of two different goal models – service goal and orchestration goal – to assure controllability across the entire IT ecosystem rather than achieving adaptation for a specific functionality. While this paper is limited in scope in terms of examples, with only a fragmentary example of the firefighting scenario introduced, in fact we have already constructed a simulation of an entire IT ecosystem for unmanned forest management which includes the firefighting scenario. Our future research plans include evaluating the proposed architecture within a simulated IT ecosystem and refining the architecture implementation by reflecting evaluation results.

References

1. Rausch, A., Muller, J.P., Niebuhr, D., Herold, S., Goltz, U.: It ecosystems: a new paradigm for engineering complex adaptive software systems. In: 2012 6th IEEE International Conference on Digital Ecosystems Technologies (DEST), pp. 1–6. IEEE Press, Campione d'Italia (2012)
2. Schneider, K., Meyer, S., Peters, M., Schliephacke, F., Mörschbach, J., Aguirre, L.: Feedback in context: supporting the evolution of IT-ecosystems. In: Ali Babar, M., Vierimaa, M., Oivo, M. (eds.) PROFES 2010. LNCS, vol. 6156, pp. 191–205. Springer, Heidelberg (2010)
3. Vromant, P., Weyns, D., Malek, S., Andersson, J.: On interacting control loops in self-adaptive systems. In: Proceedings of the 6th International Symposium on Software Engineering for Adaptive and Self-Managing Systems (SEAMS 2011), pp. 202–207. ACM, New York (2011)
4. Weyns, D., Schmerl, B., Grassi, V., Malek, S., Mirandola, R., Prehofer, C., Wuttke, J., Andersson, J., Giese, H., Göschka, K.M.: On patterns for decentralized control in self-adaptive systems. In: de Lemos, R., Giese, H., Müller, H.A., Shaw, M. (eds.) Software Engineering for Self-Adaptive Systems. LNCS, vol. 7475, pp. 76–107. Springer, Heidelberg (2013)
5. Herold, S., Klus, H., Niebuhr, D., Rausch, A.: Engineering of it ecosystems: design of ultra-large-scale software-intensive systems. In: Proceedings of the 2nd International Workshop on Ultra-large-scale Software-intensive Systems, ULSSIS 2008, pp. 49–52. ACM, New York (2008)
6. Souza, V.E.S., Lapouchnian, A., Robinson, W.N., Mylopoulos, J.: Awareness requirements for adaptive systems. In: Proceedings of the 6th International Symposium on Software Engineering for Adaptive and Self-Managing Systems (SEAMS 2011), pp. 60–69. ACM, New York (2011)
7. Sabatucci, L., Cossentino, M.: From means-end analysis to proactive means-end reasoning. In: 2015 IEEE/ACM 10th International Symposium on Software Engineering for Adaptive and Self-Managing Systems (SEAMS), pp. 2–12. IEEE Press, Florence (2015)

Contents Based Traceability Between Research Artifact and Process for R&D Projects

Jong-Won Ko, Sun-Tae Kim, Jae-Young Choi and Young-Hwa Cho

Abstract The software industry generally defines traceability as an ability to manage the relation between the requirements and other outputs from the requirements, and the documents in various forms produced over the entire cycle of the R&D Project have been managed in the form of configuration management along with the source codes as they were recognized as configuration items. However, as the management of the outputs was conducted in the file unit, it is difficult to verify consistency and integrity of the files and limited to define the relation between the targets to support traceability and the target items. This paper defined the scope and target of traceability regarding RD traceability and RD based Traceability Information Model from the Research Descriptor that defined the various forms of outputs produced over the entire cycle of the R&D Project including the execution process in the R&D Project and keywords and key sentences of the outputs. In addition, it proposed to research on traceability that escapes from the limitations of traceability on the configuration items in the file unit focusing on the outputs in the existing researches on traceability through tracking of the content of the RD and RD relation types when conducting the R&D Project and that connects to the quality of R&D Project through the relevance between traceability based on such RD relation types and SW development quality indicator.

Keywords Project management approach · Research descriptor · Software traceability

1 Introduction

The document files in various forms produced over the entire cycle of the R&D Project have been managed in the form of configuration management with source

J.-W. Ko · J.-Y. Choi(✉)· Y.-H. Cho
College of Information and Communication Engineering,
SungKyunKwan University, Suwon-si, South Korea
e-mail: {jwko0820,jaeychoi,choyh2285}@skku.edu

S.-T. Kim
Department of Software Engineering, Chonbuk National University, Jeonju-si, South Korea
e-mail: stkim@jbnu.ac.kr

© Springer Science+Business Media Singapore 2015   891
D.-S. Park et al. (eds.), *Advances in Computer Science and Ubiquitous Computing*,
Lecture Notes in Electrical Engineering 373,
DOI: 10.1007/978-981-10-0281-6_124

codes as they were recognized as configuration items in the file unit. However, due to the file unit output management, there is a limit that the researchers specializing in the area of the R&D Project are required to verify the file content after voluntarily checking it in person rather than the verification of consistency and integrity of the file content is supported from the perspective of the system. Furthermore, the existing researches failed to fully consider the research items existing in the form of various document files and SW configuration items, for example, the support for traceability among SW UML models and source codes, UI Form and test cases and the issues on traceability with the conducted R&D Process to make such outcomes.

Therefore, this paper defined the scope and target of traceability regarding the RD traceability and RD based Traceability Information Model from the Research Descriptor which defined the outputs in various forms produced over the whole period of the R&D Project including the execution process of the R&D Project and keywords and key sentences of the outputs. Also, it is aimed to ultimately improve the quality of the R&D Project by defining the relation type between RD and RD to support traceability between the R&D process and the produced outputs during the process and considering the relevance between the content of RD in executing the R&D Project and quality indicator to manage the R&D Project by tracking the RD relation type.

Chapter 2 describes the existing researches on SW traceability, Traceability Information Model and the existing researches on RD based Project Management and Chapter 3 covers the information on the proposed Research Descriptor and RD Based Traceability Information Model, the definition of the relation type between RDs and the cases of RD based Traceability. Chapter 4 gives the conclusion of the proposal of this paper and future research plans.

2   Traceability in the General Software Industry

The definition of traceability varies depending on the research areas and industries. Each field including the measurement standard of machine industry, material engineering, software engineering and food processing applies different definitions of traceability, and the most standardized form of traceability refers to the ability to connect the unique identifiable entities to the time concept and verify the relation. In general, the software industry defines traceability as an ability to manage the relation between the requirements and other outputs derived from the requirements. Verification and validation of whether the outputs produced by the requirements (design documents, all kinds of implementations (source codes, execution files), test case plans, etc.) satisfied all of the functions that were specified in the requirements specification are the activities that can back up the previously defined traceability. However, it is difficult to consider that such a general definition of traceability can be embraced from every aspect regarding software development. Traceability should be defined to apply in various ranges including traceability between different outputs as well as traceability within a certain output or traceability of the changes of outputs depending on the time change.

3   Content Based Traceability Between Research Artifact and Process for R&D Projects

This chapter describes what kind of information the Research Descriptor has, which is the key concept in the Research Descriptor Based Traceability Framework proposed in this paper, from the perspective of traceability, how the information structure is composed and what kind of relation type the information has, and shows the above with the TIM model. It also explains the RD based traceability through the cases that applied the TIM model.

3.1   Research Descriptor Traceability

The Research Descriptor generated in the R&D project has traceability to support the semantic analysis technology, similarity analysis, or semantic-based tests. The traceability for Research Descriptor can be defined as follows from four perspectives.

— Traceability from the perspective of work products : Traceability between Document RD which is the result in the form of document files Tracing of various versions for one document file and traceability for other related document files
— Traceability from the perspective of process : Tracing of relation between the generated Document RD and any tasks or activities on the project WBS
— Traceability from the perspective of the monitoring quality metrics : Tracing of Document RD that affects monitoring quality metrics
— Traceability from the perspective of software development : Tracing of components or class related to software development associated with generated Research Descriptor or source files

3.2   Research Descriptor Based Traceability Information Model

As reviewed in the relevant researches in Chapter 2, what is the information to be traced, how to define the information structures and how to set the relation between the information to be traced will be explained using the TIM model when defining the traceability through the TIM model. Therefore, this paper defined the TIM model as shown in figure 1 to define RD-based traceability in executing the R&D Project.

The TIM model defined in Picture 1 demonstrates the structure and the Compose relation between Project, Phase, Activity and Task from the process aspect among the targets to be traced first, and the Produce relation with R&D Documents produced through certain Activity and Task. Also, the relation between Activity and Activity, Task and Task can be expressed as the Depend (FS, SS, FF, SF) and Branch relation.

Also, it defined various forms of outputs produced over the entire period of the R&D Project as R&D Document; described the Section or Subsection within the Document as R&D Document Item; and defined the relation between Section and

Subsection as Compose and the relation between Document and Document and Document Item and Document Item as Refine, Depend, Branch, Evolve and Refer. Each relation type will be explained in detail in the next Section.

In addition, the relations between Software Requirement or Test case and UML Class or Code Class will be expanded into Extract or Satisfy, Verify and Implement as a target to support traceability from the SW perspective.

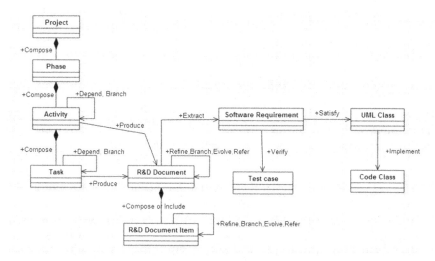

Fig. 1 Research Descriptor based Traceability Information Model

3.3   Definition of Relation Type for Research Descriptor

To define the relation type for supporting traceability, the RD relation type has been identified as follows:

To define the relation type between Document – Document / Item(Section)-Item(Section) / Activity-Activity / and Task-Task, it is required to consider how RD will support the process concept with the requirements for relation type between RDs, that is the relation type between RDs, and how to support the quality indicator of the R&D Project through the traceability between RDs. Therefore, the RD relation type was defined as follows depending on the target of each identified relation type.

a. Refine

: A relation type where a certain research topic continues to be reviewed and developed over the period of the R&D Project. There should be high similarities between the document names or item names for two Document or Docu-ment Item that can be described using the relation type.

In terms of the difference with the Evolve relation type, when the names of the documents or items are identical same, it is identified as the Evolve relation and when the names are highly similar, the relation type is defined as Refine.

b. Depend
: To change the content of a document or item, when the content of the documents or items under the relation type that should be confirmed includes all of the information of a certain document or item.

c. Branch
: When independence of the relation was stronger than correlation on two target Documents or items as in Refine or Depend, when there is relatively lower similarity between the names of documents or items.

d. Refer
: When the target document refers to the internal and external documents of the R&D Project (patent, thesis, standard, report, etc.)

e. Evolve
: The version and outputs of the target documents are revised over time, and the Evolve relation can be established between the documents before (D1-V1.0) and after (D1-V2.0) the revision.

f. Compose
: Mainly a relation type between Document and Document Item and a relation type between a document and its internal Items (A doc – A doc Item), which is also established between Section and Subsection.

Table 1 Classification of Relation type

	Relation type	Example of Relation type	
		Source	Target
Artifact RD – Artifact RD	Refine / Depend / Branch	Research plan	Research Note
		Research plan	Research Report
		Requirement Spec.	Design Document
	Compose	Research plan	Section, subsection in Research plan
	Evolve	Research plan v.1	Research plan v.2
	Refer	All Documents	the internal and external documents of the R&D Project (patent, thesis, standard, report, etc.)
Process RD – Process RD	Compose	Phase	Activity
		Activity	Task
	FS (Finish-Start)	relation of starting after finishing between Act-Act, Task-Task	
	SS (Start-Start)	A relation of concurrent starting between Act-Act, Task-Task	

Table 1 (*Continued*)

	FF (Finish-Finish)	A relation of concurrent finishing between Act-Act, Task-Task	
	SF (Start-Finish)	A relation of finishing after starting between Act-Act, Task-Task	
	Branch	An independent relation regardless of the precedence relation between the same level Processes (Act-Act or Task-Task)	
Process RD Artifact RD	Produce	Monthly Meeting	Meeting minutes
		Requirement Definition	Requirement Document

3.4 Example of Research Descriptor Based Traceability

Figure 2 demonstrates an example of the relations of quality indicators in SW development by tracing the RD relation types. As defined in ISO 9126, it is possible to guess the change rate of configuration items by tracing the Evolve relation from the baseline documents regarding the change rate of the requirements among the quality indicators in SW development or from a certain document of configuration items. In addition, it can be applied to the progress rate or requirement realization rate by tracing the Refine relation from research plan or the baseline documents about requirements.

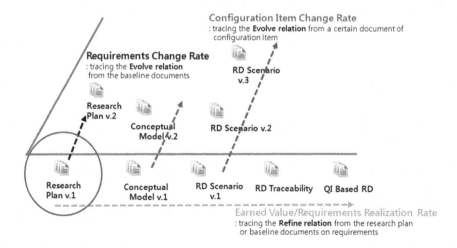

Fig. 2 Example of the Traceability between Relation Types and SW Quality Metrics

Figure 3 demonstrates the example that applied the RD relation types defined in the previous Section. It describes an example of WBS structure of the R&D Project with Process RD as the Compose relation; expresses the requirements document produced by Activity of defining requirements as the Produce relation; and the relationship between each version of requirements documents as the Evolve relation due to the revision over time. In addition, any standards, patents and reports that were referred in producing requirements document are defined as the Refer relation.

Moreover, the relationship between the analysis documents and requirements documents produced in different processes can be described as the Refine relation.

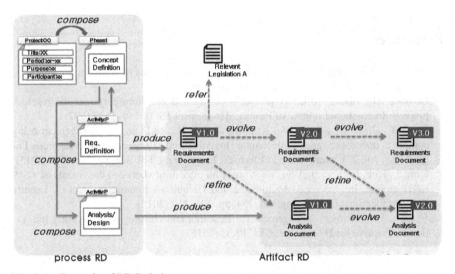

Fig. 3 An Example of RD Relation type

4   Conclusion and Further Works

As reviewed in Chapter 3, this paper defined the scope and target of traceability regarding RD traceability and the RD based Traceability Information Model from the Research Descriptor that defined various forms of outputs produced over the whole period of the R&D Project including the execution process of the R&D Project and the keywords and key sentences of the outputs. It also aimed to get out of the limitations of traceability on the configuration items in the file unit focusing on outputs in the existing researches on traceability by tracing the content and relation types of RD in conducting the R&D Project by defining the relation type of RD and RD to support traceability between the R&D Process and the produced outputs from the process. It considered the targets and perspectives that require traceability and the tracking relation types to support traceability in executing the R&D Project from various aspects, and suggested a research on traceability linked to the quality of R&D Project through the correlation between traceability based on such RD relation types and quality indicators of SW development.

Going forward, it is planned to apply the existing researches on similarity to verify the RD relation types to support traceability that was defined in this paper. Currently, researches have been being conducted on what kind of impacts the quality indicator of SW development has and how the quality indicator can be defined by tracking the quality indicator of SW development and relation types.

Acknowledgements This research was supported by the Next-Generation Information Computing Development Program through the National Research Foundation of Korea (NRF) funded by the Ministry of Science, ICT & Future Planning (NRF-2014M3C4A7 030503).

References

1. Simon, L.: Managing creative projects: an empirical synthesis of activities. International Journal of Project Management **24**, 116–126 (2006)
2. Gokhale, H., Bhatia, M.L.: A project planning and monitoring system for research projects. International Journal of Project Management **15**(3), 159–163 (1997)
3. Ko, J.W., Choi, J.Y., Kim, J.A., Kim, S., Cho, Y.H.: R&D project management using context in document: research descriptor. In: Advanced Multimedia and Ubiquitous Engineering, vol. 2. Lecture Notes in Electrical Engineering, vol. 354, pp. 203–210 (2015)
4. Kim, S., Kim, H., Kim, J., Cho, Y.: A study on traceability between documents of a SW R&D project. In: Advanced Multimedia and Ubiquitous Engineering, vol. 2. Lecture Notes in Electrical Engineering, vol. 354, pp. 203–210 (2015)
5. Vingate, L.M.: Project Management for Research and Development: Guiding Innovation for Positive R&D Outcome. CRC Press (2015)

Design and Implementation of Panoramic Vision System Based on MPEG-V

Jongseol Lee, Saim Shin, Dalwon Jang, Seong-Dong Kim, Min-Uk Kim
and Kyoungro Yoon

Abstract This paper proposes the panoramic vision system using MPEG-V specification. ISO (International Organization for Standardization) developed MPEG-V
(Moving Picture Experts Group-Virtual world) for the standardization of image-
making, which is defined many sensors and actuators. We extend MPEG-V metadata specification about RADAR detectors and multiple cameras to describe information around cars. In our implementation, the proposed usage scenarios and
metadata schema are applied to panoramic vision system, which consist of 4 cameras and OpenCV. The designed panoramic vision system can be applied to IVI
(In Vehicle Information) system or monitoring system, providing a situation environment information of the area behind the cars.

Keywords IVI · MPEG-V · Panoramic vision · Sensory effect · Metadata

1 Introduction

There is a lot of information which can be acquired in the process of driving and the
information is used for enhancing the safety. The information may be real-world
information around vehicles, and can be used to provide realistic regeneration of
real-world driving situation in virtual panoramic IVI (In-Vehicle Information) in
cars. [4,5,6] In this paper, we introduce several usage scenarios based on the newly
proposed automobile sensors, and IVI interface can provide the information
acquired from the proposed sensors by reflected in virtual world.

J. Lee · S. Shin · D. Jang · S.-D. Kim
Korea Electronics Technology Institute, Seongnam, Gyeong-gi, Korea
e-mail: {leejs,mirror,dalwon,sdkim}@keti.re.kr

J. Lee · M.-U. Kim · K. Yoon(✉)
Konkuk University, Seoul, Korea
e-mail: {minuk,yoonk}@konkuk.ac.kr

© Springer Science+Business Media Singapore 2015   899
D.-S. Park et al. (eds.), *Advances in Computer Science and Ubiquitous Computing*,
Lecture Notes in Electrical Engineering 373,
DOI: 10.1007/978-981-10-0281-6_125

ISO (International Organization for Standardization) developed MPEG-V (Moving Picture Experts Group-Virtual world) for the standardization of image-making. MPEG-V (Media context and control), published in ISO/IEC 23005, provides an architecture and specifies associated information representations to enable the interoperability between virtual worlds, e.g., digital content providers of a virtual world, (serious) gaming, simulation, and with the real world, e.g., sensors, actuators, vision and rendering, robotics[1,2,3]. MPEG-V is applicable in various business models/domains for which audiovisual contents can be associated with sensorial effects that need to be rendered on appropriate actuators and/or benefit from well-defined interaction with an associated virtual world.

2   Usage Scenarios

These scenarios assume that the automobile of the real world contains additional sensors – RADAR detectors and multiple cameras – outside of cars. In other words, the cars supporting the virtual panoramic vision equip RADAR detectors and cameras in front and back of the cars. RADAR detector is an object-detection system that uses radio waves to determine relative speed, angle of arrival, and distance. The outside cameras of cars construct a top view images which allow 360^0 overhead view of a vehicle. Hence the information acquired from the real-world can be used for safety driving or the re-construction of car accidents in virtual world.

2.1   *Virtual Panoramic IVI (In-Vehicle Information System)*

In this scenario, we also assume that the output displays for IVIs support digitalized displays. The output displays for virtual panoramic IVIs mean all kinds of output interfaces drivers usually get the necessary information while driving in cars – dashboard, head-units, side mirrors or back mirrors and so on. In usual driving situation, drivers cannot directly see the driving situation out of cars in real-world, because the real world drivers are located in cars. In order to assist drivers, RADAR detectors and multiple cameras which are located in outside of cars acquire the information about situations of driving. The virtual panoramic vision system reconstructs the real-time situation of driving in virtual world. The drivers can get 360^0 overhead information timely and correctly. In dangerous situations, the virtual panoramic IVI reconstructs the personalized IVI by constructing possible output displays in the car in order to alarm the urgent situations effectively.

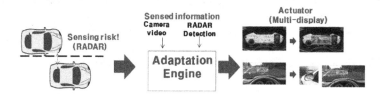

Fig. 1 Virtual panoramic IVI.

From the communication with the virtual world by the virtual panoramic vision, the driver can control the real-world driving environments more safely. The convenient driving environments in real-world can be constructed with the communications with the virtual world by virtual panoramic IVIs.

2.2 *Virtual Panoramic Black Box*

The virtual panoramic vision techniques gather visual information and the information of the adjacent objects in all directions around 360^0 overhead of a car. This information can be significantly used to understand situations when occurring accidents. With the information of the virtual panoramic vision, this virtual panoramic black box can reconstruct the past accident scenes perfectly in virtual world, it will be helpful to manage the interrupted troubles in driving such as car accidents.

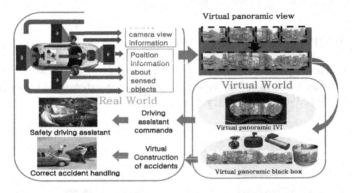

Fig. 2 Usage of virtual panoramic images

3   Extension of MPEG-V Metadata Schema for Panoramic Vision

We designed metadata type for automobile related sensors, such as RADAR sensor and Array Camera based on MPEG-V specification.

3.1 *RADAR Sensor Type*

The RADARSensorType describes sensed information with respect to an RADAR sensor. This sub clause specifies a sensor type which senses the moving or stationary target by continuously observing relative speed, angle of arrival and distance for surrounding objects.

It specifies TimeStamp, DetectedObject, Distance, Orientation, RelativeSpeed, FValWithUnitType and unit. The Distance element describes the distance of detected object. The Orientation describes the angle of detected object. RelativeSpeed describes the relative speed of detected object. FValWithUnitType is the tool for describing float value with specific unit. Unit attribute specifies the unit of the sensed value.

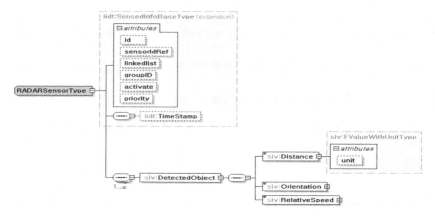

Fig. 3 RADARSensorType description

3.2 Array Camera Sensor Type

This sub clause specifies a sensor type which senses the real world by array of cameras. Array Camera sensor type is defined as an extension of Camera Sensor type of MPEG-V with additional relative location information.

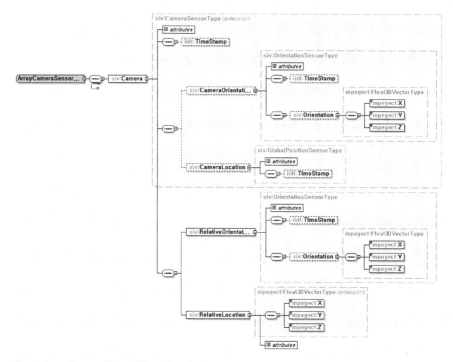

Fig. 4 ArrayCameraSonsorType description

Camera element describes the individual camera sensor. TimeStamp element describes the time that the information is acquired (sensed). RelativeOrientation element describes the relative orientation of each camera sensor. If relative orientation is specified, then CameraOrientation element in CameraSensorType is ignored. All camera sensors are relatively set on the basis of first camera sensor.

RelativeLocation element describes the relative location of each camera sensor. If relative orientation is specified, then CameraLocation element in the CameraSensorType is ignored. All camera sensors are relatively set on the basis of first camera sensor.

4   Implementation of Panoramic Vision and Metadata Generation

For panoramic vision we used 4 Logitech C920 cameras,   OpenCV2.4.9(C++)[7] and Stitching Library[8] which set ORB feature descriptor [9], BestOf2NearestMatcher, Reprojection bundle adjuster, Cylindrical Warper, Gain Compensator, and Multi-Band Blending. We can get panoramic vision videos which have 20 frames per second and 320 x 240 resolution.

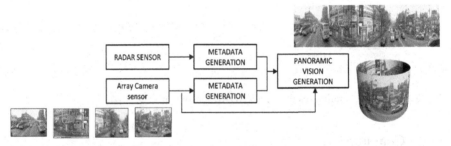

Fig. 5 Implementation of the panoramic vision system

Fig. 6. shows the description of an array of camera with the following semantics. The sensor has an ID of "ACST001" and four sub cameras each of which has an ID, such as "CID000 ~ CID003". The sensor shall be activated. The sensors shall be sensed at timestamp="6,000" where there are 100 clock ticks per second.

First camera sensor is located in (0, 0, 0). And its camera orientation is (0, 0, 0). Remaining camera sensors are located in (90, -250, 0), (0, -500, 0), (-90, -250, 0) relative positions respectively with its orientations (0, 0, 90), (0, 0, 180), (0, 0, 270). All sub cameras' orientation is only varying on z-axis to cover outside of the vehicle.

Fig. 7. shows the description of a detect object sensed by RADAR sensor with the following semantics. The sensor has an ID of "RST001". The sensor shall be activated and the distance value to the detected object is 100 meter. The sensor shall be sensed at timestamp="6,000" where there are 100 clock ticks per second. The angle of detected object with respect to the RADAR sensor shall be 30 degree, and its relative speed shall be 60 km per hour.

```
<iidl:SensedInfoList>
    <iidl:SensedInfo xsi:type="siv:ArrayCameraSensorType" id="ACST001" activate="true">
        <siv:Camera focalLength="10" aperture="2.8" id="CID000">
            <iidl:TimeStamp xsi:type="mpegvct:ClockTickTimeType" timeScale="100" pts="6000"/>
            <siv:RelativeOrientation unit="degree">
                <siv:Orientation>
                    <mpegvct:X>0</mpegvct:X>
                    <mpegvct:Y>0</mpegvct:Y>
                    <mpegvct:Z>0</mpegvct:Z>
                </siv:Orientation>
            </siv:RelativeOrientation>
            <siv:RelativeLocation unit="cm">
                <mpegvct:X>0</mpegvct:X>
                <mpegvct:Y>0</mpegvct:Y>
                <mpegvct:Z>0</mpegvct:Z>
            </siv:RelativeLocation>
        </siv:Camera>
        <siv:Camera focalLength="10" aperture="2.8" id="CID001">
            <iidl:TimeStamp xsi:type="mpegvct:ClockTickTimeType" timeScale="100" pts="6000"/>
            <siv:RelativeOrientation unit="degree">
                <siv:Orientation>
                    <mpegvct:X>0</mpegvct:X>
                    <mpegvct:Y>0</mpegvct:Y>
                    <mpegvct:Z>90</mpegvct:Z>
                </siv:Orientation>
            </siv:RelativeOrientation>
            <siv:RelativeLocation unit="cm">
                <mpegvct:X>90</mpegvct:X>
                <mpegvct:Y>-250</mpegvct:Y>
                <mpegvct:Z>0</mpegvct:Z>
            </siv:RelativeLocation>
        </siv:Camera>
```

Fig. 6 Metadata generation result using ArrayCameraSonsorType

```
<iidl:SensedInfoList>
    <iidl:SensedInfo xsi:type="siv:RADARSensorType" val-ue="100" detected="true" unit="meter" id="RST001" acti-vate="true">
        <iidl:TimeStamp xsi:type="mpegvct:ClockTickTimeType" timeScale="100" pts="6000"/>
        <siv:Distance unit="meter">100</siv:Distance>
        <siv:Orientation unit="degree">30</siv:Orientation>
        <siv:RelativeSpeed unit="kmperhour">60</siv:RelativeSpeed>
    </iidl:SensedInfo>
</iidl:SensedInfoList>
```

Fig. 7 Metadata generation result using RADARSensorType

5   Conculusion

This paper described usage scenarios and metadata schema for panoramic vision System based on the MPEG-V specification. To this end we extended MPEG-V metadata specification about RADAR detectors and multiple cameras.

We implemented panoramic vision system based on MPEG-V using OpenCV library to handle graphic images. Implementation results show a significant possibility of panoramic vision as an IVI system and a traffic management system providing situation environment information of the area behind the cars.

References

1. ISO/IEC IS 23005-1:2014 Architecture
2. ISO/IEC IS 23005-2:2014 Control Information
3. ISO/IEC IS 23005-3:2014 Sensory Information

4. Maik, V.: Automatic top-view transformation for vehicle backup rear-view camera. In: ISCE 2014 (2014)
5. Lin, Q., Han, Y., Hahn, H.: Vision-based navigation using top-view transform and beam-ray model. In: ICACSIS, pp. 371–376 (2011)
6. Szeliski, R.: Computer Vision: Algorithms and Applications. Springer (2011)
7. https://en.wikipedia.org/wiki/OpenCV
8. Brown, M., Lowe, D.: Automatic Panoramic Image Stitching using Invariant Features. International Journal of Computer Vision **74**(1), 59–73 (2007)
9. Rublee, E., Rabaud, V., Konolige, K., Bradski, G.R.: ORB: an efficient alternative to SIFT or SURF. In: ICCV 2011, pp. 2564–2571 (2011)

Computational Fluid Dynamics Analysis of the Air Damping for an Electromechanical Converter

Gong Zhang, Jimin Liang, Chaomeng Jiang, Zhipeng Zhou, Xianshuai Chen and Lanying Yu

Abstract Aiming at the air damping of the electromechanical converter under the conditions of high frequency and high speed, a computational fluid dynamics analysis of air damping is proposed for three different thrust coil frameworks based on a dynamic mesh technology. Simulation results show that with the increase of the moving frequency and speed, the air damping effect of the thrust coil framework becomes more significant. Punching holes on the end face of the thrust coil framework could greatly improve the air flow characteristics and the distribution of the pressure and velocity. Compared with proposal one, the air damping of proposal two is decreased by 81%, and proposal three is decreased by 98%. It can be seen that the air damping can be reduced by the structural optimization of electromechanical converter at the high speed working conditions.

Keywords Electromechanical converter · High speed · Air damping · Dynamic mesh

1 Introduction

An electromechanical converter, converting the electric signals into the mechanical quantities continuously and proportionally, is a key part of electric-hydraulic

G. Zhang(✉) · J. Liang · C. Jiang · X. Chen
Guangzhou Institute of Advanced Technology, Chinese Academy of Science,
Guangzhou, China
e-mail: gong_zhang@foxmail.com, {jm.liang,xs.chen}@giat.ac.cn, jiangchaomeng@163.com

C. Jiang · L. Yu
School of Mechanical Engineering, Southwest Jiaotong University, Chengdu, China
e-mail: lyyu@home.swjtu.edu.cn

Z. Zhou
DFAC Especial Vehicles Division, Wuhan, China
e-mail: luotxt@sina.com

© Springer Science+Business Media Singapore 2015   907
D.-S. Park et al. (eds.), *Advances in Computer Science and Ubiquitous Computing*,
Lecture Notes in Electrical Engineering 373,
DOI: 10.1007/978-981-10-0281-6_126

proportional control systems. At present, the moving coil electromechanical converters are receiving increased attention for use in applications requiring linear motion at high speed and high accuracy [1, 2]. The force generated is 1.5 times higher than the others of the same size [3, 4].

Conventional properties of electric-hydraulic proportional control systems could be improved by raising the response time and speed of the electromechanical converters, which is also the development trend of the electric-hydraulic proportional control technology [5, 6]. Generally speaking, the air damping of the electromechanical converters is usually ignored [7]. However, the operating speed and the structure of the electromechanical converters have great influence on the air damping. Importantly, along with the raising of speed, the air damping has a great impact on the dynamic performance of the high speed electromechanical converters while ignored previously. However, as the diversity and complexity of the structure of cavities enclosed by the electromechanical converter, the normal momentum analysis could no longer be able to evaluate the air damping of the movement process accurately and detailedly. Therefore, the flow fluid simulation method is proposed. In addition, the analysis of the interior flow fluid is necessary to the optimum design [8].

Towards this end, in this study, a comparatively investigation for the air damping characteristics of an electromechanical converter with three different thrust coil frameworks is conducted using a computational fluid dynamic (CFD) software Fluent based on a dynamic mesh technology, and the method that could reduce the air damping are obtained as well.

2 Numerical Simulation

Theoretical Basis. Assumptions of the model are as follows: considering the low speed of coil and coil framework, the air is regarded as incompressible here; the gravity and heat transfer are not taken into consideration in the process of calculation; because of the small size and complex shape of the air cavity, the interior air is deemed to in the turbulent state.

The basic control equations used to evaluate are the mass conservation equation, the Reynolds equation and the k-ε turbulent control equations [9].

Mass conservation equation can be expressed as:

$$\frac{\partial u}{\partial x} + \frac{\partial v}{\partial y} = 0 \tag{1}$$

Where u, v are the components of velocity vector in the x, y direction respectively.
And Reynolds equation could be expressed as:

$$\frac{\partial u}{\partial t} + u\frac{\partial u}{\partial x} + v\frac{\partial u}{\partial y} = -\frac{1}{\rho}\frac{\partial p}{\partial x} + \mu_{eff}(\frac{\partial^2 u}{\partial x^2} + \frac{\partial^2 u}{\partial y^2}) + \frac{\partial}{\partial x}(\mu_{eff}\frac{\partial u}{\partial x}) + \frac{\partial}{\partial y}(\mu_{eff}\frac{\partial v}{\partial x}) + F_x \tag{2}$$

$$\frac{\partial u}{\partial t}+u\frac{\partial \upsilon}{\partial x}+\upsilon\frac{\partial \upsilon}{\partial y}=-\frac{1}{\rho}\frac{\partial p}{\partial y}+\mu_{eff}(\frac{\partial^2 \upsilon}{\partial x^2}+\frac{\partial^2 \upsilon}{\partial y^2})+\frac{\partial}{\partial x}(\mu_{eff}\frac{\partial u}{\partial y})+\frac{\partial}{\partial y}(\mu_{eff}\frac{\partial \upsilon}{\partial y})+F_y \quad (3)$$

Where $\mu_{eff}=\mu+\mu_t$, $\mu=\rho C_\mu k^2/\varepsilon$, ρ is the density, μ is kinetic viscosity, P is the pressure acting on fluid microelements, F_x and F_y are the force on the microelements.

Realizable k-ε equations are as follows:

k equation can be expressed as:

$$\frac{\partial(\rho k)}{\partial t}+\frac{\partial(\rho k u_i)}{\partial x_i}=\frac{\partial}{\partial x_i}\left[(\mu+\frac{\mu_t}{\sigma_k})\frac{\partial k}{\partial x_i}\right]+G_k-\rho\varepsilon \quad (4)$$

ε equation can be written as:

$$\frac{\partial(\rho k)}{\partial t}+\frac{\partial(\rho\varepsilon u_i)}{\partial x_i}=\frac{\partial}{\partial x_i}\left[(\mu+\frac{\mu_t}{\sigma_\varepsilon})\frac{\partial\varepsilon}{\partial x_i}\right]+\rho C_1 E\varepsilon-\rho C_2\frac{\varepsilon^2}{k+\sqrt{\mu\varepsilon/\rho}} \quad (5)$$

Where μ_t is turbulent viscosity, k is turbulent kinetic energy, σ_k is the corresponding Prandtl number with k, ε is turbulent dissipation rate, σ_ε is the corresponding Prandtl number, μ_i is time averaged velocity, G_k is the turbulent kinetic energy caused by mean velocity gradient, C_1 and C_2 are the experience dates.

Physical Model. Fig.1 depicts a schematic of the electromechanical converter mainly comprised of the coil, armature, permanent magnet, thrust coil framework, isolation ring, etc. The coil is mounted on the thrust coil framework, which is joined with the spool of electric-hydraulic proportional valve, named as the coil subassembly.

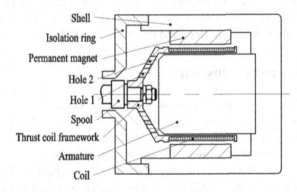

Fig. 1 Schematic of the electromechanical converter

When voltage signal is loaded on the coil, the spool moves along with the coil under the electromagnetic force generated by the coil owing to the invariable magnetic field, and then controls the port of the electric-hydraulic proportional valve. The value and direction of the electromagnetic force acted on the moving coil depend on the value and direction of the control current. The changed electromagnetic force of the coil subassembly due to the direction change of the control current, realizes the bidirectional movement of the electromechanical converter.

As showed in Fig.1, the coil subassembly moves periodically in high frequency in an airtight cavity enclosed by shell and thrust coil framework. To evaluate the air damping characteristics of the electromechanical converters with three different thrust coil frameworks parallel for convenience, 2D simplified models are used. The simplified models are described as below. Proposal one: the original thrust coil framework, without hole, as showed in Fig. 2(a). Proposal two: punch hole 1 (8×Ø4mm) on the end of the thrust coil framework, as showed in Fig. 2(b). Proposal three, punch hole 1 (8×Ø4mm) and hole 2 (15×Ø3.5mm), as showed in Fig. 2(c).

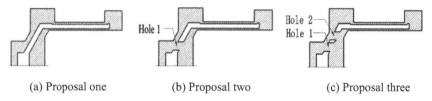

(a) Proposal one   (b) Proposal two   (c) Proposal three

Fig. 2 2D simplified model of different thrust coil frameworks

Calculation Configuration. In the moving process of coil subassembly, the calculation domain is changing, so transient model is needed when analyzing. The modeling steps are listed as follows: import the prepared CAD model into ICEM to mesh, for the better renewal of meshing, triangle mesh is applied; open the mesh file in Fluent, define the model as a 2D pressure based transient axisymmetric calculation, choose Realizable k-ε turbulence model, and adopt Smoothing and Remeshing to renew the mesh; set axis to Deforming type and the outline of coil subassembly to Rigid Body type. And the form of movement is defined by UDF; finally, set the time step size, number of time steps and max iterations before calculation.

3   Results and Discussion

The coil subassembly moves periodically in the range of -0.5mm~+0.5mm according to the Cosine rule, the amplitudes of the velocity are set as 1m/s, 2m/s and 3m/s respectively. Fig. 3 illustrates the air damping curves of three different velocities over time in the transient calculation of proposal one.

As evident in Fig. 3, with the increasing of velocity, the peak of the air damping of coil subassembly is increased accordingly. The air damping peaks are 19.6N, 8.9N, 2.2N when the velocity amplitudes are 3m/s, 2m/s, 1m/s respectively. Hence, the air damping is nearly proportional to the square of the speed, as showed in fig. 4.

Compared with the electromagnet force and the hydrodynamic force, the air damping of the electromechanical converter at the low-frequency and low-velocity motion conditions can be neglected. Nevertheless as the speed and frequency of the electromechanical converter are even higher, the impact of the air damping on the working performance becomes more significant. Consequently, it is necessary to study the characteristics of the air damping, and seek the ways to decrease it.

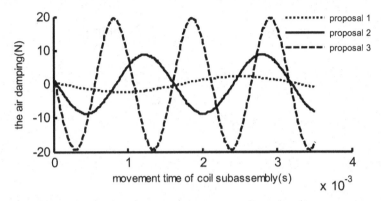

Fig. 3 The air damping of different velocities

Proposal two and proposal three are the structure optimization of proposal one through punching holes on the end of thrust coil framework. Below are the results of the comparative analysis of three different structures using the same calculation configuration.

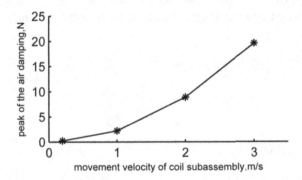

Fig. 4 The air damping peak of different speeds

Fig. 5 shows that there is a similar change law among the air damping characteristics of three proposals at different speeds. And the air damping curves of three proposals at the speed of 3m/s are chosen to evaluate. The air damping peaks are 19.6N, 3.72N and 0.34N respectively. The air damping of proposal one is significant, yet proposal three is very tiny. Compared with proposal one, the air damping of proposal two is decreased by 81%, and proposal three is decreased by 98%, which means the effect is remarkable.

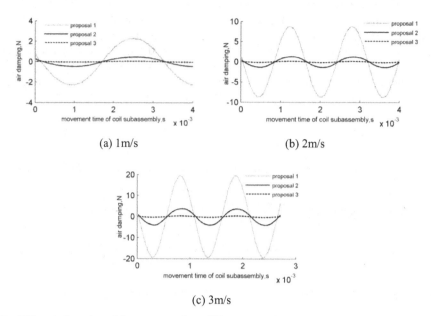

(a) 1m/s

(b) 2m/s

(c) 3m/s

Fig. 5 The air damping of three proposals at different speeds

From the graphics of the pressure contour and the velocity vector of three proposals at the limiting movement condition of 3m/s, conclusions are as follows:

The maximum positive pressure of proposal one is 3963.9Pa, the maximum negative pressure is -549.9Pa. The maximum positive pressure of proposal two is 156.9Pa, the maximum negative pressure is -456.7Pa. The maximum positive pressure of proposal three is 180Pa, the maximum negative pressure is -68.1Pa.

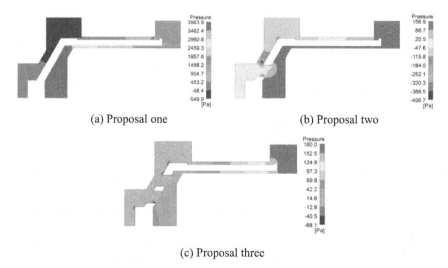

(a) Proposal one   (b) Proposal two

(c) Proposal three

Fig. 6 Pressure contour diagram of different proposals at the speed amplitude of 3m/s

And the whole pressure distribution regularity is changed. The maximum velocity of proposal one is 47.4m/s, proposal two is 22.3m/s, proposal three is 9m/s. And the whole velocity distribution regularity is changed as well. After holes are punched on the end of the thrust coil framework, the air channel increases and the gradient of pressure and velocity of the narrow cavity enclosed by the coil and the shell is decreased obviously. The distribution of pressure and velocity is improved as well as the air flow characteristics, the air damping is reduced accordingly.

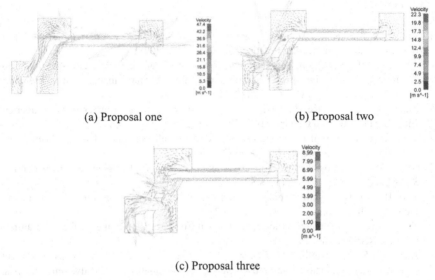

(a) Proposal one   (b) Proposal two

(c) Proposal three

Fig. 7 Velocity vector diagram of different proposals at the speed amplitude of 3m/s

4   Conclusions

This paper proposed a comparatively investigation of the air damping characteristics for an electromechanical converter based on a dynamic mesh technology, conclusions are as follows:

The air damping of the coil subassembly is proportional to the square of the speed. The comparative analysis of the three different thrust coil frameworks reveals that compared with proposal one, the air damping of proposal two is decreased by 81%, and proposal three is decreased by 98%, which means the effect is remarkable.

As detailed from the above computational fluid dynamics analysis results, the air damping of the electromechanical converter can be reduced by the structural optimization at the high speed working conditions.

Acknowledgments This research is supported by the National Natural Science Foundation of the People's Republic of China (Grant No. 51307170), the Shenzhen City Knowledge Innovation Program of China (Basic Research Project) (Grant No. JCYJ20140901003939032), and the Guangzhou City Scientific Research Projects of China (General Project) (Grant No. 201505051734437). The authors gratefully acknowledge the help of Guangzhou Institute of Advanced Technology, Chinese Academy of Sciences and School of Mechanical Engineering, Southwest Jiaotong University.

References

1. Zhao, S., Tan, K.K.: Adaptive feedforward compensation of force ripples in linear motors. Control Engineering Practice **13**, 1081–1092 (2005)
2. Abdou, G., Tereshkovich, W.: Performance evaluation of a permanent magnet brushless DC linear drive for high speed machining using finite element analysis. Finite Elements in Analysis and Design **35**, 169–188 (2000)
3. Zhang, G., Yu, L., Ke, J.: High frequency moving coil electromechanical converter. Electric Machines and Control **11**(3), 298–302 (2007). (in Chinese)
4. Tanaka, H.: History of the proportional electromagnetic solenoid. Tokyo, Japan: Japan Fluid Power System Society **31**, 50–56 (2000)
5. Sadre, M.: Electromechanical converters associated to wind turbines and their control. Solar Energy **6**(2), 119–125 (1997)
6. Yao, B., Xu, L.: Adaptive robust motion control of linear motors for precision manufacturing. Mechatronics **12**, 595–616 (2002)
7. Yu, K., Lu, YA.: Review of Electromechanical Converters. Machine Tool & Hydraulics, no. 1, 2–7 (1991). (in Chinese)
8. Amirante, R., Moscatelli, P.G., Catalano, L.A.: Evaluation of the flow forces on a direct (single stage) proportional valve by means of a computational fluid dynamic analysis. Energy Conversion & Management **48**, 942–953 (2007)
9. Wang, F.: Computational Fluid Dynamic Analysis. Tsinghua University Press, Beijing (2004)

Author Index

Printed in the United States
By Bookmasters